**ATP 3-18.1**

Army Techniques Publication
No. 3-18.1

Headquarters
Department of the Army
Washington, DC,
21 March 2019

I0092924

# Special Forces Unconventional Warfare

## Contents

*This publication supersedes TC 18-01, 28 January 2011.

# Figures

# Table

This page intentionally left blank.

# Preface

ATP 3-18.1 defines the current U.S. Army Special Forces (SF) concept of understanding, planning, and executing unconventional warfare (UW) operations. It continues the doctrinal education process that begins with JP 3-05, *Special Operations*; JP 3-05.1, *Unconventional Warfare*; ADP 3-05, *Special Operations*; ADRP 3-05, *Special Operations*; FM 3-05, *Army Special Operations*; FM 3-18, *Special Forces Operations*; and ATP 3-05.1, *Unconventional Warfare*.

ATP 3-18.1 describes UW fundamentals, activities, and considerations involved in the planning and execution of UW throughout the full range of military operations. It identifies principles of resistance and principles of UW and clearly establishes the relationship of SF to supporting resistance through UW. ATP 3-18.1 is authoritative but not directive. It serves as a guide and does not preclude SF units from developing their own standard operating procedures (SOPs) to meet their needs. It explains techniques used by units and headquarters (HQ) at the SF battalion level and below to conduct UW.

This ATP is designed to be UNCLASSIFIED in order to ensure the widest distribution possible to the appropriate Army special operations forces (ARSOF) and other interested Department of Defense (DOD) and United States Government (USG) agencies while protecting technical or operational information from automatic dissemination under the International Exchange Program or by other means. Subordinate classified appendixes are published separately.

The principal audience of this publication is commanders, staff officers, and operational personnel at the team (operational detachment–alpha [ODA]), company (operational detachment–bravo [ODB]), and battalion (operational detachment–charlie) levels. This ATP is specifically for SF Soldiers; however, it is also intended for use Army wide to improve the integration of SF into the plans and operations of other special operations forces (SOF) and conventional forces. Commanders and trainers should use this and other related publications in conjunction with command guidance and the Combined Arms Training Strategy to plan and conduct successful UW operations.

Commanders, staffs, and subordinates ensure their decisions and actions comply with applicable U.S., international, and, in some cases, host-nation laws and regulations. Commanders at all levels ensure their Soldiers operate in accordance with the law of war and the rules of engagement. (See FM 27-10.)

ATP 3-18.1 uses joint terms, where applicable. Selected joint and Army terms and definitions appear in both the glossary and the text. Terms for which ATP 3-18.1 is the proponent publication (the authority) are marked with an asterisk (*) in the glossary. Definitions for which ATP 3-18.1 is the proponent publication are **boldfaced** in the text. For other definitions shown in the text, the term is *italicized* and the number of the proponent publication follows the definition.

ATP 3-18.1 applies to the Active Army, the Army National Guard/Army National Guard of the United States, and United States Army Reserve unless otherwise stated. Unless this publication states otherwise, masculine nouns and pronouns do not refer exclusively to men.

The proponent of this publication is the U.S. Army Special Operations Center of Excellence, USAJFKSWCS. Submit comments and recommended changes on DA Form 2028 (Recommended Changes to Publications and Blank Forms) directly to Commander, U.S. Army Special Operations Center of Excellence, USAJFKSWCS, ATTN: AOJK-SFD, 3004 Ardennes Street, Stop A, Fort Bragg, NC 28310-9610.

# Acknowledgements

The copyright owners listed here have granted permission to reproduce material from their works. Other courtesy credits listed.

Citations from Steve Cheng, *Navigating Ambiguity: The Contentious World of Nonstandard Logistics* (Unpublished Essay), 2016.

Citations and images from *Total Resistance*, by Major H. von Dach, Paladin Press, 1965.

Citations and images from U.S. Government Support to Resistance Framework (DRAFT) Version 0.33, USSOCOM J-7.

Citations and images from Risa Borsykowsky, The Ayalon Institute: Kibbutzim Hill Rehovot, https://www.jewishgiftplace.com/Avalon-Institute.html.

# Introduction

ATP 3-18.1 is the Army's doctrinal foundation for UW execution at the SF battalion level and below. This publication also serves as the Army's description of how SF below the SF Group (Airborne)-level and assigned and supporting units conduct UW to larger joint and interagency audiences. ATP 3-18.1, therefore, provides doctrine directly useful to all users within the U.S. Army, but it is deliberately intended to be useful to other Services in the DOD and joint, interagency, intergovernmental, multinational audiences.

The "Resistance Profession," which is ARSOF either supporting or defeating resistance movements, is depicted in introductory figure 1. The conduct of UW is within the category of support to a resistance movement. ATP 3-18.1 identifies principles of resistance and principles of UW and clearly establishes the relationship of SF to supporting resistance through UW. ATP 3-18.1 is fundamental to the SF Soldier's understanding of resistance movements and how to support them.

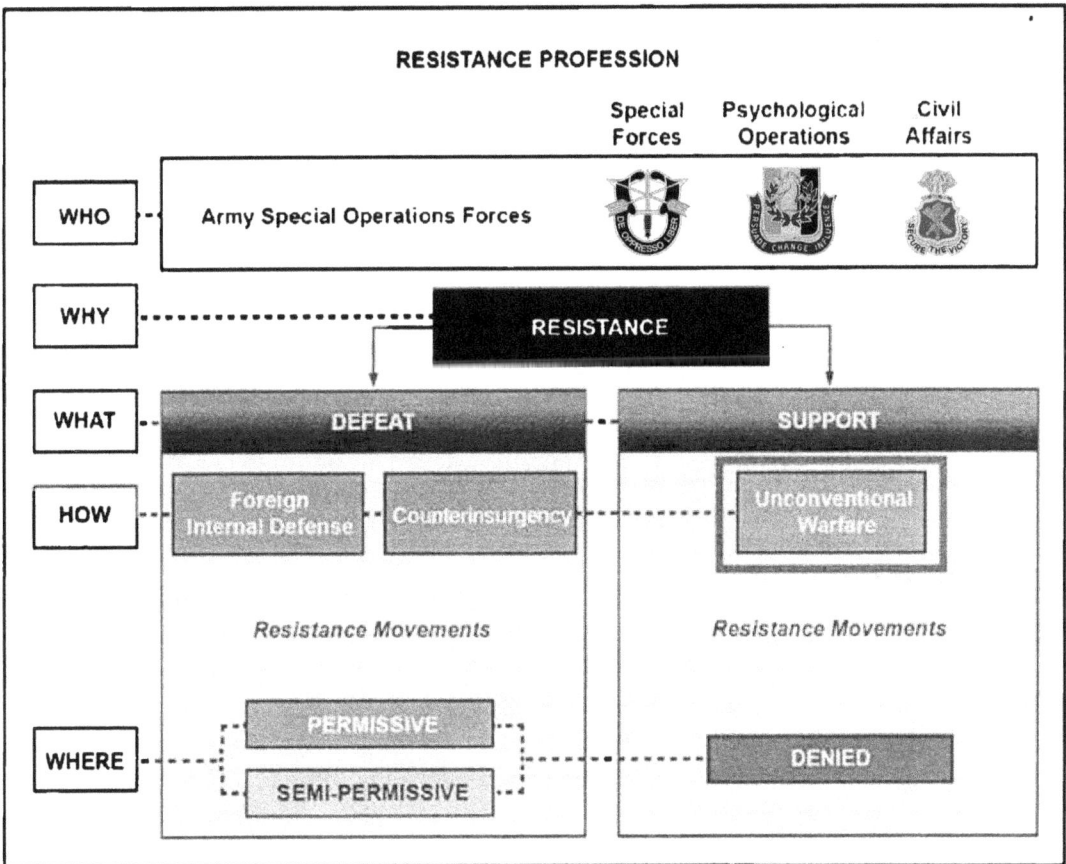

**Introductory Figure 1. Special Forces and unconventional warfare**

UW may be conducted as the main effort or as a line of effort in large-scale ground combat operations to support the conventional force. UW can be used in the competitive space to deter an adversary through coerce and disrupt activities.

ATP 3-18.1 is intended to be of practical value to SF Soldiers and commanders at the tactical level. The ATP is therefore written in significant depth and at a more basic level of detail to assist the SF Soldier's conduct of UW on the ground. However, the ATP also includes a brief articulation of larger policy and operational contexts that frame and help distinguish the narrower focus of tactical SF units.

ATP 3-18.1 contains six chapters and twenty-six appendixes that are summarized in the following paragraphs.

**Chapter 1** provides a conceptual overview of UW, sets it within the context of national policy, and discusses why UW is a viable policy option. The chapter explains the difference between coerce, disrupt, and overthrow and provides an appreciation for the criticality of considering when, and by what level of command, UW is feasible.

**Chapter 2** considers how and why populations resist; it also provides the basic components of traditional U.S. resistance theory as documented in traditional U.S. UW doctrine.

**Chapter 3** introduces the concept of support to resistance (STR) to U.S. UW doctrine, offers five identifiable types of STR, and explains UW as two of the types within that larger STR context.

**Chapter 4** outlines the basic traditional concepts of UW from the perspective of the executing unit, to include the seven-phase model of U.S.-sponsored insurgency, and it includes clarifying explanations of persistently misunderstood traditional concepts.

**Chapter 5** is a detailed discussion of the core activities that comprise UW, to include intelligence operations, preparation of the environment (PE), unconventional assisted recovery, subversion, sabotage, and guerrilla warfare.

**Chapter 6** provides practical information on supporting activities for UW.

**Appendix A** (U) Sensitive Activities Support to Unconventional Warfare (S//NF) to be published separately.

**Appendix B** (U) Unaccompanied Resistance Considerations for Unconventional Warfare (S//NF) to be published separately.

**Appendixes C–Z** present amplification of the following topics:
- (C) Intelligence Support to Unconventional Warfare.
- (D) Varieties of Resistance.
- (E) Varieties of Violent and Nonviolent Methods.
- (F) Resistance Development Models.
- (G) Theater Special Operations Command Mission Letter—Examples.
- (H) Sample Area Study Format.
- (I) Area Assessment Formats—Initial and Principal.
- (J) Special Operations Mission Planning Folder Formats.
- (K) Infiltration and Exfiltration Considerations.
- (L) Communications.
- (M) Logistics.
- (N) Caching.
- (O) Link-Up Operations.
- (P) Special Forces Resistance Advisor Considerations.
- (Q) Resistance Training Considerations.
- (R) Shadow Government Operations Model.
- (S) Area Complex Development Model.
- (T) Administrative Considerations.
- (U) Guerrilla Periods of Instruction.
- (V) Special Forces Advisor's Guide to the Combat Employment of Guerrilla Forces.
- (W) Unconventional Warfare in an Urban Environment.
- (X) Medical Aspects.
- (Y) Resistance Demobilization and Transition Considerations.
- (Z) Stay-Behind Unconventional Warfare Considerations.

Based on current doctrinal changes, certain terms for which ATP 3-18.1 is the proponent have been added or modified for purposes of this publication (introductory table 1). The glossary contains acronyms and defined terms.

**Introductory Table 1. New Army Terms**

| Term | Acronym | Remarks |
|---|---|---|
| area complex | -- | new Army term and definition |
| guerrilla warfare | -- | new Army term and definition |
| public component | -- | new Army term and definition |
| resistance area command | -- | new Army term and definition |
| resistance partner | -- | new Army term and definition |
| safe area | -- | new Army term and definition |
| stay-behind resistance operation | -- | new Army term and definition |
| stay-behind unconventional warfare | -- | new Army term and definition |
| subversive political action | -- | new Army term and definition |
| support to resistance | STR | new Army term, definition, and acronym |
| surrogate | -- | new Army term and definition |

This page intentionally left blank.

# Chapter 1

# Overview

*Unconventional warfare: Activities conducted to enable a resistance movement or insurgency to coerce, disrupt, or overthrow a government or occupying power by operating through or with an underground, auxiliary, and guerrilla force in a denied area.*

JP 3-05.1, Unconventional Warfare

---

### Unconventional Warfare Definition Breakdown

Who: Those ordered to conduct it; typically U.S. Army Special Forces, supported by joint, interagency, and sometimes multinational partners.

What: *(activities)* Intelligence operations, preparation of the environment, unconventional assisted recovery, subversion, sabotage, guerrilla warfare.

When: When ordered to conduct it.

Where: *(in a denied area)* The primary consideration is that the effects occur in the denied area.

Why: *(to enable a resistance movement or insurgency to coerce, disrupt, or overthrow a government or occupying power)*

How: *(by operating through or with an underground, auxiliary, and guerrilla force)*

---

# INTRODUCTION TO UNCONVENTIONAL WARFARE

1-1. Per JP 1, *Doctrine for the Armed Forces of the United States*, the U.S. military recognizes two forms of warfare:

- **Traditional Warfare.** Traditional warfare is characterized as a violent struggle for domination between nation-states or coalitions and alliances of nation-states. With the increasingly rare case of formally declared war, traditional warfare typically involves force-on-force military operations in which adversaries employ a variety of conventional forces and SOF against each other in all physical domains, as well as the information environment (which includes cyberspace).
- **Irregular Warfare.** Irregular warfare is characterized as a violent struggle among state and non-state actors for legitimacy and influence over the relevant population(s). In irregular warfare, a less powerful adversary seeks to disrupt or negate the military capabilities and advantages of a more powerful military force, which usually serves that nation's established government.

This dichotomy is useful for distinguishing the relevant targets of activity and the force mix most appropriate to achieve desired objectives.

1-2. UW is a form of irregular warfare, but it can also be used in support of traditional warfare. Per JP 3-05.1, there are two distinct types of UW efforts that support strategic outcomes. One type consists of UW as a supporting line of operation within the military effort of a larger military campaign. The second type of UW is employed as the strategic main effort—either as an initiative or as a response to aggression.

1-3. UW operations can also be characterized by scale. During large-scale UW, operations focus largely on military aspects of the conflict because of the eventual introduction of conventional forces. By contrast, a small-scale UW effort may be composed of compartmented small teams under the direction of a standing joint force commander (JFC) or compartmented task forces.

---

## Principles of Unconventional Warfare

1. Achieve U.S. policy objectives in denied areas through support of resistance movements and insurgencies.
2. UW involves the strategic art of indirect approach.
3. Employ UW in times, places, and circumstances where direct, overt, and overwhelming U.S. and allied power is not the best option available.
4. Effects desired from U.S. support to resistance must be feasible by the resistance, and feasible at U.S. tactical, operational, strategic, and policy levels.
5. Support actors who are typically less powerful than the state or occupying power they are resisting with all that implies.
6. Specially-trained U.S. and allied cadres, typically led by U.S. government civilian agency operators and Army Special Forces, assist, advise, accompany, and lead resistance actors to achieve mutual U.S. and resistance objectives.
7. Resistance actors are influenced, not commanded, and ultimately cooperate for their own interests.
8. UW is always a political activity in the human domain requiring U.S. operators to be agents of influence.
9. Ultimately, it is the resistance—not U.S. cadres themselves—that must achieve the desired effects.
10. The essence of UW is resistance subversion of opponent strength in all domains through a wide range of activities.
11. The activities associated with resistance and supporting resistance are timeless, but every different resistance or insurgency and every instance of UW will have unique characteristics.
12. UW cadres are typified by intelligence, superior Soldier skills, small-unit leadership excellence, stealth and situational awareness, political acumen, competence in managing complexity, self-reliance, and physical and moral courage.

*Principle*: A comprehensive and fundamental rule or an assumption of central importance that guides how an organization or function approaches and thinks about the conduct of operations (ADP 1-01).

---

1-4. UW is the core task and organizing principle for U.S. Army SF. UW capabilities encompass the skill sets required for all other SF missions. SF is specifically organized, trained, and equipped for the conduct of UW. SF is regionally oriented, language-qualified, and specifically trained to conduct UW against hostile nation-states and occupying powers to achieve U.S. goals.

# THE 21ST CENTURY GLOBAL ENVIRONMENT

1-5. Today's globally connected and networked world affects change in the strategic environment through the spread of technology into an increasing number of cultures and societies. Complicating this trend, the variety of ways many cultures and societies use technology may be significantly different in purpose and method than what is common in western democratic societies. Culture is a kind of collective mental programming based on a set of values, attitudes, beliefs, artifacts and other meaningful symbols. It represents a pattern of life adopted by a people that helps them interpret, evaluate, and communicate as members of society. Having an expert understanding of the culture and its dimensions is key to successful UW activities.

1-6. The complex global landscape faced by the USG includes vulnerable nation-states and the destabilizing activities of both adversarial nation-states and non-state actors, to include hostile neighbors and non-state actors, such as transnational terrorists and criminal enterprises. Simultaneously, global opportunities are emerging as networked populations are seeking improvements in governance, security, and economic opportunity. Power and influence are now diffusing to a greater range of actors, both state and non-state, who have not traditionally wielded it—some malign, others beneficial. Many governments are struggling to adjust to the new realities. The opportunities to identify, assess, organize, train, equip, advise, assist, and enable resistance efforts as part of a U.S. policy option is, in many ways, multiplied by this increased global complexity and destabilizing trends.

1-7.    Between states, the diffusion of technological advancements is providing greater options for both rising powers and non-state actors to pursue their interests. In some cases, this means expanding their claims of sovereignty outside of the borders and parameters currently recognized. In other cases, they are sponsoring and relying upon non-state actors to act on their behalf abroad. These states increasingly avoid direct conflict with the USG and indirectly attack the security of partner nations to undermine U.S. national interests. Within states, this diffusion of technologies and hyper-connectivity allow greater ability for beneficial, and malign, non-state actors and groups to network, organize, and challenge the legitimacy of governments. While the fundamentals of UW remain unchanged, these technological changes and shifting political methods offer the USG new opportunities of employing UW to shape new global challenges.

# UNCONVENTIONAL WARFARE AS A NATIONAL STRATEGIC POLICY OPTION

1-8.    UW is a strategic policy option available to the President of the United States. UW—combined with other operations and activities in the arsenal of U.S. instruments of national power—should always be considered an available component of the U.S. national strategy. When use of the military instrument of U.S. power is appropriate, UW provides a lower military footprint and signature to the policy makers compared to large-scale military forces. In some cases, UW is a strategic policy option main effort that can be discreetly targeted on the most high-stakes, sensitive, difficult, intractable and impenetrable problems with specifically selected, prepared, and supported forces. In other cases, UW is an attractive strategic policy option as a global economy-of-force effort in secondary theaters, or a strategic and operational military option as a supporting line of effort to larger joint force campaigns.

1-9.    UW may involve the acceptance of significant risk to trainer or advisor personnel, as well as diplomatic or political risk, but such risks may be deemed more acceptable than the risks associated with inaction on the part of the USG or the use of large-scale U.S. forces executing a major military operation or campaign. The inherent risk in UW is the time required to develop viable forces, the risks to small numbers of SOF operating with the insurgents or resistance forces, and the diplomatic or political risks depending on the desired signature and effects. Regardless of the size of the effort, the actual joint force footprint deployed into the denied area can be a small percentage of the equivalent conventional forces that would be required to create similar effects.

# SUPPORT TO RESISTANCE

1-10.    UW constitutes two of five types of STR efforts in a foreign country or area to meet USG objectives through enabling resistance movements and insurgencies. As such, UW is a specific USG policy option contained within the following larger contexts:

- Normal international political activity between states and non-state actors.
- USG efforts to achieve its objectives in the global steady-state environment using a combination of all instruments of U.S. national power.
- USG STR.

1-11.    *Support to resistance* **is a United States Government policy option to support foreign resistance actors that offers an alternative to a direct U.S. military intervention or formal political engagement in a conflict.** Such a policy decision to conduct STR may or may not involve significant conventional military operations. STR may involve a range of special operations, and sometimes, a policy decision will be made to conduct STR using the specific special operation of UW (figure 1-1).

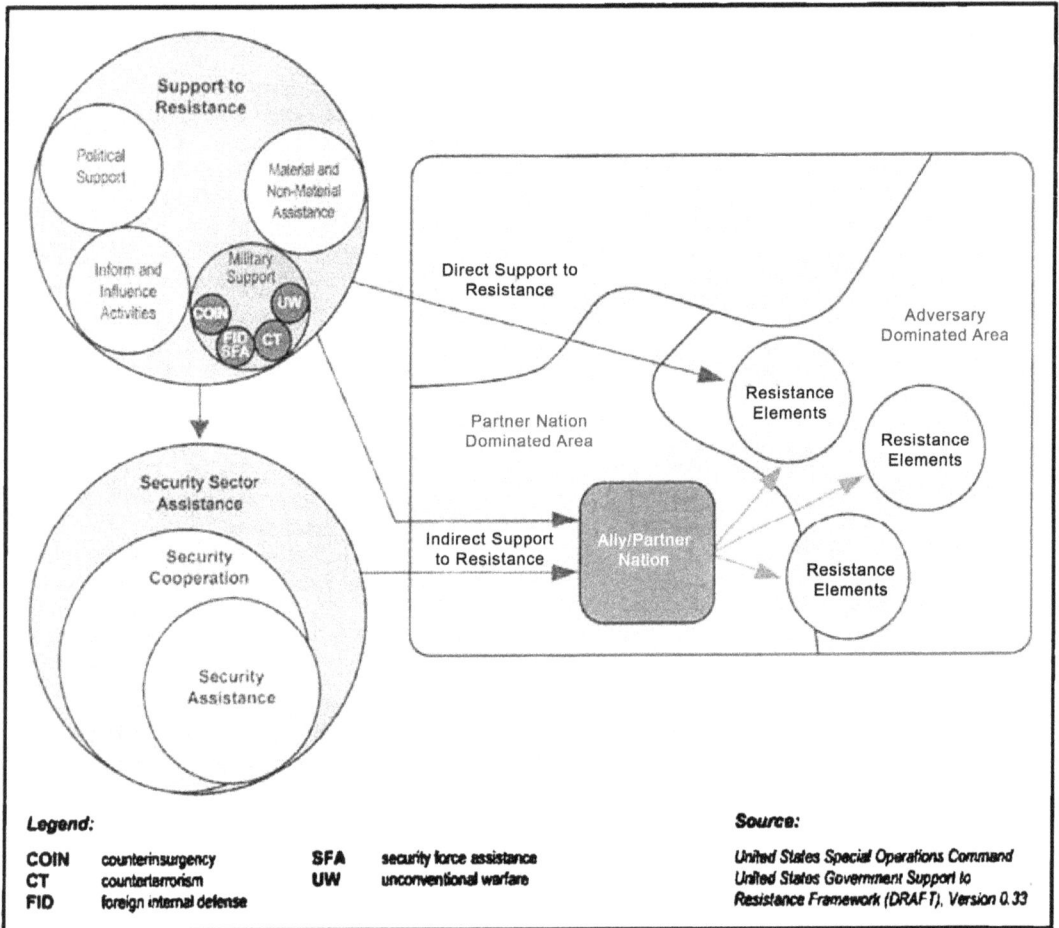

**Figure 1-1. Unconventional warfare within the context of support to resistance**

1-12.  There are many types of resistance—most of which will not be good candidates for USG support. Even those resistance efforts supported will pursue their own objectives; many of which may differ in significant ways from those of the USG. Moreover, circumstances change over time. Objectives agreed upon between the USG and any resistance partner at a specific time may subsequently diverge as conditions and perceptions change. UW is therefore a limited political arrangement between cooperating partners. It is important to note that resistance partner is not synonymous with surrogate.

1-13.  A *resistance partner* is a partner conducting resistance with whom the United States Government mutually establishes agreements to cooperate for some specified time in pursuit of mutually supporting specific objectives. A resistance partner is influenced; he is not an employee or subordinate to be commanded and controlled. A *surrogate* is someone who acts on behalf of another. A surrogate is an employee or subordinate that an employer commands and controls and for whose actions the employer bears some legal and moral responsibility. A surrogate operates on the behalf of someone else and connotes an employer-employee relationship, wherein the employee or surrogate is a compensated agent of the USG—thus imposing legal and moral responsibilities on the employer. While surrogates are likely to be employed by the USG for unilateral purposes in support of a UW campaign, the fundamental UW partner relationship with a supported resistance is one of USG influence upon cooperatives only—not control of surrogates.

1-14. UW is conducted using a modified whole-of-government approach based on mission requirements and a need to know. Most of the USG's efforts to identify and support any resistance antecedent to a policy decision to conduct UW is likely to be conducted by other USG departments and agencies—notably the Department of State (DOS) and the intelligence community. Collaboration for DOD participation in a given theater would include the appropriate combatant commands, the theater special operations command (TSOC), Service component commands, and specifically identified Service units.

# UNCONVENTIONAL WARFARE AND THE MILITARY INSTRUMENT OF NATIONAL POWER

1-15. UW is a specific mission that Congress directs the DOD to undertake as codified in law at Title 10, United States Code (USC), Section 167, *Unified Combatant Command For Special Operations Forces*. As such, UW exists as a subset within several larger contexts. The USG has many instruments of national power with which to pursue its objectives—characterized as diplomatic, information, military, economic, financial, intelligence, and law enforcement (DIMEFIL). A more comprehensive way to look at the instruments of national power is to list the departments in the Executive Branch of the USG as the Departments of—

- Agriculture.
- Commerce.
- Defense.
- Education.
- Energy.
- Health and Human Services.
- Homeland Security.
- Housing and Urban Development.
- Interior.
- Justice.
- State.
- Transportation.
- Treasury.
- Veterans Affairs.

1-16. In both views, the military instrument, department, or option is just one means of many the USG has of pursuing national objectives. However, when an STR effort includes the use of UW, the DOD, the DOS, and the intelligence community will be the primary actors.

1-17. This understanding is important to put UW in its proper contexts (figures 1-2 and 1-3, page 1-6). Resistance—as an umbrella term—exists in many times, places, and forms throughout the world. Most of those instances of resistance are of no significance to USG priorities and do not merit careful, detailed attention or study. Indeed, many nascent forms of resistance are indistinguishable from typical political turmoil and lawlessness and are practically invisible to observation. In some cases, however, specific examples of resistance throughout the world are a consideration in how the USG pursues its objectives.

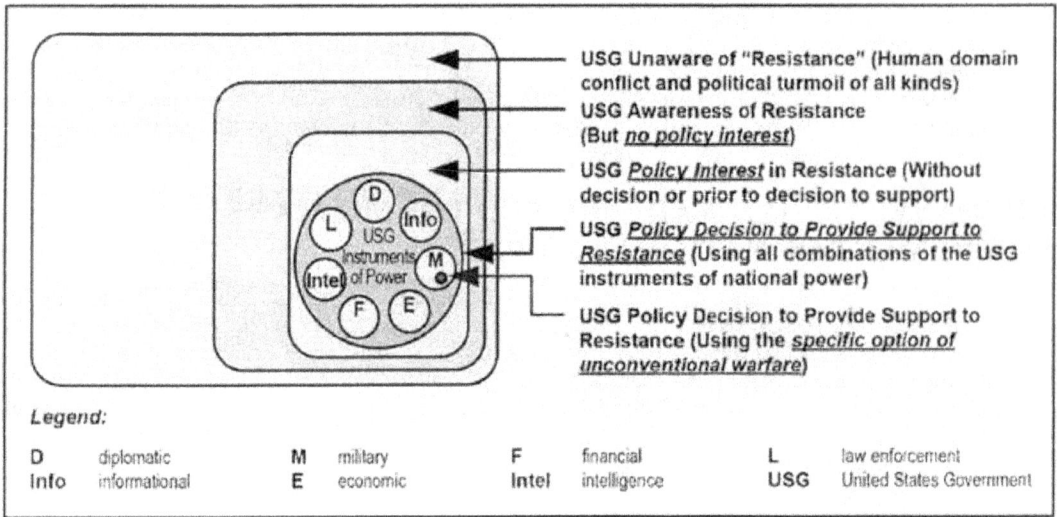

**Legend:**

| | | | | | | | |
|---|---|---|---|---|---|---|---|
| D | diplomatic | M | military | F | financial | L | law enforcement |
| Info | informational | E | economic | Intel | intelligence | USG | United States Government |

Figure 1-2. Resistance contexts

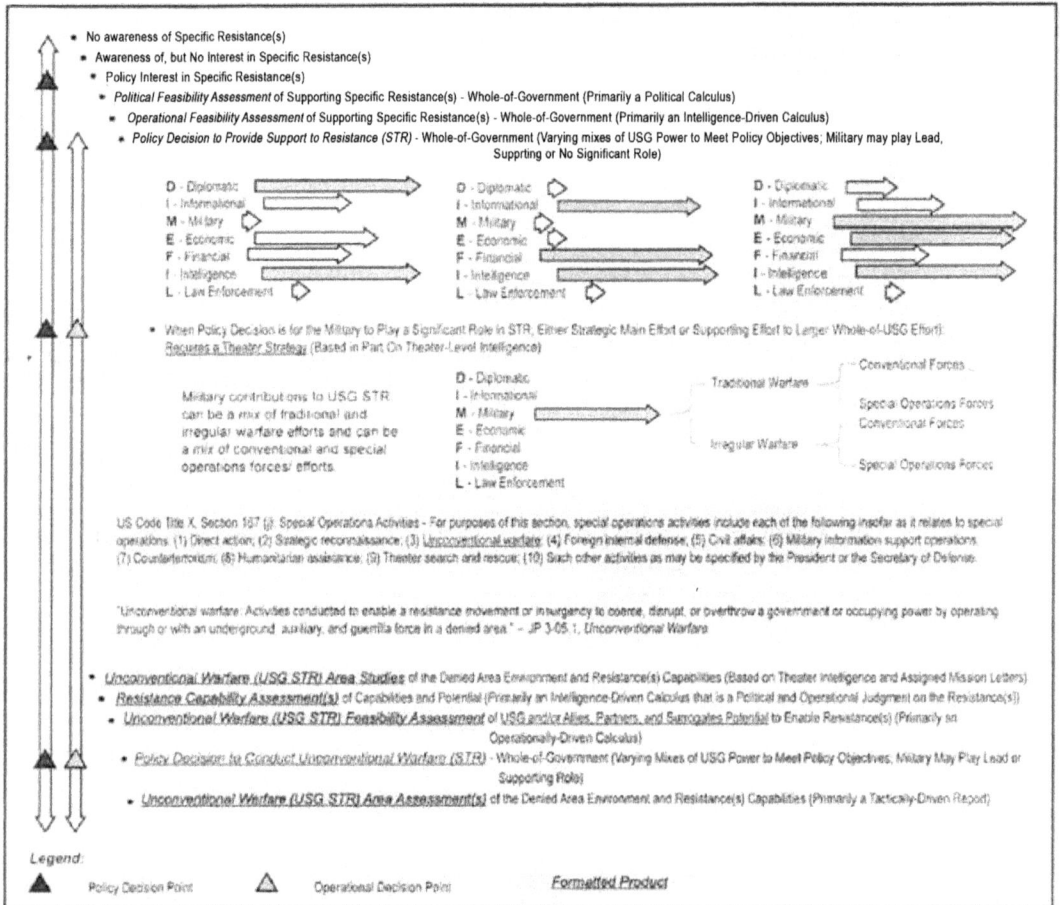

Figure 1-3. Policy interest and decisions, feasibility assessments, and support to resistance

1-18. Sometimes those resistance efforts are seen as actors that can have meaningful effects for good or ill on a given situation. The USG may perceive potential for such resistance movements to shape that situation in ways that may assist the achievement of USG objectives. The serving USG administration may decide to explore the political feasibility of rendering assistance to the resistance movement, and through a combination of routine and special activities, all of the instruments of national power or executive departments may potentially assist informing that political feasibility assessment.

1-19. Political feasibility assessments will include considerations, such as the importance and scale of USG interests, risk versus reward calculations, international law, international and domestic attitudes toward action or inaction by the USG, sensitivity to USG involvement, and many other factors. The exact mix of factors and how they are weighted in any given instance will vary with the circumstances at the time and the world view and values of the administration in power. There is no set format or immutable procedure for determining the political feasibility of any potential STR action. It is important to note that political feasibility assessments inform but critically differ from political or policy decisions to adopt a specific course of action (COA). COAs deemed politically feasible may or may not be approved.

1-20. One important component of assessing what is politically feasible is the subordinate assessment of what is operationally feasible; starting from the national strategic perspective. From the USG's overall perspective, operational feasibility is a function of all instruments of national power at the disposal of the executive. If the executive can support an actor—in this case a resistance effort—by using a less costly, intrusive, or objectionable means, it is logical to do so. For example, if giving overt political support to a resistance effort through normal diplomatic channels suffices to satisfy USG policy objectives, there may be no need to apply more costly, disagreeable, or dangerous methods. If modifications of trade policy or shrewd strategic messaging achieves USG policy objectives, this will almost always be regarded as preferable to more forceful measures. In all of these examples, the practices of normal statecraft seek to amicably achieve mutually desired goals with other international actors first and then proceed, when necessary, to more disruptive or coercive means at the USG's disposal.

# THE VARIETIES OF COERCION, DISRUPTION, AND OVERTHROW

1-21. Based on the objectives of the indigenous resistance movement and USG objectives, UW is conducted to support one of three strategic end states against an adversarial governing authority: coerce, disrupt, or overthrow. Each involves creating effects through a series of activities by the resistance and USG participants, with identified decisive points, mission parameters, and risk profiles.

## COERCE

1-22. Coercion is forcing someone or some entity to do something it would rather not do. In order to coerce a designated government or occupying power, the supported insurgency and resistance must provide adequate pressure to cause a change in actions. The measure of success is the ability to apply enough pressure from the insurgency or resistance with the desired effects and objectives. Transition from UW coerce or disrupt missions may consist of support to other continued clandestine indigenous operations, or the relationship with indigenous components could be terminated, go dormant, or be handed off to other USG departments or agencies or foreign partners. Unlike UW to overthrow, transition planning may include emphasis on preservation of the resistance. Coordination of the UW effort with significant application of other instruments of national power may be necessary to create the desired effects. Although coercion can be relatively small-scale, it can also be large-scale, such as sufficiently large-scale resistance activity in one region of a hostile nation-state that forces the belligerent government to shift large resources to pacification duties in that region. Coercion outcomes may be permanent or temporary. Transition following coercion will differ from an overthrow outcome and most likely require planning for preservation of the resistance as the U.S. support is ended or diminished. In some cases, support may simply be terminated.

1-23. In the example below, Bad Actor Country X is threatening U.S. Allied Countries A, B, C, and D with intimidation and potential cross-border invasion. A USG policy decision is made to relieve the pressure being exerted on U.S. allies by diplomatic efforts in the international arena, perhaps by favorable economic supporting efforts to the allied economies, increasing intelligence and law enforcement support and other measures. Among military options, the United States could choose to increase military support to the allies, deploy conventional forces to Country X's border, conduct shows-of-force off the coast or in the skies

- neighboring Country X, and so on. Or, a policy decision could be made to coerce Country X by forcing it to turn its attention toward other threats and away from the allies by using UW to support resistance efforts inside Country X's eastern regions (figure 1-4 and figure 1-5, page 1-9).

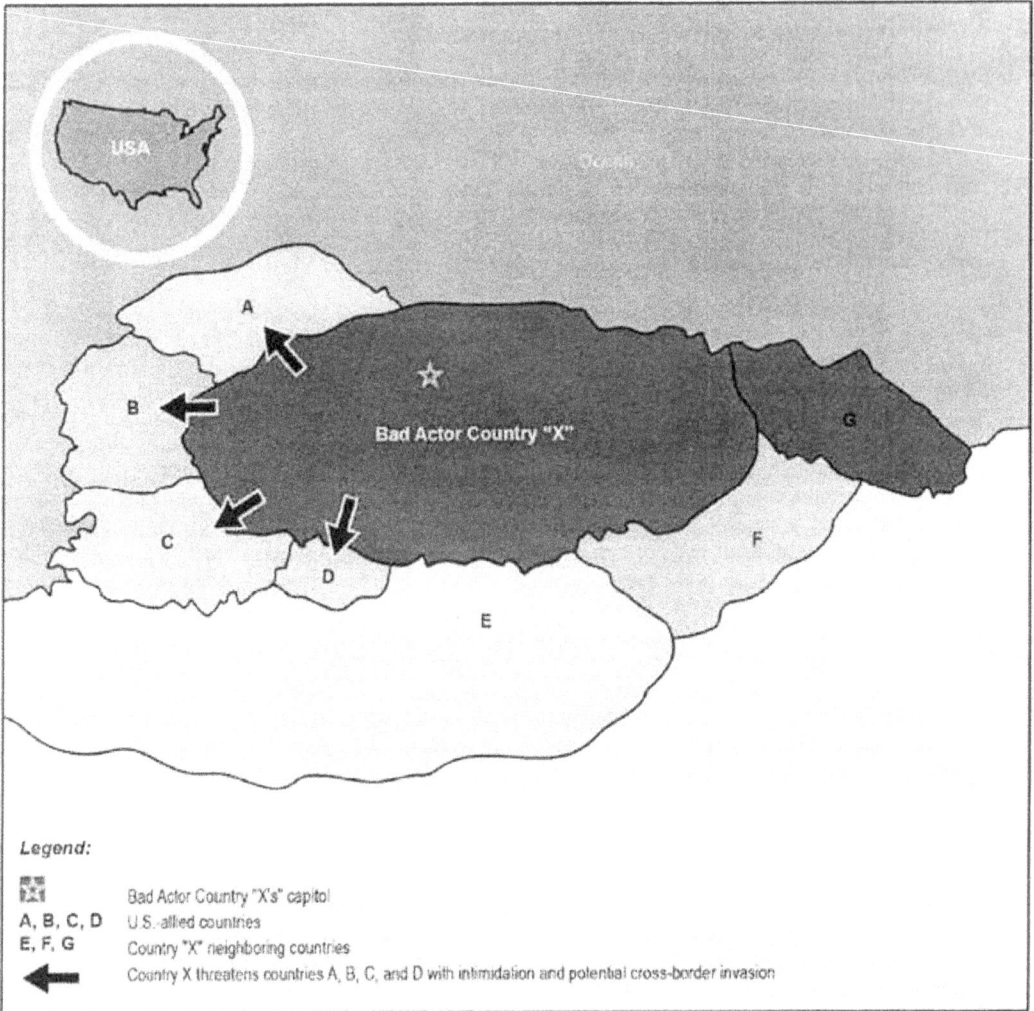

**Figure 1-4. Bad actor country threatens U.S. allies**

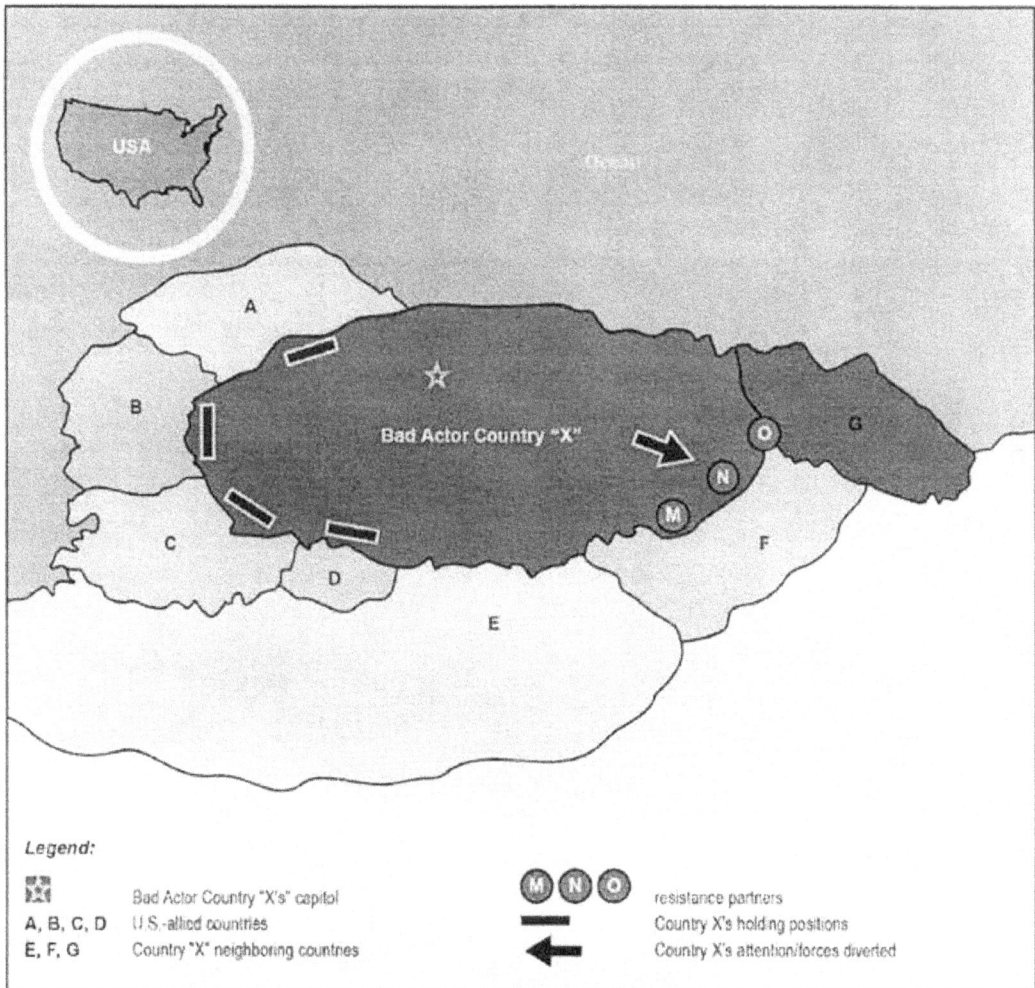

**Figure 1-5. Unconventional warfare with resistance partners coerce bad actor country to divert its attention to new threat**

## DISRUPT

1-24. The obverse of coercion is disruption, which prevents or impedes someone or some entity from doing something it would prefer to do. Although disruption can be a relatively small-scale operation, it can also be large-scale—such as coordinated regional resistance. Disruption is an integral part of traditional warfare by the USG against an enemy. Alternatively, the USG can apply UW to support a resistance against an adversarial government to support U.S. national interests and objectives. Disruption effects may be subtle, may take a long time to discover, and may have delayed, indirect, or cumulative effects. Disruption may become an option when regime change by a resistance movement or insurgency is assessed as challenging, impossible, or unnecessary but the resistance could effectively disrupt some of the adversary's significant lines of effort. Disruptive activities may be broad in nature, from sabotaging production of fuels or war materials, to attacking rear security areas and line of communications (LOC), or interdicting government protected drug trafficking. Transition following disruption will also differ from an overthrow outcome, and it will most likely require planning for preservation of the resistance as the U.S. support is ended or diminished. In some cases, support may simply be terminated. Both coercion and disruption policy option considerations include the sponsor acknowledging the risk that the insurgency or resistance may achieve the capability of overthrowing the government, independent of external support.

1-25. In the example below, Bad Actor Country X has threatened or commenced invasion of U.S. Allied Countries A, B, C, and D. In addition to using all other instruments of U.S. national power to oppose Country X's aggression, a USG policy decision is made to disrupt Country X's operations by using UW to support resistance efforts inside Country X or the territory it occupies in allied countries. Among military options, U.S. supported resistance operations may include: reducing Country X's operational and strategic materials; blocking enemy advances; reducing enemy troop strength; forcing the enemy to guard his rear areas; cutting off strategic and operational resources and communications; fomenting political upheaval in the enemy's capitol and other cities and ports; and so on (figure 1-6 and figure 1-7, page 1-11).

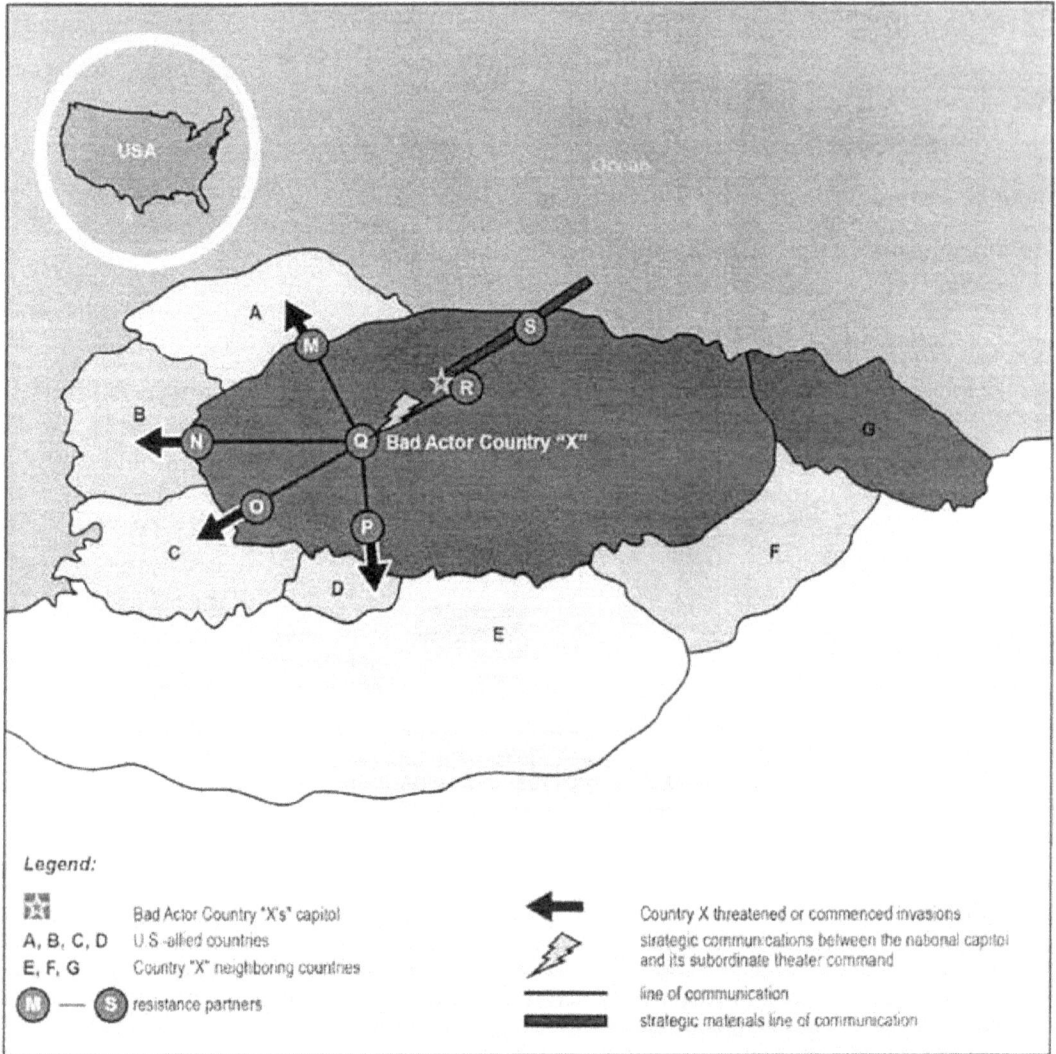

**Figure 1-6. Bad actor country invades friendly or neutral countries and unconventional warfare enables resistance partners**

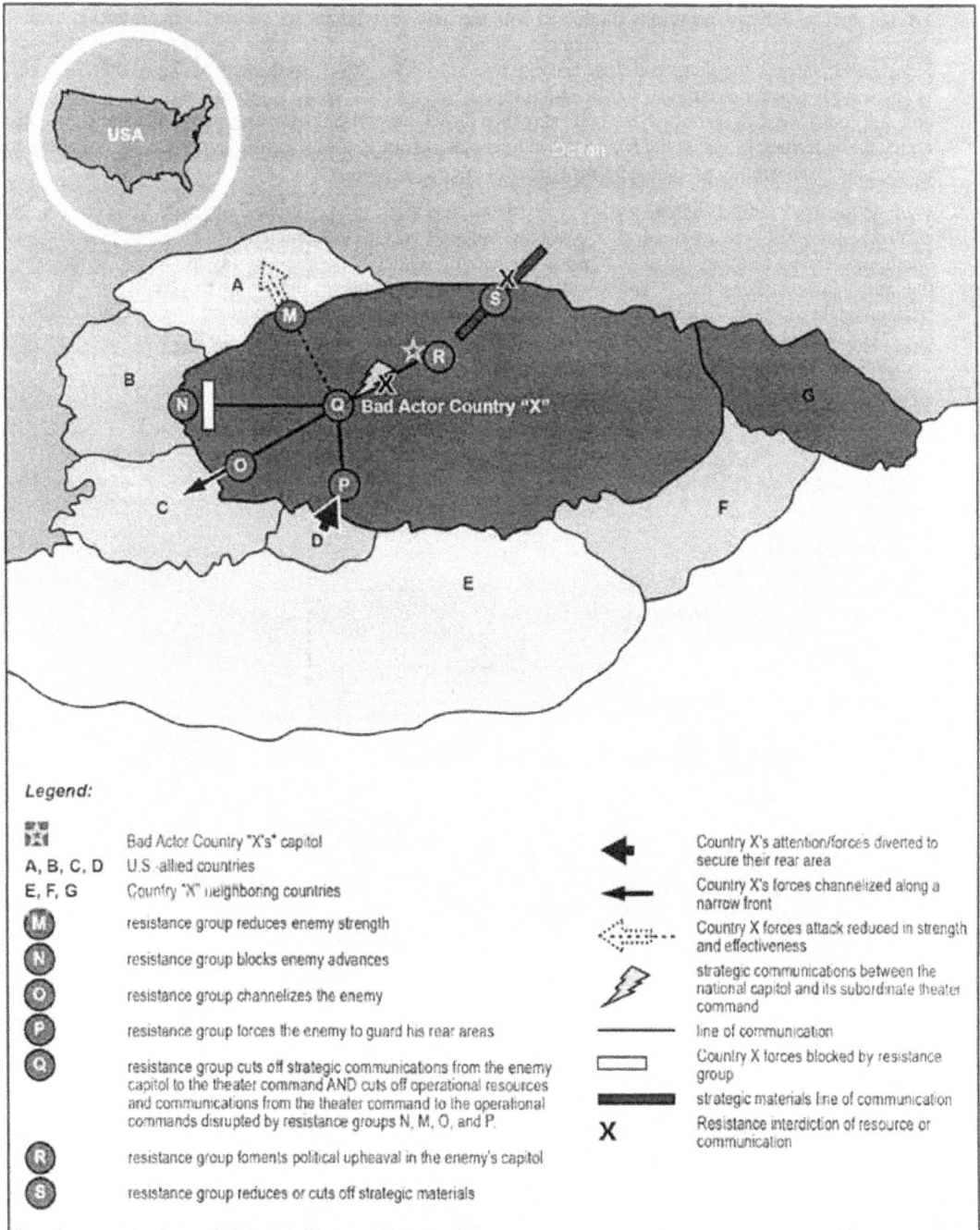

**Legend:**

| | |
|---|---|
| (capitol icon) | Bad Actor Country "X's" capitol |
| A, B, C, D | U.S.-allied countries |
| E, F, G | Country "X" neighboring countries |
| (M) | resistance group reduces enemy strength |
| (N) | resistance group blocks enemy advances |
| (O) | resistance group channelizes the enemy |
| (P) | resistance group forces the enemy to guard his rear areas |
| (Q) | resistance group cuts off strategic communications from the enemy capitol to the theater command AND cuts off operational resources and communications from the theater command to the operational commands disrupted by resistance groups N, M, O, and P. |
| (R) | resistance group foments political upheaval in the enemy's capitol |
| (S) | resistance group reduces or cuts off strategic materials |

| | |
|---|---|
| (solid arrow) | Country X's attention/forces diverted to secure their rear area |
| (wide arrow) | Country X's forces channelized along a narrow front |
| (dashed arrow) | Country X forces attack reduced in strength and effectiveness |
| (lightning bolt) | strategic communications between the national capitol and its subordinate theater command |
| ——— | line of communication |
| (open bar) | Country X forces blocked by resistance group |
| (solid bar) | strategic materials line of communication |
| X | Resistance interdiction of resource or communication |

**Figure 1-7. Resistance partners disrupt bad actor country**

## OVERTHROW

1-26. UW may be conducted to overthrow a governing authority or occupying power—typically with the intent that the supported successful resistance will eventually inherit political control over the contested area Following overthrow, U.S. support may have to rapidly transition to some form of stability operations or foreign internal defense operations conducted with the new government for security and protection against general lawlessness or former regime member attempts to regain power. From the beginning of planning, leaders and planners should consider the second- and third-order effects of desired strategic outcomes. This is particularly true for overthrow, which, if improperly planned for, could result in dangerous instability or counterproductive regimes, which the USG may then have to establish relationships with. Planners must also anticipate outcomes in which the supported indigenous partner is unsuccessful in, unable, or unwilling to assume the responsibility of, taking political control after an overthrow.

1-27. In the two examples below, a USG policy decision is made to overthrow Country X. In figure 1-8, the United States—and perhaps allied sponsor forces—support various indigenous resistance groups to conduct operations leading to the eventual collapse of the native Country X regime's ability to govern. In this case, the indigenous resistance itself—albeit with external support—achieves the decisive outcome; it is the UW strategic main effort. In figure 1-9, page 1-13, the resistance is only one supporting line of effort within a larger U.S. and multinational partner campaign to invade and topple the X regime. In this case, the combined conventional operations are likely to achieve the decisive outcome while UW supports.

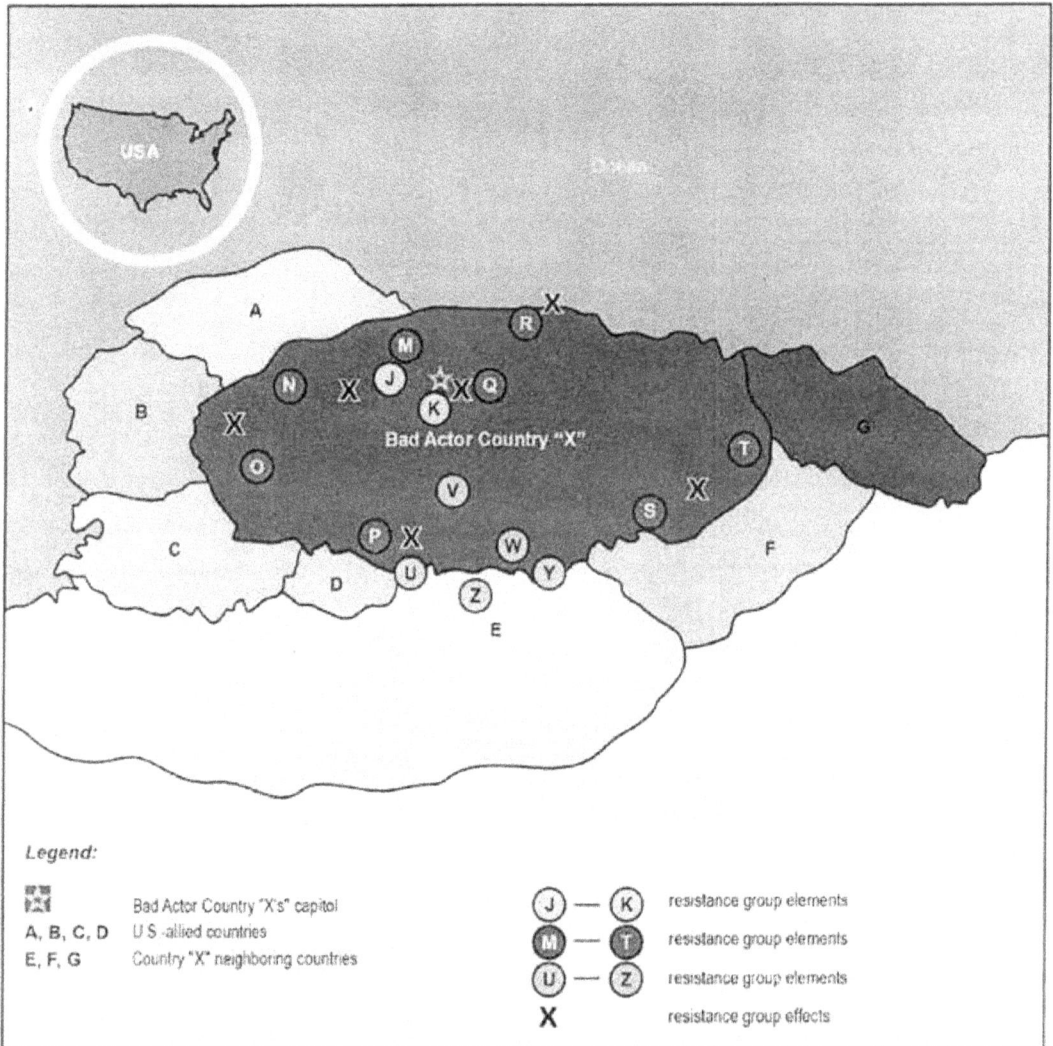

**Figure 1-8. Resistance partners enabled to overthrow by coordinated activities as an unconventional warfare strategic main effort**

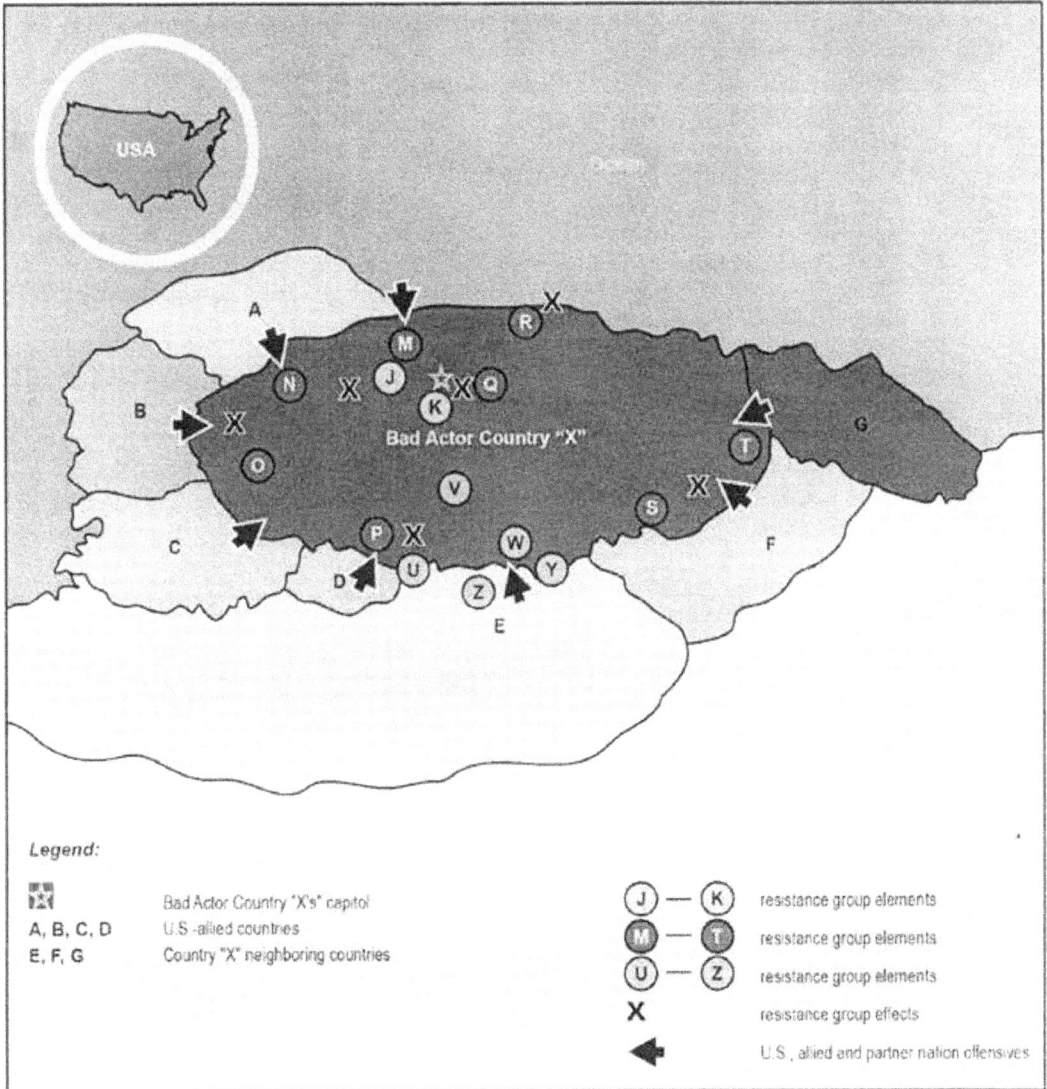

**Figure 1-9. Resistance partners enabled to overthrow as one line of operation in a larger campaign**

# END STATE FIRST—SETTING STRATEGIC, OPERATIONAL, INTERMEDIATE AND INITIAL OBJECTIVES

1-28. SF, at the battalion level and below, should understand how mission development fits into the larger contexts of a JFC's UW campaign. Most SF units executing some aspect of a campaign plan are unlikely to receive a mission to conduct UW, or to coerce, disrupt, or overthrow a government; these will be national or theater objectives.

1-29. Rather, most SF units will be tasked to achieve more narrowly defined and specific objectives; albeit often of operational or strategic significance. For example, it may be a—

- National policy objective to relieve the threat pressure on allied nations bordering Country X.
- Geographic combatant commander's (GCC's) theater objective to coerce Country X to divert its attention from threatening its neighbors on one flank by creating the perception of a priority security threat on Country X's opposite flank.
- TSOC and special operations joint task force commander's operational objective to empower indigenous resistance partners inside Country X.
- Joint special operations task force (JSOTF) commander's tactical objective to infiltrate SF teams into denied territory to link up with and enable specific resistance elements to conduct interdiction operations against Country X lines of communication.

*Note:* ATP 3-05.1 (Change 1, 25 November 2015, Chapter 5, Campaign Planning) contains a detailed, step-by-step outline for developing a UW campaign.

1-30. The special operations task force (SOTF) commander and subordinate commanders will likely have narrowly defined tactical objectives within relatively small joint special operations areas (JSOAs) within a larger joint operations area.

## DETERMINE TERMINATION CRITERIA, MILITARY END STATE, OBJECTIVES, AND INITIAL EFFECTS

1-31. Campaign design and planning translates national or theater strategic objectives into action by integrating end states, objectives, effects, and tasks among all components of the command. The commander and staff determine the appropriate termination criteria for military activity that will enable achievement of the national strategic end state. These criteria describe the conditions in-theater that must be met before conclusion of the campaign or operation, or before transition of the campaign to a supporting campaign, to enable other elements of power to achieve the national end state.

1-32. The commander and staff must then translate those criteria into a concise statement of the military end state, develop a set of objectives that will lead to achievement of the military (or theater) end state, and begin analysis to determine the effects (specific conditions) required to achieve the objectives (figure 1-10, page 1-15). As the strategic aims shift, so too must consideration of termination criteria and operational objectives. The commander may provide significant change to his guidance as a result of his reassessment of the operational environment, the associated problem definition, and the resulting adjusted operational approach. This significant change is reframing and will almost certainly result in adjusted planning guidance.

## DETERMINE DECISIVE POINTS

1-33. A decisive point is a geographic place, specific key event, system, or function that, when acted upon, allows commanders to gain a marked advantage over an adversary or contribute materially to achieving a desired effect, thus greatly influencing the outcome of an action. Developing decisive points orients on the key vulnerabilities or other critical factors identified through center-of-gravity analysis. Ideally, commanders design campaigns that attack adversary vulnerabilities at decisive points so that the results they achieve are disproportionate to the resources applied. Commanders and their staffs must determine and prioritize which vulnerabilities or capabilities, or key events, offer the best opportunity to achieve the effects on the operational environment that are needed to accomplish U.S. objectives. Some potential decisive points might include:

- In-theater ports, airfields, rail lines, or roads needed for support of UW operations are coordinated in neighboring countries.
- Pilot teams have infiltrated the operational area, linked up with, and assessed the resistance potential of indigenous groups.
- Human infrastructure has been established in support of future UW operations.
- Infiltration routes have been established.
- Indigenous resistance groups have sustained success in guerrilla operations against the regime.
- Country X's counterinsurgency operations are significantly degraded.

- International community begins to recognize resistance movement as a legitimate political actor in Country X.

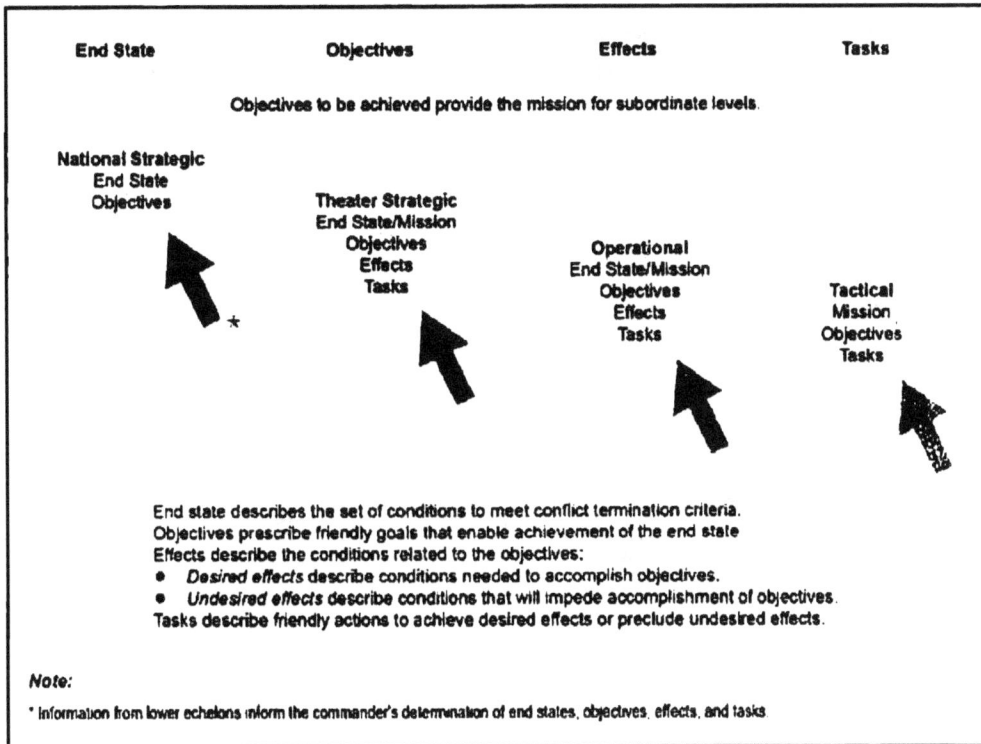

**Figure 1-10. Relationship of end state, mission, objectives, effects, and tasks**

## DEVELOP LINES OF OPERATION OR LINES OF EFFORT

1-34. Commanders may describe the operation along lines of operation or lines of effort, or both. Lines of operation or lines of effort offer a good framework for the commander to describe his visualization of a campaign. The staff refines lines of operations or lines of effort as it develops COAs, and throughout execution of the campaign, to ensure coordinated, synchronized joint action.

1-35. A line of operation is a logical line that connects actions, nodes, or decisive points related in time and purpose with an objective. A line of operation is also a physical line that defines the interior or exterior orientation of the force in relation to the enemy or that connects actions, nodes, or decisive points related in time and space to an objective. A line of effort links multiple tasks and missions using the logic of purpose—cause and effect—to focus efforts toward establishing operational and strategic conditions.

1-36. Planners may use both lines of operations and lines of effort to build their broad concept. Lines of operation portray the more traditional links among objectives, decisive points, and centers of gravity. A line of effort, however, helps planners link multiple tasks with goals, objectives, and end state conditions. Combining lines of operation and lines of effort allows planners to include nonmilitary activities in their broad concept. This combination helps commanders incorporate stability or civil support tasks that, when accomplished, help set end state conditions of the operation.

1-37. Developing lines of effort requires creative analysis and the ability to envision how potentially decisive events throughout the campaign link together. In developing and portraying lines of effort, the staff must—

- Understand and portray the critical initial conditions in the operational environment.
- Understand and portray the desired theater conditions with some sense for timing of those conditions.

- Understand and portray the theater objectives for the campaign (phased over time as needed) to achieve the desired theater conditions.
- Array decisive events (actions, functions, and so on) or locations for adversary and friendly efforts.
- Examine the decisive points and group or connect them into patterns or unifying factors.
- Collect and describe the pattern or unifying factors into lines of effort or unifying themes that run throughout the campaign. Figure 1-11 shows sample lines of effort for UW.

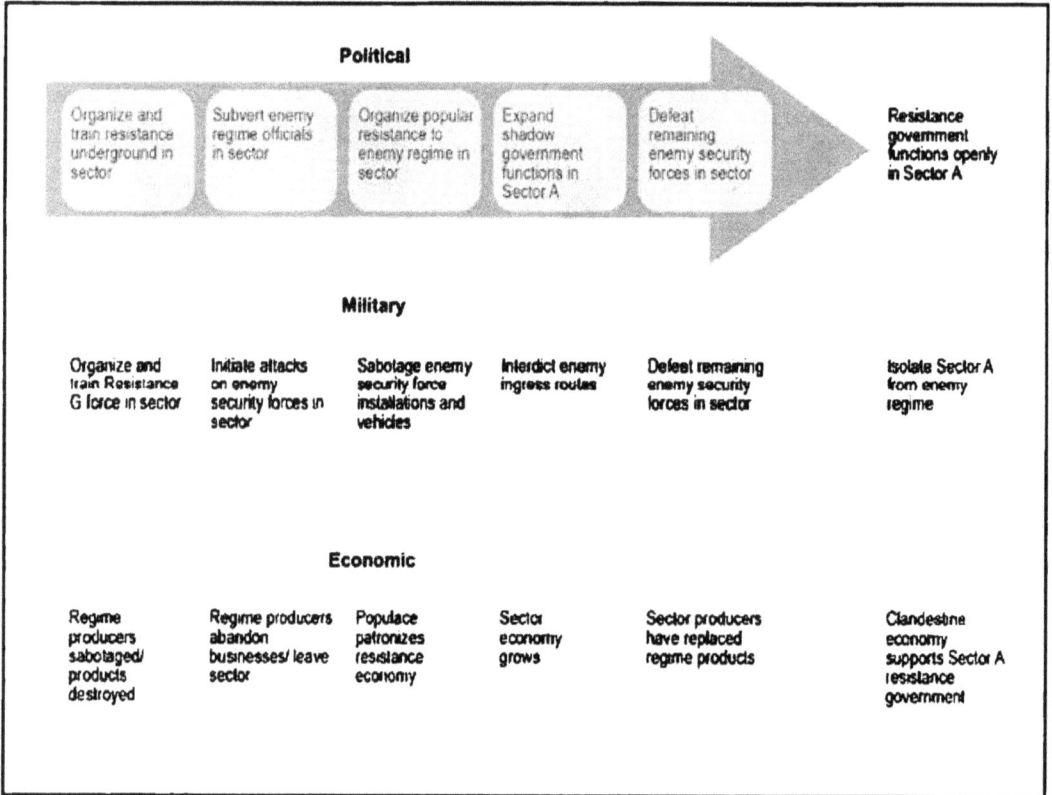

**Political**

| Organize and train resistance underground in sector | Subvert enemy regime officials in sector | Organize popular resistance to enemy regime in sector | Expand shadow government functions in Sector A | Defeat remaining enemy security forces in sector | Resistance government functions openly in Sector A |

**Military**

| Organize and train Resistance G force in sector | Initiate attacks on enemy security forces in sector | Sabotage enemy security force installations and vehicles | Interdict enemy ingress routes | Defeat remaining enemy security forces in sector | Isolate Sector A from enemy regime |

**Economic**

| Regime producers sabotaged/ products destroyed | Regime producers abandon businesses/ leave sector | Populace patronizes resistance economy | Sector economy grows | Sector producers have replaced regime products | Clandestine economy supports Sector A resistance government |

Figure 1-11. Sample lines of effort for unconventional warfare

# Chapter 2

# Fundamentals of Resistance

*Man is born free, and everywhere he is in chains.*

Jean Jacques Rousseau, *The Social Contract*

Resistance is the basic starting point for considering the utility of UW. Much of UW is an art, because every instance of resistance is different.

## WHY AND HOW POPULATIONS RESIST

2-1. Resistance generally begins with the desire of individuals to remove intolerable conditions imposed upon them by an unpopular regime or occupying power. Opposition toward the governing authority and for existing conditions that conflict with the individual's values, interests, aspirations, and way of life spread from the individual to the family, close friends, and neighbors.

2-2. Sometimes, this resistance is narrowly based along specific demographic criteria, such as religious, ethnic, socioeconomic, cultural, or other classifications. Insurgencies motivated and organized along such lines may represent only a small fraction of the populace. When that is the case, it is likely to be more difficult to organize and support a successful insurgency because the resistance is a minority of the subject population

2-3. On the other hand, sometimes an entire community may possess an obsessive hatred for the established authority, making the task of mobilization to victory much easier. This is especially true of classic resistance to occupation by an external power. An invader is likely to offend the interests of most or all demographic groups in a society. Therefore, organizing the mass of the population for resistance to occupation will almost always be easier than organizing relatively narrow factions for insurgency.

2-4. Initially, this hatred may manifest as sporadic, spontaneous nonviolent and violent acts toward authority. As the discontent grows, natural leaders—such as former military personnel, clergymen, college professors, local office holders, and neighborhood representatives—emerge to channel this discontent into organized resistance that promotes its growth. To maintain support for the cause, the population must continue to believe that the potential benefits are worth the risk of failure.

2-5. Key to transitioning from growing discontent to insurrection is the perception by a significant portion of the population that they have nothing to lose by revolting and a belief that they can succeed. In addition, there must be a spark that triggers insurrection, such as a catalyzing event that ignites popular support against the government power and a dynamic insurgent leadership that is able to exploit the situation.

*Note:* See Principles of Resistance segment on page 2-2.

## Principles of Resistance

1. Resistance is efforts by individuals or groups to resist, oppose, or overthrow an oppressor. In the context of special operations doctrine, the oppressor usually connotes an established government or occupying power.
2. Resistance is fundamentally a political activity.
3. Resistance requires a "will to resist," however derived, for at least some members of any resistance.
4. The essence of resistance is subversion of opponent strength in all domains through a wide range of activities.
5. Potential resistance activities inhabit a range from individual's and organized groups' passive, noncooperation, and demonstration nonviolent activities through a wide range and scale of violent activities.
6. Resistance activities typically involve a combination of overt, low-visibility, clandestine, and covert methods.
7. The requirement to conduct certain resistance activities clandestinely is a function of both the opponent's ability to repress and the nature of the resistance activity.
8. Resistance efforts typically must begin with small-scale clandestine organization and sporadic activity to survive.
9. Resistance strategy, methods, organization, narrative, and leadership style are subordinate to resistance objectives.
10. Every instance of resistance will have unique characteristics based on its underlying social, economic, cultural, historical, and political circumstances.
11. Not all persons useful to a resistance are members of that resistance, are witting to its true objectives, or even aware of its existence.
12. Most successful resistance efforts require some level of popular support.

*Principle*: A comprehensive and fundamental rule or an assumption of central importance that guides how an organization or function approaches and thinks about the conduct of operations (ADP 1-01).

## CLANDESTINE RESISTANCE

2-6. People who are part of the resistance but outwardly follow their normal mode of existence conduct clandestine resistance. Leadership cadres and activists conduct a mix of overt and clandestine political evangelizing, recruitment, and community organizing among the citizens' sociopolitical structures in rural villages, towns, and urban cities; within the military, police, and administrative apparatus of government; and among labor groups, educators, students, other intellectuals, the press, and similar groups. This type of resistance is organized and controlled and conducts the following activities as groups or individuals:

- Political actions, propaganda, agitation, subversion, cadre generation, and organization-building.
- Espionage, intelligence-gathering, and infiltration of government and social organizations.
- Sabotage and acts of politically driven targeted violence.
- Traffic in contraband, extortion, and other illicit fundraising activities.

*Note:* The emphasis in this section is on understanding resistance and the activities typical of resistance. Identifying, assessing, and reporting on activities the USG does not conduct or condone (such as espionage or terrorism) does not imply USG support for such activities.

2-7. For many decades, the dominant USG model for characterizing the structure of an insurgency or resistance movement—especially in its earliest phases of development—has been an iceberg.

(See Appendix F, Resistance Development Models.) Most of the structure is said to be below the surface and only the relatively small peak is visible. It is true that in building a resistance structure, insurgent leaders give principal attention to the development of a clandestine supporting infrastructure. However, this classic model asserts that only at the end of a lengthy linear process of clandestine resistance infrastructure building does the entire effort surface, to include overt guerrilla actions.

2-8. While this traditional metaphor for resistance is instructive and insightful, it is only one imperfect model of reality that cannot capture the variations of each specific case of resistance. Resistance development can vary widely based on security conditions and the governing authority's oppressive capabilities; demographics and social peculiarities; the relationship and proportion of rural to urban society and organization; the breadth and depth of indigenous popular support; external diaspora or external sponsor support; and a host of other variables. In actuality, resistance structures are a result of functions required to first survive and then advance a cause in the face of opposition. Every specific case is different. A resistance continuum exists where the extent of clandestine, discreet and overt; lethal and nonlethal; violent and peaceful; spontaneous and organized; and so on will appear and disappear. No one model fits all cases.

## OVERT RESISTANCE

2-9. Anyone who does not hide their opposition to the government is displaying open, overt resistance. As is the case with clandestine activities, no single model is capable of capturing the reality of every specific case. In fact, it is possible that the same actors can operate both openly and clandestinely. The ability to conduct open resistance is partly a function of the extent to which state oppression is constrained by external forces, explicitly permits, passively tolerates, is unaware of, or is simply unable to suppress open resistance. Some situations are so oppressive and the state security apparatus is so effective and pervasive that open resistance is rare, guarded, and brief. It is difficult, however, for a state's oppressive power to be so omniscient and omnipotent, and overt resistance will be possible at many different points on the resistance continuum. Where practicable, overt resistance can be preferable to clandestine activities as the open phenomenon of resistance may further weaken state legitimacy and authority. Overt resistance examples include: nonconformity; withholding of support; disobedience of basic laws; sabotage via work slowdowns, strikes and demonstrations; student, labor, and similar uprisings; peasant or laborer rebellions; political rallies, sabotage via physical damage to state property and infrastructure, open violence, and many more. (See Appendix E, Varieties of Violent and Nonviolent Methods.)

2-10. Overt resistance includes diverse acts of violence. The small-scale end of a violence continuum would include activities, such as projecting credible threats of violence, physical intimidation of individuals, small-scale brutalities and vandalism, single acts of sabotage, kidnappings, targeted political murders, and so on. This violence continuum expands to include ever-broader, more ambitious and more destructive activities, such as widespread acts of brutality and murder, sabotage campaigns, larger-scale violent strikes and riots, and a planned campaign of terror. In cases where the state power has significant surveillance and coercive power, the resistance actor will typically surface the role of violent actor temporarily to conduct these violent activities and then re-submerge the resistance role into a covering public role. These violent activities are certainly politically motivated acts of violence and will be characterized by the standing authority as lawlessness, criminal activity, and possible insurgent activity. Sometimes, it would be also accurate to characterize this level as small-scale guerrilla activities.

2-11. In some cases, the lack of omnipresent state control permits temporary violent actors to form in larger groups on a more frequent basis to tackle longer, harder, bigger, and more distant missions. The continuum continues to the point in which these larger groups are able to survive on a semipermanent basis—eventually requiring greater organization, regimentation and discipline aided by paramilitary structure, formal ranks and chains-of-command, uniform and insignia standardizations, and the adoption of larger-unit maneuvers and tactics. Such formations become the classic model of guerrilla warfare units. In relatively rare historical cases, such as the Chinese Resistance to Japanese occupation in World War II and the subsequent Chinese Civil War, the guerrilla forces grow into such large-scale military formations that they essentially transform into overt standing armies capable of committing to direct battle with opposing conventional armies.

*Note:* Some mistakenly believe that the guerrilla force conducts the overt aspects of resistance and the underground and auxiliary necessarily conduct everything else. This is incorrect. These component labels—which will be discussed more below—are merely categories of a model that try to represent general—not exclusive—functions. In fact, all three of these functions are blended in resistance actors and they cannot be distinctly separated. Moreover, the newly articulated public component—while primarily regarded as an overt component—is also likely to engage in a blend of overt and clandestine functions.

# DYNAMICS OF SUCCESSFUL RESISTANCE

2-12. There are several ways to understand the basic dynamics of insurgency. Army UW doctrine continuously, since at least 1998 (FM 31-20-2), has used the seven dynamics below. ATP 3-18.1 retains this traditional view to provide a simple, conceptual baseline to unify the initial thinking of those who must understand resistance. Such lists are a model imposed on reality and no single list will be accurate in every case. These dynamics are, therefore, only a starting point, and additional resistance analysis theories and analytical outlines are available in Appendix F, Resistance Development Models. However, some combination of the following dynamics are likely to transform popular discontent into an organized and effective movement.

## LEADERSHIP

2-13. A group committing random political violence is not a resistance. In a resistance, the group is committing directed and focused political violence. It requires strategic, operational, and tactical leadership to provide vision, direction, guidance, coordination, and organizational coherence. The resistance leaders must make their cause known to the people and gain popular support. Their key tasks are to break the ties between the people and the government and to establish credibility for their movement. The leaders must replace the government's legitimacy with that of their own. Their ability to serve as a catalyst that motivates and inspires others to have faith in their cause is vital to the movement's growth. Their ability to organize and willingness to distribute power across the organization is vital to the long-term success of the movement. Organizations dependent upon key charismatic personalities to provide cohesion and motivation for the movement are vulnerable to disruptions if the enemy removes or co-opts those players.

## IDEOLOGY

2-14. The resistance must have a program that justifies its actions in relation to the movement's grievances and explains what is wrong with the status quo. The most important aspect of a successful resistance is the viability of the message. It is essential that the message resonates with the people and possesses meaning to their way of life. This makes the language, culture, and geography of the masses particularly important. Ideology is an important factor in unifying the many divergent interests and goals among the resistance membership. As a common set of interrelated beliefs, values, and norms, ideology is used to manipulate and influence the behavior of individuals within the group. Ideology will serve as the rallying call for all members of the population to join the struggle. Major components of an ideology are commonly found in language, socioeconomics, ethnicity, and religion. The ideology of the resistance and the motivation of the resistor must remain linked. Once delinked, the counterinsurgent will be able to address individual grievances and negate the unity of the resistance.

## OBJECTIVES

2-15. The resistance needs to carefully choose what efforts to undertake and how they should be sequenced. Although resistance is a strategy, implementation requires intermediate objectives—specifically, strategic, operational, and tactical goals.

2-16. At the tactical level, goals are most closely correlated with direct actions on the ground. These actions, in artfully designed combination, lead to operational goals. For example, raids, ambushes, and supporting propaganda urging enemy forces to serve their tour quietly and go home alive, to consolidate into large bases away from the cities, or to stay off the roads for their own safety, achieve the goals of disrupting enemy control over territory and weakening the enemy's commitment to counterinsurgent strategy.

2-17. Operational objectives address how the resistance will progress toward its strategic goal. Examples may include:

- Attaining a level of popular and passive support in a key region.
- Preventing government occupation or use of—or transit through—a liberated area.
- Gaining international recognition or external support.

2-18. The strategic objective is the desired end state. In general, the strategic objective is to gain concessions or remove the regime in power. Typically, the strategic objective is critical to cohesion among resistance groups. It might be the movement's only clearly defined goal. Some examples of strategic goals are as follows:

- Anarchist—destructively eliminate government entirely.
- Egalitarian—impose universal equality.
- Traditionalist—resist change and return to a perceived norm.
- Pluralist—break a monopoly on political discourse.
- Secessionist—break off some section from the polity.
- Reformist—modify the application of laws and mores.
- Preservationist—safeguard valued institutions from change.
- Globalist—provoke supranational reorganization.
- Apocalyptic—act as catalyst for an envisioned end times.
- Utopian—impose a theoretical vision of man's perfection.
- Commercialist—facilitate greed through violent illegalities.

## ENVIRONMENT AND GEOGRAPHY

2-19. The environment and geography (including demographics) greatly affect the strategies and tactics of a resistance. A resistance may form its base in urban environments, rural environments, or a combination of both. By maintaining a combination of urban cells and rural bases, a resistance can often take full advantage of the benefits of both models (urban and rural) without becoming constrained by the shortcomings of either model.

2-20. A resistance located in rural areas enjoys the relative safety of remote terrain or safe havens, such as jungles or mountains. These geographical conditions make it possible for them to form larger guerrilla bands and conduct large-scale guerrilla operations. Disadvantages of a rural base may include:

- Length and speed of communications and supply lines.
- Displacement of the resistance from the populace.
- Susceptibility of resistance to conventional military counterguerrilla operations.

2-21. Urban resistance can overcome the lack of suitable restrictive terrain by operating within ethnic ghettos or enclaves within sympathetic, densely populated urban areas. These areas often create safe havens that government forces are unwilling or unable to access. This type of urban basing requires a high degree of compartmentalization, which makes it more difficult for the group to train and organize for large-scale operations. Disadvantages of an urban base may include:

- Greater susceptibility of infiltration and observation.
- Vulnerability of resistance to police actions.

## EXTERNAL SUPPORT

2-22. One important factor that could increase a resistance's chance of success is the provision of external support, and there are many examples of such external support throughout world history. Governments providing support to a resistance may or may not share beneficial interests, common ideology, or values with the resistance. However, a mutually agreed-upon set of objectives needs to be established between the internal resistance and external sponsoring partners—at least temporarily. External support can breed a degree of resistance dependency on the foreign power. However, understanding that UW involves a partner relationship is a critical requirement. Partners cooperate in a manner that recognizes that the sponsor influences rather than controls the supported partner. This differs from the relationship of an employer and his surrogate.

2-23. There are many paths to, and varieties of, external sponsorship. (See Chapter 3, Support to Resistance.) STR as U.S. policy can occur without the use of UW or even military support in the form of diplomatic, financial, informational or other means. Non-state external support may be derived from ethnic enclaves or diasporas in third-party countries in terms of political voice, money, personnel, and sanctuary. When a government-in-exile or resistance public component entities exist outside of the target area, they can be instrumental in coordinating external support for their cause. In addition, some cases of resistance have exploited or made business arrangements with narcotics traffickers, kidnappers-for-ransom, smuggling, or other external organized crime groups to raise funds and support their resistance operations. While this tactic can prove extremely effective for generating revenue, it likely imposes a counterproductive, international political cost in perceived resistance legitimacy. Examples of external support may include:

- Moral or political support in the international forum.
- Resources, such as money, weapons, food, advisors, and training.
- Sanctuary, such as secure training sites, operational bases over a border, or protection from extradition.

## PHASING AND TIMING

2-24. Army UW doctrine, since at least 1990 (FM 31-20 and FM 100-20), has used the mid-20th century Chinese, mass-population-based, three-phased model of resistance development phasing below. ATP 3-18.1 retains this traditional view to provide a simple conceptual baseline to unify the initial thinking of those who must understand resistance. This baseline example is a model imposed on reality and no single model will be accurate in every case. This model is, therefore, only a starting point, and additional theoretical and analytical resistance development models are available in Appendix F, Resistance Development Models.

2-25. However, this model has existed in Army UW doctrine for good reason. The Chinese communist insurgency survived Nationalist state oppression and concerted attempts to exterminate it in the 1920s and 1930s. The Chinese communist resistance then provided significant Chinese national opposition to Japanese Imperial World War II occupation of China in the 1930s and 1940s. After the war, the Chinese communist insurgency expanded into a peer belligerent of the Nationalist Chinese state government in the Chinese Civil War, eventually achieving the complete overthrow of the state regime and accession to complete Chinese Communist Party power in 1949. The length, breadth, depth, and discipline of the political and organizational efforts to mobilize, organize, and employ millions of people in a political-military resistance effort on this geographic and violence scale is without equal.

2-26. Successful resistance efforts pass through phases of development. Several ideas exist of how these phases should be characterized. However, the three-phase construct presented below is the Army UW doctrinal baseline because of its simplicity. From the perspective of the insurgent or resistor, Mao Zedong (Tse-Tung) referred to the three phases as: Strategic Defensive, Strategic Stalemate, and Strategic Offensive. This is the most basic articulation at the conceptual, policy, and strategic levels of how resistance develops.

2-27. Admittedly, not all insurgencies and resistance movements are mass-based protracted conflicts. (See Appendix D, Varieties of Resistance.) Not all insurgencies and resistance movements seek unconditional and total victory over mass invading armies or re-conquest of lost territories, nor do all need to experience every phase as characterized in the passage above. Mao refers to the jigsaw pattern of resistance warfare. Progression through all phases is not a requirement for success. The same resistance movement may be in different phases in separate regions of a country. Successful resistance can revert to an earlier phase when under pressure, resuming development when favorable conditions return. Activities within phases also vary by level of warfare. Movements in strategic defense, for example, will nevertheless conduct tactical, operational, and political offensive actions, and so on.

---

### The Three Stages of the Protracted War

"...[I]t can reasonably be assumed that this protracted war will pass through three stages. The first stage covers the period of the enemy's strategic offensive and our strategic defensive. The second stage will be the period of the enemy's strategic consolidation and our preparation for the counter-offensive [described below as strategic stalemate]. The third stage will be the period of our strategic counter-offensive and the enemy's strategic retreat. It is impossible to predict the concrete situation in the three stages, but certain main trends in the war may be pointed out in the light of present conditions. The objective course of events will be exceedingly rich and varied, with many twists and turns, and nobody can cast a horoscope for the Sino-Japanese war, nevertheless it is necessary for the strategic direction of the war to make a rough sketch of its trends. Although our sketch may not be in full accord with the subsequent facts and will be amended by them, it is still necessary to make it in order to give firm and purposeful strategic direction to the protracted war."

Mao Tse-Tung, *On Protracted War*, May 1938

---

*Note:* The writings of Mao were popularized in the West in part by General Samuel B. Griffith's translation of Mao's 1937, *On Guerrilla Warfare*, in 1961. In Griffith's introduction, he characterized Mao's protracted war context as three phases consisting of organization, consolidation, and preservation; progressive expansion; and decision or destruction of the enemy. Subsequently, the terms latent and incipient, guerrilla warfare, and war of movement were used in SF training at Fort Bragg throughout the 1960s, 1970s, and 1980s and found their way into official Army doctrine at least by 1990 as euphemisms for the aforementioned strategic defensive, strategic stalemate, and strategic offensive. In 2013, ATP 3-05.1, *Unconventional Warfare*, parenthetically resurrected the original terms used by Mao in parallel with the familiar doctrinal terms. This volume reverses parenthetical order of the terms to get even closer to the original meaning. Not every resistance needs to go through a so-called guerrilla warfare or war of movement phase—especially if the resistance's objectives are more limited than overthrowing the state. However, every resistance can be usefully analyzed politically and at all levels of warfare by using Mao's original three terms of strategic defense, stalemate, and offense.

## Phase I—Strategic Defensive (Latent or Incipient)

"Mao conceived this type of war as passing through a series of merging phases, the first of which is devoted to organization, consolidation, and preservation of regional base areas situated in isolated and difficult terrain. Here volunteers are trained and indoctrinated, and from here, agitators and propagandists set forth, individually or in groups of two or three, to 'persuade' and 'convince' the inhabitants of the surrounding countryside and to enlist their support. In effect, there is thus woven about each base a protective belt of sympathizers willing to supply food, recruits, and information. The pattern of the process is conspiratorial, clandestine, methodical, and progressive. Military operations will be sporadic."

Samuel B. Griffith, *Strategy, Tactics, and Logistics in Revolutionary War*,
Introduction to *On Guerrilla War* (pp. 20–21)

2-28. During this phase, the leadership of the resistance develops the clandestine supporting infrastructure upon which all future efforts will rely while attempting to avoid state scrutiny, repression, or extermination. This is almost certain to be the most protracted phase and is by far the most important. The resistance organization uses a variety of subversive techniques to prepare the population psychologically to resist. Some techniques include propaganda, demonstrations, boycotts, and sabotage. Subversive activities include lending aid, comfort, and moral support to individuals, groups, or organizations that advocate the overthrow of incumbent governments by force and violence. All willful acts that are intended to be detrimental to the best interests of the government and that do not fall into the categories of treason, sabotage, or espionage will be placed in the category of subversive activity.

2-29. Subversive activities frequently occur in an organized pattern without any major outbreak of armed violence. Activities include the following:

- Recruit, organize, and train cadre.
- Infiltrate key government organizations and civilian groups.
- Establish cellular intelligence, operational, and support networks.
- Organize or develop cooperative relationships with legitimate political action groups, youth groups, trade unions, and other front organizations. (This approach develops popular support for later political and military activities.)
- Solicit and obtain funds.
- Develop sources for external support.

2-30. The goal is to prepare or transition the population into accepting insurgent direction and future increased scope and tempo of overt actions—sometimes up to including guerrilla warfare and larger operations when appropriate and possible—and to gain the support of the local population and weaken the power of the existing government. Operationally, the resistance seeks to gain large-scale popular support by multiple tactical efforts—physically, politically, and organizationally—to convince local population organizations and neighborhoods and to expand resistance operations without the risk of compromise by the local population. At various points in resistance development, and in many different locations, if the resistance is to expand, it will become impossible for the resistance to conduct the operations it desires without the population being aware of it. The basic, classic resistance phasing model forecasts that success with all of these activities eventually will allow the resistance to expand into Phase II (Strategic Stalemate).

## Phase II—Strategic Stalemate (Guerrilla Warfare)

"In the next phase, direct action assumes an ever-increasing importance. Acts of sabotage and terrorism multiply; collaborationists and 'reactionary elements' are liquidated. Attacks are made on vulnerable military and police outposts; weak columns are ambushed. The primary purpose of these operations is to procure arms, ammunition, and other essential material, particularly medical supplies and radios. As the growing guerilla force becomes better equipped and its capabilities improve, political agents proceed with indoctrination of the inhabitants of peripheral districts soon to be absorbed into the expanding 'liberated' area.

One of the primary objectives during the first phases is to persuade as many people as possible to commit themselves to the movement, so that it gradually acquires the quality of 'mass.' Local 'home guards' or militia are formed. The militia is not primarily designed to be a mobile fighting force; it is a 'back-up' for the better-trained and better-equipped guerrillas. The home guards form an indoctrinated and partially trained reserve. They function as vigilantes. They collect information, force merchants to make 'voluntary' contributions, kidnap particularly obnoxious local landlords, and liquidate informers and collaborators. Their function is to protect the revolution."

Samuel B. Griffith, *Strategy, Tactics, and Logistics in Revolutionary War,*
*Introduction to On Guerrilla War* (p. 21)

2-31. The goal of this phase is to degrade the government's security apparatus (the military and police elements of national power) to the point in which the government is susceptible to defeat. Strategic stalemate is a generic term which connotes the resistance has grown beyond the requirements of mere survival and has developed sufficient proven, multifaceted capabilities to attack the state's power at varying times in many different ways and places. The state may continue to dominate many separate locales tactically and operationally, but it does not control everything everywhere and all of the time. In some locales, the instruments and infrastructure of state power are continually at risk from resistance attack. In some locales the resistance may dominate. A campaign of guerrilla attacks and sabotage degrade the government's military and police forces. Subversive activities continue to build and maintain support from the population Pro-resistance radio broadcasts, newspapers, and pamphlets openly challenge the control and legitimacy of established authority. Depending on perceived success and legitimacy, the resistance fighters or insurgents may achieve legal belligerent status from the international community if they meet the internationally accepted criteria.

2-32. When expanding armed actions—up to and including guerrilla warfare into larger and more capable units—is part of the resistance strategy, Phase II will see increased resistance need to gather such forces, communicate and coordinate operations, conduct training, receive logistics, rest and hide after operations, and plan future operations. Their need for intelligence collection and security also increases in Phase II. As subversive armed action elements multiply and guerrilla forces grow in size, number, and capability, so must their clandestine support mechanisms. Growth of small armed action elements may be able to remain concealed from state oppression as the popular support for the resistance continues to grow. However, standing, sizeable resistance guerrilla forces are not yet strong enough to survive government destruction during concerted counterguerrilla operations. The basic, classic resistance phasing model forecasts that while all other resistance activities continue, the resistance will attempt to achieve a rough parity between resistance armed elements' and standing guerrilla forces' destructive power versus security force combat power. This period of stalemate will have ebbs-and-flows—some favoring the state and some favoring the resistance, without either side seemingly able to secure a decisive victory. This phase may last indefinitely. If the resistance objectives are more limited than complete overthrow of the state or removal of an occupier, the conditions of stalemate may be adequate for both sides to reach a negotiated settlement. However, if the resistance objectives are total victory it may continue the struggle to the conditions of Phase III (Strategic Offensive).

## Phase III—Strategic Offensive (War of Movement)

"Following Phase I (organization, consolidation, and preservation) and Phase II (progressive expansion) comes Phase III: decision, or destruction of the enemy. It is during this period that a significant percentage of the active guerrilla force completes its transformation into an orthodox establishment capable of engaging the enemy in conventional battle. This phase may be protracted by 'negotiations.' Such negotiations are not originated by revolutionists for the purpose of arriving at amicable arrangements with the opposition. Revolutions rarely compromise; compromises are made only to further the strategic design. Negotiation, then, is undertaken for the dual purpose of gaining time to buttress a position (military, political, social, economic) and to wear down, frustrate, and harass the opponent. Few, if any, essential concessions are to be expected from the revolutionary side, whose aim is only to create conditions that will preserve the unity of the strategic line and guarantee the development of a 'victorious solution'."

Samuel B. Griffith, *Strategy, Tactics, and Logistics in Revolutionary War*, *Introduction to On Guerrilla War* (pp. 21–22)

2-33. The goal of the resistance in Phase III is to bring about the collapse of the established government or withdrawal of an occupier after weakening it through the cumulative efforts of military and subversive actions. The resistance does not necessarily need to transform into a conventional military, but it must position itself to defeat the government or occupying power. For example, the resistance might degrade the enemy's capabilities to a point that an urban uprising against the presidential palace would topple the government. Success of this tactic assumes that the resistance has subverted the ability of the state security forces to intervene. This could be achieved through a carefully coordinated campaign of political, organizational, and subversive activities in conjunction with relatively limited armed actions, rather than a decisive showdown by massive units on the battlefield. Every resistance environment and resistance organization situation is unique.

2-34. When the resistance objective is to overthrow the state or oust an occupier, someone will become responsible for the population, resources, and territory previously managed by the removed regime. In some cases, the resistance intends to take power. In other cases, the resistance may be simply the vehicle by which the road to power is cleared for others, such as a political ally, civil leadership emerging from a shadow government, or perhaps a legitimate indigenous government-in-exile. If the resistance fails to plan and execute posthostility activities, the population may lose confidence in the resistance and turn to the old government, a breakaway faction, or a splinter group of the resistance.

2-35. Failure to achieve the following objectives may cause the resistance movement to lose the opportunity it gained by initial victory and revert to an earlier phase. Based on the conditions set earlier, an effective resistance intending to seize or enable others to seize power must plan to—

- Establish an effective civil administration.
- Establish an effective military organization.
- Provide balanced social and economic development.
- Mobilize the population to support the resistance organization and the postresistance governance elements.
- Protect the population from hostile actions by former regime elements and general lawlessness.

### ORGANIZATION AND OPERATIONAL PATTERNS

2-36. The organizational and operational pattern of a given movement is similar to its order of battle. From its outset, the organization has a concept of its development based on its goals. Although there are numerous traditional models for resistance, the planner must avoid following a famous model without considering the way that model worked for its historical environment and must determine if the model is appropriate for the current problem set. (See Appendix D, Varieties of Resistance.) The structure of the organization will depend on many variables, to include the other resistance dynamics, available resources, state security threat, population distribution, and many other factors. All cases of resistance are unique and rarely follow one model exclusively. It is unlikely the structure would resemble a uniform organization, such as the military, in which all units look relatively the same. Function takes precedence over form. Factors that an analyst, or one who would develop or advise development of resistance organization, would consider include:

- Various subordinate components and their orientation.
- Commands—down to the lowest tactical level.
- Supporting infrastructure.

2-37. The organization's most important component level is the local level, in which it obtains and sustains support and manifests actions. Echelons above the local level coordinate all functions (political, military, external support, and so on). Overall command provides purpose and direction.

## SECTION 2. UNITED STATES RESISTANCE THEORY—THE BASIC MODEL OF UNDERSTANDING RESISTANCE

2-38. Reality and models are two different things. Models are imposed upon reality in order to assist in making some sense out of what is seen or expected to be seen.

2-39. Moreover, the term *component* suggests that each of what follows are separate physical entities when, in reality, they refer to degrees of openness or secrecy. The same resistor could conduct underground, auxiliary, armed (or guerrilla) force, or public component activities at different times. This is the difference between "a resistor in the underground" and "a resistor when underground." Each component has similar caveats. The reader should remember that the component models below are clear oversimplifications of a blurred reality.

# THE BASIC MODEL

2-40. The essential paramilitary and clandestine functions of resistance warfare have been explicit or implicit in Army doctrinal writings continuously since the Office of Strategic Services (OSS) manuals of World War II. In 1943, OSS *Provisional Basic Field Manual*, explicitly used the terms underground and guerrilla warfare as essential components of how the Army understood resistance. In 1958, FM 31-21, *Guerrilla Warfare and Special Forces Operations*, added the term auxiliary with specific reference to resistance auxiliary civilian organizations and auxiliary units. Since the 1961 version of FM 31-21, Army UW doctrine has continuously identified these three essential elements as the primary resistance organization components. In 2015, JP 3-05.1 added a fourth primary component of resistance—the public component— to the U.S. basic model of resistance and the UW doctrinal canon. While the functions and effects distilled in the term have always been implicit in resistance theory, ATP 3-18.1 introduces the public component term into Army UW doctrine for the first time.

*Note:* These concepts, although ancient in origin and of longstanding in Army doctrine, are merely a model. The primary purpose of the model is to educate U.S. Soldiers and others in the basic concepts of resistance and to serve the doctrinal role of establishing a foundational and unifying professional vocabulary. As such, it provides a method to analyze and organize the reality of each specific situation. It is, however, only a model. It is not reality. Each situation and each resistance is unique. The resistance may not use these labels to characterize themselves, and reality itself, resists being binned in confining language. Superimposed categories will blur.

It is important to understand and focus on the essential **function** of each component as an action—**when operating underground, when providing auxiliary support**, and **when conducting armed or public resistance activities**—rather than the physical membership of any individual resistor. Resistance members may move back and forth between these components, and any resistance member may conduct any of these functions. Remember, too, that there may be more than one resistance in a UW operational area.

2-41. The goals, objectives, and success of the resistance will determine the level of development and relationships among the components. The underground and guerrillas are politico-military entities that may conduct political acts and violent acts, which represent the ends of a spectrum between close state scrutiny and means of repression on one end and relative freedom from such scrutiny and suppression on the other end. This has a longstanding parallel spectrum in doctrine between the percentage of activity required to be done clandestinely and that which can be done more overtly. All resistance actors are likely to act in a mixture of clandestine, low-visibility, and overt manners—with a few possibly acting covertly as well. The underground is intended to operate where guerrillas generally cannot. The auxiliary represents a clandestine support structure for both the underground and guerrillas. The public component functions as an overt, political, or material support entity. The public components may negotiate with the state government or occupying power on behalf of resistance movement objectives and typically make overt appeals for domestic and international support. Public components may represent resistance strategic leadership, a subordinate organizational support element, or merely an interest section.

2-42. There are numerous additional or supplementary components included in traditional Army doctrine for resistance. These additional components are included in the order of importance to the resistance effort: leadership and command (to include the resistance area command); area complex; base population; shadow government; government-in-exile; and diaspora population.

*Note:* Remember, the following component terms and concepts are a model of the resistance itself—not UW! UW is only one specific, military form of STR. Each of these terms describes an essential or potentially helpful function necessary to conduct resistance. The resistance will have or develop and use these functions or component elements and organizations or not, will do so well or badly, and will progress or regress and ultimately meet its fate whether or not the USG has a policy interest in supporting the resistance or is even aware of its existence. The SF Soldier uses this basic, traditional Army model for analyzing the resistance structure and function and, when so ordered, organizes it or advises its improvement to the resistance partner.

# PRIMARY COMPONENTS

2-43. The primary components of the resistance model are the underground, the guerrilla or armed force, the auxiliary support to the underground and guerrilla or armed force, and the public component.

# THE UNDERGROUND

2-44. The *underground* is a cellular covert element within unconventional warfare that is compartmentalized and conducts covert or clandestine activities in areas normally denied to the auxiliary and the guerrilla force. (ADRP 3-05). The following paragraphs contain descriptions of the function, organization, membership, and the nature of operations of the underground. Figures are included from previous doctrinal sources and nondoctrinal sources for amplification.

## FUNCTION

2-45. The underground within the resistance movement or insurgency has the ability to conduct operations in areas that are often inaccessible to guerrillas, such as urban areas under tight control of the local security forces. The underground can function in these areas because it operates in a clandestine manner, allowing it to avoid elimination by state security, as depicted in figure 2-1, page 2-13. Although these are not unique to the underground, examples of typical underground functions include the following:

- Intelligence and counterintelligence networks.
- Subversive radio stations.
- Information and influence networks that control newspaper or leaflet development, rumors, night letters, graffiti, webpages, and social media sites, blogs, and postings. Resistance movements are not obliged to use U.S. terminology, and influence and information networks have traditionally been referred to as "propaganda" networks as depicted in figure 2-1, page 2-13.
- Special materiel fabrication, such as false identification, explosives, weapons, and munitions.
- Control of networks for moving personnel and logistics.
- Individuals or groups that conduct acts of sabotage in urban centers.
- Clandestine medical facilities.

2-46. The underground supports the resistance area command, auxiliary, guerrilla force, and public component based on whatever method the resistance uses to analyze factors, such as mission, enemy, terrain and weather, troops and support available–time available and civil considerations (METT-TC). These personnel commit sabotage, intelligence gathering, and acts of deception through the action arm, intelligence, supply, and personnel sections.

## ORGANIZATION

2-47. The underground organizes into compartmented cells. (See Basic Cellular Structure Models in Appendix F, Resistance Development Models.) It forms these cells within various political subdivisions of the local area or sector, such as provinces, counties, cities, towns and villages, and neighborhoods. The underground environment may be urban or rural; however, since the function of the underground is to operate where guerrillas cannot, and state repressive power is likely to be at its strongest in urban areas, the underground will likely have its greatest utility in the cities.

2-48. If a member of the underground is compromised, the information that he can reveal is limited. A command committee organizes and controls underground activities. The committee members perform duties and responsibilities based on their skills and the degree of risks they are willing to accept. Figure 2-2, page 2-14, shows a notional mass-based, party-controlled organization that explicitly shows underground (fronts penetrants), guerrilla force (military forces), and public component (mass civil organization) elements. The underground exists to function where the guerrilla force cannot. The figure provides an example that the underground would be the front organization penetrating agents and possibly—but not automatically—some leadership echelons.

**Organization of Resistance Movement**

Recruiting section
Expansion of resistance movement. Replacement of losses. Recruiting specialist

Communications section
Maintains communications with the following by means of clandestine radio station, carrier pigeons and couriers.
Remaining portions of own army.
Guerrilla units
Liaison with Allied countries if own army no longer exists.

Training section
Especially of specialists and leaders. Evaluation of "lessons learned".

Escape section
Evacuation of Allied aircrew involved in emergency landings, escaped prisoners and civilian fugitives.

Transportation group
RR employees, drivers, persons providing lodging
Liaison with guerrilla detachments which will accept these persons.

Counterfiet section
Identification papers, ration cards, money.

Sabotage section
Industrial sabotage
Post, telegraph and telephone sabotage
Railroad sabotage
Highway sabotage
Sabotage power supply
Liaison with guerrilla units who supply ammunition and explosives and state "desires" concerning targets.

Finance section
Procurement of money for the resistance movement. Supplies the various branches with the necessary funds.

Combat elements of the resistance movement will be organized at a later stage. Remain silent and wait for the moment of open uprising

Information and propaganda section
Poster section, section writing slogans on walls
Editorial staff
Secret Press
Leaflets, handbills, underground newspaper, posters.
Information squad
Instruction for population concerning appropriate behavior.
Clandestine Radio Station
Disseminates information about progress of war.
Propaganda.
Relay to
* Remainder of own army
* Guerrilla units
* Friendly foreign countries
Monitor section
Monitoring politically and militarily important news

Police section
Counter espionage
Unmasking agents and informers. Ascertain "camouflaged" collaboration.
Assassination section
To punish especially cruel officials of the enemy or important traitors.
Archive
Records of brutalities committed by the foreign dictators.
Evaluation Squad
Debriefs people who have been arrested, interrogated, imprisoned or deported by the enemy. Develops new rules of behavior toward enemy.

Notes:
1. Planned splitting up into very many sections No centralization.
2. Each member knows only the bare essentials about the organization. As a result, the entire organization cannot be compromised by the apprehension of one or a small number of individuals. Information extracted from captured resistance members will compromise only a small portion of the resistance movement, therefore the organization can recover.

Source:
H. von Dach, *Total Resistance*

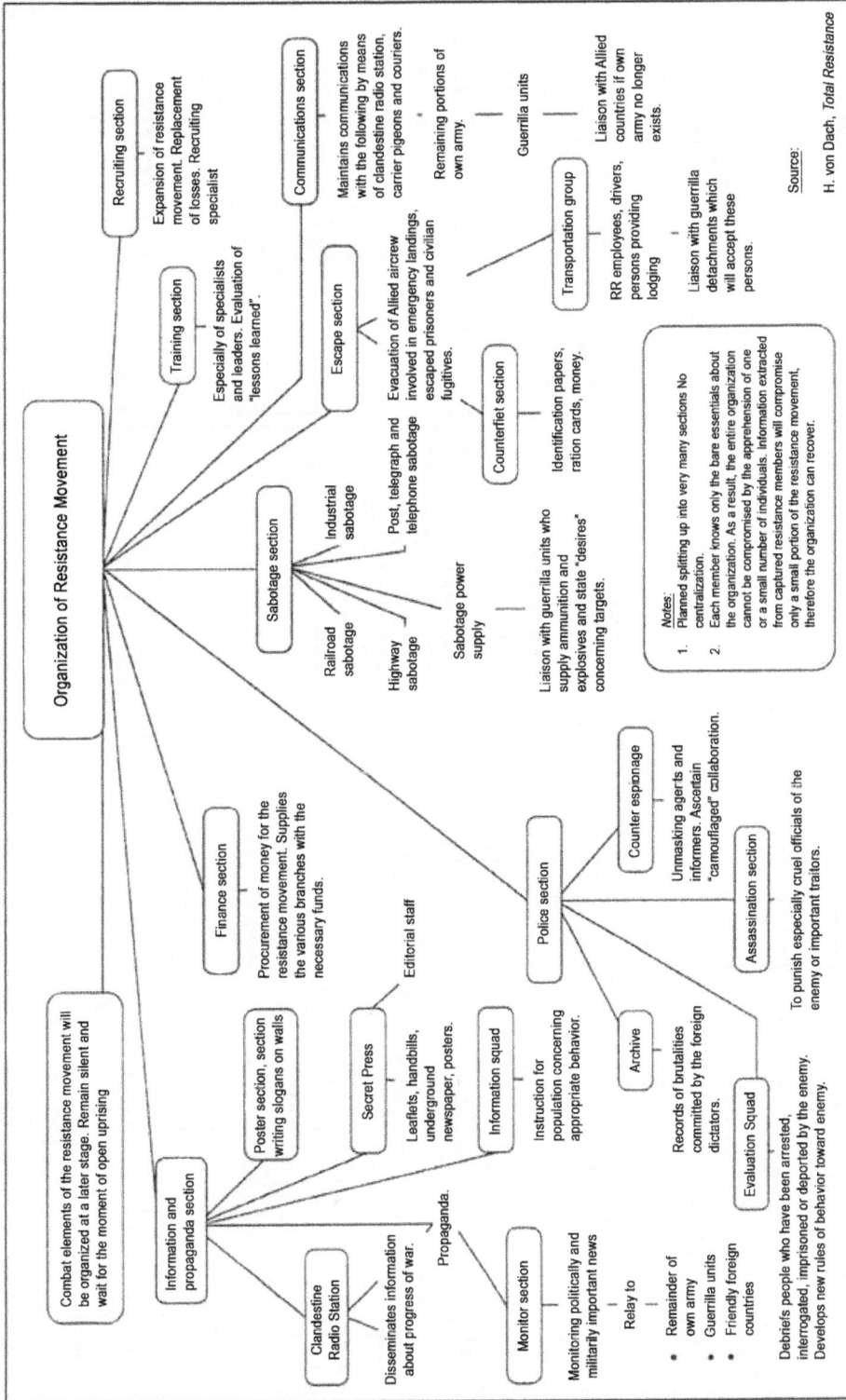

Figure 2-1. Simple, classic resistance function sketch

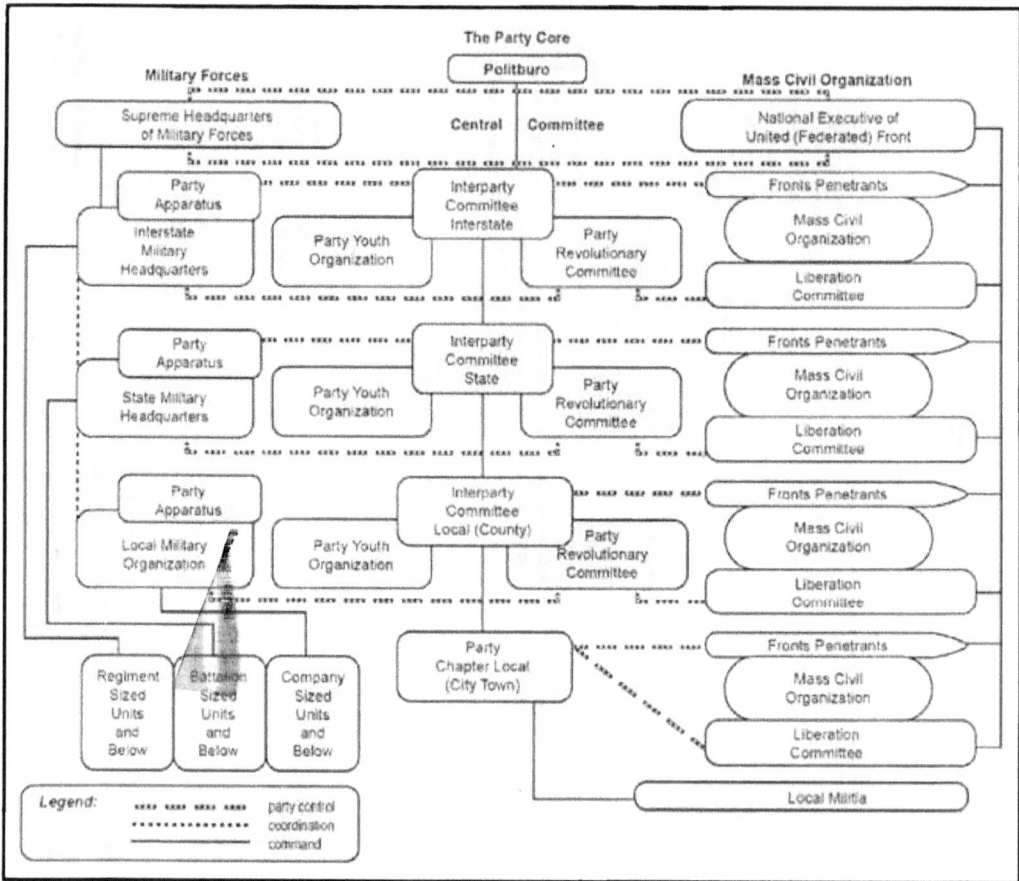

**Figure 2-2. Notional template for a mass-based, party-controlled resistance organization**

## MEMBERSHIP

2-49. Underground members normally are active members of the community, and their service is a product of their normal life or position within the community. They operate by maintaining compartmentalization and delegating most risks to their auxiliary workers. Urban members of the underground typically enable the resistance movement to affect the urban areas.

2-50. For security reasons, the size of the underground is kept as small as possible. The underground member needs to apply traditional behavior patterns to create a positive image. He is frequently prohibited from taking anything from the people without paying for it. He may even have to befriend certain segments of the population to influence them in support of the movement. He strives to conform to the normal behavior and daily activities of his neighborhood. By appearing conventional and inconspicuous, he makes it difficult for the security force to detect, identify, or find him. Without records or physical evidence, he is difficult to link to the organization. Contact and communications between members are key survival and critical points of subversive operations.

2-51. To survive and be effective, a member of the underground must exhibit many positive personality traits necessary to endure hardship, and he must be highly self-motivated. Among these traits, he or she must—

- Be stealthy, have self-discipline, and where required be technically and tactically proficient in combat skills.
- Maintain and display an inconspicuous, average citizen image toward the enemy.

- Be very cautious about confiding in others. Secrecy and dedication to the cause come before any personal benefit, allowing for greater longevity.

## NATURE OF OPERATIONS

2-52. The underground receives evaders, downed pilots, and other key people. It hides people and moves them as needed or moves them out of the JSOA entirely. The auxiliary also moves and hides people within its sector of operation. However, whereas the auxiliary is most likely to remain near its local area, the underground is more likely to be able to move people across the entire area complex. It has a system of safe houses that have been carefully selected and prepared to hide these people. It develops procedures so that people can be moved along selected routes at the best times to avoid detection. The underground moves them in false-bottom containers, to include laundry carts, fish crates, vehicles, furniture, and caskets. It also moves sensitive documents and equipment in a similar manner.

2-53. To support other operations, particularly those involving the smuggling of personnel and materials, the underground commits acts of deception to steal bona fide documents (identification cards, passes, ration coupons, money, and passports) and use them under false pretenses. If the underground cannot obtain the bona fide documents, it prepares facsimiles.

2-54. Underground operations are a mixture of clandestine, covert, low-visibility and overt. These operations have major similarities with those of both the guerrillas and the auxiliary. Some involve overt and violent actions and others are passive. The following paragraphs discuss the underground's relationship to other elements of the resistance.

2-55. Underground cells support the resistance operations of the guerrillas and auxiliary (within their sector commands) with acts of deception, sabotage, and movement of personnel and equipment. Underground cells cooperate closely with other underground counterparts in their sectors and throughout the area complex. Cooperation enables successful movement of personnel and equipment in and out of the JSOA. The underground may commit acts of violence or sabotage to interdict enemy operations and movements.

2-56. The underground may publish a resistance newsletter or newspaper promoting the resistance cause among the civilian population. In countries with a controlled media system, a private publication of this nature will be of high interest. However, the underground must anticipate enemy reprisals for possession of the paper. The enemy will probably search for the printing press. A defense against this threat is to print the paper outside the JSOA and smuggle it into the country, storing the supply in various locations.

2-57. The underground may engage in covert operations to disseminate embarrassing or incriminating information about the enemy or its officials. These revelations may inflame existing problems in sensitive areas and degrade the enemy's rapport with the civilian population or cause dissension in the enemy's ranks.

# THE GUERRILLA (OR ARMED) FORCE

## GUERRILLA FORCE

2-58. *Guerrilla warfare* consists of **military and paramilitary operations conducted in enemy-held, hostile, or denied territory by irregular, predominately indigenous, guerrilla forces to reduce the effectiveness, industrial capacity, and morale of the enemy.** In most cases, the guerrilla force is a relatively small group employing offensive tactics. They do not undertake defensive operations unless forced to prevent enemy penetration of guerrilla controlled areas or gain time for their forces to accomplish a specific mission. Guerrilla warfare is not an end in itself; it is a component of a larger resistance plan, which supports the achievement of objectives en route to an end-state political decision.

---

*Note:* Chapter 5, Section 6, Guerrilla Warfare; and Appendix V, Special Forces Advisor's Guide to the Combat Employment of Guerrilla Forces, provide more information.

---

2-59. The following paragraphs are descriptions of the function and nature of operations, the organization and membership of the guerrilla force, and a discussion about the armed force as an important alternate perspective of the guerrilla force function.

## Function and Nature of Operations

2-60. Guerrillas are irregular combatants generally operating with firepower, manpower, equipment, and logistics disadvantages compared to the security forces of the regime or occupying power. Guerrillas initially cannot hope to meet and decisively defeat a conventional unit in a pitched battle. Guerrilla units, therefore, must emphasize preservation of their military forces and attack at points most disadvantageous to the enemy. These attacks are normally conducted during periods of low visibility and directed against isolated outposts, weakly defended locations, or the moving enemy (figure 2-3, page 2-17). By recognizing their own limitations and weaknesses, guerrillas can hope for survival and eventual success. Although typically inferior to the enemy in many ways, guerrillas are often equal or superior to the enemy in the collection of information and intelligence, cover and deception, and the use of surprise to gain and maintain the initiative. The guerrilla's greatest advantage over a numerically superior and better resourced enemy is the initiative. Guerrillas exploit these advantages to compensate for their physical disadvantages.

2-61. In guerrilla warfare, the situation is always fluid. Both guerrilla units and enemy counterguerrilla forces move about on the battlefield—the former to avoid decisive engagement and annihilation and to maintain the initiative—and the latter in the attempt to find, fix, and finish the guerrillas. Guerrilla operations wear down and inflict casualties upon the enemy, cause damage to supplies and facilities, and hinder enemy operations. Because guerrilla operations are targeted against LOCs, industrial facilities, and key installations, they impede or interdict the movement of personnel and materiel, thereby seriously affecting the enemy's capability to supply, control, and communicate with its combat forces. In addition, the enemy is compelled to divert manpower and equipment to combat guerrilla activities. The success of guerrilla operations, even the fact that the guerrillas continue to exist, lowers the morale and legitimacy of the hostile regime, while maintaining the indigenous population's morale and willingness to resist.

2-62. In the early stages of UW, guerrilla forces are rarely concerned with seizing and holding terrain, primarily because they do not have the strength to do so against government countermeasures. Guerrillas attack by gaining a momentary advantage of firepower, executing their mission to capture or destroy personnel and equipment, and leaving the scene of action as rapidly as possible. Normally, the guerrillas vary their operations so that no pattern is evident. Guerrilla forces often make use of difficult terrain to aid concealment and frustrate enemy pursuit or occupation (figure 2-4, page 2-18). Guerrilla forces occupy assigned areas within which they will operate against government power, constantly move, and gradually develop into insurgent sanctuary enclaves, relatively free from enemy interference. Ultimately, these enclaves will grow in size and autonomy and remain governed by the insurgents as liberated territory. In the most successful examples of UW, the guerrillas eventually operate in close coordination and more openly with insurgent regulars or outside supporters' conventional maneuver units.

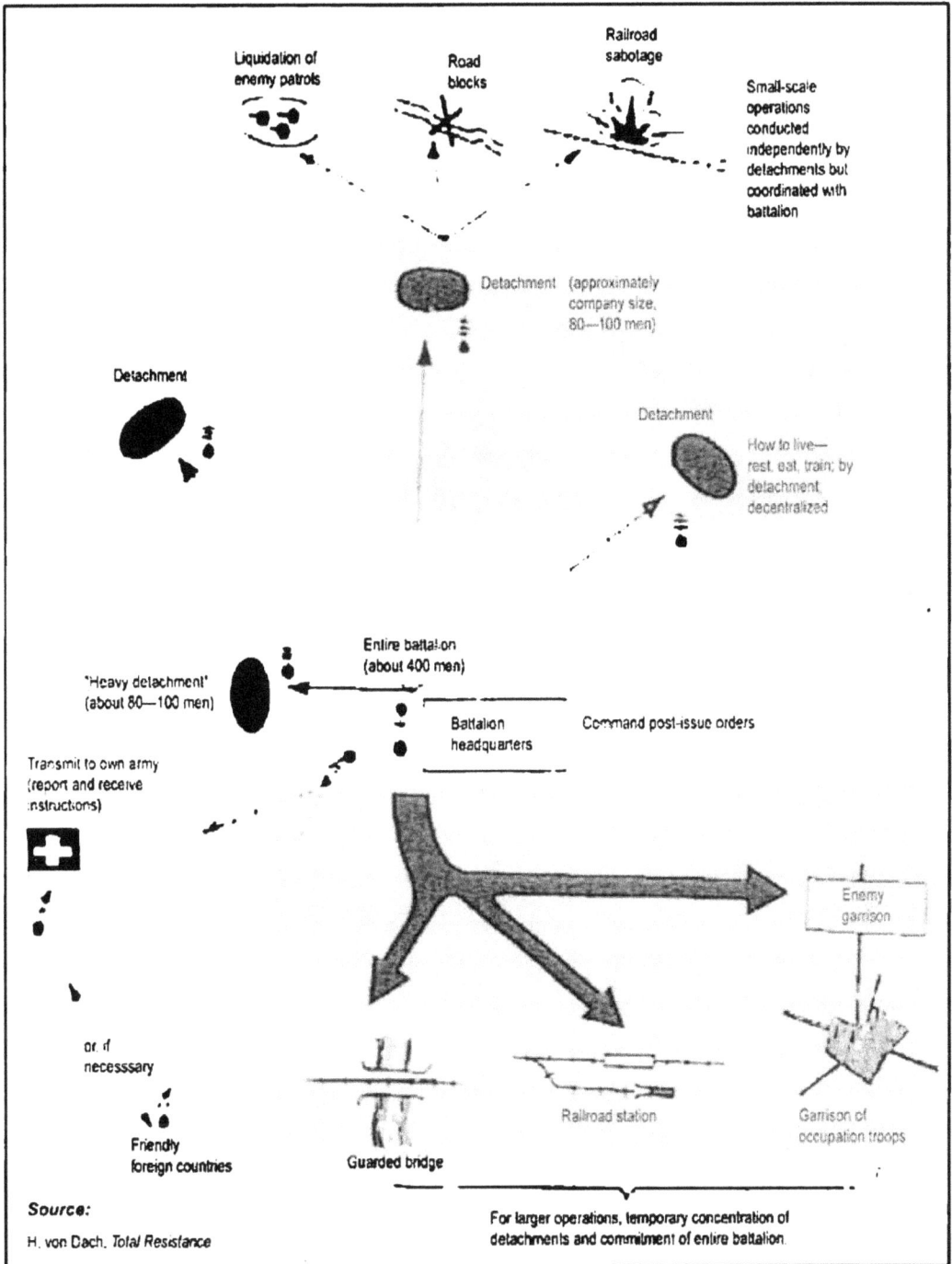

**Figure 2-3. Simple, classic guerrilla force sketch**

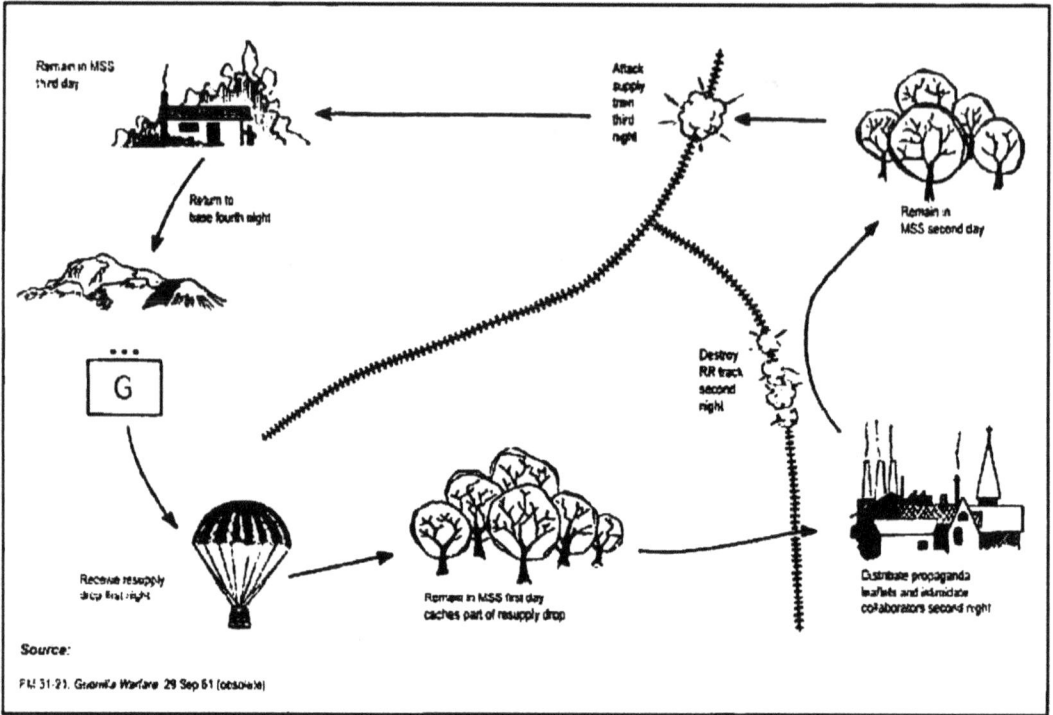

**Figure 2-4. Simple, classic guerrilla activities sketch**

## Organization and Membership

2-63. There is no common size or organizational structure to guerrilla forces, and they will vary as the situation demands. Depending on the degree of control over the local environment, the size of guerrilla elements can range anywhere from squad to brigade-size groups or larger. In the early stages of an insurgency, the guerrilla force's offensive capability might be limited to small standoff attacks. As the guerrilla force's base of support from the population grows, its ability to challenge government security forces more openly with larger-scale attacks increases. At some point in an insurgency or resistance movement, the guerrillas may achieve a degree of parity with host-nation forces in certain areas. In these cases, units may start openly fighting, rather than as guerrilla bands. In well-developed insurgencies, formerly isolated pockets of resistance activity may eventually connect and create liberated territory, possibly even linking with a friendly or sympathetic border state.

2-64. Membership of guerrilla forces will vary depending on the situation of the resistance. A nation-at-arms reacting to foreign invasion and occupation will likely have a larger aggrieved population to draw from than a more narrowly motivated insurgency. Assuming the worst case scenario of foreign invasion, the indigenous society will be fractured in many ways. Early membership of guerrilla units will likely coalesce around remnants of broken and isolated military units, militia, police, emergency services, and others with organizational training. Leadership could come from any sector of society, and it is important for previous subordinates to take the initiative and organize, train and recruit for the functions, skills, and manpower needed.

## Resisting the 'Guerrilla Force' in the Army
## Unconventional Warfare Doctrine Definition

**(The purpose of this section is to explain and clarify often-asked questions regarding the "Guerrilla Force" in the UW definition.)**

The definition for UW current since 2009 identifies its primary components as "...underground, auxiliary and guerrilla force...." As has been shown above, these three primary components have been consistent for decades in Army UW doctrine, and do provide a reasonable, adequate and basic starting foundation to understand the essential functions of resistance. However, there are problems with this current construction.

First, as discussed above, the unprecedented JP 3-05.1 (2012) for the first time included the concept of *public component*—a term which is not included with the other primary components in the current UW definition (2009). Although the new term is now an authoritative part of doctrine, the UW definition may or may not be changed to comprehensively include this new addition. Such a continuing omission would be a logical incongruity. However, the functions characterized by the new term "public component" have always been implicit or explicit in traditional UW doctrine. Therefore, just because the term is not included in the definition, does not denote that the definition or expanding body of doctrine is inadequate. For example, resistance professionals have long recognized the importance of leadership, the area command, the area complex and other terms as critical concepts to UW without requiring them to be included in a usefully succinct definition.

Second, as each resistance phenomenon – and UW potential to support that resistance – is unique, the specific characteristics of each component's contribution to that specific case will vary by the resistance dynamics mentioned above and many other factors. For example, a resistance in a large country with significant nonurban areas and widely dispersed populations may support a larger guerrilla force than a small, highly urbanized and densely populated country which likely necessitates a more clandestine approach to resistance activity armed or otherwise. Or, some cases of resistance may eschew widespread violence and only seek limited objectives and not require the classic concept of guerrilla bands conducting lethal military operations to cripple an opposing regime. By contrast, other cases of resistance may adopt or require violent methods, up-to-and-including guerrilla forces and even resistance armies. For these reasons, some readers complain that "...underground, auxiliary and guerrilla force...." [with emphasis on the "and"] is inaccurate, invalid and must be changed to "or" (or simply ignored). There are at least three primary reasons why this argument is wrong:

- **This argument either does not understand or fails to support the Army-prescribed role of doctrine.** ADP 1-01, the Army's 2014 Doctrine Primer states: "Army professionals use doctrine in two contexts: study and reflection as well as conducting (planning preparing, executing, and assessing) operations. Thus, doctrine is—and must be—both theoretical and practical. Doctrine is not a catalogue of answers to specific problems. Rather, it is a collection of fundamentals, tactics, techniques, and procedures for thinking about military problems, which operations are the most complex, and what actions best solve them. Doctrine is not <u>what</u> to think or <u>how</u> to solve specific problems." Those who claim to be "constrained by doctrine" do not understand the purpose which the Army intends for it. Is it possible that a specific resistance can be successful without a "guerrilla force?" Of course. But the "guerrilla force" concept's inclusion in doctrine provides a starting model to help professionals consider what, if any, is the nature and characteristics of the resistance's "armed force" capabilities.

- **Doctrine does not become automatically inadequate or obsolete just because it does not neatly or easily fit any specific current condition or problem set.** Doctrine is necessarily conservative, and somewhat like law, it rests on a body of precedent. In the case of Army doctrine for resistance, guerrilla and unconventional warfare, it rests on eight decades worth of precedent. Doctrine, also like law, can be changed when circumstances change and/or when it is proven to be obsolete. However, the U.S. basic theory of resistance

(continued on next page)

---

**Resisting the 'Guerrilla Force' in the Army
Unconventional Warfare Doctrine Definition (continued)**

has not changed since the United States Government started writing it into doctrine in World War II. From its beginnings, UW (or "guerrilla warfare") doctrine has identified both the guerrilla warfare and underground components as the founding primary functions of the UW concept; not simply one or the other. In fact, the essential difference between the two is only the required function of the underground to operate where the guerrilla forces cannot.

• The attempt to force a change from "A, B, and C" to "A, B, or C" is illogical because it weakens, not strengthens the explanatory power of the doctrinal concept. "A, B, and C" means that all three functions are required for the concept and the commander or planner analyzing their assigned case should anticipate seeing the specific indigenous examples of these modeled functions. For example, one should expect to find the resistance armed element (guerrilla or armed force), the element that can operate where the armed element cannot (underground), and an element that supports them both (auxiliary). "A, B, or C" means that any of the three functions can satisfy the intent of the model. For example, it would mean that the Army is claiming that UW can be conducted with just an underground, just an auxiliary, just a guerrilla force, or a combination of any two or all three. This mistakes membership in a "component"—which is merely a piece of a model—for the essential function of the component, it would overturn eight decades of Army UW doctrine precedent, and simply does not reflect reality.

## ARMED FORCE

2-65. The traditional component label guerrilla force is in-part a misnomer, because the classic concept of guerrilla warfare is too narrow to cover the entire spectrum of armed violence that could be applied by a resistance to meet its objectives. There is no distinct dividing line between armed, violent resistance in temporary, episodic, single violent political acts and permanent, uniformed armed combatant formations. Rather, there is a continuum of violence that can be used in the service of resistance goals. (See Appendix E, Varieties of Violent and Nonviolent Methods.)

2-66. One example is a drive-by shooting in which a shooter assassinates a public official in broad daylight on a city street. This is more than mere clandestine activity and—who the shooters and drivers are and what component of the U.S. resistance theory model they are in—is largely irrelevant. The action at its decisive moment is overt, and yet it is not the basic image suggested by the generic label guerrilla warfare. Another example is the political violence of the Sturmabteilung in Germany or the Freikorps in Austria and Czechoslovakia in the 1930s. Despite being explicitly inspired and led by military veterans and organized along paramilitary lines, it is still a different structure than later French, Greek, or Russian partisans in the 1940s wood line. The classic characterization of the 1960s Viet Cong as farmer by day, guerrilla by night does not neatly fit into the overt-clandestine or underground-guerrilla warfare dualities either. In addition, the 1980s mujahedeen fighter's uniform, tribal organization and leadership, and tactics were generally unmodified from traditional political and cultural reality.

2-67. SF Soldiers, therefore, need to use the doctrinal terms as intended regarding guerrilla warfare as a basic starting point for understanding the resistance's armed force. A more-or-less organized armed element is a primary function that characterizes classic understanding of resistance. What form that violent, organized function takes will vary by the specific case of resistance, by the underlying environment, by phase of resistance development, by location, and sometimes simply on resistance success.

> "The individuals of a resistance movement who band together on a military basis are the guerilla forces of a resistance movement. Unfavorable conditions may prevent a resistance movement from finding active expression in guerilla warfare, and the movement may be limited to undercover, poorly organized forms of expression. . . ."
>
> FM 31-21, *Organization and Conduct of Guerilla Warfare*,
> 5 October 1951 (pg. 3)

# THE AUXILIARY: SUPPORT TO THE UNDERGROUND AND GUERRILLA OR ARMED FORCE

2-68. The *auxiliary* is, for the purpose of unconventional warfare, the support element of the irregular organization whose organization and operations are clandestine in nature and whose members do not openly indicate their sympathy or involvement with the irregular movement (ADRP 3-05). The auxiliary refers to that portion of the population that provides active clandestine support to the guerrilla force and the underground. Members of the auxiliary are part-time volunteers that have value because of their normal position in the community.

## FUNCTION

2-69. Soldiers should not think of the auxiliary as a separate organization but as a different type of individual providing specific functions as a component within an urban underground network or guerrilla force's network. These functions can take the form of logistics, labor, or intelligence collection. Auxiliary members may not know any more than how to perform their specific function or service that supports the network or component of the organization. In many ways, auxiliary personnel assume the greatest risk. They are also the most expendable element within the insurgency. Resistance leaders sometimes use auxiliary functions to test a recruit's loyalty before exposing him to other parts of the organization. Auxiliary functions are like embryonic fluid that forms a protective layer, keeping the underground and guerrilla force alive. Specific functions include the following:

- Logistics procurement and distribution (all classes of supply).
- Labor for special materiel fabrication.
- Security and early warning for underground facilities and guerrilla bases.
- Intelligence collection.
- Recruitment.
- Communications network staff, such as couriers and messengers.
- Influence product (or propaganda) distribution.
- Safe house management.
- Logistics and personnel transport.

## ORGANIZATION AND MEMBERSHIP

2-70. The auxiliary organizes into compartmented cells. (See Basic Cellular Structure Models in Appendix F, Resistance Development Models.) It forms these cells within various political subdivisions of the local area or sector, such as provinces, counties, cities, towns and villages, and neighborhoods. The auxiliary environment may be urban or rural and since its function is to support the other components, it will be found anywhere it is needed and possible.

2-71. Active support from some of the civilian population and passive support from most of the remainder is essential to extended guerrilla operations. To ensure that both active and passive support is responsive to the resistance area command, some form of organization and control is required. Control of civilian support is accomplished primarily through the auxiliaries. Auxiliary forces compose that element of the resistance area command established to provide for and organize civilian support of the resistance movement.

2-72. Auxiliary is a term used to denote people engaged in a variety of activities. It is applied to those people who are not members of other resistance elements but who knowingly and willingly support the common cause. It includes the occasional supporter as well as the hardcore leadership. Individuals or groups who furnish support, either unwittingly or against their will, are not considered auxiliaries. Auxiliaries may be organized in groups or operate as individuals.

2-73. Auxiliary units are composed of local civilians, normally living in the smaller towns, villages, and rural areas. Unlike guerrilla units, the auxiliaries are not expected to move from place to place to conduct operations. The fact that the auxiliary forces are local and relatively static is highly desirable from the resistance area command viewpoint in that it provides support for the mobile guerrilla forces throughout most of the operational area.

2-74. Auxiliary forces normally organize to coincide with or parallel the existing political administrative divisions of the country. This method of organization ensures that each community and the surrounding countryside is the responsibility of an auxiliary unit. It is relatively simple to initiate since auxiliary commands may be established at each administrative level, for example—regional, county, district or local (communities and villages). This organization varies from country to country depending on the existing political structure. Although the auxiliary is not explicitly represented in figure 2-2 (notional template for a mass-based, party-controlled resistance organization), organization of auxiliary units can commence at any level or at several levels simultaneously and is either centralized (figure 2-5) or decentralized (figure 2-6, page 2-23).

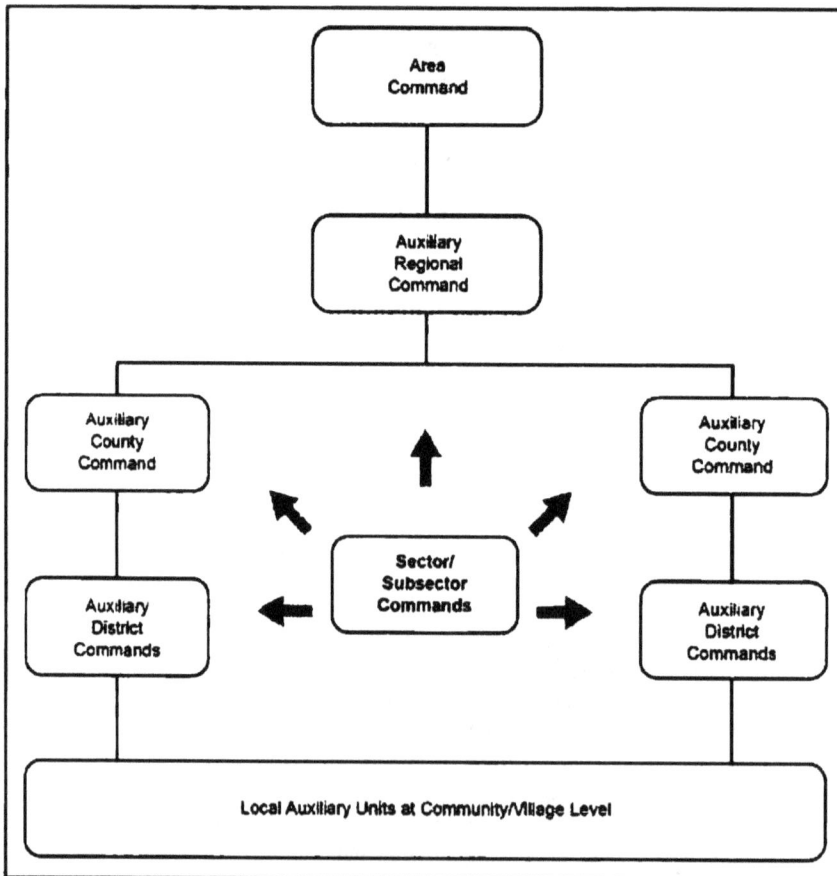

**Figure 2-5. Notional template for a centralized auxiliary structure**

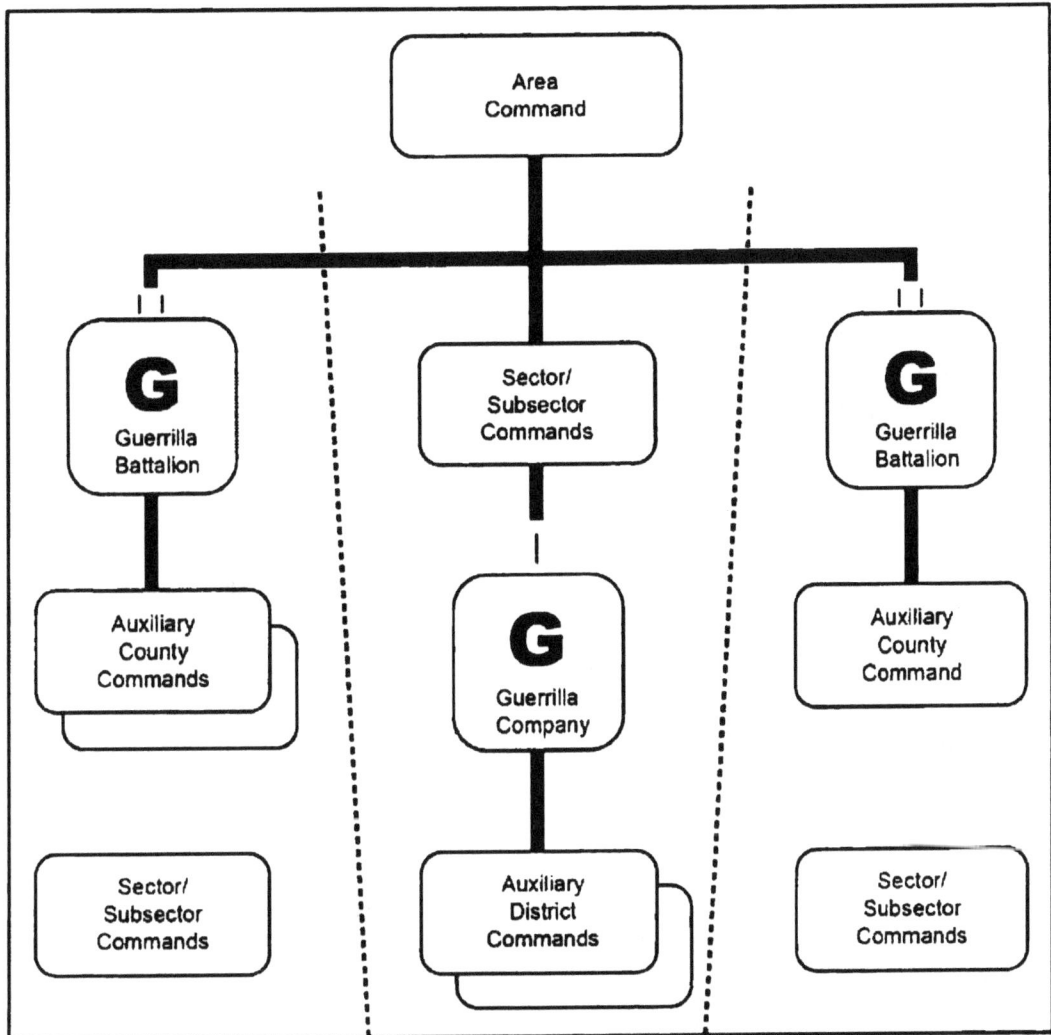

**Figure 2-6. Notional template for a decentralized auxiliary structure**

2-75. The basic organization at each level is the command committee. This committee controls and coordinates auxiliary activities within its area of responsibility. In this respect, it resembles the command group and staff of a military unit. Members of the command committee are assigned specific duties, such as supply, recruiting, transportation, communications, security, intelligence, and operations (figure 2-7, page 2-24). At the lowest level, one individual may perform two or three of these duties.

2-76. The command committee may organize civilian sympathizers into subordinate elements or employ them individually. When possible, these subordinate elements are organized functionally into a compartmented structure. However, because of a shortage of loyal personnel, it is often necessary for each subordinate auxiliary element to perform several functions.

2-77. The home guard is the paramilitary arm of the auxiliary force. Home guards are controlled by the various command committees. All auxiliary elements do not necessarily organize home guards. Home guards perform many missions for the local auxiliary force, such as tactical missions, guarding of caches, and training of recruits. Their degree of organization and training depends upon the extent of effective enemy control in the area.

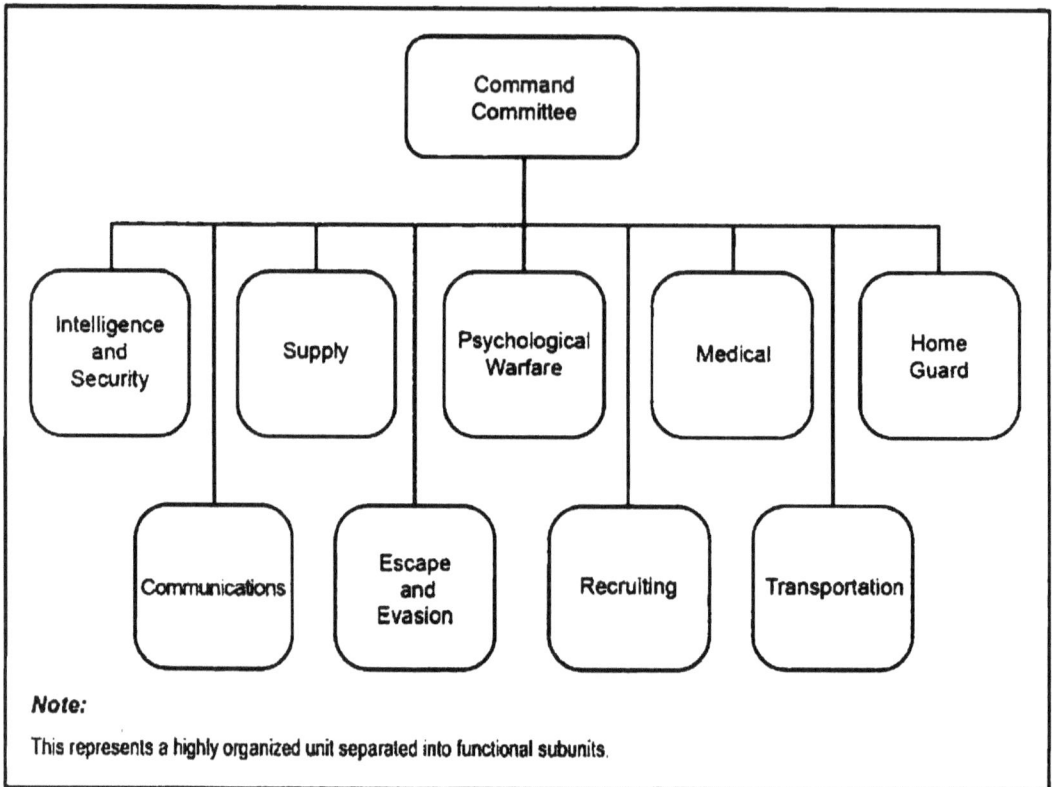

**Note:**

This represents a highly organized unit separated into functional subunits.

**Figure 2-7. Notional template for an auxiliary command structure**

## NATURE OF OPERATIONS

2-78. Auxiliary units derive their protection in two principal ways: a compartmented structure and operating undercover. While enemy counterguerrilla activities often force the guerrillas to move temporarily away from given areas, the auxiliaries survive by remaining in place and conducting their activities so as to avoid detection. Individual auxiliary members carry on their normal, day-to-day, routine while secretly carrying out the many facets of resistance action. Examples of what not to do include:

- Failing to maintain an average citizen image.
- Being repeatedly absent from work without reasonable explanations.
- Showing an unusual concern about enemy activities.
- Failing to account for missing supplies.
- Appearing unusually nervous or habitually tired.
- Confiding too freely in strangers.
- Asking questions and unusual favors of people of questionable loyalties.
- Being too eager to recruit people without adequate security checks.

2-79. Auxiliary units frequently use the passive or neutral elements of the population to provide active support to the common cause. Usually, this is done on a one-time basis because of the security risks involved in repeated use of such people. The ability of auxiliary forces to manipulate large segments of the neutral population is further enhanced by the demonstrated success of friendly forces.

2-80. The support missions discussed herein are the principal ones performed by auxiliary forces to support the resistance area command. Some of these tasks are coordinated directly with guerrilla units, while others are controlled by their own higher HQ. Normally, auxiliary units are assigned direct support missions for guerrilla units in their areas.

2-81. Typical auxiliary missions include:

- **Security and Warning.** Auxiliary units provide a physical security and warning system for guerrilla forces. They organize extensive systems of civilian sympathizers who keep enemy forces under surveillance and who warn the guerrillas of enemy moves. These civilians are selected as part of the security system because of their advantageous location, which permits them to monitor enemy movement toward guerrilla bases.

- **Intelligence.** Auxiliary units collect information to support their own operations and those of the resistance area command. The auxiliary force provides direct intelligence support to guerrilla units operating within their area of responsibility.

- **Counterintelligence.** The auxiliary unit assists the resistance area command counterintelligence effort by maintaining watch over transitory civilians, by screening recruits for guerrilla units, and by monitoring refugees and other noninhabitants of the area. Because of their intimate knowledge of local people, auxiliaries should be able to report attempts by enemy agents to infiltrate the area. They may also name those inhabitants whose loyalty to the resistance might be suspect.

- **Logistics.** The auxiliary unit supports guerrillas in all phases of logistics operations. They provide transportation or porters for the movement of supplies and equipment. Auxiliaries often care for sick and wounded guerrillas, provide medical supplies, and arrange for doctors and other medical personnel. They establish and secure caches, and they collect food, clothing, and other supplies for guerrilla units through a controlled system of levy, barter, or contribution. Sometimes auxiliaries provide essential services to guerrillas, such as repair of clothing, shoes, and certain items of equipment. Auxiliary units furnish personnel to assist at drop zones (DZs) and landing zones (LZs), and they distribute supplies throughout the area. The extent of logistics support furnished by the auxiliary force depends upon the resources of the area, the degree of influence the auxiliaries exert on the population, and the enemy activities. (See Appendix M, Logistics, for further discussion of logistics.)

- **Recruiting.** The guerrilla units depend upon the local population for recruits to replace operational losses and to expand their forces. Auxiliaries spot, screen, and recruit personnel for active guerrilla units. If recruits are provided through reliable auxiliary elements, the enemy's chances for placing agents in the guerrilla force are greatly reduced. In some instances, auxiliary units provide rudimentary training for guerrilla recruits.

- **Psychological Warfare.** A very important mission in which auxiliary units assist is psychological warfare. The spreading of rumors, leaflets, night letters, thumb drives, radios, and posters is timed with guerrilla tactical missions to deceive the enemy. Leaflets can mislead the enemy as to guerrilla intentions, capabilities, and locations. The spreading of these influence products usually involves little risk to the disseminator and is very difficult for the enemy to control.

- **Civilian Control.** To control the population and give the enemy an impression of guerrilla power, the auxiliary units establish a rudimentary legal control system. This system can control black marketing and profiteering for the benefit of the guerrilla force. Collaborators may be terrorized or eliminated by the auxiliaries. In addition, control of large numbers of refugees in the area is assumed for the guerrilla force by the auxiliary units.

- **Evasion and Escape.** Auxiliary units are ideally suited for the support of evasion and escape mechanisms. Their contact with and control over segments of the civilian population provide the area commander with a means of assisting evaders.

- **Other Missions.** Auxiliary units may be called upon to perform a number of other missions to support guerrilla operations. Examples of these include:
  - Activity in conjunction with the guerrillas against other targets. Cutting of telephone lines between an enemy installation and its reserve force prior to a guerrilla attack is an example of such support.
  - Operation of DZs or LZs.
  - Operation of courier systems between widely dispersed guerrilla units.
  - Furnishing guides to guerrilla units.
  - Under some circumstances, conducting active guerrilla operations in their areas of responsibility on a part-time basis.

# THE PUBLIC COMPONENT

2-82. The *public component* is an overt political manifestation of a resistance.

## FUNCTION

2-83. Public components are primarily responsible for negotiations with the state government or occupying power representatives on behalf of resistance movement objectives. Public components also typically make overt appeals and organize efforts for support of all kinds—from domestic and international sympathizers and stakeholders. Public components may be the strategic leadership of the resistance, or it may be merely an overt interest section wielded by such leadership located elsewhere. The public component is not synonymous with shadow government or government-in-exile. The public component may be a robust organization, or it may be only a lone spokesperson or lobbyist (figure 2-8).

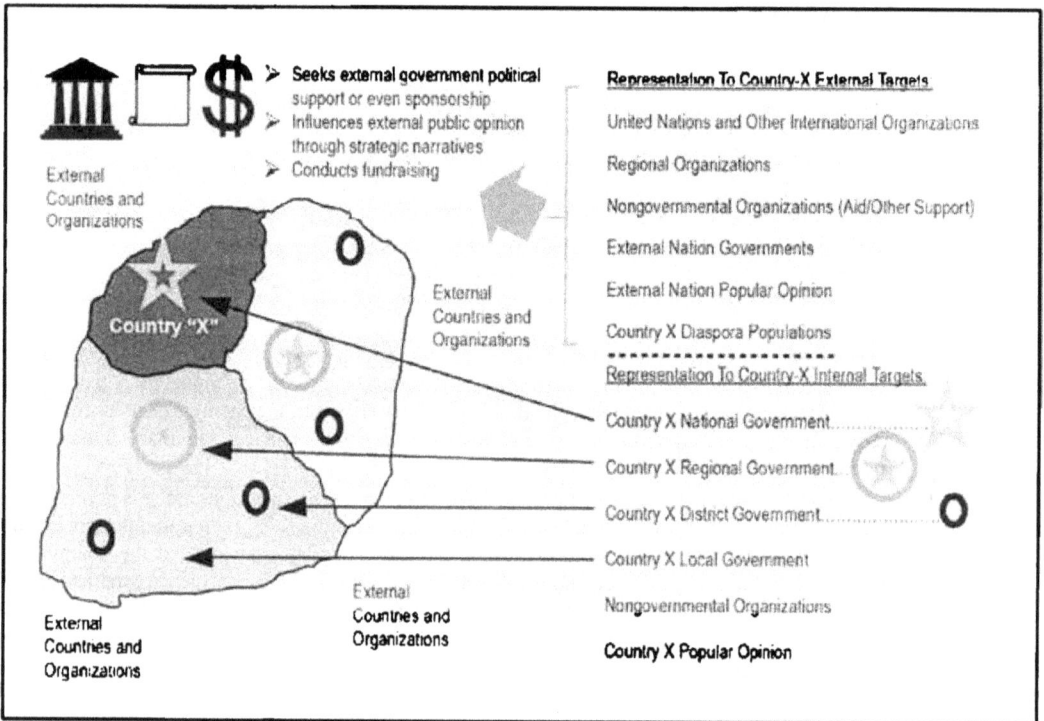

Figure 2-8. Resistance public component

## ORGANIZATION AND MEMBERSHIP

2-84. The functions of the public component will be based on the resistance's goals, perceived support requirements, and access The public component may function as a principal interface between the resistance and potential supporters, contributors, sponsors or—conversely—potential adversaries. What the resistance seeks to gain from public interaction will drive the placement and goals of each public component element. It will be limited by where those elements are permitted to exist and conduct overt activities. A resistance which seeks international or third-party support external to its area of operations (AO) will likely place public component elements in such locations. To the extent a state or occupying power will allow the continued existence of public component entities within the AO, such entities could be useful at every echelon of indigenous governance and control.

## NATURE OF OPERATIONS

2-85. Every case of resistance is unique, and the degree to which public representatives exist will vary.' In many cases, widespread popular discontent with the state or occupying power rule is represented by overt, legal political organizations, which may or may not continue as the political situation worsens. In such cases, these public attempts to create change may actually precede the formation of clandestine violent anti-regime efforts. As the state applies pressure to these public political adversaries, it is possible that the public component will modify its overt political position to maintain permission to operate openly within the state's jurisdiction. Any continuing support for the resistance will have to be clandestine or presented indirectly in such a manner as to not threaten the public component's right to operate openly.

2-86. If the state decides to suppress these public components completely, the public components may have to dissolve and go underground. In such cases, the resistance may then choose to infiltrate existing legal organizations to manipulate them as clandestine fronts for pursuing the objectives of the movements. The public components may not be located in the UW operational area at all. An alternative choice is for the public components to displace to another country where they operate openly on behalf of the resistance.

# ADDITIONAL (OR SUPPLEMENTARY) COMPONENTS

2-87. The additional (or supplementary) components of the resistance model are the leadership and command, the area complex, the base population, the shadow government, the diaspora population, and the government-in-exile.

## LEADERSHIP AND COMMAND

### Function

2-88. Leadership is not a separate type of component as much as it is a function. Strategic leadership, which conceptualizes the ends, ways, and means of the insurgent or resistance strategy, can be located anywhere—in the AO or in another country—so long as they have a means of conveying strategic guidance to insurgent elements in the AO. Some relatively small number of leaders will also likely be outside of the AO functioning as part of one or more insurgent support networks. Strategic leadership is also not a function of any particular component; it may be present in any or all of the principal components.

### Organization and Membership

2-89. Most of the organizational leadership, however, will of necessity be located inside the AO and often immersed in and influencing the population. Operational and tactical leadership is essential to the creating, buildup, and employment of all underground, auxiliary, and guerrilla elements. For this reason, most successful insurgencies and resistance movements devote much time and resources into developing trained, indoctrinated, and disciplined cadre who can magnify unifying themes, multiply adherents, expand the organization, and lead operational and tactical missions and activities.

2-90. Every resistance leadership structure will differ by organizational scope and goals and the quality of that leadership at all levels. Two examples of large-scale efforts are presented below. Figure 2-9, page 2-28, displays a notional organization with four echelons of leadership and highlights the idea of component function variety at the lowest action level. Figure 2-10, page 2-29, shows five levels of the historical organization of the French Resistance during World War II.

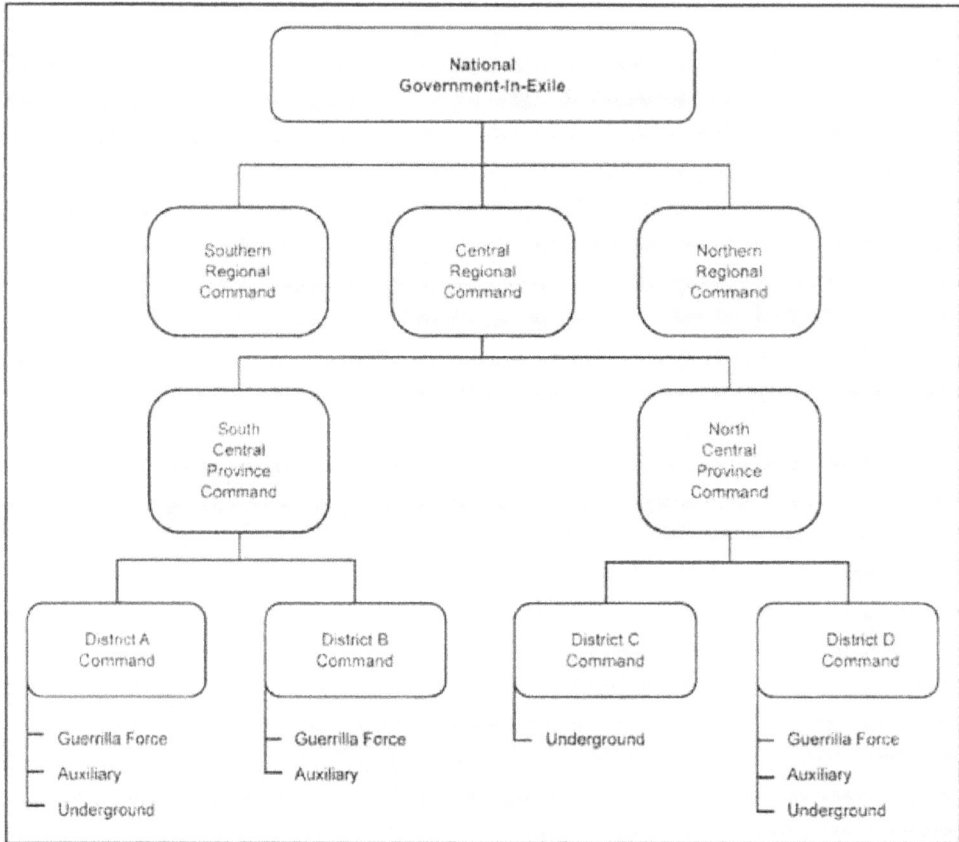

**Figure 2-9. Notional template for a resistance organization**

## Nature of Operations

2-91. Like large political or military organizations, much of the highest-level leadership may be far removed from the actions inside the AO. Therefore, a significant division exists between those providing the political and strategic leadership and support and those conducting resistance operations and tactical actions in the field. The organization chart examples (figure 2-9, above, and figure 2-10, page 2-29) represent the geographic overlay of command.

2-92. Resistance actions inside the AO will be organized in some indigenous manner, whether local leadership aligns with pre-existing social structures and authority figures or whether resistance organizational structure is tailor-made for the cause. Since 1951, traditional U.S. Army guerrilla warfare and resistance doctrine has used the term area command and such terms as sector command, district command, subsector command, and so on to model this function of resistance command inside the AO.

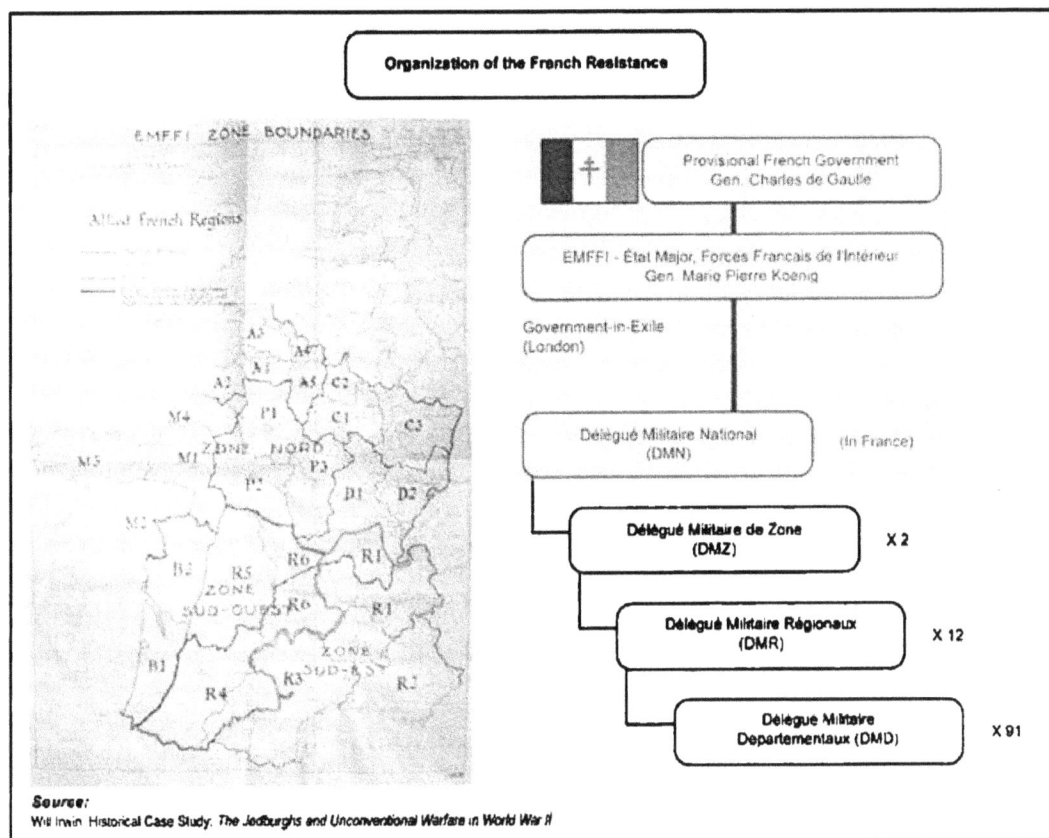

Figure 2-10. Organization of the French resistance

---

*Note: Area command* is defined in UW doctrine (ATP 3-05.1 [Change 1], 25 November 2015) as: "In unconventional warfare, the irregular organizational structure established within an · unconventional warfare operational area to command and control irregular forces advised by Army Special Forces." Because it refers to USG STR and SF interaction with the resistance, area command is currently denoted as a term of UW—not resistance per se. However, area command should be useful as a term for the model of resistance whether or not the USG makes a policy decision to conduct STR. The term for *resistance area command* will be proposed for inclusion in the DOD Dictionary of Military and Associated Terms and ADP 1-02 as: "The largest territorial resistance organization commanded by a senior resistance leader inside a defined resistance area of operations."

---

## Resistance Area Command

2-93. A *resistance area command* is **the largest territorial resistance organization commanded by a senior resistance leader inside a defined resistance area of operations.** The area commander is responsible for the command and control, direction, integration, and support of all resistance activities inside the resistance area command. The area commander may be the resistance senior political leader, may be only one member of a resistance collective senior leadership with assigned area commander duties, or may be subordinate to the resistance senior leadership. Therefore, the resistance senior leadership may or may not be co-located with the resistance area command. The area commander is likely to be assisted by a resistance area command staff, subordinate sector commanders, separate element leaders and actors, and sometimes resistance entities outside of the resistance area command's physical boundaries.

2-94. The size of the resistance area command is dependent on conditions specific to that resistance, where it is operating, and the capabilities of the opposing state or occupying power. In early stages of resistance, when survival is paramount, the functions and forces of resistance must be decentralized because of the ever-present possibility of the resistance area command being destroyed by enemy action. The area commander should be located where he can safely control the resistance movement and its activities. Flexibility, intelligence, mobility, and operations security (OPSEC) are the keys to survival and success. The area commander or his designated representatives should make frequent visits to subordinate units, both for morale enhancement and to become acquainted with the local situation. Where personal visits are not possible, the commanders should communicate with each other frequently.

2-95. The resistance area command is likely to be developed relatively early in the phasing of resistance, and resistance senior underground, guerrilla (or armed force), auxiliary, and public component leaders will conduct some centralized strategic and operational planning with the area commander. However, because of the hazards of resistance activities, they must give maximum latitude for decentralized execution to their subordinate sector, functional, and unit commanders. This policy provides latitude for subordinate commanders to determine the how to in planning and executing their missions. Mutual confidence, cohesion, and trust must exist between the area commander and his subordinate sector, functional, and unit commanders to be effective.

2-96. There are no rigid patterns for the structure and function of a resistance area command. The resistance area command is compartmented, but it should include representatives from all elements of the resistance movement. Resistance area command meetings will include a command group of resistance leaders representing the primary components. The meetings may include shadow government leaders, government-in-exile representatives, or select and vetted external actor representatives, when available. The current enemy situation is the basis for an area command or sector meeting.

2-97. Suitable locations for the site should be in a secure and isolated location, either rural or urban. When indigenous or enemy government-imposed population control measures are in effect, it may be easier for meetings to be held in rural or guerrilla-controlled territory. Outer and inner security zones should be formed. The outer security ring consists of observers who are strategically placed away from the meeting place to observe avenues of approach and to provide early warning by cell phone, landlines, handheld radios, or other means. The inner security zones surround the meeting place itself and consist of enough personnel to allow for the escape of the resistance leadership in case of compromise. Before the meeting convenes, those present must decide on the actions to take in case of a compromise. They must consider escape plans and routes. A security element, usually members of the inner security zones, serves as a rear guard while key personnel, equipment, and documents are removed and quickly evacuated.

2-98. During the initial organization of a resistance area command, the intelligence section of the staff is given special emphasis. Throughout all phases of the organization of a resistance force, the intelligence net is expanded progressively until the intelligence requirements (IRs) for the resistance area command HQ can be fulfilled. The functions of a resistance area command intelligence section are to—

- Collect, record, evaluate, and interpret information of value to the resistance, generally and the guerrilla (or armed) forces, specifically. It distributes the resulting intelligence to the area commander and staff and to higher and lower commands.
- Organize, supervise, and coordinate, together with the operations section, special intelligence teams (airfield surveillance, air warning, and coast watcher).
- Plan and supervise the procurement and distribution of maps, charts, photographs, and other materiel for intelligence purposes.
- Recommend intelligence and counterintelligence policies.
- Collect and distribute information on evasion and recovery, to include instructions for downed aircrews, evaders, and escapees.
- Establish liaison with the intelligence staffs and lower commands.
- Provide intelligence personnel for duty at lower commands.
- Conduct training to carry out intelligence functions.

2-99. The intelligence section should include deployable intelligence forward detachments. When it is anticipated that enemy pressure will forces the resistance area command HQ to move to alternate locations, forward intelligence detachments can set up and begin operations in the prepared alternate site before the rear echelon moves. During large-scale, multielement resistance operations, the forward intelligence detachment can provide an intelligence section for an advance command post, organized to direct the operations of two or more sector commands and disparate resistance elements. Guerrilla units produce most of the intelligence needed for their own security and for local operations against the enemy, and they must gather information required by higher HQ. Although these subordinate units are supervised by the resistance area command rear intelligence section, intelligence detachments and information collectors should be decentralized as far as practicable to subordinate sector commands.

## Sector Commands

2-100.   Large resistance area commands may establish subordinate sector commands. Sectors are formed to simplify operations and to provide a mechanism to promote centralized planning and decentralized execution. If the resistance area command is subdivided into sector commands, its component units are the subordinate sector commands. The sector command is the command element of the multicomponent resistance effort in a given sector. The same factors that define the boundaries of resistance area commands define the boundaries of sectors. The sector command performs the same functions as the resistance area command, except within the limits of its own boundaries. Although the U.S. Army basic model of resistance only expects one resistance area command per resistance movement, there is no limit to the number of sector commands that can be established under the resistance area command. Moreover, sectors can be further divided into an unlimited number of subsectors as determined by the needs of resistance.

## Critical Cadres

2-101.   Cadres play a critical role in resistance organizational development and control, operational effectiveness and, in some cases, resistance political control. Cadres play an important role in receiving initial resistance training and indoctrination and then in turn expanding the organization by disseminating the content to others. This force multiplier role is critical, when resistance efforts seek to expand the scope of their capabilities from small initial efforts, and when dissemination of resistance know-how must occur quickly and efficiently.

"In each situation the Special Forces commander decides which of the following training systems will be the most beneficial to resistance force personnel: (1) centralized training, (2) decentralized training, (3) individual or on-the-job training, or (4) specialized training for selected personnel. In most instances one or all of the systems noted above will be used. To expedite training programs and to provide effective instruction to dispersed units, a centralized training course may be designed and presented to selected resistance personnel, who in turn act as instructor cadres to dispersed units. When this occurs, a decentralized training system is put into effect."

ST 31-201, *Special Forces Operations*, November 1978,

Chapter 5, Section V: Training of the Resistance Force (p. 5-23)

2-102.   Cadres are also important to large resistance efforts requiring the mobilization and control of large resistance forces and huge populations. Not all resistance efforts are mass population-based, protracted warfare. However, the organizational power and political control and force multiplication potential of deliberate cadre development and employment was obvious to the Communist Chinese.

2-103.   Cadres are also a critical concept in the analysis and operational employment of resistance efforts at a cellular level. Figure 2-11, page 2-32, shows the importance of considering access to various social entities in expanding and safeguarding a resistance organization. Each specific individual proceeding from the primary cell cadre is himself a key cadre based on the conditions permitting interface with the social entities approached.

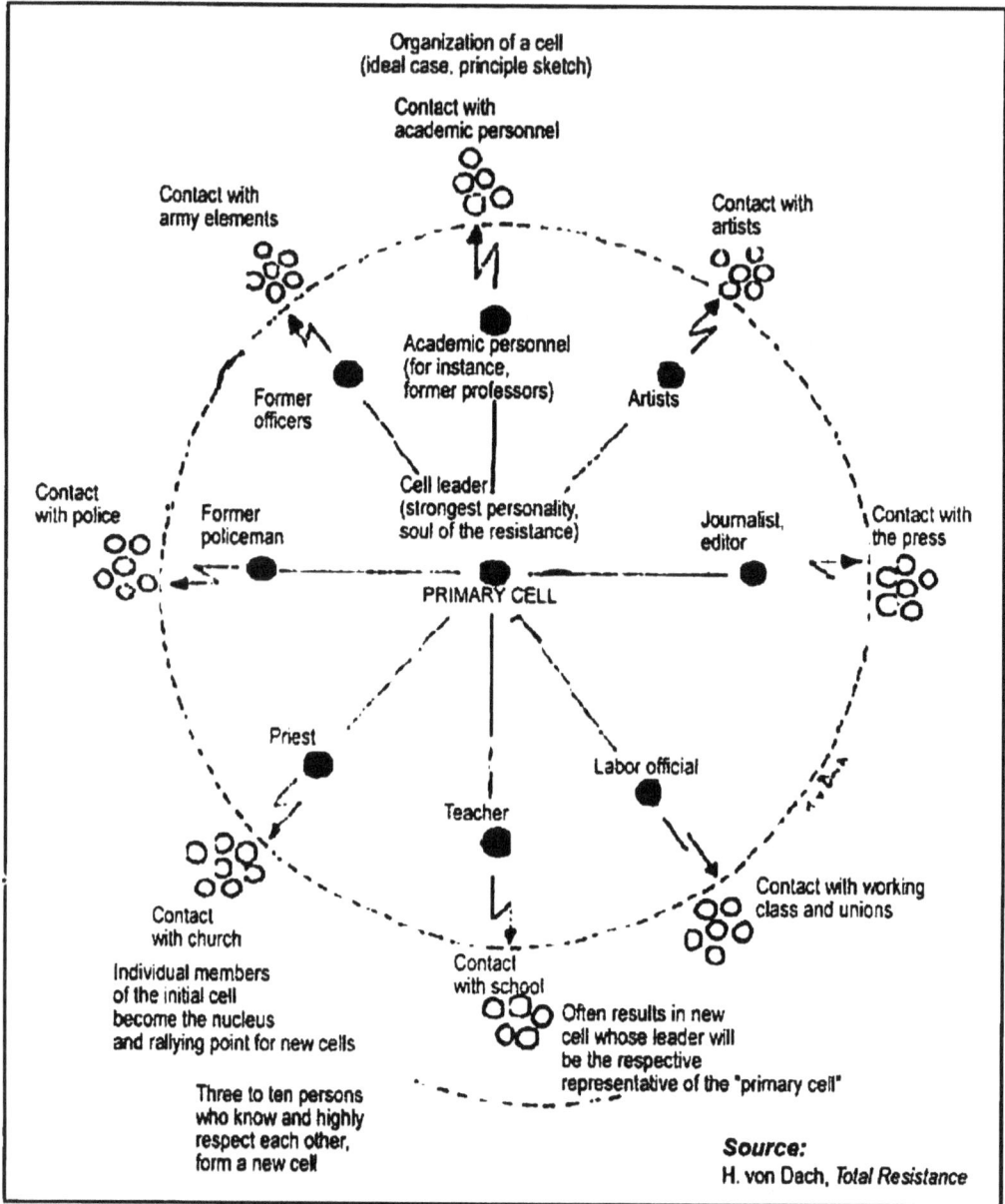

**Figure 2-11. Critical cadres example of cellular organization**

2-104. Cadres can also play an essential role in establishing and maintaining political control of resistance and other organizations. In figure 2-12, page 2-33, the triangles represent a parallel party-controlled structure embedded into resistance and outside organizations. These highly disciplined political cadres ensure the party line is adopted and adhered to by the leaders of all echelons of the subordinate organization. They serve as a loyal feedback loop reporting on those subordinates to the party's senior leadership so those who deviate from expected attitudes and performance can be corrected or eliminated and replaced. Organization and unit operational and tactical decisions will be guided, monitored, and when necessary vetoed, by the political officer cadres at every level, sacrificing initiative and freedom for party control. Although the example shown is a classic communist model, analysis of the presence and effectiveness of this political cadre principle is applicable to resistance efforts of all kinds and sizes.

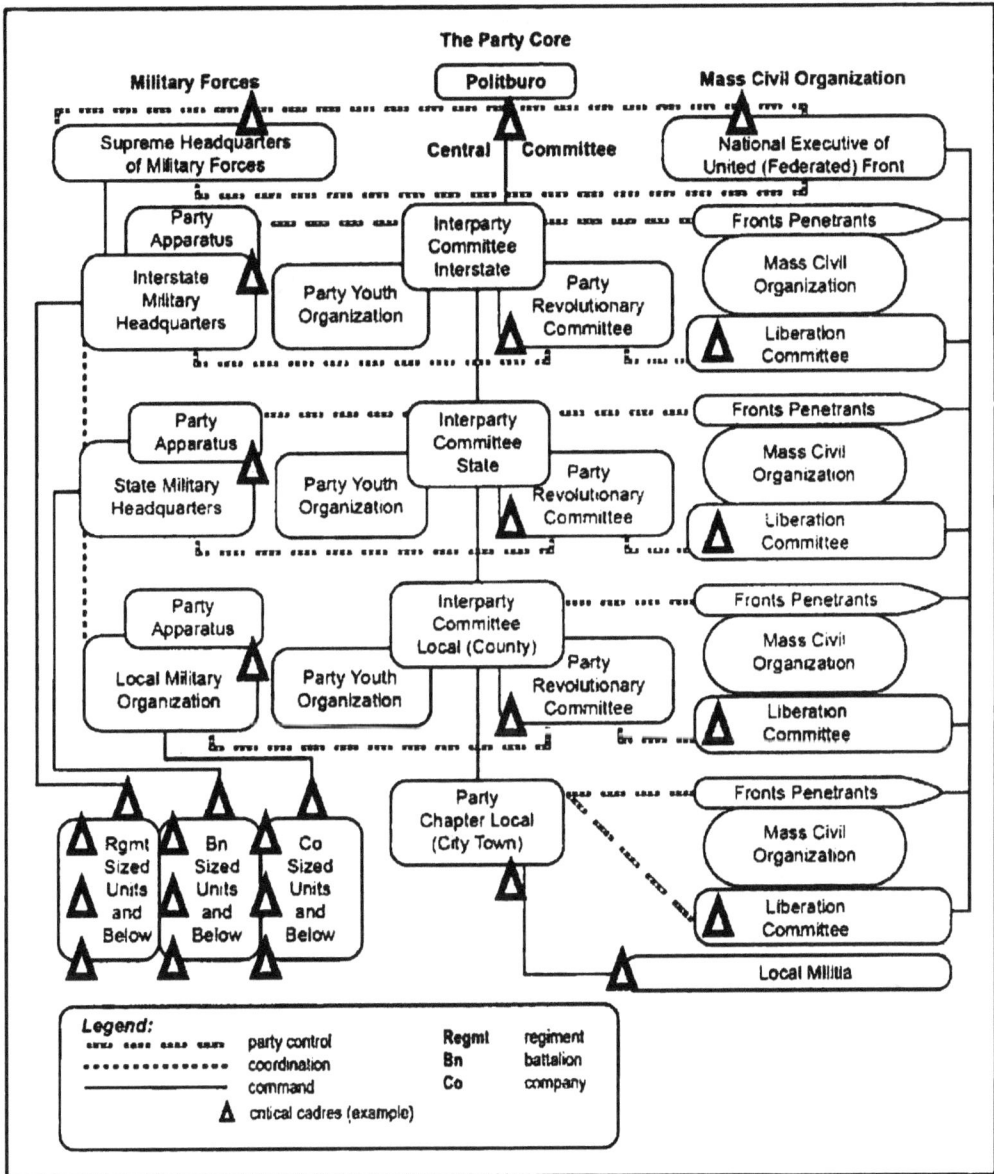

**Figure 2-12. Critical cadres example of political control**

"The Chinese Communist Party is a party leading a great revolutionary struggle in a nation several hundred million strong, and it cannot fulfil its historic task without a large number of leading cadres who combine ability with political integrity. In the last seventeen years our Party has trained a good many competent leaders, so that we have a framework of cadres in military, political, cultural, Party and mass work....

**(continued on next page)**

---

**(continued)**

...We have the responsibility for organizing and training them and for taking good care and making proper use of them. Cadres are a decisive factor, once the political line is determined. Therefore, it is our fighting task to train large numbers of new cadres in a planned way.... We must know how to judge cadres. . . .

We must know how to use cadres well. In the final analysis, leadership involves two main responsibilities: to work out ideas, and to use cadres well. Such things as drawing up plans, making decisions, and giving orders and directives, are all in the category of 'working out ideas.' To put the ideas into practice, we must weld the cadres together and encourage them to go into action; this comes into the category of 'using the cadres well'. . .

The criterion the Communist Party should apply in its cadres policy is whether or not a cadre is resolute in carrying out the Party line, keeps to Party discipline, has close ties with the masses, has the ability to find his bearings independently, and is active, hardworking and unselfish. . . ."

Mao Tse-Tung, The Role of the Chinese Communist Party in the National War,
In *Selected Works*, Vol. II, October 1938 (p. 202)

---

## AREA COMPLEX

### Function

2-105.   An *area complex* is **a clandestine, dispersed network of facilities to support resistance activities in a given area designed to achieve security, control, dispersion, and flexibility.** The area complex is contested territory or an area that contains clandestine supporting infrastructure. It is not liberated territory. It represents the resistance's AO. Resistance elements can maintain their clandestine infrastructure in the area complex. The clandestine infrastructure provides the resistance with a measure of freedom of movement and support. These areas overlap with areas under the control of the government or occupying military. These areas can eventually transform into liberated areas if the enemy's ability to challenge the resistance forces degrades to a level of parity with the guerrilla forces. To support resistance activities, an area complex must include a security system, guerrilla bases, communications, logistics, medical facilities, and a series of networks capable of moving personnel and supplies. The area complex may consist of friendly villages, towns, or portions of urban areas under guerrilla military or political control. Figure 2-13, page 2-35, illustrates some of the functions and entities typical of an area complex. (See Appendix M, Logistics, for more details on the logistics functions of an area complex.)

### Organization, Membership, and Nature of Operations

2-106.   The area complex will include the supporting facilities, functions, and materiel available to the resistance. It will include guerrilla bases and resistance support networks, to include security, logistics, transportation, medical, communications, information and propaganda, recruitment, intelligence and counterintelligence, and finance. The auxiliary will play a leading role in establishing, developing, stocking, concealing, maintaining, guarding, retrieving, and transporting materiel support in parts of the area complex that requires clandestine activity to succeed. The auxiliary will also be the primary component involved in movements of materiel, personnel, casualties, and evaders, and it will play a significant role in intelligence and security networks. However, the area complex reflects the level of permissiveness in the denied area, generally.

2-107.   In large, lightly guarded, and relatively remote rural areas, the facilities, functions, and stores available to the resistance might be accessed with the same level of security and awareness expected of a combat patrol in enemy territory—or perhaps the security awareness of bootleggers, drug producers, or poachers. In large sectors of undergoverned, sporadically policed megacities, large quantities of facilities, functions, and material can be accessed and transported at the same level of security and awareness expected of organized crime's racketeering and gang violence activities. Conversely, in areas where state surveillance and oppression is constant, thorough, multilayered, and quick responding, the most carefully concealed methods will be required to access tightly-controlled facilities, functions, and materials. (See Appendix S, Area Complex Development Model, for more details.)

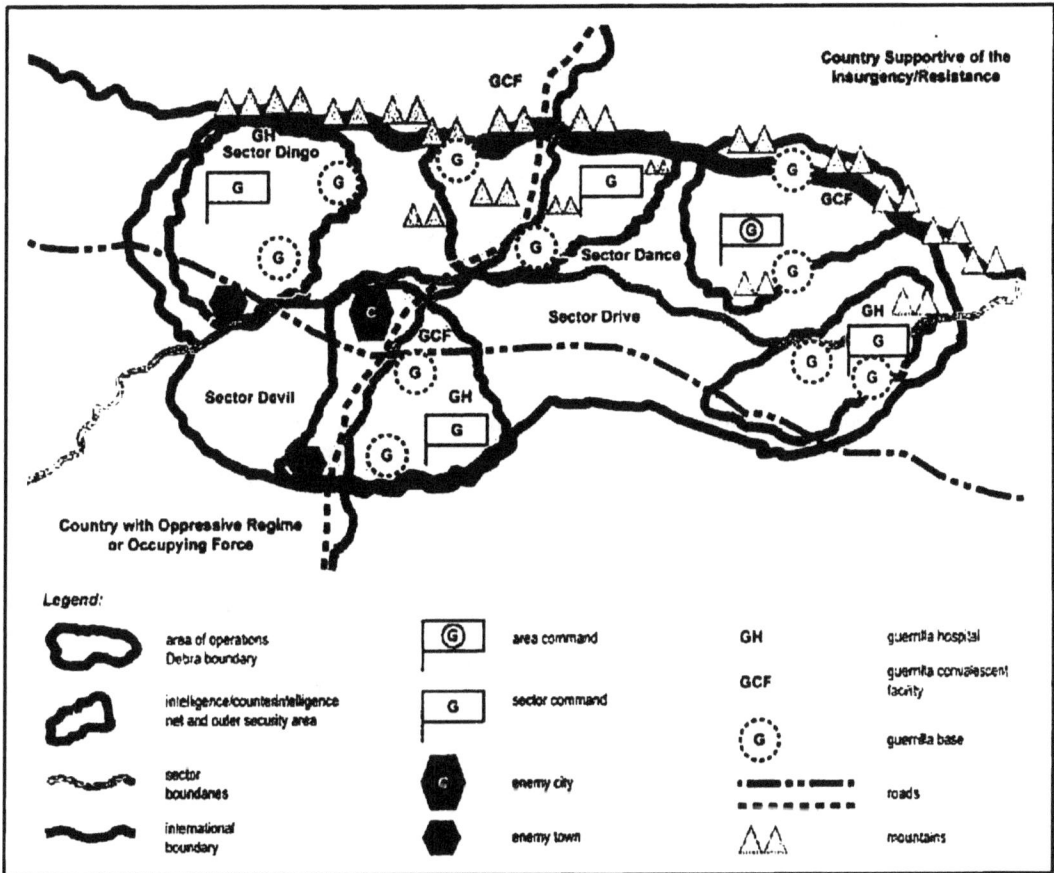

**Figure 2-13. Joint special operations area Debra area complex sketch**

## BASE POPULATION

### Function

2-108. The nature and involvement of the base indigenous population in the resistance AO must be a consideration in analyzing resistance; it can be considered a supplementary component of resistance in some cases. The role played by the indigenous population—if any—will differ based on several variables. Variables will include: the motivating grievances of the resistance, how widely and deeply such grievances may be shared by the larger population and its demographic subsets, cultural qualities, such as respect for or fear of authorities and attitudes toward using violence for political effects, apathy, resistance stated and perceived goals, resistance real and perceived effectiveness in meeting its goals, prospects for individual gain through supporting a resistance, and many other factors. The base population will be a critical factor in mass-based resistance approaches, but it may be deemed less important in elite-vanguard resistance approaches like Bolshevism or coups-d'états.

### Organization, Membership, and Nature of Operations

2-109. The communist method—especially since Mao—recognized the population as the raw material of revolution. Infiltration of popular institutions and organizations that could mobilize popular support was the key to influencing, recruiting, and manipulating ever-expanding popular groups. Insurgent leaders, therefore, tasked elements of the underground with infiltrating civil institutions and manipulating popular grievances and overt indigenous political activities to support insurgent objectives. Many of these activities, such as strikes, labor unrest, food riots, and so on, could be effective methods of combating and weakening the adversary government, without being directly associated with the insurgent effort. (Section 4, Subversion, in Chapter 5, Core Activities, discusses such subversive techniques in more detail.)

2-110.   Communist insurgents of the 20th century referred to the general population as the mass base—a hyperbolic image which magnified assumptions of popular support. The participants in such mass-popular activities may be unaware of their manipulation by and for the insurgent movement. Therefore, it would be inaccurate to consider these participants a direct part of the resistance organization. However, resistance manipulation of such popular organizations, fronts, and groups can be a powerful, indirect political and psychological line of operation against the state.

## SHADOW GOVERNMENT

### Function

2-111.   The shadow government consists of all organizations and key individuals that provide goods or services that parallel or supplant the functions and control of the enemy state government. A shadow government operates at all levels and across the entire broad spectrum of society; it does not refer only to senior indigenous political leaders.

### Organization, Membership, and Nature of Operations

2-112.   Shadow government key individuals and organizations may come from any of the four primary components of resistance, or they may come from outside of the resistance organization. Members of a shadow government may originate as resistance organization cadre who deliberately set up services and functions to support the population whose allegiance and support the resistance seeks. In other cases, those same resistance cadre may infiltrate and co-opt organizations which already exist, bending those organization's activities in support of the resistance. Alternatively, key individuals who attempt to fill popular needs in the community left unaddressed by state hostility, disregard, incompetence, or absence may find common cause with the goals of the resistance. Shadow government leaders may be unaligned with the resistance, but the services provided inadvertently separate the local communities from the state government in ways useful to the resistance.

2-113.   The shadow government operates in the resistance AO in a denied area. A shadow government may or may not have a relationship with a government-in-exile, should one exist. In fact, any government-in-exile may or may not be aware of the existence and capabilities of a shadow government. In some cases, the shadow government may have opposing objectives from a government-in-exile, or the resistance, or both. A shadow government should not be assumed to be a monolith; there may be many disparate shadow governments in separate regions and neighborhoods—the attempted unification of which may present a significant political task for all political actors in the AO seeking their allegiance.

2-114.   A parallel structure of government can undermine public confidence in the hostile regime and secure population support for passive resistance efforts. A population dependent on a shadow government will be compelled to comply with passive resistance. Ideally, the shadow government can perform normal governmental functions in a clandestine manner and synchronize those functions with the resistance movement. The shadow government is critical because it exercises a degree of control, supervision, and accountability over the population at all levels, including district, village, city or province, and it further discredits and delegitimizes the existing government. These functions provide basic human necessities that sustain population groups and usually have the effect of knitting communities together into political entities. Who controls each of these functions—whether it is the government, the resistance, or a third party—wields influence, which may be turned into operational political capabilities. (See Appendix R, Shadow Government Operations Model, for more information.)

## DIASPORA POPULATION

### Function

2-115. Ethnic enclaves or diasporas in third-party countries can provide significant support in terms of political voice, money, personnel, and sanctuary. Some segments of diaspora populations may have loyalties to the resistance that can be voluntarily enlisted from outside the denied area. Other segments of the diaspora population might be exploited involuntarily. Some diasporas may be wealthy and well-integrated into their host societies. Many, however, will be impoverished, desperate, and possibly ostracized from their host nations' societies and vulnerable to intimidation.

2-116. In addition, the value of information diaspora population members may provide about conditions in the denied area could degrade over time. Diaspora members selected for reinsertion into denied areas may have missed changes to the environment and their inappropriate responses to challenge could put them at higher risk of compromise. Moreover, those remaining in the denied area are likely to distrust those who disappeared and then later reappear. The longer the diaspora member was away, the more this will apply.

### Organization, Membership, and Nature of Operations

2-117. Financial networks can be a critical aspect of supporting, sustaining, and resourcing operations, and financial support can sometimes be drawn from diaspora populations. Typical third-country sources include voluntary donations by supporters, tithes conflating religious duty with support for political activities, and legitimate money-producing front enterprises. Examples of involuntary domestic sources include collecting mandatory union dues and association fees that are then funneled into resistance coffers, revolutionary taxes imposed directly on the populace, extortion from legitimate neutral businesses, and outright criminal activity, such as bank robbing or drug trafficking. The USG would not support or condone such activities, but Soldiers need to acknowledge that some resistance entities may employ these techniques.

2-118. With freedom to operate outside of state or occupying power oppression, resistance operatives within the diaspora populations living and working outside of the denied area can leverage international sources of support. Diaspora actors can seek direct donations from sponsoring or sympathetic states, sympathetic nongovernmental organizations (NGOs), and purpose-built fundraising front groups and legitimate business enterprises. Covert financial support can be arranged through money laundering by banks and other legitimate businesses or series of front groups. In the information age, digital transference of enormous sums is often easily achieved. Meanwhile, ancient methods of illicit cross-border activity (such as smuggling) are as relevant today as ever.

2-119. Some historical examples of such resistance diaspora activities are as follows:
- The Liberation Tigers of Tamil Eelam of Sri Lanka were adept at international fundraising, particularly in Europe and Canada.
- The New Peoples' Army and the Provisional Irish Republican Army drew support from the ethnic Filipino and Irish diasporas respectively, particularly in America.
- The Palestine Liberation Organization placed an emphasis on careful coordination of propaganda explanations of their operations in external media for the purpose of creating international sympathy.
- The Palestine Liberation Organization, Hezbollah, and al-Qaeda have all been skilled in the goal of drawing support from the larger Muslim world and Europe by careful manipulation of messages and appeals to religious obligation.

## GOVERNMENT-IN-EXILE

### Function

2-120. A government-in-exile is a government displaced from its country of origin, yet remains recognized as a legitimate sovereign authority of a nation. A government-in-exile will normally take up sanctuary in a nearby allied or friendly nation-state. Governments-in-exile should be expected to seek international support for the nation-state entity it represents but does not currently control, and it should be expected to seek reinstatement of its own power.

## Organization, Membership, and Nature of Operations

2-121. A government-in-exile is likely to come into existence due to foreign occupation of its country or when the regime is removed from power by indigenous opponents. The classic examples are the World War II governments that displaced to London upon Nazi occupation of their homelands. These governments-in-exile were seen as legitimate, considered allies of the victorious powers, and were supported in most cases to regain their previous status. Exiled old regimes which have lost power because of indigenous power struggles may or may not enjoy such legitimacy. In some cases, such old regimes have no realistic hope of regaining power, international recognized legitimacy may be transferred to the new regime, and the personalities of the old regime will fade away. In other cases, the old regime claimants may linger on, plotting to regain power with or without support from the international community of states. Finally, some governments-in-exile may be cynical creations of a sponsoring power, hoping to install a puppet government in another country.

2-122. Some governments-in-exile, therefore, may be legitimate representatives as seen by both the international community and the indigenous populations they represent. This would represent the ideal situation if that government seeks to lead, support, or follow-on from any resistance movement in the former country it controlled. However, not all cases are like this. A standing government-in-exile may initiate a resistance movement and believe it is in control, or the resistance movement may arise spontaneously and not acknowledge government-in-exile control. In fact, the resistance leadership may have different political goals than the exiled government and may or may not welcome that government's participation, once power in the denied area is achieved. Even when working together, the government-in-exile may lose control of or be an ineffective leader of resistance on the ground. Those who fought for victory may not welcome back those who sat on the sidelines in relative safety.

# Chapter 3

# Support to Resistance

*He that would make his own liberty secure, must guard even his enemy from oppression;*
*for if he violates this duty, he establishes a precedent that will reach to himself.*

Thomas Paine, *Dissertation on First Principles of Government*, 23 December 1791

STR is a USG policy option. An STR approach offers an alternative to a direct U.S. military intervention or formal political engagement in a conflict. STR can function as an economy of force action, enabled by a small U.S. footprint approach and requiring a modest investment of resources by relying primarily on indigenous personnel and other means. STR is inherently flexible from the U.S. perspective; the United States does not commit its national prestige to the same extent as the employment of U.S. forces, providing a sound option for the pursuit of limited objectives. Even when the United States intervenes directly in a conflict, resistance elements can complement U.S. military forces.

The United States may have difficulty identifying a legitimate resistance movement to support. Broad differences may exist between the United States and possible resistance elements regarding ends, ways, and means. Accurate all-source intelligence is needed to assess the resistance and understand who is who in the environment before selecting a suitable partner. The United States will have limited control and influence over the resistance. Even if there is initial agreement in many areas, interests will likely diverge over time. In some instances, known support from the United States can undermine the legitimacy of a resistance in the eyes of the local people. Providing STR can have unanticipated consequences over the long term and may inadvertently contribute to regional instability, even after the conclusion of the USG STR campaign.

## SECTION 1. FUNDAMENTALS OF SUPPORTING RESISTANCE

3-1    STR does not automatically equate to the use of the military instrument of USG power. Any combination of the DIMEFIL can be used for STR. The United States should be alert to the possibility of supporting nonviolent resistance movements that may arise in the environment, with non-defense USG departments or agencies leading assistance efforts. It is generally preferable that local friendly actors retain primary ownership of the problem, and STR may in some instances offer a means to this end.

3-2.    STR also provides an alternative to confront adversary non-state actors in ungoverned or undergoverned spaces—providing U.S. policymakers with an option when the host nation is incapable or unwilling to oppose an occupying force. The United States may find that the international community, in some situations, does not accept an STR approach. Without an early and decisive commitment to an STR strategy, the approach may falter. Inadequate efforts to prepare the environment and to plan for the postconflict transition can be major challenges. With the wrong resistance partner, STR can create a major stabilization problem. In each instance, the United States must weigh the risk of action against the dangers of inaction.

3-3.    Given the civil or popular nature of resistance movements, STR generally may include nongovernmental entities, although friendly foreign governments may sponsor and participate in resistance activities. In some instances, the USG may work with a partner nation to enable resistance activities, possibly as part of an overall deterrence or national defense strategy and via the use of security cooperation tools and programs. The vestiges of a vanquished, displaced, or exiled friendly government administration could form the nucleus of a resistance. STR may include the provision of political support and efforts to deliver nonlethal and lethal aid. It may also encompass a range of activities to organize, train, and equip resistance elements in permissive, uncertain, or hostile environments. The provision of advisory assistance as part of STR would signal a high level of U.S. commitment to a resistance organization and movement. In certain circumstances, U.S. leaders may direct the Armed Forces to conduct operations in support of, or in coordination with, resistance elements. STR may play a key role in confronting adversary state and non-state actors who employ conventional and irregular capabilities as part of hybrid strategies. As these actors take ground, an indigenous resistance may arise to oppose them within the occupied territories.

---

### Key Support to Resistance Terms and Definitions

It is important that SF Soldiers, others in the resistance community of interest, and anyone else using this publication use terms correctly. As stated in ADP 1-01, *Doctrine Primer*, 2 September 2014, pg. 1-3, para. 1-9, basic functions of doctrine include: "provide a common frame of reference and cultural perspective," and "provide a common professional language." Use terms correctly. Educate others to do so.

### Resistance Partner Versus Surrogate

There is a critical difference between these terms. Neither is specifically defined in Joint and service doctrine, although this volume introduces such definitions for its own clarity and proposes each be added to the joint lexicon. The term "partner" itself is described many places in joint and Army doctrine; often vaguely as "multinational partner," "foreign partner," "partner nation" and other usages.

> A **resistance partner** specifically is: "A 'partner' conducting resistance with whom the USG mutually establishes agreements to cooperate for some specified time in pursuit of mutually supporting specific objectives. A resistance partner is influenced, not commanded and controlled."

> A **surrogate** specifically is: "Someone who acts on behalf of another. A surrogate is an employee or subordinate that an employer commands and controls and for whose actions the employer bears some legal and moral responsibility."

Joint and Army doctrine uses the term "surrogate," but only sparingly, and only in the context of personal recovery; a unilateral, contractual arrangement entirely consistent with the above distinction. Do not use these terms interchangeably.

STR (and its subset of UW—see Chapter 4 of this publication) requires at least one resistance partner. Although one can have (perhaps many) surrogates operating in support of a STR mission, at least one partner has to be involved. Otherwise, you are doing special (or conventional) operations but not STR.

Finally, while the terms *surrogate* and *proxy* are used in international relations taxonomy as a policy perspective, such usage does not obviate the importance of the distinction made above. The emphasis for both policy decisions and planning should be on lowering decision makers' and practitioners' expectations of "control" over a resistance partner." As the USSOCOM STR Framework Guide (DRAFT) Version 0.33, 13 October 2016, page 15, states: "While a state sponsor may seek to use the resistance organization as its surrogate, actual control will often be limited. As interests fall out of alignment, alliances may well weaken or dissolve over time."

---

3-4. STR involves the synchronized planning and execution of a series of complex activities, requiring close and ongoing collaboration across USG departments and agencies (figure 3-1, page 3-3). Cooperation with a range of partners in the environment is also essential. From the U.S. perspective, STR may be the main effort or a supporting activity within a comprehensive USG campaign. The goal of a USG STR effort may be to coerce, disrupt, or overthrow an adversary government or occupying power as part of a broad DIMEFIL approach. The following paragraphs provide a description and details regarding key STR efforts.

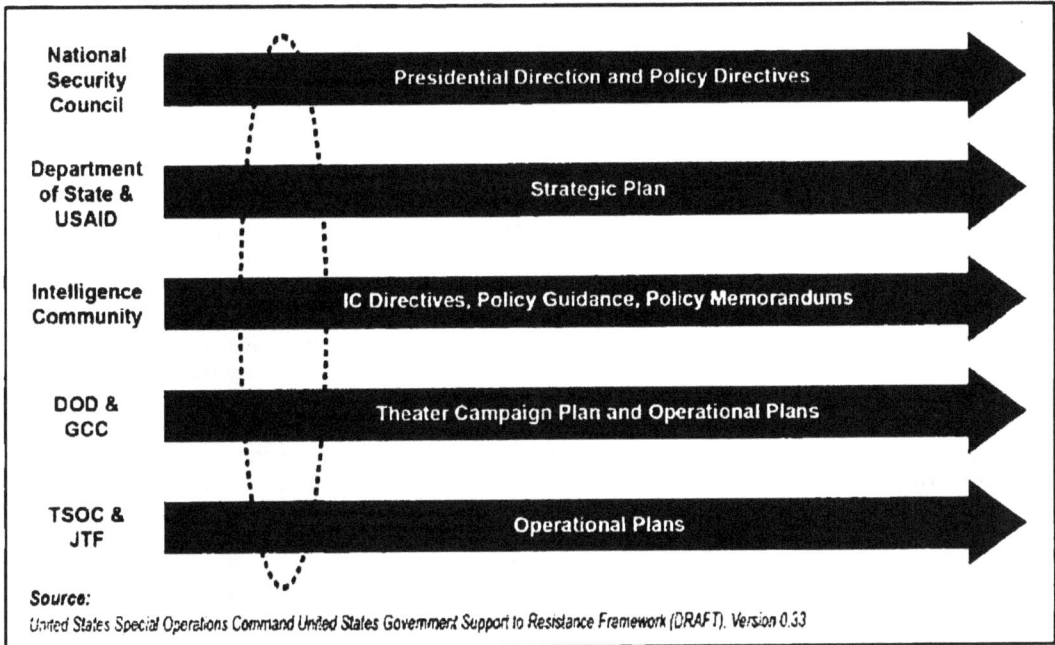

**Figure 3-1. Collaborative planning across United States Government agencies**

# SUPPORT TO RESISTANCE APPROACH

3-5.  To succeed in an STR campaign, the USG and its partners will need to understand the environment and assess resistance potential, plan the STR effort, prepare the environment and execute STR, and transition to a desired postconflict state.

> *Note:* USSOCOM, *U.S. Government Support to Resistance Framework (DRAFT) Version 0.33,* provides more detailed information on STR.

## UNDERSTAND THE ENVIRONMENT AND ASSESS RESISTANCE POTENTIAL

3-6.  Before embarking on an STR campaign, the USG and its partners must assess the resistance potential in the area of interest. These efforts include activities to ascertain a potential resistance group's political objectives, capabilities, and possible limitations in its ability to generate support in the environment. Efforts to obtain necessary insights will require a combination of intelligence activities and engagement with partners. The USG must assess the practicability of developing a unified messaging campaign with the resistance. It is essential for the USG to identify and understand key resistance and regime groups, leaders, and associated networks, as well as their priorities and influences on their behavior. Insight into the grievances, goals, worldview, and ideology of the resistance—and an assessment of its capabilities for political action, internal governance, and military operations—are exceedingly important. The USG must clearly identify the resistance elements the United States would and would not support and clearly understand the second and third order effects of each policy decision.

3-7.  Detailed intelligence, operational, and policy assessments will determine if STR is a viable option. The USG must attempt to anticipate the potential outcomes of conducting STR and impacts across a range of U.S. and international stakeholders. The USG needs to foresee an adversary's potential countermeasures and responses to STR efforts, as well as the possible economic and social consequences. The United States must assess which international partners are willing to participate in the STR campaign and the extent of their participation, such as authorities, national caveats, and so on. A detailed analysis of costs, benefits, and risks is necessary to evaluate all options. The legal implications of conducting STR may vary from one situation to another. For example, supporting a resistance within a friendly foreign country against a hostile occupying power would have a more solid legal foundation than an STR effort against a government that is antagonistic toward the United States, but has not engaged in any aggression.

## PLAN THE SUPPORT TO RESISTANCE EFFORT

3-8.   STR requires an effective political strategy to enhance the legitimacy of the resistance and generate support in the environment. This strategy will guide lethal and nonlethal efforts and enable the resistance to consolidate gains and achieve enduring outcomes. STR efforts must account for a resistance group's stage of development. Some groups may be in a formative stage and incapable for some time of engaging in action effectively. USG representatives must be prepared to coordinate strategy issues and develop information sharing arrangements with resistance leaders. Elements of the USG will have the necessary authorities to assess resistance potential. However, a Presidential decision is necessary prior to initiating assistance efforts, particularly in providing aid to an armed resistance group. An existing or new Presidential finding is necessary to conduct covert action. Early efforts to secure operational authorities to prepare the environment and conduct STR are critically important. STR efforts must consider the range of USG capabilities (DIMEFIL) to both strengthen the resistance and to isolate or weaken the adversary regime or occupying power. The development of command and control structures and collaboration or deconfliction mechanisms deserves careful consideration. USG departments and agencies must have a shared understanding of the continuum of actions that might form part of the STR campaign. Planners should identify measures of performance and measures of effectiveness with which to assess the progress of STR.

## PREPARE THE ENVIRONMENT AND EXECUTE SUPPORT TO RESISTANCE

3-9.   Preparatory activities may not necessarily lead to execution. PE is an umbrella term for operations and activities conducted by selectively trained special operations and paramilitary forces to develop an environment for potential future operations. These preparatory activities may include information PE and the development of local support and organizations to conduct resistance operations. U.S. SOF will typically advise an armed resistance movement to organize an underground, an auxiliary, and a guerrilla force. When conditions are appropriate and when directed by the President, USG departments and agencies conduct STR to coerce, disrupt, or overthrow an adversary regime or occupying power. The STR campaign will often include the range of STR activities described below. Continual assessment of the STR campaign and its impact on the behavior of relevant actors will provide necessary information to adjust efforts during execution. This statement builds on the definition of PE provided in JP 3-05. Specifically, the text above adds mention of paramilitary forces

## TRANSITION TO A DESIRED POSTCONFLICT STATE

3-10.   The USG must have a vision of the desired postconflict state at the start of the STR campaign and an understanding of how it will transition following a conclusion of hostilities. Changes in the lead USG department or agency can be anticipated during the transition stage. A compelling vision of the postconflict state will help generate support, build unity of effort, and sustain resolve.

# SUPPORT TO RESISTANCE ACTIVITIES

3-11.   STR will require a range of multidisciplinary activities, requiring close collaboration across USG departments and agencies. The following paragraphs outline key efforts.

## ENABLE POLITICAL ACTIVITIES

3-12.   The resistance must have a vision of the type of political order it seeks to facilitate once the adversary regime or occupying power is no longer in control if overthrow is the STR objective. Resistance leaders must also determine how they will mobilize the population and sustain resolve over a possibly protracted campaign. In some instances, support to a civil or nonviolent resistance movement may be capable of furthering or accomplishing U.S. goals without the occurrence of an armed struggle. Other political activities may include initiatives to develop a unified opposition front, a government-in-exile, or a shadow government—that may lay the ground work for a new or rehabilitated political process. The USG may also support the resistance in dispatching envoys and even diplomatic missions to communicate its message in other countries. Related initiatives may include the formation of a friends group or an international coalition that is sympathetic and supportive to the resistance. The USG must engage various resistance groups and facilitate dialogue among them to develop unity of effort. Contact with resistance elements will assist the USG in determining the type of sponsorship that is necessary, verifying that support is legally appropriate, and developing a strategy to enable assistance efforts. In some cases, the USG may determine that it is beneficial to work with various civil society groups, which may include:

- Developing or fostering community cohesion and resiliency.
- Partnering with independent media outlets.
- Assisting those who provide political training.

## SHAPE THE INFORMATION ENVIRONMENT

3-13. The USG and its partners must identify key audiences and determine the most effective communications means to reach them. In some situations, crucial audiences may include individuals and groups outside the conflict area and in neighboring countries. It is vital to work with resistance elements to develop an effective narrative and create desired effects that will further shared objectives. In some instances, USG personnel will need to assist the resistance to work with local media, identify networks that will convey effective messaging to key audiences, and counter the adversary narrative and propaganda. Outreach to international audiences and the use of local or nontraditional information capabilities should be key considerations. The role of cyberspace capabilities may vary based on the extent of Internet penetration in the conflict area. Information efforts must anticipate unintended consequences and scenarios that may arise during the STR campaign and develop contingency messaging that will mitigate adverse effects. The USG and the resistance should work together to align words, deeds, and images. In this instance, as in others, integrated USG action across departments and agencies is essential.

## PROVIDE MATERIAL AND NONMATERIAL ASSISTANCE

3-14. The USG must tailor its support to the needs of the resistance and the demands of each particular situation, assessing the level of assistance necessary to sustain the relationship with the resistance and to achieve shared objectives. The USG and its partners should provide support in such a manner that it enhances the legitimacy and credibility of the resistance. The local people should feel that the United States is helping them in their struggle, but not dictating terms. U.S. allies and international partners may be in a position to contribute to assistance efforts. When providing lethal aid, the USG must verify that recipients understand and comply with the laws of war. In some instances, the nonattribution of U.S. aid will be a concern. The United States and its partners must plan for a scalable approach, which can ramp up support as the resistance grows in strength, expands its areas of influence, and accelerates its operational tempo. Friendly forces might not grasp the extent of the resistance potential at the start of a campaign. Efforts to provide material and nonmaterial assistance to a resistance organization may include initiatives to—

- Provide nonlethal aid, such as—
  - Medical supplies.
  - Money.
  - Clothing.
  - Food.
- Furnish lethal aid, such as—
  - Small arms.
  - Antitank weapons.
  - Antiaircraft missiles.
  - Ammunition.
- Share intelligence that has been approved for foreign disclosure.
- Organize, train, and advise resistance elements, which may include land, air, and maritime or riverine components.
- Facilitate public order, governance, and essential services in liberated areas.

- Build internal and external (international) communications capabilities.
- Enable humanitarian assistance.
- Establish screening mechanism and counterintelligence procedures to prevent the adversary from infiltrating the resistance.

## CONDUCT MILITARY AND PARAMILITARY OPERATIONS

3-15. USG departments and agencies must identify and obtain operational authorities prior to the conduct of military or paramilitary operations. The resistance should engage in progressively sophisticated activities as its capabilities improve and opportunities in the environment arise. Military and paramilitary efforts in support of the resistance may potentially include efforts to—

- Provide intelligence support.
- Provide indirect support, such as training and exercises outside the conflict zone.
- Provide direct support without involving combat, such as—
  - Military information support operations (MISO).
  - Logistics.
  - Transportation.
- Conduct combat operations, such as—
  - UW operations.
  - No-fly zone operations.
  - Maritime blockades.
  - Air support.
  - Combined arms operations.
- Defend sanctuary areas where resistance forces can train and refit.

## SECTION 2. CLARITY: INTELLIGENCE, FEASIBILITY, AND INITIAL CONTACT

3-16. STR has a lineage in U.S. Army doctrine dating back to the World War II "classic period" with the OSS field manuals. For over 70 years, doctrinal tradition has continuously provided the basic concepts and character of, and actors involved in, STR. As mentioned previously, guerrilla warfare and underground operations are foundational ideas of this concept—generalized at different times under the overarching label of special operations, strategic services, guerrilla warfare, UW, or STR. Other central STR activities since the classic period have included intelligence operations, escape and evasion or unconventional assisted recovery, psychological or morale operations and preparation, subversion, sabotage, and network development. Recognition of the importance of these activities and the consistent inclusion over the decades is one of the strengths of Army doctrine on resistance.

3-17. Over the same decades, however, and in one manual revision after another, there have been aspects of these concepts that have remained vague or inconsistent. Part of the confusion is that from the earliest OSS manuals in 1943 up to and including the unprecedented JP 3-05.1, *Unconventional Warfare*, in 2012, and innovative nondoctrinal concepts, such as the USSOCOM *USG Framework for STR* (DRAFT) 2017, these doctrine and concepts have been articulated in military publications from a predominantly military perspective. It is important to remember that all these activities are subordinate to policy. The USG can use any combination of the instruments of national power, and the military instrument may often play only a supporting role in any specific instance of STR. (See figure 1-3, page 1-6, Policy Interest and Decisions, Feasibility Assessments and Support to Resistance Contexts.) Three of the most confusing examples of important STR concepts include the intelligence aspects of STR; the criticality of the feasibility assessment; and the varieties of preparation, initial contact, and infiltration included for decades as the first three steps in the seven phases of the UW model. All three of these ideas are related to each other. The purpose of this section is to add clarity to these interrelated concepts in Army doctrine.

# INTELLIGENCE ASPECTS OF SUPPORT TO RESISTANCE

3-18. The OSS manuals of World War II identified intelligence as an important activity necessary to conduct STR. All subsequent STR doctrine (regardless of what it was titled) has recognized intelligence gathering as a core STR activity.

---

### Functions of the Secret Intelligence Branch

"The principal function of the Secret Intelligence Branch is to collect and evaluate secret intelligence and to disseminate such intelligence to appropriate branches of OSS and to military and other authorized agencies. Supplementary functions are: to establish and maintain direct liaison with Allied secret intelligence agencies; and to obtain information from underground groups by direct contact or other means."

FM 5, *Secret Intelligence Field Manual: Strategic Services (Provisional)*

Office of Strategic Services, 22 March 1944

---

3-19. However, there are at least two key differences between OSS era intelligence and that of the early 21st century. The first is that the OSS predated the establishment of the Central Intelligence Agency in 1947 and the subsequent multidecade development of the expansive U.S. national intelligence establishment. which now includes 17 separate agencies. The Armed Services always had intelligence sections that were mostly focused on the relatively narrow needs of each Service, and information and intelligence gathering has always been a part of diplomacy. However, the immediate needs of world war, on an unprecedented scale, highlighted the need for high-risk, on-the-ground intelligence activities that the OSS Secret Intelligence Branch was established to provide.

3-20. When the OSS was deactivated at the end of the war, the intelligence capabilities were split between the national secret IRs in the Central Intelligence Agency and the more narrowly focused military requirements in the Armed Services. Since the 1952 establishment of SF to conduct guerrilla warfare and UW, this once-unified capability for special operations plus both military intelligence and secret intelligence was now bifurcated. For the majority of the UW or guerrilla warfare (STR) doctrine since World War II, the need for intelligence at all levels was vaguely articulated as something the USG would have to accomplish generically without a detailed explanation of specific USG department and agency responsibilities for that intelligence. On one end of an STR-intelligence spectrum, the Central Intelligence Agency was responsible for national-level intelligence and, on occasion, specific operational activities, such as the establishment of early relationships with potential resistance partners. On the other end of the STR intelligence spectrum, SF was understood to be provided pre-existing USG-resistance partner relationships, and while STR—like UW—always had potential strategic utility, the majority of SF intelligence activities focused on intelligence support to operational and tactical operations.

3-21. This longstanding assumption in doctrine remains generally accurate in both law and practice. Short of war declared by the U.S. Congress or specific exception by Presidential finding, the Central Intelligence Agency is the only entity authorized in peacetime to conduct covert actions.

3-22. However, this is not the entire story. Instability throughout the world sees stable nation-states quickly dissolve into misgoverned, undergoverned or ungoverned areas. In addition, allied or partner nations can become adversaries; and resistance potential may appear during the execution of other types of operations. SF are regionally oriented, culturally experienced, and frequently engaged with foreign partners overseas. For all of these reasons, SF routinely maintains the capability to develop resistance potential with strategic, operational, and tactical utility during any stage of resistance development or stage of the USG policy decision process to explicitly conduct STR.

---

*1.7

(a) THE CENTRAL INTELLIGENCE AGENCY. The Director of the Central Intelligence Agency shall:

(1) Collect (including through clandestine means), analyze, produce, and disseminate foreign intelligence and counterintelligence;

(2) Conduct counterintelligence activities without assuming or performing any internal security functions within the United States;

(3) Conduct administrative and technical support activities within and outside the United States as necessary for cover and proprietary arrangements;

(4) Conduct covert action activities approved by the President. No agency except the Central Intelligence Agency (or the Armed Forces of the United States in time of war declared by the Congress or during any period covered by a report from the President to the Congress consistent with the War Powers Resolution, Public Law 93-148) may conduct any covert action activity unless the President determines that another agency is more likely to achieve a particular objective;

(5) Conduct foreign intelligence liaison relationships with intelligence or security services of foreign governments or international organizations consistent with section 1.3(b)(4) of this order;

(6) Under the direction and guidance of the Director, and in accordance with section 1.3(b)(4) of this order, coordinate the implementation of intelligence and counterintelligence relationships between elements of the Intelligence Community and the intelligence or security services of foreign governments or international organizations; and

(7) Perform such other functions and duties related to intelligence as the Director may direct."

Executive Order 12333, *United States Intelligence Activities*,
4 December 1981. Section 1.7(a),
Amended by Executive Orders 13284 (2003), 13355 (2004), and 13470 (2008)

---

3-23. The second key difference from the OSS era is the gigantic amount of information available and the capacity of the USG, partners, and NGOs capacity to capture, gather, transmit, and process such information. The ubiquity and interconnectedness of 21st century communications and the relative transparency of most societies compared to the mid-20th century is an epochal change. On one hand, this means that the ease of gathering information useful to policy and operational decision making is unprecedented. Therefore, the information readily available to support USG STR decision making specifically is available at higher policy echelons than previously. This has the effect of relegating the needs of intelligence to either the most sensitive and narrowly focused of national missions or to the on-the-ground requirements of operational and tactical survival and success. For example, sometimes the best solution for a specific example of STR may be targeted, covert funding through routine commercial channels based on expert understanding of international finances. This is probably not an STR example requiring military participation. Another example of STR is a psychological assessment of specific audiences and targeted strategic messaging from outside the denied area, which results in adversary policy changes that may meet the STR with little or no requirement for intelligence. There are many potential examples to illustrate the effect of information age intelligence on the artful application of STR

# CRITICALITY OF THE FEASIBILITY ASSESSMENT

3-24. The changes to information and intelligence also meaningfully highlight differences in what feasibility means at different echelons of command in ways not adequately addressed in previous, traditional STR (UW) doctrine. There are at least four different levels at which feasibility to conduct STR has meaning. The political, operational, resistance capability, and the feasibility to conduct UW (as a subset of STR).

## POLITICAL FEASIBILITY OF UNITED STATES SPONSORSHIP

3-25. When there is policy interest in possibly conducting STR by the USG, the routine intelligence processes and executive departments will provide strategic information and national policy-level advice to national decision makers. The political feasibility assessments at these levels will include national, nonmilitary instruments of USG power. There is no set format, but some of the considerations likely to be entertained at this level of assessment and decision making include these questions:

- What effects are the USG trying to achieve? What could or must be done through STR to achieve them?
- How might STR alter the strategic military-political balance of power?
- What are the significant risks of adopting a STR policy domestically and internationally?
- What intended effects might trigger unwanted escalation by our adversaries? What second and third order effects might do the same?
- Who would be involved? Who else could help? Who would have to be informed and why? Who must it be concealed from?
- What assets are in place to affect the outcome now? What can be brought to bear and what does that entail?
- How might STR help set conditions for international crisis resolution on terms favorable to the United States or allies without the need for an overt U.S. conventional force commitment?
- What are the coordination and execution requirements for envisioned STR options?
- What mix of overt, low-visibility, clandestine, and covert activities will be required?
- What is the probability of STR success? What does success look like?
- What is the probability of STR failure? What do we stand to lose if this all goes badly?
- Who should be the lead agency?
- What are our range of options?

3-26. The policy questions above are not specific to any particular executive department or agency and this national security council-level consideration could lead to any combination of USG-wide efforts. However, the military participants at this level will attempt to answer the questions above, both as advice to general policy and through the specific lens of military capabilities.

## OPERATIONAL FEASIBILITY OF SUPPORT TO RESISTANCE

3-27. When policy interest advances to the point where STR is envisioned involving the military instrument of power, the GCC may direct the conduct of feasibility assessments at theater strategic, operational, and tactical levels. These GCC and inter-GCC considerations will include the entire range of conventional and special operations capabilities available. Sometimes, the best option for STR may be a conventional aircraft flyover, publicized naval patrols, or a maneuver show-of-force on a key border. One should not automatically assume that the STR feasibility query at this point is a UW feasibility assessment. There is no set format, but some of the considerations likely to be entertained at this level of assessment and decision making include the policy questions above, as well as:

- What is the mission?
- What are the objectives?
- What effects are the USG trying to achieve? What could the GCCs do to achieve them?
- What is the current adversary correlation-of-forces with our own in theater? How might STR alter the strategic military-political balance of power?
- To what extent is support for this specific resistance already outlined in current operation plans (OPLANs), if at all?
- What forces do we currently have available to meet different policy options?
- What are the significant military risks of adopting a STR policy?
- What intended effects might trigger unwanted military escalation by our adversaries? What second and third order effects might do the same?
- Who would be involved? Who else could help? Who would have to be informed and why? Who must it be concealed from?
- What assets are in place to affect the outcome now? What can be brought to bear and what does that entail?

- What are the coordination and execution requirements for envisioned STR options?
- In which options are we the main effort and in which are we the supporting effort?
- What is the right mix of conventional forces and SOF?
- What mix of overt, low-visibility, clandestine, and covert activities will be required?
- What is the probability of STR success? What does success look like?
- What is the probability of STR failure? What do we stand to lose if this all goes badly?
- What are our range of military options?
- What do we know about the resistance?
- Do we have relationships with anyone in the resistance already?
- What is our access to the resistance?

3-28. The joint GCC staff will consider the questions above from the perspective of the entire theater and primarily through a conventional force capability first. However, as the military aspects of STR have traditionally been associated with special operations, the GCC will task the TSOC to provide a STR feasibility assessment from a special operations core missions' capability perspective. This may or may not include, or proceed to, a UW feasibility assessment. Five types of STR—two traditional UW and three non-UW STR—are outlined in Section 3 below.

## RESISTANCE CAPABILITY ASSESSMENT

3-29. Senior U.S. leaders are likely to emphasize U.S. processes and inputs to mission success in STR; however, STR involves support to a resistance, and it is important that STR planners emphasize knowledge of the resistance itself. The resistance is not a subordinate echelon of the USG and U.S. commanders that can be commanded and controlled. The resistance is a partner that has a mind of its own, may have uncertain loyalties and tenacity, and has objectives that may or may not align well with the USG's objectives, and must be influenced as a cooperative partner. Moreover, the United States should not assume that it knows all necessary relevant details on the ground inside the denied area, at the beginning of an STR feasibility process. On the contrary, getting enough detail on the resistance to make reasonable policy option recommendations on the feasibility of STR can be a daunting and dangerous challenge.

> *Note:* Once again, there is no set, absolute format for a resistance capability assessment, and each specific case of resistance requires its own evaluation. Traditional Army STR (UW) doctrine since at least FM 31-21 (1958) has continuously used the area assessment as a guiding format for evaluation of the resistance. (See Appendix H, Sample Area Study Format; and Appendix I, Area Assessment Formats—Initial and Principal.) ATP 3-05.1—Appendix E, Revolutionary and Insurgent Warfare Analysis, and Appendix F, Unconventional Warfare Campaign Analysis—provides useful resources for resistance assessment. Portions of several collective tasks in TC 18-01.1, TC 18-01.2, and TC 18-01.3 (Conduct Premission Activities; Integrate Military Information Support Operations; Integrate Civil Affairs Operations; Develop an Area Command Group and Staff; Develop the Area Complex; and Assess and Advise on the Coordination of Shadow Government Functions) provide useful lenses through which to focus detailed analysis, as do the network models in Appendix A, Sample Planning Charts, in each of the referenced TCs.

3-30. Another start point for the resistance capability assessment is a list drawn from the dynamics and components addressed in Chapter 2:

- What is the detailed, outline history of the resistance to date?
- What are the political, strategic, operational, and tactical objectives of the resistance?
  - To what extent are these objectives deliberately stated by the resistance or inferred by others?
  - How do openly stated objectives differ from underlying, clandestine objectives?
  - How do these objectives differ for different audiences?
  - How capable is the resistance in meeting these objectives?

- What ideology does the resistance espouse?
  - How are narratives shaped to message this ideology?
  - How does the resistance deliver the messages to target audiences?
  - How effective are these messages in achieving the effects desired by the resistance?
- What strategic, operational, and tactical approaches does the resistance use to meet its objectives?
- If the resistance approach is conducted in phases, how are those phases understood, structured, and implemented?
- How is the resistance organized?
  - How does the organizational structure enable the resistance to carry out operations?
  - What are the resistance's organizational strengths and weaknesses?
  - What is the resistance equivalent to the underground, guerrilla (or armed) force, the auxiliary, and the public component?
  - What are the functions, membership, organization, and nature of operations of each of these components?
  - What is the resistance internal support structure?
  - What is the resistance equivalent to the area complex? How is it organized? Who controls, maintains, and secures it? Who adds to or draws from it?
  - What methods does the resistance employ to solicit or seize internal support?
- What is the structure of the leadership?
  - Who provides the overall policy leadership for the resistance? Where are they located?
  - Who provides the strategic, operational, and tactical leadership? Where are they located?
  - What is the resistance equivalent to the resistance area command?
  - What is the chain of command? Who are the key leadership nodes critical to resistance functions?
  - What is the resistance equivalent of sector commands?
  - What is the legitimacy of the resistance leadership as perceived by the other resistance leaders? By the indigenous populace who supports the resistance? By any external supporters? By other state and non-state actors?
  - How is organizational discipline maintained? Who is responsible for enforcing it?
  - How does the organization expand? Who is responsible for multiplying its force and capabilities?
  - Does the resistance use cadres? If so, how?
  - What are the resistance leadership interfaces, if any, with any shadow government leadership and mass population base leadership?
- What is the underlying environmental, geographic, political, and demographic contexts of the denied area in which the resistance operates?
  - What is the mix of urban and rural areas of operation in the denied area? How does this mix affect how the resistance operates?
  - What are the economic and social conditions in the denied area? How do they affect how the resistance operates?
  - What are the underlying political conditions which determine who supports the resistance, supports the state or occupying power, supports other rivals for power, and who remains neutral and why?
- What sources of external support are available to the resistance?
  - What sources of external support does the resistance source from its own organization?
  - What external supporters exist? What do they provide? What is their relationship to the resistance?
  - In addition to material support, who sponsors the resistance from external locations?

- Does the resistance have a government-in-exile? If so, where is it located? What legitimacy does it have? What does it provide the resistance? How are communications between the external location and the denied area achieved?
- What methods do the resistance employ to solicit external support?
- What is the state or occupying power's capability to suppress the resistance?
  - What countermeasures does the resistance employ to provide for its own survival and security?
  - What is the range of overt, low-visibility, clandestine, and covert activities necessary for resistance survival and effectiveness?
  - What is the spectrum of nonlethal, to increasingly lethal, capabilities used successfully by the resistance?
  - What is the scale of each resistance's nonlethal and lethal capability?
  - What is the current correlation-of-forces between the resistance and the adversary state, occupying power, and any other rivals for power in the denied area?

3-31. Every resistance and potential resistance will be different. In some cases, the USG will have large amounts of detailed, specific information and long experience with a resistance. In such cases, answering resistance capability assessment questions like the ones above will be relatively easy, and information gaps can be easily filled by routine USG intelligence community collection efforts and other standard observation and reporting. However, such information gathering for assessment will differ by each unique instance of resistance, the sensitivity of USG interest, and the level of inquiry. For example, although much may be known, or can be gathered, about a specific resistance at a political and perhaps strategic level and records of resistance phenomena can be gathered and analyzed, the on-the-ground operational and tactical details of the resistance may be difficult to determine. A range of USG DIMEFIL capabilities can be employed to fill those more elusive information gaps. In some cases, the best option will be to insert operatives dedicated to close preparation of the denied area environment to answer such resistance capability questions.

## FEASIBILITY TO CONDUCT UNCONVENTIONAL WARFARE

3-32. When the USG has determined a policy interest in one or more specific resistances, the initial political and operational feasibility assessments may determine sufficient promise to continue pursuit of that policy interest. The USG considers the results of initial resistance capability assessments and decides on a policy of STR. In those cases in which other DIMEFIL solutions to STR are inadequate, a feasibility to conduct UW assessment may be undertaken. What may appear straightforward and relatively simple at the policy and senior command level, may be risky and difficult at the operator level. Although no one should expect to have complete information and every question answered, it is critical that the UW feasibility assessment be conducted with as much fidelity as possible.

### The Criticality of the Feasibility Assessment

3-33. Planning remains limited until leadership validates certain assumptions. If operations proceed without a proper feasibility assessment, the likelihood of unintended consequences is high. To gain an accurate picture, operational personnel need to meet with indigenous personnel who represent the resistance forces. This meeting may take place inside the denied territory, in the United States, or in a third-party nation. Although meeting representatives in the United States or a third-party nation is safer for an assessment team, it also provides a less reliable assessment of potential capabilities and may be politically damaging if a covert or clandestine operation is exposed. Participation of all components is preferred to enable an accurate assessment of all potential resistance capabilities.

3-34. The assessment analyzes the feasibility, acceptability, and adequacy of a mission. This is an assessment based on METT-TC to determine if the necessary means and resources are available to meet mission requirements. It also addresses whether the potential gain or desired effect outweighs or otherwise justifies the potential losses or cost. In addition, the assessment determines if achieving the desired objectives would accomplish the desired effects.

3-35. While determining the feasibility of a potential campaign, planners must have clear objectives, a desired end state, and knowledge of exactly what level of support is available and acceptable. Without these specifics, negotiations with potential resistance forces are futile. If planners determine conditions are unfavorable during the assessment, then they need to consider any measures that could transform the current situation into a more favorable one. For example, can the USG—

- Persuade a potential resistance group to cease unacceptable tactics or behavior?
- Persuade a coalition to accept a specific resistance group's participation under certain conditions?
- Degrade the enemy's control over the population?
- Bolster the will of the population to resist?
- Achieve desired objectives within the given time constraints?

3-36. SF can actively engage their resistance counterparts to encourage adherence to international norms of behavior and law. They can also change attitudes and beliefs about other groups participating in the resistance effort as part of unity and cohesion building. SF may leverage Psychological Operations (PSYOP) and Civil Affairs (CA) enablers to help engage and influence their partners. In circumstances where the resistance has sanctuary outside of denied territory, CA elements and PSYOP units may be able to link up directly to provide support.

3-37. Planners need to be careful of attempting to overcome a potential resistance shortcoming by creating resistance or guerrilla forces that are not indigenous. Historically, the United States has not had success creating and transplanting these types of resistance forces to an operational environment without an existing clandestine infrastructure that connects the local population to the foreign forces.

3-38. Expatriates are a valuable resource, particularly in regions where the culture is largely unfamiliar or alien to a planner's frame of reference. However, planners should carefully ensure the individual's claims are valid. An expatriate's influence in a given country can be inversely proportional to the length of time he has been away from his former homeland. Although there are many reasons an expatriate might exaggerate his influence in a region and attempt to exploit the situation in his favor, he may be legitimately surprised to find his own assessment of his influence to be grossly inaccurate. During normal peacetime conditions, a person can spend years away from a country and expect to maintain their contacts and influence. This period significantly shrinks under the pressures of a harsh regime or occupying force.

## Feasibility Analysis

3-39. UW feasibility assessments are a continuation of all previous USG assessments for the specific purpose of UW with an emphasis on operational- and tactical-level requirements to achieve effects inside a specified denied area. The focus is to evaluate proposed mission accomplishment, using available resources within the plan's contemplated time frame. Measures to enhance feasibility include adjusting the concept plan (CONPLAN), ensuring sufficiency of resources and capabilities, and maintaining options and reserves. Representative questions that would be included in the TSOC's evaluation are likely to include:

- What are the GCC's objectives and commander's intent for this proposed UW plan?
- What are our limitations and constraints?
- What are the METT-TC factors?
- Is the USG in contact with or can it make contact with individuals representing the resistance potential in an area?
- Specifically, what assets do we have that are already in contact with the resistance inside the denied area?
- What assets are critical to develop inside the denied area in order to conduct STR operations? How long would it take to establish, mature, and have them reach maximum effectiveness?
- What do we know and what do we still need to learn about resistance groups in the operational area?
- What is the resistance's state of development?
- What resistance capabilities could be enhanced by U.S. support?
- What objectives have the USG and senior resistance leadership mutually agreed to?
- What means of support have been agreed to?
- What is the appropriate composition and strength of a U.S. advisory effort to meet USG objectives?

- What are the operational options for delivering support?
- Will the environment geographically and demographically support resistance operations?
- Is the enemy effectively in control of the population?
- What are the appropriate locations for U.S. personnel to be located outside of the denied area to render support to the resistance?
- What are the appropriate locations for U.S. personnel to be located inside of the denied area to render support to the resistance?
- What are the appropriate locations for U.S. personnel to be co-located with the resistance inside and outside the denied area?
- Where, how, and when could we infiltrate U.S. advisors?
- What staging bases are required to support UW operations for mission command, intelligence, infiltration, sustained logistics support, and so on?

### Complete In-Progress Review—Final

3-40. Once the feasibility of STR (UW) is assessed and reported, a draft OPLAN may be created. Once the plan is completed, the GCC submits it with the associated time-phased force and deployment data file to the joint staff for review. The joint planning and execution community reviews the plan for:
- **Adequacy.** Does the plan satisfy the mission and comply with guidance provided?
- **Feasibility.** Are the required resources available in the time frames anticipated?
- **Acceptability.** Are the anticipated operations proportional and worth the anticipated costs? Is it politically supportable?
- **Completeness.** Does the plan include all required parts?
- **Compliance.** Does the plan align with current U.S. and international law and joint procedures?

## AREA ASSESSMENTS

3-41. Two additional levels of assessment remain for STR (UW). One is subsequent to an approved operation order (OPORD) by the Secretary of Defense that results in the GCC directing detailed mission planning by subordinate units. The second would be used subsequent to an execute order and executing unit infiltration into the denied area. Both use the same area assessment format and emphasize tactical and operational details inside the denied area. (See Appendix I, Area Assessment Formats—Initial and Principal.) During mission planning, the format focuses the executing unit on necessary details for mission accomplishment. The unit will attempt to fill in the format using all information accumulated in ongoing area studies conducted by them or others and provided by the military intelligence detachment, the mission brief received by higher, special operations mission planning folders (SOMPFs) and relevant target intelligence packages (TIPs), if any, and ongoing requests for information (RFIs). Once the unit is infiltrated into the denied area, it immediately reports data included in the initial assessment. The principal assessment is then continuously updated for the duration of the unit's mission to support all subsequent activities and operations.

# VARIETIES OF PREPARATION, INITIAL CONTACT, AND INFILTRATION

## BACKGROUND CONTEXT

3-42. The ideas of preparation, initial contact, and infiltration have been key parts of the Army STR—such as OSS, guerrilla warfare, and UW—tradition since World War II. All three classic STR concepts exist in OSS manuals in an as-yet-unconnected context (see text excerpts from various OSS manuals below). The earliest example of preparation from FM 2 is clearly referring to psychological preparation of varied target audiences to increase chances of allied success through inspiring resistance among the subject population and to degrade the morale and effectiveness of the axis and their collaborators.

"22. FREEDOM STATIONS.

Various types of *cover* are used by freedom stations. They may pose as: the organ of a subversive (freedom) group within an enemy area; a regular official station of the enemy; the special organ of an official enemy group (Army, Navy, or party); the organ of an anonymous or private group within enemy territory, not openly subversive. Some freedom stations have no very definite cover at all beyond that of being a freedom station run by a colorful personality, the question of identity being left as a mystery.

a. USES OF FREEDOM STATIONS. Freedom station transmissions can be used for the following purposes:

(1) To spread demoralizing rumors among enemy soldiers and civilians;

(2) To encourage 'patriotic' resistant groups within enemy territory by acting as their spokesman;

(3) To stimulate and direct sabotage and subversive activity;

(4) To make preparations for, and direct, popular uprisings and pro-Allied activity on D-day;

(5) To terrorize collaborationists and Axis officials by black-balling, giving the impression of a vast and powerful underground;

(6) To divide the enemy group from group, or nation from nation, by spreading divisive stories, by posing as a representative of one group or nation and condemning, insulting, ridiculing another group or nation;

(7) To heckle enemy broadcasts."

FM 2, *Morale Operations Field Manual: Strategic Services (Provisional)*,
Office of Strategic Services,
26 January 1943, p. 22, para. 22

3-43. Initial contact, as represented in FM 4, refers to gaining contact and establishing good relations with resistance groups.

"SECTION X—THE SELECTION OF SPECIAL OPERATIONS TASKS AND MISSIONS...

39. TYPES OF TASKS

In sabotage and in contact with and support of resistance groups there is a large field of possible SO [special operations] tasks, including:

a. ORGANIZATIONAL TASKS—the recruiting of agents, gaining contact with and establishing good relations with such groups, assisting in their training, organization, leadership and supply."

FM 4, *Special Operations Field Manual: Strategic Services* (Provisional),
Office of Strategic Services
23 February 1944, p.16, para. 39a

3-44. Surprisingly, very little verbiage is dedicated to an explicit discussion of infiltration in the OSS series manuals, but it is inherently necessary to the types of STR the OSS executed. The term infiltration does appear in FM 6 in the required training curriculum section.

"10. CURRICULUM

a. Members of Operational Groups should receive adequate training in the following subjects:

(1) Map study, including map sketching map-and-compass problems, direction-finding by field expedients, study of aerial photos.

(2) Scouting and patrolling, including instruction and practice in use of physical cover, reconnaissance, signalling, infiltration. . . ."

FM 6, *Operational Groups Field Manual: Strategic Services* (Provisional), Office of Strategic Services, 25 April 1944, p.11, para.10a

3-45. These three terms—preparation, initial contact, and infiltration—along with the other foundational concepts of OSS-era STR—would be incorporated into the longstanding Army-phased model of U.S. STR, which eventually became known and is still used today as *The Seven Phases of UW*. The first sequential use of the six phases of STR in Army doctrine and nonconsecutive seventh phase of STR in Army doctrine was from as early as 1951, as seen in the following excerpt. Therefore, the coalescing doctrinal model of the seven phases is already present in an Army (STR) manual which predates the establishment of the Army SF itself. By 1965, the essential seven phase model had taken shape and has remained consistent in Army STR (guerrilla warfare and UW) doctrine to the present day.

"b. Organizing and conducting guerilla forces behind enemy lines may be expected to involve staff studies, tentative operation plans, and operation plans for six phases. They are—
(1) Psychological preparation.
(2) Initial contact.
(3) Infiltration of operational groups.
(4) Organization and partial supply.
(5) Logistical build-up.
(6) Exploitation.

c. In many cases, the first three phases must be accomplished to obtain enough information on which to base estimates and plans for the operations involved in the last three phases.

d. The theater special forces commander issues directives to place in effect his plan for a proposed operation. These directives normally cover a complete operation and include tentative plans to allow lower echelons and supporting agencies time to prepare. Tentative operations plans are amended kept up-to-date and placed in effect by supplemental directives and orders of, the special forces commander."

FM 31-21, *Organization and Conduct of Guerrilla Warfare*, 5 October 1951, p. 38, para. 37b–d

"CHAPTER 35. DEMOBILIZING GUERILLA FORCES

224. GENERAL

a. As the tide of battle rolls beyond an area, the purpose for which a guerilla force was organized and its ability to assist regular forces cease to exist. When this point is reached, the force should be demobilized without delay. . . ."

FM 31-21, *Organization and Conduct of Guerrilla Warfare*, 5 October 1951, p. 228, para. 224

3-46. There have been some minor changes over the decades. Phase six has morphed from exploitation, to combat employment, to employment reflecting an appreciation for both the partner—versus subordinate or surrogate—role of the resistance and the more holistic range of employment capabilities that a resistance can achieve beyond the emphasis on guerrilla warfare as the decisive phase. Phase seven has changed from demobilization to transition for some of the same reasons. Depending on the ultimate policy goal of the STR (UW), the resistance may transition to the new, victorious regime resulting in the armed resistance still being demobilized, or alternatively incorporated into the new regime's security service, or a combination of both. When the STR (UW) policy goal is more limited, such as for coerce or disrupt objectives, transition may mean to disband, go dormant, change methods, change sponsors, proceed to a simple termination of activities, and so on.

3-47. In addition, the STR founding concept of psychological preparation of target audiences remains valid and important presently. However, the importance of preparation STR (UW) doctrine is much broader than just that former singular, specific application. The FM 31-21 (1958) had eight instances of preparation but did not mention psychological preparation at all. In subsequent decades during which UW phase one was normalized as psychological preparation, the overwhelming preponderance of preparation usages nevertheless referred to other forms of preparation, such as unit and mission. For example, while FM 31-20 (1971) continued to state UW phase one as psychological preparation, over 93-percent of the uses of preparation did not refer to phase one. In fact, the 1971 manual explicitly identified multiple levels and types of preparation—from strategic to tactical—and varied by command levels, staff sections, and executing units. In 2003, FM 3-05.201, *Special Forces Unconventional Warfare Operations*, acknowledged this long-recognized reality and changed UW phase one to the current preparation.

---

### Phases in Development of Guerrilla Warfare

¨34. Phases in Development of Guerrilla Warfare

a. General.

(1) Throughout the development of guerrilla warfare, the importance of security and intelligence procedures must be emphasized.

(2) The cycle of a U.S.-sponsored resistance movement consists of seven phases of development. These phases do not necessarily occur in the sequence indicated and several phases may occur simultaneously.

b. Phase I, Psychological Preparation. Psychological operations are initiated well in advance to prepare the local populace to receive U.S. personnel. This requires a well-coordinated psychological operations effort executed at the JUWTF level.

c. Phase II, Initial Contact.

(1) Before Special Forces detachments can be introduced into a guerrilla warfare operational area, assets must be developed in the area. Initial contact provides for the development of assets and their subsequent reception of, and assistance to, the infiltrated operational detachment.

(2) It is advantageous to have an asset exfiltrated from the operational area to the SFOB to assist in briefing the operational detachments and to accompany the detachment during infiltration. The lack of an asset does not necessarily preclude the infiltration of the Special Forces detachment.

d. Phase III, Infiltration. The entry of the Special Forces operational detachment, with accompanying supplies, into the enemy or enemy-controlled area, to include establishing communications with the SFOB. . . .

e. Phase IV, Organization. Area assessment is initiated. Organization and development of the guerrilla warfare operational area begins immediately after infiltration. . . .

f. Phase V, Build Up. The guerrilla force is expanded. The organizational structure, including various clandestine support activities, is compartmented to enhance security and improve operational potential. All activities increase in scope and intensity to include the capability of the guerrilla force to support theater objectives. . . .

**(continued on next page)**

---

> **Phases in Development of Guerrilla Warfare (continued)**
>
> g. Phase VI, Combat Employment. The guerrilla force is committed in consonance with unconventional warfare objectives of the unified commander. This phase continues until linkup occurs between conventional and guerrilla forces, or hostilities cease. . . .
>
> h. Phase VII, Demobilization. Once the ultimate objectives for the activation and organization of the guerrilla force have been achieved, demobilization of the guerrilla force should be accomplished. . . ."
>
> FM 31-21, Special Forces Operations—U.S. Army Doctrine,<br>5 June 1965, pp. 61-62, para. 34

## CURRENT APPLICATION

3-48. ATP 3-18.1 consciously modifies the concepts of preparation, initial contact, and infiltration in doctrine once again by differentiating different types and levels appropriate to these terms. This modification is necessary based on the previous section's doctrinal explanation. There has been a widespread—but mistaken—conflation of the relatively recent doctrinal term *preparation of the environment* with "preparation" as phase one of the UW model. Moreover, there are many varieties of what it means to prepare generally.

> **The Seven Phases of Unconventional Warfare**
>
> "2-31. The seven phases of U.S.-sponsored insurgency are a conceptual template that planners use to aid understanding of a UW campaign effort. As a template, it merely serves as a guide for planning and execution. Not all phases will necessarily take place nor do they have to be executed in a rigid linear order. For example, U.S. forces may come to assist a large, previously established insurgent/resistance force with significant prior combat experience. In such a case, the organization, buildup, and employment phases may already be completed or ongoing before SOF ever infiltrate the JSOA. The seven phases are a model, not a checklist. With the exception of Military Information Support (MIS) forces, no other SOF will mirror the seven phases, but may support a specific portion or phase of the larger UW campaign. Regardless, operational personnel should understand how their efforts integrate with and contribute to the overall campaign plan."
>
> ATP 3-05.1, Unconventional Warfare (Change 1),<br>25 November 2015, p. 2-8, para 2-31

3-49. There are many different levels at which preparation, initial contact, and infiltration have useful but dissimilar meanings. These levels include at least:

- Steady state existing between nation-states.
- USG policy interest in potential STR.
- USG holistic DIMEFIL STR activities at strategic, operational, and tactical levels of activity exclusive of the military instrument.
- U.S. military STR activities at strategic, operational, and tactical levels of activity (exclusive of UW).
- U.S. military STR (UW) activities at strategic, operational, and tactical levels of activity (inclusive of UW) prior to activities inside the denied area.
- U.S. military STR (UW) activities at strategic, operational, and tactical levels of activity (inclusive of UW) during and after initiation of USG activities inside the denied area.

### Preparation

3-50. Preparation can be any deliberate, premeditated effect imposed on a target audience, object, or process. At the highest policy and strategic levels, examples might include deliberate strategic messaging through U.S. Embassies, military shows-of-force, manipulation of currency exchange rates, or a wide range of other

USG DIMEFIL capabilities. Use of these capabilities represents the art of statecraft. At lower, tactical levels examples of preparation could mean psychological preparation of specific target audiences as part of a larger JFC's campaign plan; deliberate establishment of supporting human and physical infrastructure for potential follow-on missions; unit training to meet mission-essential task list (METL) requirements; coordination of prearranged communications security measures; packaging of emergency supply bundles; or an endless variety of military STR (UW) capabilities and techniques, tactics, and procedures. Use of these sometimes represents the art of STR (UW) and sometimes merely SOPs.

## Initial Contact

3-51. Initial contact also has a wide variety of meanings depending on many variables. Examples include: echelon of USG activity, significance and authority of interlocutors at each level, linear temporal point at which such contacts are made, direct versus indirect contact, physical human face-to-face contact versus virtual-only contact, direct contact between two principles versus contact filtered through human or technical cutouts, initial contact between two already acquainted parties which represents not an interpersonal personalities change but a change in bargaining positions, and so on. Like the confusing vagueness of traditional doctrine's explanation of psychological preparation, initial contact has also been insufficiently articulated. Figure 3-2, page 3-20, shows one representative—not all-inclusive—possible way of acknowledging the wide range of meaning for initial contact.

## Infiltration

3-52. Consistent with the layout of preparation and initial contact above, infiltration also requires a modern relook of long-standing doctrinal representations and community assumptions. Infiltration is not perceived only from the national policy and strategic perspective as the first infiltration affected during the STR (UW) campaign. From the national perspective, there will often be such a first incidence of infiltration according to a named plan, but there are at least three critical shortcomings of this interpretation:

- Every USG and allied entity that is part of a STR (UW) plan and is intended at some point to cross a threshold into a denied area will conduct infiltration. To limit the idea of infiltration to only that of the first incidence—from the perspective of the strategic plan—is inaccurate and needlessly confusing. Every unit that will enter the denied area in UW conducts infiltration.
- Much of the modern world is already interconnected digitally and by other nonhuman, nontangible modes of communication. When parts of a STR (UW) plan include remote, cyber, or other virtual activities not allowing or not requiring a physical presence in the denied area, the idea of infiltration is profoundly modified. Not only are there many instances of infiltration, there are also many varieties.
- The high degree of already existing global, transnational, transcultural interconnectedness sometimes means that infiltration is unnecessary. In cases where operatives, partners, and surrogates already located in a denied area and supporting a STR (UW) plan simply commence activities, infiltration may be unnecessary. Similarly, when STR (UW) effects in a denied area from a remote location involve a keystroke, breaking of a circuit, propagation of airwaves, and so on, traditional notions of infiltration are inapt.

## Conclusion

3-53. The above discussion highlights the variety of applications of standard English terms and specific professional positions on semantic meaning. The many examples above are all reasonable and appropriate. However, ATP 3-18.1 is written for SF units tasked with preparing for and executing UW. In Chapter 4, Fundamentals of Unconventional Warfare, the primary use of the seven phase model of UW will be interpreted as a model from the tactical and operational perspective of the executing unit—not primarily from the national policy and strategic levels nor emphasizing conceptual, academic speculations. The start point for this perspective will be the operational detachment's receipt of a UW mission. All SF units preparing for and executing UW conduct all seven phases of the model. The interpretation by some Soldiers that certain units do some phases of the traditional seven phase UW model and other units do the remaining phases of the UW model has been—and herein is explicitly identified as—invalid.

| | Known /Emerges Spontaneously | Unknown /Emerges Spontaneously | Known or Unknown / Solicited |
|---|---|---|---|
| **STRATEGIC** | Routine, KNOWN, OFFICIAL STATE (& POSSIBLY INTERSTATE) government contact with access to a "denied area" emerges as a "resistance" figure (national political/strategic level) [Example: disgruntled regime cabinet minister or military general signals displeasure/resistance potential against targeted regime].<br><br>Routine, KNOWN, UNOFFICIAL/NON-STATE (& POSSIBLY INTERSTATE) contact with access to a "denied area" emerges as a "resistance" figure (national political/strategic level) [Example: transnational religious leader, global business executive; leading cultural figure or NGO manager]. | UNKNOWN, OFFICIAL STATE (& POSSIBLY INTERSTATE) government contact with access to a "denied area" emerges as a "resistance" figure (national political/strategic level) [Example: disgruntled regime cabinet minister or military general signals displeasure/resistance potential against targeted regime].<br><br>UNKNOWN, UNOFFICIAL/NON-STATE (& POSSIBLY INTERSTATE) contact with access to a "denied area" emerges as a "resistance" figure (national political/strategic level) [Example: trans-national religious leader; global business executive; leading cultural figure; or NGO manager]. | KNOWN or UNKNOWN, OFFICIAL STATE (& POSSIBLY INTERSTATE) government contact with access to a "denied area" is solicited as a "resistance" figure (national political/strategic level) [Example: disgruntled regime cabinet minister or military general signals displeasure/resistance potential against targeted regime].<br><br>KNOWN or UNKNOWN, UNOFFICIAL/NON-STATE (& POSSIBLY INTERSTATE) contact with access to a "denied area" is solicited as a "resistance" figure (national political/strategic level) [Example: Trans-national religious leader; global business executive; leading cultural figure or NGO manager]. |
| **OPERATIONAL** | Routine, KNOWN, OFFICIAL SUB-STATE government contact with access to a "denied area" emerges as a "resistance" figure (sub-national political/operational level) [Example: disgruntled regime provincial governor, big city mayor, army or corps commander, regional police, fire or emergency medical chief signals displeasure/resistance potential against targeted regime].<br><br>Routine, KNOWN UNOFFICIAL SUB-STATE contact with access to a "denied area" emerges as a "resistance" figure (sub-national political/operational level) [Example: provincial-level religious leader; regional business executive; big landowner; regional cultural figure; tribal chief; provincial-impact NGO leader]. | UNKNOWN, OFFICIAL SUB-STATE government contact with access to a "denied area" emerges as a "resistance" figure (sub-national political/operational level) [Example: disgruntled regime provincial governor; big city mayor; army or corps commander; regional police, fire or emergency medical chief signals displeasure/resistance potential against targeted regime].<br><br>UNKNOWN, UNOFFICIAL/SUB-STATE contact with access to a "denied area" emerges as a "resistance" figure (sub-national political/operational level) [Example: provincial level religious religious leader; regional business executive; big landowner; regional cultural figure; tribal chief; provincial-impact NGO manager]. | KNOWN or UNKNOWN, OFFICIAL SUB-STATE government contact with access to a "denied area" is solicited as a "resistance" figure (sub-national political/operational level) [Example: disgruntled regime provincial governor; big city mayor; army or corps commander, regional police, fire or emergency medical chief signals displeasure/resistance potential against targeted regime].<br><br>KNOWN or UNKNOWN, UNOFFICIAL/SUB-STATE contact with access to a "denied area" is solicited as a "resistance" figure (sub-national political/operational level) [Example: provincial-level religious leader; regional business executive; big landowner, regional cultural figure or provincial-impact NGO manager]. |
| **TACTICAL** | Routine, KNOWN, OFFICIAL SUB-REGIONAL government contact with access to a "denied area" emerges as a "resistance" figure (sub-regional political/tactical level) [Example: disgruntled regime small city or town mayor or village headman; division, regiment or lower military commander; local official; local official service provider signals displeasure/resistance potential against targeted regime].<br><br>Routine, KNOWN UNOFFICIAL SUB-REGIONAL contact with access to a "denied area" emerges as a "resistance" figure (sub-regional political/tactical level0 [Example: district or parish religious leader; local business leader; local cultural figure or local-impact NGO manager; union official; local gangster chief; outlaw leader]. | UNKNOWN, OFFICIAL SUB-REGIONAL government contact with access to a "denied area" emerges as a "resistance" figure (sub-regional political/tactical level) [Example: disgruntled regime small city or town mayor or village headman; division, regiment or lower military commander; local official; local official service provider signals displeasure/resistance potential against targeted regime].<br><br>UNKNOWN, UNOFFICIAL/SUB-REGIONAL contact with access to a "denied area" emerges as a "resistance" figure (sub-regional political/tactical level) [Example: district or parish religious leader; local business leader; local cultural figure or local-impact NGO manager; union official; local gangster chief; outlaw leader]. | KNOWN or UNKNOWN, OFFICIAL SUB-REGIONAL government contact with access to a "denied area" is solicited as a "resistance" figure (sub-regional political/tactical level) [Example: disgruntled regime small city or town mayor or village headman; division, regiment or lower military commander; local official; local official service provider signals displeasure/resistance potential against targeted regime].<br><br>KNOWN or UNKNOWN, UNOFFICIAL/SUB-REGIONAL contact with access to a "denied area" is solicited as a "resistance" figure (sub-regional political/tactical level) [Example: district or parish religious leader; local business leader; local cultural figure or local-impact NGO manager; union official; local gangster chief; outlaw leader]. |

**Figure 3-2. Varieties of initial contact**

## SECTION 3. VARIETIES OF SUPPORT TO RESISTANCE

3-54. There are five primary types of STR, as discussed in the following paragraphs. The distinctions between them are most important at the policy level, where the political purpose for conducting the specific STR is decided. Although the policy objectives and limitations will shape commanders' possible operational approaches, the vast majority of tactical activities and operations will be the same for operators on the ground.

3-55. The USG may make a policy decision to provide support to a resistance through a combination of U.S. national instruments of power. Sometimes that support may include the military option to train, advise, assist, and support that resistance with military advisors in, or adjacent to, the denied area where the resistance is taking place. In some cases, assistance to the resistance effort inside the denied area can be rendered remotely and sometimes from great distances through digital, virtual, or other disembodied channels.

# SUPPORT TO RESISTANCE TYPE ONE: RESISTANCE AGAINST AN OCCUPYING POWER (CLASSIC UNCONVENTIONAL WARFARE)

3-56. In type one STR (figure 3-3), *country e* invades and occupies all or part of friendly *country c*. Indigenous citizens of *country c* then resist the occupation. This STR type represents a classic form of UW. Its purpose is to assist the indigenous resistance to defeat and expel the foreign invaders, thus allowing the citizens to regain control and reestablish the sovereignty of their own country. Resistance to external invasion is most likely to anger most every demographic category and socioeconomic strata in the occupied country, making the task of organizing resistance by the indigenous actors themselves and STR by external actors easier. An example of this type would be U.S. support to the French Maquis resisting German occupation of their country in World War II.

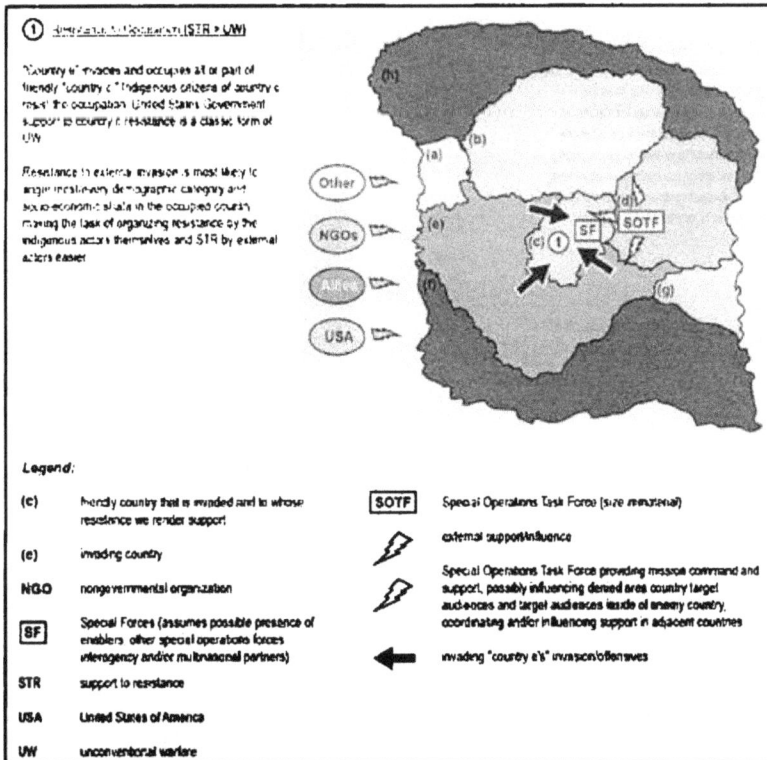

**Figure 3-3. Support to resistance type one: resistance to occupation (classic unconventional warfare)**

3-57. Planning for stay-behind resistance is not a completely different type of STR but a variation of this one. Preparations made to conduct future resistance against possible invasion can be an integral part of a nation's defense planning. Support to such allied nations' resistance efforts is a classic form of UW. (See Appendix Z, Stay-Behind Unconventional Warfare Considerations.)

# SUPPORT TO RESISTANCE TYPE TWO: INSURGENCIES AGAINST A SOVEREIGN STATE GOVERNMENT (CLASSIC UNCONVENTIONAL WARFARE)

3-58. In type two STR (figure 3-4), factions within *country a* commit violence and other acts of subversion for political objectives. These insurgent factions represent resistance against the authority of the standing government. USG support to insurgency is a classic form of UW. Its purpose is to assist an insurgency to coerce, disrupt, or overthrow the current regime so that the insurgency has the opportunity to influence and institute change or take power and see a new government formed. Unlike broad-based resistance to occupying power, insurgencies are more likely to be organized and motivated along specific demographic, ideological, religious, and other categories and may be concentrated in specific socioeconomic strata, much narrower than the population as a whole. This fact is likely to make the task of organizing resistance by the indigenous actors themselves and STR by external actors much more difficult relative to resistance to occupation. An example of this type would be U.S. support to Tibetan insurgents in China during the early 1950s.

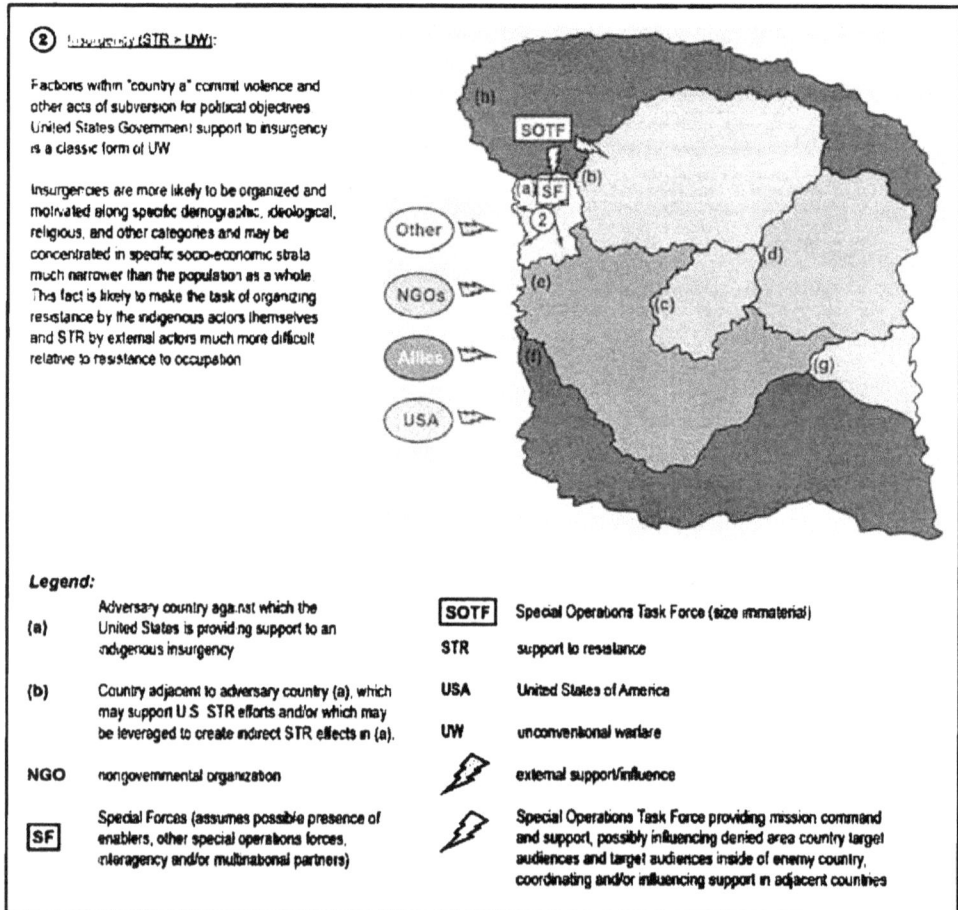

Figure 3-4. Support to resistance type two: insurgency (classic unconventional warfare)

# SUPPORT TO RESISTANCE TYPE THREE: INDIGENOUS RESISTANCE ELEMENTS IN SUPPORT OF FOREIGN INTERNAL DEFENSE

3-59. In type three STR (figure 3-5), sovereign nations whose national rulers and administrations are recognized by the international community as the legitimate government nevertheless have portions of their territory under the physical control of non-state actors. When the degree of such non-state actors control rises to the level of providing state-like functions to the populace, such as governance, law enforcement, provision of basic state services like health care or sanitation, and so on, it is treated similar to an occupying power. In this case, resistance are those elements oppressed by the non-state actor governing authority willing to resist against that non-state actor's control. An example of this type would be U.S. support to Iraqis in Mosul resisting Islamic State in Iraq and Syria control of northern Iraq in 2017.

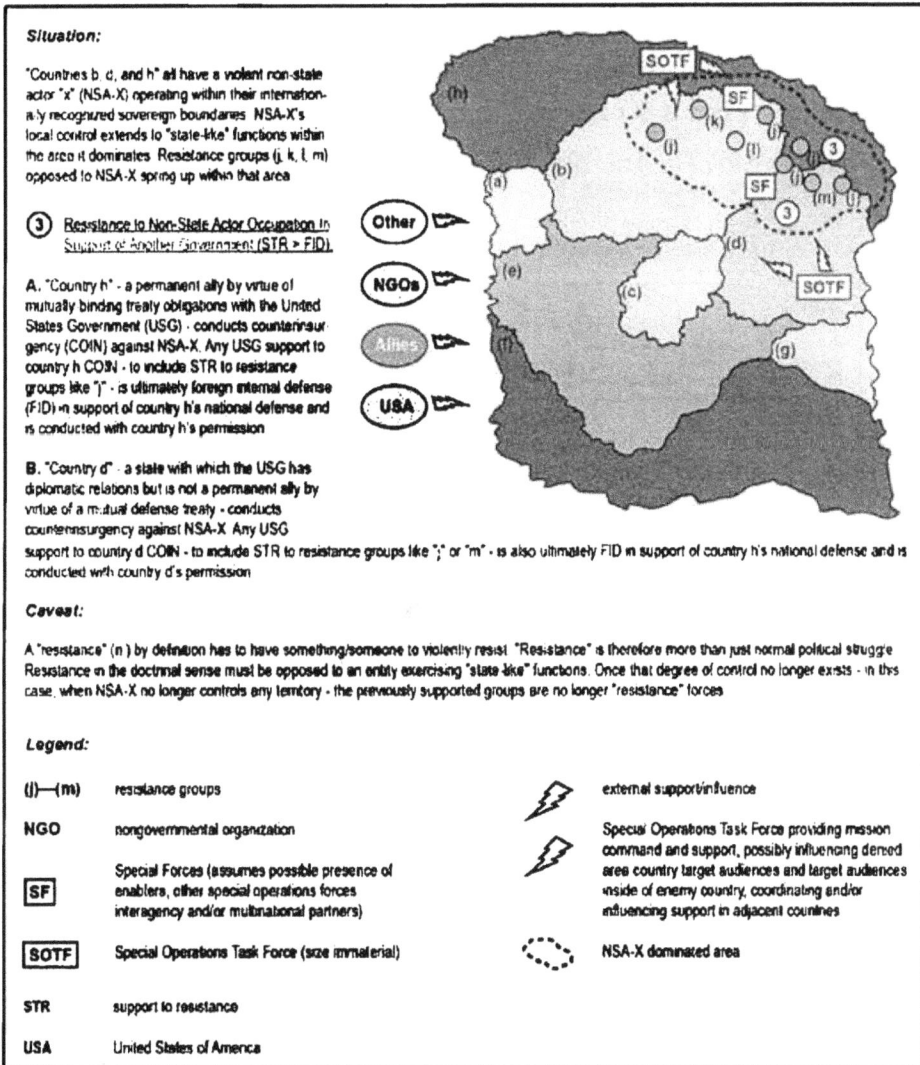

**Situation:**

"Countries b, d, and h" all have a violent non-state actor "x" (NSA-X) operating within their internationally recognized sovereign boundaries. NSA-X's local control extends to "state-like" functions within the area it dominates. Resistance groups (j, k, l, m) opposed to NSA-X spring up within that area.

③ Resistance to Non-State Actor Occupation in Support of Another Government (STR > FID).

A. "Country h" - a permanent ally by virtue of mutually binding treaty obligations with the United States Government (USG) - conducts counterinsurgency (COIN) against NSA-X. Any USG support to country h COIN - to include STR to resistance groups like "j" - is ultimately foreign internal defense (FID) in support of country h's national defense and is conducted with country h's permission.

B. "Country d" - a state with which the USG has diplomatic relations but is not a permanent ally by virtue of a mutual defense treaty - conducts counterinsurgency against NSA-X. Any USG support to country d COIN - to include STR to resistance groups like "j" or "m" - is also ultimately FID in support of country h's national defense and is conducted with country d's permission.

**Caveat:**

A "resistance" (n) by definition has to have something/someone to violently resist. "Resistance" is therefore more than just normal political struggle. Resistance in the doctrinal sense must be opposed to an entity exercising "state-like" functions. Once that degree of control no longer exists - in this case, when NSA-X no longer controls any territory - the previously supported groups are no longer "resistance" forces.

**Legend:**

| | |
|---|---|
| (j)—(m) | resistance groups |
| NGO | nongovernmental organization |
| SF | Special Forces (assumes possible presence of enablers, other special operations forces interagency and/or multinational partners) |
| SOTF | Special Operations Task Force (size immaterial) |
| STR | support to resistance |
| USA | United States of America |

external support/influence

Special Operations Task Force providing mission command and support, possibly influencing denied area country target audiences and target audiences inside of enemy country, coordinating and/or influencing support in adjacent countries

NSA-X dominated area

**Figure 3-5. Support to resistance type three: resistance to non-state actor occupation in support of foreign internal defense**

3-60. STR types three and four differ by whether or not the USG is supporting resistance in cooperation with the internationally recognized national regime that owns the sovereign territory in which the non-state actor occupier is located. The purpose of type three STR is defeat of the non-state actor to ultimately support the host-nation government's counterinsurgency, counterterrorist, and broader national security objectives. Because USG STR is a policy option to support another sovereign government, the USG STR in this case is ultimately conducted on behalf of foreign internal defense. With the benefit of international legitimacy on the side of the sovereign government, political support for the STR effort from the international community should be easier to obtain. It should be noted that when the non-state actor is degraded to the point where it no longer is able to control territory and provide state-like functions to a populace, it would no longer be meaningful to regard opponents of non-state actor remnants as resistance; therefore, STR ceases at that point in lieu of other internal defense options.

# SUPPORT TO RESISTANCE TYPE FOUR: INDIGENOUS RESISTANCE ELEMENTS IN SUPPORT OF COUNTERTERRORISM, COUNTERPROLIFERATION, OR STABILITY OPERATIONS

3-61. The purpose of type four STR (figure 3-6) is defeat of the non-state actor for USG objectives despite noncooperation or lack of permission from the national government in whose sovereign territory the operations take place. An example of this type would be U.S. support to Syrians resisting Islamic State in Iraq and Syria control of northern and eastern Syria from 2011 to 2017.

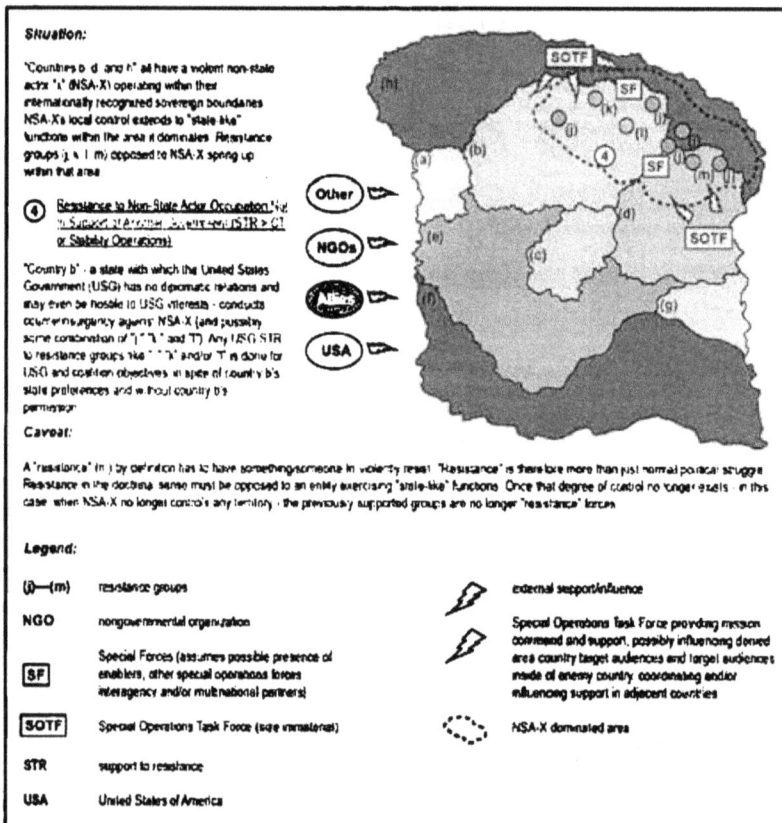

**Figure 3-6. Support to resistance type four: resistance to non-state actor occupation in support of counterterrorism, counterproliferation, or stability operations**

3-62. In STR type four, USG objectives may include counterterrorism, counterproliferation, and support to regional stability or similar objectives. Because such objectives are usually transnational in nature, the USG may be supported by multinational and other international partners. Once again, it should be noted that when the non-state actor is degraded to the point where it no longer is able to control territory and provide state-like functions to a populace, it would no longer be meaningful to regard opponents of non-state actor remnants as resistance; therefore, STR ceases at that point in lieu of other operations. Effective achievement of STR objectives and elimination of the non-state actor's ability to control territory to the level of providing state-like functions will probably be to the advantage of the government that is the sovereign authority over that territory. Sovereign governments should be expected to occupy the vacuum created by the degradation of the non-state actor that had once controlled some of its national territory.

# SUPPORT TO RESISTANCE TYPE FIVE: INDIGENOUS RESISTANCE ELEMENTS IN A CONTESTED, UNGOVERNED SPACE

3-63. This is the weakest and least likely case of supporting a resistance, and this situation is most likely to be confused with merely supporting surrogates intervening in foreign politics (figure 3-7). An example of this type would be U.S. support to factions in the Second Libyan Civil War from 2014 to 2018.

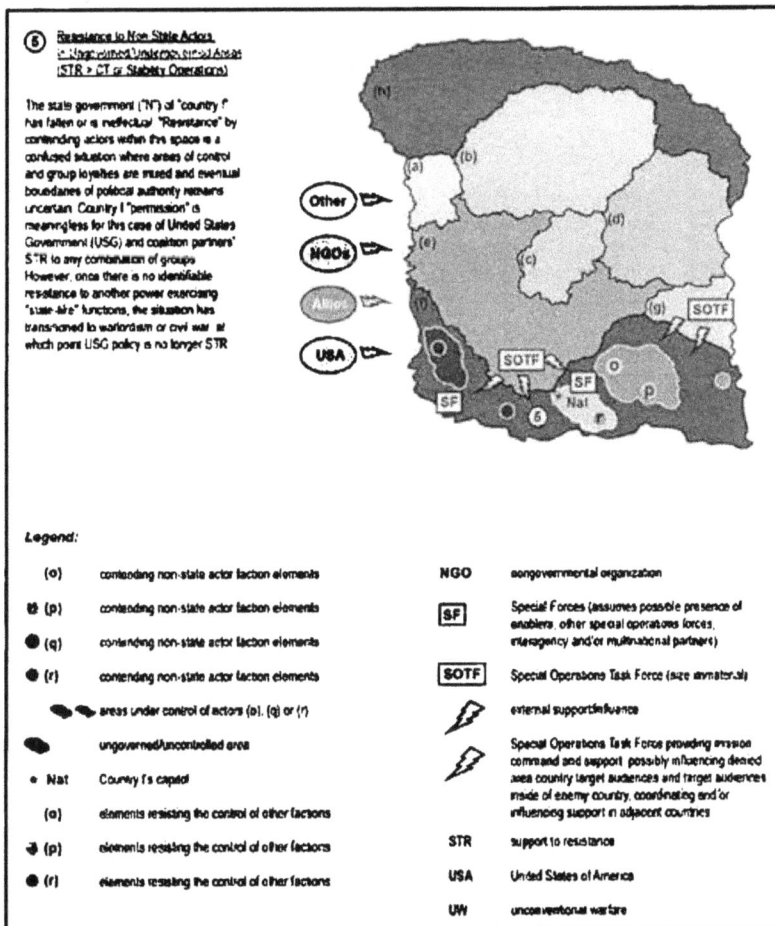

**Figure 3-7. Support to resistance type five: resistance to non-state actors in ungoverned or undergoverned areas**

3-64. In this type of STR, a previous sovereign government has fallen or is so ineffectual that a state of civil war exists between actors contesting over misgoverned, undergoverned, or ungoverned space. Therefore, no internationally recognized, legitimate national government exists. If a national government did exist and the USG supported it, USG efforts—including any STR—would be in the context of foreign internal defense in support of an ally and might rise to the level of including major combat operations. If a national government existed that was opposed by the USG, the USG could choose to conduct major combat operations against the regime or, at a lower level of commitment, conduct STR within the area of that government's control. However, such STR against an opponent regime would be STR type two (support to insurgency).

3-65. The purpose of type five STR is the defeat of one or more non-state actors who exercise some state-like functions over a populace to meet USG objectives. This STR provides greater stability through elimination of the most serious threats and promote consolidation of opposing factions. STR is not support to the non-state actors themselves; it is support to resistors within the area of control of such non-state actors. Once again, it should be noted that when the non-state actor is degraded to the point where it no longer is able to control territory and provide state-like functions to a populace, it would no longer be meaningful to regard opponents of non-state actor remnants as resistance; therefore, STR ceases at that point in lieu of other operations. Effective achievement of STR objectives and elimination of the non-state actor's ability to control territory to the level of providing state-like functions will probably be to the advantage of other actors contesting over the same space. Other actors—sometimes to include intrusive external power seeking to exploit an advantage—should be expected to occupy the vacuum created by the degradation of the non-state actor that had once controlled some territory.

3-66. All five STR types are combined in figure 3-8, page 3-27.

**(1)** Resistance to Occupation (STR > UW):

"Country e" invades and occupies all or part of friendly "country c." Indigenous citizens of country c resist the occupation. United States Government (USG) support to country c resistance is a classic form of UW. Resistance to external invasion is most likely to anger most/every demographic category and socio-economic strata in the occupied country, making the task of organizing resistance by the indigenous actors themselves and STR by external actors easier.

**(2)** Insurgency (STR > UW):

Factions within "country a" commit violence and other acts of subversion for political objectives. USG support to insurgency is a classic form of UW. Insurgencies are more likely to be organized and motivated along specific demographic, ideological, religious, and other categories and may be concentrated in specific socio-economic strata much narrower than the population as a whole. This fact is likely to make the task of organizing resistance by the indigenous actors themselves and STR by external actors much more difficult relative to resistance to occupation.

**Situation:**

"Countries b, d, and h" all have a violent non-state actor "x" (NSA-X) operating within their internationally-recognized sovereign boundaries. NSA-X's local control extends to "state-like" functions within the area it dominates. Resistance groups (j, k, l, m) opposed to NSA-X spring up within that area.

**(3)** Resistance to Non-State Actor Occupation In Support of Another Government (STR > FID):

A. "Country h" - a permanent ally by virtue of mutually-binding treaty obligations with the USG - conducts counterinsurgency (COIN) against NSA-X. Any USG support to country h COIN - to include STR to resistance groups like "j" - is ultimately foreign internal defense (FID) in support of country h's national defense and is conducted with country h's permission.

B. "Country d" - a state with which the USG has diplomatic relations but is not a permanent ally by virtue of a mutual defense treaty - conducts COIN against NSA-X. Any USG support to country d COIN - to include STR to resistance groups like "j" or "m" - is also ultimately FID in support of country h's national defense and is conducted with country d's permission.

**(4)** Resistance to Non-State Actor Occupation Not In Support of Another Government (STR > CT or Stability Operations)

"Country b" - a state with which the USG has no diplomatic relations and may even be hostile to USG interests - conducts COIN against NSA-X (and possibly some combination of "j," "k," and "l"). Any USG STR to resistance groups like "j," "k," and/or "l" is done for USG and coalition objectives in spite of country b's state preferences and without country b's permission.

**Caveat:**

A "resistance" (n ) by definition has to have something/someone to violently resist. "Resistance" is therefore more than just normal political struggle." Resistance in the doctrinal sense must be opposed to an entity exercising "state-like" functions. Once that degree of control no longer exists - in this case, when NSA-X no longer controls any territory - the previously supported groups are no longer "resistance" forces.

**(5)** Resistance to Non-State Actors in Ungoverned/Undergoverned Areas (STR > CT or Stability Operations)

The state government ("N") of "country f" has fallen or is ineffectual. "Resistance" by contending actors within this space is a confused situation where areas of control and group loyalties are mixed and eventual boundaries of political authority remains uncertain. Country f "permission" is meaningless for this case of USG and coalition partners' STR to any combination of groups. However, once there is no identifiable resistance to another power exercising "state-like" functions, the situation has transitioned to warlordism or civil war - at which point USG policy is no longer STR.

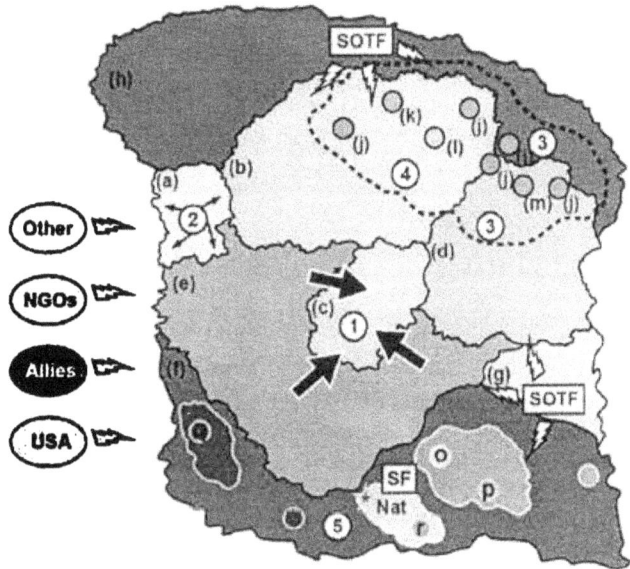

**Figure 3-8. Five types of support to resistance**

This page intentionally left blank.

## Chapter 4

# Fundamentals of Unconventional Warfare

*The start point for this perspective will be the operational detachment's receipt of a UW mission. All SF units preparing for and executing UW conduct all seven phases of the model. The interpretation by some Soldiers that certain units do some phases of the traditional seven phase UW model and other units do the remaining phases of the UW model has been—and herein is explicitly identified as—invalid.*

ATP 3-18.1, *Special Forces Unconventional Warfare*

## SECTION 1. PREPARATIONS PRECEDING OPERATIONAL DETACHMENT ISOLATION FOR AN UNCONVENTIONAL WARFARE MISSION

## MISSION CONTEXT

4-1. The SF conduct of UW has two basic contexts: first, the congressional mandate to prepare for it, and second, when an order is given to execute it. The U.S. Congress directs DOD to maintain a capability for UW in 10 USC 167. DOD designates USSOCOM as the responsible DOD command for UW. USSOCOM designates USASOC as the lead service component for UW doctrine, organization, training, materiel, leadership and education, personnel, and facilities in USSOCOM Directive 525-89, *(U) Unconventional Warfare (S//NF)*. USASOC provides forces selected, trained, and equipped to conduct UW to GCCs U.S. Army SF Soldiers are specifically selected, trained, and educated to shape foreign, political, and military environments by working with and through host nations, regional partners, and indigenous populations. UW is the primary mission that most distinguishes SF (FM 3-18).

4-2. UW has an approved joint and Army definition, which is characterized in detail by approved joint and Army doctrine. Commander USSOCOM, as advised by the Commander of USASOC, can initiate the process for a change, when necessary, to the definition through a joint service doctrine process. Meanwhile, joint and Army doctrine is periodically revised through a structured, systematic process to maintain its traditional, essential, and enduring nature but to change as needed to address new requirements. UW is not whatever one believes it is—UW is specified in current official doctrine by the joint force and the Army.

4-3. The USG uses combinations of its DIMEFIL instruments of U.S national power to pursue USG objectives in a competitive world. Most of those USG DIMEFIL activities are routine and follow openly acknowledged, peaceful, and well-established procedures, and the military instrument plays a supporting role to the other instruments of USG power. Sometimes, the USG will make a policy decision to conduct STR The military may play no role, some supporting role, or a leading role in this STR policy decision. Military participation in STR will involve a mix of conventional force operations and special operations, and it will represent varying mixes of traditional and irregular warfare.

4-4. After multiple levels of political and operational feasibility assessments, a policy decision to conduct UW may be made. In a few rare cases, the President of the United States will issue a finding that directs the Central Intelligence Agency to conduct covert action as the lead agency for UW, with the military in a limited supporting role. In a majority of cases, the GCC will task subordinate units to begin planning for UW This mission planning will be guided by standing approved GCC theater OPLANs, TSOC mission letters, and command guidance specific to the particular mission. It will be further informed by all available information and intelligence provided by higher echelons to the subordinate units. Although UW can have significant strategic effects and international political implications—and the operational detachment must always be mindful of operational effects—the specific mission planning for each prospective executing unit will be narrowed down to a tactical focus. In the majority of approved cases, UW missions will be carried out upon receipt of a Secretary of Defense-approved execute order. (See Appendix A, (U) Special Activities Support to Unconventional Warfare (S//NF), published separately; Appendix C, Intelligence Support to Unconventional Warfare; Appendix G, Theater Special Operations Command Mission Letter—Examples; Appendix H, Sample Area Study Format; and Appendix J, Special Operations Mission Planning Folder Formats, for more information.)

# MISSION PLANNING

4-5.  SF commanders conduct both deliberate and time-sensitive mission planning and targeting. They receive their missions as a result of the joint strategic planning process. Strategies, policies, and SF missions are subject to continual change. This section describes how SF missions are identified and how SF battalions, companies, and ODAs conduct mission analysis, feasibility assessments, and mission planning to perform UW missions.

## MISSION ANALYSIS

4-6.  In UW mission analysis, the battalion staff analysts view all sources of mission letters and taskings. They set priorities for resources and efforts through a clear statement of the battalion commander's intent and concept of operation. Mission analysis provides the basis for the battalion METL preparation and, with the guidance in the mission letter, drives unit training for all the operational detachments.

4-7.  Home station mission analysis of UW follows a specific application of the deliberate planning process. This process assumes the SF battalion has more than one OPLAN it may have to execute. The deliberate planning process allows staff analysts to allocate resources and set priorities of effort. The analysts develop and refine draft METLs based on METT-TC planning. The bulk of the work on area studies is completed before beginning specific deliberate planning.

---

*Note:* GTA 31-01-003, *Detachment Mission Planning Guide*, outlines specific ODA mission planning procedures, to include deliberate and time-sensitive (crisis action) planning.

---

4-8.  Deliberate (peacetime) planning is based on projected political and military situations that are applicable to a UW environment. The objective of this process is to develop an OPLAN with CONPLANs to provide for flexible execution. A crisis is fluid by nature and involves dynamic events, making flexible planning a priority. Deliberate planning supports time-sensitive (crisis-action) planning by anticipating potential crises and developing CONPLANs that assist in the rapid development and selection of a COA.

## DELIBERATE PLANNING PROCESS

4-9.  The deliberate planning process begins with receipt of the tasking order (TASKORD) and TIP (figure 4-1, page 4-3). The process will ultimately result in a completed SOMPF. A mission letter and TIP are needed to begin a feasibility assessment or develop a plan of execution. Each TIP contains the operational intent of the TSOC and group commanders. It also contains a specification of premission constraints and considerations that will define parameters for the assessment and planning process. The group commander and staff review the TIP and SOMPF to determine the general shortcomings and requirements inherent in the TASKORD. They then assign a battalion to assess the mission's feasibility and to begin planning the mission. Although the group passes the action to the battalion for more detailed analysis, group staff elements must continue their own analysis and coordinate the known and anticipated shortfalls.

## FEASIBILITY ASSESSMENT PROCESS

4-10.  Upon receipt of the TASKORD, the SF battalion begins the feasibility assessment process. This assessment determines the ability of the ODA to perform the mission and the abilities of the battalion or ODB to support the mission. The warfighting functions and special operations imperatives are the guides to use for conducting a feasibility assessment. The process requires direct involvement of the battalion or SOTF staff, the ODB staff, and the ODA. Each has specific degrees of concern and areas of responsibility that, when addressed, will accurately reflect mission feasibility. Reviewing these items ensures unity of effort of all units in the JSOA and avoids conflict among friendly units. When determining mission feasibility for a preconflict mission assignment at the home station, planners must forward detailed mission shortfalls in training or resource requirements that are beyond the battalion's capability to fix to the SF Group (Airborne)—such as school quotas and mission specific equipment. Once completed, the feasibility assessment must clearly show whether mission feasibility is conditional upon the resolution of identified shortfalls. The battalion commander translates training deficiencies into additions to the operational detachment METL for training. He then tasks his staff elements to coordinate the resolution of materiel or personnel shortfalls.

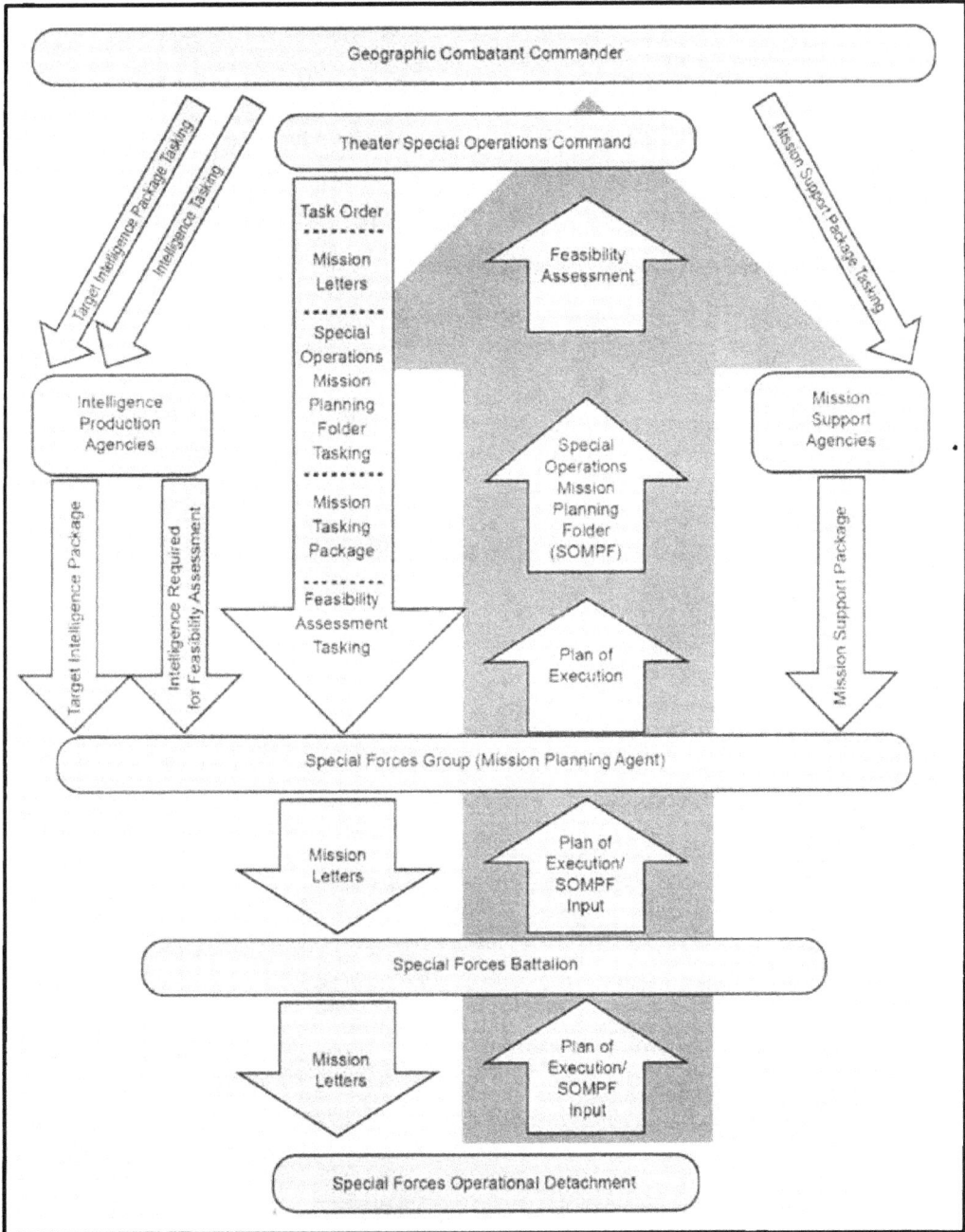

**Figure 4-1. Deliberate planning process**

## BATTALION FEASIBILITY ASSESSMENT ACTIVITIES

4-11. The battalion commander and his staff consider the specific mission requirements in the context of overall group and battalion plans. They must assess the ability of the battalion or SOTF to prepare, support, sustain, command, and control elements conducting UW once they are committed to an operation. The SF battalion has two principal roles in the feasibility assessment process: to guide and support the operational detachment's efforts. If the operational detachment's and the battalion staff's assessments conflict, the battalion commander makes the final determination. The battalion operations staff serves as the focal point for coordinating infiltration and exfiltration feasibilities. Operational detachments and battalion staffs must consider all methods of infiltration and exfiltration during the feasibility assessment. Infeasible or inappropriate methods of infiltration or exfiltration are excluded from future consideration. An example of when a mission becomes infeasible might be when all potential infiltration and exfiltration methods are infeasible. The battalion or SOTF staff forwards to the tasked ODA and ODB any known constraints, available information, and the commander's intent, which become their basis for assessment.

## COMPANY FEASIBILITY ASSESSMENT ACTIVITIES

4-12. A battalion typically gives missions directly to each of its 3 ODBs and 18 ODAs. This tasking does not imply that the company HQ does not participate in tasking its organic ODAs. Unless there are security reasons for compartmentalizing the mission, the company receives the TASKORD from the battalion commander and staff. The battalion then tasks one of its ODAs or ODBs and supervises its assessment activities The SF battalion commander has three reasons to pass the TASKORD through the company commander. First, the ODB (company) manages the day-to-day activities of the ODA and ODB. Second, operational experience of the SF company HQ normally exceeds that of the ODA; the company staff mentors the ODA. Third, after deployment, the most common missions for an ODB will be that it is independently deployed into the JSOA or deployed to support the ODA as an advanced operations base or as a special operations command and control element (SOCCE). For UW, the company's feasibility assessment concerns include:

- Communications means and schedules.
- Emergency exfiltration requirements.
- Emergency resupply considerations.
- Preliminary contact and link-up plans.

4-13. The ODA or ODB performing feasibility assessments for specific missions does not develop plans at this time. Either detachment can determine mission requirements and examine applicable tactics, techniques, and procedures. It identifies interagency coordination, required resources, training beyond its organic capability, and augmentation needs. It must include these requirements as conditions in the final feasibility assessment. Feasibility assessments are stated as either:

- Affirmative—trained, equipped, and routinely supported.
- Conditional—we can do it, but the following conditions or requirements exist.
- Negative—we cannot accomplish this mission because of a clearly stated reason.

## SUPPORTING AGENCY FEASIBILITY ASSESSMENT ACTIVITIES

4-14. Aviation assets conduct feasibility assessments for aircraft availability and compatibility with mission guidelines and routes for infiltration and exfiltration. Technical augmenters conduct feasibility assessments to ensure compatibility with the mission. This feasibility assessment should include:

- Capability to execute the selected infiltration and exfiltration.
- Emergency resupply.
- Fire support.
- Other mission-specific skills necessary for mission success.

## FIRE SUPPORT ASSESSMENT

4-15. Joint fires, close air support, naval gunfire, fire support types, availability, limitations, and requirements need to be identified early in the mission planning sequence. During planning, the commander:

- Establishes restricted fire areas, no-fire areas, and fire support coordination lines as needed.
- Develops fire support coordination measures that protect deployed operational detachments and are not unnecessarily restrictive on the conventional or supporting forces.

## POSTFEASIBILITY ASSESSMENT ACTIVITIES

4-16. Once a mission is determined feasible or conditionally feasible, the SOTF staff or operational detachment anticipates an actual TASKORD. The battalion or SOTF staff immediately identifies requirements for schooling, training, and equipment.

4-17. Making changes to the modified tables of organization and equipment or task-organizing early to meet mission requirements are examples of long-term fixes that should begin during the feasibility assessment. The battalion staff and the operational detachment retain all documentation and working papers from the feasibility assessment since they form the basis for subsequent plan of execution development.

## PLAN OF EXECUTION DEVELOPMENT

4-18. The plan of execution is the final element of the SOMPF. It shows how the ODA intends to carry out the assigned mission. Specific mission preparation begins when the battalion receives the tasking to prepare a SOMPF to accomplish a tasked mission. This tasking takes the form of a SOMPF shell containing all elements developed to that point. Appendix J provides information on SOMPF contents. Options for conducting specific mission planning may include:

- Activating the SOTF for training and isolating ODAs so that they may accomplish their detailed planning.
- Conducting planning as part of the routine training day.

## BATTALION ACTIVITIES

4-19. The battalion commander and staff plan for support of the overall operation. In addition, the battalion has activities that support operational detachment plan of execution development. Building upon the analysis conducted during the feasibility assessment, the battalion staff uses the military decisionmaking process as described in JP 5-0, ADP 5-0, and FM 6-0. For planning that is not time-critical, the staff prepares formal written estimates that identify anticipated mission-specific requirements of the operational detachments. They also address planning options available to the tasked operational detachment. For example, the signal section addresses available communications capabilities and procedures for transmitting data without undue risk of compromise. At the battalion level, the commander directs the decision-making process toward—

- Developing estimates to be used in preparation of the battalion order and operational detachment planning guidance.
- Developing the commander's intent into a feasible concept of operations.
- Planning for SOTF activities required to support the employed operational detachments.

## Battalion Order

4-20. The battalion planning process produces the battalion order. The battalion order encompasses all missions that the battalion's operational detachments will accomplish. ODAs and ODBs will normally receive information on their JSOA only.

4-21. Once completed, the coordinated battalion order or plan becomes a supporting plan to the group or combined JSOTF OPLAN. The revised or new mission letter normally requires the staff to review previous home station analysis and coordination.

## Commander's Intent

4-22. The commander's intent serves to communicate guidance in the absence of specific orders. The commander's intent must be stated clearly. It must be brief and to the point. The commander must include a separate statement of intent for each specific TASKORD. The intent directs mission concept development at the operational detachment level. The order must contain sufficient detail to guide the operational detachment commander's mission analysis and decision-making process, without limiting the operational detachment's flexibility during the planning options.

## Mission Letter, Operation Plan, and Special Operations Mission Planning Folder

4-23. The battalion commander issues a mission letter to the ODAs and ODBs. He then provides the battalion OPLAN and SOMPF to the ODA. The ODA needs time to review the information before receiving the mission briefing from the battalion commander and staff. The mission briefing consists of battalion staff officers and key noncommissioned officers and mission-specific operation cell members if required. They summarize their specific areas and elaborate on data or changes in the OPLAN that impact on the ODA's mission. The ODA does not leave the briefing until it understands the commander's intent and has asked all the questions generated by its review of the battalion OPLAN and SOMPF. The battalion staff answers all questions concerning the mission, for example, infiltration platforms, support equipment availability, and rehearsal areas available to the operational detachment.

## COMPANY ACTIVITIES

4-24. The ODB may also prepare for its own follow-on area command mission in the future. When the ODB is not compartmentalized from the ODA receiving its mission, the SF company commander and staff monitor the mission briefing with the ODA. This practice ensures the ODB—

- Is available to the ODA during the planning process.
- Understands the training and resource requirements related to the mission.
- Understands the missions it is supporting or controlling, if tasked to act as an advanced operations base or as SOTF augmentation.

## SECTION 2. SEVEN PHASES OF UNCONVENTIONAL WARFARE AS CONDUCTED BY SPECIAL FORCES DETACHMENTS

---

### The Seven Phases of Unconventional Warfare

"2-31 The seven phases of U.S.-sponsored insurgency are a conceptual template that planners use to aid understanding of a UW campaign effort. As a template, it merely serves as a guide for planning and execution. Not all phases will necessarily take place nor do they have to be executed in a rigid linear order. For example, U.S. forces may come to assist a large, previously established insurgent/resistance force with significant prior combat experience. In such a case, the organization, buildup, and employment phases may already be completed or ongoing before SOF ever infiltrate the JSOA. The seven phases are a model, not a checklist. . "

ATP 3-05.1, *Unconventional Warfare (Change 1)*, 25 November 2015, p. 2-8, para. 2-31

---

# PHASE I—UNIT MISSION PREPARATION UPON RECEIPT OF AN ORDER TO CONDUCT UNCONVENTIONAL WARFARE

4-25. Phase one, unit mission preparation, encompasses several groups of activities that occur at different times. These groups are preisolation mission preparation, deliberate mission planning conducted during isolation, and post approval or preinfiltration mission preparations.

## MISSION PREPARATION

4-26. Mission preparation must predate isolation and the deliberate planning process. During mission preparation, specific mission employment is not required. Operational detachments tasked to conduct a specific type of mission or use a specific means of infiltration continue to support their METL through scheduled training. Mission preparation, which is typically done at the unit's home station, includes METL, individual training, and mission-specific training.

4-27. Based on the group commander's training guidance, the battalion commander assigns the missions and approves the operational detachment's draft METL that supports the assigned missions. The company commander must plan, conduct, and evaluate company and operational detachment training to support this guidance and the approved METL for the mission.

## Premission Training

4-28. The operational detachment commander ranks the tasks that need training. Since there will never be enough time to train in every area, he focuses on the METL tasks that are essential for mission completion. He emphasizes building and maintaining proficiency in those tasks that have not been performed to standard and sustaining proficiency in those tasks that are most difficult. He will rank those tasks during training meetings with the ODAs and ODBs.

4-29. Once the tasks for training are selected, the ODA commander builds a training schedule and plans on those tasks. He provides the previous training requirements to the battalion commander. The battalion commander approves the list of tasks to be trained, and then the ODA commander includes them in the training schedule.

4-30. The company commander coordinates the support and resource requirements with the battalion S-3 well before the scheduled training to allow S-3 personnel sufficient time for coordination. He ensures that tasks, conditions, and standards are enforced and conducted according to the principles outlined in ADP 7-0, *Training*.

4-31. In UW, the operational detachment must know the operational environment. The operational detachment often deploys to the actual AO, even when the activities in the potential JSOA are not directly related to the TASKORD. The opportunity to survey the climatic, geographic, cultural, and other environmental factors must not be lost. Where it is impossible to deploy to the actual area and conduct offset training, the S-3 coordinates with the battalion S-2 to identify accessible locations for training that replicate each detachment's AO. The operational detachment trains to exercise its mission plans as carefully and realistically as possible, while always maintaining OPSEC. One of the most difficult aspects of mission training is replicating the conditions of a denied area. The S-3 coordinates with other units to provide realistic and aggressive opposing forces field tactical training whenever possible. The hazards and austere conditions of a denied area also highlight the requirement and importance of broad detachment cross-training of skills to maximize resilience and redundancy.

## Intelligence

4-32. The battalion S-2 is responsible for all intelligence-related matters. He is responsible for keeping the commander and his staff informed on all UW security matters, to include hostile and indigenous force activities. Because of the nature of UW, the battalion S-2 must play a key role in mission success. All mission planners will understand the goals of priority intelligence requirements (PIRs) and IRs. The S-2 must ensure the collection plan PIRs and IRs on which the mission is based have not been nor can be satisfied by other sources. If the S-2 identifies other mission capable sources, he informs the battalion commander through the S-3, and then the support is requested. The SF group commander determines minimum essential preparation tasks. He then modifies the deliberate planning process to do those tasks in the time available. The SF group commander must inform the TSOC and 1st Special Forces Command when he cannot accomplish these tasks with an acceptable degree of risk for mission success.

## Target Intelligence Packets

4-33. Intelligence is perishable over time. The battalion S-2 has the primary responsibility for keeping the intelligence database current. The S-2 requests additional information from higher and adjacent HQ to meet and monitor PIRs and IRs. The S-2 section conveys changes to the situation and to TIPs that affect mission accomplishment for the appropriate operational detachment. The S-2 section must provide liaison with local intelligence forces to maintain the command's security. The S-2, with S-5 input, also provides current situation updates and intelligence summaries to the operational detachment and responds to intelligence-related questions and requests.

## Area Studies

4-34. At the company and operational detachment level, the country area study is the primary tool for tracking intelligence over time. The operational detachment continually updates the area study at the home station. The battalion S-2 provides the database for updating the study by using all available sources, to include all DOD systems, as well as open-source information and publicly available information. (See Appendix H, Sample Area Study Format.)

## Personnel

4-35. The S-1 section provides for and coordinates personnel service support. Its primary combat duties focus on strength management, casualty management, and replacement operations. Other responsibilities include mail, awards, pay, Uniformed Code of Military Justice, enemy prisoners of war (EPWs), and Soldier readiness processing. The section takes part in the operations order process by developing administrative annex material, preparing personnel estimates, and recommending replacement priorities. In the personnel service support coordination role, the S-1 serves as the focal point for personnel, administration, financial, religious, medical, public affairs, and legal activities.

## Logistics and Communications

4-36. All levels of command review the UW-specific requirements for logistics. The S-4 must redistribute the available supplies and equipment within the battalion and make inventory adjustments. In addition, the S-4 makes preparations for all classes of external supply, to include automatic, on-call, and emergency. In UW, communications requirements are usually long term and unique. The signal section must review the compatibility and type of equipment to ensure all requirements can be met. Battery inventory is particularly critical. The annual budget must include projections for training requirements. Where nonstandard equipment is used, the signal officer and his staff must identify reliable internal sources for prescribed load list repair parts and batteries. (See Appendix L, Communications; Appendix M, Logistics; and Appendix N, Caching.)

## Pre-Employment Preparation

4-37. The S-3 must periodically review and update mission SOMPFs. This review and update should take place whenever:
- There is a significant change in the situation (intelligence).
- The supported plan or basic tasking has major revisions.
- Personnel turnover affects mission readiness.

4-38. After the battalion has been alerted to execute a mission, it activates the SOTF. Normally, the SOTF will operate from a forward secure area, but it may also operate from the home station. The deliberate planning process and isolation procedures apply to the UW mission.

## Time-Sensitive (Crisis-Action) Planning Process

4-39. The time-sensitive (crisis-action) planning process continues the sequence begun during the deliberate planning process with receipt of the mission letter and continues with receipt of the TASKORD and TIP. Deliberate planning presumes that the operational detachment has completed a SOMPF for the mission and conducted its mission preparation at the home station. Time-sensitive tasks, such as reviewing the current intelligence summary for changes and confirming infiltration and exfiltration means, are critical and timely activities. Failure to allow the operational detachment to complete these activities in the 96-hour time-sensitive planning process decreases the probability of mission success.

## Emerging Missions

4-40. Emerging missions are those other missions for which a requirement was not anticipated and no TIP or SOMPF has been prepared. The time-sensitive planning process for such missions involves a 96-hour cycle. At the SOTF, the operational detachment uses the first day to prepare orders and may dedicate the last three days to the deliberate planning process, rehearsals, inspections, and rest during the operational detachment isolation. As with any such procedural guide, the time frames are approximate and planners can adjust them as required. The 96-hour time frame obviously does not permit the type of meticulous, METL-driven mission preparation described in the previous paragraphs of this chapter. If, however, the battalion has done a thorough job of mission analysis for its assigned JSOA, the general conditions of the emerging mission will parallel other missions for which it has previously prepared its operational detachments. The JSOA may be similar (if not the same) as will the threat and other factors of the operational environment. The battalion and company commanders must make every effort to assign the emerging mission to the operational detachment that has most closely prepared for the similar operation. During time sensitive planning, staffs must anticipate mission requirements and staff-to-staff coordination. Planners must consider the transportation and information needs commonly requested for the type target being considered. Some specific areas that are particularly critical to time-sensitive planning are communications, intelligence, operations, map coverage, logistics, and infiltration and exfiltration methods.

4-41. Significant planning cannot begin without adequate and current map coverage. The SOTF S-2 must supply current maps to the operational detachment. Quality imagery will significantly enhance the probability of mission success. Detailed planning for infiltration and exfiltration is critical. Compromise of the operational detachment on infiltration can indicate the mission will be a failure. Since infiltration and exfiltration usually involve joint external assets, coordination is more difficult in this short time period. Face-to-face coordination between the operational detachment and the supporting asset (for example, pilot, boat operator, or ship captain) is crucial.

4-42. Supplies from the SOTF support center are usually the only logistics support available for unanticipated emerging missions. This situation highlights the need for the battalion S-4 to plan and deploy with a large assortment of equipment and supplies rather than just the minimum required for the preplanned missions.

## SPECIAL FORCES OPERATIONAL DETACHMENT DELIBERATE PLANNING PROCESS

4-43. Deliberate UW planning during isolation prepares the ODA for a hypothetical crisis based on the most current intelligence. This assumption allows for the preplanned use of resources and personnel projected to be available when the plan becomes effective. The assumptions make it improbable that any CONPLAN that the ODA has implemented will be totally usable without modifications. The detailed analysis and coordination that was accomplished during the time available for deliberate planning will expedite effective decision making as the crisis unfolds. At this time, assumptions and projections are replaced with hard data. Once the SOTF commander selects prospective ODAs to execute the mission, each detachment moves into an isolation facility, receives the OPORD and mission brief, and begins its mission preparation as follows

*Note:* The following paragraphs provide an overview of mission planning during Phase 1 ODA Mission Preparation for UW. For the detailed steps of the military decisionmaking process, see GTA 31-01-003, *Detachment Mission Planning Guide*, Chapter 2.

## Step 1: Receive the Mission

4-44. The operational detachment receives the battalion staff's planning guidance. This guidance may come in a mission letter from the group during normal peacetime operations, or it may come in the form of an OPORD or TASKORD and be provided a SOMPF, if available. The company passes command guidance to the ODA. The operational detachment commander reviews this planning guidance and activates the staff sections within the ODA. Unit SOPs designate ODA members for each staff section according to their military occupational specialty and assigned staff responsibilities. Each operational detachment member reviews his portion of the OPORD. When review of the OPORD is complete, the operational detachment discusses the battalion commander's intent. Each member voices his concerns and develops questions to be answered in the mission briefing to the team. The operational detachment develops RFIs and puts them in writing. These RFIs address unanswered questions and unclear points in the commander's intent. These RFIs are forwarded to the battalion staff and a mission briefing is scheduled. The battalion staff presents the mission briefing, during which it answers as many RFIs as possible and provides a working status on the others. The operational detachment questions each staff member as required. The ODA commander ensures the perception of the commander's intent is correct through face-to-face discussion with the battalion commander.

## Step 2: Exchange Information

4-45. After the mission briefing is complete and all questions have been answered or noted for further research and coordination, the operational detachment commander conducts a mission analysis session with the operational detachment. This session ensures operational detachment members understand PIR, rules of engagement, intelligence indicators, legal and political constraints, and the role of CA and PSYOP. In this session, the operational detachment reviews all available information to ensure that all operational detachment members agree on what has been presented. Operational detachment members with specialized skills or experience (either for the mission or in the JSOA) provide information on their unique perspectives or requirements. If confusion over information or interpretations of information exists, the operational detachment develops and forwards additional RFIs to the battalion for clarification.

## Step 3: Restate Commander's Intent and Produce Planning Guidance

4-46. After exchanging information, the operational detachment commander, assistant detachment commander (warrant officer), and the operational detachment operations sergeant (or company operations warrant officer or sergeant major if an ODB) meet to develop the restated mission and intent to produce OPORD planning guidance. Here, the operational detachment leaders:

- Review the specified and implied tasks from the OPORD.
- Review the battalion mission statement and commander's intent.
- Consider the information received to date.
- Review the feasible infiltration and exfiltration means.

4-47. Based on this information, the operational detachment leaders review and develop a comprehensive list of specified and implied tasks, and the commander develops a restated mission. The wording of the task in the OPORD does not need to be rearranged. The restated mission specifically identifies the task as the *what* of the *who, what, when, where,* and *why.* Based on the knowledge of the skills, capabilities, current and achievable standards of training of the operational detachment, and available resources, the operational detachment leaders develop numerous COAs for consideration.

4-48. Planning responsibilities different from, or not covered by, unit SOPs are also clearly stated. When the operational detachment leaders have completed the analysis, they pass the findings out to the entire operational detachment. Normally, information is given verbally; however, written guidance or training aids (for example, flip charts) are preferable where guidance is complicated. As a minimum, the operational detachment leaders must usually present three COAs in writing.

## Step 4: Prepare Staff Estimates

4-49. Based on the planning guidance and unit SOPs, the operational detachment members prepare estimates for their areas of responsibility. These estimates are not limited to those of the traditional staff areas of responsibility. PSYOP, as well as civil-military operations (CMO), estimates will always be prepared for UW operations. At the home station, operational detachment's planning (where time is not a factor) includes written estimates. Written estimates provide needed continuity for replacements when operational detachment members rotate. Staff estimates must identify support requirements for each COA. Once completed, the detachment commander briefs the staff estimates to the operational detachment. The briefing serves the purpose of exchanging information between operational detachment members.

## Step 5: Prepare Commander's Estimate and Decision

4-50. After the staff estimate briefings, the operational detachment commander reassembles the operational detachment leaders and, with their assistance, prepares the commander's estimate. A critical portion of this step is finding and weighing the factors to be used in evaluating the COAs. The operational detachment leaders select the specific factors for the mission based on the commander's intent and the specified and implied tasks. The commander's estimate must include specific factors for each of the warfighting functions. The special operations imperatives should be translated into specific factors related to the mission. The commander's estimate is prepared in written form. Based on his estimate, the commander decides which COA the operational detachment will plan to execute. The product of this step is a statement of the operational detachment commander's intent and a concept of the operation. The operational detachment commander briefs them to the entire operational detachment. This briefing serves to answer any questions the operational detachment may have and serves as the operational detachment murder board for the concept. After briefing the battalion, the commander schedules an informal mission concept briefing with the operational detachment and the battalion commander.

## Step 6: Present Mission Concept Briefing

4-51. The mission concept briefing is an informal briefing presented to the battalion commander to ensure the planning efforts meet his intent. This briefing allows approval of the operational detachment's concept of the operation before expending time in detailed planning. The goal of the briefing is to obtain the commander's approval for a COA. Mission concept briefings have no specific format but normally include the following information:

- Mission.
- Higher commander's intent.
- Concept of the operation.
- COAs considered.
- Factors used to evaluate the COA, risk assessment, and METT-TC.
- Specific UW tasks.
- Task-organization (to include requests for required attachments).
- Infiltration, contact, and exfiltration means.
- Identification of external support required (such as interagency approval of special activities and nonstandard equipment).
- General statement of the commander's concept of operations.
- Mission-essential personnel and equipment problems not previously addressed.

*Note:* ATP 5-19, *Risk Management*, provides doctrinal guidance on managing risks within the conduct of operations.

4-52. Visual aids should be those that the operational detachment is already using for mission planning. The mission concept briefing should not be so formal that the event hinders the operational detachment's planning activities. The battalion commander approves the concept, modifies it, or directs the operational detachment to return to Step 3, providing additional guidance to clarify his mission intent. Key battalion staff members may accompany the battalion commander to the mission concept briefing and familiarize themselves with the approved mission concept.

4-53. Based on the approved concept, the battalion staff anticipates the operational detachment's support requirements. For example, if the briefing calls for high altitude-low opening infiltration of a 12-man operational detachment, the S-4 does not wait for a support request for high altitude-low opening air items for the infiltration. The selected concept drives IRs. Because of the generally repetitious information requirements, intelligence personnel can anticipate and deliver many of their requirements. However, the operational detachment is still responsible for all details of the UW operation.

4-54. Approval of the mission concept ends the concept development phase for this planning session. Responsibility for the concept now rests with the battalion commander, and the operational detachment continues with their planning.

## Step 7: Orders Production—Complete the Plan

4-55. Upon approval of the mission concept, the operational detachment leaders produce the body of the plan in an OPORD format. The written plan specifies taskings to subordinate elements and individual members of the operational detachment. Annexes are not included at this point; they are produced during the detailed planning phase. The plan must be in keeping with the battalion commander's guidance and understood by all members of the operational detachment. When these criteria are satisfied, the battalion commander approves the plan after completion of the briefback.

## Step 8: Conduct Detailed Planning

4-56. The operational detachment organizes for planning the same as it would for mission execution. When the mission does not require the entire operational detachment, nonessential members help subordinate elements in the mission planning process. Compartmentalization within the operational detachment is generally counterproductive. Each operational detachment member or element completes a detailed plan for the execution of assigned tasks. Members of the operational detachment brief their respective areas, while other members provide an appropriate critique. The battalion staff should be available to provide the same service with the added advantage of greater experience and objectivity.

> *Note:* GTA 31-01-003, *Detachment Mission Planning Guide*, provides detailed information on SF operational detachment isolation duties and responsibilities.

4-57. In UW, certain aspects of the mission may be beyond the experience of any of the planners. In these circumstances, rehearsals are excellent mission planning tools. New or unfamiliar employment techniques may be tested by realistic rehearsals of portions of the plan during its development. Often, walking through an action will reveal the need for changes to the plan. In any case, before an operational detachment briefs a plan of execution to its battalion commander, operational detachment members should physically confirm the viability of the plan under the most realistic circumstances possible. Formats for specific subelements of the detailed plan that support the plan of execution vary with the mission. Some portions of the plan of execution may require annexes; others only require reference to a larger section. The plan of execution should relate how the exceptional activity contributes to collecting and reporting requirements or how it enhances the survivability of the operational detachment. Alternate plans are also considered. The plan of execution lists all mission-essential equipment and accounts for the disposition of that equipment in operational detachment packing and resupply plans. The operational detachment leaders ensure that all supporting subplans are consistent and mutually supportive. They also must ensure that specialized equipment used for one activity can also be used for another. Upon completion of all supporting subplans, the operational detachment leaders supervise the preparation of the formal plan of execution. It includes all the annexes, notes, narratives, and graphics essential to conduct the mission.

## Step 9: Conduct Briefback

4-58. The briefback serves a distinct purpose. Although the plan of execution details what an ODA intends to accomplish, the briefback explains to the battalion commander how the ODA will execute the assigned tasks. All information is contained in the commander's folder and must be able to stand alone. The ODA is now ready to accomplish its mission. The battalion commander's last effective influence on these activities is through guidance given at the briefback.

4-59. The briefback format is driven by the plan, not the reverse. Several mission briefback formats are available that provide general guidance. There is no best format or checklist for UW. The ODA prepares the briefback using the completed detailed plan. Existing formats are useful as a means to organize the presentation (in general terms) and as a checklist to look for obvious areas that were overlooked during planning. Briefback preparation often reveals gaps in planning.

4-60. In adapting the format, the following principles apply:
- The format must provide a detailed description of the activities of each ODA member throughout the execution of the mission. This description provides a mental picture of the operation for the commander receiving the briefing.
- The briefing uses visual aids only if they help to clarify the briefing.
- Briefers must avoid constant reference to the commander's folder as it denies the staff access to necessary information—use charts instead.
- The briefing must provide the commander with adequate information to judge the efficacy of the plan.

4-61. The ODA presents its briefback in the isolation facility where the planning was conducted. This practice enhances OPSEC and reduces both administrative and support requirements. Each ODA member briefs his own responsibilities. The staff members focus questions on areas where they did not hear adequate information to judge the completeness or viability of the plan. However, all staff elements should have thoroughly coordinated their input during the mission-planning phase. The traditional habit of quizzing ODA members' memory of mission-essential details is appropriate for this briefback since mission preparation and training will continue. The purpose of the briefback is for the commander and staff to judge the merits of the plan. For example, if the battalion communications representative hears reference to a communications system that he cannot personally verify as available, he questions the ODA on the availability of that system. If he hears discussion on the use of an antenna that appears inappropriate, he confirms the reason for that selection. The briefback is intended to show any weaknesses in the plan while they can still be corrected. If the ODA cannot justify any action, no matter how minute, that action needs to be reconsidered, corrected, and then implemented.

4-62. Once the battalion commander is confident the plan is workable and is delegated the authority by the group commander, he approves the plan of execution. If further work is still required, he gives specific guidance and returns the ODA to the planning phase. The commander determines the extent of revision and whether another full briefback is required. As a minimum, he should require the appropriate staff officer to personally brief him on any changes. Once the plan is approved, the ODA commander is responsible only for the preparation of the ODA and execution of its mission. The battalion commander assumes responsibility for the viability of the plan. He should withhold final approval of plans until all support requests are confirmed and the TASKORD authority approves the plan of execution.

## Step 10: Obtain Plan of Execution Approval

4-63. Once the plan of execution is completed and approved by the battalion commander, the S-3 forwards it through the SF group to the tasking agency. The tasking agency then compiles the SOMPF by obtaining the mission support package and other supporting documents. The S-2 or S-3 secures the basic folder and returns a copy to the mission planning agent. This procedure constitutes plan of execution approval.

## DETAILED PREINFILTRATION MISSION PREPARATIONS

4-64. After developing the OPORD, the operational detachment members continue detailed preparations for the mission. OPSEC measures remain important. Operational detachment members obtain supplies, equipment, and training materials and prepare for infiltration and mission accomplishment. Operational detachment members hone their military skills through mission rehearsals, conduct extensive area and language orientation, obtain input from various staff elements, and may receive an asset debriefing that also enhances detailed planning. If no member of the operational detachment has had previous contact in any form with the resistance, and the asset debriefed is a member of the resistance verified by the battalion, this meeting would constitute the operational detachment's initial contact with the resistance. In most cases, an asset provided for debriefing by the battalion will not be a member of the resistance. (See Phase Two—Initial Contact, below.)

4-65. The operational detachment staff sections follow up on previous requests for additional resources and support not already delivered. They contact supporting agencies according to established procedures to find the status of their requests. Operational detachment staff sections consider alternate COAs when supporting agencies fail to provide the required resources and support. They route functional area information requests to the other staff sections through the operational detachment S-3 and route IRs through the operational detachment S-2. Operational detachment staff sections modify previously developed estimates and plans according to the latest information available. They also update, through the operational detachment S-3, the commander's critical information requirement list according to the latest information available and their needs for additional commander's critical information requirements arising from modified estimates and plans. The actions of the operational detachment staff members are listed below.

*Note:* The operational detachment consults TC 18-01.1, *Unconventional Warfare Mission Planning Guides for the Operational Detachment Level—Alpha Level,* for assistance in detailed mission preparation.

## Commander

4-66. The commander commands and controls the operational detachment. He ensures the operational detachment completes mission preparation according to the higher commander's OPORD or OPLAN and the operational detachment's OPORD. The commander ensures all operational detachment members know and understand the operational detachment's OPORD. He approves tentative changes to the OPORD if the changes satisfy requirements according to his intent, identify mission objectives, and follow the higher commander's OPLAN or OPORD. The commander ensures all legal questions have been clarified and all operational plans are according to applicable legal guidance and directives. The commander is overall responsible for command of the operational detachment's preparation for all planned STR activities.

## Assistant Detachment Commander

4-67. The assistant detachment commander ensures mission preparation is according to the event time plan. He supervises and directs all staff functions acting as the operational detachment chief of staff. He is responsible for coordinating all staff efforts, identifying critical preparatory planning shortfalls, and evaluating all regional implications of the mission. He assists and supervises in the areas of operations, intelligence, CA, and PSYOP. He assists the commander with overall operational detachment staff function preparations for STR activities.

## S-1 (Senior Medical Sergeant)

4-68. The operational detachment S-1 ensures preparation for movement with Soldier readiness processing and the requirements are met according to current Army personnel processing regulations and the unit's SOP. He follows up on all previous personnel services requests according to the operational detachment OPORD. He informs the commander of any problems in the administrative preparation of the operational detachment for infiltration. He establishes procedures for personnel accountability and strength reporting in support of the operational detachment. He makes preparations for resistance administrative support appropriate to the specific mission requirements. (See Appendix T, Administrative Considerations.)

## S-2 (Intelligence Sergeant)

4-69. The operational detachment S-2 supervises operational detachment requests for, and dissemination of, intelligence according to the commander's PIRs and IRs and the information collection plans. He updates both the enemy and the prospective resistance partners' situations using the latest available information and intelligence. The S-2 informs the entire operational detachment of changes in the situation that will affect planned mission execution. He monitors the implementation of the operational detachment's information collection plans, to include updating the commander's PIRs and IRs, conducting an area assessment, and requesting additional intelligence support. The S-2 also monitors the operational detachment's OPSEC measures and plans to ensure they effectively counter the anticipated threat according to the current situation. He then recommends appropriate actions. He makes preparations for resistance intelligence and counterintelligence support and any unilateral U.S. intelligence needs appropriate to the specific mission requirements. (See Appendix C, Intelligence Support to Unconventional Warfare; Appendix H, Sample Area Study Format; and Appendix I, Area Assessment Formats—Initial and Principal.)

## S-3 (Operations Sergeant)

4-70. The operational detachment S-3 leads all mission preparations according to the event time plan. He oversees OPSEC as the operational detachment continues mission preparations with members of the supporting higher staff. The S-3 ensures deception, PSYOP, CA, fires and other support needed in the JSOA is coordinated with the operational detachment and incorporated into the plan. He incorporates any approved changes from higher echelons into the operational detachment's OPORD and disseminates all approved OPORD changes to the operational detachment. He assigns mission tasks to operational detachment members. The S-3 ensures operational detachment predeployment training and rehearsals are according to the operational detachment's OPORD, the resistance training program of instruction in the language spoken within the JSOA, the unit's SOP, and the time event plan. He ensures weapons are test fired. The operations sergeant prepares all scheme of maneuver requirements necessary to achieve the mission from departure of the isolation area, to staging areas, through infiltration of and movement in a denied area, actions on the objective, exfiltration and recovery to friendly control. The operations sergeant assigns the preparation, approval, and delivery of the SF evasion and recovery plan. He oversees all operational detachment preparations for comprehensive resistance support appropriate to the specific mission requirements. (See Appendix K, Infiltration and Exfiltration Considerations; and Appendix V, Special Forces Advisor's Guide to the Combat Employment of Guerrilla Forces.)

## S-4 (Senior Engineer Sergeant)

4-71. The operational detachment S-4 inspects, accounts for, and ensures the serviceability of required supplies and equipment (less medical and communications supplies and equipment) in the operational detachment's custody. He stockpiles additional required supplies and equipment according to the operational detachment's OPORD, to include the basic load of ammunition to support follow-on or contingency missions. The S-4 follows up all previous requests for additional logistics resources or support according to the operational detachment's OPORD. He informs the assistant detachment commander of any logistics problems in preparing the operational detachment for infiltration. He makes preparations for resistance logistics support appropriate to the specific mission requirements. (See Appendix M, Logistics; and Appendix N, Caching.)

## S-5 (Operational Detachment Intelligence and Operations Sergeant)

4-72. The operational detachment S-5 position is established by operational detachment SOPs or assigned by the operations sergeant. He analyzes the civil component of the JSOA using areas, structures, capabilities, organizations, people, and events (ASCOPE) and ensures civil considerations and CA team activities are integrated into the plans and activities of all other staff sections. He updates the civil situation using the latest available information. He also analyzes the psychological themes and delivery methods directed in the battalion order and ensures operational detachment preparation to employ or leverage them. He assists the intelligence sergeant in identifying target audiences in the operational detachment's assigned AO and ensures military information support considerations and PSYOP team activities are integrated into the plans and activities of all other staff sections. The operational detachment S-5 monitors the activities of the operational detachment and anticipates future requirements for CMO and MISO based on current and planned operations, as well as the results of completed operations. He makes preparations for resistance CA and MISO appropriate to the specific mission requirements.

> *Note:* TC 18-01.1 (Appendix G, Military Information Support Operations; and Appendix H, Civil Affairs Support) provides additional information.

## S-6 (Communications Sergeants)

4-73. The operational detachment signal staff obtains the cryptographic materials, signal operating instructions (SOI), and additional related materials according to the higher commander's OPLAN or OPORD and the operational detachment's OPORD. They conduct inspections to account for required communications equipment. The signal staff conducts function tests of communications equipment and systems as required by the higher commander's OPORD or OPLAN, the operational detachment's OPORD, and the unit's SOP. The signal staff informs the assistant detachment commander of any communications problems in preparing for the operational detachment's mission. They make preparations for resistance communications support appropriate to the specific mission requirements. (See Appendix L, Communications.)

### Medical (Operational Detachment Medical Sergeants)

4-74. The operational detachment medical staff ensures the operational detachment members meet medical and dental Soldier readiness processing requirements according to the unit's SOP. They ensure that the operational detachment's required immunizations are current according to AR 40-562, *Immunizations and Chemoprophylaxis for the Prevention of Infectious Diseases*, and that the unit's SOP contains information published by the Center for Disease Control and the World Health Organization. This information comes from the Armed Forces Medical Intelligence Center. The medical staff also ensures the operational detachment members receive any required medical or dental treatment identified during Soldier readiness processing. They follow up on all previous requests for additional medical resources or support according to the operational detachment's OPORD. The medical staff accounts for and inspects the serviceability of the required medical supplies and equipment. They make preparations for operational detachment medical sustainability in the AO. They also inform assistant detachment commander of any health services problems in preparing for the operational detachment's mission. They make preparations for resistance medical support appropriate to the specific mission requirements. (See Appendix X, Medical Aspects.)

### Instructors (All)

4-75. Each operational detachment instructor prepares his training materials in the resistance force's language, if possible. He prepares a lesson outline for each period of instruction according to the program of instruction, the tentative training schedule, and other assigned classes. The instructor includes in each lesson outline, as a minimum, the task, conditions, standards, performance measures, required training aids, doctrinal publications, and handouts. He prepares for the use of translators when required. He makes preparations for resistance cadre building procedures in his field of expertise as appropriate to the specific mission requirements. (See Appendix Q, Resistance Training Considerations; and Appendix U, Guerrilla Periods of Instruction.)

# PHASE II—UNIT INITIAL CONTACT WITH SPECIFIED RESISTANCE ASSETS DURING MISSION PREPARATION OR EXECUTION

## VARIETIES OF INITIAL CONTACT BY LINEAR TIME AND ECHELON

4-76. In the large majority of cases, an operational detachment receiving a mission to conduct UW will not be the first USG representative to have made initial contact with a member of a resistance that the USG has decided to support. The previous chapters discussed in detail how traditional UW doctrine in previous decades blended—and often confused—immediate tactical activities with larger conceptual ideas. In reality, initial contact has multiple meanings, and there can be many distinct instances of so-called initial contact related to linear events and decision process development timelines, to include political negotiation and decision echelons, differences in military or other USG national instruments of power organizational involvement, military echelon, planners versus executors, and so on. Moreover, the same element may have more than one type of initial contact. Phase two is articulated below from the perspective of the operational detachment that has received a mission and is preparing to conduct UW.

### Higher to Lower United States Government Echelon Resistance Introduction

4-77. Most cases of initial contact with a resistance will be an introduction between echelons. Traditionally, this is most often interpreted as the operational detachment being directed to use a contact plan provided by higher to execute, after infiltration into a denied area, and ideally an asset knowledgeable of that same denied area to debrief for information while still in isolation.

## Prearranged Contact Plan

4-78. The operational detachment in isolation, conducting mission planning and preparation for a UW mission, should receive a contact plan between the operational detachment and the resistance provided to it during receipt of the higher commander's order. This implies that previous, higher echelon USG mission analysis had concluded that support to a resistance was at least potentially feasible. This further implies that a USG policy decision was made to attempt cooperation with a resistance entity of some kind, and at least rudimentary agreements were made between the USG at some level and some member or agent of that resistance in order to put the operational detachment in contact with that same resistance. In most cases, the details leading to those decisions included USG or allied initial contacts with the resistance at a level higher than the operational detachment. Those contacts may have come through diplomatic, commercial, social, or other channels. Previous contacts may have been developed and handed off by allies or other third parties, or may have been deliberately developed by U.S. intelligence organizations or some other military element. Only in the rarest of cases would an operational detachment be given a mission to contact a resistance in which no other USG entity, ally, or agent had ever engaged with before at any time, place, or level.

4-79. Traditionally, the operational detachment mission will include physical infiltration into a denied area and physical linkup with a designated resistance. In such cases, this is the most meaningful instance of initial contact for that operational detachment. Satisfaction with this instance of initial contact is based on the immediate operational and security success, resulting from the operational detachment's successful movement in a denied area from the infiltration location without fatal compromise, and achieving face-to-face linkup with indigenous actors at a time and place and in a manner designated in the mutually prearranged contact plan. This represents the baseline meaning of initial contact for an operational detachment; however, there are other meaningful varieties of initial contact.

4-80. It is possible, and even likely, that the elements making the first contact with the operational detachment in a denied area are not significant resistance organization decision makers. It may be days, weeks, or months before the operational detachment is escorted to, or permitted to, meet with any significant resistance leadership. Until such a meeting occurs, the meaning of initial contact is very shallow, and it awaits a more significant initial contact with a resistance leader authorized to negotiate. In the modern age of virtual reality and nonphysical relationships, initial contact could take on an entirely different meaning. It is possible that unit initial contact with the representatives of the resistance will be virtual prior to any physical contact. Beyond that, it is also conceivable that both initial and subsequent contact with the resistance will be limited to such virtual interactions. (See Appendix B, (U) Unaccompanied Resistance Considerations for Unconventional Warfare (S//NF), published separately.) In addition, the operational detachment may be provided an asset for debriefing prior to infiltration. In most cases, an asset provided for debriefing by the battalion will not be a member of the resistance. However, if no member of the operational detachment has had previous contact in any form with the resistance, and the asset debriefed is a member of the resistance verified by the battalion, this meeting would constitute the operational detachment's initial contact with the resistance.

## Asset Debriefing

4-81. The purpose of an asset debriefing is to provide the operational detachments firsthand and timely information that will allow them to prepare and plan a more detailed mission. The asset should have comprehensive background knowledge of the objective area. The preferred asset is indigenous to the JSOA or AO and is the designated representative of the area commander. The asset has been searched and screened to validate and vet his authenticity. In addition, he will have:

- Recently exfiltrated from the area.
- Been thoroughly interrogated by trained and certified personnel for any information and intelligence.
- Volunteered or have been recruited to assist in the SF mission.

4-82. The SF debriefer should quickly establish and maintain rapport, gain the asset's confidence, and question him as a peer, not as an enemy prisoner. Information needed from an asset should include, but not be limited to, the following:

- Current background information on the asset (experience, history).
- His method and reasons for assistance to the United States.
- The resistance force goals (past, present, and future).
- Local and regional combat and support power of the enemy security force.
- General and specific JSOA intelligence needed for initial contact, infiltration, operations, sustainment, and exfiltration.

## Pilot Team Considerations

4-83. Traditional UW doctrine typically conflated initial contact as a STR concept with the tactical event of an operational detachment's—however constituted—initial contact with resistance members on the ground, and lumped both dissimilar ideas into the same UW model (Phase Two–Initial Contact). That self-inflicted doctrinal confusion was compounded further by adding the specific mission of conducting a resistance feasibility assessment with a vanguard element known as a pilot team. All four of these ideas are separate, and beginning with this publication, these ideas are expressly articulated as such.

### Initial Contact Above Operational Detachment Level

4-84. As discussed before, the STR concept of initial contact can have many occurrences with many actors and USG echelons before a decision to conduct the specific UW type of STR is made. In fact, initial contact with a resistance may not result in a decision to conduct UW at all. Therefore, the traditional UW concept of the pilot team is not necessarily to make overall initial contact for the USG.

### Initial Contact at Operational Detachment Level During Unconventional Warfare

4-85. As discussed before, the baseline meaning of initial contact for an operational detachment conducting UW is the initial meeting with the resistance at linkup, inside the denied area at the end of infiltration. This meaning of initial contact applies to any operational detachment directed to conduct a UW mission. In certain cases, an operational detachment may be explicitly designated a pilot team.

### Pilot Team as Feasibility Assessment Team

4-86. Pilot teams have two different potential functions: as primarily a feasibility assessment team or as primarily a vanguard team for a larger planned force. The former traditional assessment role is, by definition, a mission prior to approval to conduct UW; it is PE—not UW per se. The traditional mission of a pilot team as carried forward in ATP 3-05.1 is still valid.

4-87. A pilot team with the above traditional mission may be ordered to exfiltrate, pending further policy decisions on whether or not the USG will eventually conduct STR (and in some cases, UW specifically), may be ordered to initiate the conduct of UW at that moment, or it may be ordered to operate as a vanguard element for a larger follow-on force. It is important to acknowledge, however, that until the pilot team is given an order to initiate UW—it is not conducting UW; it is conducting PE.

### Pilot Team as Vanguard of a Larger Force

4-88. By contrast, an operational detachment conducting a UW mission in a denied area may be designated a pilot team with the primary mission of making the initial on-the-ground contact with the resistance, for the purpose of facilitating the reception, staging, onward movement, and integration of planned follow-on forces. A pilot team with a vanguard force primary mission is not primarily doing PE.

> "A pilot team is a deliberately structured composite organization comprised of Special Forces operational detachment members, with likely augmentation by interagency or other skilled personnel, designed to infiltrate a designated area to conduct sensitive preparation of the environment activities and assess the potential to conduct unconventional warfare in support of U.S. objectives. If possible, planners may exfiltrate a resistance representative, referred to as an asset, from the operational area to brief the pilot team during its planning phase and possibly to accompany the team during their infiltration into the operational area and linkup with resistance forces.
>
> The pilot team conducts detailed area assessments to expand its understanding of the operational environment, particularly human terrain analysis. This analysis provides information on the degree of support for the UW effort among the local populations...."
>
> ATP 3-05.1, *Unconventional Warfare (Change 1)*, 25 November 2015,
> p. 2-9, para. 2-36 and 2-37

### Lower to Higher United States Government Echelon Resistance Introduction

4-89. There are exceptions to the typical pattern of higher-to-lower introduction of resistance initial contacts to operational detachments, however. Rather than a preconceived and carefully considered policy decision initiated at senior USG policy levels and introduced downwards, it is possible that initial contact with potential resistance partners will be made at tactical levels and transmitted upward.

### Chance Contact

4-90. During the conduct of other missions, U.S. forces may be approached by self-identified resistance members or their agents to solicit U.S. support for their cause. This initial contact of potential resistance members is unplanned, and it is therefore outside of the standard seven-phase model and traditional understanding of U.S. STR (UW). Such contacts should be reported. Although initially unintended, this chance contact may begin the feasibility assessment process and eventually result in its own UW mission.

### Deliberate Contact

4-91. Although the exception rather than the rule, it is possible that previously unidentified and uncontacted population groups will be approached to solicit their potential interest in fomenting resistance or accepting external sponsor support. This, too, represents a departure from the traditional standard seven-phase model and traditional understanding of U.S. STR (UW). In such a case of no prior contact of any kind, the mission is not UW per se, but rather, it is the creation of initial contact itself. This differs from a pilot team operation in which case it is assumed that contact with the resistance of some kind has been made previously at some level.

### VARIETIES OF INITIAL CONTACT BY TYPE

4-92. Like the confusing vagueness of traditional doctrine's explanation of psychological preparation, initial contact has also been insufficiently articulated. Initial contact has a wide variety of meanings by the type of contact it represents. Some examples include:

- Echelon of USG activity.
- Significance and authority of interlocutors at each level.
- Linear temporal point at which such contacts are made.
- Direct versus indirect contact.
- Physical human face-to-face contact versus virtual-only contact.
- Direct contact between two principles versus contact filtered through human or technical cutouts.
- Initial contact between two already acquainted parties, which represents not an unprecedented interpersonal meeting event but rather a change in political interests or bargaining positions.

# PHASE III—UNIT INFILTRATION OF THE DENIED AREA, INITIAL ASSESSMENT, AND LINKUP WITH THE RESISTANCE

## INFILTRATION

4-93. The operational detachment with a UW mission infiltrates the denied area according to the approved OPORD to link up with a designated resistance. Infiltration into the denied area is not the UW mission itself; it is merely a phase of the operation. Infiltration begins when the operational detachment leaves the last friendly departure or launch point, continues through a passage into the denied area, places the operational detachment on the ground in the denied area, and continues through movement in the denied area until link up with the resistance is achieved. In terms of the seven phase model of UW, at the point of linkup with the resistance the infiltration phase is over.

*Note:* Although there is no substitute for the ground truth veracity of SF operational detachments in the denied area working shoulder-to-shoulder with resistance partners, in the modern age there will be some cases where physical infiltration of the SF operational detachment into the denied area is not even required, may be improbable or even counterproductive. In such cases, successfully supporting a resistance still fits the definition of UW if the mutually desired effects can be achieved inside the denied area. However, the standard and most likely template for conducting UW—even in the modern age—still requires the SF operational detachment's stealthy physical entry into the denied area.

4-94. Successful infiltration into the JSOA requires detailed planning and preparation at the joint level. Several methods are considered based on METT-TC. (See Appendix K, Infiltration and Exfiltration Considerations, for additional information.) With the exception of cases where infiltration or exfiltration mechanisms into or out of denied areas may be under the sole control of the resistance, there is nothing about special operations infiltration that is specific to UW only. The basic activities of infiltration into denied areas will include specialized movement techniques, use of caches, evasion planning, and resistance link-up procedures. UW does, however, traditionally use a highly detailed form of information reporting called the area assessment, which begins immediately upon infiltration of the operational detachment into a denied area.

## MOVE IN A DENIED ENVIRONMENT

4-95. Movement in a denied area is a basic expected requirement of many kinds of special operations and even some conventional force operations. Movement in a denied area during UW differs from most conventional operations based on the purpose of the UW mission—assisting, advising, and accompanying resistance forces. Whereas conventional forces use movement to contact for deliberately engaging the enemy and developing the situation into a larger event, the operational detachment seeks to avoid the enemy at all times—except when it can be assured of having the initiative. The conventional force assumes that units in contact with the enemy will be augmented by additional and larger maneuver forces and supported by fires with increasing destructive power. The operational detachment in the denied area has no such expectations. Even though long-range, conventional reconnaissance forces can be deliberately placed into enemy-controlled areas, such forces typically do not go deep into such territory and do not stay long. By contrast, the operational detachment conducting UW may be placed very deep in enemy territory for the duration of a conflict, with little or no recourse to friendly direct support. The operational detachment moving in a denied area while conducting a UW mission does not win or survive by its own firepower.

## Movement Techniques

4-96. An operational detachment conducting UW in a denied area is extraordinarily vulnerable to destruction if spotted and engaged by enemy forces, which enjoy advantages of size, firepower, maneuverability, communications, and so on. As a basic principle, the operational detachment must prioritize remaining undetected during movement and eluding enemy engagements. This is especially true during the highly vulnerable period between initial insertions onto the ground in the denied area prior to linkup with the resistance. All of the techniques of stealthy, dismounted patrolling must be mastered and employed.

*Note:* ATP 3-21.8, *Infantry Platoon and Squad*, Chapter 6, Patrols and Patrolling; and TC 3-21.76, *Ranger Handbook*, provide additional information.)

## Caches

4-97.  An operational detachment conducting UW in a denied area is beyond routine conventional sources of supply. Since most UW missions may involve long periods of austere conditions and self-reliance, judicious use of hidden caches is an integral part of the operational detachment's long-term sustainment plan. The infiltration period of initial insertion into a denied area prior to linkup with the resistance offers an opportunity to emplace emergency caches known only to the operational detachment, higher echelon, and those with a need to know. (See Appendix N, Caching.)

## Evasion

4-98.  An operational detachment conducting UW in a denied area is a high risk activity. Starting with infiltration, numerous events throughout the course of a UW mission could trigger the operational detachment entering into evasion. Examples of such events include: compromise at the time and place of initial insertion, compromise while en route to link up with the resistance, catastrophic defeat of the resistance by the enemy postlinkup, catastrophic loss of rapport with the resistance, orders by higher command, or other reasons. Careful and detailed evasion planning is essential to maximizing the operational detachment's probability of survival when under duress. (See ATP 3-18.72, *(U) Special Forces Personnel Recovery (S//NF)*, for more information on evasion planning.)

## Linkup With the Resistance

4-99.  Linkup between two lethal forces is always a potentially dangerous operation. Such linkup is typically conducted at night or during periods of limited visibility, often between forces that are unknown and mutually unfamiliar. The chance of being mistaken for an enemy increases tensions on all sides. These tensions are increased during UW initial linkup between the operational detachment and the resistance link-up element. In a denied area, the operational detachment will assume that every other actor it encounters is potentially hostile. Finding the one friendly resistance actor in this environment is looking for the needle in the haystack. Operational detachment and resistance partners likely speak different languages, have uncommon and unfamiliar appearances, and may interpret body language and interactive responses differently. All of these additional challenges make the operational detachment resistance link-up event more dangerous. This puts a premium on receiving good, quality information; obtaining mutually agreed-upon detailed link-up procedures and signals; and conducting rehearsal of those procedures during mission preparation, prior to infiltration. Where possible, operational detachment linkup with the resistance can be aided by infiltrating a resistance member known to the receiving party as a trusted guide for both parties. (See Appendix O, Link-Up Operations.)

## AREA ASSESSMENT

4-100.  The operational detachment members and the commander begin an area assessment immediately after entry into the JSOA. This assessment is the collection of special information and serves as the commander's estimate of the situation. The area assessment is a continuous process that confirms, corrects, refutes, or adds to previous intelligence acquired before commitment. The area assessment also serves as a basis for changing premission operational and logistics plans. There are no fixed formulas for conducting an area assessment. Each commander has to decide for himself what should be included and what conclusions may be drawn from the information he collects.

4-101.  When making an area assessment, the operational detachment commander considers all the major factors involved, including the enemy situation, security measures, and the many aspects of the civil component as defined by ASCOPE. The operational detachment should only disseminate new information and intelligence that differs significantly from the intelligence received before commitment. An area assessment is either initial or principal, depending on the urgency involved. (See Appendix I, Area Assessment Formats—Initial and Principal.)

## Initial Assessment

4-102. Initial assessment includes those requirements deemed essential to the operational detachment immediately following infiltration. These requirements must be satisfied as soon as possible after the team arrives in the AO. Much of this initial assessment may be transmitted in the initial entry report (ANGUS) or situation report (CYRIL). Figure 4-2 provides examples of these report formats.

---

Proword: **ANGUS**
(1) Mission designator and ODA codename.
(2) Location (6-digit UTM coordinates).
(3) Casualties. Code name of personnel who are unable to continue the mission using code words listed below:

UNCLE: Killed in Action
FROST: Wounded in Action (explain).
SPARK: Captured or Missing in Action.
(4) DTG contact/linkup made with friendly elements.
(5) Strength of FID/guerrilla force.
(6) Location of MSS/patrol pase (6-digit UTM coordinates). If different from line 1.
(7) DTG surveillance established on target/area.
(8) Additional information.

Proword: **CYRIL**
(1) Location of outstation(s) (6-digit UTM coordinates).
(2) Guerrilla/FID strength.
(3) Condition of detachment (excellent, good, fair, poor).
(4) Enemy contact information (DTG, causality/status).
(5) Major activities since last report (what, where, and when).

Include post PSYOP and TACAIR, and so on results in this paragraph. Also include non-nuclear destruction of targets.
(6) Projected major activities (what, where, and when).
(7) Additional information.

*Note:* Avoid lengthy descriptions, but furnish sufficient detail to allow the formulation of supporting or contingency plans at the base.

---

**Figure 4-2. ANGUS and CYRIL examples**

## Principal Assessment

4-103. Principal assessment forms the basis for all other subsequent UW activities in the JSOA. It is a continuous operation and includes those efforts that support the continued planning and conduct of operations. It should be transmitted using the format planned during isolation. This format may be abbreviated by deleting information already confirmed. This report should include new or changed information.

---

**Area Assessment in an Urban Environment**

Decades of traditional UW doctrine have been focused primarily on rural areas for understandable reasons. Prior to the late 20th century, a majority of the world's population lived in rural areas. In addition, the centers of enemy state or occupying power surveillance and repressive strength were concentrated in cities as key state infrastructure nodes. As a matter of sheer survival, resistance efforts tended to form and operate from inaccessible, usually rural areas.

Today, a majority of world population lives in urbanized areas and that demographic trend is increasing. Area once considered physically remote and inaccessible have increasingly been integrated into a modernized grid. Rural, underdeveloped and sparsely populated area is shrinking in relative terms. Moreover, technological advances have increased the surveillance potential of opponents to unforeseen levels. It is more difficult than ever before to remain hidden in physical hinterlands. Finally, resistance is a human activity. Organizing and assisting human population groups to achieve political ends by contesting state power usually requires being where the people are. That increasingly is in the cities.

What makes urban guerrilla warfare so different from rural guerrilla warfare or conventional military contest is the presence of a large audience to the struggle. Unlike rural guerrillas, urban guerrillas cannot withdraw to some remote jungle where they are safe from observation and attack They must be able to live in the midst of hundreds of witnesses and potential informers. Although the principle of *area assessment* remains the same, the conditions of assessing urban environments is different. Appendix W, Unconventional Warfare in an Urban Environment, provides information for the planning and conduct of UW in an urban environment.

---

# PHASE IV—UNIT SUPPORT TO RESISTANCE ORGANIZATION, CONFIRMATION OF MUTUAL OBJECTIVES, PRINCIPAL ASSESSMENT, AND RESISTANCE CADRE DEVELOPMENT

4-104. After infiltration, the major task facing the operational detachment is to support organizing the resistance. This is the most critical phase of the entire seven-phase UW model. Correct, holistic operational detachment identification and assessment of the resistance movement's capabilities, and sound operational detachment advice and assistance to improve and expand future resistance capabilities is all rooted in the decisions made during this phase. By contrast, hasty or ill-advised operational detachment recommendations could prove fatal to the USG's desire to continue policy support of the resistance, the resistance's desire to cooperate with the United States, or to the resistance and operational detachment themselves.

## OBJECTIVES

4-105. The initial task of the operational detachment is to reaffirm the agreements made between the USG and the resistance leadership prior to arrival of the operational detachment in the denied area. To the senior leadership in the area command, the operational detachment is the direct representative of the USG.

### United States Government Objectives

4-106. The USG provides external support of a resistance ultimately to meet USG policy goals. The cost-benefit analysis, which underlies the STR policy will be different for every specific situation. The analysis is likely to change as circumstances and larger political contexts change. The resistance has a destiny to either succeed and survive, or fail and die, regardless of whether the United States provides support to it or not. The operational detachment must remember that it is always, ultimately, a matter of USG strategic policy interests that determines the extent and duration of any such sponsorship. In addition, contingencies to ensure the survival of the operational detachment as tactical circumstances change on the ground, may require the operational detachment to break off physical contact with the resistance, regardless of an enduring sponsorship commitment.

## Confirmation of Mutually Agreed Upon United States Government Resistance Partner Objectives

4-107. Competent and mature resistance leadership will understand that the USG is providing support to them for USG interests. The operational detachment does not need to belabor this truth. The operational detachment should, however, emphasize the goals that the USG and the resistance have in common. Sometimes, resistance culture, history, and common values will make continuing commitment to mutual goals relatively intuitive and easy. These objectives may be short-range or long-range, tactical, operational, strategic, or political in varying combinations. At other times, many cultural, historical, and value challenges will make commitment to mutual objectives difficult to achieve and sustain. Potential or initial resistance partners may violate human rights in ways that cannot be accepted by the USG, forcing the USG to withdraw sponsorship. Resistance partners may share strong mutual goals at some levels of war but diverge significantly at others. For example, the immediate destruction of enemies at a tactical level may be mutually desirable, but such effects may result in a new form of resistance-led government which is anathema to the USG.

4-108. Ultimately, the resistance is a partner of the USG which must be influenced and negotiated with—not a surrogate which can be directed—and resistance leadership will act in their own interests. The operational detachment monitors the fluid relationship and reports changes as needed.

## Build Rapport

4-109. The operational detachment's mission is not to build rapport. Its mission is to achieve the mission objectives described above. However, rapport is a critical skill that will enable the operational detachment to build a sense of trust. The operational detachment members must be diplomats, as well as military advisers, and establish a good working relationship with the resistance organization. This relationship develops from acceptance of the USG conditions necessary for U.S. sponsorship in balance with the resistance leaderships' satisfaction that the relationship continues to benefit the resistance. Honesty and demonstrable support play a measureable role in gaining and sustaining rapport. Shared hardship and heroism and qualitative human personality traits play an unquantifiable role. In a lasting partner relationship, the clear foundational message of rapport will almost always be that the operational detachment and resistance are united with a common goal against a common enemy. (See Appendix P, Special Forces Resistance Advisor Considerations.)

## DEVELOPMENT OF THE JOINT SPECIAL OPERATIONS AREA

4-110. Once mutual USG-resistance objectives have been confirmed, and initial rapport has been established, the operational detachment must identify, assess, and provide advice to the resistance leadership on the basic structure of the resistance organization. The longstanding and enduring U.S. model of resistance, along with the accompanying list of successful resistance dynamics, provide the operational detachment the basic template to proceed.

> *Note:* See TC 18-01.1, especially task #16 (Develop an Area Command Group and Staff), task #17 (Organize Forces), and task #18 (Develop the Area Complex), for detailed lists of what structures the SF advisor should expect to find (or intend to develop) in a resistance.

## Identify

4-111. The operational detachment must do its best to identify the structure and functions of the resistance as they actually are beginning immediately upon linkup. Through a thorough preinfiltration area study, accurate and recent intelligence updates, and recent knowledgeable asset debriefs, the operational detachment will learn as much as is known about the resistance prior to infiltration. From infiltration on, the operational detachment will remain observant—acutely aware of everything in its operational environment—to collect information and fill out and update the area assessments and other formats. Aiding the operational detachment is the longstanding U.S. model of resistance, which outlines the rudimentary functions necessary for resistance survival, control, and effectiveness. However, that model is only a template. Every situation will differ, requiring the operational detachment to identify the actual structure and functions on the ground before them. This is not as straightforward as it might seem.

4-112. First, the resistance is unlikely to use the terms in U.S. doctrine to describe itself, especially in the beginning of the relationship. This is why the operational detachment must identify the functions the resistance uses to then identify the form of the resistance structure. This requirement also partly explains why deep area orientation and mission preparation aids the operational detachment's understanding of the foreign culture they are now immersed in.

4-113. Second, the ability to identify is largely dependent on the ability to first observe. There are several reasons why this observation will be a challenge. How much the resistance allows the operational detachment access to the resistance is dependent in part on how much the resistance leadership trusts the distant USG, generally, and the operational detachment in its presence, specifically. It may take a long time for the resistance to trust the operational detachment sufficient to expose its structure and capabilities. In some cases, that level of trust may never come. In most cases, such resistance self-exposure will be incremental. It may take a very long time for the operational detachment to get a somewhat complete sketch of the overall resistance organization. In all cases, the operational detachment should expect that the resistance will withhold at least some of its own secrets permanently. Even the most stalwart allies withhold some information from each other. To expect that the resistance would ever voluntarily give up all of its clandestine mechanisms, sources of indigenous power, secret political benefactors, and emergency sanctuaries is a naive assumption. Finally, even the most cooperative and forthcoming resistance partner cannot escort the operational detachment to every resistance location at the same time. The operational detachment is a small element that cannot be everywhere at once. The larger the scale of the resistance, the size of the denied area territory, and the more effective the state or occupying power's repressive capabilities are, the more this will be true. While the operational detachment will likely be involved in select resistance tactical actions, the operational detachment most often will have to work at the resistance operational and strategic levels, work through area and sector commands, and may have to trust the version of the resistance painted for them by the resistance leadership. The operational detachment will do its best to verify the ground truth of the entire resistance.

## Assess

4-114. The challenges of identifying the resistance ground-truth, mentioned above, will shape the accuracy of the operational detachment assessments. Nevertheless, despite the challenges and the fact that every situation is different, the operational detachment has the advantage of the U.S. model of resistance as the basic template for the required functions of a resistance. The operational detachment should seek to find the local resistance equivalent of each function listed in U.S. UW doctrine (at a minimum). Cooperative resistance counterparts should be able to answer these questions, and hundreds similar, such as—

- What mechanisms do you have for acquiring and reporting intelligence in Urban Area X?
- How does leadership communicate in a timely manner with dispersed elements throughout the country?
- How do you coordinate the actions and activities of your different elements?
- How do you sustain your efforts? Where are stores maintained?
- How do you maintain security?
- What successful tactical armed actions have you taken? What were the immediate and secondary results?

4-115. Sometimes, the operational detachment will have to infer the answers based on incomplete information. Sometimes, the operational detachment will have to use its judgment and experience to correctly assess the reality of a situation, despite faulty or misleading information provided by the resistance.

4-116. There will be different kinds of assessments. The principal assessment is the standard format for collecting and updating everything about the resistance for transmission to the JFC. Assessments will cover all aspects pertinent to the resistance at all levels of war and comprehensive of all essential functions. Personality assessments of key leaders, population attitudes to the resistance, and prospects for successful achievement of U.S.-only and U.S.-resistance mutual goals will be among the most important. Most assessments will be for U.S. use only. Some—particularly those which improve and strengthen the resistance—will be deliberately shared with the resistance partners.

## Develop

4-117. Once the resistance is assessed, the operational detachment and resistance leadership craft a plan for resistance development within the parameters of the original USG-resistance leadership agreement. Once again, the longstanding U.S. model of resistance is the basic template for expected structure and functions that the operational detachment assists the resistance to achieve. Every situation will be different, and the development needs of the resistance will vary, depending on how large, developed, and effective it already is; the structural and functional requirements the resistance needs based on its objectives; the strength of the enemy opponent and stage of development; whether the resistance is primarily urban or rural based; and many other factors. Development of a large, already combat experienced, well-led and well-resourced resistance will differ markedly from the nascent effort merely struggling to survive. The organization phase of the traditional seven-phase UW template is intended to consider the latter example—the hardest possible case of a fledgling resistance that needs every kind of advice and assistance.

## Basic Command

4-118. The resistance leadership structure and the chain of command must be outlined. Use of the resistance area command and sector command structures represent a geographically based structure. Political leadership outside this command structure should also be identified with the venues they operate in and audiences they might influence. Any government-in-exile or displaced political leadership that legitimately speaks for the resistance should be identified as well. Declared key leaders should be cross-referenced with essential functions to identify any critical function gaps.

## Basic Organizational Structure

4-119. The operational detachment uses the generic core and supplementary component categories provided in U.S. UW doctrine—cross-referenced with resistance objectives—resources available, and a realistic development timeline to advise on the development needs of the resistance organizational structure. Generally speaking, mere survival of the resistance is the first priority, so structures set up to ensure security and intelligence will likely be first. Sustainment and procurement structures, messaging capabilities, and possibly recruitment will probably be next. To the extent that political messaging and popular support is an important aspect of resistance strategy, both overt and clandestine political organization structures must be created or co-opted. To the extent the resistance desires and requires structural growth of any kind to meet its objectives, critical cadre development mechanisms will also be an early structural consideration. Action units—capable of low-level political agitation and demonstrations and modest armed forces capable of low-level violent activities—will be early structural considerations. Larger units, capable of larger actions will come later.

## Basic Support Structure

4-120. The operational detachment uses the sustainment network lists and diagrams in U.S. UW doctrine regarding the area complex as a start point for advising creation, improvement, and future expansion of all essential supporting functions. These organizational considerations are a combination of physical infrastructure—caches, safe houses, training areas, medical facilities, equipment, and so on—and human infrastructure—the people who provide the functions. A majority of the organizational considerations of the core resistance component auxiliary will be relevant here.

## Critical Cadres

4-121. The operational detachment itself is an elite U.S. Army cadre designed, in large part, to multiply resistance capabilities through experienced advice and leadership in core functions. Traditionally, and especially during the height of the 20th century Cold War, the expectation was that the standard 12-man ODA was capable of training, advising, and assisting an indigenous guerrilla light infantry battalion, to include supporting the underground and auxiliaries, the sector or area command elements, and the development of the area complex. Twelve men were intended to advise or lead a 450–600 resistance force along with all associated clandestine command and supporting mechanisms in a denied area. This meant that the ODA occupied (or advised) the key positions of that battalion. In detail, this meant the ODA commander functioned as a light infantry battalion commander, the ODA operations sergeant functioned as a light infantry battalion operations officer, and the rest of the ODA's warrant officers and noncommissioned officers acted as key battalion staff or light infantry company commanders. This U.S. cadre function provided experienced (and unified) advice and good examples to the indigenous leadership, until such time as those indigenous leaders were experienced enough to operate on their own. The same function is critical to resistance organic expansion. There are a finite number of U.S. advisors available to any given situation, and ultimately the success or failure of the resistance depends primarily on their own indigenous efforts. For those resistance movements which desire and require physical growth, the identification, recruitment, and training of critical cadres, as part of a well-conceived development plan, is a vital organizational requirement.

# PHASE V—UNIT SUPPORT TO RESISTANCE BUILDUP AND EXPANSION OF RESISTANCE CAPABILITIES

4-122.    The traditional seven-phase model of UW includes this phase to acknowledge that, in most cases, a supported resistance is almost always likely to expand in size and capabilities from some previous, less-developed period in the past—especially after the anticipated improvements provided by U.S. support and advice. This is a basic concept big idea of the model; there is no fixed, phase five-specific list of tasks that can only be done in this phase. The key concept is expansion in all resistance structures and functions.

4-123.    There are, however, a couple of qualifications to the above paragraph. Buildup is based on the classic mid-20th century notion of UW as STR that always has as its goal the complete overthrow of the state or occupying power. Since this is conceived as a large undertaking, it is also assumed that the resistance structure necessary to meet that effort will also always have to be very large. This template was true for the successful Chinese Communist revolution. It was not as true, however, for the successful Bolshevik revolution. Therefore, the degree of buildup required will vary according to the specific conditions of the resistance. Moreover, the 21st century definition of UW identifies coerce and disrupt outcomes as potential objectives of UW; in a majority of cases were the resistance seeks more limited objectives, large-scale buildup may not be required. Buildup or expansion may be much more limited and could be very specific to limited target audiences and capabilities. Finally, regardless of the overall resistance objectives, buildup and expansion is not a monolithic effort conducted in perfect unison across the entire resistance and denied area. Buildup and expansion is almost certain to be episodic, unevenly achieved, and in some cases may actually retract in some locales, while it increases in others. Sometimes, the resistance itself may choose to disband or decrease efforts in some local areas while expanding generally.

4-124.    One additional caution is important for the operational detachment to consider in its advice to the resistance during this phase. There is a tradeoff between operational effectiveness and security. Expansion and buildup should not be attempted haphazardly; it should be done deliberately with much forethought and as a sequenced step in a well-outlined resistance plan. If the resistance grows too fast and too obviously and if resistance elements grow beyond their ability to provide for their own local security, competent and powerful enemy security forces will destroy them.

# PHASE VI—UNIT ADVICE AND SUPPORT TO THE EMPLOYMENT OF RESISTANCE CAPABILITIES

4-125.    Inclusion of this phase in the traditional seven-phase model of UW originates from the classic World War II UW example in the European Theater of Operations in which resistance efforts were one line-of-effort supporting the overall conventional campaigns. Although supported resistance elements engaged in continual harassment of enemy forces and provided routine intelligence and personnel recovery support throughout the war, the larger supreme Allied command intention was to prepare the resistance for the culminating big event of support to the Allied landings on D-Day in 1944. (See buildup and expansion in phase five.) In decades past, U.S. UW doctrine even labeled this phase combat employment to characterize this culmination of efforts.

4-126.    This phase is also a basic concept big idea of the UW model. Although the very basic idea of developing resistance capabilities to the point of desired effectiveness is valid—especially so for those instances of support to large-scale operations and unconditional overthrow objectives—this phase is also somewhat of a misnomer. The UW seven-phase model is just a model. Employment is not limited to only combat. Employment is not something that happens only when all previous phases have been completed and is not necessarily done only after all preparations are complete and executed all-at-once, on command, across the entire theater. The resistance does not hold its fire until a command is given to commence resistance phase six. The reality of resistance is much messier. Employment can be nonlethal political and organizational activities conducted very early in the life of the resistance and without which the ability to conduct combat later on might never be reached. Resistance employment is likely to include very small-scale acts of violence early on. Employment of resistance information and influence activities or agitation or street demonstrations by public component front groups is no less resistance employment than a guerrilla force setting an ambush of an enemy supply convoy in a mountain pass. All of chapter 5, *Core Unconventional Warfare Activities*, discusses in detail scores of representative employment activities of resistance. (See Appendix P, Special Forces Resistance Advisor Considerations; Appendix R, Shadow Government Operations Model; and Appendix V, Special Forces Advisor's Guide to the Combat Employment of Guerrilla Forces.)

# PHASE VII—UNIT TRANSITION FROM UNCONVENTIONAL WARFARE MISSION RESPONSIBILITIES TO FOLLOW-ON MISSIONS

4-127. In decades past, UW doctrine discussed phase seven as demobilization because the classic origins of the UW model was focused primarily on the guerrilla forces and their transition from irregular combatants to disarmed, postwar peaceful civilians—nonthreatening to the new government after cessation of hostilities. In more recent decades, the Army better appreciated that the guerrilla force was only one aspect of resistance; all of the multifaceted considerations of a resistant society had to be considered and, where possible, assisted to transition from hostilities to the postconflict reality. (Appendix Y, Resistance Demobilization/Transition Considerations, discusses the requirements of this phase in detail.)

4-128. There are two characteristics of the transition phase that merit additional comment. First, for decades in UW doctrine, transition (or demobilization) assumed that the objective of UW was always assumed to be overthrow of an enemy state or removal of an occupying power. This unconditional and unequivocal result demanded complete redirections and reintegration of an entire subgroup into a new larger society—a society that was also classically assumed to be under the benevolent control or supervision of allied victors. Bad guys out; good guys in. These assumptions are invalid when the objective of the UW effort is coerce or disrupt, when there is no resistance or U.S. intention to proceed to complete overthrow. If the objectives, expenditures, scale, risk accepted, and time frame are all limited, if the enemy state or occupying power will still be in power at the conclusion of the UW effort, the transition will be correspondingly limited as well. Transition in such circumstances will not mean transition of resistance into a new society. Transition will mean change in the supporting relationship between the resistance and the U.S. temporary sponsor.

4-129. It is possible that transition of resistance forces could still mean the reintegration of the resistance as loyal members of the still existing state regime. This is conceivable when a negotiated settlement between the resistance and the (now former enemy) state is possible. When negotiated settlement and peaceful integration are not possible with the still enemy state or occupying power, transition will likely mean that the U.S. support is stopped altogether. Alternatively, the United States may continue support but at a lower level for decreased objectives or changed strategies, such as to maintain the resistance in a dormant state or to support it as a nonlethal-only political actor, respectively. It is also possible that support of the resistance could be handed over to an allied or third-party sponsor.

4-130. The second critical UW transition consideration is that it assumes there is an articulated end state envisioned for the UW effort. From a purely theoretical view, some argue that there is no such thing as an end state because human relations—and political interaction at all levels—never end. If and when the USG ceases all support to the resistance and presence in the denied area, the indigenous population will continue to exist and in some sense problems never end. While that is generally true, the idea of an end state is still necessary and useful, because it attempts to articulate the objective of the UW operation. As mentioned in chapter 1, it is critical that every UW feasibility assessment and OPLAN be based on a clear end state. The final phase of transition must be considered at the very beginning of planning and assessment of a UW effort. Planners and commanders must envision the end state they intend the UW effort to transition into.

# Chapter 5

# Core Unconventional Warfare Activities

*The shifting from political struggle to armed struggle was a very great change that required a long period of preparation. If insurrection is said to be an art, the main content of this art is to know how to give to the struggle forms appropriate to the political situation at each stage.*

Vo Nguyen Giap, *Peoples' War, Peoples' Army*, 1962

5-1. The five UW core activities of subversion, sabotage, guerrilla warfare, personnel recovery, and intelligence operations have been consistently present in Army UW and guerrilla warfare doctrine, dating back to World War II. These five core UW activities plus the derivative new core activity of PE constitute the six core activities likely to be present in any specific example of UW. It is important to note that the logic does not work in reverse. Each activity also has utility outside of UW. Just because one may be conducting intelligence operations or sabotage or PE, for example, does not mean one is necessarily conducting UW.

5-2. This publication helps SF Soldiers identify and assess techniques a resistance may use, and it represents the core capabilities that the SF Soldier conducting a UW mission should be expected to advise others on or lead himself. The examples below are a representative sample of what has been used historically by many actors throughout world history. Presentation of subversive techniques in this publication is not a statement that U.S. forces necessarily conduct or support any one of them, nor should such U.S. conduct or support be inferred. The United States does not conduct or condone terrorism.

## SECTION 1. INTELLIGENCE OPERATIONS

5-3. Insurgencies and resistance movements rely on accurate and timely intelligence for survival and effectiveness like any other political actor with the will to use violence. The resistance area command is not, however, primarily an intelligence agency but a military force. The intelligence system of the operational area is primarily geared to support the command itself. The primary intelligence interests of the resistance area command will be obtaining and acting on intelligence that supports its own insurgent tactical operations and political activities. However, when an insurgency or resistance movement gains U.S. support as part of a UW campaign, such intelligence, as can be collected by the indigenous resistance forces, can be of tremendous value to the combatant commander. Definitions for intelligence terms include:

- **Intelligence Operations.** *Intelligence operations* are the tasks undertaken by military intelligence units through the intelligence disciplines to obtain information to satisfy validated requirements (ADP 2-0). These requirements are normally specified in the information collection plan. Intelligence operations collect information about the intent, activities, and capabilities of threats and relevant aspects of the operational environment to support commanders' decision making. Intelligence operations, like reconnaissance, surveillance, and security operations, are shaping operations used by the commander for decisive action.

- **Intelligence Preparation of the Battlefield.** *Intelligence preparation of the battlefield (IPB)* is the systematic process of analyzing the mission variables of enemy, terrain, weather, and civil considerations in an area of interest to determine their effect on operations (ATP 2-01.3). The IPB process considers all threat capabilities within and across each domain within the unit's AO and area of interest and the relevant aspects of the information environment. The other staff sections assist the intelligence staff in developing the IPB products required for planning. The IPB process consists of these four steps:

- Define the operational environment.
- Describe the environmental effects on operations.
- Evaluate the threat.
- Determine threat COAs.

5-4.  IPB starts immediately upon receipt of the mission, is refined throughout planning, and is updated to support subsequent operational planning. (ATP 2-01.3 provides details of the IPB process.)

*Note:* Like PE, UW intelligence operations are highly sensitive and current examples will almost always be classified. Therefore, the examples given in this section rely on the classic era of UW—World War II—to illustrate some of the constituent ideas. Although dated, these examples provide insight into the types of intelligence operations that have been and could again be conducted. Presentation of intelligence operations techniques in this publication is not a statement that U.S. forces necessarily conduct or support any one of them, nor should such U.S. conduct or support be inferred.

# TACTICAL INTELLIGENCE

5-5.  Tactical intelligence, which allows the insurgent or resistance organization to survive and to take action against the opponent is the priority. Insurgent intelligence assists associated guerrilla forces by providing valuable data about the enemy and the area of impending combat. This may include the number of enemy troops, their deployment, their unit designations, the nature of their arms and equipment, the location of their supply depots, the placement of their minefields, the pattern and routine of their patrols, the morale of the troops, and various topographical factors, such as swamps and ravines, that govern access to enemy emplacements. This information will be used by the insurgent or resistance organization itself for its own tactical planning. Every member of a resistance is an information provider.

5-6.  The location of the area command in enemy-controlled territory also makes available to the theater commander an additional means of developing intelligence generally unavailable to other theater forces. Some examples include:

- Finely detailed and timely threat characteristics intelligence data.
- Indigenous information crucial to the development of effective insurgent psychological warfare activities.
- Information of political, sociological, and economic intelligence value at a resolution and vantage otherwise unavailable.
- Precise, accurate, and timely target information for tactical and strategic air forces plus poststrike information.
- Intelligence data to support specific tactical operations, such as airborne, amphibious, or armored operations.

5-7.  Sometimes this information is obtained directly by underground personnel by visual observation of the targets. For example, members of the French resistance reconnoitered German coastal defenses in the preparation for the Allied invasion of France in June 1944. Such data may also be collected by the local populace, or popular antennae, as these sources are described in one Vietminh manual. The Vietminh used children playing near French fortifications as a source of information on troop arrivals and departures, the guard system, and other pertinent details, which aided the guerrillas in planning attacks—all of which were easily observable by untrained children.

5-8.  In UW, intelligence training and guidance by the sponsoring power can improve the quality of such information because most undergrounds initially lack the personal experience in this type of work. For example, the World War II Jedburgh teams in France trained resistance workers in their intelligence surveys and helped to coordinate their collection efforts. Likewise, Red Army personnel were assigned to the Soviet partisans to direct their activities. In the modern context, U.S. Army SF and other governmental agencies can provide this kind of training and guidance.

5-9.   A Soviet guidebook distributed to partisans for use in regions under German occupation instructed that, "If you happen to encounter troops...do not show that you observe the enemy...ascertain the colour of their headgear, their collar braid, and the figures on their shoulder straps. If they have questioned the inhabitants about something, try to find out what the Fascists have asked. . . ."

5-10.   The guidebook also gave tips about ascertaining enemy intentions, "...if an attack is planned, trucks will arrive loaded and depart empty; if the enemy intends to retreat, fuel and foodstuffs will be removed. roads and bridges will be demolished, telephone wires will be removed, and trains and trucks will arrive empty and depart full."

# INTELLIGENCE IN SUPPORT OF SABOTAGE

5-11.   Reconnoitering transportation and communications facilities prior to sabotage attacks occupied much of the time of French resistance persons. Often working closely with Allied advisors, these people surveyed targets earmarked for sabotage on D-day. In reconnoitering a bridge, for example, resistance members looked for such factors as the guard system covering the bridge. If a number of permanent troops were evident, a step to eliminate them had to be included in the sabotage plan. When there was only an occasional patrol, the resistance would time an attack to avoid the patrol. Observers also noted and reported the characteristics of bridge construction so that the size of the explosives could be calculated.

5-12.   By determining the schedule of enemy train movements, saboteurs were able to destroy stretches of railroad track while it was in use, thereby compounding the wreckage and complicating repair work. Danish railroad saboteurs had an elaborate system to provide this information. Throughout Jutland, underground members were stationed near major terminals to note the departures of enemy troop trains. Whenever one was seen, the observer telephoned prearranged code phrases to the sabotage cell in the town next on the railroad line. Members of this cell then proceeded to predetermined spots on the tracks to lay their mines. With this advance notice, the mines could be placed at the last moment, preventing detection by patrolling guards. The train delayed by sabotage might eventually reach the next stop, but observers there would be waiting to repeat the process. Using these observers along a train's route, the resistance was sometimes able to slow a train's progress by days, or even weeks.

5-13.   Production facilities were also surveyed by undergrounds in preparation for sabotage attacks. When possible, underground personnel were often aided in planning factory sabotage by outside intelligence experts, for these were best qualified to make the necessary technical judgments. It is a problem in itself to determine just which components in a plant should be incapacitated. Prior to the blowing up of a Norwegian heavy-water plant being operated by the Germans during World War II, the preliminary reconnaissance was done by a special operations executive agent who parachuted into Norway. Details about the factory's equipment were obtained from a Norwegian scientist in London. Other data, perhaps about the guard system and access to the equipment, apparently were supplied by underground workers in the plant.

# INTELLIGENCE FOCUSED ON SCIENTIFIC AND MILITARY SECRETS

5-14.   Secret scientific and military data can be obtained by recruitment of employees in scientific and military installations, or by simple observation elsewhere. An example of the latter was the valuable data about the N-2 rockets obtained by the Danish resistance. During the summer of 1943, fishermen near the island of Bornholm began to report the rash of unidentified objects in the sea; they were recorded by the resistance leader on the island. In August, the island's police commissioner notified the underground leader of the crash of a flying craft in a nearby field. The two men rushed to the scene before the Germans and found the wreckage of what was clearly a new kind of aircraft. The only identification mark was a number: V1-83. The men took photographs immediately before the arrival of the German investigators. From the skid marks, the underground leader was able to determine that the device had come from the southwest. From the pictures, the underground chief drew a complete sketch of the weapon. This sketch, the photographs, and the notations as to the direction from which the missile had come were sent by courier to England, providing the British with perhaps their first technical data on the new German rocket.

## POLITICAL INTELLIGENCE

5-15. Underground agents also collect political information for analysis by the intelligence analyst cell. They note the statements and activities of persons to determine who favors the regime, so that these persons may be closely watched or eliminated if their actions seriously threaten the underground. In Belgium during World War II, the Movement National Beige kept files on collaborators and campaigned by threatening phone calls and letters to dissuade these individuals from working with the enemy. If this failed, the collaborators were often executed. The list of collaborators was never made public in order to keep concealed the extent of cooperation with the enemy.

5-16. In wartime, the underground also notes the morale of the enemy soldiers. The Polish Home Army systematically collected data on German troops by reading their mail. There were too few Germans to handle all of the postal work, so many Poles were employed. These workers would open letters and photograph the contents before sending them on. From these letters, a fairly good estimate could be made of the enemy's morale.

---

### The Value of Insurgent/Resistor Intelligence
### in Unconventional Warfare

"There is no doubt that most partisan action inflicted damage upon the opposing forces. Some of the damage was severe.... [However,] their second great contribution was in the field of intelligence.... [It] cannot be doubted that the partisans served well as field intelligence, especially after Army intelligence officers had been seconded to all partisan staffs in 1943. The scope was wide—the partisans were everywhere—their location was ideal—behind the enemy's front—and their instructions were detailed—in the Field Service Regulations, the Partisan Handbook, the Guide Book for Partisans, and so on.

We can be almost certain that again and again Russian attacks were mounted in those areas which partisan reports had indicated as vulnerable. The Russians during the war became expert in attacking the enemy's weakest points: the small front line gaps in the winter of 1941-2, the front held by German satellite troops at the beginning of the Stalingrad battle; and if there was neither gap nor satellite, it was almost always the seam between two enemy formations which the Red Army selected for its breakthrough attempts.... There was only one source which could consistently direct the Red Army against the weakest link of the enemy front, and this task...was entrusted to the partisans.

We are of course better informed about the value of French partisan intelligence. 'In fact, the day the battle (in France) began,' says General De Gaulle, 'all the German troop emplacements, bases, depots, landing fields and command posts were precisely known, the striking force and equipment counted, the defense works photographed, the minefields spotted.... Thanks to all the information furnished by the French resistance, the Allies were in a position to see into the enemy's hand and strike with telling effect.'

These words speak for themselves; no finer testimonial could be given."

Otto Heilbrunn, *Partisan Warfare* (1962)

---

## ADDITIONAL CONSIDERATIONS

5-17. The underground organization and many of its activities are based upon a fail-safe principle so that it is organized and if one element fails, the consequences on the total organization will be minimal. This is especially true for intelligence cells. Almost all clandestine organizations that are susceptible to compromise by security forces have parallel organizational units and networks of units. In every case, the underground attempts to have a backup unit that can perform the same duties as the primary unit if the latter is compromised. It usually takes a long time to establish a unit or net and the underground must plan for contingencies, such as the compromise of the primary unit or increased government security measures. Thus, the organizational expansion of undergrounds is usually in a lateral direction by duplicating units and functions. The decentralization extends to all functions.

5-18. If the cell operates as an intelligence unit, its members may never come in contact with each other. The agent usually gathers information and transmits it to the cell leader through a courier or mail drop. The intelligence cell leader may have several agents, but the agents never contact each other and only contact the cell leader through intermediaries. The underground usually does not jeopardize intelligence units by demanding that they perform sabotage as well; sabotage operations may draw attention to individuals and compromise their usefulness as intelligence agents.

5-19. Some resistance intelligence tasks call for the recruitment of persons with access to important information. Gathering intelligence is a major activity of an underground, and to do it effectively, an underground contacts and develops sources in all areas of interest. The range of persons so utilized may be wide, varying from a peasant woman living near an enemy convoy route to a person in a sensitive governmental bureau. Immediate access to valuable information is not the only criterion for selection. Sometimes the underground may recruit individuals to provide information that is valueless in order to keep them on a string, in the hope that they may advance in their profession and be in a position to supply useful information someday. If time permits, it may be better to wait for someone to work his way up into a position of trust rather than to seek to recruit someone in a high position with the attendant risk of failure or compromise.

5-20. The unusual factors of large-scale, illegal immigration and the special political status of the mandate in Palestine created a unique situation conducive to underground operations. Drawing on former members of World War II undergrounds, the Haganah developed an effective intelligence network. It also transcended national boundaries and used the sympathy of other nations for the refugees to aid the movement.

## SECTION 2. PREPARATION OF THE ENVIRONMENT

5-21. *Preparation of the environment* is an umbrella term for operations and activities conducted by selectively trained special operations forces to develop an environment for potential future special operations (JP 3-05). During UW, PE occurs in a dual capacity—

- That conducted by U.S. forces to operate on the periphery of and within a designated UW operational area.
- The preparation of the denied area by resistance elements for the conduct of resistance operations and activities.

*Note:* ATP 3-18.20 and ATP 3-18.16, *(U) Preparation of the Environment (S//NF)*, provide detailed information on PE.

5-22. U.S. forces conduct PE during UW to develop knowledge of the operational environment, to establish networks and physical infrastructure, and for general target development in support of future and ongoing SOF operations. Resistance forces focus on developing infrastructure that supports survivability and recoverability of the overall resistance effort. Much of what a resistance does to survive is based on careful observations of the adversary and coupled with organization and tactics that outmaneuver and outwit hostile security forces and their mechanisms. The development of courier networks, for example, establishes a means by which decentralized elements communicate to coordinate and execute resistance activities. Both human networks and accompanying physical infrastructure are required to operate a courier network. Specially selected, trained, and compartmented individuals provide the service, and clandestine sites that facilitate the exchange of communication material are carefully selected and safeguarded. Although this example represents just one facet of resistance PE, it does characterize its essence.

5-23. As resistance PE develops, other functions are added to establish and expand capabilities that support intelligence collection, escape and evasion of underground workers, subversion and sabotage activities, and guerrilla warfare. Resistance PE establishes the basis of the area complex, and it is critical for the security and expansion of the organization. When U.S. forces partner with resistance elements, a compilation of PE capabilities occurs. For example, safe areas identified as void of enemy air defense capabilities become potential territories for allied reinforcement delivery through DZs and LZs.

5-24. Resistance PE physical infrastructure development involves the consideration of topographic and hydrographic terrain in both rural and urban settings, environmental conditions, man-made structures, facilities, transportation capabilities, and lines of communication. The distances associated with these elements that influence the deployment and employment of forces are also considered.

5-25. Caching is one aspect of infrastructure development that supports the entire spectrum of resistance activity. Cached supplies meet emergency needs where personnel are barred from their normal supply sources by sudden developments. Caching solves supply problems of long-term operations conducted far from a secure base. Caching also provides for anticipated needs of wartime operations in areas likely to be overrun by the enemy.

5-26. To facilitate resistance intelligence operations, infrastructure is developed that emphasizes security and communications. This particular infrastructure supports clandestine meeting sites, recruitment, training, material development, and the controlled distribution of items, such as documentation, uniforms, or special equipment. Resistance infrastructure associated with intelligence operations may include:

- Safe houses for clandestine meetings.
- "Joe Houses" for locations to allow intelligence collectors a cool down period.
- Safe areas and hide sites to facilitate escape when under duress. A *safe area* is **a designated area in hostile or denied territory that offers a reasonable chance for a resistance organization to conduct clandestine activities without compromise (ATP 3-18.1, proposed for inclusion in ADP 1-02).**
- Cache sites for storing supplies and equipment.
- Key terrain for courier operations or impersonal contacts.

5-27. Resistance evasion and escape infrastructure is developed to counter enemy compromise of individuals, cells, or networks. Enemy surveillance is instituted through local and state agencies, informants, penetration agents, moles, and collaborators, and it can trigger search and seizures, arrests, and interrogation activities. Evasion and escape infrastructure provides the means for the displacement of resistance personnel to safer areas where they can resume their resistance work. These networks can also expand to reach safety in areas under resistance control or even sympathetic neutral countries.

5-28. Acts of subversion pitch the resistance organization against the enemy with the intent of undermining their institutions and infrastructure on every level. At the grass roots level, infrastructure useful for subversive resistance activity might include community centers, youth shelters, rehabilitation centers, donation centers, bookstores, cafés, taverns, and social clubs. At these locations, resisters engage with the local population by word-of-mouth or through social media to influence subversive activities, such as boycotts, demonstrations, walk-outs, or strikes.

5-29. To affect activities above the local level, resistance subversive agents also infiltrate key enemy institutions, posing as compliant workers or low level supervisors. When established in the work environment, they compete against their peers and eventually rise through the ranks to first echelon managers or staff. Once in a position of influence, these agents gradually exercise degrees of control over the organization. In doing so, civilian sector industrial complexes, law enforcement, military units, or branches of government can be subverted to diminish the enemy's ability to mass-produce, maintain order, or govern effectively.

5-30. Resistance sabotage activities require access to and use of infrastructure that relies on absolute secrecy for survival. To prepare for sabotage, the resistance needs clandestine infrastructure in which it can plan, train, construct devices, and rehearse for operations. During execution, additional infrastructure is required to support the clandestine movement into and out of the operational area. In addition, locations to pre-position and make final preparations within striking distance of the target, as well as sites that support post-strike regress and concealment, are required. Infrastructure typically associated with sabotage operations includes safe sites, safe thoroughfares, caches, mission support sites (MSSs), and hide sites.

5-31. Infrastructure necessary for successful guerrilla force operations is organized to withstand hostile reaction to armed resistance and sustain combat operations. Much of the early phase of guerrilla activity involves the conduct of political work, small-scale attacks, and sabotage. As the size and capability of the guerrilla force increases, attacks are launched from guerrilla base areas. The location of guerrilla units within a base area is not fixed, and only in extreme circumstances do they defend it. A system of alternate unit locations is established to enable guerrillas to avoid and survive counterguerrilla campaigns. Rugged mountains, swamps, and forest areas provide relatively secure guerrilla base areas.

5-32. The timely warning of enemy approach and rapid movement to predetermined alternate locations are the essential elements of an effective security system. Local guerrilla forces and the civilian support infrastructure are organized and developed in an outer security zone. This area serves multiple purposes in addition to security, such as providing a primary source of recruits and medical support for the guerrilla forces. From this area, medical personnel establish casualty collection points and systems to evacuate the wounded to guerrilla bases or civilian care facilities during combat operations.

5-33. During the conduct of guerrilla force operations, the area commander locates his HQ where he can directly influence organization and operations in the most important sectors of his area. His HQ is in a secure area where access is limited or uncontrolled by the enemy. The surrounding terrain should not favor large-scale, enemy-mounted, or dismounted operations. Logistics concerns and health conditions in the area are further considerations. The area commander also selects two or more alternate sites and prepares them for emergency use. (See Appendix M, Logistics, and Appendix S, Area Complex Development Model, for further details and examples of resistance area complex development.)

# SECTION 3. UNCONVENTIONAL ASSISTED RECOVERY

5-34. As a UW activity, SOF conduct unconventional assisted recovery unilaterally with indigenous or surrogate personnel or with other governmental agencies, employing compartmented tactics, techniques, and procedures. The military aspects of unconventional assisted recovery require sensitive methods for which SOF are specifically organized, trained, and equipped. In the conduct of unconventional assisted recovery, SOF may be deployed into a JSOA before strike operations to support recovery operations. The intent of unconventional assisted recovery is to merge U.S. and resistance recovery mechanisms to bring isolated personnel into contact with—and ultimately into the custody of—a recovery force as soon as possible and then move them to an area where exfiltration to definitive USG control can occur.

5-35. Unconventional assisted recovery is highly sensitive, and such operations may be covert or clandestine. All of the tactics, techniques, and procedures pertinent to U.S. Army SF interaction with indigenous forces in UW—particularly those methods which empower insurgent or resistant undergrounds—are useful to unassisted recovery mechanisms. In concert with indigenous partners, procedures and supporting assets are developed to achieve specified personnel recovery tasks of: locate, contact, authenticate, support, move, or exfiltrate U.S. military and other designated personnel to friendly control. The artful design of these procedures and assets results in the creation of unconventional assisted recovery mechanisms.

*Note:* ATP 3-18.20; ATP 3-18.72, *Special Forces Personnel Recovery*; ATP 3-05.71, *(U) Army Special Operations Forces Resistance and Escape (C)*; and FM 3-05.70, *Survival*, provide additional information on unconventional assisted recovery.

5-36. For a resistance organization, the primary purpose of personnel recovery is to develop evasion and escape networks to preserve its members, cells, and networks from enemy compromise. Resistance evasion and escape networks assist evaders in reaching safety in areas under resistance control or in friendly or sympathetic neutral countries. As an ancillary function, these networks also serve to move information, material, and equipment into and out of enemy territory.

5-37. Evasion and escape networks are established for resisters in imminent danger of being exposed or being hunted by the enemy, families of captured resistance members, refugees, and stranded personnel, such as escaped resistance members.

5-38. The resistance organizes escape and evasion networks into cellular, chain-like operations consisting of clandestine routes and safe houses. Multiple operational cells are used for controlling, contacting, hiding, passing, and evacuating the resistance members. The control cell—

- Consists of three to five members.
- Plans operations.
- Coordinates the procurement of supplies, documents, clothing, and money needed by those traversing the net.
- Controls the communications system.

- Establishes and monitors security measures and procedures.
- Controls and coordinates the activities of the other cells in the network.

5-39. Under the direction of the control cell, a system of safe houses is established. Safe houses allow a pause during movement through the network and the ability of personnel to hide for limited or extended periods. The safe house is the focal point of the resistance evasion and escape network, and each area develops its own chain. Through coordination with other areas, the network is extended and routes are established. Routes may encompass the immediate country, as well as extend into adjacent countries, leading toward a final destination or exfiltration site.

5-40. Resistance evasion and escape networks are divided into two separate categories of unassisted and assisted. Unassisted evasion is the most dangerous and relies on the skills of the evader to provide his own security, movement, and refuge. Resistance assisted evasion and escape provides the best chance of success and is typically divided into three distinct segments: contact, hiding, and passing.

5-41. An evader or escapee's best chance for success will result from making contact with local resistance elements when being actively hunted by the enemy. Once contact is established, he is introduced to the evasion and escape mechanism. In addition to evaders or escapees, couriers, liaisons, and officials can make use of the mechanism when required and when permitted. These trusted individuals are provided with addresses and passwords necessary to gain entry. Others, such as refugees or underground members of destroyed networks, fleeing enemy authorities usually do not have such information. For them, making contact is an extremely dangerous undertaking. To aid such individuals, resistance lookout cells are established to spot those fleeing while they are on the move and attempt to bring them into the network.

5-42. Between each segment in the evasion and escape line, personnel are hidden in safe houses while the next stage of their journey is coordinated. Usually, these safe houses are within one day's travel of each other. Accommodations may vary, but they are always controlled by members of the resistance. Homes of underground members are sometimes used in emergency situations, but this compounds the risk already confronting the worker and greatly increases the risk of compromise to the organization.

5-43. Some situations require hiding personnel for short periods with only enough time for them to rest, eat, and prepare for the next leg. Directions are also given only to the next stop. This reduces the number of safe houses that could be compromised in the event of capture and interrogation. Other safe houses may be needed for long durations, especially when persons sought by the enemy may require more or less permanent lodging. This situation is typically experienced when individuals are unable to escape either from the country or take refuge in hidden rural camps. Many may be forced to stay at home, hiding under the cloak of false documents.

5-44. Sick or wounded personnel also require special medical treatment and may need to hide for extended periods. In such cases, makeshift care facilities or hospitals where members of the staff are sympathetic to the resistance can be of great value. Camps may also be established in forests or mountainous areas to accommodate personnel. Such camps, however, are usually totally dependent on the underground, guerrilla units, or local inhabitants for food and clothing.

5-45. Passing is the moving of personnel from safe house to safe house through the evasion and escape line until they reach their destination or the planned exfiltration site. Passing is also another dangerous part of the journey as personnel are constantly exposed to enemy control measures. Enemy checkpoints are often randomly established to search and question personnel designed to expose evaders as well as inhibit clandestine resistance activity. Border checkpoints pose special problems as border police are usually well trained and are generally the most difficult to deceive. Identification cards, passports, and other required documents are usually carefully checked. Those whose documents appear suspicious are usually detained and questioned.

5-46. Various methods are used to pass personnel through the line—such as cut-out rendezvous and concealed movement. Through the use of cut-out rendezvous, personnel are taken from one section in the line by a guide who then leaves them at a prearranged site. At a designated time later, another guide from the next section arrives and collects them. Guides never meet and there is no direct contact between segments. Concealed movement involves selecting clandestine routes or concealing personnel during movement. All measures of concealment may be employed, such as the use of an ambulance or hearse to hide personnel under the guise of injury or death.

5-47. Interested readers may also gain insight into the potential range of activities possible in unconventional assisted recovery mechanism design by looking at legacy vignettes. Traditionally, personnel recovery has often been referred to as escape and evasion or escape and recovery. During World War II and the Cold War, for example, military, civilian, and indigenous special operatives set up escape and evasion networks (or nets) or evasion and recovery nets. There are many historical examples available; below is one example.

---

### Example Vignettes/Concepts of Escape and Evasion Networks
### From Previous Eras

Throughout the 20th century, "escape and evasion nets" essentially consisted of escape routes, hideouts or "safe houses;" and some secretive organization of shadowy figures to assist the evader through the network in enemy country until he could be delivered to areas under friendly control

There were three general categories of safe houses: the temporary stopover, the emergency hideout, and the permanent refuge. Couriers and traveling agents used the temporary stopover to facilitate travel. Escapees and persons in danger also used temporary safe houses along escape routes for food, rest, and directions to the next stopover. An operative who became suddenly ill, wounded, or sought by the police could use an emergency hideout. Such safe houses were typically private homes of loyal and reliable persons who were supporters of, but not identified with, the underground movement. Other facilities have also been used as safe houses. Algerian physicians loyal to the Front de Liberation Nationale (FLN) hid evaders as patients in the Algiers Municipal Hospital. The permanent or long-term safe house was often an isolated farm or cabin, a distant encampment, or a location in a nearby nation sympathetic to the underground movement.

In the 1930s, the Soviet Comintern utilized extensive auxiliary offices and bases for their agents abroad. The Seamen's and Port Worker's International, for instance, controlled seamen's clubs in every major port in the Western Hemisphere. These clubs served as reporting and relocation bases for agents operating in or traveling through the country. Personnel at these auxiliary bases arranged contacts, passports, cover addresses, and funds for agents. When an agent lost contact with his organization, he simply reported to the nearest auxiliary base for food, shelter, funds, and instructions.

In planning riots and demonstrations, the Viet Cong pre-established safe zones in sections of the city where they would store weapons and assemble agitators. They identified shopkeepers and homeowners willing to provide shelter for the demonstrators. The agitators hid in these safe houses until the police had completed their postdemonstration search. Afterwards, the agitators would withdraw from the city by predetermined routes.

The safe houses along an escape-and-evasion network would usually be placed only within one day's travel or each other. The person maintaining the safe house seldom engaged in any other subversive activity that might have drawn attention to him. The underground would have supplied him with extra food, clothes, and any identification papers or documents needed to conduct his portion of the clandestine effort. Each person in the route knew how to reach only the next link, and no one person knew the identity or location of every link. Guides generally escorted the escapee from one link to the next. The guides would meet at a prearranged spot halfway between the two safe houses and neither guide would know the location of the other's safe house.

The Viet Cong infiltration process from North Vietnam to South Vietnam provides an illustration of safe house and fail-safe concepts. After completing training in the North, Communist infiltrators were trucked to the Laotian border just above the demarcation line where they rested for several days before beginning their move southward. An infiltration group usually numbered 40 to 50 men, but once they reached the border they broke up into smaller groups.

(continued on next page)

---

---

**Example Vignettes/Concepts of Escape and Evasion Networks
From Previous Eras (continued)**

Each man carried a three- to five-day supply of food, a first-aid packet, hammock, mosquito netting, and similar items. No one was permitted to carry personal papers, letters, or photographs that might be used by the enemy to identify him. The infiltration routes along the Laos-South Vietnam border included way stations. A chain of local guides led the units along a network of secret trails. Each guide knew only his own way station and conveyed troops to the next way station just as the network conveys escapees between safe houses. Conversation was discouraged in transit and only the leader of the group was permitted to speak with the guide. In this manner, the network maintained a degree of security and contained damage if one guide defected or was captured.

During World War II, underground escape-and-evasion nets devised some unusual techniques to pass escapees beyond checkpoints. Police members of the net would handcuff escapees and pass them through the checkpoint as prisoners. Underground members also hid escapees in maternity homes until their passage through the escape route could be secured. In one incident, an escapee was passed through a checkpoint by placing him in an ambulance and having him feign insanity.

---

# SECTION 4. SUBVERSION

5-48. A hostile government or occupying power relies on some critical, minimum degree of military infrastructure and economic, psychological, and political strength and morale. Insurgencies and resistance movements attempt to undermine the human component of such strength and morale by employing techniques of subversion. Of all UW core activities, the subversive efforts of U.S.-supported insurgencies and resistance movements is the single, most strategically valuable activity because it provides discreet methods of influencing an opponent's behavior without resorting to more overt, large-scale, and unilateral U.S. actions. U.S. support to subversion conducted by indigenous actors is the very heart of UW. Subversion, subversive political action, and psychological action are defined as follows:

- **Subversion.** *Subversion* is actions designed to undermine the military, economic, psychological, or political strength or morale of a governing authority (JP 3-24).
- **Subversive Political Action.** A *subversive political action* is a planned series of activities designed to accomplish political objectives by influencing, dominating, or displacing individuals or groups who are so placed as to affect the decisions and actions of another government (ATP 3-18.1, proposed for inclusion in ADP 1-02).
- **Psychological Action.** *Psychological action* is the lethal and nonlethal actions planned, coordinated, and conducted to produce a psychological effect in a foreign individual, group, or population (FM 3-53).

*Note:* The detailed knowledge required to understand the vulnerabilities of a nation's political, economic, military, and social organizations, and the sophisticated analysis of target audience susceptibilities is fundamental to the successful conduct of subversion and UW. Early inclusion of, and continuous coordination with, PSYOP and CA Soldiers in UW campaign planning is critical.

5-49. Typically, insurgencies are won by a combination of military and political means. Much of the political leverage involved in settlements is derived from the psychological and practical effects of mobilizing the populace against the government. Insurgent subversion strategy is to separate the existing government from its basis of power by capturing the institutional supports upon which it rests, by eroding mass support for the government, and by overtaxing internal security forces with problems of unrest. Insurgencies may not be able to automatically seize power through fomenting disorder, but the undermining of regime power and the creation of a security and governance vacuum is a prerequisite to the insurgency establishing its own control. When obedience to law breaks down among a populace, a tense, highly emotional state develops, which gives the underground a chance to channel dissatisfactions. Insurgent subversion methods and tactics involve the psychological objectives of creating social disorganization and the political objectives of creating alternative organizations to contend for power. Psychological objectives may include:

- Creating social disorganization and conditions of uncertainty. The resultant unrest and confusion are used to conceal underground operations. A characteristic of this kind of social confusion is a condition of general apathy among a large segment of the populace and an unwillingness to help either side. This indifference benefits the resistance underground; apathetic people do not cooperate by supporting government programs, and they seldom volunteer the information and intelligence necessary for detecting underground elements and operations.
- Creating doubt and suspicion of government and government officials. This focuses attention and grievances on the ineffectiveness of government.
- Crystallizing attitudes and organizing dissident elements to resist government action and policies.

5-50. Subversive influence activities not only affect the operational environment in which populations reside, but also the informational environment. The real world or the facts are relatively unimportant in subversion operations; the shaping of attitudes, values, and beliefs are vital, however. Underground organizations align their appeals with a society's recognized, accepted values, such as independence and land for the landless. Members of the population who already accept widely held values are more easily persuaded to accept an insurgency which espouses them. Insurgents also reward those who are loyal and punish any who oppose them. Riots and passive resistance provide strong social pressure to influence the undecided or uncommitted.

5-51. Subversive influence activities are conducted in a variety of forms, such as—

- Psychological actions.
- Mass media.
- Interpersonal communications.
- Theatrical performances.
- Programs for local civic improvement.
- Audio, visual, and audiovisual messages.

5-52. Although the themes, targets, objectives, and other informational guidance used throughout the operation is determined at the highest echelon of the organization, successful execution of influence activities depends in large part upon the training, experience, and ingenuity of the operators at the local level.

5-53. In attempting to shape popular attitudes, values, beliefs, and behavior, and to develop support for subversion activities, influential messages and actions are directed at specific target audiences or groups. Occupational, religious, ethnic, and other social groups are often identified, and messages and actions are tailored to be effective for a particular group. The purpose of underground information activities may be to obtain support from neutral and uncommitted key groups and individuals, to raise morale and reinforce existing attitudes and beliefs among underground members and their supporters, to undermine the legitimacy and credibility of the existing government, and to lower the morale of government forces and personnel.

5-54. Underground movements tailor emotional and other types of appeals to various segments of society. For example, groups that are reluctant to take up arms against the government may be susceptible to emotional appeals and directed into passive measures. Religious or pacifist groups, women, children, and the elderly may also possibly be mobilized for passive resistance. In organizing demonstrations and riots, attention is given to selecting groups most likely to respond to the agitator's call to action—student groups, dissatisfied labor union members, and groups with known grievances. Insurgencies and resistance movements can use subversive techniques to manipulate crowds and civil disturbances for the purpose of advancing the overall insurgent strategy.

5-55. Resistance organizations deliver influential messages to audiences while agitation typically consists of interpersonal and task-specific communication directed toward relatively small, selected audiences. Messages and action (agitation) are interdependent and complementary. Typically, mass communications build support for the cause of the movement by articulating emotions associated with conditions (grievances) through messages, and agitation operationalizes those messages by associating the movement with actions being taken to solve or address those grievances on behalf of the people affected.

# MESSAGES

5-56. Persuasive messages attempt to create feelings of doubt and uncertainty about future events and to promote perceptions of crisis. Care is taken to differentiate between the government and the people and attribute blame or fault to the government. Message audiences are generally divided between broad, general ones based on demographics, region, and so on, and those crafted for more local reception.

5-57. Messages typically used by resistance organizations consist of three types: informative, directive, and persuasive. When the message captures people's attention, they will be receptive to the persuasive content of the insurgent's message.

5-58. In order to ensure continuity and consistency, selected messages are provided at regular intervals to all resistance information elements. In addition, insurgencies and resistance movements may have their own doctrine and procedures that must be trained and reinforced. The larger, more mass-based, and more multinational an insurgency is, the more this will be true. Moreover, such instructional content may be used for execute orders to dispersed resistance cells. Previous insurgencies have used the double language routine in which disguised instructions are embedded in routine media broadcasts.

5-59. Persuasive messages are designed to bolster the morale of the insurgents, to undermine confidence in the government and its policies among audiences, and to win active supporters or at least sympathizers to the movement. The messages tend to be phrased in highly emotional terms.

## GENERAL THEMES

5-60. Several devices are used to justify the movement through a consensual validation—creating the appearance of majority approval. Resistance themes tend to stress the legitimacy as well as the reality of insurgent power vis-à-vis the opposing regime. Some examples include:
- Insurgents speak for the people.
- The opposing regime is ruthless, engaging in unwarranted and unprincipled aggression to exploit the populace.
- The opposing regime is arrogant and contemptuous of the indigenous population.
- The opposing regime is deceitful in all actions and statements; nothing done by the regime is as it seems.
- The power of the opposing regime is overstated; it is not to be feared.
- The victory of the insurgency is inevitable.
- External support enjoyed by the insurgency is framed as widespread and legitimate international support, whereas regime external support proves collusion with foreign interests, betrayal of local interest, and lack of regime popular support.
- In some cases, the prevailing theme is the effectiveness of individual initiatives taken against the government, whereas other themes stress the helplessness of individuals who ought to bind themselves to the collective strength of the insurgent movement.

## LOCAL APPEALS

5-61. Appeals can facilitate insurgent organization building or at least encourage acquiescence with insurgent activities. Some local target audiences may be urbane and politically motivated enough to be persuaded by theoretical arguments. These, however, are the exception. Most target audiences will be initially reluctant to participate in anti-regime activities and will be most persuaded by local issues of grievance and security. Local appeals are characterized by—
- Avoiding theoretical arguments versus demonstrated provision of improving local goods and services.
- Couching allegiance in terms of who will eventually win and who will lose rather than consideration of who in the long run is right or wrong.
- Exploiting family and community ties and local preferences and prejudices.
- Motivating and serving the populace (insurgent emphasis) versus eliminating insurgent fighters and repressing the populace (regime emphasis).

● Demonstrating regime impotence and absence versus insurgent effectiveness and sustained presence (action propaganda).

# MASS MEDIA

5-62. There are several techniques of message delivery. The following paragraphs discuss these techniques.

## TELEVISION

5-63. The television has been an important means of mass message delivery—especially in the developed world—for well over half a century. As globalization occurs, the reach of television expands concurrently. Wikipedia estimated that the number of television sets worldwide in 2012 was approximately 1.4 billion, or 1 television set for every 5 people on earth. Other than cell phones, television is the largest source of unified messaging technology available. Moreover, television content transmission is expanding due to the increasingly widespread use of satellite signals replacing earth-bound line-of-sight transmissions.

## SOCIAL MEDIA

5-64. The widespread use of cell phone and Internet technology allows agitation on an unprecedented scale. These social media are much more responsive than a leaflet, radio, or newspaper and other mass media, and when signals are unblocked by the government, they allow real-time organization across wide distances and among disparate groups. Statista 2018 states: "In 2019 the number of mobile phone users is forecast to reach 4.68 billion."

## NEWSPAPERS

5-65. Newspapers are important messaging tools. Newspapers that are centrally controlled help spread insurgent information and ideas uniformly and so are instrumental in tying its readers into a close mental community; they become a mental rallying point (RP)—they spread the party line. However, the importance of such clandestine newspapers extends beyond the obvious function of influence. Production and distribution of subversive newspapers is an organizational challenge that contributes to resistance organization building.

> "Arranging for and organizing the speedy and proper delivery of literature, leaflets, proclamations, etc., training a network of agents for this purpose, means performing the greater part of the work of preparing for future demonstrations or an uprising."
>
> Vladimir Ilyich Lenin, *What is to be Done?*
> Collected Works, Vol 5, 1961

5-66. Undergrounds not only produce their own newspapers but also use existing legitimate newspapers for their own ends. Press criticism of certain aspects of local authority can provide fertile ground for implantation of rumors about public officials targeted for subversion. Insurgent undergrounds will seek to infiltrate established media entities to influence their coverage choices and alter their editorial positions.

## RADIO

5-67. Radio broadcasts have the obvious advantage of simultaneously reaching a large number of persons over a considerable range of territory. The same coverage by newspaper takes longer and is much more dangerous for the publishers, distributors, and recipients. Radio broadcasts can be made abroad, providing challenges to enemy efforts to shut them down. Likewise, foreign-based insurgent radio stations can claim to be broadcasting from within the liberated territory—a claim which is both embarrassing to the government and difficult to disprove.

### INTERPERSONAL COMMUNICATIONS

5-68. Possibly the oldest method of passing information is also one of the most effective: word-of-mouth interpersonal communications. Resistance movements or insurgencies operating in rural, agrarian, or low-literacy societies can effectively influence and shape local attitudes through word-of-mouth transmission.

5-69. In some cases, travelers, such as merchants, itinerant storytellers, religious figures, or dedicated and professional resistance cadres can make circuits of remote areas. Many remote villagers are likely to be both illiterate and lacking outside sources of information. Such persons are vulnerable to influence by skilled subversives. The remoteness of outside information is easily discredited or reinterpreted toward insurgent goals by agents who physically visit the remote area as part of a larger coordinated plan of continuous persuasion.

5-70. Some mass-based insurgencies will develop village infrastructures with dedicated, local political agents to control access to information. Such persons may control the only portable radio, cell phone, or newspaper in the village. The agitator, to ensure a maximum degree of credibility, tries to bar other external sources of information from his target group or area. This can be done through the confiscation of radios, threats to rural newspaper distributors, and impairment of government access to the area. In effect, the political agent seeks to develop a captive audience in order to facilitate his job of influencing attitudes and behavior and generating popular support for the underground movement.

# AGITATION

5-71. Agitation is essential for creating mass support. It takes more than a presentation of information to a group of people for them to accept a cause and be persuaded to support it. Exposure to information does not imply absorption of it. There are psychological, as well as physical barriers that inhibit the flow of information and ideas. There may be general apathy in which a large portion of the population is unfamiliar and unconcerned with particular events. Another barrier is the phenomenon of selective exposure: a tendency to hear only information conforming to individual tastes, biases, and existing attitudes. There is also selective interpretation: information understood only in terms of current attitudes. Frequently, individuals who do alter their attitudes as a result of new information do so only within the context of their prior attitudes.

5-72. The task of the agitator lies in overcoming these barriers by developing and delivering messages that are credible and meaningful. Agitators must reach the indifferent, must blend messages with the existing attitudes of target groups, and must make the resultant attitude change that can be exploited to elicit mass action. The agitator must remove any complacency that exists among a group of people, intensify their unrest, and channel the unrest to suit the purposes of the underground.

# PASSIVE RESISTANCE

5-73. People frequently overestimate the effectiveness, centrality, and importance of guerrilla warfare to UW, and they tend to underestimate the critical role of subversive populace organization and the time required to do such organizing. Resistance movements or insurgencies whose strategy relies on widespread involvement of the population must be able to mobilize the population to engage in widespread passive resistance. In UW, it is primarily operatives of the underground who organize and direct passive resistance efforts to persuade the ordinary citizens to carry them out.

### OBJECTIVES

5-74. Passive resistance implies a large, unarmed group whose activities capitalize upon social norms, customs, and taboos in order to provoke action by security forces that will serve to alienate large segments of public opinion from the government or its agents. If the government does not respond to the passive resisters' actions, the resisters will immobilize the processes of public order and safety and seriously challenge the writ of government.

5-75. Passive resistance rests on the basic thesis that governments and social organizations, even when they possess instruments of physical force, depend upon the voluntary assistance and cooperation of great numbers of individuals. Therefore, the passive resistance method of opposing an established power structure is to persuade as many persons as possible to refuse to cooperate with it, including members of governmental organizations, such as security and the military.

5-76. The principal tactic used to induce noncooperation with the government can be described as persuasion through suffering. One of the persistent myths of passive resistance is that persuasion through suffering aims only to persuade the governing opponent by forcing it to experience a guilty change of heart and a sense of remorse. This conception of the role of suffering makes the fundamental error of presuming that only two actors are involved in the process of passive resistance—the government and the resisters. Actually, passive resistance operates within a framework involving three actors: the government, the resisters, and the critically important larger public audience (the population in general). More than anything else, the objective of passive resistance is to create situations that will engage public opinion and increase popular involvement· in resistance activities, and to wear down and frustrate the overall effectiveness of the established power structure.

5-77. When the passive resister suffers at the hand of the government, the insurgent seeks to shape the interpretation of that suffering as a demonstration of resister integrity, commitment, and courage, while simultaneously demonstrating the injustice, cruelty, and tyranny of the government. The essential function of such suffering is comparable to the emotional mob interaction that takes place between a martyr and a crowd

5-78. If the passive resister provokes a response from the security forces or government that can be made to seem unjust or unfair, the resister's charges of tyranny and persecution are confirmed. Should the government fail to act, it abdicates its control over the population, over the enforcement of law, and over the maintenance of order. This dilemma thrusts upon a government the initiative, and also the responsibility, for uninvited conflict with unarmed citizens. Either way, the position of the passive resister is legitimized, the effectiveness and morale of government security forces is degraded, and the insurgency's subversive political strength is increased. The ideal situation is when popular passive resistance is self-generating, self-multiplying, and ultimately victorious with relatively little violence as in India in the 1940s or in both Poland and the Philippines in the 1980s.

## TECHNIQUES

5-79. Actions of passive resistance may range from small isolated challenges to specific laws to complete disregard of governmental authority. In all cases, passive resistance is a two-edged sword—it seeks to lower the morale and effectiveness of the governing regime or occupying power, while at the same time raising the morale of the populace and creating a feeling of defiance and unity that could be channeled later into more significant resistance activity. The techniques of passive resistance can be classified into three general types: attention-getting devices, noncooperation, and civil disobedience.

### Attention-Getting Devices

5-80. Passive resistance in the early stages usually takes the form of actions calculated to gain attention, provide persuasive messages for the cause, or be a nuisance to government forces. Attention-getting devices include demonstrations, mass meetings, picketing, and the creation of symbols. Demonstrations and picketing help advertise the resistance campaign and educate the larger public to the issues at stake. Such activities provide information and agitation for both internal and external consumption.

5-81. Many insurgencies have used mass demonstrations and general strikes—notably during the Iranian and Bolshevik revolutions—to provoke regime repression and to bring attention to the cause. World War II resistance efforts provide several examples of both nuisance campaigns (spitting on occupiers, cold-shoulder ostracism, deliberate rudeness, deliberate joke-and-ridicule campaigns, whispering rumor campaigns, and false reporting and over-reporting of bogus threats) and creation of symbols (the aforementioned martyrs, the cross of Lorraine, graffiti slogans, or the continued and very public presence of the Danish king as a symbol of continued Danish identity under Nazi occupation). Possibly, the most dramatic example of symbol creation was the very effective self-immolations in Vietnam in the 1960s and before the Arab Spring in 2010 and 2011.

## Noncooperation

5-82. Techniques of noncooperation call for a passive resister to perform normal activities in a slightly contrived way, but not so that police or the government can accuse him of breaking ordinary laws. Such activities as slowdowns, boycotts of all kinds, and various forms of disassociation from government are all examples of noncooperation. Other examples include falsification of blueprints so that structures are built improperly; deliberate errors in adjustment of machine tools and precision instruments that create flaws in end items; shipping mail or parcels to wrong addresses, shipping incorrect orders, or forgetting to include items in the shipment; feigned sickness which leaves gaps in the workforce; feigned fear which provides a rationale for both poor work performance and attendance and taxes the health system; and many other examples. Such techniques of noncooperation can also be considered sabotage.

5-83. Noncooperation is a principal tool of passive resistance and has been shown to be most effective in disrupting the normal processes of society and severely hampering and challenging the writ of a government—all in a way that is difficult for the government and its security forces to challenge. Many individuals altering their normal behavior only slightly can add up to a society behaving most abnormally.

## Civil Disobedience

5-84. Mass participation in deliberately unlawful acts—generally misdemeanors—constitutes civil disobedience. This is perhaps the most extreme weapon of passive resistance. The boundary between misdemeanors and serious crimes can be considered the dividing line between nonviolent and violent resistance. Some examples of civil disobedience include:
- Breaking specific laws, such as—
  - Tax laws (not paying taxes).
  - Traffic laws (disrupting traffic).
  - Restrictive public laws that prohibit meetings, publications, and free speech.
- Certain kinds of strikes and walkouts.
- Resignations en masse.
- Minor destruction of public or private property.

5-85. Civil disobedience is a powerful technique, but to be effective it must be exercised by large numbers. There is a calculated risk involved: the breach of law automatically justifies and involves punishment by the government and security forces. However, the more massive the scale on which civil disobedience is organized, the less profitable it is for the government to carry out sanctions.

5-86. Organizers of passive resistance are selective about the laws that are to be broken. The laws should be related in some manner to the issues being protested or the demands being made. In summary, the underlying consideration in most passive resistance techniques is whether they serve to legitimize the position of the passive resister, while alienating or challenging the government.

## ORGANIZATION

5-87. Obviously, the success of passive resistance rests largely on its ability to secure widespread compliance within the society. A government cannot be robbed of the popular support upon which it depends if only a few individuals act. A boycott, for example, requires participation by great numbers.

5-88. Organization is of critical importance to passive resistance. Although a few individuals can launch a passive resistance movement, in order to succeed they must be joined by thousands whose participation is strategically channeled. Modern technology has made the organization of large numbers of sufficiently motivated resisters in a short period of time much easier than in previous eras. Given proper cadre planning and coordination, the summoning and mustering of resistance flash mobs at strategic locations can have a potentially decisive psychological and political impact. Support to the indigenous underground in developing, growing, and providing both tactical and strategic employment of such popular resistance organizations is an early, continuous, and critical focus of U.S. Army SF conducting UW. Once again, the detailed social, demographic, and psychological expertise of U.S. Army CA and PSYOP units is invaluable in planning for such passive resistance and for identifying what forces and factors induce people to mobilize for passive resistance.

### Normative Factors

5-89. One method by which leaders of passive resistance movements secure widespread compliance is by clothing their movement and techniques in the beliefs, values, and norms of society—those things people accept without question. For example, when a minority secular regime like the Shah's or a minority secular occupier like the Soviets attempted to force social change on majority conservative Islamic societies in Iran and Afghanistan, such initiatives provided a ready-made focus for organizing popular resistance. They offended the traditional and dominant norms.

5-90. An opposite example would be constriction or repression of individual liberties in a society accustomed to enjoying them. The American Revolution and all of the multinational resistance to Nazi occupation in World War II were, in part, reactions to lost liberties. Other examples include the nearly universal disgust at the abuse of women, children, the elderly, beloved church leaders, and so on.

### Consensual Validation

5-91. The technique of consensual validation—in which the simultaneous occurrence of events creates a sense of their validity—is often used to make public opinion coalesce. For example, if demonstrations take place at the same time in diverse parts of a country, the cause which they uphold appears to be valid simply because a variety of persons are involved. A minority group can organize a multitude of front organizations, so that seemingly widely separated and diverse organizations simultaneously espouse the same cause and give the impression that a large body of opinion is represented. Passive resistance organizers effectively use the psychology of consensual validation to rally public opinion.

### Religious Factors

5-92. Rare or extraordinary factors, such as charisma play an important part in mobilizing public opinion. Obviously, exceptional resistance leaders like Gandhi and Martin Luther King provided probably indispensable charismatic power to mobilizing public sentiment. Polish-born Pope John Paul played a significant role in solidarity's success, as did Ayatollah Khomeini in Iran, and the self-immolating Buddhist monks in Vietnam. In all three cases, the difficult-to-quantify factor of religious authority contributed to mobilizing passive resistance.

*Note:* GTA 41-01-005, *Religious Factors Analysis*, includes more information on conducting analysis to better understand religious factors in planning resistance.

### Pressure for Conformity

5-93. The same techniques used by passive resisters against the government can be used to ensure widespread social compliance within the resistance movement. Ostracism is frequently used to apply pressure on individuals not participating in the passive resistance campaign. Instances of organized ostracism of collaborators were common from occupied territories in Europe and Asia during World War II through resistance to colonial powers throughout Africa in the late 20th century.

## COMMUNICATION AND PERSUASION

5-94. As noted earlier, the first phase of passive resistance is characterized by a period of attention-getting information activities and psychological actions: parades, demonstrations, posters, newspapers, and other forms of communication—either clandestine or open. Once the resistance movement is launched, there must be a continuing means of spreading the word. No movement can operate without some form of communication between the leaders and the led.

## TRAINING

5-95. Once organizational steps are taken to secure widespread social compliance, an effort must be made to instruct and train passive resisters. The idea is to erect a mental barbed-wire fence between resisters and authority. This instruction often takes the form of codes of dos and don'ts. Many undergrounds have found that it is easier to tell people what not to do—than what to do.

5-96. Training is particularly critical when positive, not just negative, actions are desired. Noncooperation and civil disobedience are positive acts that necessarily involve training, organization, and solidarity on the part of the resisters, whether they operate in the open or clandestinely.

# PARALLEL GOVERNMENT STRUCTURE

5-97. One method that is frequently used to both undermine public confidence in a government and to secure population support for passive resistance efforts is the establishment of parallel structures of government. If a population depends upon an underground-sponsored government, it will be compelled to comply with the underground's passive resistance program and withdraw its support from the regular government.

5-98. The techniques and societal values capitalized upon to undermine popular support of the government also serve the positive function of solidifying public opinion around a larger sense of community and national identification. Examples of constructing parallel governments include Hezbollah in Lebanon, the Naga in India, the Fuerzas Armada Revolucionarias de Colombia in Colombia, and the classic communist template from China to Cuba to Vietnam.

# SUBVERSIVE MANIPULATION OF CROWDS, RIOTS, AND DEMONSTRATIONS

5-99. The next step—more dramatic, sometimes violent, and notoriously difficult to control—is the wielding of popular support as a weapon. Crowds can be subversively manipulated to conduct peaceful mass political gatherings, aggressive demonstrations, or riots that effectively support the insurgency or resistance. Popular unrest in the form of strikes, riots, and demonstrations usually have limited goals, such as better working conditions, relief from food or other shortages, or demands for limited social change. Typically, these activities are relatively spontaneous, narrowly focused, and short-lived; a single peaceful mass political demonstration or riot does not make an insurgency.

5-100. The goal of the insurgent underground, however, is to coerce, disrupt, or overthrow the government and to shape government behavior or seize power. The insurgent does this by either exploiting such spontaneous grievances in the service of the resistance movement, or by deliberately creating such demonstrations. The subversive manipulation of crowds and civil disturbances involves a relatively small number of insurgent underground members who try to guide and direct legitimate protests. They attempt to direct the crowd toward emotional issues and arouse them against authority. The emotional perceptions and beliefs of the crowds that participate in civil disturbances often do not coincide with objective reality, and the individuals involved, very often, do not realize that their grievances are being manipulated by the insurgency or resistance movement in politically subversive ways.

5-101. Subversively manipulated civil disturbances may be considered as having four phases:
- The precrowd phase.
- The crowd phase.
- The civil disturbance phase.
- The postcivil disturbance phase.

## PRECROWD PHASE

5-102. In the preparation or precrowd phase, the underground elements are primarily concerned with building an organization. Training the network of agents necessary for the rapid and appropriate delivery of informative and persuasive messages accounts for the greater part of the work required to prepare for a demonstration or uprising. This step is a necessary precursor to a strike or demonstration organization.

5-103. Selected individuals are given special training in the manipulation of crowds. They are taught how to build barricades and conduct street fighting, how to mobilize blocks in the city and workers in plants, how to develop a local strike into a general strike and a general strike into a city uprising, and how to coordinate these into a national uprising. Sometimes, outside specialists can be brought in to direct the training activities. They are instructed in the potential for nonviolent demonstrations and specific techniques for sit-ins, no-shows, silent mass gatherings, marches, public mourning for dead compatriots, suspension of civic and social activities, economic boycotts, refusal to pay taxes or fees, traffic shutdowns, and many other forms of resistance that provoke government reaction.

5-104.    Some sort of planning on the part of the underground must take place. It may vary from rudimentary to highly sophisticated. Underground agents are instructed to infiltrate target groups by joining formal organizations, clubs, or any association that gives them access to such audiences.

5-105.    The next step is the selection of a population target. It is chosen primarily for its potential to achieve the desired psychological or political objective. Furthermore, any group that is not susceptible to manipulation, at least after some preparation, is not considered a viable target. Groups identified by their common interests (for example, ethnic minorities, laborers, farmers, educators, youth, the unemployed, and so on) offer great potential for covert manipulation because attention can be centered on issues affecting them rather than on philosophical or ideological arguments.

5-106.    The desired change in the attitudes and behavior of the members of the target group is usually accomplished through the delivery of carefully crafted messages and executing psychological actions—the desired effects of which are intended to increase anxiety and emotional stress. Word of mouth, radio, telephone, leaflets, and other messaging have been effectively used. Pistols, rifles, materials for making Molotov cocktails and explosives, and other weapons, such as clubs and lengths of pipe, and also handbills, signs, armbands, and banners, must be acquired and stored.

5-107.    Arrangements for members of the underground group to flee the area must be completed. These consist primarily of establishing routes of escape containing safe houses or other hiding places. Safe zones are established with householders and shopkeepers where demonstrators may seek cover when fleeing from the police.

5-108.    In places where demonstrations or strikes can be planned in advance, the underground mounts a campaign directed at preconditioning target groups. Chosen themes are constantly emphasized in messages and actions. By concentrating on local and specific grievances, a group is conditioned to phrases and slogans to which its members may later react under conditions of emotional stress.

## CROWD PHASE

5-109.    The indispensable element in civil disturbances is the crowd—not just any crowd, but a crowd made up of individuals who have been conditioned either by subversive manipulation or by other events. There are several ways to assemble a crowd. Cell members infiltrate mass organizations so that strikes or mass meetings can be changed into armed demonstrations. There are built-in sanctions within labor unions or other disciplined organizations that can be used to punish members who do not comply with the decisions of the organization. Therefore, if the infiltrated union or organization calls for a strike or demonstration, its members can be brought into a particular place at a particular time. Student groups are highly volatile on many social issues and can be induced to participate in demonstrations for the sheer excitement. Some factors of crowd formation and subversive manipulation include:

- **Informal Gatherings.** Demonstrations can be brought about at parades, street parties, dances, or during normal rush-hour periods.
- **Hired Demonstrators.** Organizers bring in paid outside agents and ruffians to expand crowds and accelerate radicalization.
- **The Precipitating Event.** The precipitating event which results in the formation of a crowd depends, for a great deal of its effectiveness, upon communication, especially upon exploitation of the event. Organizers associate the event with existing issues or exploit the natural distortion which accompanies word-of-mouth communication. The precipitating issue or event can be a martyred individual, a report of police brutality, or a symbolic act, such as the desecration of a flag.

- **Mob Management Techniques.** Mob management has several general characteristics, to include:
  - Leader elements are organized into an external command—well-removed from the activity and which can observe the demonstration (and outside reaction)—and an internal command located in the crowd, which is responsible for directing the demonstration.
  - Bodyguards who surround and shield the internal command from the police and facilitate their escape, if necessary.
  - Messengers who carry orders between the internal and external commands.
  - Shock guards who are armed with weapons and act only as reinforcements or are able to create diversionary violence, allowing resistance leaders to escape if they become engaged by the police.
  - Banner carriers who switch from banners expressing general grievances to those reflecting direct insurgent propaganda at the appropriate time.
  - Cheering sections consisting of special demonstrators who rehearse the slogans and chants and the order in which they are to be raised.
  - Media handlers who ensure that sympathetic media are well-placed to capture dramatic and inflammatory images useful to insurgent propaganda.
- **The Agitator.** After the crowd has been formed, the agitator assumes a significant role. His function is to enflame smoldering resentments of his listeners through emotional appeals and then provide a justification for their resistant actions. The agitator in the crowd plays upon the audience's suspicion of things they do not understand and manipulates whatever grievances motivate that specific crowd (material deprivation, repressive laws, inequality, nationalist or religious redress, martyred comrades, or other perceived injustices). The agitator then further enflames the crowd by painting a mental picture of dangers, disasters, insecurity, uncertainty, and hopelessness. Critically, the agitator must then identify the other—the bogeyman or oppressor responsible for the injustices and the conditions and who must now be opposed by that particular crowd. Where the conditions for insurgency or resistance are latent and incipient, the agitator rarely needs to invent grievances, merely exploit them. Even when issues are contrived, the skilled agitator rarely justifies his facts, but relies on vague appeals, generalities, and emotional themes.
- **The Disturbance Leader.** After the crowd has been emotionally aroused, some event must set it in motion. Often it begins its riotous activity by following a leader who merely shouts, "Follow me! Let's go!" The insurgent internal command assumes the leadership role if an emergent leader does not arise spontaneously. The event which sets the crowd in motion may, like the precipitating incident that brought them together, be either factual or fabricated. In the case of nonviolent protest or strike, the disturbance leader will act to calm the crowd in the face of opposition from the security apparatus or opposing groups. The disturbance leader will ensure that heavy-handed repression by the government is witnessed and recorded so that it can be exploited to build local, national, and international support.
- **Commands, Signals, and Communications.** Small groups have face-to-face communication and interaction. In large groups, however, communication comes through second- and third-hand sources. Insurgent media, preplanned distributions, cell phones or tactical radio, social media devices, and so on, all aid the magnification of insurgent command and control. A relatively new political phenomenon is the virtual crowd, which magnifies for the viewer the perceived size and activities of others engaging in behaviors the agitator wants to encourage. Such mobilization and actions may be real or fake. Either way, this technique presents the agitator with a significant opportunity to proselytize and preprogram wide audiences who are then subject to rapid mobilization and exploitation. Very large actual crowds—or many crowds dispersed over a large area—can also be susceptible to rumors. This was truer before the widespread use of social media. However, social and other electronic media can be blocked, rendering the crowd again vulnerable to word-of-mouth influence. Moreover, skilled operatives can use social and other electronic media to spread rumor campaigns. Rumors can be useful to mobilize collective action because they float on generalized beliefs rather than on demonstrable facts, and they tend to lend perceived substance to generalized beliefs. Where such generalized beliefs can be identified early, or even pre-planted into mob consciousness, agile insurgent use of timely rumors can spark desired crowd reactions.

## CIVIL DISTURBANCE PHASE

5-110.   Some factors of shaping and controlling crowds once they are rioting include:

- **Maintaining Emotional Excitement.** Once the destructive action of the crowd is under way, the agitator tries to maintain the level of emotional excitement. This can be accomplished in various ways. Cheerleaders can chant rhythmic and inspiring phrases or songs. Slogans can be displayed and banners unfurled. Booster incidents—the most universal being the looting of stores and shops—can be created or capitalized upon. Bank holdups and kidnappings are also carried out during the chaos. Other acts, such as the verbal abuse and stoning of police, not only permit the individual to release aggression and hostility against the symbols of authority but also increase the emotional involvement. The use of Molotov cocktails or similar dramatic destructive events also ratchet up crowd excitement.
- **Expanding the Violence.** Agitators attempt to capitalize upon the contagious effect of civil disturbances by spreading the violence and creating new incidents in nearby areas. Attacks upon symbols of authority, such as police stations and the offices of local officials, increase the intensity of the disorder. If possible, radio stations, newspapers, water, and power services are seized. Newspapers and radios spread the rumors, and control of water and power plants spreads social disorganization and fear.
- **Countering Police Activity.** Police and army counter-riot tactics are studied by the insurgent planners so that steps can be taken to circumvent them. Routes usually taken by internal security forces are blocked with barricades, overturned vehicles, and debris. Attacks upon police stations and their communications systems serve to disrupt police countermeasures. Cadres are usually guarded by strong-arm squads and avoid confrontation with the police so they will not be jailed. Appeals are made to army or police units not to attack their own countrymen, and cadres appeal to security force shame if or when police activities cause civilian casualties, especially among the very young, old, women, and children. Media handlers or internal insurgent media will be pre-positioned to record footage of any use of government force, and the most sophisticated movements will have cadres in place to shape, package, and distribute messages supportive of the insurgent's goals.
- **Employing Nonviolent Civil Disturbance.** Organizers should also consider nonviolent civil disturbances as an option for achieving intermediate and long-term objectives. Nonviolent actions can achieve many of the same goals as violent disturbances, and often with longer-lasting effects. These demonstrations can create a spectacle for media attention, raise public awareness of common grievances, disrupt the normal order of civic life, affect the economy, inspire recruitment through a show of strength and persistence, and enable the committed to exercise their revolutionary zeal, in a way that builds their confidence and strengthens resolve without exposing them to mass repression. Nonviolent action makes the movement more appealing to those who are sitting on the fence, presents a more sympathetic profile in the media, discourages security-force crackdown while creating an opportunity for the resistance to exploit atrocities, and builds support for the movement by exposing the excessive use of force by government security forces.

## POSTCIVIL DISTURBANCE PHASE

5-111.   The civil disturbance and its effects should be exploited using traditional, local, and international media; social media; and cellular and Internet communication mechanisms. The resistance movement should control the release of photographs, interviews, and videos surrounding the event to build internal and external support. After a civil disturbance has subsided, underground elements use a variety of means to exploit the situation. One way of maintaining the interest and emotional involvement of the population is a 24-hour general strike. Workers, especially those in key industries and utilities, are encouraged to protest against the government by staying away from work 24 hours. This is time enough to interrupt vital utilities and affect the entire population. Individuals and property owners are faced with the dilemma of going about their normal routine and facing violence or staying home for a day.

5-112.   To turn specific issues into grievances against the government, the underground makes appeals to all workers to join in a united front against the government. Cells in various factories, districts, zones, and businesses demand that their organizations support the strike in the form of a united front. Factory cells incite union members to stay off the job. They seek union sponsorship.

5-113.   In order to demonstrate the uncompromising position of the government, the demands against it are usually vague and impossible to meet. Original issues, such as higher wages or repeal of a sales tax, are now changed to antigovernment demands. A call is made for the release of political prisoners, and the police and army are asked to join the rioters. It is customary to insist on nothing less than the complete overthrow of the existing government. These demands can be articulated in protest meetings that keep the public aroused and involved. Committees are formed in every village or city to protest government action. Every attempt is made to get notable and respected citizens to lend their names to the protest.

## SECTION 5. SABOTAGE

5-114.The terms sabotage and subversion have distinct military definitions, but in common English usage they are frequently used interchangeably. Sabotage is defined as an act or acts with intent to injure, interfere with, or obstruct the national defense of a country by willfully injuring or destroying, or attempting to injure or destroy, any national defense or war materiel, premises, or utilities, to include human and natural resources. Sabotage is technically a component of subversion because it consists of actions that do contribute to the undermining of the military, economic, psychological, or political strength or morale of a governing authority. However, subversion generally connotes the actions directed at human beings and meant to undermine the sources of political power, whereas sabotage generally connotes actions directed at physical things and processes and meant to undermine the sources of material power. Nevertheless, there will continue to be instances such as noncooperation with authorities, which are equally understood as both subversion and sabotage.

# GENERAL AND STRATEGIC SABOTAGE CHARACTERISTICS

5-115.   Sabotage is an attempt to damage the resources of the government's war effort—military and economic organizations; industrial, food, and commodities production; and public morale and law and order. In most specific instances of general (or simple) sabotage, the material damage inflicted on the enemy by sabotage is relatively small. However, general sabotage has the effect of cumulatively degrading all forms of enemy strength. Carl von Clausewitz indicated that war includes friction which retards, slows, and frustrates war efforts. Sabotage is like creating man-made friction. General sabotage disrupts normal flows of production, wears everything out faster, and makes everything operate less optimally. The effect of generalized sabotage is to increase enemy expenditure of money, time, and manpower, while hurting his morale and cohesiveness.

5-116.   In some cases, however, a single event of strategic sabotage may have significant strategic or operational impact disproportionate to the means employed. A small team of Norwegian underground and special operatives, for example, may have prevented the Nazis from developing the atomic bomb with obvious war-winning ramifications.

5-117.   Sabotage can be low-tech or high-tech. In the Norwegian example, the strategic sabotage mission relied on good intelligence and preparation work by the underground and the daring of the saboteurs. Nevertheless, the explosive charges used and the means of physically emplacing them on the critical equipment were low-tech activities. Today, the ability to surreptitiously emplace malware through the Internet is not only a superlative use of indirect and disproportionate means, it can use cutting-edge technology to achieve strategic sabotage results. Stuxnet, an electronic worm released into the computer networks of the Iranian nuclear enrichment program, is an example of a precision cyber weapon. Stuxnet was designed to target a specific system configuration known to be associated with specialized industrial control processes. Most experts agree that Stuxnet was a product of a state-sponsored computer network operations program; however, the source has remained untraceable. The Stuxnet worm was confirmed by Iranian officials to have affected centrifuges used in the nuclear development process.

## Operations GROUSE, FRESHMAN, and GUNNERSIDE

During World War II, the Allies decided to remove the heavy water supply and destroy the Vemork heavy water plant in Telemark, Norway, in order to inhibit the Nazi development of nuclear weapons. Between 1940 and 1944, a sequence of sabotage actions by the Norwegian resistance movement, as well as Allied bombing, ensured the destruction of the plant and the loss of the heavy water produced.

In Operation GROUSE, the British Special Operations Executive (SOE) successfully placed four Norwegian nationals as an advance team in the region of the Hardanger Plateau above the plant. Later in 1942, the unsuccessful Operation FRESHMAN was mounted by British paratroopers. They were to rendezvous with the Norwegians of Operation GROUSE and proceed to Vemork. This attempt failed when the military gliders crashed short of their destination, as did one of the tugs, a Halifax bomber. The other Halifax bomber returned to base, but all the other participants were killed in the crashes or captured, interrogated, and executed by the Gestapo.

In 1943, a team of SOE-trained Norwegian commandos succeeded in destroying the production facility with a second attempt, Operation GUNNERSIDE. The Norwegian advance team collected intelligence on vital elements of the plant's operation, such as the number, hours of change, and behavioral pattern of the German guards; the layout of the plant; and routes of ingress and egress. The underground made contact with the plant's chief engineer, a Norwegian, who gave them information about the floor plans and the location of critical machinery. In order to plan and rehearse their attack, they built an exact model of the heavy water plant in England.

After nearly a year's planning, an 11-man team infiltrated the area for the actual attack. The team divided into two groups: blocking and demolition. The time for the strike was set for 12:30 a.m. to ensure that the off-duty guards would be asleep and to allow the team 5 hours of darkness to escape. In the actual attack, the blocking unit forced the entrance and covered the German guard barracks, while the demolition unit entered the plant itself through a cable tunnel and set the charge at a predetermined location. Although the explosion was so small it did not arouse the German garrison, it destroyed key apparatus in the plant and 3,000 pounds of heavy water, nearly half a year's production.

These actions were followed by Allied bombing raids. The Germans eventually elected to cease operations and remove the remaining heavy water to Germany. Norwegian resistance forces subsequently sank the ferry *SF Hydro* on Lake Tinnsjo, preventing the heavy water from being removed.

After the war, the SOE judged Operation GUNNERSIDE to be the most successful act of strategic sabotage in all of World War II. The Nazis did not win the race to develop the atomic bomb.

5-118.   Sabotage can have several purposes in a UW campaign. Planners should obviously seek to degrade the enemy regime or occupying power's resources generally and also identify strategic targets when possible. However, there are other advantages of conducting sabotage:

- Whereas UW is largely about organizing and channeling popular resistance and because anyone can be trained to conduct simple acts of sabotage, sabotage provides a relatively low-cost and low-risk method for the indigenous populace to participate in resistance.
- Such participation is an act of war, which means the UW operators are force-multiplying the combatants.
- Such participation is an easy way to build "esprit de resistance" amongst the population and correspondingly weaken enemy morale and inculcate fear in the security forces. Moreover, sabotage indicates popular resistance to the regime or occupying power in a way difficult to pin on any one individual or group.
- Simple sabotage most often uses natural, ordinary materials and is typically conducted by average citizens with routine access to targets. It is thus an indirect method of "waging warfare" through the populace. By contrast, strategic sabotage may require highly specialized saboteurs (such as the British commandos in Norway) or highly sophisticated techniques or technology (such as computer hacker denial-of-service attacks or the surreptitious insertion of the Stuxnet worm)
- The cumulative effects of general sabotage in a given area can have strategically significant results (for example, the interdiction of railways, roadways, and communications throughout Normandy

before and during the Allied landings, or the massive and widespread timed failure of imported computer chips integral to indigenous defense systems prior to a conventional strike). This economy-of-force quality of sabotage is a fundamental characteristic of UW's utility when it is conducted in support of major operations.

# TECHNIQUES AND SELECTED CONSIDERATIONS

5-119. Sabotage techniques are only limited by imagination, opportunity, and acceptable risk level. For obvious safety reasons, a detailed list of specific techniques will not be provided here. However, generalized techniques and selected considerations of employing sabotage follow.

5-120. General or nuisance sabotage is closely related to passive resistance in that it requires neither trained sabotage teams nor carefully selected targets. Sabotage acts in this category usually express individual resistance and take the form of noncooperation, such as deliberate slowdowns on factory production lines, or harassment, such as telephoned bomb threats that force the evacuation and search of buildings and plants.

5-121. Noncooperation sabotage was used extensively in occupied Europe during the German occupation during World War II. Workers slowed their pace of production, went on strike, and refused to help Germans apprehend rebel patriots; postal workers intercepted letters addressed to the Gestapo. The underground pressured doctors into signing medical certificates stating that certain key people were unable to work. During the Cold War, Soviet, Polish, and other Iron Curtain populations used the same techniques against the communist state. Today, strikes by Egyptian air traffic controllers, state workers in Thailand, or food service workers in Greece effectively use coordinated work stoppages to aggravate authorities, burden the economy, and tie down government security forces.

5-122. In World War II, the underground distributed camouflaged pamphlets of sabotage techniques to prompt patriots to sabotage the German occupation. Techniques included methods to slow down production in factories. Sharp metal objects were put in the streets to puncture automobile tires. This technique, which virtually halted traffic, was very effective because the only people who had automobiles were German officials and collaborators. Other techniques included failing to lubricate machines according to maintenance schedules, hiding repair parts, and dropping tools and other foreign objects into moving parts. Today's more complex technology and reliance on electronic and computerized components and digitized transmission provides new government vulnerabilities for the saboteur. Highly sophisticated machinery relies on tighter tolerances that correspondingly make them easier to miscalibrate or break. Ubiquitous electronic components are vulnerable to electromagnetic pulses or power stoppages. The digital commons has opened up unprecedented vulnerabilities to the skilled digital saboteur.

5-123. Where destruction is involved, the traditional and timeless weapons of the citizen-saboteur are salt, nails, candles, pebbles, thread, or any other materials he might normally be expected to possess as a householder or as a worker in his particular occupation; they appear to be innocent. His arsenal is the kitchen shelf, the trash pile, his own usual kit of tools and supplies; they excite no suspicion. The targets of his sabotage are usually objects to which he has normal and inconspicuous access in everyday life. In the 21st century, commonplace items that can be weaponized for sabotage now also include the cell phone, the remote control, the personal computer, and ordinary use of airplanes, trains, and automobiles.

5-124. A different type of simple sabotage requires no destructive tools whatsoever and produces physical damage, if any, by highly indirect means. It is based on universal human error opportunities to make faulty decisions, to adopt a noncooperative attitude, and to induce others to follow suit. Making a faulty decision may be simply a matter of placing tools in one spot instead of another. A noncooperative attitude may involve nothing more than creating an unpleasant situation among one's fellow workers, engaging in bickering, or displaying surliness and stupidity. The continuously expanding interconnection of human activities through social media have exponentially enabled and accelerated the potential for such antisocial disruption effects, and empowered trained cadres to rapidly mobilize such efforts.

5-125. Motivating the saboteur to begin and continue to conduct sabotage over sustained periods is a special problem. Simple sabotage is often an act which the citizen performs according to his own initiative and inclination; usually, he cannot be closely controlled. The will to conduct sabotage is usually counterintuitive. He must be persuaded that he is acting in self-defense against the enemy or retaliating against the enemy for other acts of destruction. On the other hand, not all potential saboteurs are necessarily motivated by patriotic

5-132. Nevertheless, while the above statement is true, it is also true that almost all UW endeavors requiring an overthrow outcome will not succeed without conducting effective guerrilla warfare (or at least coordinated acts of armed force)—the results of which are integrated into a well-conceived overall resistance campaign plan. UW is warfare and assumes violence. Some—sometimes much—of such violence is armed propaganda, political murder, and terrorism. Some is the small warfare known as guerrilla warfare. The following are key guerrilla warfare terms:

- **Guerrilla.** A *guerrilla* is an irregular, predominantly indigenous member of a guerrilla force organized similar to military concepts and structure in order to conduct military and paramilitary operations in enemy-held, hostile, or denied territory. Although a guerrilla and guerrilla forces can exist independent of an insurgency, guerrillas normally operate in covert and overt resistance operations of an insurgency (ATP 3-05.1).
- **Guerrilla Force.** A *guerrilla force* is a group of irregular, predominantly indigenous personnel organized along military lines to conduct military and paramilitary operations in enemy-held, hostile, or denied territory (JP 3-05).
- **Guerrilla Warfare.** *Guerrilla warfare* consists of military and paramilitary operations conducted in enemy-held, hostile, or denied territory by irregular, predominantly indigenous, guerrilla forces to reduce the effectiveness, industrial capacity, and morale of the enemy (ATP 3-18.1, proposed for inclusion in ADP 1-02).

# CHARACTERISTICS OF GUERRILLA WARFARE

5-133. Guerrilla warfare comprises combat operations conducted in enemy-held territory by predominantly indigenous forces on a military or paramilitary basis to reduce the effectiveness, industrial capacity, and morale of the enemy. Guerrilla operations are conducted by relatively small groups employing offensive tactics. In UW as conceived by U.S. Army SF, guerrilla warfare is not an end in itself; it is a component of a larger UW campaign plan that supports the achievement of intermediate objectives en route to an end state political decision.

# NATURE OF GUERRILLA WARFARE

> "The strategy is one to ten[;] the tactics are ten to one."
> Li Tso-Peng Commentary on Mao's *On Protracted War*

5-134. Guerrillas are irregular soldiers generally operating with firepower, manpower, equipment, and logistics disadvantages compared to the security forces of the regime or occupying power. Guerrillas initially cannot hope to meet and decisively defeat a conventional unit in a pitched battle. Guerrilla units, therefore, must emphasize preservation of their military forces and attack at points most disadvantageous to the enemy. These attacks are normally conducted during periods of low visibility and are directed against isolated outposts, weakly defended locations, or the moving enemy. By recognizing his own limitations and weaknesses, the guerrilla can hope for survival and eventual success. Although typically inferior to the enemy in many ways, the guerrilla is often equal or superior to the enemy in the collection of information and intelligence, cover and deception, and the use of surprise to gain and maintain the initiative. The guerrilla exploits these advantages to compensate for his physical disadvantages.

5-135. In guerrilla warfare, the situation is always fluid. Both guerrilla units and enemy counterguerrilla forces move about on the battlefield—the former to avoid decisive engagement and annihilation and to maintain the initiative, and the latter in the attempt to find, fix, and finish the guerrillas. Guerrilla operations wear down and inflict casualties upon the enemy, cause damage to supplies and facilities, and hinder enemy operations. Because guerrilla operations are primarily directed against lines of communications, industrial facilities, and key installations, they impede or interdict the movement of men and materiel and seriously affect the enemy's capability to supply, control, and communicate with his combat forces. In addition, the enemy is compelled to divert manpower and equipment to combat guerrilla activities. The success of guerrilla operations—even the fact that the guerrillas continue to exist—lowers enemy morale and prestige and maintains the morale and will to resist of the indigenous population.

5-136.   In the early stages of UW, guerrilla forces are rarely concerned with seizing and holding terrain, primarily because they do not have the strength to do so against government countermeasures. The guerrilla attacks by gaining a momentary advantage of firepower, executes his mission to capture or destroy personnel and equipment, and leaves the scene of action as rapidly as possible. Normally, the guerrilla varies his operations so that no pattern is evident. Guerrilla forces often make use of difficult and inaccessible terrain to aid concealment and frustrate enemy pursuit or occupation. Guerrilla forces will typically he assigned areas of responsibility within which they will operate against government power, constantly move, and gradually develop into insurgent sanctuary enclaves, relatively free from enemy interference. Ultimately, these enclaves will grow in size and autonomy and will be governed by the insurgents as liberated territory. In the most successful examples of UW, the guerrillas eventually operate in close coordination and more openly with insurgent regular field armies (for example, China in World War II, the Chinese Civil War, and Vietnam) or outside supporters' conventional maneuver units.

# THREE TYPES OF GUERRILLA WARFARE MISSIONS: SUMMARIZED

5-137.   Guerrilla warfare missions are generally of three basic types, which are summarized below:

- **Guerrilla Warfare Missions in Support of the Theater Commander's Unconventional Warfare Campaign.** UW may represent the entire U.S. strategic initiative or response to an opponent. In these cases, U.S. overt support will be limited. U.S. large-scale intervention is not anticipated and the main effort is the UW campaign itself. U.S. operations will be mostly limited to low-visibility, clandestine, and covert activities. Initially, indigenous emphasis is on the tactical techniques of guerrilla warfare for their own survival; only afterwards is the emphasis on their operational usefulness to the UW campaign plan. These operations assume the least amount of U.S. direct control. U.S. Army SF and other U.S. operatives typically provide combat advice and assistance.

- **Guerrilla Warfare Missions to Assist Conventional Forces Engaged in Combat Operations.** UW may represent a line of operation in support of large-scale U.S. intervention involving major operations. Although the indigenous tactical techniques of guerrilla warfare remain the same, the emphasis in this case is their operational usefulness to conventional major operations. These operations assume a greater degree of U.S. direct control, possibly with U.S. tactical commanders exercising operational control of guerrilla forces. U.S. Army SF also function as a liaison for the tactical commander's more direct authority.

- **Guerrilla Warfare Missions Conducted After Linkup With Friendly Forces.** Missions may be assigned to guerrilla forces after linkup with friendly conventional forces. If combat operations continue against a major opponent still in control of denied territory, such missions may constitute continuing UW in new areas of operation. In such a case, U.S. Army SF and other U.S. operatives are very likely to continue providing combat advice, assistance, and liaison. If combat against major opponents has culminated in U.S. or coalition victory, such missions may represent some stage of Phase VII (Transition) whereby the indigenous guerrilla forces' status has changed to an adjunct of the new governing authority's security posture—something more like light infantry, scouts, or constabulary. This would probably represent an end to UW operations, and subsequent activities would likely represent support to a foreign internal defense mission in support of the new government. The authorities of the new host-nation government, the degree of any direct U.S. operational control, and the continued involvement by U.S. Army SF or other U.S. operatives would then vary by specific circumstances.

5-138.   Guerrilla warfare may be conducted as part of a standalone UW initiative in which the guerrilla force is the predominant or only friendly armed force in the AO. Alternatively, forces conducting guerrilla warfare may be a small adjunct to a conventional force main effort when UW is only one line of operation in major combat operations. Thirdly, in some cases, guerrilla warfare may continue after linkup with friendly forces has occurred. The three types of guerrilla warfare missions are expanded in the following main paragraphs.

# GUERRILLA WARFARE MISSIONS IN SUPPORT OF THE THEATER COMMANDER'S UNCONVENTIONAL WARFARE CAMPAIGN

5-139. Guerrilla forces conduct missions and activities supportive of the UW campaign plan These missions will have a range of tactical, operational, and strategic objectives; will have both immediate and long-term effects on the enemy; and will have direct military and indirect political and psychological implications. These missions and activities comprise the basic techniques of guerrilla warfare. They consist of the following:

- Offensive guerrilla combat operations are comprised of the raid and ambush. These techniques are the fundamental tactics of all guerrilla operations throughout the history of warfare.
- Interdiction is a major operational technique and emphasis in UW and is therefore a basic component of guerrilla warfare. Interdiction—properly planned and employed as part of a comprehensive UW campaign plan—is probably the single most important guerrilla warfare activity that guerrilla forces conduct.
- Psychological actions may be executed by UW indigenous partners who are, in turn, supported by U.S. PSYOP units that advise and assist in the development and employment of guerrilla information capabilities. All operations have a psychological impact and all should be conducted in a manner that will enhance organization and favorable influence of designated target audiences. The psychological impact of guerrilla operations may, in fact, be more significant—individually or cumulatively—than the material combat results. Therefore, guerrilla force activities affect influence efforts. Nevertheless, influence activities are part of the larger context of the UW campaign plan and will be managed primarily by elements other than the guerrilla forces themselves.
- Intelligence tasks are a basic Soldier responsibility. Guerrilla forces will conduct reconnaissance and surveillance missions as they would other combat patrols and will report observed tactical information as a matter of SOPs. Like PSYOP, indigenous conduct of specific and special intelligence activities will be conducted primarily by members of the underground and auxiliary.
- Evasion-and-escape mechanisms are developed to assist in the recovery of separated friendly personnel. Although guerrilla units assist evasion-and-escape activities, such operations are likewise conducted primarily by the underground and auxiliary.
- Defensive combat operations are almost a misnomer because the overwhelming majority of guerrilla force defensive activities are passive. Especially in the early periods of a resistance or insurgency, guerrilla forces are at a marked strength disadvantage compared to government or occupying power security forces. In most cases, guerrilla elements forced to defend themselves from security force attack connotes loss of security and loss of the initiative. In most such cases, the guerrilla force is likely to be destroyed; therefore, guerrillas defend themselves primarily through low-visibility or high-mobility elusiveness.

*Note:* UW core activities are broader than just guerrilla warfare. Guerrilla intelligence units may conduct PE. Guerrilla actions will contribute to subversion of the enemy, and those actions will almost certainly include sabotage. Moreover, guerrilla forces support evasion as part of unconventional assisted recovery and contribute to resistance intelligence operations. These core activities are discussed separately from guerrilla warfare.

## OFFENSIVE COMBAT OPERATIONS

5-140. Raids and ambushes are the principal offensive techniques of the guerrilla force. Raids and ambushes may be combined with other actions, such as mining and sniping, or these latter actions may be conducted independently. All such tactical actions can be integrated into operational plans to interdict enemy lines of communications, key areas, military installations, and industrial facilities. Operational plans will usually drive the selection of tactical objectives. Operational plans must routinely include consideration of second- and third-order effects of any combat actions on both the enemy and the local population.

5-141. Careful and detailed planning is a prerequisite for effective guerrilla combat operations. Depending on the resistance movement, the area command will typically issue general operational guidance, and dispersed units will use maximum tactical latitude—often because there is no alternative—to achieve the command's objectives. Prior to initiating combat operations, a detailed intelligence collection effort is made in the projected objective area and, where possible, the target or objective area will be kept under surveillance up to the time of attack. While the resistance may be weak overall strategically, the individual guerrilla force units must choose tactical targets that they can seal off, overwhelm, and dominate for a short specific period. That same guerrilla force's relative weakness mandates an emphasis on tactical surprise on initiation and rapid preplanned withdrawal along alternate routes when concluded.

5-142.   Tactical objective selection is very much a function of guerrilla force capabilities and limitations. Guerrilla forces, by definition, are relatively weak and under-equipped for communications and mobility. If guerrilla forces are widely dispersed and are to converge on an objective, yet have inadequate means of communications, unforeseen contingencies and reactive responses must be carefully considered. Likewise, if lack of mobility would put withdrawing guerrilla forces at a disadvantage to pursuing forces, countervailing measures must be implemented. Engagement by superior enemy reaction forces could result in unacceptable risk to the force. These risks represent serious concerns that supporting SF teams can help mitigate.

5-143.   Tactical objective selection and the tactical concept of the operation will also be a function of desirability as assessed by the criticality, accessibility, recuperability, vulnerability, effect, and recognizability (CARVER) matrix. Once targets are selected, control measures, such as zones of action, axes of advance, and limits of advance, and supporting efforts, such as MSSs, are established. (An MSS is a preselected area used as a temporary base or stopover point. The MSS is used to increase the operational range within the JSOA.) Premission training and rehearsals are conducted. SF Soldiers are particularly effective in advising and assisting with such planning, training, and execution considerations.

## Raids

5-144.   A raid is an operation to temporarily seize an area in order to secure information, confuse an adversary, capture personnel or equipment, or destroy a capability culminating with a planned withdrawal. Raids are conducted by guerrilla units to destroy or damage supplies, equipment, or installations (such as command posts, communications facilities, depots, radar sites, and so on); capture supplies, equipment, and key personnel; or cause casualties among the enemy and his supporters. Other effects of raids are to draw attention away from other operations, keep the enemy off balance, and force him to deploy additional units to protect his rear areas. Raids—like all other combat operations—can also be used as confidence targets and training events for the guerrilla raid force itself.

5-145.   The size of the raid force depends upon the mission, nature, and location of the target and the enemy situation. The raid force may vary from a single squad attacking a police checkpoint or unprotected rail lines, to a multibattalion raid attacking a large garrison or supply depot. Regardless of size, the raid force consists of three basic elements: assault, support, and security. The assault element is organized and trained to accomplish the objectives of the raid. It consists of a main action group to execute the major task of the raid mission (such as blow up a transformer, capture a key individual, or assault a garrison) and may include personnel detailed to execute special tasks (such as place explosives, handle hostages, or breach perimeters). The support element supports the assault element by concentrating high-volume, direct and any indirect fires. The security element supports the raid by preventing the enemy from reinforcing or escaping. In additional, the security element covers the withdrawal of the assault element and acts as a rear guard for the raid force. The size and positioning of the security element depends upon the enemy's capability to intervene in or react to the operation.

5-146.   Movement to and withdrawal from the objective are key raid considerations and an intensive intelligence effort precedes each operation. Insurgent intelligence cells, auxiliary members, and guerrilla force reconnaissance elements conduct reconnaissance of the routes to the target and, if possible, of the target itself. Local auxiliary sources may be required to furnish guides, transports, or MSSs. Surveillance of the target is continuous up to the time of the attack. The raid force commander exercises extreme caution to deny the enemy any indications of the impending operation.

5-147.   Movement to the objective area is planned and conducted to allow the raid force to approach the target undetected. Movement may be over single or multiple routes. The preselected route or routes terminate in or near one or more MSSs. During movement, every effort is made to avoid contact with the enemy. Upon reaching the MSS, security groups are deployed and final coordination takes place prior to movement to the attack position (figure 5-1).

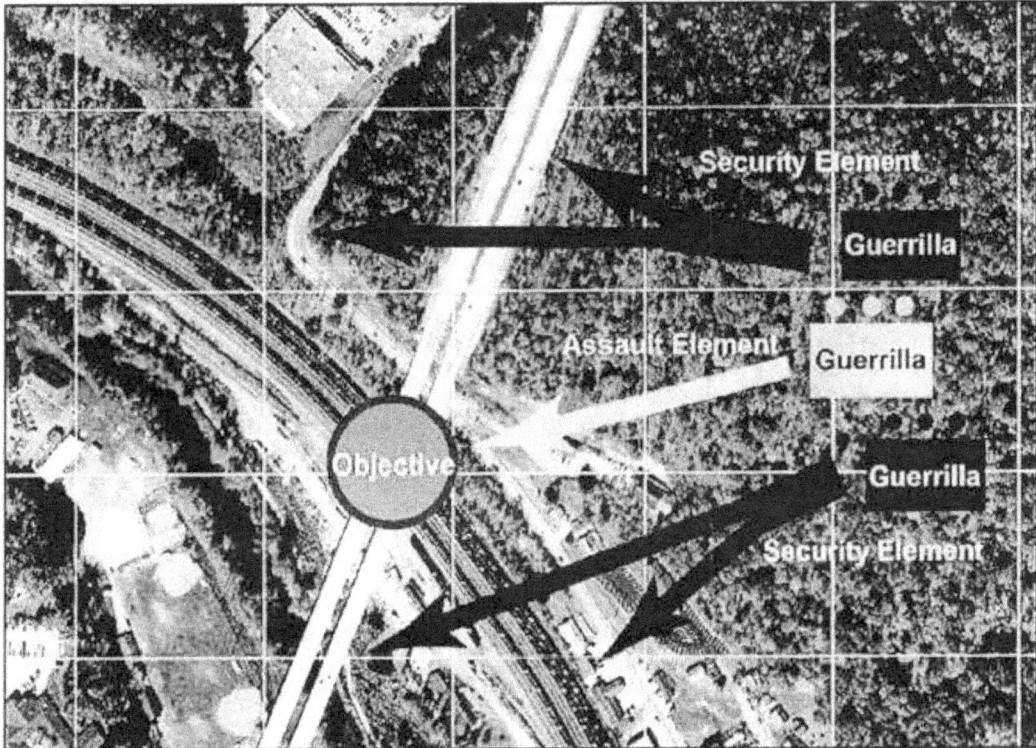

**Figure 5-1. Movement to an objective**

5-148.   Withdrawal is accomplished in a manner designed to achieve maximum deception to the enemy and to facilitate further action by the raid force. The various elements of the raiding force withdraw, in order, over predetermined routes through a series of RPs. Frequently, the raid force disperses into smaller units, withdraws in different directions, and reassembles at a later time and at a predesignated place to conduct other operations. Elements of the raid force may conduct further operations, such as an ambush of the pursuing enemy force, during the withdrawal.

5-149.   Should the enemy organize a close pursuit of the main body, the security element assists by fire and movement, distracting the enemy and slowing him down. Elements of the raiding force that are closely pursued by the enemy do not attempt to reach the initial RP, but on their own initiative, they lead the enemy away from the remainder of the force and attempt to lose the enemy by evasive action over difficult terrain. If the situation permits, an attempt is made to reestablish contact with the raid force at other RPs or to continue to the base area as a separate group. When necessary, the raiding force separates into small groups or even individually to evade close pursuit by the enemy (figure 5-2, page 5-31).

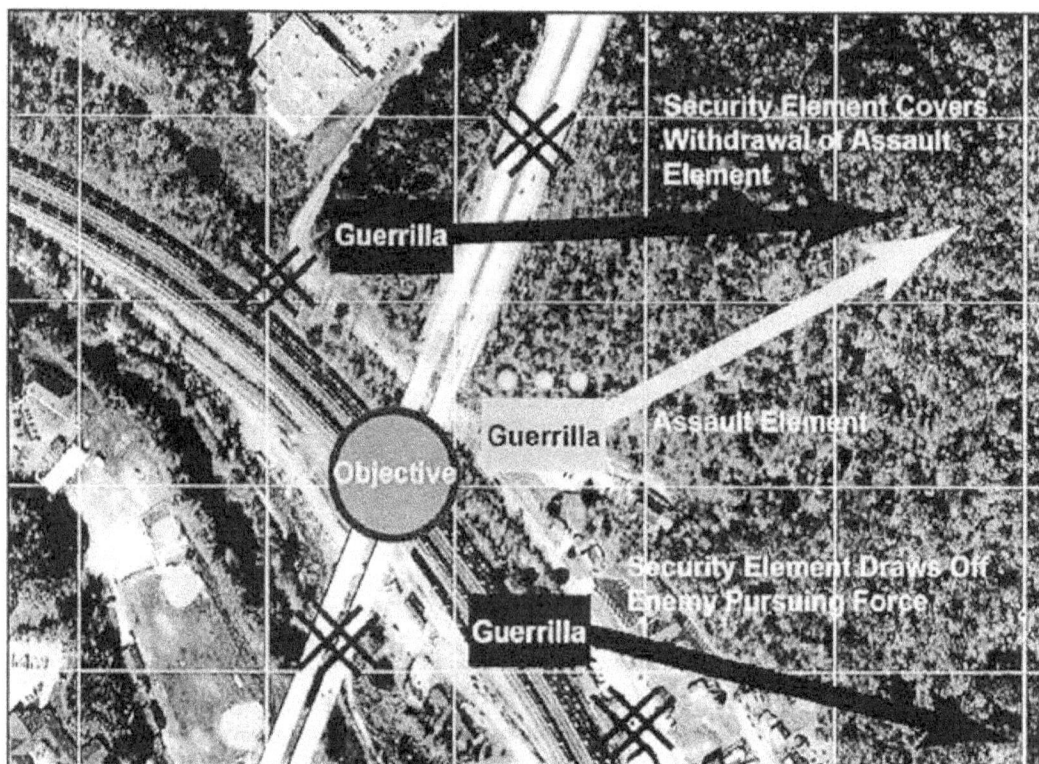

**Figure 5-2. Withdrawal from an objective**

## Ambushes

5-150. An ambush is a form of attack by fire or other destructive means from concealed positions on a moving or temporarily halted enemy. In an ambush, the enemy sets the time, and the attacker sets the place. Ambushes are conducted to destroy or capture personnel and supplies, harass and demoralize the enemy, delay or block movement of personnel and supplies, and canalize enemy movement by making certain routes useless for traffic. The result usually is concentration of the majority of movements to principal roads and railroads where targets are more vulnerable to attack by other theater forces. As in all guerrilla force combat operations, the area command will issue general operational objectives, which may then be best achieved by selected tactical ambush operations.

5-151. Like the raid force, the ambush force is organized into assault, support, and security elements. The assault element conducts the main attack against the ambush target, which includes halting the column, killing or capturing personnel, recovering supplies and equipment, and destroying unwanted vehicles or supplies that cannot be moved. The support element supports the assault element by concentrating high-volume direct and any indirect fires. The security force isolates the ambush site using fires, demolitions, or roadblocks.

5-152. Preparation for an ambush is similar to that of a raid except that selection of the ambush site is an additional key consideration. Planning factors include these questions:

- Will the operation be a single ambush against one column or an area ambush, or a series of ambushes against one or more routes of communication? If it will be a series of ambushes, how are they to be sequenced to achieve maximum effectiveness?
- What are the probable sizes, strength, and composition of the enemy force that is to be ambushed, formations likely to be used, and enemy reinforcement capability and reaction times?

5-153. Other factors may include:
- Finding terrain along the route favorable for an ambush is a key consideration; natural conditions should provide concealment and physical security, enhance the destructive effectiveness of the attacker, and channel and contain the target into the kill zone.
- Terrain considerations include finding unobserved routes of approach and withdrawal to and from the objective.
- Timing of the ambush should provide maximum advantage to the attacker and maximum disadvantage to the target. Most ambushes are conducted during periods of limited visibility.

5-154. Intelligence preparation will be similar to those for a raid with one key exception—effort must be expended to establish enemy traffic patterns and strengths. However, analysis and prioritization of these patterns should already be a central part of operational targeting. Like the use of raids, the value of ambushes is not limited only to their tactical results; their primary function is to contribute to the imposition of operational effects on the enemy. Movement to and from the objective, training and rehearsals, MSS preparations, and security measures will be essentially the same as for raids. Given enough damage consistently, the enemy may even abandon a route—and access to the terrain it runs into—entirely. This can be a huge operational achievement—providing an additional liberated area, as well as a potentially strategic psychological victory.

5-155. There are a few special ambush situations to consider. Most ambushes are conducted relatively close to the enemy kill zone. Ambushes typically involve members of the assault element, briefly occupying the kill zone to capture prisoners or to capture and destroy material (a near ambush). However, if the purpose of an ambush is not to destroy but merely to harass, the ambush site selection and engagement methods will have to be modified. Such far ambushes have the insurgent advantage of putting more distance and possible terrain features between the ambush force and enemy forces.

5-156. Ambushes against columns protected by armored vehicles depend upon the type and location of armored vehicles in a column and the weapons of the ambush force. If possible, armored vehicles are destroyed or disabled by fire of antitank weapons, land mines, improvised explosive devices, Molotov cocktails, or by throwing hand grenades into open hatches. An effort is made to immobilize armored vehicles at a point where they are unable to give protection to the rest of the convoy and block the route of other supporting vehicles.

5-157. Ambushes against moving railroad trains may be subjected to harassing fire, but the most effective ambush involves derailing the train. The locomotive should be derailed on a down grade, at a sharp curve, or on a high bridge. This causes most of the cars to overturn and results in extensive casualties among passengers. It is desirable to derail trains so that the wreckage remains on the tracks to delay traffic for longer periods of time. Fire is directed on the exits of overturned coaches and designated groups armed with automatic weapons rush forward to assault any coaches that are still standing. Other groups take supplies from freight cars and then set fire to the train. Rails are removed from the track at some distance from the ambush site in each direction to delay the arrival of reinforcements by train. In planning the ambush of a train, guerrilla forces must remember that the enemy may include armored railroad cars in the train for its protection and that important trains may be preceded by advance guard locomotives or inspection cars to check the track.

5-158. Waterway traffic like barges, ships, and other craft may be ambushed in a manner similar to a vehicular column. The ambush party may be able to mine the waterway and thus stop traffic. If mining is not feasible, fire delivered by direct-fire heavy weapons can damage or sink the craft. Fire should be directed at engine room spaces, the waterline, and the bridge. Recovery of supplies may be possible if the craft is beached on the banks of the waterway or grounded in shallow water.

## INTERDICTION

5-159. UW forces use interdiction as the primary means of accomplishing operational objectives. Interdiction is designed to prevent or hinder, by any means, enemy use of an area or route. Interdiction is the cumulative effect of numerous, smaller offensive operations, such as raids, ambushes, mining, and sniping. Enemy areas or routes that offer the most vulnerable and lucrative targets for interdiction are industrial facilities, military installations, and lines of communication.

5-160.   The results of planned interdiction programs include:

- Effectively interfering with the movement of enemy personnel, supplies, equipment, and raw material.
- Destroying storage and production facilities.
- Destroying military installations.
- Frustrating the enemy's ability to accurately locate guerrilla bases by analyzing guerrilla operations.
- Causing the enemy to overestimate the strength and support of the guerrilla force.
- Demoralizing the enemy, lessening his will to fight, and modifying his movement, access, support, and control patterns.

5-161.   Interdiction targets should usually be selected based on the achievement of operational objectives supporting the overall UW campaign plan. Therefore, suitable targets for interdiction are typically facilities and material used by the regime or occupying power to maintain political power and enable security forces. Major interdiction target systems include:

- **Transportation.** This includes railroads, highways, waterways, and airways, to include physical networks, controlling facilities, and automotive machines and stock.
- **Communication.** This includes telephone, telegraph, radio, television, and computer production and broadcast infrastructure; public address systems and venues; and newspaper, magazine, and book production and distribution networks.
- **Industry.** This includes anything supporting government or occupier power, but especially manufacturing facilities for weapons, aircraft, vehicles, ammunition, and shipping, to include support and distribution infrastructure.
- **Power.** This includes electric, nuclear, chemical, hydro, wind, solar, and so on; and generation and processing facilities and distribution networks.
- **Fuel.** This includes gas, oil, diesel, and so on; and generation and processing facilities and distribution networks.
- **Military.** This includes military installations and personnel and all relevant supporting infrastructure.
- **Human Nodes.** This includes key human nodes in the formal and informal state civilian power structures and key popular or non-state influence brokers or collaborators.

---

*Note:* Raids and ambushes are critical components of interdiction campaigns without being subordinate components of interdiction. Although UW planners should always put tactical actions into the context of operational and strategic plans—and raids and ambushes are common techniques contributing to campaign interdiction plans—raids and ambushes are sometimes justified by themselves. For example, a strategic raid or ambush that results in the killing or capture of a strategic enemy leader, the destruction of weapons of mass destruction in enemy hands, or the liberation of large numbers of friendly personnel from enemy captivity are valuable combat operations, regardless of any broader interdiction implications.

---

5-162.   There are a few special considerations when planning interdiction campaigns:

- Interdiction is both an action (operation) and an effect created by a combination of insurgent or resistance actions. Commanders and UW planners should remember that all methods can contribute to an interdiction plan, including acts of sabotage and subversion by elements of the underground and auxiliary.
- Actions should be directed against the primary and alternate critical elements of each target system; the quality of political, military, economic, social, information, infrastructure, physical environment, time (PMESII-PT), CARVER, and other methods of target analysis is important.
- Commander's intent is particularly important in interdiction operations because effects achieved are more important in interdiction than literal tactical results achieved. For example, destruction of a bridge (literal interdiction result) to interdict stream crossings (commander's intent) does little good if a usable ford is nearby.

5-163. While much of target analysis is concerned with selecting targets to destroy or incapacitate, another critical function is to identify which targets are irreplaceable or would require too much investment of time and resources to reconstruct either by or for the new government. If an infrastructure target is identified as necessary to disrupt, primacy must be given to determining what part of the system can be most easily destroyed or removed and then quickly replaced by friendly forces or entities. This reverse CARVER principle is based on the fundamental logic that UW ties much of its success to the populace support and governmental function within the UW operational area. Although destruction of key civil infrastructure can immediately achieve desired effects, it ultimately sets conditions for long-term difficulty or failure.

## Mining (Including the Use of Improvised Explosive Devices) and Sniping

5-164. Mining affords the area commander a means of interdicting enemy routes of communication and key areas with little expenditure of manpower. Mines allow the user to move away from the mined site before the enemy activates them. The planned use of mines as an interdiction technique also has a demoralizing effect on enemy morale. Mines may be employed in conjunction with other operations, such as raids, ambushes and sniping, or used alone. When used alone they are emplaced along routes of communication or known enemy approaches within an area at a time when traffic is light. This allows personnel emplacing the mines to complete the task without undue interference and then make good their escape. Mines can be used to cover the withdrawal of a raiding or ambush force to slow enemy pursuit. Mines can also be emplaced around enemy installations to cause casualties to sentinels and patrols.

5-165. Sniping is another interdiction technique. It is economical in the use of personnel and has a demoralizing effect on enemy forces. A few trained snipers can cause casualties among enemy personnel, deny or hinder use of certain routes, and require the enemy to employ a disproportionate number of troops to drive off the snipers. Snipers may operate to cover a mined area, as part of a raiding or ambush force, or by themselves. Snipers can operate as small as a two-man element or as part of a larger sniper observer element with an MSS.

## Railroad Systems

5-166. Railroads present one of the most profitable and easily accessible target systems for attack by guerrilla forces. Most nations rely heavily upon railroad traffic for movement of heavy military equipment. Therefore, guerrilla interdiction attacks against rails can have far-reaching operational and strategic effects. Railroads are characterized by open stretches of track and vulnerable switches, repair facilities, power plants and substations, and coal and water supplies providing unlimited opportunities for attack. Rail crews are subject to interdiction through intimidation.

5-167 Railroad tracks are easily attacked by guerrilla units because it is almost impossible to guard long stretches of track effectively Lightly armed, mobile guerrilla units can inflict heavy damage on tracks. Attacks on open tracks use fewer explosives than attacks on other railroad installations and components. A small guerrilla unit can destroy a considerable amount of railroad track in one night. It is possible for a small group of guerrillas working regularly to keep a single track out of operation permanently. Attacks on tracks should cover a wide area. Telegraph and telephone lines along the railroad are cut simultaneously

5-168. Critical equipment and facilities may be harder to attack because they are usually guarded. However, careful mission planning, CARVER analysis, and skilled execution can still make repair facilities, reserve stocks of equipment, railroad cranes, and other critical items vulnerable. Often, the underground or auxiliary will have better access to rail facilities than will guerrillas.

5-169. Rolling stock may be simultaneously attacked with track interdiction. Demolition of tracks at the time when trains are passing can increase the damage to the tracks and track bed, result in captured supplies, kill and wound enemy personnel, or liberate prisoners. Bridges, tunnels, and narrow railway passes, although usually well-guarded, are critical points for train attacks. Trains moving through areas menaced by guerrillas move slowly and are guarded. Attacks on guarded trains require well-trained and well-armed guerrillas. Heavy weapons and explosives are useful in attacking guarded trains.

5-170. Early interdiction of railroads after hostile occupation may interfere with the enemy's offensive momentum and may forestall large-scale deportation of civilian populations. Cumulatively, the primary effect of interdiction of railroads is disruption of the enemy's flow of supplies, movement of troops, and industrial production. Disruption of orderly processes may result in the accumulation of sizable targets jammed up at rail terminals, junctions, and marshalling yards. These targets then make lucrative targets susceptible to attack by other units using high-explosive, stand-off weapons. Interdiction will increasingly deplete reserves of repair materials, will increase the burden upon enemy security forces and repair crews, and may force transfer to more costly and ponderous highway traffic.

## Highway Systems

5-171.   Highways are less vulnerable targets than railroads. Damage inflicted is more easily repaired, and repairs require fewer, critical materials and less skilled labor. Bridges, underpasses, tunnels, cuts, and culverts are vulnerable points on road networks. Sections of road that may be destroyed by flooding from adjacent rivers, canals, or lakes are also vulnerable. In addition, a road may be interdicted by causing rock slides or landslides.

5-172.   Since highways have fewer vulnerable spots, it is likely that these points will be heavily defended. If so, it is better to concentrate on attacking enemy convoys and columns using the highways. In the initial stages of hostilities, small bridges, tunnels, and levees may be insufficiently protected. As guerrilla attacks increase in frequency and effect, enemy security forces increase protection of these likely guerrilla targets.

5-173.   Where the roads cannot be destroyed, traffic is interrupted by real and dummy mines. Ambushes are conducted when suitable terrain is available. Long-range fires from positions away from roads disrupt enemy traffic. Points for interdiction are selected in areas where the enemy cannot easily reestablish movement by making a short detour.

## Waterway Systems

5-174.   Like railroads, waterways typically carry heavy and big-bulk cargos, the interdiction of which could seriously impact the enemy's economy and military sustainability. The most vulnerable portions of waterway systems are electrical installations, dams, and locks, which are usually well guarded. The destruction of these installations can disrupt traffic effectively for long periods. Although they are more recuperable, signal lights, beacons, and channel markers can be effectively attacked. Sinking vessels in restricted channels by floating mines, limpets, or fire from heavy-caliber weapons may be effective in blocking waterway traffic. Dropping bridges into the waterway, creating slides where the channel is narrow, and destroying levees all hinder ship movement on waterways. Personnel who operate the waterway facilities, such as pilots and lock operators, may be eliminated. These personnel are not easily replaced and their loss will affect operation of the waterway. Mines and demolitions charges may be placed at strategic points on the waterway.

## Airway Systems

5-175.   Airways are interdicted by attacking those facilities that support air movement. Air terminals; communications systems; navigational systems; petroleum, oils, and lubricants dumps; maintenance facilities; and key personnel are targets for attack. Since air traffic is dependent upon fuel, lubricants, spare parts, and maintenance tools, the lines of communications and installations providing these items are attacked. The specific conditions of the operational environment will indicate the criticality of the regime's or occupying power's reliance on airways.

## Communications Systems

5-176.   Wire and fiber optic cable communications, wireless devices and transponders, and ground-based components of satellite relay systems are all potentially vulnerable to guerrilla attack. However, destruction of a single axis of a wire or cable or the destruction of a single relay tower or ground installation will seldom result in the complete loss of adversary communications. Alternate routing is normally available and such rerouting often occurs automatically and almost instantaneously. The degree of such communications system redundancy will vary depending on the technology level in the JSOA. However, even in systems with significant redundancy, the destruction of any portion of the system tends to overload the remaining facilities, thereby degrading the effectiveness of the adversary's system and taxing his repair and maintenance assets. Moreover, multiple simultaneous attacks, carefully planned and sequenced series of attacks, or persistent cumulative series of attacks may be able to stop communications completely, even if only temporarily or in a small, targeted area.

5-177. Cables and wires can be cut and wire poles can be felled. Relay towers and masts can be brought down or the satellite transceivers can be destroyed. Underground cable often runs through concrete conduits and requires more time to destroy. Repair of cables can be delayed by removing a section of the cable. Destruction of physical Internet server sites, telephone central offices and repeater stations, or the supporting power facilities causes greater damage and takes longer to repair than cutting cables or dropping towers. However, radio, television stations, and Internet server sites may be located in well-protected areas and difficult to attack. Destruction of the relay towers and masts, antenna sites, or the transmission lines and cables is usually easier to accomplish than destruction of the receiver or transmitter station. Finally, high-technology developments may allow a virtual interdiction of adversary communications systems. However, this is more likely to be done by the underground or support elements in third-party countries than by the guerrilla force itself.

## Power Systems

5-178. Power lines are vulnerable to attack much in the same manner as wire communications. Large transmission towers often require demolitions for destruction. Critical points in any power system are the transformer stations. If these stations are not accessible to attacks by guerrilla units, long-range fire from small- or large-caliber weapons may disrupt their operations. Power-producing plants and steam-generating plants may be too heavily guarded for raid operations. To disable them, guerrilla forces should concentrate on interdicting the fuel supply or other support infrastructure.

## Water Supply Systems

5-179. The disruption of water lines supplying industries can often be profitably accomplished; water supplies generally are conducted through underground pipelines, and may be destroyed with explosive charges. Raids against reservoir facilities and purification plants are also feasible, but the possible effects upon the civilian population must be considered.

## Fuel Supply Systems

5-180. Petroleum and natural gases for an industrial area are usually supplied by pipelines; damage to lines inflicted by rupture and ignition of fuel is considerably greater than damage inflicted on water lines. Large storage tanks at either end of a pipeline are highly vulnerable to weapons fire, especially when using incendiary projectiles. Contaminating agents may be injected into pipelines or fuel tanks.

## DEFENSIVE OPERATIONS

5-181. Guerrilla operations are primarily offensive in nature. Guerrilla units, with their relatively light weapons and equipment, are normally inferior in strength and firepower to organized enemy forces. They should not, therefore, undertake defensive operations unless forced to do so or in support of special operations conducted by other theater forces. When the enemy attacks, guerrillas defend themselves by movement and dispersion, by withdrawals, or by creating diversions. Whenever possible, defensive operations are accompanied by offensive actions against the enemy's flanks and rear.

## Passive Measures

5-182. Most guerrilla defense is passive. Guerrilla commanders typically make use of distant, inhospitable, and rugged terrain to impede the enemy's ability to find, fix, and pursue them. Guerrilla forces rely heavily on intelligence measures that provide advance warning of impending large-scale counterguerrilla operations. For this reason, coordination with auxiliary inner and outer security rings and underground intelligence cells is a critical early and ongoing requirement in UW. Selected conditions or indicators that might indicate impending enemy offensives in guerrilla operational areas include:

- Advent of suitable weather for extensive field operations.
- Arrival of new enemy commanders.
- Any change in the conventional battle situation, which releases additional troops for counterguerrilla operations. Such changes include enemy victories over allied conventional forces, a lull in active operations, and a reduction in the size of the battle area.

- Increase in the size of local garrisons or the arrival of new units in the area, especially if these are combat troops or troops with special counterguerrilla capabilities, such as radio direction finding units; chemical, biological, and radiological units; rotary-winged aircraft; or mountain, airborne, or reconnaissance troops.
- Extension of enemy outposts, increased patrolling, and aerial reconnaissance.
- Increased enemy intelligence effort against the guerrillas.

5-183. Upon receiving information that indicates the enemy is planning a counterguerrilla campaign, the commander should increase his own intelligence effort, determine the disposition and preparedness of his subordinate units, and review plans to meet the anticipated enemy action. Guerrilla forces must always be prepared to displace.

## Active Measures

5-184. To divert the enemy's attention, the commander may have diversionary activities initiated in other areas. Likewise, he may intensify his operations against enemy lines of communications and installations. Full utilization of underground and auxiliary capabilities assists in diversionary measures. In preparing to meet enemy offensive action, key installations within a guerrilla base are moved to an alternate base; essential records and supplies are transferred to new locations, while those less essential are destroyed or cached in dispersed locations. In the event that the commander receives positive intelligence about the enemy's plans for a major counterguerrilla operation, he may decide to evacuate his bases without delay.

5-185. The commander may decide to delay and harass the advancing enemy. The object is to make the attack so expensive that the enemy will terminate operations. First, security activities on the periphery as well as within any bases are accelerated. Maximum utilization is made of the defensive characteristics of the terrain. Ambushes are positioned to inflict maximum casualties and delay, and antipersonnel mines are employed extensively to harass the enemy. As the enemy overruns various strongpoints, the defenders withdraw to successive defensive positions to delay and harass again. When the situation permits, they may disperse, pass through the line of encirclement, and initiate attacks on the enemy's flanks, rear, and supply lines. If the enemy is determined to continue his offensive, the guerrilla forces should disengage and evacuate the area, dispersing into small elements that will rally elsewhere at predetermined locations. Under no circumstance does the guerrilla force allow itself to become so engaged that it loses its freedom of action and permit enemy forces to encircle and destroy it.

5-186. A counterinsurgent's encircling maneuver is the greatest danger to guerrilla forces because it prevents the guerrillas from maneuvering. Once the counterinsurgent has succeeded in encircling a guerrilla force, he may adopt one of several possible COAs. The simplest is to have his troops close in from all sides, forcing the guerrillas back until they are trapped in a small area which is then assaulted. Differences in terrain make it almost impossible for counterinsurgent troops to advance at an equal rate all around the perimeter, thus creating the possibility of gaps between individuals and units. In other cases, the counterinsurgent may decide to break down the original circle into a number of pockets that will then be cleared one by one. In this situation, the guerrillas must either break out or escape through gaps that may appear as counterinsurgent forces are maneuvering into new positions.

5-187. Perhaps the most difficult situation for guerrillas to counter with is an assault after encirclement has been accomplished. In this maneuver, counterinsurgent forces on one side of the encircled area either dig in or use natural obstacles to block all possible escape routes, while other counterinsurgent forces on the opposite side of the encirclement advance, driving the guerrillas against the fixed positions. As the advance continues, counterinsurgent forces that were on the remaining two sides are formed into mobile reserves to deal with any breakouts.

5-188. A guerrilla commander must be constantly on the alert for indications of encirclement. When he receives indications that an encircling movement is in progress, such as the appearance of enemy forces from two or three directions, the guerrilla commander immediately maneuvers his forces to escape—while enemy lines are still thin and spread out and coordination between advancing units is not yet well established. Records and surplus equipment are either cached or destroyed. Thus, the guerrilla force either escapes the encirclement or places itself in a more favorable position to meet it. If, for some reason, escape is not initially accomplished, movement to a ridgeline is recommended. The ridgeline affords observation, commanding ground, and allows movement in several directions. The guerrillas wait on this high ground until periods of low visibility or other favorable opportunity for a breakthrough attempt to occur. If gaps between enemy units exist, combat detachments seize and hold the flanks of the escape route. When no gaps exist in enemy lines, these detachments attack to create and protect an escape channel. The breakthrough is timed to occur during periods of poor visibility, free from enemy observation and accurate fire. The breakout can be aided by guerrilla diversionary attacks at other points of the encircling perimeter.

5-189. If the breakout is successful, the guerrilla force should increase the tempo of its operations whenever possible, thus raising guerrilla morale and making the enemy cautious in the future about leaving his bases to attack the guerrilla areas. If the breakout attempt is unsuccessful, the commander divides his force into small groups and instructs them to infiltrate through the enemy lines at night or to hide in the area until the enemy leaves. This action should be taken only as a last resort, as it means the force will be inoperative for a period of time and the morale of the unit may be adversely affected. Reassembly instructions are announced before the groups disperse.

# GUERRILLA WARFARE MISSIONS TO ASSIST CONVENTIONAL FORCES ENGAGED IN COMBAT OPERATIONS

5-190. When the JSOA exists within and in support of a joint operational area, operational control of the guerrilla forces concerned may be passed to the joint task force commander. This happens when UW represents a line of operation in support of large-scale U.S. intervention involving major operations. The classic example of UW—World War II in both European and Pacific Theaters—represents missions in support of a larger conventional fight. Guerrilla operations are potentially of great importance in supporting tactical objectives.

5-191. Guerrilla forces can attack lines of communications, enemy supply depots and material, and isolated security forces to degrade the enemy's strength generally, and they are particularly useful prior to and during invasions. Guerrilla forces can expect missions that directly assist combat operations of friendly tactical units. Since guerrilla forces occupy denied and enemy-controlled space with local area expertise, they can be particularly valuable in providing timely and detailed intelligence. Therefore, guerrilla forces can be employed as a reconnaissance and security force. They can mount diversionary attacks to support friendly operations elsewhere. Although guerrillas typically cannot hold and defend for long against strong opposition, they can temporarily seize key terrain in the enemy's rear area, such as key bridges, tunnels, defiles, dams, or other installations.

5-192. Although primarily of value in support of the tactical offense, guerrilla warfare can also assist friendly forces engaged in defensive operations, primarily through small-scale supporting offense actions. As major operations expand, evasion-and-escape operations expand to handle larger numbers of friendly personnel who may find themselves as evaders. Psychologically, the impact of friendly conventional forces' success is magnified by intensified UW activity, and MISO should be crafted to exploit this. Finally, as conventional forces liberate previously enemy-controlled areas, linkup between friendly tactical commands and guerrilla forces usually takes place.

5-193. Tactical commanders who employ guerrilla forces must carefully consider their capabilities when assigning them operational tasks. Guerrilla units are usually organized as light paramilitary formations, and they typically lack support structure, mobility, and communications. Typical guerrilla forces are composed of irregulars without formal military training and experience. Commanders should assume that guerrilla forces are not comparable to similarly sized conventional forces. Assignment of missions to guerrilla units should take advantage of their light-infantry characteristics and area knowledge. U.S. Army SF personnel working with the guerrillas are best suited to recommend to the tactical commander appropriate tasks for guerrilla forces.

5-194. The severest limitations common to guerrilla forces when employed with friendly tactical units is their shortage of adequate voice communications equipment and transportation. This is particularly true when guerrilla units are operating with a mobile force in a penetration, envelopment, or exploitation. For this reason, guerrilla units have a slower reaction time in terrain favoring a high degree of mechanical mobility. Conventional commanders may overcome this disadvantage by providing the necessary equipment or using the guerrilla force on an area basis. Another special consideration is the forward presence of guerrilla units behind the enemy's line of control. It is imperative that tactical commanders have SF liaison personnel on their staff to ensure establishment and adherence to no-fire areas where guerrilla forces operate.

## SUPPORT OF GROUND OFFENSIVE OPERATIONS

5-195.   As the JFC's area of influence overlaps the UW operational area, guerrilla units shift to operations planned to produce immediate effects on enemy combat forces. Initially, these activities are directed against enemy communication zones and army support troops and installations. As the distance between guerrilla and conventional forces decreases, guerrilla attacks have greater influence on the enemy combat capability. Guerrilla operations support penetrations and envelopments and are particularly effective during exploitation and pursuit.

5-196.   Because of the high density of enemy combat troops in the immediate battle area, guerrillas can give little direct assistance to friendly forces in the initial phases of a penetration. Guerrilla forces can best support the attack by isolating, or assisting in the seizure of, the decisive objective. Guerrilla forces hinder or prevent movement of enemy reserves, interrupt supply of combat elements, and attack enemy command and communications facilities, fire support means, and airfields. Locations of critical installations and units that the guerrillas cannot effectively deal with are reported to the tactical commander for attack. As friendly forces near the decisive objective, guerrilla units direct their operations toward isolating the objective from enemy reserves. In some instances, guerrilla forces may be able to seize and hold the objective or key approaches to it for a limited time pending linkup with the conventional force.

5-197.   Guerrilla units assist an enveloping force in much the same way as in a penetration. Guerrillas can conduct diversionary attacks to assist other forces' cover and deception plans. As in the penetration, guerrillas hinder movement of reserves, disrupt supply, attack command and communications installations, and reduce the effectiveness of enemy fire support. They may assist in containment of bypassed enemy units. They attempt to isolate the objective of the enveloping force. They may seize and hold critical terrain, such as bridges, defiles, and tunnels, to prevent enemy destruction. They may perform screening missions to the front and flanks or be a security element to fill gaps between dispersed units of the enveloping force. If used in a reconnaissance or security role, guerrilla units operate on an area basis; that is, they perform their security or screening role within a specified area during the time the enveloping force passes through the area. Guerrilla units usually do not possess the transportation or communications to accompany mobile forces.

5-198.   As friendly tactical units pass from a successful penetration or envelopment to the exploitation of their gains, guerrilla operations increase in effectiveness. As the enemy attempts to reconstitute an organized defense or withdraw to new positions, he should be attacked at every opportunity by guerrilla forces. Enemy units previously assigned to rear area security duties are likely to be recommitted to attempts to restore a defensive position elsewhere. This may increase the vulnerability of rear area installations to guerrilla attack. Guerrilla forces can also assist in containing bypassed enemy units, can round up stragglers and prisoners, may be able to seize control of areas not occupied by the exploiting force, and contribute to the general enemy demoralization caused by the exploitation and subsequent pursuit. As linkup with the exploiting conventional force is accomplished, guerrilla forces and their underground and auxiliary counterparts can contribute to consolidation and transition activities.

5-199.   Throughout coordinated use of guerrilla forces in support of major operations, operational control of the guerrilla force is retained at the level best able to coordinate the actions of the operation. SF working with guerrilla forces provide an excellent method of liaison and coordination with conventional forces. However, not all guerrilla forces will necessarily have SF or other coalition advisors. In such cases, as linkup becomes imminent, guerrilla units nearest the attacking force may be attached to or placed under the operational control of that force. Concurrent with linkup, responsibility for administrative support of the guerrilla force is passed to the tactical command.

## SUPPORT OF AIRBORNE OPERATIONS

5-200.   Guerrilla forces, by virtue of their location in enemy-controlled areas, can materially assist conventional or other forces engaged in airborne operations. They support airborne forces during the assault phase and subsequent operations. They may also be employed in conjunction with airborne raids and area interdiction operations.

5-201.   Initially, guerrilla forces can provide selected current intelligence of the objective area upon which the airborne force commander bases his plans. Immediately prior to the assault, guerrilla units may be able to secure DZs and LZs, seize objectives within the airhead line, and occupy reconnaissance and security positions. Concurrent with landing of the assault echelon, guerrillas can conduct reconnaissance and security missions, provide guides and information, interdict approaches into the objective area, control areas between separate airheads and dispersed units, attack enemy reserve units and installations, and conduct diversionary attacks as a part of the cover and deception plan. In addition, the guerrilla forces, auxiliary, and underground may be able to help control civilians within the objective area.

5-202.   Correct timing of guerrilla operations with the airborne assault is essential. If committed prematurely, guerrilla forces may nullify the surprise effect of the operation and, in turn, be destroyed by the enemy. Conversely, if committed too late, the desired effects of the guerrilla force employment may never be realized. U.S. Army SF operating with UW forces are particularly effective in coordinating such guerrilla force activities with conventional operations.

5-203.   As the assault phase of an airborne operation passes into the defensive or offensive phase, guerrilla forces continue to exert pressure on the enemy forces in the vicinity of the objective area. Guerrillas continue to provide up-to-date information on enemy moves and disposition. Attacks are directed against enemy units attempting to contain or destroy the airborne force, thus requiring the enemy to fight in more than one direction. Airborne forces that have an exploitation mission may employ recovered guerrilla units in reconnaissance and security roles as guides and to assist in control of void areas between dispersed units. If the airborne force is to be withdrawn, the guerrillas can assist to cover the withdrawal by diversionary operations conducted in the rear of enemy forces.

5-204.   Guerrilla forces assist airborne raids in a similar fashion as they do the assault phase of an airborne operation. They provide information and guides, perform reconnaissance and security missions, and divert enemy forces during the withdrawal of the raiding force. An additional factor to consider before using guerrilla forces to support an airborne raid is the undesirable effect of enemy reaction on resistance organizations after withdrawal of the raiding force.

5-205.   Airborne units are seldom committed to Normandy-like guerrilla warfare areas to conduct interdiction operations if the guerrilla force has the capability to conduct such operations. However, in areas where no effective resistance exists, or when supporting major invasions, airborne forces may be committed to conduct interdiction operations. Whatever guerrilla forces are located in areas selected for airborne interdiction, they assist the airborne force to conduct their operations. They provide current, ground truth information and guides, conduct reconnaissance and security missions, control the civilian population, assist in collecting supplies, and generally aid the airborne force commander in making the transition from conventional operations to guerrilla operations. SF detachments with experience in the operational area can prove particularly valuable in assisting the coordination of large airborne force guerrilla operations with those of the indigenous guerrilla force.

## SUPPORT OF AMPHIBIOUS OPERATIONS

5-206.   Guerrilla forces can also support conventional forces engaged in amphibious operations. Typical support activities include conducting operations to hinder or deny the enemy approach to the beachhead; seizing and holding all or a portion of the beach head; assisting airborne operations, which may be part of or complementing the amphibious assault; and conducting cover and deception operations to deceive the enemy as to the location of the actual beachhead. Guerrilla forces with adequate communications connectivity can provide current intelligence or spot for naval fires. Coastal defenses, obstacles, and radars would be typical targets in support of such an amphibious assault.

5-207.   Guerrilla forces operating within the objective area will probably be assigned to the operational control of the amphibious task force commander when he becomes responsible for the objective area. Such operational control will then probably pass to the landing force commander when he assumes responsibility for operation ashore. Depending on the situation, guerrilla forces may be advised and assisted by U.S. SOF. Such forces could be U.S. Army SF, but if they have been operating in the littoral areas of the UW area for some time, it is likely they have been developed and will be advised by U.S. naval or marine SOF conducting a UW mission. This is an example of the inherently joint nature of UW and highlights the imperative of effective joint planning and coordination. As in guerrilla force support to airborne operations, proper timing is critical to operational effectiveness and guerrilla force survival.

## LINK-UP OPERATIONS

5-208.   Most offensive operations in which guerrilla forces assist tactical commands involve a juncture between elements of the two forces. Normally during link-up operations, the guerrilla force is the stationary force, and the conventional unit is the link-up force. However, not all guerrilla forces in an operational area are involved in linkup with tactical units. Some guerrilla units may be assigned missions assisting tactical commands where the requirements of the operation preclude physical juncture. For example, during a raid or area interdiction operation by airborne forces or when conducting operations as part of a cover and deception plan for an amphibious force, it is often undesirable to link up all guerrilla units with the attacking units.

5-209.   Linkup, particularly between dissimilar combined combat organizations, can be dangerous. Therefore, careful coordination is required to mitigate risk of incident. Operational control of guerrilla forces supporting conventional operations should be made clear prior to linkup. The mechanics of linking up must not be confused by questions of who the guerrillas are to answer to. As in most other cases of operational synchronization, the attachment of U.S. Army SF to the guerrilla force, prior to linking up, provides an invaluable conduit for command, control, communications, and coordination.

5-210.   The value of the trust SF build with their counterparts can be crucial in facilitating adherence to successful link-up procedures. Alternate, but less certain, methods of liaison preparatory to linkup include coordination through the underground, temporary exfiltration then re-infiltration of a guerrilla liaison team, or infiltration of conventional force liaison elements.

5-211.   Linkup with guerrilla forces should not be assumed to be a one-time, one-place event. Guerrilla forces are typically dispersed throughout the operational area. Linkup may denote an entire series of relatively small events. Therefore, not only should coordination be made with the area command to have UW forces rally at predesignated points to facilitate linkups, but also the conventional force commander must issue guidance for small-scale and unexpected link-up opportunity SOPs.

5-212.   Fire control lines and no-fire areas must be established to protect both the link-up force and the guerrilla forces from each other's fires. Guerrilla unit dispersion and the fact that civilian elements are a part of the UW force make these control measures particularly important. Communications coordination will be the key to successful linkup and minimizing of incidents. Here again, embedded SF units have the right communications equipment to provide liaison to the larger guerrilla forces. For smaller elements or individual combatants wishing to link up with conventional forces, turn-in procedures and near-recognition or identification measures should be disseminated through available joint force and resistance media channels.

# GUERRILLA FORCE MISSIONS CONDUCTED AFTER LINKUP WITH FRIENDLY FORCES

5-213.   In the event control of guerrilla forces is retained by the United States, missions may be assigned guerrilla forces after linkup with friendly forces has been accomplished. Depending on the political arrangements made with any resistance government-in-exile, operational control of guerrilla forces may be passed to tactical commanders. They may be directly attached to a SOTF or JSOTF, combined JSOTF, or conventional division-sized element. To avoid misuse of such light paramilitary forces and to prepare for the challenges of amicable transition to postcombat activities, any attached SF detachments should remain with guerrilla units during this period. Conventional commanders must not overlook the new support requirements of these nonconventional forces.

## TRANSITION CONSIDERATIONS

5-214.   Most guerrilla forces are irregular, paramilitary, and not suitable for standard conventional operations. However, properly coordinated guerrilla forces may be ideal for conducting reconnaissance and security missions, screening the flanks of friendly forces, patrolling void areas between dispersed units, and providing guides. Some guerrilla forces may be used for light, static security duties in rear areas. Particularly capable and reliable guerrilla forces may be useful in counterguerrilla operations against enemy dissidents or newly coined resistors. The theater commander should be mindful of the role these guerrilla fighters may play in the challenges of transition to the newly formed indigenous government, and utilize, protect, or prepare them for that purpose.

## CONVENTIONAL COMBAT OPERATIONS

5-215. In some rare cases, properly trained and equipped guerrilla units can be employed as conventional combat units. Normally, they would require additional combat and logistics support, such as armor, artillery, and transportation. A period of retraining and reequipping is usually necessary prior to commitment to combat. When so employed, they should be commanded by their own officers with assistance from SF advisors.

## RECONNAISSANCE AND SECURITY MISSIONS

5-216. Because of their familiarity with the terrain and people in their operational areas, guerrilla forces possess a unique capability in a reconnaissance and security role. However, their lack of vehicular mobility and voice communications equipment are limitations on their employment with mobile forces.

5-217. The normal method of employment in reconnaissance and security missions is to assign guerrilla units an area of responsibility within which the guerrilla forces patrol difficult terrain and gaps between tactical units, establish roadblocks and observation posts, screen flanks, provide guides to conventional units, and seek out enemy agents and stragglers.

## REAR AREA SECURITY

5-218. Guerrilla forces may be assigned rear area security missions. Such assignments should be based on their familiarity with and knowledge of the assigned area and populace. Typical areas to be secured include logistics and administrative installations, supply depots, airfields, pipelines, rail yards, ports, and tactical-unit train areas. Guerrilla forces on rear area security may also patrol difficult terrain that contains bypassed enemy units or stragglers; police towns and cities; and guard lines of communications, such as railroads, highways, telecommunications systems, and canals. When provided with appropriate transportation, guerrilla units may be employed as a mobile security force reserve. Selected guerrilla, auxiliary, and underground elements may be effectively used in support of civil censorship operations conducted throughout the controlled area.

## COUNTERGUERRILLA OPERATIONS

5-219. Guerrilla forces are adapted by experience and training for use in counterguerrilla operations. Their knowledge of guerrilla techniques, the language, terrain, and population are important capabilities that can be exploited by conventional commanders engaged in counterguerrilla operations. Guerrilla forces may provide the principal sources of information and intelligence about dissident elements opposing friendly forces. They have the capability of moving in difficult terrain and locating guerrilla bands. They detect guerrilla supporters in villages and towns and implement control measures in unfriendly areas. When properly organized and supported, guerrilla forces may be made completely responsible for counterguerrilla operations in selected areas.

## INDIGENOUS SUPPORT TO CIVIL AFFAIRS OPERATIONS

5-220. Because of their specific area knowledge, cultural preparation, and linguistic capabilities, guerrilla forces or selected civilian support elements may be assigned to assist CA units. Guerrilla forces can—

- Assist in refugee collection and control duties.
- Assist in civil police duties.
- Help establish civil government.
- Recruit labor.
- Furnish or locate technicians to operate public utilities.
- Guard key installations and public buildings.

- Assist in the review and censorship of material for dissemination through public media facilities.
- Assist in restoring the area to some semblance of normality.

5-221. Shrewd commanders will anticipate the utility of transitioning guerrilla forces from a combat experience into adjuncts of new government authority, and they will plan for such forces' reintegration, organization, compensation, and accountability.

# Chapter 6

# Supporting and Related Unconventional Warfare Activities and Legal Considerations

*The whole is greater than the sum of its parts.* [paraphrased]

*Aristotle's Metaphysics*

Several enabling operations, partner forces and entities, and considerations and supporting activities must be considered when planning for UW.

## MILITARY INFORMATION SUPPORT OPERATIONS

6-1.   MISO—and much more broadly, the full range of U.S. information and influence activities to shape perceptions—will always play a central role in successful UW. The earliest practical synchronization of efforts between the TSOC, GCC, and the chief of mission in the area is essential. MISO must be involved at the earliest stages of planning and should prepare the indigenous population prior to the insertion of SOF. The MISO portion of the influence plan should include desired effects, proposed actions, end state objectives, supporting objectives, and the requirements for each phase of the campaign. The MISO plan should also outline key messages and themes for different audiences during each phase. MISO will be fundamental to every phase of a UW campaign.

> "In guerrilla warfare, more than in any other type of military effort, the psychological activities should be simultaneous with the military ones, in order to achieve the objectives desired."
>
> CIA, Tayacan, *Psychological Operations in Guerrilla Warfare*, pp. 7–8.

6-2.   One of the most salient characteristics of UW conducted in the 21st century, compared to previous eras, is the widespread use of electronic devices worldwide. This development creates additional opportunities to shape the operational environment through both direct influence operations to individuals and exploiting vulnerabilities in regime information architectures. UW planners must consider exploiting the full range of recent developments in the cyberspace environment. Cyber infrastructure questions will include:

- What portion of the population has Internet access?
- What other telecommunications media exist?
- How effective are the target regime's censors and controls?
- What are their weaknesses and vulnerabilities?

6-3.   In addition, the state of development of such useful technology will often outpace the doctrine and procedures intended to exploit it. Planners must integrate cyber warfare subject-matter experts into unconventional campaign planning at the earliest practical opportunity, both to assist in developing exploitable cyberspace and electronic warfare options and to identify authority and approval gaps. United States Strategic Command, United States Cyber Command, other DOD organizations, or other government agencies will be typical partners in modern UW planning.

6-4.   UW operations may be part of a limited regional contingency or conducted as part of a major theater war. In either event, the nature of UW distinguishes it as a type of psychological combat or contest of wills, making MISO a vital component of any UW operation. PSYOP forces are integrated with other SOF conducting core, supporting, and related UW activities. UW operations are usually conducted through low-visibility, covert, or clandestine means that entail a high degree of political risk. PSYOP forces are integrated in all phases of UW as a capability essential for success.

*Note:* ATP 3-05.1 provides additional information on UW operations.

6-5. USG information activities focus on the psychological PE. These activities seek to influence populations to provide information of intelligence value necessary for the conduct of operations and to build support for USG-sponsored forces in the UW operational area. During UW, PSYOP influence activities are conducted to undermine the legitimacy of occupying forces or hostile regimes, while simultaneously building the legitimacy and credibility of U.S.-sponsored resistance movements. Interagency or intergovernmental support may also be conducted concurrently with influence activities to bolster regional support for USG activities and objectives and to undermine any support the occupying power or hostile regime may have.

6-6. During UW operations, PSYOP forces—

- Create, reinforce, or sustain those attitudes held by the population that cause them to act in a manner beneficial to their own and toward U.S. objectives.
- Inform key target audiences of conditions within the UW operational area to build support for the resistance movements and insurgent forces.
- Facilitate recruitment of local support personnel and combatants—by using messages through key local leaders to communicate and contrast the goals of the resistance movement with that of the enemy government or occupying power.
- Publicize the efforts of resistance movements and insurgent forces to build the credibility of movement leaders.
- Shape the operational environment for the introduction of U.S. forces.

6-7. PSYOP forces conduct operations from inside or outside the UW operational area through approved programs and, when the mission dictates, infiltrate into the area with SF operational detachments to begin building an indigenous information capability (see figure 6-1, page 6-3). Initial efforts center on assessing the ability of resistance and insurgent forces to conduct influence activities and determining what capability is required.

6-8. PSYOP forces conduct influence activities initially, but the primary objective is to organize, train, and equip local forces to establish or expand an indigenous influence capability. Instructing resistance members and insurgents in the law of war is an essential element in this training, and it is vital to maintaining credibility of words and deeds to validate the resistance movement. PSYOP forces—

- Advise leaders on the psychological effects of operations.
- Train indigenous members to apply influence activities.
- Assist indigenous irregular forces in the conduct of influence activities to effectively convey words that validate deeds.
- Assess conditions in the UW operational area and recommend actions to facilitate influence activities.
- Provide truthful accounts of operations to capitalize on the psychological effects of successful operations, while minimizing the effects of unsuccessful operations.
- Recommend targets to resistance and insurgent leaders based on potential psychological effect.

6-9. Principle target audiences for PSYOP forces focus on the enemy military forces, civilian collaborators, the civilian population, and the resistance. PSYOP forces focus on enemy soldiers' frustrations to lower their morale and reduce their effectiveness by inducing feelings of inadequacy, insecurity, and fear, with the intention of causing them to consider desertion or defection. Collaborators are typically targeted with isolation programs designed to instill doubt and fear in conjunction with a positive political action program. PSYOP forces' objectives of targeting the civilian population focus on persuasion and reassurance that resistance goals are attainable, and that the United States is in support of the same political and social objectives. PSYOP forces also ensure that specific goals for the resistance are reinterpreted and reemphasized continually—particularly during periods of hostility.

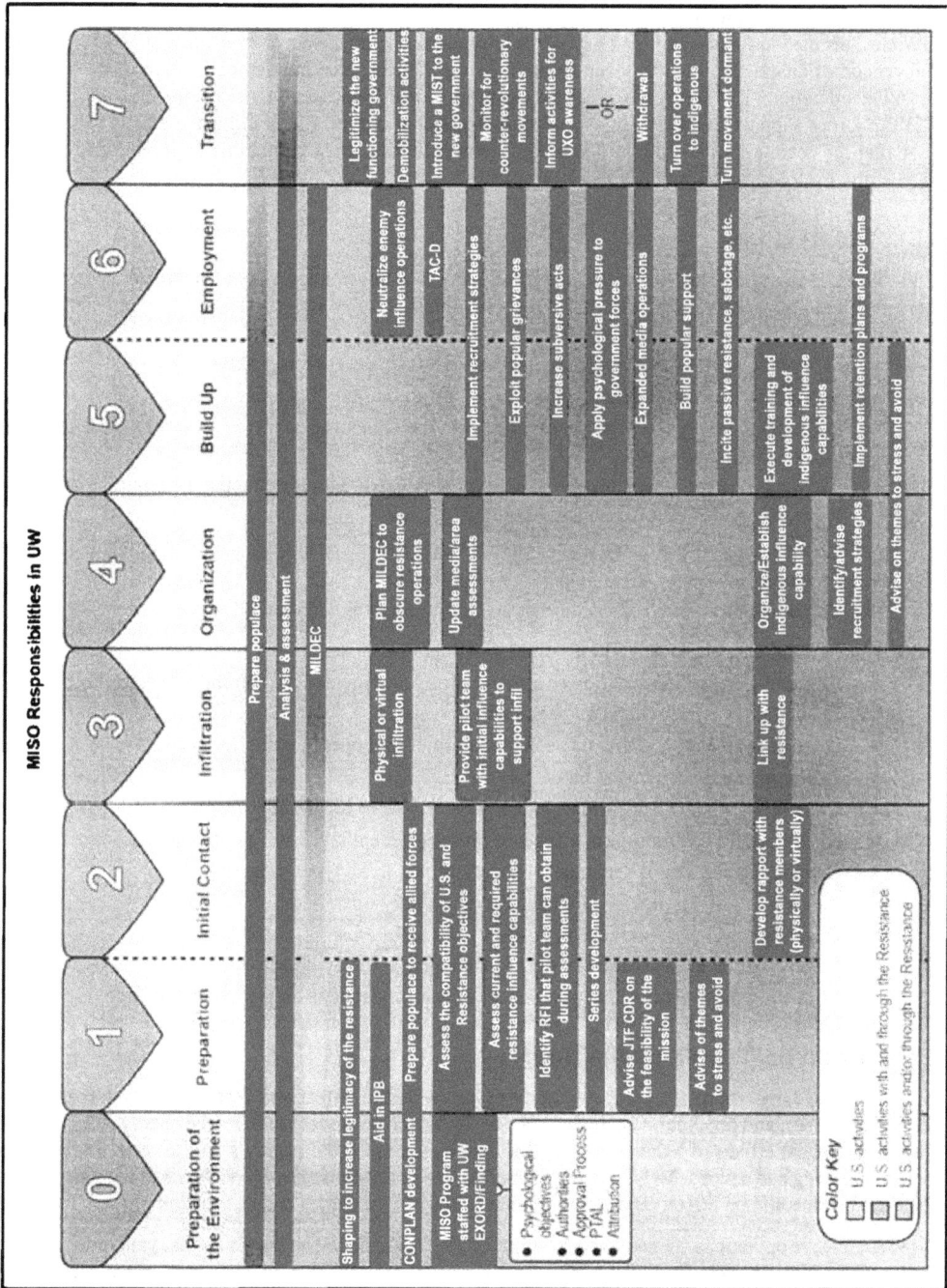

**Figure 6-1. Military information support operations responsibilities in unconventional warfare**

*Note:* The information in figure 6-1 is also reflected in USAJFKSWCS Memorandum, *Primer–Military Information Support Operations in Unconventional Warfare*, 28 May 2015.

6-10. The transition at the end of a UW operation is critical with respect to psychological effect. Successful UW elevates the legitimacy and credibility of a group seeking to subvert or overthrow an existing authority. Subversion of an existing authority can create a power vacuum, making demobilization of the resistance a sensitive operation. PSYOP forces support demobilization processes to accelerate the return to peace. PSYOP forces support DOD, interagency, and international organizations influence efforts to stabilize the area after the cessation of hostilities. These long-term activities assist in building legitimate and sustainable democratic institutions.

## COMBAT OPERATIONS

6-11. PSYOP forces conduct influence activities and other information related activities, such as military deception and senior leader engagement, throughout all phases of an operation. Prior to commencing combat operations, PSYOP forces are employed to assist with—
- Preventing armed involvement by neutral or hostile neighboring states.
- Preventing interference by civilians during future military operations.
- Degrading enemy combat power through the encouragement of desertion and creation of dissension in the ranks.
- Planning military deception to reduce risks associated with introduction of forces.
- Preparing the operational environment for U.S. and coalition combat operations.
- Creating tension and stress within enemy target audiences in anticipation of combat.
- Reducing the enemy target audience will to resist.
- Degrading an enemy's decision-making abilities and operational effectiveness.

6-12. During combat operations, PSYOP forces are employed to assist with—
- Degrading enemy combat power through the encouragement of surrender and malingering.
- Reducing the enemy target audience will to resist.
- Degrading an enemy's decision-making abilities and operational effectiveness.
- Exploiting successes on the battlefield.
- Creating the element of surprise and combat initiative for friendly forces.
- Exploiting the psychological effects of military deception activities on enemy leaders and forces.
- Conducting successful personnel recovery efforts in all types of operations.
- Preventing interference by civilians during military operations.
- Degrading and destroying enemy assets and capabilities.
- Protecting key infrastructure critical to achieving military objectives.
- Mitigating the unintended consequences of military actions on the target audience.

## OFFENSIVE AND DEFENSIVE OPERATIONS

6-13. Offensive operations are combat operations conducted to defeat and destroy enemy forces and seize terrain, resources, and population centers. They impose the commander's will on the enemy. MISO reinforce the psychological effects of lethal and nonlethal actions to reduce the enemy's will to resist and affect their decision making and actions. MISO increase the effects of shock caused by the violence of the attack, degrade the enemy's operational effectiveness, and minimize civilian casualties and destruction of the infrastructure.

6-14. Defensive operations are combat operations conducted to defeat an enemy attack, gain time, economize forces, and develop conditions favorable for offensive and stability operations. To support defensive operations, PSYOP forces conduct influence activities and military deception to reduce the enemy's combat power, reduce the enemy's ability to conduct offensive operations, and delay the advance of enemy units. By supporting the defense, PSYOP units continue to actively shape the information environment to shift to offensive operations.

## STABILITY OPERATIONS

6-15. During stability operations, the U.S. military conducts various tasks and activities abroad and in coordination with other USG elements of power to maintain or reestablish a safe and secure environment, and to provide essential governmental services, emergency infrastructure reconstruction, and humanitarian relief. The ability of the host nation and USG to convey intent and information to the local populace is central to achieving a stable, restored, and secure environment. Employing PSYOP forces to interface with and engage local populations is vital to gaining their trust and confidence—a necessary step in stabilizing an AO.

6-16. Given the wide visibility and global media coverage of military activities, the coordination and synchronization of actions to establish and amplify the commander's narrative are essential to gaining and maintaining credibility and transparency. Credibility and transparency encourage accurate media reporting.

6-17. PSYOP forces typically provide the preponderance of the commander's influence capability and contribute vital support to lethal actions—using deliberately crafted and targeted messages and psychological actions for influence and effect. Psychological actions are lethal and nonlethal actions planned, coordinated, and conducted to produce a psychological effect in a foreign individual, group, or population. PSYOP forces are able to influence target audiences prior to, during, and after military actions. They provide the commander the ability to convey intent and vital information to selected target audiences and populations. Most often, the understanding, cooperation, and support of the local populace for U.S. efforts are necessary for achieving mission success.

6-18. The focus of stability operations is to improve conditions for the civilian populace, the host-nation government, and legitimate private institutions. There are a number of military and nonmilitary objectives for USG and host-nation efforts. The objectives include public safety and security, effective governance, reconstruction, and restoration of basic services. When security is critical, PSYOP forces direct populations to comply with the U.S. and host-nation government's security and safety measures. During USG or host-nation efforts to meet the critical needs of the people, PSYOP forces are often called upon to inform local populations about the locations of food, water, shelter, and medical care. During stability operations, actions and messages directed at selected target audiences and populations contribute to—

- Acceptance of friendly forces in the operational area.
- Rejection of enemy presence and activities by the local populace in stabilized areas.
- Legitimacy of U.S. efforts and military activities in the operational area.
- Legitimacy and viability of the host-nation government.
- Isolation of adversary, enemy, and criminal elements capable of destabilizing the area

# CIVIL AFFAIRS OPERATIONS

6-19. CA forces support UW through the execution of CA operations, which is critical to the planning and execution of UW campaigns. When integrated throughout all phases of UW planning and execution, CA forces provide the capability to analyze the civil component's strengths and vulnerabilities as applicable, to both the existing regime and to the resistance movement. CA operations also provide a comprehensive approach toward assisting the resistance in legitimacy and transitional governance, from the initial resistance movement through transition, to an emergent stable government. CA forces also are able to assist in developing broader CMO efforts in support of the resistance.

6-20. CA forces are typically sought for their unique capabilities with regards to identifying and mitigating the underlying causes of instability in order to create a stable environment. This same analysis, however, can be used by the resistance to identify and degrade those identified strengths and bonds of the existing regime into vulnerabilities, resulting in its continued de-legitimization. This, in turn, creates legitimacy opportunities for the movement. Separately, within their sphere of control and influence, the resistance can consolidate legitimacy and initial governance by using CA assessments, strengthening civil vulnerabilities, and cementing a bond with the greater population. Methods to realign the legitimacy of power should consider the timeliness required to restore essential services and strengthen the bonds between the population and the resistance movement or new government upon the collapse of the old regime.

6-21. When UW results in overthrow and CA forces support transition to a new government, CA forces conduct support to civil administration (SCA). SCA is assistance given to a governing body or civil structure of a foreign country, whether by assisting an established government or interim civilian authority or supporting a reconstructed government. SCA occurs when military forces support the DOS in the implementation of interim civil authority or U.S. foreign policy in support of host-nation internal defense and development. SCA supports the U.S. diplomatic, informational, military and economic instruments of national power abroad through executing tasks affiliated with cooperative security, theater security cooperation, and foreign internal defense as a function of stability operations and irregular warfare.

6-22. Through SCA, CA forces can support a "shadow government or government-in-exile" to plan for and administer civil government in the areas of rule of law, economic stability, infrastructure, governance, public health and welfare, and public education and information. SCA is the systematic application of specialized skills for assessing and advising on the development of stability and governance. When conducting civil reconnaissance and civil engagement to develop civil consideration data, CA forces assigned to special operations CA formations collaborate with CA military government specialists within United States Army Reserve CA formations to formulate governance and stability lines of effort for the resistance.

6-23. The resistance gains legitimacy and transitional governance by addressing grievances and providing essential services to create a civil strength or bond with the population. As the regime becomes severely degraded or collapses, CA will assist with addressing the remaining civil vulnerabilities to create stability for the emergent government, and inclusively, across the indigenous population.

6-24. It is critical to note that figure 6-2, page 6-7, is not intended to represent a linear process, but it rather represents the relationship between civil strengths and civil vulnerabilities as they pertain to civil instability for the current regime, civil opportunities for the resistance, and stability for a new government. The left side depicts CA operations and corresponding resistance actions that increase the de-legitimization of the current regime, through degradation of civil strengths with the population, and attacking civil vulnerabilities to provide legitimacy opportunities for the resistance.

6-25. All CA operations may support UW, although the most important role of CA operations is facilitating the swift transition of power from the resistance forces to a legitimate government after the cessation of hostilities. CA forces may also assist detachments in planning and executing UW operations by—

- Advising detachments in cultural, political, and economic considerations within the JSOA.
- Assessing impacts of proposed missions to the local populace.
- Advising detachments on development of resistance organizations and the expansion of the JSOA in gaining and maintaining popular support.
- Assisting partisan forces to develop auxiliaries and to conduct populace and resources control operations.
- Assisting detachments in integrating with other government agencies (for example, the DOS and the U.S. Agency for International Development).
- Advising and assisting detachments in planning, coordination, and establishment of dislocated civilian camps (key recruitment source).
- Advising and planning measures to gain support of the UW force's civilian populace.
- Planning mobilization of popular support to the UW campaign.
- Analyzing impacts of resistance on indigenous populations and institutions and centers of gravity through CA inputs to IPB.
- Providing the supported commander with critical elements of civil information to improve situational awareness and understanding within the battlefield.
- Assessing the knowledge, skills, abilities, and attitudes of individuals among the local populace and resistance organizations to identify potential posthostility social, political, and economic leaders.
- Advising detachment and partisan forces on development of civil administration within the JSOA as a legitimate government begins to operate.

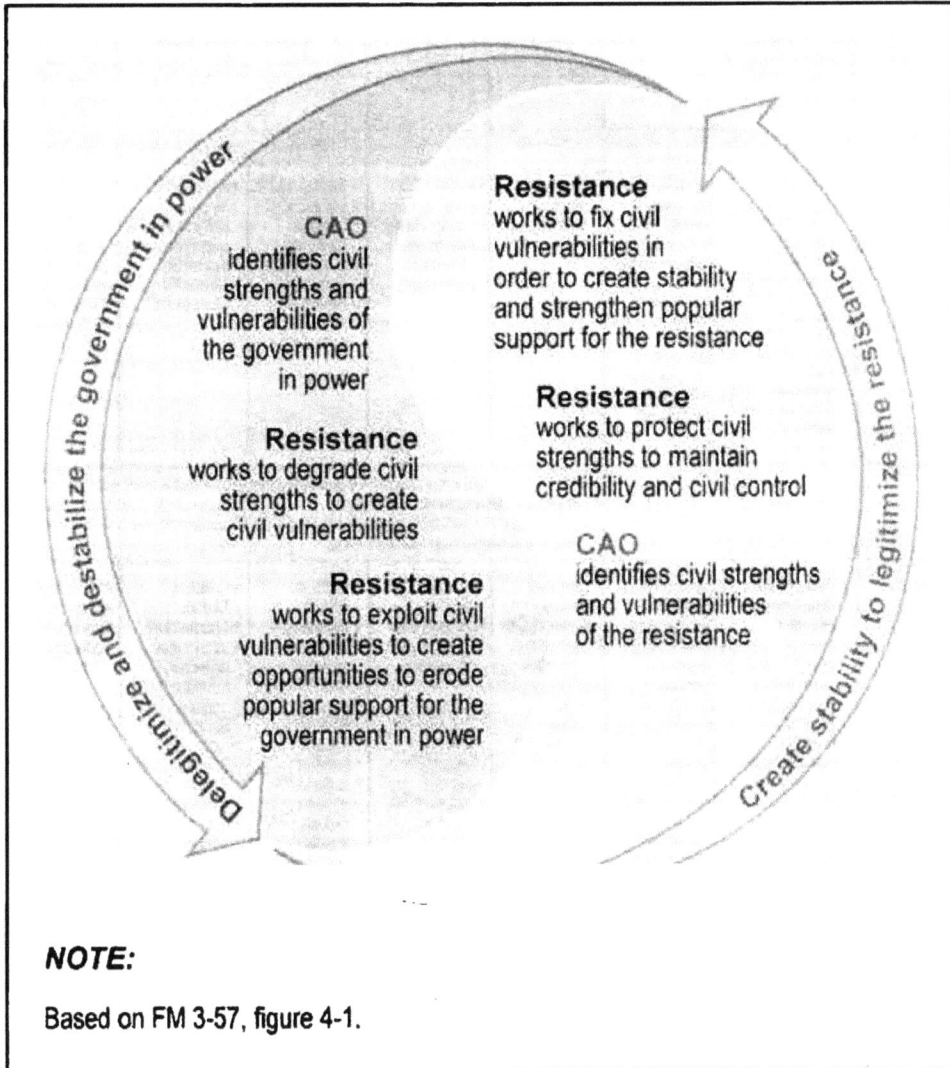

**CAO**
identifies civil
strengths and
vulnerabilities of
the government
in power

**Resistance**
works to degrade civil
strengths to create
civil vulnerabilities

**Resistance**
works to exploit civil
vulnerabilities to create
opportunities to erode
popular support for the
government in power

**Resistance**
works to fix civil
vulnerabilities in
order to create stability
and strengthen popular
support for the resistance

**Resistance**
works to protect civil
strengths to maintain
credibility and civil control

**CAO**
identifies civil strengths
and vulnerabilities
of the resistance

Delegitimize and destabilize the government in power

Create stability to legitimize the resistance

**NOTE:**

Based on FM 3-57, figure 4-1.

**Figure 6-2. Civil Affairs operations support to unconventional warfare**

6-26. CMO are inherent to UW (figure 6-3, page 6-8). CA forces are capable of providing support to all seven phases of a UW campaign: preparation, initial contact, infiltration, organization, buildup, combat employment, and transition. The UW environment contains military and civilian components that are scattered and intertwined within the JSOA. Although the detachment generally focuses its efforts on the military aspect of an insurgency, it must also consider the nonmilitary aspects of the JSOA. Natural, routine, planned, or unpredictable indigenous activities may hinder or help the activities of the guerrilla force during all phases of a U.S.-sponsored insurgency.

6-27. CMO are the commander's activities that establish, maintain, influence, or exploit relations between military forces (including guerrilla or insurgent), government, NGOs, and the indigenous populations and institutions in the JSOA. CMO facilitate other military operations and consolidate and achieve U.S. objectives. In CMO, military forces may perform activities and functions that are normally the responsibility of local, regional, or national government. These activities will occur before, during, or after other military actions. They may also occur, if directed, in the absence of other military operations in a UW environment.

| Phase | I Preparation | II Initial Contact | III Infiltration | IV Organization | V Buildup | VI Employment | VII Transition |
|---|---|---|---|---|---|---|---|
| Who | CAPT/TCAPT/ J-9/G-9/S-9 | CAT | CAT | CAT/CMOC | CAT/CMOC | CAT/CMOC | CAPT/CAPT/ CMOC |
| Where | CONUS/OCONUS | JSOTF/SOTF/AOB | SOTF/AOB | SOTF/AOB | SOTF/AOB | SOTF/AOB | SOTF/AOB |
| General Support | • Develop commander's intent and end state for CAO/CMO<br>• Provide CAO inputs to IPOE<br>• Identify MOEs<br>• Advise SF commander<br>• Initiate transition/ disengagement concept planning | • Integrate with pilot team planning cell<br>• Provide politico-military civil assessments<br>• Provide CAO analysis and evaluation of pilot team assessments<br>• Advise SF commander | • May insert with pilot team or SF detachment | • Validate CA inputs to IPB<br>• Conduct key IPI engagements<br>• Advise SF commander | • Validate MOEs<br>• Monitor effects<br>• Facilitate resistance force buildup<br>• Refine CA inputs to IPB<br>• Advise SF commander | • Monitor MOEs<br>• Monitor/ mitigate impacts of populatuinon operations<br>• Advise SF commander | • Conduct RSOI USAR<br>• Provide liaison<br>• Legitimize post-hostilities institutions<br>• Advise SF commander |
| | | | NOTE 1: CA infiltration of a denied area is nonstandard. Any CA infiltration of a denied area will be mission dependant.<br>NOTE 2: UW resulting in limited coerce or disrupt objectives will also have a limited transition. CA's role may also be limited. | | | | |
| CAOSupport | • Prepare area study/running estimates<br>• Identify strengths and vulnerabilities of established government<br>• Identify strengths and vulnerabilities of resistance force<br>• Conduct in-depth civil consideration analysis and evaluation | • Conduct CR/CE<br>• Conduct initial assessment<br>• Conduct CIM<br>• Update civil information collection plan<br>• Identify key leaders from IPI that may influence CAO/CMO plan<br>• Identify international organizations/ NGOs and other resources | • Continue CR/CE<br>• Conduct CIM<br>• Update civil information collection plan<br>• Support friendly infiltration of the denied area | • Continue CR/CE<br>• Conduct CIM<br>• Update civil information collection plan<br>• Establish CMOC<br>• Support FHA activities<br>• Support FA activities<br>• Support PRC activities<br>• Initiate SCA | • Continue CR/CE<br>• Conduct CIM<br>• Update civil information collection plan<br>• Support CMOC operations<br>• Support FHA activities<br>• Support FA activities<br>• Support PRC activities<br>• Support DC operations<br>• Advise and assist IPV resistance<br>• Support network development | • Continue CR/CE<br>• Conduct CIM<br>• Update civil information collection plan<br>• Support CMOC operations<br>• Support FHA activities<br>• Support FA activities<br>• Support PRC activities<br>• Support DC operations<br>• Deconflict international organizations / NGO operations | • Provide SCA<br>• Conduct CIM<br>• Update civil information colection plan |

Stability Operations (When UW resultis in overthrow of adversary)

***Legend:***

| | |
|---|---|
| **AOB** | advanced operations base |
| **CA** | Civil Affairs |
| **CAO** | Civil Affairs operations |
| **CAPT** | Civil Affairs planning team |
| **CAT** | Civil Affairs team |
| **CE** | civil engagement |
| **CIM** | civil information management |
| **CMO** | civil-military operations |
| **CMOC** | civil-military operations center |
| **CMSE** | civil-military support element |
| **CONUS** | continental United States |
| **CR** | civil reconnaissance |
| **DC** | dislocated civilian |
| **FA** | foreign assistance |
| **FHA** | foreign humanitarian assistance |
| **G-9** | assistant chief of staff, Civil Affairs operations |
| **IPB** | intelligence preparation of the battlefield |
| **IPI** | indigenous populations institutions |
| **IPOE** | intelligence preparation of the operational environment |
| **J-9** | civil-military operations directorate of a joint staff |
| **JSOTF** | joint special operations task force |
| **MOE** | measure of effectiveness |
| **NGO** | nongovernmental organization |
| **OCONUS** | outside the continental United States |
| **PRC** | populace and resources control |
| **RSOI** | reception, staging, onward movement, and integration |
| **S-9** | battalion or brigade Civil Affairs operations staff officer |
| **SCA** | security cooperation assistance |
| **SF** | Special Forces |
| **SOTF** | special operations task force |
| **TCAPT** | theater Civil Affairs planning team |
| **USAR** | United States Army Reserve |
| **UW** | unconventional warfare |

**Figure 6-3. Civil Affairs support to unconventional warfare by phase**

6-28. The intent of U.S. military UW operations is to exploit a hostile power's political, military, economic, and psychological vulnerability by developing and sustaining resistance forces to accomplish U.S. strategic objectives. CMO planners and CA forces are well equipped to assist detachments in developing the internal and external factors that make up the operational environment of UW operations, and in achieving the support or neutrality of various segments of society within or influencing the JSOA.

6-29. Some internal factors that CA forces and CMO planners contribute may include:

- Strengths, weaknesses, vulnerabilities, functions, and center of gravity within and influencing the JSOA.
- Interrelationships between the centers of gravity and key members of the indigenous populations and institutions within and influencing the JSOA using ASCOPE.
- Goals and motivating factors for key civil sector factors.
- Relationships with USG, other government agencies, international organizations, and NGOs within and influencing the JSOA.

6-30. Some external factors that CA forces and CMO planners contribute may include:

- Scope and limitations of each agency's influence and programs.
- Legal and political restrictions and considerations on SF activities.
- Civil sources and assistance available to SF to further ensure mission accomplishment.
- Intent and goals of NGOs and other key civilian indigenous populations and institutions in the JSOA.
- Status of relationships of indigenous populations and institutions, international organizations, and NGOs with representatives of the USG.
- Intent and goals of international organizations (for example, United Nations, African Union, and North Atlantic Treaty Organization).

# CONVENTIONAL FORCES SUPPORT

6-31. There is no set template or linear checklist for the potential roles of conventional forces support in UW. (See Chapter 1, Overview; and Chapter 2, Fundamentals of Resistance, to review the context of STR and the variables that will shape each unique application of UW.) When conventional forces and SOF are employed in the same joint operating area as part of the JFC's campaign plan, the JFC synchronizes the effects of each in mutually supporting ways that achieve the overall campaign plan objectives. Sometimes conventional forces and SOF physically link up on the battlefield. Most of the time such link up is unnecessary or unattainable. In both cases, SOF advisers, assigned to support indigenous resistance or insurgent forces, provide the most significant tactical and operational conduit between conventional forces commanders and indigenous resistance or insurgent leaders.

6-32. When UW is a supporting line of operation to a larger campaign, it is much more likely that the JFC will incorporate UW effects into a supporting effort for the conventional campaign rather than the other way around. However, coordinated air or naval transport, conventional forces fires that disrupt or conventional forces maneuvers that distract enemy counterinsurgent activities, strategic messaging, bases of support, and so on, are some examples of how conventional forces could support operational detachments conducting UW.

# JOINT, INTERAGENCY, AND INTERORGANIZATIONAL SUPPORT

6-33. As a part of joint unified action, SOF operators must increasingly function in joint, interorganizational and multinational environments and must do so even at lower, operator-level echelons. UW is inherently an operation involving multiple agencies, organizations, and stakeholders, and operators must be able to coordinate actions with a wide array of potential partners and stakeholders. UW practitioners must understand the requirements of other agencies, services, or nations to accomplish the mission. U.S. Army SF routinely operate as part of a joint, combined, and multinational special operations team under the guidance of civilian agencies, and they are well prepared to support any combination of operations. Proactive liaison is essential to successful integration.

---
*Note:* JP 3-05.1 and ATP 3-05.1 provide additional information.

---

## JOINT

6-34. By definition, UW is conducted among populations on land, and the U.S. Army SF is the unique unit specifically designated to conduct UW under the proponency of USASOC. However, according to USSOCOM Directive 525-89, that does not mean that other components do not conduct or support UW. Marine, Navy, and Air Force component participation in UW campaigns is determined by the specific conditions pertaining to the situation. Naval Special Operations Command; United States Marine Corps Forces, Special Operations Command; and Air Force Special Operations Command all have core or derived capabilities to conduct or support a UW effort.

## INTERAGENCY COORDINATION

6-35. All U.S. official international interactions, including support to a UW campaign, involve DOS and intelligence community participation—most crucially in the Phase 0 preparations in the operational area prior to U.S. military involvement. Most other USG agencies (for example, the Departments of Treasury, Justice, or Commerce) can also potentially play a role in support of a UW campaign. Figure 6-4 shows a notional template of joint and interagency coordination.

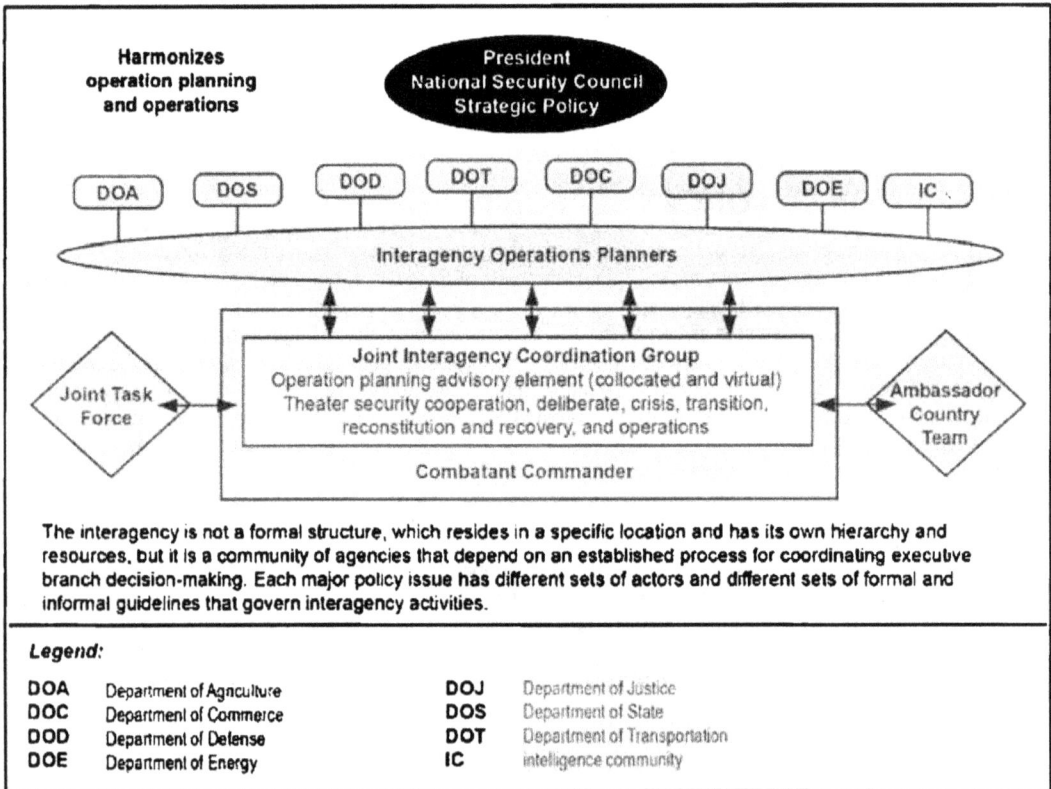

**Figure 6-4. Joint interagency coordination**

## Standard Tactical Examples

6-36. The degree to which operational detachment activities inside a denied area will include interagency partners will vary depending on the specific situation. The baseline UW example is an operational detachment that has already benefitted from interagency feasibility assessments, coordination, and operational support, then typically infiltrates through joint platforms by itself into a denied area. In some cases, other agencies with particular expertise or special capabilities may meet the operational detachment upon infiltration, link up with them sometime after infiltration, or will infiltrate with it. In relatively rarer cases, the operational detachment will infiltrate as part of an interagency team with interagency partners in the lead. In those cases in which UW is conducted to overthrow a state or occupying power, Phase VII (transition) is by definition a gradual handover of lead responsibilities to—and therefore close cooperation on the ground with—interagency partners in liberated territory.

## Joint Interagency Task Force

6-37. Joint interagency task forces (JIATFs) are formal organizations usually chartered by the DOD and one or more civilian agencies and guided by a memorandum of agreement or other founding legal documents that define the roles, responsibilities, and relationships of the JIATF members. The JIATF is staffed and led by personnel from multiple agencies under a single commander or director. JIATFs may be separate elements under the JFC, or they may be subordinate to a functional component command, a JSOTF, or a staff section (such as the operations directorate of a joint staff [J-3]). JIATF members can coordinate with the country team, their home agencies, joint interagency coordination groups in the area of interest, and other JIATFs to defeat complex networks. Because they use more than the military instrument of national power, JIATFs can develop and drive creative nonlethal solutions and policy actions to accomplish their mission.

---

*Note:* JP 3-05.1 provides additional information.

---

6-38. Coordinated and synchronized civilian and military efforts are essential to the conduct of successful UW operations. UW commanders and planners should collaborate with other USG agencies as appropriate in a whole-of-government approach to plan, prepare for, and conduct UW operations. As part of this effort, the JFC should support the development, implementation, and operations for Phase VII (Transition) early in the planning phase, aimed at unity of effort in rebuilding basic infrastructure; developing local governance structures; fostering security, economic stability, and development; and building indigenous capacity to support this infrastructure.

## U.S. Country Team

6-39. All steady-state activities that are undertaken to support a host-nation government are managed through the elements of the U.S. country team, led by the chief of mission. The U.S. country team is the primary interagency coordinating structure that is the focal point for unified action during steady-state activities. As permanently established interagency organizations, with deep reservoirs of local knowledge based on long interaction with the host-nation government and population, country teams represent important resources during stability operations. However, during the conduct of UW, a U.S. country team may not be present in the country where UW efforts are to be focused. Coordination with the surrounding country teams will be necessary to support those UW efforts. Figure 6-5, page 6-12, depicts the country team.

## Interorganizational

6-40. A wide variety of actors outside of USG control can potentially affect the successful conduct of UW. Identifying these actors, assessing their potential impact in each phase and on every aspect of the UW campaign, and creating an integrated approach with U.S. joint and interagency partners to influence these actors is an important responsibility of commanders and staffs.

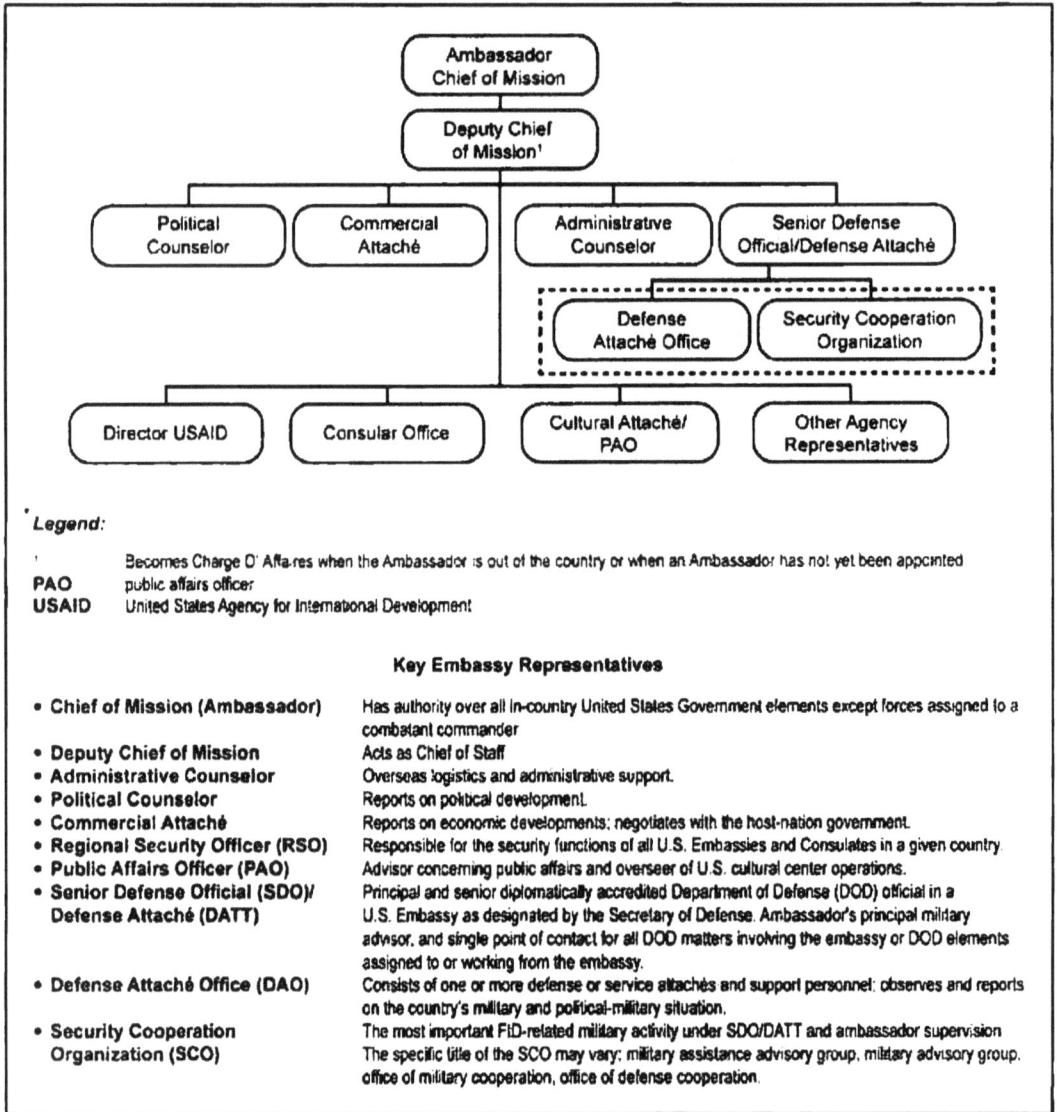

**Legend:**

| | |
|---|---|
| | Becomes Charge D' Affaires when the Ambassador is out of the country or when an Ambassador has not yet been appointed |
| PAO | public affairs officer |
| USAID | United States Agency for International Development |

**Key Embassy Representatives**

- **Chief of Mission (Ambassador)** — Has authority over all in-country United States Government elements except forces assigned to a combatant commander
- **Deputy Chief of Mission** — Acts as Chief of Staff
- **Administrative Counselor** — Overseas logistics and administrative support.
- **Political Counselor** — Reports on political development.
- **Commercial Attaché** — Reports on economic developments; negotiates with the host-nation government.
- **Regional Security Officer (RSO)** — Responsible for the security functions of all U.S. Embassies and Consulates in a given country.
- **Public Affairs Officer (PAO)** — Advisor concerning public affairs and overseer of U.S. cultural center operations.
- **Senior Defense Official (SDO)/ Defense Attaché (DATT)** — Principal and senior diplomatically accredited Department of Defense (DOD) official in a U.S. Embassy as designated by the Secretary of Defense. Ambassador's principal military advisor, and single point of contact for all DOD matters involving the embassy or DOD elements assigned to or working from the embassy.
- **Defense Attaché Office (DAO)** — Consists of one or more defense or service attachés and support personnel; observes and reports on the country's military and political-military situation.
- **Security Cooperation Organization (SCO)** — The most important FID-related military activity under SDO/DATT and ambassador supervision The specific title of the SCO may vary: military assistance advisory group, military advisory group, office of military cooperation, office of defense cooperation.

**Figure 6-5. United States Embassy country team**

## Coalition Nations

6-41. Coalitions are temporary agreements to cooperate between different sovereign state governments. Each government represents different populations, sensibilities, and laws that sometimes conflict with coalition plans and with other coalition partner nation objectives. Political resolve rarely is consistently robust across a coalition. U.S. policy makers should be cognizant of the difficulties some coalition members face in maintaining popular support for their participation.

## Neighboring Countries

6-42. Many insurgencies depend on safe havens and relatively secure support areas in countries adjacent to the target nation. This requires an accurate assessment of the status and level of support these neighboring governments provide. Sanctuary may be given willingly or may be beyond control of the government there.

The U.S. sponsor must also negotiate the permissible level of friendly coalition activities within these third-party nations. Significant support to UW core, supporting, and related activities in the UW operational area may be conducted from such adjoining areas. Recognition of such U.S. sponsor support may be sensitive and require extra effort to remain clandestine or covert. In addition, the populations of neighboring countries should be assessed for potential support to the UW effort.

## Diaspora Communities

6-43. Ethnic and other unique characteristics shared between insurgent groups in the UW operational area and outside groups provide multiple avenues to potentially support the UW effort.

## Intergovernmental Organizations

6-44. International organizations form when two or more national governments sign a multilateral treaty to form such a body and finance its operations; they possess legal personality in international law and their staffs enjoy diplomatic status. Most international organizations are regionally focused and, as such, when international organizations member states could be adversely affected by an insurgency in their region, the organization may act collectively to provide (or deny) legitimacy, sanctuary, and support to insurgents. International organizations can also play an important role in humanitarian assistance and development relevant to the larger context of the UW campaign.

## Nongovernmental Organizations

6-45. NGOs are private, self-governing, nonprofit organizations. Their activities (a direct function of the interests of their donors) are very diverse, but they include interests, such as education, health care, environmental protection, human rights, conflict resolution, and similar issues. Some NGOs are witting, implementing partners for U.S. foreign activities, but these are relatively rare. However, in order to secure freedom of movement, including access to uncertain environments, most NGOs generally strive to be independent, politically neutral, and needs-driven organizations. Consequently, they often try to minimize contact with uniformed military personnel or other governmental actors, who are seeking humanitarian space in which to operate.

6-46. Despite their best efforts, NGOs may not be seen as neutral by either the target state or some insurgents. In some cases, NGO's reliance on indigenous government resources—military and otherwise—for transportation and protection make them susceptible to insurgent targeting. However, the access provided by such collaboration may have potential benefits to the insurgent cause and their U.S. sponsors. In other cases, NGOs may oppose the government or occupying power and actively forge links with insurgents. In addition, as an independent and often credible source of ground truth about the areas in which they work, they are an important source of information to many interested parties. Each case of NGO cooperation and utility must be judged separately.

6-47. Fraternal organizations, societies, clubs, trade guilds, and lodges many times have a long history of indigenous membership and activities in an area. Such organizations can be deeply embedded in a population's social fabric through business or other ties, enabling their activities to be viewed from simply benign to very beneficial in nature. Fraternal organizations may have an advantage in being perceived as truly local in origin, grass roots in approach, yet viably connected. This may be a distinctly different light than which larger traditional NGOs with blatant agendas may be viewed.

## Multinational Corporations

6-48. Multinational corporations usually become indirectly involved in USG foreign activities when their corporate interests (financial interests, foreign-based personnel, or infrastructure) are threatened or when a financial advantage is perceived. UW planners do not always have the ability to influence the activities of multinational corporations in an affected country, but they may find that their interests and potential access may complement the UW effort.

## Media

6-49. Media is a key actor in a successful UW campaign information strategy. The USG is accustomed to interaction with Western media groups in campaigns dominated by conventional forces and large-scale overt U.S. involvement. However, gaining cooperation or assuring constructive foreign media reporting in a low-visibility, long-duration, and potentially highly sensitive UW campaign requires more finesse. MISO should therefore be involved in assessing and shaping foreign media participation from the outset of UW planning.

# LOGISTICS CONSIDERATIONS

6-50. UW logistics support is different from support to other types of operations and even other types of special operations. This difference derives from the requirement to assist indigenous forces in denied areas typically characterized by a politically sensitive and physically challenging environment and usually over long time periods. Theater sustainment architecture will typically be undeveloped, especially when UW is used as a strategic initiative or response without planned U.S. major operations.

6-51. Operational detachments' UW supply requirements are a function of the specifics of the mission. Sustainment planning for any operation involves the identification of requirements, the provision of organic capabilities to meet those requirements, the subsequent identification of shortfalls, and the consideration of leveraged or acquired options to mitigate those shortfalls. This is the conventional logistics estimate model and is applicable for the UW concept of support planning as well. However, because of OPSEC requirements, the constraints and limitations inherent to UW, and the unusual support requirements often encountered in supplying irregular elements, challenges exist in the unconventional logistics arena that requires increased diligence in the planning processes. Sometimes, a resort to nonstandard, logistics problem solving will be required. (See Appendix M, Logistics, for detailed discussion of operational detachment-level logistics considerations.)

*Note:* ATP 3-05.1, Chapter 4, Supporting Activities and Legal Considerations, provides further discussion of theater-level support to UW.

6-52. Since UW is support to a resistance in a denied area, it is likely to be critical that an area complex is developed which can sustain the resistance. The difficulty of merely surviving in the denied area will require clandestine methods and a pattern of distributed and redundant networks that shape the amount, type and accessibility of logistics that can be made available.

*Note:* TC 18-01.1, Task 18: Develop the Area Complex, provides a detailed list of factors necessary for resistance sustainment.

# LEGAL CONSIDERATIONS

6-53. UW often inhabits the gray area of political engagement in which it blends into warfare. It is crucial that Soldiers planning and executing UW pay careful attention to legal considerations and respect for authorities. All ARSOF operations will comply with U.S. and international law, national policy, DOD directives, ARs, whether SF operations are conducted during war or stability operations.

## EXECUTION AUTHORITIES AND SENSITIVITIES

6-54. Military operations require an instrument authorizing their execution. The senior policy document that describes the activities and operations of UW is USSOCOM Directive 525-89. This directive conforms to U.S. statutes, the law of land warfare, and rules of engagement issued by competent legal authorities. Because of its varied preparatory activities, range of operational environments, operational modes, and extended nature, the complexity associated with a UW campaign may require a multitude of authoritative instruments—each with its own unique constraints, restrictions, limitations, content, processes, and offices—wherein lies legal authority to approve its execution.

6-55. In order to be successful, UW campaigns depend on the secrecy of the campaign's objectives; participants and relationships; locations; capabilities; timing and synchronization; and the tactics, techniques and procedures to be used. The intent is to protect sensitive information but not impede the planning and execution of UW campaigns.

## COVERT ACTION

6-56. No agency, except the CIA—or the Armed Forces in time of war—may conduct any covert action activity unless the President determines that another agency is more likely to achieve a particular objective. USC Title 50 (50 USC), Section 3093, *Presidential Approval and Reporting of Covert Actions*, defines *covert action* as an activity or activities of the United States Government to influence political, economic, or military conditions abroad, where it is intended that the role of the United States Government will not be apparent or acknowledged publicly, but does not include:

- Activities the primary purpose of which is to acquire intelligence, traditional counterintelligence activities, traditional activities to improve or maintain the operational security of United States Government programs, or administrative activities.
- Traditional diplomatic or military activities or routine support to such activities.
- Traditional law enforcement activities conducted by United States Government law enforcement agencies or routine support to such activities.
- Activities to provide routine support to the overt activities (other than those activities described [above]) of other United States Government agencies abroad.

6-57. Section 3093(a) of 50 USC provides Presidential authority and limitations in regard to the issuance of findings authorizing covert action and documentation and oversight requirements. The President may not authorize the conduct of a covert action by departments, agencies, or entities of the USG, unless he determines such an action is necessary to support identifiable foreign policy objectives of the USG, and that such action is important to the national security of the United States. This determination shall be set forth in a finding that shall meet each of the following conditions:

- Each finding shall be reduced to a written finding as soon as possible (but in no event more than 48 hours after the decision is made).
- Unless immediate action by the USG is required and time does not permit the preparation of a written finding, a finding may not authorize or sanction a covert action, or any aspect of any such action, which has already occurred.
- Each finding shall specify each department, agency, or entities of the USG authorized to fund or otherwise participate in any significant way in such action. Any employee, contractor, or contract agent of a department, agency, or entity of the USG—other than the CIA— directed to participate in any way in a covert action shall be subject either to the policies and regulations of the CIA, or to written policies or regulations adopted by such department, agency, or entity, to govern such participation.
- Each finding shall specify whether it is contemplated that any third party (which is not an element of, or a contractor or contract agent of, the USG, or is not otherwise subject to USG policies and regulations) will be used to fund or otherwise participate in any significant way in the covert action concerned, or be used to undertake the covert action concerned on behalf of the United States.
- A finding may not authorize any action that would violate the Constitution or any U.S. statute.

6-58. Section 3093(b) of 50 USC requires the congressional intelligence committees to be kept fully and currently informed of all covert actions. Any covert action finding shall be reported to the committees, as soon as possible after such approval, and before the initiation of the covert action authorized by the finding.

## PREPARATION OF THE ENVIRONMENT

6-59. The broad nature of PE activities precludes identification of specific authorization requirements, but the requirement to obtain authorization as prescribed within applicable orders, laws, and regulations is common to all. Coordination procedures for PE vary widely, but the TSOC's responsibility for the development, coordination, and implementation of PE activities with the respective country teams, regional interagency, and GCC counterparts is also common to all.

*Note:* USSOCOM Directives 525-16, *(U) Preparation of the Environment (S//NF),* and 525-5, *(U) Advanced Special Operations Techniques (S//NF),* provide additional information on PE.

6-60. Military intelligence operations and activities that are authorized by 10 USC include:
- Human intelligence (HUMINT).
- Counterintelligence.
- Signals intelligence.
- Tagging, tracking, and locating operations.
- Close-in persistent surveillance.
- Intrusive reconnaissance and surveillance.
- Measurement and signatures intelligence.
- Several classified activities or operations that support PE. (Such 10 USC military intelligence operations and activities are executed under the combatant command's 10 USC authority.)

6-61. The DOD may also conduct intelligence operations and activities under authority of 50 USC in support of national IRs. Such operations are conducted under authority granted to the Secretary of Defense under 50 USC. Such operations are approved for execution by the Secretary of Defense following the GCC's or USSOCOM commander's endorsement and favorable Director of Central Intelligence Agency directive coordination by the Director, Central Intelligence Agency; Director, National Security Agency; and pertinent chiefs of mission, chiefs of station, and the Undersecretary of Defense for Intelligence.

## GENERAL LEGAL PRINCIPLES

6-62. Seven basic legal principles forged from the Hague and Geneva Conventions, the International Declaration of Human Rights, and the customary laws of war govern all U.S. military operations. They are as follows:
- Observance of fundamental human rights will recognize the dignity and worth of the individual and the fundamental freedom of all without distinction as to race, sex, language, or religion. Human rights violations will not be tolerated. As with violations of the law of war, U.S. Soldiers will report human rights violations when they become aware of them
- Civilians shall be treated humanely and may not be used to shield military operations.
- EPWs and civilian detainees will be treated humanely and according to the provisions of the Geneva Conventions.
- U.S. Soldiers are entitled to similar humane treatment should they fall into the enemy's hands.
- Orders to commit war crimes must be disobeyed.
- Soldiers who violate the law of war will be held responsible for their actions. Superiors who order violations of the law of war are criminally and personally responsible for such orders, as are subordinates who carry out the orders.
- Weapons, munitions, and techniques calculated to cause unnecessary pain and suffering are forbidden.

## ABDUCTIONS

6-63. Abduction is the forcible, nonconsensual removal of a person by one state from the territory of another state. American law enforcement officials refer to such abductions as arrests. To be acceptable under international law, abduction must satisfy more exacting standards than the availability of an arrest warrant issued by the state responsible for the action. Abduction may be considered criminal conduct in the state from which a person was taken. In addition, regardless of whether the state from which an individual was taken is hostile to the abducting state, abductions constitute an extremely sensitive operation that can significantly affect international relations.

6-64. The United States reserves the right to engage in nonconsensual abductions for three specific reasons:

- If a state for internal political reasons may be unwilling to extradite a target or give its public consent to the target's removal. Unofficially, the state may be prepared to have the target removed without granting formal consent and may even offer some cooperation in carrying out the action.
- When the target is an extremely dangerous individual and is accused of grave violations of international law, such as air piracy.
- To prevent terrorists, other dangerous individuals, and their state supporters from assuming they are safe from such unilateral action.

6-65. The legality of abduction by a government or its agents is heavily dependent on the facts of each circumstance. Consulting with a legal advisor is strongly encouraged prior to executing any mission-related abduction.

## ATTACKS ON TERRORISTS AND TERRORIST CAMPS

6-66. The United States recognizes and strongly supports the principle that a state, subject to continuing terrorist attacks, may respond with appropriate use of force to actively defend against further attacks. This policy complies with the inherent right of self-defense recognized in the United Nations Charter.

## USE OF OTHER FORCE

6-67. Presidential Executive Order 12333 states: "No person employed by or acting on behalf of the United States Government shall engage in, or conspire to engage in, assassination." Article 23B of the Hague Conventions of 1907 essentially prohibits wartime assassination, outlawing the treacherous wounding or killing of the enemy, or offering a reward for an enemy dead or alive. Neither Executive Order nor the Hague Conventions prohibit attacks on individual soldiers or officers of the enemy, whether in the zone of hostilities, occupied territory, or elsewhere. An individual combatant can be targeted lawfully whether he or she is directly involved in combat, providing support, or acting as a staff planner.

## TREATMENT OF AN ENEMY PRISONER OF WAR

6-68. The operational detachment captures a prisoner and is now at an immediate disadvantage. One or two men must guard the EPWs, which reduces the strength of the operational detachment and impacts on its capability to complete the mission. EPWs will hamper the movement of the operational detachment and increase the likelihood of the operational detachment's compromise.

6-69. The operational detachment must treat the EPWs humanely and afford them all rights and privileges required under the Geneva Convention for the Treatment of Prisoners of War. EPWs will not be targeted, even if their presence jeopardizes the mission. The operational detachment has four options with regard to EPWs:

- Take the individuals into custody and evacuate them later.
- Take the individuals into custody, hold them, and later leave them where they are likely to be found. While in U.S. custody, they must be treated according to the Geneva Conventions.
- Take the individuals into custody and later release custody to an allied power. The allied power must be a signatory to the Geneva Conventions and willing to treat the EPWs as required by the Geneva Conventions.
- Take the individuals into custody; release them near the point of capture.

## USE OF THE ENEMY'S UNIFORM

6-70. The law of war generally prohibits the use of enemy uniforms in combat. However, DOD Law of War Manual, Section 5.23, *Use of Enemy Flags, Insignia, and Military Uniforms*, indicates that the use of enemy uniforms outside of combat as part of a ruse or to avoid detection is permissible. Use of enemy uniforms to avoid detection cannot be to close with the enemy to kill or wound, but they can be used to infiltrate. In addition, using enemy uniforms as a necessity and not to deceive the enemy is also permissible (for example, cold weather uniform items). In that instance, enemy insignias and ranks should be removed. Finally, prisoners of war are permitted to wear enemy uniforms to facilitate escape and to evade capture. EPWs should not engage in combat while in the enemy's uniform, or else their status may convert to spies or saboteurs (See the DOD Law of War Manual for further discussion.)

## GROUP JUDGE ADVOCATE

6-71. Judge advocates should be consulted throughout the operations process to ensure proposed plans and decisions are legally sound and the commander is presented with thoroughly vetted options. In some instances, legal review of unit activity is required. Regardless of the requirement, UW operations present unique legal issues, so it is highly encouraged that commanders and staff seek legal input concerning operations.

6-72. The group judge advocate assigned to the unit provides legal advice to the group commander as required. SF missions may be politically sensitive, particularly in stability operations, and present multiple, complex legal issues and questions. During mission planning and execution, the commander must ensure his units operate within the scope of the law of war and U.S. law (such as security assistance and intelligence statutes), international treaties, and host-nation support agreements. Failure to comply with legal and policy requirements could cause embarrassment and even criminal investigation and prosecution for the commander and his staff.

6-73. The judge advocate provides legal advice to the commander and staff. The group or battalion judge advocates in the SF group know the applicable laws, policy, and regulations associated with the mission of the client. The judge advocate must have a comprehensive understanding of the force structure, missions, doctrine, and tactics of the units he/she advises. The group or battalion judge advocate makes themselves readily available to commanders within their organization and other leaders who seek legal advice. They integrate with the staff to provide legal advice to planning and command post operations and ensure operations are executed within the scope of relevant laws, policies, and regulations. Group and battalion judge advocates will consult with judge advocates in their technical chain and other legal experts to ensure competent legal advice is provided to the commander and staff.

# SENSITIVE ACTIVITIES SUPPORT TO SPECIAL OPERATIONS

6-74. Sensitive activities are programs that restrict personnel access, such as alternative compensatory control measures; sensitive support to other Federal agencies; clandestine or covert operational or intelligence activities; sensitive research, development, acquisition, or contracting activities; special activities; and other activities excluded from normal staff review and oversight because of restrictions on access to information.

6-75. Sensitive activities support many different kinds of both SOF and conventional forces operations. Sensitive activities may be used in support of any special operations core mission. Sensitive activities may be used to support UW; it is not equivalent to UW. Sensitive activities is not a type or phase of UW. (See Appendix A, (U) Sensitive Activities Support to Unconventional Warfare (S//NF), published separately.)

> *Note:* AR 380-381, *Special Access Programs (SAPS) and Sensitive Activities,* and ATP 3-18.20 provides more information on sensitive activities.

## Appendix A

# (U) Sensitive Activities Support
# to Unconventional Warfare (S//NF)

## INTRODUCTION

A-1.  In Unconventional Warfare (UW), success hinges on maintaining secrecy, flexibility, and operational security while supporting resistance movements that seek to coerce, disrupt, or overthrow established authorities. Sensitive activities (SA) play an essential role in ensuring that UW operations can unfold effectively and discreetly within hostile or denied territories. These activities involve a range of classified and compartmentalized actions that shield UW missions from adversarial detection, safeguard sensitive information, and protect personnel and resources from exposure.

A-2.  Sensitive activities encompass multiple operational facets, including clandestine intelligence collection, covert logistical support, and alternative compensatory control measures (ACCMs) that restrict personnel access to critical information. They also provide discreet support to other federal agencies, conduct covert training and advising of indigenous forces, and implement advanced counterintelligence measures. Collectively, these activities enable UW missions to achieve their objectives without compromising security or disclosing the identity of supporting entities.

A-3.  Sensitive activities function as force multipliers in UW, securing and amplifying the effects of each phase—from the initial Operational Preparation of the Environment (OPE) to the decisive stages of combat or political action. While sensitive activities are integral to UW, they are neither a type of UW nor an independent phase. Instead, they are enablers that operate in conjunction with all stages of UW, providing critical infrastructure, intelligence, and logistical support to sustain resistance operations.

## ROLE OF SENSITIVE ACTIVITIES IN UW OPERATIONS

### 1. OPERATIONAL PREPARATION OF THE ENVIRONMENT (OPE)

A-4.  Operational Preparation of the Environment (OPE) is the foundational stage in UW, setting the stage for future resistance activities. During this phase, covert and secure operations are conducted to assess the operational environment, establish networks, and develop resources that will be instrumental in later stages. Sensitive activities are pivotal during OPE as they provide the infrastructure and security needed to prepare for full-scale UW missions.

A-5.  Sensitive activities contribute to OPE through clandestine intelligence gathering, establishing covert communications channels, and conducting reconnaissance. By limiting access through ACCMs and other secure access measures, sensitive activities ensure that only essential personnel are privy to the mission details, minimizing the risk of compromise.

This initial setup phase often involves assessing local attitudes, power dynamics, and potential allies within the region, which helps identify the most promising pathways for supporting or inciting resistance.

A-6.  For example, sensitive activities may involve deploying covert intelligence operatives who are embedded within local populations to gather information on enemy positions, capabilities, and morale. They may also establish secure supply caches, communication networks, and logistical channels that will later support larger resistance operations. Sensitive activities in this phase are designed to prepare the ground covertly, creating a secure environment that enables UW operatives to operate efficiently and effectively when the mission moves into more active phases.

## INTELLIGENCE COLLECTION AND ANALYSIS

A-7   Intelligence is the cornerstone of any successful UW operation, as it informs strategy, decision-making, and tactical planning. In denied or hostile areas, traditional intelligence-gathering methods may not be feasible or safe, making sensitive activities indispensable for obtaining reliable intelligence on the enemy and the environment. Sensitive activities use a range of clandestine and covert methods to gather, analyze, and secure intelligence without compromising operational security.

A-8   Through covert intelligence collection, sensitive activities provide critical insights into enemy capabilities, supply lines, infrastructure, and weaknesses. This intelligence is essential not only for identifying high-value targets but also for understanding the social, political, and cultural context in which the UW mission will operate. Sensitive activities help gather intelligence through human intelligence (HUMINT) sources embedded in the local population, signals intelligence (SIGINT) from intercepted communications, and other classified methods tailored to the mission's specific needs.

A-9   Intelligence collected via sensitive activities informs not only the strategic decisions made at higher levels but also the tactical choices made by field operatives. For instance, if intelligence indicates a strong enemy presence in a particular area, UW operatives can adjust their routes or alter their logistical plans to avoid detection. Additionally, intelligence gathered through sensitive activities often includes details on local sentiments and support levels, enabling UW operatives to tailor their outreach and influence efforts to resonate with the local population.

## LOGISTICAL SUPPORT IN DENIED AREAS

A-10   Logistical support in denied areas is a complex and high-risk endeavor, requiring careful planning and secure execution. Sensitive activities provide the means to establish and maintain covert supply chains that are essential for sustaining resistance efforts in areas where conventional military logistics are unavailable or impractical. These covert logistical networks ensure that personnel, weapons, medical supplies, and communication equipment reach their destinations without alerting enemy forces.

A-11   Sensitive activities manage logistics through a combination of covert transport routes, hidden cache sites, and decentralized stockpiles. These logistical operations often rely on local support networks, trusted intermediaries, and commercial cover to disguise the movement of supplies. For instance, sensitive activities might use civilian vehicles, rural transportation routes, or even livestock-based transport in rural environments to transport equipment under the radar of enemy patrols.

A-12  One of the key benefits of sensitive activities in this context is their ability to establish redundant logistical networks, meaning that if one supply route is compromised, others remain available. This redundancy is critical in high-risk UW environments where adaptability and flexibility are paramount. By building logistical depth through sensitive activities, UW operatives can sustain long-term operations without relying on conventional military support, thus enhancing the autonomy and resilience of the resistance movement.

## TRAINING AND ADVISING INDIGENOUS FORCES

A-13  Training and advising indigenous resistance forces is a cornerstone of UW. Through discreet training programs, sensitive activities build the operational capacity of these forces while maintaining deniability for external sponsors. This support ranges from basic military training to specialized tactics, sabotage, intelligence gathering, and operational planning.

A-14 Sensitive activities use covert training locations, controlled access measures, and carefully selected trainers to ensure that the presence of external advisors remains concealed. These training sessions empower resistance fighters to operate independently, using guerrilla tactics and unconventional strategies that are tailored to their specific environment. By integrating covert advisory support, sensitive activities help indigenous forces become self-sufficient, enabling them to continue the resistance even in the absence of direct external support.

A-15  In some cases, sensitive activities may provide advanced training in areas such as psychological operations, counterintelligence, or clandestine communication techniques. These skills are crucial for indigenous forces operating in contested areas where both physical and informational battles are fought. By equipping resistance forces with these capabilities, sensitive activities enhance the movement's ability to carry out sophisticated operations without relying on continuous oversight from external entities.

## PSYCHOLOGICAL OPERATIONS (PSYOP) AND INFLUENCE CAMPAIGNS

A-16  Psychological operations (PSYOP) play a pivotal role in UW, shaping the perceptions, beliefs, and behaviors of key audiences, including the local population, enemy forces, and neutral observers. Sensitive activities support PSYOP by developing covert influence campaigns designed to undermine enemy morale, foster support for the resistance, and promote narratives that align with the UW mission's objectives.

A-17  Through secure communication channels and discreet influence networks, sensitive activities enable operatives to disseminate information that appears organic to the local environment. This can involve using local influencers, creating anonymous or pseudonymous messaging channels, or staging events that demonstrate the strength and resolve of the resistance. By keeping the source of these messages hidden, sensitive activities help protect the resistance movement's credibility, allowing it to build genuine local support.

A-18  One example of PSYOP supported by sensitive activities might be the dissemination of pamphlets, social media posts, or radio broadcasts that highlight enemy atrocities, local grievances, or the resilience of the resistance. These messages are tailored to resonate with local sentiments and create psychological pressure on enemy forces, ultimately making them more susceptible to defection or demoralization. Sensitive activities ensure that these operations remain covert, preventing the enemy from identifying and countering the influence campaign.

## COUNTERINTELLIGENCE MEASURES

A-19    Counterintelligence is essential for protecting UW operations from espionage, infiltration, and sabotage by enemy intelligence services. Sensitive activities implement rigorous counterintelligence measures to safeguard personnel, plans, and assets from hostile interference. These measures include vetting all participants, securing communication channels, and conducting regular security assessments to identify and mitigate potential risks.

A-20  Sensitive activities are responsible for establishing secure counterintelligence protocols that restrict access to mission-critical information, ensuring that only trusted individuals have access to operational details. This compartmentalization is particularly important in denied or contested areas, where the risk of infiltration is high. Counterintelligence measures may also involve conducting surveillance on key locations, monitoring for unusual activity, and employing deception tactics to confuse and mislead enemy intelligence operatives.

A-21    Through these protective measures, sensitive activities enhance the resilience and security of UW missions, enabling operatives to conduct operations with reduced risk of compromise. By establishing and maintaining a robust counterintelligence posture, sensitive activities safeguard both the operational integrity of UW missions and the safety of personnel involved.

# COORDINATION OF SENSITIVE ACTIVITIES IN UW

A-22  Sensitive activities in UW require seamless coordination across multiple levels and entities to ensure that all operations are aligned with overarching strategic objectives. Effective coordination in this context entails interagency collaboration, stringent operational security, adherence to legal and policy guidelines, clear command structures, and comprehensive risk management.

A-23  **Interagency Collaboration:** Sensitive activities often involve working closely with other federal agencies, such as intelligence and law enforcement bodies, to access specialized capabilities and intelligence resources. This collaboration enhances the effectiveness of UW by leveraging each agency's unique strengths, whether in HUMINT collection, cyber capabilities, or strategic analysis.

A-24  **Operational Security (OPSEC):** Sensitive activities prioritize operational security through strict access controls, compartmentalization of information, and encrypted communication channels. OPSEC measures protect the integrity of sensitive operations, prevent unauthorized disclosure, and shield personnel from hostile surveillance.

A-25    **Legal Compliance and Oversight:** Sensitive activities in UW must adhere to domestic laws, international standards, and established policies to ensure the legitimacy and accountability of operations. Compliance with legal frameworks is essential to avoid unintended consequences, prevent diplomatic fallout, and uphold ethical standards in covert operations. Sensitive activities operate within authorized parameters, ensuring that all actions align with U.S. and international law, especially when collaborating with other federal entities or engaging with foreign allies. Legal oversight ensures that sensitive activities contribute to UW in a way that reinforces rather than jeopardizes the overarching mission.

A-26 **Command and Control**. Clear command structures and well-defined communication channels are critical for coordinating sensitive activities with broader UW objectives. Sensitive activities must integrate smoothly with conventional military operations and other aspects of UW to ensure that all efforts work toward unified strategic outcomes. This coordination often requires establishing dedicated command posts or secure communication networks to streamline decision-making and reduce delays in sensitive, time-critical operations.

A-27 **Risk Management**: Sensitive activities involve inherent risks, including potential exposure, compromise of assets, or mission failure. Rigorous risk assessment and mitigation procedures are employed to address these challenges, with contingency plans and alternative courses of action prepared in advance. Sensitive activities frequently use layered security measures, allowing operatives to respond flexibly to shifting threats and environmental conditions. Effective risk management protects personnel and assets while enabling the continuation of UW operations, even under high-stakes circumstances.

A-28 By addressing these essential elements, sensitive activities support UW in a cohesive, legally sound, and highly secure manner, making it possible to achieve strategic goals without compromising the safety of personnel or the integrity of the mission.

# CHALLENGES AND FUTURE CONSIDERATIONS FOR SENSITIVE ACTIVITIES IN UW

A-29 Although sensitive activities are indispensable in supporting UW, they come with their own set of challenges and complexities. Managing these challenges is essential to ensure the long-term effectiveness and adaptability of UW operations.

## BALANCING SECRECY AND OPERATIONAL TRANSPARENCY

A-30 One of the central challenges of sensitive activities in UW is the need to maintain strict secrecy while also ensuring that stakeholders have sufficient insight to make informed decisions. Excessive secrecy can limit information sharing, leading to coordination issues and reduced situational awareness among operatives. Conversely, too much transparency risks exposing sensitive operations to hostile forces. Sensitive activities must strike a careful balance, leveraging compartmentalization and need-to-know principles to protect essential information while providing operational leaders with the intelligence and resources necessary to make strategic choices.

## RESOURCE ALLOCATION AND LOGISTICAL CONSTRAINTS

A-31 Sensitive activities often require specialized resources, from high-security communication devices to covert transportation networks, all of which must be managed within logistical constraints. The allocation of these resources is particularly challenging in denied areas where resupply may be infrequent or impossible. Budgetary limitations, environmental factors, and unexpected disruptions may all impact the availability of crucial resources. Effective planning and prioritization are essential to ensure that sensitive activities have access to the tools they need to support UW, even in resource-constrained environments.

## ETHICAL AND LEGAL IMPLICATIONS

A-32  Sensitive activities operate in a complex ethical and legal landscape, where covert operations can raise questions about legality, transparency, and accountability. Issues such as the use of clandestine methods, working with foreign allies, and operating in denied areas can present legal and ethical dilemmas. Ensuring that sensitive activities are conducted within the bounds of the law is vital for protecting the legitimacy of UW missions and avoiding unintended diplomatic repercussions. Ethical considerations also come into play when dealing with local populations, whose lives may be impacted by these operations.

## INTERAGENCY COORDINATION AND COMMUNICATION

A-33  UW operations involving sensitive activities require seamless interagency coordination to pool resources, align strategies, and maintain unified objectives. However, differences in agency priorities, procedures, and security protocols can complicate this coordination. Sensitive activities require dedicated communication channels and clearly defined roles to facilitate smooth interagency cooperation. When multiple agencies are involved, sensitive activities benefit from a central coordinating authority to ensure that each component supports the mission in a cohesive manner.

## ADAPTING TO EMERGING THREATS AND TECHNOLOGICAL CHANGES

A-34  Sensitive activities must remain adaptable in the face of emerging threats and rapidly evolving technology. Cybersecurity, for instance, has become a major concern as hostile actors increasingly employ cyberattacks and digital espionage to undermine UW operations. Sensitive activities must evolve to address cyber vulnerabilities, developing resilient communication systems and leveraging cybersecurity measures to protect classified data. Advances in artificial intelligence, surveillance technologies, and digital misinformation campaigns also require sensitive activities to continually adapt their tactics and strategies to stay ahead of adversaries.

## MAINTAINING OPERATIONAL SECURITY IN HOSTILE ENVIRONMENTS

A-35  In UW, maintaining operational security is crucial for preventing adversaries from gaining insight into sensitive activities. Hostile environments, however, present unique challenges to security. Environmental factors, local alliances, and shifting political landscapes can expose sensitive operations to unexpected risks. Sensitive activities require robust security protocols, flexible contingency plans, and adaptable threat responses to maintain operational security under challenging conditions.

# FUTURE DIRECTIONS FOR SENSITIVE ACTIVITIES IN UW

A-36 As unconventional warfare continues to evolve, sensitive activities will need to expand and adapt to address new threats, enhance existing capabilities, and align with future UW strategies. Anticipated future directions include the integration of advanced technologies, a focus on hybrid warfare, and increased collaboration with emerging partners in both governmental and non-governmental sectors.

## INTEGRATION OF ARTIFICIAL INTELLIGENCE AND AUTONOMOUS SYSTEMS

A-37 With advancements in artificial intelligence (AI) and machine learning, sensitive activities are likely to incorporate AI-driven solutions to enhance intelligence analysis, automate repetitive tasks, and improve decision-making in real time. AI can help in pattern recognition, behavioral prediction, and large-scale data analysis, providing operatives with deeper insights into enemy strategies and local conditions. Autonomous systems, such as drones, can also support sensitive activities by conducting reconnaissance, monitoring logistics routes, and even delivering supplies covertly. The integration of these technologies will allow sensitive activities to increase efficiency and operational effectiveness while reducing risk to personnel.

## FOCUS ON CYBERSECURITY AND COUNTER-CYBER OPERATIONS

A-38 As digital threats grow, sensitive activities will place a greater emphasis on cybersecurity, ensuring that communication networks, data storage, and intelligence-sharing platforms remain secure from cyber infiltration. Future sensitive activities will likely develop dedicated counter-cyber units to monitor and respond to digital threats. This focus on cybersecurity will not only protect sensitive information but also provide operatives with tools to disrupt enemy networks, control misinformation, and defend resistance groups against cyberattacks.

## STRENGTHENING PARTNERSHIPS WITH ALLIED AND LOCAL FORCES

A-39 The future of sensitive activities will likely involve deeper partnerships with allied countries and local non-governmental organizations to enhance UW efforts. By expanding relationships with trusted allies and leveraging local resources, sensitive activities can extend their reach, acquire specialized knowledge, and gain additional support without expanding their physical footprint. Local partnerships offer unique insights into cultural dynamics, while international allies may provide additional technological and logistical resources that are otherwise difficult to access in denied areas.

# CONCLUSION

A-40   Sensitive activities represent a fundamental enabler of unconventional warfare, providing secure, covert, and flexible support that enhances the resilience, effectiveness, and security of UW missions. Through intelligence gathering, logistical management, counterintelligence, psychological operations, and covert training, sensitive activities shield UW operations from exposure and amplify their impact within hostile environments. By coordinating across multiple agencies, adhering to strict legal and ethical standards, and adapting to emerging threats, sensitive activities ensure that unconventional warfare missions remain viable and effective in today's complex security landscape.

A-41 As unconventional warfare continues to adapt to global shifts in power, technology, and resources, sensitive activities will evolve alongside it. Future sensitive activities will integrate advanced technologies, address new challenges in cybersecurity, operate in extremely hostile areas, and collaborate more deeply with local and international partners. These adaptations will help sensitive activities meet the demands of future conflicts, maintaining the secrecy, flexibility, and impact that make them indispensable to unconventional warfare operations.

A-42 Through the ongoing refinement and expansion of sensitive activities, the U.S. military can ensure that its unconventional warfare efforts remain strategically aligned, ethically sound, and operationally secure. Sensitive activities provide the covert foundation on which effective resistance movements and insurgencies can build, enabling unconventional warfare to support national security objectives in even the most challenging environments.

## Appendix B

# (U) Unaccompanied Resistance Considerations for Unconventional Warfare (S//NF)

## INTRODUCTION

B-1  Unaccompanied resistance in Unconventional Warfare (UW) describes resistance movements that operate independently, relying on indigenous leadership, resources, and strategies without direct support or active presence from external military advisors or forces. These movements are often grassroots efforts, born out of the local population's aspirations, grievances, and determination to resist occupying powers or oppressive regimes. In UW contexts, unaccompanied resistance is critical as it highlights a self-sustained, decentralized approach that maintains operational security, reduces external footprint, and enhances the legitimacy of the resistance among local populations.

B-2  For effective UW, understanding unaccompanied resistance requires exploring the operational challenges and structural needs that enable these movements to succeed autonomously. This appendix outlines key considerations for unaccompanied resistance, from leadership and operational security to resource acquisition and adaptability, providing a comprehensive look at the unique characteristics and strategic imperatives of autonomous resistance operations.

## AUTONOMOUS ORGANIZATION AND LEADERSHIP

B-3  At the heart of unaccompanied resistance lies the necessity for strong, self-reliant leadership structures and organizational frameworks. Without external oversight or continuous support, these movements must be able to organize, sustain, and adapt independently. Autonomous organization emphasizes leadership rooted in local credibility, cultural alignment, and the ability to inspire and mobilize resistance fighters under a shared cause.

### DECENTRALIZED LEADERSHIP

B-4  In unaccompanied resistance, leadership often adopts a decentralized approach, empowering small, independent cells or units that can operate autonomously without constant coordination. This setup allows the movement to maintain resilience, as the loss or capture of a single cell does not compromise the entire resistance.

### GRASSROOTS CREDIBILITY AND ALIGNMENT WITH LOCAL ASPIRATIONS

B-5  Leaders within unaccompanied resistance movements typically emerge from the local population, embodying the community's values, grievances, and aspirations. This alignment ensures that the movement's objectives resonate with the broader population, fostering a sense of shared purpose and mutual support.

## STRATEGIC DECISION-MAKING

B-6 Autonomous leaders must be capable of making strategic decisions that align with the movement's long-term objectives. This involves assessing risks, prioritizing operations, and making resource allocations with limited input or resources from external actors. Leadership in unaccompanied resistance requires a mix of tactical acumen, adaptability, and the ability to balance immediate needs with overarching goals.

# OPERATIONAL SECURITY AND SECRECY

B-7 In the absence of direct external support, unaccompanied resistance movements must place a high priority on operational security (OPSEC) and secrecy. Maintaining low visibility and avoiding detection are essential for the survival and efficacy of these movements, which operate within environments that are often hostile and heavily monitored by occupying or controlling forces.

## CLANDESTINE OPERATIONS

B-8 Unaccompanied resistance movements frequently operate in clandestine ways, using cover identities, safe houses, and encrypted communication channels to avoid detection. These measures prevent the compromise of personnel and operational plans, which is crucial in high-risk environments.

## CELL-BASED ORGANIZATION AND LEADERLESS RESISTANCE

B-9 Many unaccompanied resistance movements adopt a "leaderless resistance" structure, where small, independently operating cells act without centralized command. This approach reduces the likelihood of infiltration, as the capture of one cell does not jeopardize others or provide comprehensive intelligence on the broader movement.

## COUNTERINTELLIGENCE MEASURES

B-10 Vigilant counterintelligence practices are crucial for protecting unaccompanied resistance efforts. This includes rigorous vetting of new members, surveillance awareness training, and counter-surveillance techniques to monitor and counteract enemy intelligence efforts.

## OPERATIONAL DISCIPLINE AND INFORMATION COMPARTMENTALIZATION

B-11 Unaccompanied resistance cells compartmentalize information, sharing operational details strictly on a need-to-know basis. This approach limits the risk of a single operative compromising the movement's broader strategy, contributing to the movement's overall resilience.

# RESOURCE ACQUISITION AND SUSTAINABILITY

B-12 Unlike externally supported UW operations, unaccompanied resistance movements must secure resources—such as weapons, food, medical supplies, and intelligence—independently. The sustainability of these movements relies on their ability to acquire, manage, and distribute resources without continuous external assistance.

## SELF-SUFFICIENCY IN SUPPLY CHAINS

B-13 Unaccompanied resistance groups often establish clandestine supply networks that leverage local support. They may utilize underground economies, black markets, or sympathetic local suppliers who provide the necessary resources discreetly.

## IMPROVISATION AND RESOURCEFULNESS

B-14 Autonomous resistance movements frequently resort to improvisation, repurposing civilian materials for military use, creating makeshift weapons, or using alternative means to secure essential supplies. This ingenuity is crucial in denied areas where access to conventional military resources is limited.

## DEPENDENCE ON LOCAL SUPPORT NETWORKS

B-15 Local sympathizers and communities play a critical role in providing logistical support. By embedding within the population, resistance groups can tap into local knowledge, shelter, and supplies, reinforcing their operational sustainability without revealing their dependence on community resources.

## FINANCIAL INDEPENDENCE AND ALTERNATIVE FUNDING

B-16 To sustain operations, unaccompanied resistance groups often engage in alternative funding methods, including fundraising from diaspora communities, illicit trade, or revenue-generating activities that are difficult to trace. Financial independence reduces vulnerability to external influence and ensures the movement's long-term viability.

# TRAINING AND CAPACITY BUILDING

B-17 In the absence of external trainers, unaccompanied resistance movements must develop internal training programs to enhance their operational effectiveness. This self-contained approach to capacity building involves training operatives in guerrilla tactics, intelligence gathering, and sabotage techniques, fostering a high level of self-sufficiency.

## INTERNAL TRAINING PROGRAMS

B-18 Resistance cells create internal training regimens for new recruits, covering skills essential for clandestine and guerrilla operations. Training focuses on low-cost, high-impact tactics suited to limited-resource environments, ensuring that operatives are equipped to operate effectively with minimal equipment.

## FOCUS ON TACTICAL FLEXIBILITY

B-19  The training emphasizes flexibility and adaptability, preparing fighters to operate in dynamic environments. Resistance fighters learn to adjust their tactics in response to changing conditions, ensuring they remain unpredictable to enemy forces.

## LOCAL KNOWLEDGE AND CULTURAL FAMILIARITY

B-20  Training in unaccompanied resistance emphasizes the importance of local knowledge, cultural awareness, and language skills. Familiarity with the cultural and social nuances of the operating area helps operatives blend in, gather intelligence, and interact with civilians in a way that avoids arousing suspicion.

## CROSS-TRAINING FOR MULTIPLE ROLES

B-21  Given the limited personnel and resources, individuals within resistance groups are often cross-trained to fulfill multiple roles, from intelligence gathering to combat operations. This versatility enhances operational resilience and enables the group to function even with minimal manpower.

# COMMUNICATION AND COORDINATION

B-22  Effective communication within and between resistance cells is critical for maintaining cohesion and synchronizing operations. Unaccompanied resistance groups must implement secure, covert communication methods that facilitate coordination without compromising security or exposing leadership structures.

## SECURE AND COVERT COMMUNICATION CHANNELS

B-23  Unaccompanied resistance groups often use encrypted, low-profile communication methods, such as dead drops, encrypted digital messages, and face-to-face meetings in safe environments. These channels allow for secure communication without drawing attention to the group's movements.

## OPERATIONAL CODES AND SIGNALS

B-24  To reduce the risk of interception, resistance groups rely on coded language, pre-arranged signals, and simplified communication protocols. This minimizes the need for lengthy transmissions, lowering the likelihood of detection by enemy forces.

## REDUNDANCY AND COMMUNICATION RESILIENCE

B-25  Resistance movements implement redundant communication methods, allowing for backup systems in case one channel is compromised. For example, if digital communications are disrupted, operatives may switch to physical message relay systems or alternate drop locations to ensure continuity.

## COORDINATION ACROSS DECENTRALIZED UNITS

B-26 Although cells operate independently, unaccompanied resistance groups may employ centralized communication hubs or regional coordinators who act as intermediaries. These individuals relay critical information while maintaining compartmentalization, ensuring that the overall movement remains coordinated.

# CULTURAL AND SOCIAL INTEGRATION

B-27 Successful unaccompanied resistance movements are deeply embedded within the cultural and social fabric of their local environment. This integration ensures that the movement is supported and protected by the local population, providing a steady recruitment pool, operational security, and social legitimacy.

## ALIGNMENT WITH LOCAL GRIEVANCES

B-28 Effective resistance movements resonate with local grievances and broader social issues, positioning themselves as champions of the community's needs and rights. By aligning their objectives with those of the local population, resistance groups build social credibility and support.

## USE OF CULTURAL SYMBOLS AND NARRATIVES

B-29 Resistance groups often employ cultural symbols, narratives, and historical references that resonate with the population, reinforcing the movement's legitimacy. This strategy fosters a sense of shared identity and helps attract recruits who identify with the movement's goals.

## BUILDING TRUST AND RAPPORT WITH CIVILIANS

B-30 Unaccompanied resistance groups prioritize building trust and rapport with local civilians, who may provide intelligence, shelter, or logistical support. This relationship minimizes the risk of betrayal and ensures that the population views the movement as a legitimate and protective force rather than an external disruptor.

## CULTURAL SENSITIVITY IN OPERATIONS

B-31 By respecting local customs, traditions, and norms, resistance operatives reduce the risk of alienating civilians. Cultural sensitivity enhances operational security, as locals are less likely to disclose information about the movement's activities to occupying forces.

# ADAPTABILITY AND RESILIENCE

B-32 Operating without external support requires a high degree of adaptability and resilience. Unaccompanied resistance movements must be capable of adjusting their tactics, strategies, and organizational structures in response to dynamic threats, enemy actions, and changing environmental conditions.

## FLUID TACTICS AND STRATEGIES

B-33 Adaptability is key to the survival of unaccompanied resistance. These groups use guerrilla tactics that can be modified or abandoned in response to enemy action allowing them to stay one step ahead of more conventional or heavily resourced adversaries. By remaining flexible in their approach, unaccompanied resistance movements can exploit opportunities or withdraw strategically, adapting quickly to new challenges.

## RESILIENCE THROUGH REDUNDANCY

B-34 To ensure continuity in the face of losses or setbacks, resistance movements often create redundant structures, such as alternative communication networks, backup supply caches, and parallel leadership hierarchies. This redundancy enables the movement to recover swiftly if one element is compromised, ensuring that the loss of a single component does not cripple the entire operation.

## IMPROVISATION IN RESOURCE USE

B-35 Unaccompanied resistance groups are known for their ability to repurpose everyday materials for operational use. Whether constructing makeshift weapons, creating improvised explosive devices, or utilizing local resources for shelter and camouflage, their capacity for improvisation makes them formidable despite limited resources.

## ADAPTIVE LEARNING AND TACTICAL EVOLUTION

B-36 Unaccompanied resistance movements continually analyze their operational environment, learning from both their successes and setbacks. They adapt tactics based on the enemy's responses, evolving their approaches to remain unpredictable. This process of adaptive learning, combined with their decentralized structure, makes these movements resilient and difficult to counter with conventional military strategies.

# PSYCHOLOGICAL OPERATIONS AND INFORMATION WARFARE

B-37 Psychological operations (PSYOP) and information warfare are essential tools for unaccompanied resistance movements seeking to influence public opinion, weaken enemy resolve, and maintain morale within their ranks. These operations allow resistance movements to compete in the informational domain, which can be as important as physical confrontations.

## LEVERAGING LOCAL NARRATIVES AND SYMBOLISM

B-38 Unaccompanied resistance movements often harness powerful cultural narratives and historical grievances that resonate with the local population. By framing their cause as a continuation of longstanding struggles, they enhance their credibility and attract support from individuals who identify with the movement's objectives.

## COUNTERING ENEMY PROPAGANDA

B-39  Resistance groups actively work to counter the enemy's propaganda, which may portray them as criminals or terrorists. They use grassroots communication methods, such as word-of-mouth networks, underground publications, and trusted community figures, to convey their messages and counteract enemy narratives.

## MAINTAINING MORALE AND COHESION AMONG OPERATIVES

B-40  Unaccompanied resistance movements use PSYOP internally to reinforce a sense of purpose, solidarity, and resilience among their members. Regular communication of small successes, symbolic victories, and inspirational stories helps maintain morale even during periods of hardship or setbacks.

## USE OF COVERT INFORMATION CHANNELS

B-41  In many cases, resistance movements will develop covert means of spreading their messages, such as encrypted messaging platforms, underground newspapers, and graffiti. These methods allow them to reach a broad audience without risking direct confrontation with the enemy, creating an "information shadow" that supports resistance efforts while avoiding detection.

# LEGAL AND ETHICAL CONSIDERATIONS

B-42  Operating without external oversight places the responsibility of adhering to legal and ethical standards squarely on the leadership of unaccompanied resistance movements. While their primary focus is often on survival and operational success, adherence to ethical principles can influence their legitimacy and support among the local population.

## AVOIDANCE OF CIVILIAN CASUALTIES

B-43  Unaccompanied resistance movements generally aim to minimize harm to civilians, as collateral damage can erode local support and delegitimize the movement. By prioritizing civilian safety, they can maintain the moral high ground and appeal to the sympathies of the population.

## ADHERENCE TO RULES OF ENGAGEMENT

B-44  While not bound by the same formal rules as state actors, many resistance groups adopt their own codes of conduct, which dictate acceptable behavior during operations. Such rules may prohibit acts of theft, violence against non-combatants, and other actions that could harm the movement's reputation.

# STRATEGIC AUTONOMY AND LONG-TERM SUSTAINABILITY

B-45 For unaccompanied resistance movements, long-term sustainability depends on maintaining strategic autonomy and developing robust internal structures that can endure beyond immediate operational objectives. This autonomy not only reduces vulnerability to external influence but also fosters a sense of ownership and commitment within the movement.

## INDEPENDENT DECISION-MAKING

B-46 The independence of unaccompanied resistance movements means they must make decisions that are fully aligned with their local context and long-term vision. This decision-making process is often collective, involving key leaders or representatives from different regions, which reinforces the legitimacy of the movement.

## BUILDING RESILIENT INSTITUTIONS

B-47 Over time, successful resistance movements may evolve from decentralized cells into more structured organizations, with formalized institutions that handle recruitment, resource allocation, intelligence, and communication. These institutions ensure continuity, creating a foundation for governance and stability in the event of a successful insurgency or political transition.

## EVOLUTION TOWARDS POLITICAL GOALS

B-48 While immediate objectives may focus on military resistance, many unaccompanied resistance movements aim to transition toward political legitimacy. This shift involves developing political wings, engaging in dialogues, and preparing for potential negotiations. Strategic autonomy allows these movements to adapt to political opportunities without compromising their principles or objectives.

## LOCAL AND REGIONAL ALLIANCES

B-49 Sustaining an unaccompanied resistance effort often requires establishing alliances with local or regional actors who share similar goals. These alliances provide mutual support, create opportunities for resource sharing, and enhance the movement's ability to withstand external pressure. However, maintaining these alliances requires balancing shared interests with the movement's core values and mission.

# CONCLUSION

B-50 Unaccompanied resistance movements represent a distinct and resilient form of unconventional warfare, characterized by autonomy, adaptability, and a deep connection to local populations. Operating without external military advisors or logistical support, these movements rely on indigenous leadership, internal resourcefulness, and the commitment of their members to navigate complex and often hostile environments. While the lack of direct support can present challenges in terms of resources, training, and intelligence, it also affords these movements a level of strategic autonomy that allows them to adapt swiftly and align closely with the aspirations of the communities they serve.

B-51    By understanding the unique considerations associated with unaccompanied resistance, unconventional warfare strategists can better appreciate the strengths, limitations, and complexities of autonomous resistance operations. These insights are crucial for assessing the feasibility of unaccompanied resistance efforts, whether as part of a larger UW campaign or as standalone movements seeking to enact change within their societies.

B-52  In modern conflict environments, unaccompanied resistance continues to demonstrate the power of grassroots movements in achieving strategic objectives, often by leveraging the very constraints that initially seem to limit them. Through decentralized leadership, covert operations, and an unwavering commitment to their cause, unaccompanied resistance groups illustrate how self-sustained, locally driven efforts can challenge and resist powerful adversaries, ultimately reshaping the political and social landscapes in which they operate.

This page intentionally left blank.

# Appendix C

# Intelligence Support to Unconventional Warfare

This appendix gathers various frameworks for population and resistance analysis from a range of intelligence and special operations doctrine sources. It is intended to provide several similar tools that provide slightly different intelligence analysis perspectives on the role of human activity considerations for UW.

## UNCONVENTIONAL WARFARE CAMPAIGN ANALYSIS

C-1. This framework is based on the widely used PMESII-PT systems analysis approach. It is useful for analysis of an entire target country's system of systems.

*Note:* ATP 3-05.1 (Appendix F, Unconventional Warfare Campaign Analysis) provides additional information.

### POLITICAL SYSTEM POINTS OF ANALYSIS

C-2. The political system points of analysis include the following:
- Leadership (national, regional, local, and core).
- Local worker's parties.
- Secret police.
- Informants.
- Security apparatus.
- Detention camps.
- Alliances and external support.
- Regime control of national resource systems.
- Domestic image of omnipotence, omnipresence, and infallibility.

C-3. Political analysis of a foreign country begins with an assessment of the basic principles of government, governmental operations, foreign policy, political parties, pressure groups, electoral procedures, subversive movements, as well as criminal and terrorist organizations. It then analyzes the distribution of political power—whether it is a democracy, an oligarchy, a dictatorship, or has political power devolved to multiple interest groups, such as tribes, clans, or gangs. Analysis must focus on determining how the political system really operates, not the way it is supposed to operate

### BASIC GOVERNMENTAL PRINCIPLES

C-4. The starting point of political analysis is the formal political structure and procedure of a foreign nation. Analysts must evaluate—
- Constitutional and legal systems.
- Legal position of the legislative, judicial, and executive branches.
- Civil and religious rights of the people.
- People's national devotion to constitutional and legal procedures.

### GOVERNMENTAL OPERATIONS

C-5. Governments are evaluated to determine their efficiency, integrity, and stability. Information about how the government actually operates or changes its method of operation gives the intelligence user clues about the probable future of a political system. When assessing governmental operations, analysts should consider the following:

- Marked inefficiency and corruption, which differs from past patterns, may indicate an impending change in government.
- Continued inefficiency and corruption may indicate popular apathy or a populace unable to effect change.
- Increased restrictions on the electoral process and on the basic social and political rights of the people may mean the government is growing less sure of its position and survivability.

## FOREIGN POLICY

C-6. Analysis of a target country's foreign policy addresses the country's public and private stance toward the United States, foreign policy goals and objectives, regional role, and alliances. Analysts gather foreign policy data from various sources, to include:

- Diplomatic and military personnel.
- Technical collection systems.
- Official foreign government statements.
- Press releases.
- Public opinion polls.
- International businessmen.
- Academic analyses.

## POLITICAL PARTIES

C-7. Analysts study special interest parties and groups (for example, labor, religious, ethnic, and industry) to evaluate their—

- Aims.
- Programs.
- Degree of popular support.
- Financial backing.
- Leadership.
- Electoral procedures.

## PRESSURE GROUPS

C-8. With few exceptions, most states have some type of formal or informal pressure groups. Examples include political parties, associations, religious or ethnic organizations, labor unions, and even illegal organizations (for example, a banned political party). The analyst must identify these pressure groups and their aims, methods, relative power, sources of support, and leadership. Pressure groups may have international connections and, in some cases, may be almost entirely controlled from outside the country.

## ELECTORAL PROCEDURES

C-9. Elections range from staged shows of limited intelligence significance to a means of peaceful, organized, and scheduled revolution. In addition to the parties, personalities, and policies, the intelligence analyst must consider the circumstances surrounding the actual balloting process and changes from the historical norm.

## SUBVERSIVE MOVEMENTS

C-10. In many countries there are clandestine organizations or guerrilla groups whose intention is to overthrow or destroy the existing government. When analysts report on subversive movements, they should address—

- Organizational size.
- Character of membership.
- Power base within the society.
- Doctrine or beliefs system.
- Affiliated organizations.
- Key figures.
- Funding.
- Methods of operation.

## CRIMINAL AND TERRORIST ORGANIZATIONS

C-11. Criminal organizations in some countries are so powerful that they influence or dominate national governments. Analysts must examine the organization's influence or forceful methods of control. Most terrorist organizations are small, or short-lived, and not attached to any government. Analysts should determine if external factors or even the area's government assists the terrorist group.

## POLITICAL SYSTEM QUESTIONS

C-12. The following political system questions pertain to the national political structure:

- What is the type of governmental system in place?
  - Where does it draw its legitimacy from?
  - Are the sectors stable or in transition?
  - Does the electoral process affect them?
  - Where do they draw their power?
  - What is the source of their knowledge and intellectual income?
  - Who are the leaders?
  - Where do they draw their power from?
  - Does a core bureaucracy staff them?
  - Are they associated with governmental departments or agencies?
  - Who are the key leaders?
  - How are they linked within the power network?
- Are the governmental departments or agencies stable or in transition?
  - Are new departments or agencies being created?
  - If so, what is the cause of this transition (societal, cultural, educational, technical, and economic)?
  - What is the source of its workforce?
  - Who are the leaders?
  - Is it staffed by a core bureaucracy?
  - What skill level?
  - Interagency and departmental dependencies?
  - External dependencies (societal, cultural, and educational)?

C-13. The following questions pertain to the national political demographics structure:

- Ethnic and religious groups having political power:
  - Are these groups regionalized?
  - How do they exercise political power?

- What is their legislative representation?
- Is there a paramilitary structure?
- How do these ethnic and religious groups wield power within urban society?
- How do these ethnic and religious groups wield power within rural society?
- Political parties:
  - What are the political parties?
  - Are they externally or internally supported?
  - Are they associated with ethnic, religious, or cultural groups?
  - Who are their leaders? Who are their allies?
  - What is their political opposition? Who are their allies?
- Political action groups:
  - Where do they draw their power (societal, cultural, technical, economic)?
  - Where do they draw their intellectual capital?
  - What is the source of their leadership?
  - What is their knowledge?
  - What are their external elements?
  - What are the expatriate communities?
  - What is their relationship with the government?

C-14. The following questions pertain to the regional political relationships:
- Regional—nonadversarial and adversarial?
- How are relations maintained—through economics, religion, culture, ideology, or common needs?
- International—Nonadversarial and adversarial?
- How are relations maintained—through economics, religion, culture, ideology, or common needs?
- Potential allies during a conflict: national resolve to engage in conflict?
- Military resolve to engage in politically motivated action?

C-15. Other considerations include the following:
- Public confidence in government and in society.
- Factionalism or regionalism within the governmental structure.
- Challenges faced by the government.
- Political effects caused by organized groups.
- Government political response to group pressures.
- Political effects upon internal and external security—relates to military.
- Government response to diplomatic overtures.
- National economic goals affecting the political structure.
- Police mechanisms.

## MILITARY SYSTEM POINTS OF ANALYSIS

C-16. The military system points of analysis include:
- Leadership.
- Intelligence, to include—
  - Signals intelligence.
  - HUMINT.
- Logistics.
- Command and control.
- Electronic warfare.
- Civil defense.
- Underground facilities.

- Power ventilation access.
- Mobilization.
- Training.
- Stockpiles.
- Army—
  - Artillery.
  - Long-range missile systems.
  - Infantry.
  - Armor.
  - Engineers.
  - Mobility.
  - Mine-clearing command and control.
  - Bridging.
  - Countermobility.
  - Obstacles.
  - Survivability.
- Navy—
  - Surface capabilities.
  - Subsurface (submarine).
  - Remote control vehicles.
  - Mine-laying submarines.
  - SOF platforms.
  - Patrol fleet anti-ship missiles.
  - Coastal defenses.
  - Radar capabilities.
- Air Force—
  - Air-to-ground capabilities.
  - Fixed-wing platforms.
  - Rotary-wing platforms.
  - Air-defense capabilities.
  - Radar and integrated air defense system.
  - Precision munitions capabilities.
  - Bases (runways, refuel capabilities, ramp space).
- Communications: missiles (theater, ballistic, or space).
- Weapons of mass destruction (research, production, storage, and delivery).
- Industrial or technical base (for production and repair of advanced equipment).

## ANALYSIS OF THE ADVERSARY'S MILITARY

C-17. The analysis of the adversary's military will focus on its leadership, capabilities, dispositions, and morale or commitment to its government, to include:

- Key military leadership, including their training and previous experience in senior leadership.
- Installations and facilities of a military significance (both primary and secondary purposes).
- Infrastructure in place to support identified installations and force structure.
- Military units, including personnel and chain of command.
- Assigned equipment.
- Current and projected weapons system capabilities.

## MILITARY SYSTEM QUESTIONS

C-18. The following military system questions pertain to the military environment:

- Will the national leadership use military means to achieve objectives?
- Does the leadership intend to forge or enhance military ties with another state that poses a threat to regional security or U.S. interests?
- Does the leadership intend to enhance national military capabilities in a way that could be regionally destabilizing?
- Are the national leader's political goals a cause for concern?
- Key leadership—residence, office, wartime command post, telephone, email, political patronage, religious affiliations, ethnic affiliations, personal assets, nonmilitary activities, influences?
- Soldiers—ethnic or religious composition by region of regular forces and elite forces, pay, training, morale, benefits, gripes or issues?
- Capabilities, to include:
    - Equipment imports—what, from whom, where based, points of entry?
    - Support (spare parts, maintenance, and operational training)?
    - Indigenous production and assembly?
    - Raw materials and natural resources?
    - Supply—production, movement, and storage?
    - Days of supply on-hand of key supplies (ammunition and petroleum, oils, and lubricants)?
- Transportation—road capacity, primary lines of communication, and organic transportation assets, to include:
    - Rail (same as roads)?
    - Water (inland or intracoastal)?
    - Bridges (classification, construction materials, length, bypass)?
    - Tunnels (height or width restrictions, bypass)?
- Organizations, to include:
    - Garrison locations, brigade or larger combat, battalion or larger logistics and sustainment?
    - Naval port facilities and home stations?
- Airfields, to include:
    - Fixed fields?
    - Home station?
    - Associated dispersal and highway strips?
    - Number and type aircraft at base?
- Reconnaissance and surveillance, to include:
    - Assets and capabilities by echelon?
    - National-level or controlled assets?
    - Associated ground stations and downlinks?
    - Centralized processing and dissemination facilities?
    - Center of excellence and HQ for each intelligence discipline?
    - Commercial sources for imagery, dissemination capability, mapping, other?
- Military communications, to include:
    - Fixed facilities?
    - Mobile capabilities?
    - Relay and retransmission sites?
    - Commercial access?
- Integrated air defense, to include:
    - Early warning?
    - Target acquisition and tracking and guidance?

- Fixed launch sites?
- Mobile air defense assets?
- Centralized command and control?
- Airfields associated with counterair assets?
- Airborne warning aircraft (for example, Airborne Warning and Control System)?
- Electrical power requirements?
- Theater ballistic missile or coastal defense missiles, to include:
  - Fixed launch sites?
  - Mobile assets?
  - Meteorological stations supporting?
  - Command and control decision makers?
  - Target acquisition?
  - Target guidance and terminal guidance?
  - Power requirements?
- Weapons of mass effects capabilities, to include:
  - Number and type?
  - Production, assembly, storage, delivery means?
  - Imports required—source and mode of transport?
  - Command and control decision maker?
  - Command and control rivalries—personal and interservice?
  - Decision making—dissemination or transmission means (direct or through chain of command)?
- Special capabilities, to include:
  - SOF?
  - Weapons of mass destruction?
  - Theater ballistic missile?
  - HUMINT?
  - Submarines?
- Military situation, to include:
  - Under what conditions does the military execute its missions?
  - Is there internal conflict within the military that could destabilize this country?
  - Are there emerging or increasing rivalries or factionalism within the military?
  - Are there emerging or increasing power struggles within the military?
  - Is there deteriorating morale or increasing dissension within the ranks or in the officer corps?
- Civil-military relations, to include:
  - How loyal is the military to the current regime?
  - Are there cultural or religious factors that might cause frictions and dissension?
  - Are there changes or developments in civil military relations that could destabilize the country?
- Government-military relations, to include:
  - Will the senior military leadership support and defend the government against internal resistance and insurgency?
  - What factors might cause a loss of confidence or support?
  - What factors might cause a military coup to occur?

- Civil-military conflict:
  - Is there increasing conflict between the civilian and military leaders?
  - Is there a difference in views between junior and senior leaders toward service to the government? To the peoples or constitution?
  - Is there increasing civil-military conflict over constitutional or legal matters?
- Sociomilitary conflict, to include:
  - Are there growing tensions or conflicts in sociomilitary relations that could destabilize the country?
  - Is the military assuming a new internal security role or increasing its involvement in internal security affairs?
- Military activities, to include:
  - Are military operations or activities having an increasingly adverse impact on society?
  - Is the military involved in criminal activity that is contributing to increased tensions or conflict between the military and the public?
- External military threat, to include:
  - Is an external military threat emerging or increasing?
  - Is an adversary engaging in or increasing limited or covert military action?
  - Is an adversary preparing to engage in conventional military action against this country?
  - Is an adversary trying to acquire or in the process of deploying weapons of mass destruction or advanced weapons?
- Operational status or capability, to include:
  - Are there changes or developments in the military's operational status or capabilities that suggest pending military action?
  - Is there unusual change or a sudden increase in activity levels or patterns?
  - Are there changes or developments in personnel status?
  - Are there significant changes or developments in force capabilities?

## ECONOMIC SYSTEM POINTS OF ANALYSIS

C-19. The economic system points of analysis include the following:
- Industry.
- Financial.
- Distribution humanitarian aid.
- Currency.
- Arms exports.
- Corruption or linkages.
- Black market agriculture.
- Drug crops and trafficking.
- Mining.
- Natural resource areas or production.
- Foreign investment.
- Trade linkages.

## ANALYSIS OF THE ADVERSARY'S ECONOMY

C-20. Analysis focuses on all aspects of the adversary's economy that have the potential for exploitation. Among these are industrial production, agriculture, services, and armament production. Concentration will be on those elements of the economy that are factors in foreign trade and factors on the internal economy—that can have an impact on the political decision-making process and popular support for the government. Both the official and underground (black market) economies must be examined.

C-21. Concentration will be on the adversary and the regional and global countries with which it has its major trade and exchange linkages. Certain specific nations and regional economic alliances could be highly dependent upon adversary exports, and the impact upon these must be considered. The focus will be on critical elements of the trading partners that may be exploited and not their economy as a whole. In the economic system, a great deal of information is available from open sources. The initial task is to develop a baseline of information on the adversary's economy, such as gross domestic product, growth rates, unemployment rates, money supply, economic plans, inflation, and national debt. Analysis may include sources of national wealth:

- Natural resources.
- Products (agriculture and manufacturing).
- Foreign aid.
- Foreign trade.
- Import or export.
- Trading partners.
- Domestic consumption.
- Management of the economy.
- Government role.
- Private sector role.
- Corruption.
- Slush funds, leaders' bank accounts.
- Counterfeiting.

## ECONOMIC SYSTEM QUESTIONS

C-22. The following questions pertain to the economic system:
- What are the key indicators of the economic health of the countries of interest?
- Which external factors have the most impact upon the economy?
- What areas of the economy are most susceptible to foreign influences and exploitation?
- What is the impact of foreign economic assistance? What would be the impact of its reduction or removal?
- What percentage of the economy should be classified as "black or gray market"?
  - Can the activities in this sector be quantified?
  - Can the activities in this sector be influenced?
- What are the governmental rules on foreign investment? Who do they favor?
- Which nations have the most to gain or lose from damage to, or a collapse of, the economy?
- What are the most likely areas of economic growth?
- Will there be growth in the private sector share of the economy? Who would benefit the most from this change?
- How effective will be steps to diversify the economy?
- What is the inflation rate? To what extent will steps to control inflation be successful?
- Will government subsidies of selected products for domestic use continue? What would be the impact of their reduction or removal?
- What is the anticipated trend in demand for foreign (particularly U.S.) currency?
- What is the prognosis for food production?
  - Are they dependent on imports?
  - Will rationing of essential goods continue?
  - Which items are most likely to be rationed?
- How will demographic factors (for example, birth rate, adult or child ratio, or rural migration to urban areas) affect the economy in the future?
- What is the impact of the drug trade on the overall economy? Regional economies?

- Will imports of military spending or hardware increase?
  - Who are the most likely suppliers?
  - Will these be cash transactions or will a barter system be established?
- What is this nation's standing within the International Monetary Fund and World Bank?
- Is trade with European Union member nations expected to increase? If so, in what specific areas?
- Have any key members of the economic sector leadership been educated in the West or China? If so, have they maintained contacts with their former colleagues?
- Are changes to the current system of state-owned monopolies anticipated? If so, what will be the impact?
- What are the key industries of the state(s)?
- What are the major import or export commodities?
- What is the trade balance? Is this a strength or vulnerability?
- What is the labor situation (for example, unemployment statistics, labor sources, unions, and so on)?
- Who or what are the key government economic leaders or agencies?
- Who are the principal business leaders in the country?

## SOCIAL SYSTEM POINTS OF ANALYSIS

C-23. The social system points of analysis include the following:

- Culture or system.
- Personality.
- History.
- Religion command, control, and communications.
- Family ties or tribal linkages.
- Organized crime.
- Impact of local traditions.
- Families (traditional or influential controlling major decisions).

C-24. Analysis must study the way people, particularly the key leadership and natural leaders, organize their day-to-day living, including the study of groups within society, their composition, organization, purposes and habits, and the role of individuals in society. For intelligence purposes, analysts study seven sociological factors. The detailed list should be viewed as a guide for developing the necessary information to develop the sociological systems summary for the target countries.

## POPULATION

C-25. Intelligence data derived from censuses and sample surveys describe the size, distribution, and characteristics of the population, including rate of change. Most countries now conduct censuses and publish detailed data. Analysts use censuses and surveys to evaluate an area's population in terms of—

- Location.
- Growth rates.
- Age and sex.
- Structure.
- Labor force.
- Military manpower.
- Migration.

## CHARACTERISTICS OF THE PEOPLE

C-26. Analysts study social characteristics to determine people's contribution to national cohesion or national disintegration. Social characteristics evaluated by analysts include:

- Social stratification.
- Number and distribution of languages.
- Prejudices.
- Formal and informal organizations.
- Traditions.
- Taboos.
- Nonpolitical or religious groupings and tribal or clan organization idiosyncrasies.
- Social mobility.

## PUBLIC OPINION

C-27. Key indicators of a society's goals may be found in the attitudes expressed by significant segments of the population on questions of national interest. Opinions may vary from near unanimity to a nearly uniform scattering of opinion over a wide spectrum. Analysts should sample minority opinions, especially of groups capable of pressuring the government.

## EDUCATION

C-28. Analysts concentrate on the general character of education and on the quality of elementary through graduate and professional schools. Data collected for these studies include:

- Education expenditures.
- Relationship between education and other social and political characteristics.
- Education levels among the various components of society.
- Numbers of students studying abroad.
- Extent to which foreign languages are taught.
- Subjects taught in schools.

## RELIGION

C-29. Religious beliefs may be a potentially dangerous friction factor for deployed U.S. personnel; this was experienced in the Middle East with Fundamentalist Islamic sects. Understanding those friction factors is essential to mission accomplishment and the protection of friendly forces. Analysts evaluate data collected on an area's religions, which includes—

- Types.
- Size of denominations.
- Growth or decline rates.
- Cooperative or confrontational relationships between religions, the people they represent, and the government.
- Ways the government deals with religious organizations.
- Roles religious groups play in the national decision-making process.
- Religious traditions and taboos.

## PUBLIC WELFARE

C-30. To evaluate the general health of a population, analysts must identify—

- Health delivery systems.
- Governmental and informal welfare systems.
- Social services provided.
- Living conditions.

- Social insurance.
- Social problems that affect national strength and stability (for example, divorce rate, slums, drug use, or crime) and methods of coping with these problems.

## NARCOTICS AND TERRORISM TOLERANCE

C-31. A population's level of tolerance for narcotics and terrorist activities depends on the relations between these organizations and the population as a whole. Analysts should determine if the tolerance is a result of the huge sums of money that traffickers pump into the economy or a result of traffickers' use of force. Terrorists may be accepted and even supported by the local populace if they are perceived to be working for the good of the local people. The intelligence analyst must evaluate the way these organizations operate.

## SOURCES

C-32. Because of the nature of the social focus area, the preponderance of information is envisioned to be open source. The initial task is to develop a baseline of information on the target nation. Basic data will be collected and analyzed. Numerous studies, sponsored by the USG, as well as academic treatises, are available. A more difficult problem will be making the essential linkages within the sociological area and with other focus areas, particularly political and economic.

## SOCIAL SYSTEM QUESTIONS

C-33. The following questions pertain to the social system:
- What are the general perceptions of social stability?
- Who are the population's most respected figures? Why are they so respected? How do they maintain the public focus?
- What are the government's most effective tools for influencing the masses?
- What dominant areas of society are emerging and causing instability or areas of conflict?
  - Are any of these areas linked to political factors?
  - Are there ethnic or racial factors?
- What are the predominant economic areas that are contributing to, promoting, or exacerbating social instability?
- How can interrelationships be established between religious and ethnic minorities in the country of interest? How can these relationships be effectively manipulated to affect a desired outcome?
- What are perceptions of public safety primarily attached to?
  - How is the level of violence defined by society?
  - What elements may make it appear excessive?
- What psychological effects do an increased level of violence have on a person's notion of safety?
- What are the effects of increased criminal activity on the family, the town, the region, and nationally?
- How can the coalition increase the psychological perception that the global economy is surpassing the country of interest?
- How can the coalition stimulate the notion that the government is failing to provide for basic elements, or is slow to produce results?
- What are the adverse effects of increased organized criminal activity upon society by industrial component?
  - Is there white collar or financial crime?
  - Are there issues with drugs and drug smuggling?
- Proliferation of weapons: What are the types of weapons and to whom are they going?
- Gang-related activity: Is there a predominant ethnic group asserting themselves in this arena, and are they utilizing any particularly violent tactics to assert themselves?

- What are the significant effects of increased public health problems?
  - What public health issues have increased?
  - How effective is the government?
- How extensive is the division of wealth between ethnic and religious groups and what is their potential for promoting tension or conflict?
- What are the effects of environmental problems having on society?
- What are the key groups adversely affected by increasing poverty rates?
- What primary tools are used by the government for influencing the masses?
  - How do the masses validate information obtained by the government?
  - Do they feel they need to validate information?
- Who are the key opposition leaders?
  - How do they influence the masses?
  - How are they funded and by whom are they primarily funded?
- Who are the key opposition groups?
  - How do they influence the masses?
  - How are they funded and by whom are they primarily funded?
  - Are there any common themes to unite them or areas that may divide them?
- How do opposition groups recruit?
  - Do they target a specific social group?
  - Is there a hierarchical structure?
  - How are members dismissed from the ranks?
- How do these groups affect one another?
  - How do they affect similar groups in neighboring countries?
  - Do they have external support?
- What are each faction's mechanisms for influencing the others?
  - How do they communicate officially and unofficially?
  - What factions are armed?
  - Where do they get their weapons?
- Are acts of civil disobedience increasing?
  - Is the level of violence employed by the government to quell civil disobedience increasing?
  - Are acts of vigilantism on the rise?
  - How are disturbances quelled?
  - What tools are brought to bear?
- What consumer goods are most valued by the country of interest's populace?
  - Who controls supply?
  - How are they networked?
  - Is there any increase in a particular product?
- What are the "hot button" issues dividing the various factions of the society?
  - What networks and mediums can be used to subvert and confuse each faction?
  - What are the capabilities of regional allies to polarize these factions?
- How are rumors spread most effectively?
- What is the social perception of the military's ability to meet that threat?
  - What is the state's ability to meet the threat?
  - What is the state's ability to provide overall security in a micro or macro context?

- How are troops conscripted?
  - What are the incentives for service?
  - What unofficial groups or associations exist within the military?
  - How do they recruit or dismiss people?
- Is criminal behavior increasing within the military?
- What types of criminal activity occur within the military?
- What is the hierarchal structure of the military?
  - Is there a dominant ethnic group assuming more leadership roles?
  - What ethnic groups stay the most connected in the military?
  - Which groups are more apt to include outsiders?
- Which ethnic and religious minorities feel the most repressed?
  - How do they express their discontent?
  - Do any organizations exist to channel their feelings?
  - How responsive do they feel the government is to their issues?
- How does the population view outside assistance?
  - How likely is the government to ask for assistance?
  - How is the need for assistance determined?
- How are relief organizations viewed within the country?
  - Are they busy?
  - How effective are they at solving problems and meeting the needs of those they serve?
- Are there problems with immigrant flows?
- How are refugees treated?
- What consumer goods are in short supply?
  - How are those goods brought to market?
  - Who controls the flow of such goods?
  - Is there a dominant ethnic group controlling the flow?
  - How effective is the black market in producing hard-to-obtain goods?
- What goods dominate the black market?
  - Who are the primary producers and end receivers of goods?
  - Is there a particular group emerging as the leader of the black market?
- How are minority laborers networked with minority leaders?
  - What are the links between labor groups and minority activists?
  - What ethnic groups compose the majority of the skilled labor force?
  - How is skilled labor kept from going abroad?

## INFRASTRUCTURE SYSTEM POINTS OF ANALYSIS

C-34. The infrastructure system points of analysis include the following:
- Transportation, to include:
  - Rail.
  - Trains.
  - Bridges.
  - Tunnels.
  - Switches.
  - Road.
  - Ship or boat.
  - Dam locks.
  - Air.

- Communications, to include:
  - Military networks.
  - Radio telephone.
  - Teletype fiber satellite.
  - Visual.
  - Civilian.
  - Radio telephone.
  - Television speakers.
  - Signs.
- Energy or power, to include:
  - Coal.
  - Petroleum, oils, and lubricants.
  - Hydro.
  - Nuclear.
  - Water.

C-35. Infrastructure analysis focuses on the quality and depth of the physical structures that support the people and industry of the state. In developed countries, it is the underlying foundation or basic systems of a nation-state, generally physical in nature and supporting or used by other entities (for example, roads, telephone systems, and public schools).

## INFRASTRUCTURE SYSTEM QUESTIONS

C-36. The following questions pertain to the infrastructure system:

- Lines of communications, to include:
  - Where are the key ports, airfields, rail terminals, roads, railroads, inland waterways located?
  - Where are key bridges, tunnels, switching yards, scheduling or control facilities, depots or loading stations, switching yards, and so on?
- Electrical power, to include:
  - Where are power plants, transformer stations, and relay and power transmission lines located?
  - Where are the key substations, switching stations, and line junctures?
- Potable water, to include:
  - Where are the water treatment plants, wells, desalination, bottling plants, and pumping stations?
  - Where are the key pumping stations, control valves, and distribution line junctures?
- Telecommunications, to include:
  - What are the location and architecture of the domestic telephone system, cable, fiber-optic, microwave, Internet, and cell phone networks and satellite stations?
  - Where are the key control points and junctures?
- Petroleum and gas, to include:
  - Where are the gas and petroleum fields, gathering sites, pumping stations, storage areas, refineries and distribution lines?
  - Where are the key pumping stations, control valves, and distribution junctures?
- Broadcast media, to include:
  - What are the location, frequency, power, and radius of effective range (coverage) of the amplitude modulation or frequency modulation radio and television stations?
  - Where are the studios, antenna, and relay towers located?
  - How are they powered?
  - Where are the key control points and junctures?

- Public health, to include:
    - What are the locations of the hospitals and clinics?
    - Are they adequately staffed, supplied, and equipped?
    - Is the equipment well maintained?
    - Is the staff well trained?
    - Do they depend on foreign or domestic sources for their supplies, medications, and spare equipment parts?
    - Where are the key control points and junctures?
- Schools, to include:
    - What are the locations of the public, private, and religious primary and secondary schools and universities?
    - Where are the key control points and junctures?
- Public transportation, to include:
    - What are the public transportation (bus, streetcar, taxi) routes?
    - Where are the key control points and junctures?
- Sewage collection and treatment, to include:
    - Where are the collection systems, pumping stations, treatment facilities, and discharge areas located?
    - Where are the key control points and junctures?

## OTHER COMMON INFRASTRUCTURE QUESTIONS

C-37. Other common infrastructure questions include the following:
- How are key facilities linked (physically, electronically, and so on.)?
- What are the key nodes?
    - Where are they?
    - Where are the disabling yet nonlethal or nondestructive infrastructure nodes?
    - What are their alternates?
    - How are they linked to the key facilities and each other?
- Are there indigenous capabilities?
    - What indigenous capabilities could be used?
    - How are they linked and organized?
- What are the critical nodes?
    - What is the security surrounding the nodes?
    - What is the security posture at these facilities?
- Who controls the forces?
    - How are security forces, police, or paramilitary networked?
    - What training do they receive?
    - What is their level of proficiency?
    - Are they augmented as alert status (national or local) changes?
- What are the ground, naval, or air defense capabilities at or near these facilities?
    - How are they networked?
    - What groups are likely to conduct industrial sabotage?
    - How are they tasked, linked, supported?
- Who owns and who controls the infrastructure?
- Who owns or controls the aforementioned entities?
- Is ownership by private, corporate, or governmental entities?
- What organizations have regulatory oversight or control?
- What is the capability to repair damage to the system and restore it to service?

- Is maintenance and repair an integral part of the organization?
- What are their capabilities and limitations?
- Which contractors are normally used and for what purpose?
- Are repair or restore materials readily available or is there a long lead time for critical supplies and components?
- Who are the key engineering contractors for these facilities?
- Can or will they share plans, blueprints, schematics, and so on?
- What would be the second-order effects of influencing the infrastructure?

## INFORMATION SYSTEM POINTS OF ANALYSIS

C-38. The information system points of analysis include the following:
- Education.
- Propaganda: inside country; outside country.
- Newspaper or magazine information technologies.
- Radio.
- Television.
- Internet.
- Informal transmissions (word of mouth or rumor).
- Cyberspace information systems.

## ANALYSIS OF INFORMATION SYSTEMS

C-39. Analysis of information systems and operations include the following:
- Telecommunications capabilities and level of sophistication, teledensity rates, radio and television broadcast coverage, including television, landline, cellular, Internet, radio, and so on.
- Interconnectivity of communications via integrated services digital network, fiber optic, satellite, and microwave.
- Primary nodes and trunks of telecommunications infrastructure, including government, nongovernment, citizen, and military use of information operations.
- Knowledge of the country of interest key leaders' style and decision-making habits, advisors' perceptions, and cultural influences.
- Understanding governmental use of media influence, public affairs, and CA interrelationships
- Knowledge of military, NGOs, and law enforcement interrelationships.
- Understanding of effects on adversary under psychological, computer network attack and defense, electronic warfare, and space operations.
- Locations and purpose of physical infrastructure of communications and broadcast towers, cables, and supporting operations centers included within the infrastructure focus.
- Development of and use of computer network operating systems, information technology industry skill sets, and software applications.
- Media affiliations, perceptions, and sympathies, to include censorship and self-censorship in news and entertainment print and broadcast industries.

## INFORMATION SYSTEM QUESTIONS

C-40. The following questions pertain to the information system:
- How effective are the country of interest's network defense capabilities?
- What reactions could be expected following an incident?
- What recovery procedures are routinely exercised?
- What is the organizational structure of the telecommunications industry?
- How effective is the country of interest at managing physical security of infrastructure and implementing network security practices?

- What interrelationships exist between civil law enforcement, military, commercial, and nongovernmental agencies that would enhance the country of interest's response to an emergency?
- What redundancies exist within the country of interest's network to eliminate or reduce network down time?
  - Cellular, satellite, landline, power back up?
  - How effective is their exchange, backbone, architecture in providing redundancies?
- What would cause a slowdown of the country of interest's network?
- In what ways can the effect be localized (geographic, logic, by agency, and so on)?
- What bandwidth issues exist within the country of interest's communications industry?
- How well, and in what ways, does the government manage its allocation?
- What type of OPSEC practices does the country of interest routinely exhibit to deny exploitation?
- In what ways have military, civil, or corporate operations centers improved their practices or tactics in keeping with the country of interest's technological improvements?
- Do they rely more heavily on computers, cellular, or networks than in the past?
- What are the indicators, if they exist, that the country of interest has developed a more focused vision and strategic plan for using technology than it had in the late 1990s?
- What effect has technology had on productivity, transportation, logistics, and so on, in government, commerce, corporate, and private sectors?
- How does the country of interest perceive their use of technology from a governmental perspective?
  - From the citizens' perspective?
  - Military?
  - Business?
  - Legal?
  - Law enforcement?
  - NGOs?
- What is known about the country of interest's assessment of Blue network vulnerabilities and defense measures?
- Do regional and neighboring countries or satellite broadcasts (television, radio, and Internet) have an audience in the country of interest's population?
- Which broadcasts are popular with citizens and what are the audience's demographic and statistic data?
- What programs or broadcasts are popular with minority political parties, resistance movements, academia, and so on?
- What is the topology design the country of interest networks utilize?
- Which exchanges and trunks are collocated within government-controlled facilities?
- Are government-commercial partnerships used to provide network services?
- What is known of current and planned technology projects, such as—
  - Fiber-optic cabling?
  - Integrated services digital network access expansion?
  - Satellite leases and launches?
- What is the operational status and capability of country of interest's low-earth orbit satellites?
- What Internet domains are accessible to the population?
- Is reliable language interpretation software available?
- What licenses do the government require for Web hosting?
- What governmental directives address network security in supporting national security objectives?
- What messages might be effective in the country of interest?
- What themes are prevalent in the media?

- What advances in communications technology have enabled improvements in military hardware employment?
- How is telecommunications technology used in law enforcement operations?
- To what degree and direction are telecommunications infrastructure investments impacting military readiness?
- What is the state of international telecommunications connectivity to the country of interest?
- Which current telecommunications and Internet security operations have been exercised?
- Is there a national crisis action plan?
- What practices and policies do the government use in monitoring information-related media (television, radio, Internet, and so on)?
- What enforcement methods have been employed?
- Which print media and online content do citizens turn to for news or entertainment?
- Do censorship policies or self-censorship trends exist in the country of interest?
- Is there a market and distribution pipeline for recorded or intercepted news or entertainment programs?
- In what ways does law enforcement interact in this market?
- What is known about the country of interest's network operating systems?
- What information technology skill sets are known to be in high demand?
- Is there a prevalence of—
  - Software piracy?
  - Counterfeiting?
  - Drug smuggling?
  - Organized crime?
  - Identity theft?

## PHYSICAL ENVIRONMENT POINTS OF ANALYSIS

C-41. The physical environment points of analysis include the following:
- Terrain, to include:
  - Observation and fields of fire.
  - Avenues of approach.
  - Key terrain.
  - Obstacles.
  - Cover and concealment.
  - Land forms.
  - Vegetation.
  - Terrain complexity.
  - Mobility classification.
- Natural hazards.
- Climate.
- Weather, to include:
  - Precipitation.
  - High-temperature heat index.
  - Low-temperature wind chill.
  - Wind.
  - Visibility.
  - Cloud cover.
  - Relative humidity.

C-42. Physical environment questions include the following:

- What are the natural features in the area?
- What operational impact do rivers have?
- What operational impact do mountains have?
- What are the man-made features in the area?
- What is the operational impact of cities?
- What is the operational impact of airfields?
- What is the operational impact of bridges?

## TIME POINTS OF ANALYSIS

C-43. The time points of analysis include the following:

- Cultural perception of time.
- Information offset.
- Tactical exploitation of time.
- Key dates, time periods, or events.

C-44. Time questions include the following:

- What are the time schedules of operationally significant adversary activities?
- How does the environment in the area impact friendly time schedules to assemble, deploy, and maneuver units in relationship to the adversary?

# CIVIL CONSIDERATIONS ANALYSIS

C-45. This framework is based on the widely used ASCOPE civil considerations analysis approach. Civil considerations encompass the man-made infrastructure, civilian institutions, and attitudes and activities of the civilian leaders, populations, and organizations within an AO and provide insight into how these elements influence military operations.

---

*Note:* ATP 2-01 3, *Intelligence Preparation of the Battlefield,* (Table 4-4, PMESII and ASCOPE Examples; and Chapter 4, Describe Environmental Effects on Operations) provides additional information on PMESII and ASCOPE.

---

## PMESII-PT AND ASCOPE GRAPH

C-46. Figure C-1, pages C-21 and C-22, cross-references the ASCOPE characteristics with the PMESII-PT operational variables. An understanding of civil considerations—the ability to analyze their impact on operations—enhances several aspects of mission planning: among them, the selection of objectives; location, movement, and control of forces; use of weapons; and protection measures.

| | AREAS | STRUCTURES | CAPABILITIES | ORGANIZATIONS | PEOPLE | EVENTS |
|---|---|---|---|---|---|---|
| **POLITICAL** | • Enclaves<br>• Municipalities<br>• Provinces<br>• Districts<br>• Political districts<br>• Voting<br>• National boundaries<br>• Party affiliation areas<br>• Shadow government influence areas | • Courts (court houses, mobile courts)<br>• Government centers<br>• Provincial/district centers<br>• Meeting halls<br>• Polling sites | • Public administration<br>  • Civil authority practices and rights<br>  • Political system<br>  • Political stability<br>  • Political traditions<br>  • Standards and effectiveness<br>• Executive<br>  • Administration<br>  • Policies<br>  • Powers<br>  • Organization<br>• Legislative<br>  • Administration<br>  • Policies<br>  • Powers<br>  • Organization<br>• Judicial/Legal<br>  • Administration<br>  • Capacity<br>  • Policies<br>  • Civil and criminal codes<br>  • Powers<br>  • Organization<br>  • Law enforcement<br>• Dispute resolution, grievances<br>  • Local leadership<br>  • Degrees of legitimacy | • Major political parties<br>  • Formal<br>  • Informal<br>• Nongovernment organizations<br>• Host government<br>• Insurgent group affiliations<br>• Court system<br>• Covert political power<br>• Partnerships: foreign | • United Nations representatives<br>• Political leaders<br>• Governors<br>• Councils<br>• Elders<br>• Community leaders<br>• Paramilitary members<br>• Judges<br>• Prosecutors | • Elections<br>• Council meetings<br>• Speeches (significant)<br>• Security and military training sessions<br>• Significant trials<br>• Distribution of power<br>• Political motivation<br>• Treaties<br>• Will |
| **MILITARY** | • Area of influence<br>• Area of interest<br>• Area of operation<br>• Safe havens or sanctuary<br>• Multinational/local nation bases<br>• Historic ambush/improvised explosive device sites/insurgent base | • Bases<br>• Headquarters (police)<br>• Known leader houses/businesses | • Doctrine<br>• Organization<br>• Training<br>• Materiel<br>• Leadership<br>• Personnel and manpower<br>• Facilities<br>• History<br>• Nature of civil-military relationships<br>• Resource constraints<br>• Local security forces<br>• Quick reaction force<br>• Insurgent strength<br>• Enemy recruiting | • Host nation forces present<br>• Insurgent groups present and networks<br>• Multinational forces<br>• Paramilitary organizations<br>• Terrorists<br>• Multinational forces present<br>• Fraternal organizations<br>• Civic organizations | • Key leaders<br>• Multinational, insurgent, military | • Combat<br>• Historical<br>• Noncombat<br>• Kinetic events<br>• Unit reliefs<br>• Loss of leadership |
| **ECONOMIC** | • Commercial<br>• Fishery<br>• Forestry<br>• Industrial<br>• Livestock dealers<br>• Markets<br>• Mining<br>• Movement of goods/services<br>• Smuggling routes<br>• Trade routes<br>• Black market areas | • Banking<br>• Fuel<br>  • Distribution<br>  • Refining<br>  • Source<br>• Industrial plants<br>• Manufacturing<br>• Mining<br>• Warehousing<br>• Markets<br>• Silos, granaries, warehouses<br>• Farms/ranches<br>• Auto repair shops | • Fiscal<br>  • Access to banks<br>  • Currency<br>  • Monetary policy<br>  • Ability to withstand drought<br>  • Black market<br>  • Energy<br>  • Imports/exports<br>  • External support/aid<br>• Food<br>  • Distributing<br>  • Marketing<br>  • Production<br>  • Processing<br>  • Rationing<br>  • Security<br>  • Storing<br>  • Transporting<br>• Inflation<br>• Market prices<br>• Raw materials<br>• Tariffs | • Banks<br>• Business organizations<br>• Cooperatives<br>• Economic nongovernment organizations<br>• Guilds<br>• Labor unions<br>• Major illicit industries<br>• Large landholders<br>• Volunteer groups | • Bankers<br>• Employers/Employees<br>• Labor occupations<br>• Consumption patterns<br>• Unemployment rate<br>• Underemployment rate (if this exists)<br>• Job lines<br>• Landholders<br>• Merchants<br>• Money lenders<br>• Black marketers<br>• Gang members<br>• Smuggling chain | • Drought, harvest, yield, domestic animals, livestock (cattle, sheep) and market cycles<br>• Labor migrations events<br>• Market days<br>• Payday<br>• Business openings<br>• Loss of business |

Figure C-1. PMESII and ASCOPE examples

| | AREAS | STRUCTURES | CAPABILITIES | ORGANIZATIONS | PEOPLE | EVENTS |
|---|---|---|---|---|---|---|
| SOCIAL | • Refugee camps<br>• Enclaves<br>  • Ethnic<br>  • Religious<br>  • Social<br>  • Tribal<br>• Families or clans<br>• Neighborhoods<br>• Boundaries of influence<br>• School districts<br>• Parks<br>• Traditional picnic areas<br>• Markets<br>• Outdoor religious sites | • Clubs<br>• Jails<br>• Historical buildings/ houses<br>• Libraries<br>• Religious locations<br>• Schools/universities<br>• Stadiums<br>• Cemeteries<br>• Bars and tea shops<br>• Social gathering places (meeting places)<br>• Restaurants | • Medical<br>  • Traditional<br>  • Modern<br>• Social networks<br>• Academic<br>• Strength of tribal/ village traditional structures<br>• Judicial | • Clan<br>• Community councils and organizations<br>• School councils<br>• Criminal organizations<br>• Familial<br>• Patriotic/service organizations<br>• Religious groups<br>• Tribes | • Community leaders, councils, and their members<br>• Education<br>• Ethnicity/racial<br>  • Biases<br>  • Dominant group<br>  • Percentages<br>  • Role in conflict<br>• Key figures<br>  • Criminals<br>  • Entertainment<br>  • Religious leaders<br>  • Chiefs/elders<br>• Language/dialects<br>• Vulnerable populations<br>• Displaced persons<br>• Sports<br>• Influential families<br>• Migration patterns<br>• Culture<br>  • Artifacts<br>  • Behaviors<br>  • Customs<br>  • Shared beliefs/ values | • Celebrations<br>• Civil disturbances<br>• National holidays<br>• Religious holidays and observance days<br>• Food line<br>• Weddings<br>• Birthdays<br>• Funerals<br>• Sports events<br>• Market days<br>• Family gatherings<br>• History: Major wars/conflicts |
| INFORMATION | • Broadcast average area (newspaper, radio, television)<br>• Word of mouth<br>• Gathering points<br>• Graffiti<br>• Posters | • Communications:<br>  • Lines<br>  • Towers (cell, radio, television)<br>• Internet service:<br>  • Satellite<br>  • Hard wire<br>  • Cafes<br>• Cellular phone<br>• Postal service<br>• Print shops<br>• Telephone<br>• Television stations<br>• Radio stations | • Availability of electronic media<br>• Indigenous communications networks<br>• Internet access<br>• Intelligence networks<br>• Printed material:<br>  • Journals<br>  • Newspapers<br>  • Flyers<br>• Propaganda mechanisms<br>• Radio<br>• Television<br>• Social media<br>• Literacy rate<br>• Word of mouth | • Media groups and news organizations<br>• Religious groups<br>• Insurgent inform and influence activities groups<br>• Government groups<br>• Public relations and advertising agencies | • Decision makers<br>• Media personalities<br>• Media groups and news organizations<br>• Community leaders<br>• Elders<br>• Heads of families | • Disruption of services<br>• Censorship<br>• Religious observance days<br>• Publishing dates<br>• Inform and influence activities / campaigns<br>• Project openings |
| INFRASTRUCTURE | • Commercial<br>• Industrial<br>• Residential<br>• Rural<br>• Urban<br>• Road systems<br>• Power grids<br>• Irrigation networks<br>• Water tables | • Emergency shelters<br>• Energy:<br>  • Distribution system<br>  • Electrical lines<br>  • Natural gas<br>  • Power plants<br>• Medical:<br>  • Hospitals<br>  • Veterinary clinics<br>• Public buildings<br>• Transportation:<br>  • Airfields<br>  • Bridges<br>  • Bus stations<br>  • Ports and harbors<br>  • Railroads<br>  • Roadways<br>  • Subways<br>• Waste distribution, storage and treatment:<br>  • Dams<br>  • Sewage<br>  • Solid<br>• Construction sites | • Construction<br>• Clean water<br>• Communication systems<br>• Law enforcement<br>• Firefighting<br>• Medical:<br>  • Basic<br>  • Intensive<br>  • Urgent<br>• Sanitation<br>• Maintain roads, dams, irrigation, sewage systems<br>• Environmental management | • Construction companies:<br>  • Government<br>  • Contract | • Builders<br>• Road contractors<br>• Local development councils | • Scheduled maintenance (road/bridge constructions)<br>• Natural, man-made disasters<br>• Well digging<br>• Community center construction<br>• School construction |

**Figure C-1. PMESII and ASCOPE examples (continued)**

## ENVIRONMENTAL EFFECTS ON OPERATIONS DESCRIPTIONS

C-47. Due to the complexity and volume of data involving civil considerations, there is no simple model for presenting civil considerations analysis. Rather, it comprises a series of intelligence products composed of overlays and assessments. The six characteristics of ASCOPE are discussed below.

## AREAS

C-48. Key civilian areas are localities or aspects of the terrain within an AO that often are not militarily significant. This characteristic approaches terrain analysis (observation and fields of fire, avenues of approach, key terrain, obstacles, and cover concealment) from a civilian perspective. The intelligence staff analyzes key civilian areas in terms of how they may affect the missions of friendly forces, as well as how friendly military operations may affect these areas. Examples of key civilian areas are—

- Areas defined by political boundaries, such as districts within a city or municipalities within a region.
- Locations of government centers.
- Social, political, religious, or criminal enclaves.
- Ethnic or Sectarian fault lines.
- Agricultural and mining regions.
- Trade routes.
- Possible sites for the temporary settlement of displaced civilians or other civil functions.

C-49. The intelligence staff maintains this information in the civil considerations data file and constructs an areas overlay to aid in planning.

## STRUCTURES

C-50. Existing structures can have various degrees of significance. Analyzing structures involves determining how the location, functions, capabilities, and consequences of its use can support or hinder the operation. Using a structure for military purposes often competes with civilian requirements. Commanders must carefully weigh the expected military benefits against costs to the community that have to be addressed in the future. Commanders also need to consider the significance of the structure in providing stability to the AO. Certain structures are critical in providing a state of normalcy back to the community and need to be maintained or restored quickly. The possibility of repaying locals for the use of shared facilities or building more of the same facilities, time and cost permitting, should also be considered. The following are examples of structures:

- Military bases.
- Police stations.
- Jails.
- Courtrooms.
- Political offices.
- Electrical power plants and substations.
- Petroleum, oils, and lubricants refineries.
- Dams.
- Water and sewage treatment and distribution facilities.
- Communications stations and networks.
- Bridges and tunnels.
- Warehouses.
- Airports and bus terminals.
- Universities and schools.

C-51. Other structures are cultural sites that international law or other agreements generally protect, for example:

- Religious structures.
- National libraries and archives.
- Hospitals and medical clinics.
- Monuments.
- Works of art.
- Archaeological sites.
- Scientific buildings.
- Museums.
- Crops, livestock, and irrigation works.
- United Nations Educational, Scientific and Cultural Organizations—designated World Heritage Sites.

C-52. The intelligence staff maintains this information in the civil considerations data file and, because of the amount of information in these categories, constructs multiple overlays for structures to aid in planning.

## CAPABILITIES

C-53. Commanders and staffs analyze capabilities from different levels. They view capabilities in terms of those required to save, sustain, or enhance life, in that priority. Capabilities can refer to the ability of local authorities—those of the host nation, aggressor nation, or some other body—to provide a populace with key functions or services, such as—

- Public administration.
- Public safety.
- Emergency services.
- Technology.
- Basic necessities (food, water, medical availability).

C-54. Capabilities include those areas in which the populace may need help after combat operations or major operations, such as public works and utilities, public health, economics, and commerce. Capabilities also refer to resources and services that can be contracted to support the military mission, such as interpreters, laundry services, construction materials, and equipment.

C-55. The intelligence staff maintains this information in the civil considerations data file and constructs a structures overlay to aid in planning.

## ORGANIZATION

C-56. IPB considers the organization dimension (such as the nonmilitary groups or institutions) and political influence and their impacts in the AO. Organizations influence and interact with the populace, friendly forces, the threat or adversary, and each other. An important aspect of civil considerations is the political dimension of the local population and their expectations relative to threat or adversary and friendly operations.

C-57. Political structures and processes enjoy varying degrees of legitimacy with populations from local through international levels. Formally constituted authorities and informal or covert political powers strongly influence events. Political leaders can use ideas, beliefs, violence, and other actions to enhance their power and control over people, territory, and resources. There are many sources of political interest. These may include charismatic leadership; indigenous security institutions; and religious, ethnic, or economic factors. Political opposition groups or parties also affect the situation. Each may cooperate differently with U.S. or multinational forces.

C-58. Understanding the political circumstances helps commanders and staffs recognize key organizations and determine their aims and capabilities. Understanding political implications requires analyzing all relevant partnerships—political, economic, military, religious, and cultural. This analysis captures the presence and significance of external organizations and other groups, including groups united by a common cause.

Examples include private security organizations, transnational corporations, and NGOs that provide humanitarian assistance.

C-59. Political analysis must include an assessment of varying political interests and the threat's·or adversaries' political decisive point or center of gravity and will. Will is the primary intangible factor; it motivates participants to sacrifice to persevere against obstacles. Understanding what motivates key groups (for example, political, military, and insurgent) helps commanders understand the groups' goals and willingness to sacrifice to achieve their ends.

C-60. Organizations are nonmilitary groups or institutions in the AO. They influence and interact with the populace, the force, and each other. They generally have a hierarchical structure, defined goals, established operations, fixed facilities or meeting places, and a means of financial or logistics support. Some organizations may be indigenous to the area. These organizations may include:

- Religious organizations.
- Fraternal organizations.
- Patriotic or service organizations.
- Labor unions.
- Criminal organizations.
- Community watch groups.
- Political groups.
- Agencies, boards, committees, commissions (local and regional, councils).
- Multinational corporations.
- International organizations, such as United Nations agencies.
- Other host-nation government agencies (foreign version of Department of Education, United States Agency for International Development).
- NGOs, such as the International Committee of the Red Cross.

---

*Note:* The other host-nation government agencies designated above are separate from organizations within the threat capability (military, intelligence, police, paramilitary), such as the CIA.

---

C-61. To enhance their situational awareness, commanders must remain familiar with organizations operating in their AOs, such as local organizations that understand the political dimension of the population. Situational awareness includes having knowledge of how the activities of different organizations may affect military operations and how military operations may affect these organizations' activities. From this, commanders can determine how organizations and military forces can work together toward common goals when necessary.

C-62. In almost every case, military forces have more resources than civilian organizations. However, civilian organizations may possess specialized capabilities that they may be willing to share with military forces. Commanders do not command civilian organizations in their AOs. However, some operations require achieving unity of effort between them and the force. These situations require commanders to influence the leaders of these organizations through persuasion. They produce constructive results by the force of argument and the example of their actions.

C-63. The intelligence staff maintains this information in the civil considerations data file and, because of the amount of information in these categories, constructs multiple overlays for organizations to aid in planning.

## PEOPLE

C-64. The use of the general term *people* describes nonmilitary personnel encountered by military forces. The term includes all civilians within an AO, as well as those outside the AO whose actions, opinions, or political influence can affect the mission. Individually or collectively, people can affect a military operation positively, negatively, or neutrally. In stability tasks, Army or Marine Corps forces work closely with civilians of all types. Understanding the sociocultural factors of the people in the AO is a critical component of understanding the operational environment.

C-65. There can be many different kinds of people living and operating in and around an AO. As with organizations, people may be indigenous or introduced from outside the AO. An analysis of people will identify them by their various capabilities, needs, and intentions. It is useful to separate people into distinct categories, such as demographically, social and political groups, and target audiences. (A target audience list can be obtained from the PSYOP element or the information operations officer or G-3.) When analyzing people, commanders consider historical, cultural, ethnic, political, economic, religious, and humanitarian factors. They also identify the key communicators and the formal and informal processes used to influence people.

C-66. The languages used in the region will have a huge impact on operations. The staff identifies the languages and dialects used within the AO so that language training can be scheduled, communication aids such as phrase cards can be developed, and translators can be contracted. Translators will be crucial for collecting intelligence, interacting with local citizens and community leaders, and developing products.

C-67 Another aspect of language involves the transliteration guide not written using the English alphabet. This will have an impact on all intelligence operations, to include collection, analysis, dissemination, and targeting. In countries that do not use the English alphabet, a theaterwide standard should be set for spelling names. Without a spelling standard, it can be difficult to conduct effective analysis. In addition, insurgents or criminals may be released from custody if their names are misidentified. To overcome these problems, there must be one spelling standard for a theater. Because of the interagency nature of counterinsurgency operations, the standard must be agreed upon by non-defense agencies. Intelligence staffs should also know family naming conventions in places like the Middle East where various cultures do not use an individual's surname and family name.

C-68. Another major consideration when analyzing people is religion. Religion has shaped almost every conflict of the past, and there are indicators that its influence will only grow. The staff considers the following when incorporating religion in planning. They—

- Know when religious traditions will be affected by the mission and try to determine how religion will affect the mission.
- Know when religious figures have influenced social transformations in a negative or positive way.
- Attempt to understand all parties, no matter how violent or exclusive.
- Consult military religious affairs subject-matter experts, such as chaplains, priests, rabbis, imams, or other religious teachers as appropriate for detailed insight on religious attitudes.

C-69. Religion has the ability to shape the operational environment. Religion can add a higher intensity, severity, brutality, and lethality to conflicts than almost any other factor. Religion can motivate the masses quickly and inexpensively.

C-70. Part of the analysis of people is identifying cultural terms and conditions. Cultural terms and conditions describe both American and foreign ways of thought and behavior. Understanding culture gives insight into the motives and intent of nearly every person or group in the operational environment: friend, enemy, or other. This insight, in turn, allows commanders and staffs to allocate resources, outmaneuver opponents, alleviate friction, and reduce the fog of war. The study of culture for military operations is not an academic exercise and, therefore, requires specific military guidelines and definition. The analyst sets aside personal bias and judgment and examines the cultural group dispassionately, basing the analysis purely on facts. The military studies broad categories of cultural factors, such as—

- Social structure.
- Behavioral patterns.
- Perceptions.
- Religious beliefs.
- Tribal relationships.
- Behavioral taboos.
- Centers of authority.
- Lifestyles.
- Social history.
- Gender norms and roles.

C-71. Culture is studied in order to give insights into the way people think, the reasons for their beliefs and perceptions, and what kind of behavior they can be expected to display in given situations. Because cultures are constantly shifting, the study of culture is an enduring task that requires historical perspective, as well as the collection and analysis of current information.

C-72. The intelligence staff maintains this information in the civil considerations data file and, because of the amount of information in these categories, constructs multiple overlays for people to aid in planning.

## EVENTS

C-73. Events are routine, cyclical, planned, or spontaneous activities that significantly affect organizations, people, and military operations. Examples include:
- National and religious holidays.
- Agricultural crop or livestock and market cycles.
- Elections.
- Civil disturbances.
- Celebrations.
- Natural phenomenon (such as monsoon, seasonal floods and droughts, volcanic and seismic activity, natural disasters).
- Man-made disasters.

C-74. Examples of events precipitated by military forces include combat operations or major operations, congested road networks, security restrictions, and economic infrastructure disruption or stimulus. Once significant events are determined, it is important to template the events and to analyze them for their political, economic, psychological, environmental, and legal implications.

# INFORMATION ENVIRONMENT

C-75. This framework is used by PSYOP units to understand the dimensions of the information environment. Figure C-2, page C-28, depicts the relationship of the information environment to information activities.

---

*Note:* Figure 6-1, Chapter 6, Intelligence of FM 3-53, *Military Information Support Operations,* provides additional information.

---

C-76. MISO are conducted within the information environment to influence the attitudes, values, and beliefs of selected target audiences and individuals to affect their decision-making process and subsequent behavior. The information environment is where information is collected, processed, disseminated, or acted upon. It includes leaders, decision makers, individuals, and organizations. It is also where humans observe, orient, decide, and act upon information, and it is therefore the principal environment for decision making.

C-77. The information environment consists of the physical, informational, and cognitive dimensions. The physical dimension is the tangible world. The physical dimension includes geography and terrain, weather, infrastructure, and populations. The informational dimension is where information is created, and where it exists. Information is collected, processed, and communicated in this dimension. Significant characteristics of the informational dimension include quality (accuracy, relevancy, and timeliness), content (means and format of information exchange), and communication flow (internal information sharing, reaching, receiving, and comprehension). The cognitive dimension is where perceptions, awareness, beliefs, and values reside (individual and collective consciousness). It is also where decisions are made. Significant characteristics of the cognitive dimension are memory, attention, and planning that result in subsequent mental and physical performance. PSYOP units conduct activities in the physical and informational dimensions to affect the cognitive. These activities are information-based and driven.

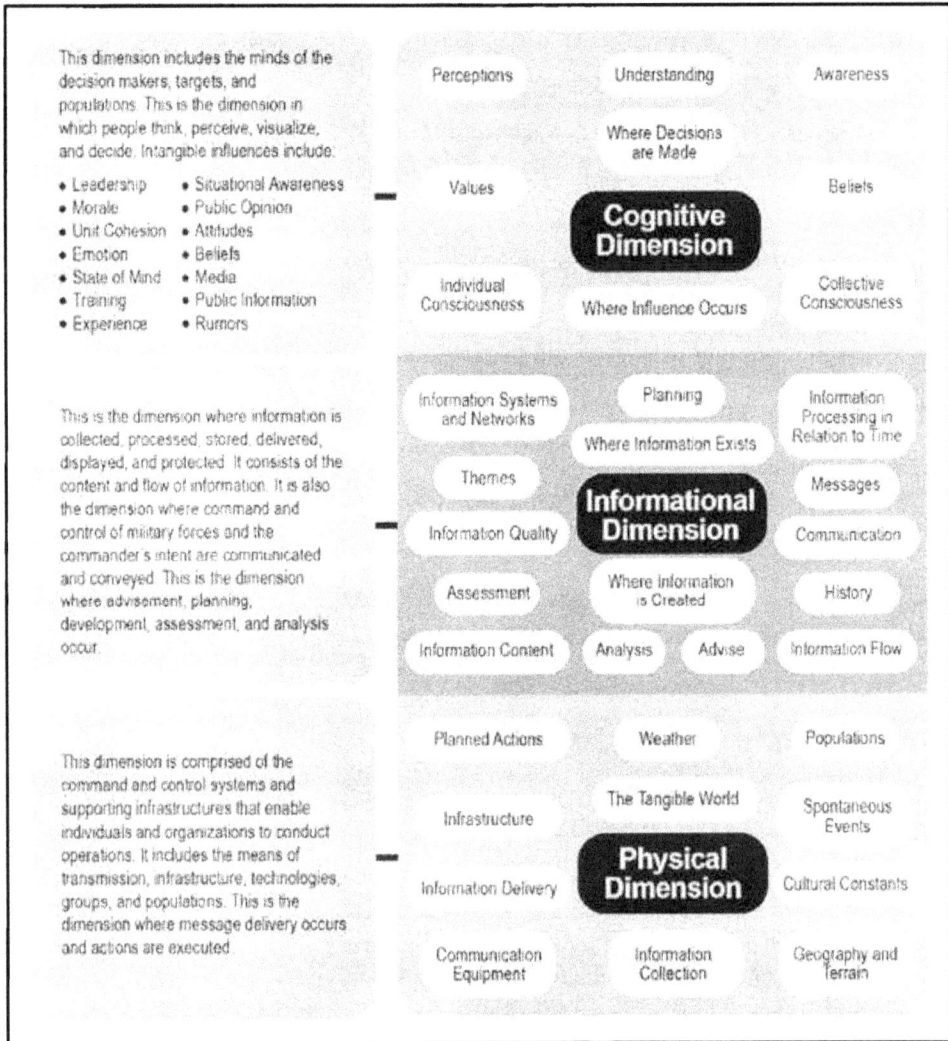

This dimension includes the minds of the decision makers, targets, and populations. This is the dimension in which people think, perceive, visualize, and decide. Intangible influences include:

- Leadership
- Situational Awareness
- Morale
- Public Opinion
- Unit Cohesion
- Attitudes
- Emotion
- Beliefs
- State of Mind
- Media
- Training
- Public Information
- Experience
- Rumors

Perceptions   Understanding   Awareness

Where Decisions are Made

Values   Beliefs

**Cognitive Dimension**

Individual Consciousness   Where Influence Occurs   Collective Consciousness

This is the dimension where information is collected, processed, stored, delivered, displayed, and protected. It consists of the content and flow of information. It is also the dimension where command and control of military forces and the commander's intent are communicated and conveyed. This is the dimension where advisement, planning, development, assessment, and analysis occur.

Information Systems and Networks   Planning   Information Processing in Relation to Time

Where Information Exists

Themes   Messages

**Informational Dimension**

Information Quality   Communication

Assessment   Where Information is Created   History

Information Content   Analysis   Advise   Information Flow

This dimension is comprised of the command and control systems and supporting infrastructures that enable individuals and organizations to conduct operations. It includes the means of transmission, infrastructure, technologies, groups, and populations. This is the dimension where message delivery occurs and actions are executed.

Planned Actions   Weather   Populations

The Tangible World

Infrastructure   Spontaneous Events

**Physical Dimension**

Information Delivery   Cultural Constants

Communication Equipment   Information Collection   Geography and Terrain

**Figure C-2. Relationship of the information environment to information activities**

# INTELLIGENCE INDICATORS IN DECISIVE ACTION

C-78. This framework is based on tables of standardized intelligence indicators. It is useful for recognizing probable, operationally significant occurrences indicated by phenomena perceived in the AO. The tables below are a select human factors-centric extract from a more complete list of indicator tables in ATP 2-33.4, *Intelligence Analysis*.

*Note:* ATP 2-33.4, Appendix B, Indicators in Decisive Action, provides more information on indicators in decisive action.

## SELECT HUMAN FACTORS—CENTRIC INDICATORS

C-79. The activities that reveal the intended threat COA are called indicators. An indicator is an activity or lack of activity that confirms or denies the action or event specified in an IR. Intelligence analysts develop indicators. Because the use of indicators is such an important part of determining threat COAs, it is imperative that all-source intelligence analysts carefully review all indicators. The figures in this appendix, although exemplary and not all inclusive, identify the different types of indicators, as well as applicable activities. These brief, select examples are designed to provide a starting point for more in-depth specific analysis for an operation. Development and refinement of indicators is an important activity that links all-source analysis to planning requirements and assessing collection.

## INDICATOR EXAMPLES

C-80. The following figures list various activities and explanations of indicators:
- Figure C-3, pages C-29 and C-30—Offensive indicators.
- Figure C-4, pages C-30 and C-31—Defensive indicators.
- Figure C-5, page C-31—Delaying indicators.
- Figure C-6, page C-32—Withdrawal indicators.
- Figure C-7, pages C-32 and C-33—Population indicators – Aggressive Behavior.
- Figure C-8, pages C-33 and C-34—Population indicators – Directed Behavior.
- Figure C-9, pages C-34 and C-35—Propaganda indicators.
- Figure C-10, pages C-35 and C-36—Commodities indicators.
- Figure C-11, page C-36—Environment-related indicators.
- Figure C-12, page C-37—Improvised explosive device indicators, observables, and signatures.
- Figure C-13, page C-37—Tribe or clan indicators.

| Activity | Explanation |
|---|---|
| Massing of maneuver elements, armor, artillery, and logistics support. | May indicate the main effort by weakening areas of secondary importance. |
| Deployment of combat elements on a relatively narrow frontage (not forced by terrain). | May provide maximum combat power at the point of attack by reducing frontages. Likely threat decisive effort. |
| Massing of indirect fire support assets | May indicate initiation of main effort. |
| Extensive artillery preparation of up to 50 minutes in duration or longer. | Initiates preparation preceding an attack. |
| Dispersal of tanks and self-propelled artillery to forward units. | Can indicate formation of combined arms assault formations with tanks accompanying the leading maneuver elements and artillery following in-bounds. |
| Surface-to-surface missile units located forward. | Provides depth to threat offensive tasks; places friendly support and unassigned areas in range. May also indicate, when employed alone, harassing or special weapons (chemical) delivery. |
| Antiaircraft artillery and mobile surface-to-surface missiles located well forward with maneuver elements | Provides increased protection to massed forces before attack. extends air defense umbrella forward as units advance. |
| Demonstrations and feints. | May precede an attack; may deceive actual point of attack |
| Establishment and strengthening of counterreconnaissance screen. | Protects assembly areas and forces as they prepare for attack. May be effort to prevent friendly forces from seeing attack preparations. |
| Concentration of mass toward one or both flanks within the forward area. | May indicate intent for single or double envelopment. particularly if massing units are armor heavy. |
| Increased patrolling or ground reconnaissance. | May indicate efforts to gather detailed intelligence regarding friendly dispositions prior to attack. |
| Command posts located well forward; mobile command posts identified. | Indicates preparation to command an offensive task from as far forward as possible. |

**Figure C-3. Offensive indicators**

| Activity | Explanation |
|---|---|
| Movement of noncombatants from the area of operations. | Indicates preparation for rapid forward advance of troops and follow-on forces. |
| Extensive conduct of drills and rehearsals in unassigned areas. | Often indicates major attacks, particularly against fortified positions or strongly defended natural or man-made barriers, which require rehearsal of specialized tactics and skills. |
| Cessation of drills and rehearsals. | Rehearsals are completed and the unit is preparing for offensive tasks. |
| Increased activity in supply, maintenance, and motor transport areas. | May indicate movement of additional forces to the front to sustain a major attack. Stocking of sustainment items, such as ammunition and medical supplies before an attack. |
| Increased aerial reconnaissance (including unmanned aircraft systems). | Threat effort to collect further intelligence on friendly dispositions or defensive positions. |
| Establishment or securing of forward arming and refueling points, auxiliary airfields, or activation of inactive airfields. | Preparation for increased sorties for aircraft and faster turnaround time and aviation sustainment. Indicates preparation, to support offensive tasks with aircraft as far forward as possible. |
| Clearing lanes through own obstacles. | Facilitates forward movement and grouping of assault units, particularly at night and usually immediately precedes an attack |
| Reconnaissance, marking, and destruction of defending force s obstacles. | Indicates where assaults will occur |
| Gap-crossing equipment (swimming vehicles, bridging, ferries, and assault boats) located in forward areas (provided large water obstacle or gap) | Expect a substantial effort to cross a water obstacle during a main attack |
| Staging of airborne, air assault, or special forces with transportation assets, such as transport aircraft or helicopters. | Airborne or air assault operations will likely indicate efforts to attack friendly commands, communications, or sustainment nodes. May indicate a main effort in which airborne forces will link with ground maneuver forces. |
| Increased signals traffic or radio silence. | May indicate intent to conduct offensive tasks, however, increased traffic may be an attempt to deceive. Radio silence denies information derived from signals intelligence. |
| Signals intelligence and electronic warfare assets located forward. | Provides electronic attack and surveillance support for the attack |

**Figure C-3. Offensive indicators (continued)**

| Activity | Explanation |
|---|---|
| Preparation of battalion and company defensive areas consisting of company and platoon strong points. | Indicates intent for holding terrain with defense in-depth, normally supported by armored counterattack forces. |
| Extensive preparation of field fortifications, obstacles, and minefields. | Indicates strong positional defense. |
| Attachment of additional antitank assets to frontline defensive positions. | Indicates intent to contest friendly armor in forward positions, and attempts to attrite and channel friendly armor into engagement areas for armor counterattack forces. |
| Formation of antitank strong points in-depth along avenues of approach | May allow penetration of friendly armor into engagement areas. Will engage armor in-depth. |
| Preparation of alternate artillery positions. | Increases survivability of artillery in the defense. Indicates great effort to support main defensive area with artillery—no withdrawal of maneuver forces from main defense unless defeated. |

**Figure C-4. Defensive indicators**

| Activity | Explanation |
|---|---|
| Concentration of armor units in assembly areas in the rear of the main defensive area. | Indicates holding armor units in reserve for possible counterattack or counteroffensive tasks. |
| Presence of concentrated antitank reserves. | Provides quick reaction capability against armor penetrations of the main defense. |
| Displacement of sustainment and medical units toward the rear area. | Facilitates defensive repositioning, maneuver, and counterattacks (support units are not in the way) |
| Prestocking of ammunition, supplies, and engineer or pioneer equipment in forward positions. | Reduces the burden on sustainment support during the battle, reduces vulnerability of interdiction of supplies, and ensures strong points can survive for reasonable periods if bypassed or cut off by advancing forces. |
| Increased depth from the forward line of troops of artillery and surface-to-surface missile units. | Allows continued employment of artillery during maneuver defense without significant rearward displacement. |
| Increased use of landline communications—often with corresponding decrease in radio traffic | Implies intent to remain in position because landlines are less vulnerable to electronic warfare and provide more secure communications. |
| Presence of dummy positions, command posts, and weapons. | Complicates friendly targeting and analysis. Deceives attacking force of actual defensive positions and strength |
| Air defense more concentrated in one particular area. | Indicates location of numerous high-value targets, such as armor, sustainment, artillery, or command posts |

Figure C-4. Defensive indicators (continued)

| Activity | Explanation |
|---|---|
| Withdrawal from defensive positions before becoming heavily engaged. | Indicates delaying action to avoid decisive engagements. |
| Numerous local counterattacks with limited objectives; counterattacks broken off before position is restored | Assists disengaging units in contact, rather than an attack to restore position. |
| Units bounding rearward to new defensive positions, while another force begins or continues to engage. | Indicates units conducting local withdrawals to new positions. Usually an effort to preserve the defending force and trade space for time. |
| Maximum firepower located forward; firing initiated at long ranges. | Intent to inflict casualties, thus slowing advance of attacking force and providing sufficient volume of fire to avoid decisive engagements. Allows for time to disengage and reposition defending forces. |
| Extremely large unit frontages compared to usual defensive positions. | Indicates delaying action to economize force, allowing larger formations to withdraw. |
| Chemical or biological weapons in forward areas. Reports of threat in chemical protective clothing while handling munitions. | Indicates possible chemical munitions use. Chemically contaminated areas cause significant delays to attacking forces. |
| Identification of dummy positions and minefields. | Indicates defending force using economy of force. Causes advancing force to determine if mines are live or inert. |

Figure C-5. Delaying indicators

| Activity | Explanation |
|---|---|
| Systematic destruction of bridges, communications facilities, and other assets. | Denies advancing force the use of infrastructure and installations in withdrawal areas. |
| Establishment of a covering force or rear guard. | Covers withdrawal of main body; usually consists of a subelement of the main force; usually only the rear guard element engages attacking forces. |
| Increased rearward movement at night, particularly during inclement weather | Attempt to avoid contact with the attacking unit in order to preserve the force and its combat power |
| Minimal presence of sustainment and medical units. | Withdrawal of nonessential sustainment and medical assets. It may also indicate the inability to move depots and dumps. |
| Establishing and marking withdrawal routes and traffic control points. | Facilitates rapid movement of forces to the rear. Indicates attempt to preserve force by conducting an organized and rapid withdrawal. |
| Preparation of new defensive positions beyond supporting range of present positions. | Indicates an attempt to establish new positions along suitable terrain before the arrival of deliberately withdrawn forces. |
| Increased engineer activity and stockpiling of explosives in threat rear area near bridges, tunnels, or built-up areas | Mobility operations facilitate a withdrawal by maintaining lines of communications for own forces. Demolition preparation indicates likely destruction of infrastructure in front of attacking force. |
| Rearward movement of long-range artillery. | Positions long-range artillery in subsequent defensive positions in order to support withdrawal with indirect fire. |
| Activation of command posts well removed (beyond usual norms) from the present battle area. Positioning of command posts along route of withdrawal. | Establishes command nodes in the new position and along route of march in order to control movement and arrival of forces. |

**Figure C-6. Withdrawal indicators**

| Indicators of Aggressive Behavior Within the Population |
|---|
| Identification of agitators, insurgents, and militias or criminal organizations, as well as their supporters and sympathizers, who suddenly appear in, or move from, an area. |
| New faces or unknown people in a rural community. |
| Unusual gatherings among the population |
| Disruption of normal social patterns. |
| Mass migration from urban to rural locations or from rural to urban locations. |
| Massing of combatants of competing power groups. |
| Increase in the size of embassy or consulate staffs from a country or countries that support indigenous disaffected groups, particularly those hostile to the United States or the current intervention. |
| Increase in neighboring countries of staff and activities at embassies or consulates of countries associated with supporting indigenous disaffected groups. |
| Lack of children playing outside in neighborhoods. |
| Increased travel by suspected subversives or leaders of competing power bases to countries hostile to the United States or opposed to the current intervention. |
| Influx of opposition, resident, and expatriate leaders into the area of operations. |
| Reports of opposition's disaffected indigenous population receiving military training in foreign countries |
| Increase of visitors, such as tourists, technicians, business persons, religious leaders, and officials from groups or countries hostile to the United States or opposed to the current intervention |

**Figure C-7. Population indicators—aggressive behavior**

| Indicators of Aggressive Behavior Within the Population |
| --- |
| Close connections between diplomatic personnel of hostile countries and local opposition groups. |
| Communications between opposition groups and external supporters. |
| Increase of disaffected youth gatherings, such as student protests or demonstrations. |
| Establishment of organizations of unexplained origin and with unclear or nebulous aims. |
| Establishment of new organizations that replace an existing organizational structure with identical aims. |
| Appearance of many new members in existing organizations, such as labor unions. |
| Infiltration of student organizations by known agitators. |
| Appearance of new organizations stressing grievances or interests of repressed or minority groups. |
| Reports of large donations to new or revamped organizations. |
| Reports of payment to locals or engaging in subversive or hostile activities. |
| Reports of the formation of opposition paramilitary or militia organizations. |
| Reports of lists of targets for planned opposition attacks. |
| Appearance of professional agitators in gatherings or demonstrations that result in violence. |
| Evidence of paid and armed demonstrators' participation in riots or violent protests. |
| Significant increase in thefts, armed robberies, and violent crime in rural areas; increase in bank robberies in urban areas. |
| Surveillance of host-nation government or mission force facilities and personnel. |

Figure C-7. Population indicators—aggressive behavior (continued)

| Opposition-Directed Aggressive Behavior Within the Population |
| --- |
| Refusal of population to pay, or unusual difficulty in collecting rent, taxes, or loan payments. |
| Trends of demonstrated hostility toward government forces or the mission force. |
| Unexplained disappearance of the population or avoidance of certain areas. |
| Unexplained disappearance or relocation of children and adolescents. |
| Reported incidents of attempted recruitment to join new movements or underground organizations. |
| Criminals and disaffected youth who appear to be acting with, and for, the opposition. |
| Reports of extortion and other coercion by opposition elements to obtain financial support. |
| Use of fear tactics to coerce, control, or influence the local population. |

| Activities Directed Against the Government or Mission Force Within the Population |
| --- |
| Failure of police and informer nets to report accurate information, which may indicate sources are actively supporting opposition elements or the sources are intimidated. |
| Decreasing success of government law enforcement or military infiltration of opposition or disaffected organizations. |
| Assassination or disappearance of government intelligence sources. |
| Reports of attempts to bribe or blackmail government officials, law enforcement employees, or mission personnel. |
| Classified information leaked to the media. |
| Sudden affluence of certain government and law enforcement personnel. |
| Recurring failure of government or mission force raids on suspected opposition organizations or illegal activities apparently due to forewarning. |
| Increased hostile or illegal activity against the government, its law enforcement and military organizations, foreigners, minority groups, or competing political, ethnic, linguistic, or religious groups. |
| Demonstrations against government forces, minority groups, or foreigners designed to instigate violent confrontations with government or mission forces |
| Increased antigovernment or mission force rhetoric in local media. |

Figure C-8. Population indicators—directed behavior

| Activities Directed Against the Government or Mission Force Within the Population (continued) |
|---|
| Occurrence of strikes or work force walkouts in critical industries or geographic areas intended to cast doubt on the government's ability to maintain order and provide security and services to the people. |
| Unexplained loss, destruction, or forgery of government identification cards and passports. |
| Recurring unexplained disruption of public utilities |
| Reports of terrorist acts or extortion attempts against local government leaders and business persons. |
| Murder or kidnapping of government, military, and law enforcement officials or mission force personnel. |
| Closing of schools. |
| Reports of attempts to obtain classified information from government officials, government offices, or mission personnel. |

**Figure C-8. Population indicators—directed behavior (continued)**

| General Indicators of Adversary Information Activities |
|---|
| Dissident propaganda from unidentified sources. |
| Increase in the number of entertainers with a political message. |
| Increase of political themes in religious services. |
| Increase in appeals directed at intensifying general ethnic or religious unrest in countries where ethnic or religious competition exists. |
| Increase of agitation on issues for which there is no identified movement or organization. |
| Renewed activity by dissident or opposition organizations thought to be defunct or dormant. |
| Circulation of petitions advocating opposition or dissident demands. |
| Appearance of opposition slogans and pronouncements by word-of-mouth, graffiti, posters, leaflets, and other means |
| Propaganda linking local ethnic groups with those in neighboring countries or regions |
| Clandestine radio broadcasts intended to appeal to those with special grievances or to underprivileged ethnic groups |
| Use of bullhorns, truck-mounted loudspeakers, and other public address equipment in spontaneous demonstrations. |
| Presence of nonmedia photographers among demonstrators. |
| Dissident propaganda from unidentified sources. |

| Adversary Information Activities Directed Against the Established Government |
|---|
| Attempts to discredit or ridicule national or public officials |
| Attempts to discredit the judicial and law enforcement system |
| Characterization of government projects and plans. |
| Radio and Internet propaganda from foreign countries that is aimed at the target country's population and accuses the target country's government of failure to meet the people's needs |

| Adversary Information Activities Directed Against the Mission Force and Host-Nation Military and Law Enforcement |
|---|
| Spreading accusations that the host-nation military and police are corrupt and out of touch with the people |
| Spreading accusations that mission force personnel will introduce customs or attitudes that are in opposition to local cultural or religious beliefs |
| Character assassinations of mission military and law enforcement officials. |
| Demands to remove strong anti-position or anti-crime military and law enforcement leaders from office |
| Calls for the population to cease cooperating with the mission force or host-nation military and law enforcement |

**Figure C-9. Adversary information indicators**

| Adversary Information Activities Directed Against the Mission Force and Host-Nation Military and Law Enforcement (continued) |
|---|
| Widespread hostile media coverage of even minor criminal violations or incidents involving mission force personnel. |
| Accusations of brutality or ineffectiveness or claims that mission or government forces initiated violence following confrontations. |
| Publication of photographs portraying repressive and violent acts by mission force or government forces |
| Refusal of business persons and shop owners to conduct business with mission force personnel. |

| Adversary Information Activities Directed Against the Education System |
|---|
| Appearance of questionable doctrine and teachings in the educational system. Creation of ethnic, tribal, religious, or other interest group schools outside the government educational system, which propagate opposition themes and teachings. |
| Charges that the educational system is only training youth to do the government's bidding. |
| Student unrest manifested by new organizations, proclamations, demonstrations, and strikes against authority |

**Figure C-9. Adversary information indicators (continued)**

| Indicators of Negative Food-Related Activities |
|---|
| Diversion of crops or meat from markets. |
| Unexplained shortages of food supplies when there are no reports of natural causes. |
| Increased reports of foodstuffs pilfering. |
| Sudden increase in food prices possibly indicating an opposition-levied tax. |
| Spot shortages of foodstuffs in regions or neighborhoods associated with a minority group or weaker competing interest group, while food supplies are generally plentiful in other areas. Conversely, sudden local shortages of foodstuffs in rural areas may indicate the existence of an armed opposition group in that region |
| Sudden increase of meat in markets, possibly indicating slaughtered livestock because of a lack of fodder to sustain them. |
| Appearance of emergency relief supplies for sale in black markets possibly indicating diversion from a starving population. |
| Appearance of relief supplies for sale in normal markets in a country or region recently suffering from large-scale hunger, which may indicate the severity of the food crisis is diminishing. |

| Indicators of Negative Arms and Ammunition-Related Activities |
|---|
| Increased loss or theft of weapons from military and police forces. |
| Discovery of arms, ammunition, and explosives being clandestinely manufactured, transported, or cached. |
| Attacks on patrols resulting in the loss of weapons and ammunition |
| Increased purchase of surplus military goods. |
| Sudden increase in prices for arms and ammunitions on the open market. |
| Reports of large arms shipments destined for neighboring countries, but not intended for that government |
| Reports of known arms traffickers establishing contacts with opposition elements. |
| Increase in armed robberies. |
| Reports of thefts or sudden shortages of chemicals, which could be used in the clandestine manufacture of explosives. |
| Reports of large open-market purchases of explosives-related chemicals without an identifiable industrial use. |
| Appearance of manufactured or smuggled arms from noncontiguous foreign countries. |

**Figure C-10. Commodities indicators**

| Indicators of Negative Clothing-Related Activities |
|---|
| Unusual, systematic purchase or theft of clothing materials that could be used for the manufacture of uniforms or footwear |
| Unusual scarcity of clothing or material used in the manufacture of clothing or footwear |
| Distribution of clothing to underprivileged or minority classes by organizations of recent or suspect origin |
| Discovery of caches of uniforms and footwear or materials that could be used to manufacture uniforms and footwear |
| Increase of males in the streets wearing military style clothing or distinctive markings. |

| Indicators of Negative Medicine-Related Activities |
|---|
| Large-scale purchasing or theft of drugs and medicines or the herbs used to manufacture local remedies |
| Scarcity of drugs and medicinal supplies on the open or black markets. |
| Diversion of medical aid donations |
| Discovery of caches or medical supplies |

| Indicators of Negative Communications-Related Activities |
|---|
| Increase in the purchase and use of radios |
| Discovery of caches of communications equipment |
| Unusual increase in amateur radio or cellular telephone communications traffic. |

Figure C-10. Commodities indicators (continued)

| Indicators of Suspicious Rural Activities |
|---|
| Evidence of increased foot traffic in the area. |
| Increased travel within and into remote or isolated areas |
| Unexplained trails and cold campsites. |
| Establishment of new, unexplained agricultural areas or recently cleared fields. |
| Unusual smoke, possibly indicating the presence of a campsite or a form of communication. |
| Concentration of dead foliage in an area, possibly indicating use of camouflage. |
| Presence of foot traps, spikes, booby traps, or improvised mines along routes and trails. |

| Indicators of Suspicious Urban Activities |
|---|
| Apartments, houses, or buildings being rented, but not lived in as homes. |
| Slogans written on walls, bridges, and streets. |
| Defacement of government and mission force information signs. |
| Sabotage of electrical power network and pollution of urban area's water supply. |
| Terrorist acts against physical targets, such as bridges, dams, airfields, or buildings |
| Change of residence of suspected agitators or opposition leaders. |
| Discovery of message dead drops. |
| Increased smuggling of currency, gold, gems, narcotics, medical supplies, and arms into urban centers. |
| Appearance of abnormal amounts of counterfeit currency. |
| Increase in bank robberies. |
| Work stoppages or slowdowns in essential industries |
| Marked decline in product quality in essential industries. |
| Marked increase in equipment failures in essential industries. |
| Unexplained explosions in essential utilities and industries |
| Establishment of roadblocks or barricades around neighborhoods associated with opposition elements |
| Attempts to disrupt public transport through sabotage. |
| Malicious damage to industrial products or factory machinery. |

Figure C-11. Environment-related indicators

| Indicators of Basic Improvised Explosive Device Indicators, Observables, and Signatures |
| --- |
| Vehicles following convoys for a long distance and then pulling off to the side of the road. |
| Dead animals along the roadways. |
| Freshly dug holes along the roadway (possible future improvised explosive device report) |
| New dirt or gravel piles. |
| Obstacles in roadway used to channel the convoy. |
| Personnel on overpasses. |
| Signal with flares or city lights (turned off or on) as convoy approaches. |
| Absence of the ordinary children in the area or merchants at a market. |
| **Key Indicators, Observables, and Signatures (Indicating Something is About to Happen)** |
| Dramatic changes in population from one block to the next |
| Dramatic changes in illumination (lights) from one area to the next during hours of limited visibility |
| Absence of children when normally present. |
| Identification of markings indicated in intelligence reports of an improvised explosive device site. |
| New dirt or gravel piles. |

**Figure C-12. Improvised explosive device indicators, observables, and signatures**

| Indicator | Information Objective |
| --- | --- |
| Local Conflict Casualty | What groups (tribes, clans) are local rivals? |
| | How intense is the rivalry? |
| | What are the relative strengths or external alliances of rival groups? |
| | U.S. presence or host government? |
| | Do locals normally carry arms? |
| | Does group rivalry parallel rivalry within the host government? |
| Recurrence of Same Clan Name Among Detainees | Is the clan native to the area? If so, where does the clan reside, in which villages? |
| | How big is the clan and how many male adults? |
| | Who is the acknowledged chief? |
| | Do or did any members of the clan have positions in former regime? Who are they? Do any of them have access to arms or ammunition? Where is their cache or source? |
| | Do any of them provide training to other relatives? |
| | What are the usual economics of the clan? |
| | Can the clan exploit these activities to gain arms or facilitate or conceal their operations? |
| | Has any relative been killed by U.S or multinational forces? If so, is there a current mood of blood vengeance within the clan? |
| | Which mosques do clan members attend? Do any of them follow a particular doctrine? Is that doctrine radical or moderate? If radical, do the imams encourage hostility to the U.S. presence? |
| | Has the clan offered protection to any strangers or foreigners? Are these people recent arrivals or long-term residents? What is the identity and agenda or business of such people? |

**Figure C-13. Tribe or clan indicators**

# CRIMINAL AND NONCOMBATANT ACTOR ANALYSIS

C-81. This framework concerns criminal and noncombatant actors that will be present in the UW operational area.

> *Note:* The sections are excerpted from Chapter 4, Criminals, and Chapter 5, Noncombatants,
> TC 7-100.3, *Irregular Opposing Forces*.

## CRIMINAL CHARACTERISTICS

C-82. Criminal elements exist at every level of society and in every operational environment. Their presence, whatever their level of capabilities, adds to the complexity of any operational environment. They may be intertwined with irregular forces and possibly with regular military or paramilitary forces of a nation-state. However, they may also pursue their criminal activities independent of such other actors.

C-83. Some individuals, groups, and activities are criminal or illegal only because they violate laws established by a recognized governing authority. Others may violate moral or ethical standards of a given society or of the international community. In some operational environments, the threat is more criminal than military or paramilitary in nature. Insurgents, guerrillas, or other armed groups often use or mimic established criminal enterprises and practices to move contraband, raise funds, or otherwise further their goals and objectives.

C-84. Criminal activity is a category of violence that is enmeshed in the daily life of most people in both urban and rural areas. However, criminal activity thrives in areas where there is instability and lack of government control or law enforcement. The actions of insurgents and guerrillas further erode stability and effective governance, creating more opportunities for criminal pursuits. Sometimes, given those opportunities, insurgent or guerrillas themselves turn to crime—either to sustain themselves or for personal profit. It may be difficult to distinguish crime from ethnic feuds, ideological and theological extremism, or other elements of a culture that incite insurgency or guerrilla warfare.

C-85. Governing authorities often characterize insurgents and guerrillas as "bandits." The reason for this is that their activities in opposing the governing authority and sustaining themselves are illegal (from the government perspective). Acts of subversion may be against the law (that is, criminal) even if not violent.

## CRIMINAL ACTIVITIES

C-86. Criminals use many and varied tactics and techniques. Some of these methods overlap with one another. The activities typically include an objective to make fiscal profit or achieve influence.

### Security

C-87. Security is crucial for criminal organizations. These organizations may use the highest degree of sophistication available to conduct intelligence collection and counterintelligence activities. These activities are a priority and can be well funded. Intelligence sources may extend to high levels within government and law enforcement agencies. The local populace may willingly provide ample intelligence collection, counterintelligence, and security support. Intelligence and security can also be the result of bribery, extortion, or coercion.

C-88. Most members of criminal organizations are capable of protecting themselves and their assets. Typically, they carry small-caliber weapons, such as handguns, pistols, rifles, and shotguns. They are lightly armed out of necessity or convenience, not for lack of resources. When greater force of arms is necessary to control people, protect vital resources, or obtain information, these organizations typically have members who can use heavier arms, such as machineguns and assault weapons. Large criminal organizations may hire private security contractors to conduct surveillance, provide personal security for leaders, or guard key facilities.

## Theft

C-89. Theft is the taking of another person's property without that person's permission or consent with the intent to deprive the rightful owner of it. Thus, theft is an overarching term that covers various crimes against property such as burglary, embezzlement, larceny, looting, robbery, and fraud. (See also identity theft and intellectual property theft, both under Cyber Crime.)

## Fraud

C-90. A fraud is an intentional deception made for personal gain or to damage another individual or entity. Defrauding people or entities of money or valuables is a common purpose of fraud. Fraud can include:
- Ponzi schemes.
- Insider trading.
- Embezzlement.
- Insurance scams.
- Money laundering.
- Forgery.

C-91. Fraud can also include cybercrimes, such as—
- Identity theft.
- Copyright infringement.

## Racketeering

C-92. A racket is a fraudulent scheme, enterprise, or activity. It is usually an illegitimate business made workable by bribery or intimidation. A racketeer is one who extorts money or advantages by threats of violence, by blackmail, or by unlawful interference with business or employment. Therefore, racketeering overlaps with bribery, intimidation, and extortion.

C-93. In racketeering, a criminal organization typically creates or perpetuates a problem or the perception of a problem for which it then offers a solution, for a fee. The intent is to engender continual patronage. In the traditional example of a protection racket, the racketeer informs a store owner that a substantial monthly fee will be required in exchange for protection. The protection provided takes the form of the absence of damage inflicted upon the store or its employees by the racket itself. Another example is malicious software (malware) that pretends to be detecting spyware or other infections on a computer and offers to download a cleaning utility for a fee. In actuality, the distributor of the malware is also the maker of the cleaning utility. This aspect of racketeering overlaps with the category of cybercrime, discussed below. In addition to protection rackets, racketeering can also involve numbers rackets or illegal lottery.

C-94. Racketeering can also include seemingly legitimate businesses that provide a front for illegal activity, such as buying and selling illegal merchandise. Loan sharking is another form of racketeering in which a criminal offers loans at extremely high interest rates to borrowers who cannot qualify for loans from legitimate sources. The debtor faces violence or other criminal means to cause harm if the debt is not paid.

C-95. Black markets fall under racketeering as illegal businesses. These can include traded goods and services, such as biological organs, transportation providers, illegal drugs, prostitution, weaponry, alcohol, tobacco, currency, and fuel. Black markets tend to flourish in most countries during wartime, because of rationing and impact commodities, such as food, fuel, rubber, and metal.

## Gambling

C-96. In most societies, the act of gambling is in itself not illegal. What becomes illegal is an illicit business based on gambling (also called gaming). It is illegal because either:
- It is conducted in a fraudulent manner.
- The governing authority has outlawed it.

- The gambling enterprise fails to share all its profits with the governing authority. Gambling in the form of a numbers racket or illegal lottery is a form of racketeering. Internet gambling is a form of cybercrime.

C-97. Many types of gambling have been, or still are, illegal in some places. Hence, criminals may be the only operators of some games. Even when a governing authority legalizes some gambling ventures, it may be difficult to keep criminals from becoming involved because of the huge potential for profits. Besides gaming operations recognized and (sometimes poorly) regulated by the state, there are many other gambling opportunities in which criminals may be involved.

C-98. Gambling is often associated with other types of crime. Operators of legal gambling establishments may receive kickbacks for allowing money laundering. Gambling operations earn large amounts of cash and present particular opportunities for skimming and money laundering. Aside from skimming profits from a legalized gambling enterprise, skimming can also occur with the granting of credit, which can lead into loan sharking. In particular, problem gamblers who need money for gambling may easily fall prey to loan sharks. Compulsive gambling can also lead to crimes, such as embezzlement, robbery, check forgery, and fraud.

## Prostitution

C-99. Prostitution is the practice of indulging in sexual relations for money. As a criminal enterprise, it also involves a criminal or criminal organization that profits from the commercial sex act. Prostitutes may also profit from the venture, although the criminal organization gets its cut. However, prostitution may be linked to human trafficking, in which case the criminals get all the profits.

## Extortion

C-100. Extortion is the act of obtaining money, materiel, information, or support by force or intimidation. Criminal organizations use extortion to obtain information or cooperation or to protect its members. Examples of extortion include:

- Intimidating politicians to vote in a manner favorable to the criminal organization.
- Intimidating judges to free an organization member.
- Forcing a farmer to grow drug-producing crops
- Extorting money from local businesses in exchange for protection, which means not harming the business or its members.
- Using death threats against an individual or his family to cause him to provide information or resources.
- Intimidating other people not to take action against the criminal organization.
- Using information warfare methods to create and maintain fear caused by extortion.

## Bribery

C-101. Bribery is giving money or other favors to influence someone. Criminals give money to people in power who make or influence decisions. For example, bribes to law-enforcement officials can cause them to have their patrols avoid a criminal organization's transit routes. If the organization is unable to bribe someone, it employs harsher methods, such as extortion, assassination, or murder, to gain cooperation.

## Arson

C-102. Arson is maliciously burning another person's dwelling, structures, or property. This may be done as punishment for noncompliance with internal rules of criminal organizations. It may be the result of not paying the criminal organization for protection. It may also be done on a for-hire basis or as a means to inflict terror on a targeted person or group.

## Hijacking

C-103. Hijacking is stealing or commandeering a conveyance. Criminals may conduct a hijacking to produce a spectacular hostage situation. Sometimes criminals may hijack a conveyance as a means of escape; in that case, the criminals may eliminate any unneeded people and materiel, such as hostages or baggage.

## Kidnapping

C-104. Kidnapping is an abduction or transportation of a person or group by force. The person is kept in false imprisonment (confinement without legal authority). This may be done for ransom or in furtherance of another crime (such as human trafficking or hostage taking). This type of crime has become very popular with criminal organizations, and the methods vary by region. Kidnapping flourishes particularly in fragile or failed states and regions in conflict, as drug traffickers and other criminal organizations fill the vacuum left by governing authority.

C-105. The risk in kidnapping is relatively lower than in hostage taking. This is primarily because the criminals take the kidnapped victim to a location controlled by the criminal organization. The criminals then make demands and are willing to hold a victim for a long time, if necessary.

## Hostage Taking

C-106. Hostage taking is typically an overt seizure of a person or persons to gain publicity, concessions, or ransom. Unlike kidnapping, in which usually a prominent individual is taken, the hostages are usually not well-known figures. Criminals attempt to hold hostages in a neutral or friendly area. The planning and conduct of a hostage taking are similar to those of a kidnapping or hijacking. However, criminals may also take hostages as an expedient measure when they have difficulty exiting a crime scene.

## Murder

C-107. Murder is the unlawful killing of another human being without justification or excuse. Murder is perhaps the single most serious criminal offense. Criminal organizations use murder as an enforcement tool and as a method of generating revenue in murder-for-hire schemes.

## Assassination

C-108. An assassination is the murder (usually of a prominent person) by a sudden or secret attack. It is usually prompted by religious, ideological, political, or military motives, but it may be done for payment. Criminals and criminal organizations may use an assassination:
- For monetary gain.
- To exert terror.
- To display power.
- To exact revenge on a public official.
- To eliminate people they cannot intimidate.
- To punish people who have left the criminal organization.

## Maiming

C-109. Maiming is a deliberate act to mutilate, disfigure, or severely wound a person so as to cause lasting damage. Maiming can involve assault and battery on a person with intent to inflict serious injury. Other methods include:
- Male castration.
- Female genital mutilation.
- Burning or branding.
- Forcible tattooing.
- Cutting off limbs.
- Removal of tongue, eyes, or ears.
- Throwing a corrosive acid or alkali to cause blindness or scarring.

C-110. The person maimed is an outward sign of the criminal organization's power and control. A criminal organization often uses or threatens maiming:

- To enforce order within the organization.
- To collect debts.
- For extortion.
- As a for-hire operation to generate revenue.

## Smuggling

C-111. Smuggling is the clandestine transportation of illegal goods or persons. It usually involves illegal movement across an international border. There are various motivations to smuggle. These include participation in illegal trade, illegal immigration or emigration, and tax evasion. Smuggling is often related to trafficking in persons, drugs, or arms.

## Trafficking

C-112. Trafficking is the transportation of goods or persons for the purpose of making a profit. Criminals conduct illegal trafficking. Human trafficking (trafficking in persons) is the second largest criminal activity in the world—followed by drug trafficking and arms trafficking.

### *Human Trafficking*

C-113. Human trafficking is the recruitment, transportation, transfer, harboring, or receipt of persons (including children) for the purpose of exploitation. It involves the threat or use of force, coercion, abduction, fraud, deception, or abuse. Human trafficking may take two forms:

- Sex trafficking, in which a person is induced to perform a commercial sex act.
- The recruitment, harboring, transportation, provision, or obtaining of a person for labor or services, for the purpose of subjection to involuntary servitude, debt bondage, or slavery. Criminals choose to traffic human beings because, unlike other commodities, people can be used repeatedly and because human trafficking requires little in terms of capital investment.

C-114. Human trafficking is not the same as people smuggling. A smuggler may facilitate illegal entry into a country for a fee, but on arrival at their destination, the smuggled person is free. The trafficking victim is coerced in some way and is further exploited after arrival, in order to derive profits. Victims do not agree to be trafficked; they are tricked, lured by false promises, or forced into it. Traffickers control their victims by coercive tactics including deception, fraud, intimidation, isolation, physical threats and use of force, debt bondage, or even force-feeding drugs.

### *Drug Trafficking*

C-115. The illegal drug trade is a black market consisting of production, distribution, packaging, and sale of illegal psychoactive substances. The legality or illegality of the black markets purveying the drug trade is relative to geographic location. The drug-producing countries may be inclined to tolerate the drug traffickers because of bribery or the effect on the country's economy. Drugs often cross international borders in order to reach the best paying customers. The massive profits inherent to the drug trade serve to extend its reach. The social consequences of drug trade include crime, violence, and social unrest.

### *Arms Trafficking*

C-116. Arms trafficking involves illicit transfers of arms, ammunition, and associated materials. Criminal organizations may be involved in two types of arms trafficking:

- Small-scale transactions by individuals or small firms that deliberately transfer arms to illicit recipients.
- Higher-value or more difficult illicit shipments of arms involving corrupt officials, brokers, or middle men motivated mainly by profit.

C-117. Arms trafficking is driven by a variety of clients, which include:
- Embargoed governments.
- Armed groups involved in war, banditry, terrorism, or insurgency.
- Criminals and criminal organizations.
- Citizens who cannot obtain guns legally.

C-118. Some arms and ammunition may come from illicit arms manufacturers. However, the source of a large proportion of illicit conventional arms is government disposals of surplus arms or thefts from insecure government stockpiles. Governments themselves may deliberately facilitate covert flows of arms to their proxies or allies or to embargoed or suspect destinations for profit.

C-119. Arms trafficking is widespread in regions of political turmoil. However, it is not limited to such areas. Most arms trafficking occurs at the regional or local level. Among the most common forms are numerous shipments of small numbers of weapons that, over time, result in the accumulation of large numbers of illicit weapons. While individual transactions occur on a small scale and do not draw attention, the sum total of weapons trafficked is large.

## Cybercrime

C-120. Cybercrimes are offenses targeting or using information technology. This includes computers, computer networks, and other telecommunications networks, such as the Internet (chat rooms, emails, notice boards, and groups) and mobile phones. Such crimes may threaten, not only individuals and groups but also, a nation's security and financial health. Cybercrimes can facilitate a variety of other criminal activities, including money laundering, extortion, fraud, racketeering, gambling, smuggling, and trafficking.

C-121. Criminals exploit the speed, convenience, and anonymity that modern technologies offer in order to commit a diverse range of criminal activities. The global nature of the Internet allows criminals to commit almost any online, illegal activity anywhere in the world. In the past, cybercrime has been committed by individuals or small groups of individuals. However, an emerging trend is for criminal networks and criminally minded technology professionals to work together and pool their resources and expertise.

C-122. Cybercrimes include network intrusions, hacking attacks, malicious software, and account takeovers, leading to significant data breaches affecting every sector of the world economy. Advances in computer and telecommunications technology and greater access to personal information via the Internet have created a virtual marketplace for transnational cyber criminals to share stolen information and criminal methodologies. The increasing level of collaboration among cyber criminals raises the level of potential harm to individuals, companies, and governments. Members of online forums discuss cybercrime topics of interest. Criminal purveyors buy, sell, and trade:
- Malicious software.
- Hacking services.
- Spamming devices.
- Personal identification information.
- Credit and debit card data.
- Bank account information.
- Brokerage account information.
- Counterfeit identity documents.
- Other forms of contraband.

C-123. Cybercrime falls into two broad categories:
- Computer crimes.
- Intellectual property crimes.

## Computer Crimes

C-124. Computer crime refers to any crime that involves a computer or computer network. The computer or network may be the target, or it may be used in the commission of a crime, the primary target of which is independent of the computer network or device.

### Crimes Targeting Computers

C-125. Criminals can cause damage to computers in many ways. For example, an unauthorized intruder can send commands that delete files or shut the computer down. Intruders can initiate a denial-of-service attack that floods the victim's computer with useless information and prevents legitimate users from accessing it. A virus or worm can use up all of the available communications bandwidth on an agency or corporate network, making it unavailable to employees. When a virus or worm penetrates a computer's security, it can delete files, crash the computer, install malicious software, or do other things that impair the computer's integrity.

C-126. Extortion threats involving damage to a computer are a high-technology variation of old-fashioned extortion. The threat need not be sent electronically. Intruders may threaten, unless their demands are met, to:

- Penetrate a system and encrypt or delete a database (erasing or corrupting data or programs).
- Distribute denial-of-service attacks that would shut down (or slow down) the victim's computers.
- Steal confidential data.

C-127. Crimes may involve intercepting or interfering with communications through the use of electronic, mechanical, or other devices. This applies to electronic communications, as well as oral and wire communications via common carrier transmissions. This may involve—

- Spyware or intruders using packet sniffers.
- Persons improperly cloning email accounts.
- Other surreptitious collection of communications from a victim's computer.

C-128. The crime may extend to the unauthorized disclosure of the contents of an illegally intercepted communication or using such information for other criminal purposes. Such crimes may also injure or destroy various types of communications operated or controlled by a government or used for a state's military or paramilitary functions.

### Crimes Using Computers

C-129. Crimes using computers or facilitated by computers include:

- Unauthorized access to computers (even government computers and perhaps law enforcement).
- Unauthorized access to stored communications (including email, social networking data, and voicemail).

C-130. Two of the most common uses are for:

- Fraud (computer fraud and wire fraud).
- Identity theft.

C-131. **Computer fraud** is the use of information technology to commit fraud. Data or information or even the use of a computer can be regarded as a thing of value. Criminals may devise various schemes or artifices to obtain money, property, goods, or services of measurable value. These include:

- Accessing a computer without authorization to obtain:
  - Commercial data.
  - Information contained in a financial record of a financial institution.
  - Information contained in a file of a consumer reporting agency on a customer.
  - Information from any government department or agency.
  - Information from a protected computer.
- Accessing without authorization a government, commercial, or personal computer and affecting the use of the owner's operation of the computer.
- Accessing a protected computer with the intent to defraud and thereby obtain anything of value.

- Causing the transmission of a program, information, code, or command that causes damage to a computer system, personal injury, or a threat to public health or safety
- Trafficking in passwords or similar information through which a computer may be accessed without authorization.
- Manipulating market data for criminal purposes.

C-132. Computer fraud can also fall under the more general category of wire fraud. In what is considered **wire fraud**, criminals may use fraudulent pretenses, representations, or promises transmitted by various means of telecommunication, including:

- Wire.
- Radio.
- Television.
- Mobile phone.
- Facsimile.
- Telex.
- Modem.
- Internet.

C-133. Money mule schemes could be categorized as computer or wire fraud. In such schemes, criminals trick people into moving money for them. Through Internet phishing, criminals can steal money from unsuspecting people by accessing their accounts. The phishers then face the problem of moving the large sums of money acquired from the victims to their own accounts, often in other countries, without attracting suspicion. Traditionally, drug smugglers have used people (referred to as mules) willing to carry small amounts of drugs across borders for them for a price. In a variant of this scheme, phishers recruit innocent people in the countries where their victims reside by offering them a lucrative job using their home computers. Those recruited are usually unaware that these are not legitimate business opportunities. Thus, phishers can dupe a large number of individuals (called money mules) into accepting relatively small amounts of money stolen from people's accounts and transferring the funds to the phishers in return for a commission.

C-134. **Identity theft** is almost always committed to facilitate other crimes. It involves unauthorized transfer, possession, or use of a means of identification of another person with the intent to commit, or to aid or abet, or in connection with, any unlawful activity. Most commonly, it involves the misuse of another individual's personal identifying information for fraudulent purposes. With relatively little effort, an identity thief can use this information to take over existing credit accounts, create new accounts in the victim's name, or even evade law enforcement after the commission of a violent crime. Identity thieves also sell personal information online to the highest bidder, often resulting in the stolen information being used by a number of different perpetrators. Personal information can also be obtained for the purpose of blackmail or other forms of extortion.

C-135. Criminals can obtain credit and debit card numbers by hacking into the wireless computer networks of major retailers. Network intrusions can compromise the privacy of individuals if data about them or their transactions resides on the victim network. Once inside the networks, criminals install "sniffer" programs that capture card numbers, as well as passwords and account information. After collecting the data, they conceal it in encrypted computer servers that they control. Then they sell the credit and debit card numbers through online transactions to other criminals. In addition to these "carders," there are "phishers," who obtain the same type of information via fraudulent emails.

C-136. The unlawful use of identification information can involve access device fraud. Such fraud can include any device, number, or other means of account access that can be used to obtain money, goods, services, or other things of value. It can also involve unlawful access to buildings or facilities.

## Intellectual Property Crimes

C-137. Intellectual property crimes involve theft of material protected by copyright, trademark, patent, or trade-secret designation. Such intellectual property is vital to local, national, and international economies.

C-138. The interconnected global economy creates unprecedented business opportunities to market and sell intellectual property worldwide. Geographic borders present no impediment to international distribution channels. If the product cannot be immediately downloaded to a computer, it can be shipped and arrive by next day air. However, the same technology that benefits rights-holders and consumers also benefits intellectual property thieves who are seeking to make a fast, low-risk profit. In addition, trafficking in counterfeit merchandise also generates large profits.

C-139. The most egregious violators are large-scale criminal networks and transnational criminal organizations, whose conduct threatens, not only intellectual property owners but also, the economy of nation-states. Because many violations of intellectual property rights involve no loss of tangible property and do not require direct contact with the rights-holder, the owner often does not even know that he is a victim for some time.

C-140. Intellectual property crimes can overlap with computer crimes, especially in the following areas:
- Unauthorized access of a computer to obtain information.
- Mail or wire fraud (can include the Internet)
- Devices to intercept communications.

C-141. Unauthorized obtainment of information or electronic media covered by copyright, trademark, patent, or trade-secret designation robs the rights-holders of their ideas, inventions, and creative expressions. Such theft is facilitated by digital technologies and Internet file sharing networks.

C-142. Although fraud schemes can involve copyrighted works, it is not mail or wire fraud unless there is evidence of any misrepresentation or scheme to defraud. Mail and wire fraud may exist, even if the perpetrator tells his direct purchasers that his goods were counterfeit, as long as he and his direct purchasers intended to defraud the direct purchaser's customers. Wire fraud can include the Internet.

C-143. Intellectual property crimes may involve intercepting and acquiring the contents of communications through the use of electronic, mechanical, or other devices. Criminals may then market the contents of an illegally intercepted communication, including intellectual property.

## Money Laundering

C-144. A criminal organization conducts money-laundering activities to transfer funds into the legitimate international financial system. Because of the legal restrictions levied by a governing authority, the organization must have a way of transferring "dirty" (illegally earned) money into "clean" money. The organization smuggles some currency back to its country of origin. However, large sums of foreign currency are not feasible for the organization because it must use the legal currency of its country to make transactions. For example, a farmer cannot receive payment for his drug crops in the foreign currency because he would be unable to use it.

C-145. A more productive way to transfer funds into the legitimate financial system is to operate through front companies. Front companies are legitimate businesses that provide a means to launder money. The criminal organization establishes its own front companies or approaches legal companies to act as intermediaries. Some front companies, such as an import or export business, may operate for the sole purpose of laundering money. Other companies (possibly targets of extortion) operate as profit-making activities and launder money as a service to the organization. Criminal organizations operate front companies in their own country, as well as in other countries.

C-146. Personnel involved in money laundering can include accountants, bankers, tellers, and couriers. Some are willing participants in the money-laundering process and accept bribes for their services. Extortion and intimidation keep unwilling members active in the process. Members of money-laundering organizations also establish and operate front companies.

C-147. Some members, such as accountants and bankers, perform their normal functions as in a legal business. However, they may conduct illegal acts on behalf of the organization. Because of banking regulations, members must conduct activities that do not draw attention to themselves. As an example, couriers deposit a small enough level of funds in bank accounts to avoid banking regulations. However, not all countries have such regulations, making it easier to launder large sums of money.

C-148. Couriers conduct many transactions with financial institutions. They travel from bank to bank, making deposits or converting money into checks and money orders to prepare for smuggling activities. Money-laundering personnel may also use electronic fund transfers, which make tracking of illegal transactions much more difficult, especially when the country has lax banking laws.

## Civic Actions

C-149. Criminal organizations conduct programs of patronage under the guise of civic actions. These programs are only indirectly intended to benefit the general populace. Rather, the main intent of the criminal organizations is to gain and maintain support, reward their supporters, and facilitate their continued activities. They may build a school, improve a road, or supply food and medicine. These projects benefit the local population because they improve the people's quality of life and may improve their standard of living. For example, road building makes it easier to transport goods to market and creates jobs. Some of the jobs may be temporary, such as those in construction, while others may be permanent, such as those in education and clinics. Through its propaganda efforts, the criminal organization ensures that the population knows who is making the improvements.

C-150. Sometimes the criminal organization cooperates with an insurgent force to conduct civic actions. They do not cooperate for ideological reasons. Their sole purpose is to inhibit the governing authority's ability to affect their activities by gaining the inherent security a grateful public can provide.

## Information Warfare

C-151. Criminal organizations may have the resources to conduct a variety of information warfare activities. However, their focus is often on a well-orchestrated perception management (propaganda) effort. This can be a powerful tool for intimidating enemies (governing authority and criminal competitors) and encouraging support of the organization's money-making efforts. The use of perception management techniques can ensure that the population knows who is making improvements in the local environment, allowing the organization to take credit for other benefits it provides to the local population.

> *Note:* TC 7-100.3 (Appendix A, Information Warfare) provides a detailed discussion on information warfare.)

C-152. The criminal organization can also use counterpropaganda to spin events against the governing authority. For example, the government may burn a farmer's drug-producing crop and then broadcast announcements that the action is a direct result of the farmer's involvement with an illegal activity. However, the criminal organization can counter this by instilling the idea that the governing authority does not have programs to help the farmer make profits legally, and reminding the populace of the new medical clinic it built.

C-153. Various elements of information warfare can also be instrumental in other criminal activities. For instance, perception management and deception are related to fraud, including computer fraud. Information attack and computer warfare are parts of cybercrime.

## Terrorism

C-154. Terrorism is always illegal, whether conducted by insurgents, guerrillas, or criminals. When conducted by regular military forces or other agencies of a nation-state (perhaps in the form of ethnic cleansing or confiscation of private property), other states can consider it illegal by international standards. For criminal organizations, terrorism becomes an action against an adverse governing authority, a criminal competitor, or an innocent populace.

> *Note:* TC 7-100.3, Chapter 6, Terrorism, provides more information on terrorism. Many of the tactics and techniques used by criminals and criminal organizations are the same as addressed for terrorism in TC 7-100.3. The only difference is that criminals may use these as means to obtain profits rather than just to instill fear or coerce governments and societies.

NONCOMBATANT CHARACTERISTICS

C-155. A host of noncombatants add complexity to any operational environment. The irregular opposing forces attempt to manipulate these noncombatants in ways that support its goals and objectives. Many noncombatants are completely innocent of any involvement with the irregular opposing forces. However, the irregular opposing forces will seek the advantage of operating within a relevant population of noncombatants whose allegiance or support it can sway in its favor. This can include clandestine yet willing active support (us combatants), as well as coerced support, support through passive or sympathetic measures, or unknowing or unwitting support by noncombatants.

C-156. Noncombatants are persons not actively participating in combat or actively supporting of any of the forces involved in combat. They can be either armed or unarmed. Figure C-14 shows examples of these two basic types of noncombatants that can be manipulated by the irregular opposing forces. These examples are not all inclusive, and some of the example entities can be either armed or unarmed.

*Note:* TC 7-100.3 (Chapter 5) provides additional details on various armed and unarmed combatants.

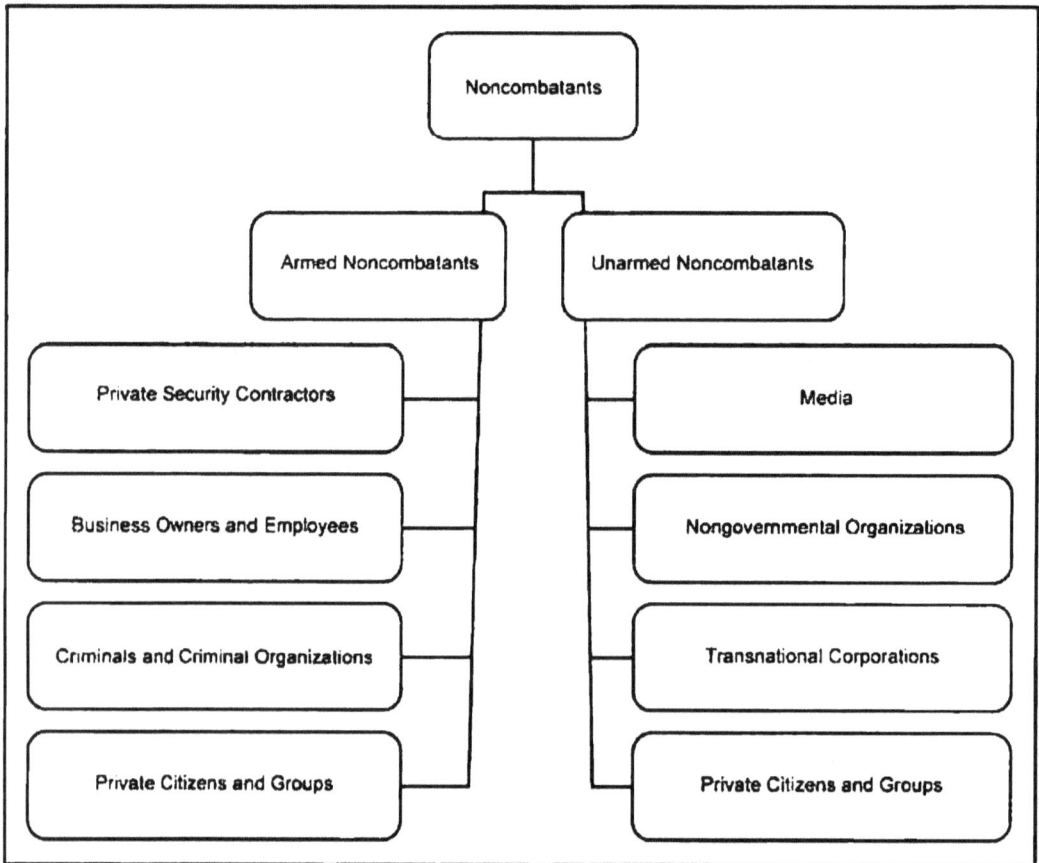

**Figure C-14. Armed and unarmed noncombatant examples**

C-157. Aside from military and paramilitary forces, the civilian population of a nation or region is often the single most important aspect of an operational environment. This situation can be further complicated by the presence of other noncombatants who are not indigenous to the country or region. From a U.S. viewpoint, the status of noncombatants is typically friendly, neutral, or unknown. Conversely, the noncombatants would view U.S. or local governing authority forces as friendly or neutral in regard to themselves.

# CULTURE AND CULTURAL GEOGRAPHY ANALYSIS

C-158. This framework addresses cultural considerations of the operational area.

*Note:* The cultural considerations are excerpted and summarized from FM 3-24, *Insurgencies and Countering Insurgencies* (Chapter 3, Culture).

## WHY STUDY CULTURE?

C-159. More than any other factor, culture informs and influences the worldview. In other words, culture influences perceptions, understandings, and interpretations of events. To be successful in interacting with the local population to gain information on the enemy or to understand their requirements, military members must do more than learn a few basic facts or "do's and do nots." They must understand the way that their actions can change the situation for the local population (both positively and negatively) and the resulting perceptions of the population toward those actions. Many people believe they view their world accurately, in a logical, rational, unbiased way.

C-160. However, people filter what they see and experience according to their beliefs and worldview. Information and experiences that do not match what they believe to be true about the world are frequently rejected or distorted to fit the way they believe the world should work. Mirror imaging is dangerous because it leads to thinking that U.S. assumptions about a problem and its solution are shared by the operational area population and multinational partners, rather than employing perspective taking and looking at the problem from the population's perspective. The four fundamental aspects of culture, when planning and executing military operations, are:
- Culture influences how people view their world.
- Culture is holistic.
- Culture is learned and shared.
- Culture is created by people and can and does change.

C-161. There are several key considerations for assessing a culture and cultural geography. These include:
- All cultures have a unique and interdependent connection with their physical surroundings.
  - A culture's physical surroundings (terrain, resources, weather, and access to transportation networks) dictate their range of choices.
  - The culture's choices in turn affect and change their physical surroundings (terracing, agriculture and cultivation, infrastructure building).

*Note:* This interdependence is subtle and varies from culture to culture and place to place. We may inadvertently change the environment if we neglect careful study.

- The economy of a culture may be formal, informal, or mixed.
- Social structure (outer dimension of how a culture organizes and bestows power and authority) and identity (inner dimension) of how a person conceives of themselves) may be in flux.
- Leaders and key influencers (may be formal or informal).
- Beliefs, values, norms, and symbols (asking about a peoples' history is one of the best ways to understand their beliefs, values, norms, and symbols.
- Religion can be a force (for good or evil).

## ORGANIZING TO UNDERSTAND CULTURE

C-162. There are three important methods for trying to understand the cultural element of the operational area:
- **Green Cell.** A green cell is an ad hoc working group, consisting of individuals with a diversity of education and experience capable of identifying and considering the perspective of the population, the host-nation government, and other stakeholders within an operational environment. Ideally, a green cell is composed of individuals with cultural expertise across all warfighting functions. If a whole-of-government approach is used, experts from other government agencies, such as the DOS,

should be included in the cell. The green cell cooperates closely with the other members of the planning staff so that cultural factors are considered throughout the range of military operations. The green cell also interfaces and coordinates with joint and interagency groups, drawing upon the collective knowledge and experience of the DOS, NGOs, and foreign resources.

- **Cultural Advisor.** The cultural advisor is a concept developed and employed in Operation ENDURING FREEDOM in Afghanistan. Cultural advisors are the principal subject-matter experts on culture and planning related to their designated geographic region of expertise, serving as the cultural and language advisors to the commander. The cultural advisor is a special staff officer for the commander and a member of the planning staff. This person, not only serves on the planning staff but also, deploys and serves as an ongoing advisor to senior leaders while they are in theater, if needed. The advantage of having a cultural advisor on staff is that this advisor can often help explain to the commander what the advisor sees on the ground in the AO. A foreign area officer or a CA Soldier may be a good selection for a cultural advisor. Both can provide an understanding of the host nation and its specific regional, religious, and ethnic differences, and they may have foreign language skills. As a result, the commander can adjust operations in response to a culturally challenging environment.

- **Human Terrain System.** The human terrain system provides tactical- to strategic-level support to commanders. The human terrain system conducts field research and analysis of the local population—to determine the civil considerations in order to help commanders better understand the operational environment from the population's perspective—and to assess how actions will potentially impact and be perceived by the local population. A human terrain team typically consists of a team leader, one or two social scientists, and one research manager. When manning demands permit, human terrain team personnel are recruited and deployed to promote engagement with elements of the population that typically have restricted access.

C-163. There is an important caveat when creating cultural cells in the staff. Civil considerations should not be relegated to only a narrow cell compartmented from other staff functions. Human and cultural considerations analysis should be integrated into routine staff fusion procedures.

# STABILITY AND INSTABILITY ANALYSIS

C-164. This framework is based on the analysis supporting stability operations, and it may initially seem counterintuitive for the analyst of resistance. Actually this is a fruitful framework of analysis. The classic objective of UW is to overthrow a government. That objective inherently involves some greater or lesser degree of destabilization. Therefore, analyzing what leads to stability, but applying that understanding in reverse, suits the purpose of UW when its objective is to overthrow a state government or occupying power. However, once overthrow is achieved, the transition phase will likely see the United States in support of indigenous entities attempting restabilization of that same society. Focusing on stability factors from both perspectives—resistance and adversary government—is necessary. Moreover, initial efforts at UW planning should include carefully considered endstate objectives—something stability analysis will inform. Finally, when the objective of UW is much more limited—to merely coerce or disrupt—the careful consideration of destabilizing some sectors of the target, while shielding others from instability, will be a critical consideration requiring stability analysis.

*Note:* The following sections were derived from FM 3-07 (Chapter 4, Stability Assessment Frameworks), and ATP 3-07.5 (Chapters 2–6 and Appendix B, Assessments and the District Stability Framework).

## DISTRICT STABILITY FRAMEWORK

C-165. To increase the effectiveness of stability missions, the United States Agency for International Development developed the District Stability Framework (USAID) (DSF). The DSF was designed to guide and support stabilization efforts by helping civilians and military organizations identify the causes of instability, develop activities to diminish or mitigate them, and evaluate the effectiveness of the activities in fostering stability at the tactical- or operational-level. The DSF can be used to create local stabilization plans and provide data for the Interagency Conflict Assessment Framework, which has a strategic (country or regional) focus. The DSF supports unity of effort by providing partners with a common framework to—

- Understand an operational environment from a stability-focused perspective.
- Maintain focus on the local population and its perceptions.
- Identify the sources of instability in a specific local area.
- Design activities that specifically address the identified sources of instability.
- Monitor and evaluate activity outputs and impacts, as well as changes in overall stability.

*Note:* ATP 3-07.5 provides a more detailed description of the District Stability Framework.

C-166. The DSF helps overcome many of the challenges to successful operations focused on stability. It can help—

- Keep military formations focused on the center of gravity for operations focused on stability—the population and its perceptions.
- Provide a common operational picture for all interagency teams in an AO. By focusing on sources of instability, partner organizations can focus their varied resources and expertise on shared priorities.
- Prioritize activities based on their importance to the local populace and their relevance to the overarching mission of stabilizing the area.
- Enhance continuity between military formations. Units can easily pass DSF data along from one unit to the next, establishing a clear baseline that identifies sources of instability and the steps taken to mitigate them.
- Empower tactical-level formations by giving them hard data useful for decision making at their level and for influencing decisions at higher levels.
- Identify measures of performance and measures of effectiveness for unit activities rather than simply tracking measures of performance.
- Track indicators of overall stability by assessing whether an area is becoming more stable.
- Identify issues that matter most to the population; the DSF helps identify information themes that resonate with the population.

## PRIMARY STABILITY TASKS

C-167. Refer to figure C-15, page C-52, when reading the characteristics of the primary stability tasks and stability principles.

### Conditions to Establish Civil Security

C-168. Of the six primary stability tasks, *establish civil security* aligns closest with normal Army capabilities and responsibilities. Throughout the phases of each stability task (initial response, transformation, and fostering sustainability), military forces will likely be the supported or executive agent, even when other nonmilitary partners have an extensive presence. Civil security requires five necessary conditions:

- **Cessation of Large-Scale Violence.** Stability requires the cessation of large-scale violence. Large-scale armed conflict has come to a halt. Military forces have separated and are monitoring warring parties, implemented a peace agreement, or have managed violent belligerents.
- **Public Order.** Stability requires public order. Military forces establish public order by enforcing laws equitably; protecting the lives, property, freedoms, and rights of individuals; reducing criminal and politically motivated violence to a minimum; and pursuing, arresting, and detaining criminal elements (from looters and rioters to leaders of organized crime networks). Military forces also improve the cleanliness and order of important public places.
- **Legitimate State Monopoly Over the Means of Violence.** Stability requires legitimate state monopoly over the means of violence. Military forces identify, disarm, and demobilize major

illegal armed groups. They have also vetted and retrained security forces so those forces can operate lawfully in a professional and accountable manner under a legitimate governing authority.

- **Physical Protection.** Stability requires physical protection. Political leaders, ex-combatants, and the general population are free from fear from grave threats to physical safety. Refugees and internally dislocated persons can return home without fear of retributive violence. Military forces protect women and children from undue violence. Military forces also protect key historical or cultural sites and critical infrastructure from attack.
- **Territorial Security.** Stability requires territorial security. People and goods can freely move throughout the country and across borders without fear of harm to life and limb. Military forces protect the country from invasion and secure borders from infiltration by insurgent or terrorist elements and illicit trafficking of arms, narcotics, and humans.

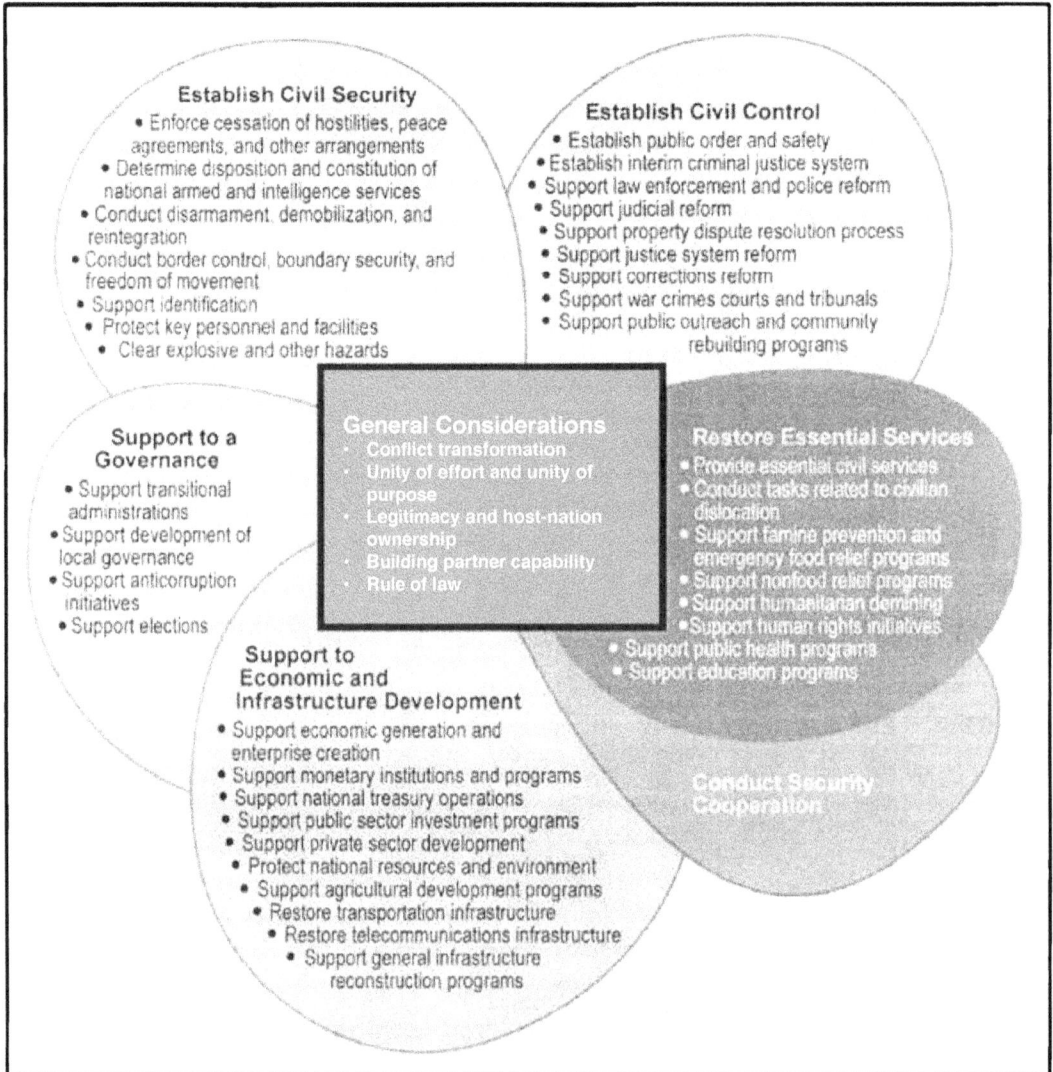

**Figure C-15. Primary stability tasks and stability principles**

## Conditions to Establish Civil Control

C-169.  Civil control ensures that citizens live in a safe society in which individuals and groups do not take the law into their own hands. Rather, they respect the decisions of and adhere to the rule of law. Rule of law enables a populace to have equal access to a self-sustaining justice system, consistent with international human rights standards, and to apply access equally. This is a long-term process conducted by civilian entities. Nevertheless, Army units take initial actions to begin establishing some level of civil control in public order and safety. Initially, Army units are the only authorities capable of implementing some level of civil control and will likely be involved in building host-nation capacity. Army units quickly transfer the lead for these efforts to the U.S. country team. Civil control includes the following necessary conditions.

- **Just Legal Frameworks.** Stability requires just legal frameworks. Laws are consistent with human rights norms, drafted in a transparent way, publicly promulgated, and ensure the separation of powers. Laws are fair, responsive to the needs and realities of the host nation, and benefit the entire population, not just powerful elites.
- **Public Order.** Stability requires public order. Unlawful activity, such as criminal and political violence and intimidation, is reduced to an acceptable minimum. Enforcers of the law pursue, arrest, and detain perpetrators, and the local population can move freely about the country without fear of undue violence.
- **Accountability to the Law.** Stability requires accountability to the law. Laws hold the population and public officials, including military officials, legally responsible for their actions through legitimate processes, norms, structures, and sanctions. Accountability is achieved both horizontally (through state institutions overseeing one another) and vertically (by citizens overseeing the actions of the state).
- **Access to Justice.** Stability requires access to justice. People, including those from marginalized groups, can seek and obtain remedy of grievances through justice institutions.
- **Culture of Lawfulness.** Stability requires culture of lawfulness. The population generally follows the law and has a desire to use the justice system to address their grievances. Most people believe that formal laws are a fundamental part of justice and that the rule of law enhances their lives and society in general.

C-170.  Military forces can establish these conditions by performing subordinate tasks during all three phases of stability. Initial response tasks focus on establishing interim mechanisms for the restoration of rule of law. During the initial response phase, Army units have to execute tasks on their own because little or no host-nation capability exists to establish order. In other cases, host-nation security forces can maintain order and require little Army unit involvement. Transformation tasks develop justice and corrections systems that meet international standards of fairness, including the treatment of detainees, and that include viable processes for redress and reconciliation. In the transformation phase, host-nation police forces and, potentially, international organizations begin to contribute and Army units focus more on security force assistance, particularly on the systems required to professionalize the host-nation security forces. Fostering sustainability tasks emphasizes the process of shifting control of the judiciary and corrections systems to host-nation personnel. In every fostering sustainability phase, Army units transfer responsibility to the host nation, as well as monitor and report. They also transition to a steady-state posture focused on advisory duties and security cooperation. The host nation assumes complete responsibility for civil control through its own justice institutions.

## Conditions to Restore Essential Services

C-171.  Normally, the host-nation government and civilian relief agencies work best to restore and develop essential services. In most cases, local, international, and U.S. civilian agencies arrived in-country long before U.S. forces. However, when these organizations are neither established nor have the necessary capacity, Army units execute these tasks until the other organizations can. These organizations advise Army units regarding the best way to perform these tasks. They aim to return to preconflict or predisaster situations. Sometimes, conditions are so poor that Army units develop, fundamentally transform, or provide more equitably the basic levels of essential services.

C-172. The provision of essential services includes several necessary conditions:
- Access to and delivery of basic needs and services.
- Access to and delivery of education.
- Return and resettlement of dislocated civilians, including refugees and internally displaced persons.
- Social reconstruction.

C-173. Military forces can establish these conditions by performing subordinate tasks during all three phases of stability. These tasks occur during all three phases of stability operations. During the initial response phase, Army units take the lead in providing for the population's immediate critical needs, supporting and enabling other actors as they become operational. In some situations, Army units provide minimal assistance since the other actors are already well established. The transformation phase occurs once the crisis is past and sufficient capacity begins to grow. This phase establishes the foundation for long-term development and resolves root causes of conflict that lead to famine, dislocated civilians, refugee flows, and human trafficking. In the fostering sustainability phase, the host nation makes the efforts permanent by institutionalizing positive change in society, ensuring it has the means to sustain progress. If the situation in the host-nation regresses, the host nation may re-employ earlier stability methods.

## Conditions to Support to Governance

C-174. Stability efforts often occur in failed or fragile state situations or when the host-nation government has difficulty resolving stabilization and reconstruction challenges. In some cases, host-nation governments can function and exercise their sovereign responsibility. In these cases, international actors, including Army units, support host-nation government authorities. However, if the host-nation government is dysfunctional or absent, international law obligates intervening military forces to provide interim governance as a transitional military authority until the host nation establishes a responsible civilian authority.

C-175. Good governance includes the following necessary conditions:
- Provision of essential services.
- Stewardship of state resources.
- Political moderation and accountability.
- Civic participation and empowerment.

## Conditions to Support Economic and Infrastructure Development

C-176. In postconflict and fragile states, host-nation actors, other USG agencies, intergovernmental organizations, and civilian relief agencies often have the best qualifications to lead efforts to restore and help develop host-nation economic capabilities. However, if security considerations or other factors make these actors incapable of assisting initially, Army units assist host-nation actors to begin the process of achieving sustainable economic development. Ultimately, the goal is to establish conditions so that the host nation can generate its own revenues and not rely upon outside aid.

C-177. Preserving assets—such as production facilities, hospitals, universities, existing companies, and markets—dramatically reduces the time required to re-establish a sustainable level of economic activity. Many assets survive an initial conflict but subsequently wither in the instability of the aftermath.

C-178. Army units first gain and maintain comprehensive situational awareness. Units work with host-nation officials and others to gather and continuously update the information needed to accurately assess the host nation. This assessment includes the status of the host-nation economy, infrastructure, civil society, and local communities. This initial socioeconomic and infrastructure assessment forms the basis for developing and implementing economic and infrastructure development strategies that establish the following conditions:
- Employment generation.
- Macroeconomic stabilization.
- Market economy sustainability.
- Control over the illicit economy and economic-based threats to peace
- Functioning civil societal infrastructure and local community development

## DISTRICT STABILITY FRAMEWORK

C-179.    The DSF is a tactical-level analysis, planning, and programming tool specifically created to guide and support stabilization efforts. The DSF supports the assessments by helping users identify local sources of instability and design programs and activities to address these sources. It reflects stability efforts, counterinsurgency, and international development best practices by emphasizing the local population's perspectives, development principles, and measuring impact—not just output. The DSF supports unity of effort by providing implementers from various organizations with a common framework to—

- Understand operational environments from a stability-focused perspective.
- Maintain focus on the local population and its perceptions.
- Identify the sources of instability in a specific local area.
- Design activities that specifically address the identified sources of instability.
- Monitor and evaluate activity outputs and impacts, as well as changes in overall stability.

C-180.    The DSF has four basic steps (figure C-16). To maximize effectiveness, all relevant partners and organizations in the area ideally stay involved in the entire process, participating in an inclusive stability working group. The four basic steps consist of situational awareness, analysis, design, and monitoring and evaluation. The DSF requires population-centric and stability-oriented situational awareness by examining an operational environment, the cultural environment, stability and instability dynamics, and local perceptions. In the second step, units analyze information from situational awareness to identify and prioritize the sources of instability in a given local area. In the third step, units use design to identify and refine proposed activities to diminish the sources of instability against a series of stabilization fundamentals, design principles, and prioritization criteria. When monitoring and evaluating, units measure effort and achievements on three levels: output (activity completion), impact (effects achieved by individual activities), and overall stability. Units then use insights from this step to adjust and develop future stabilization activities.

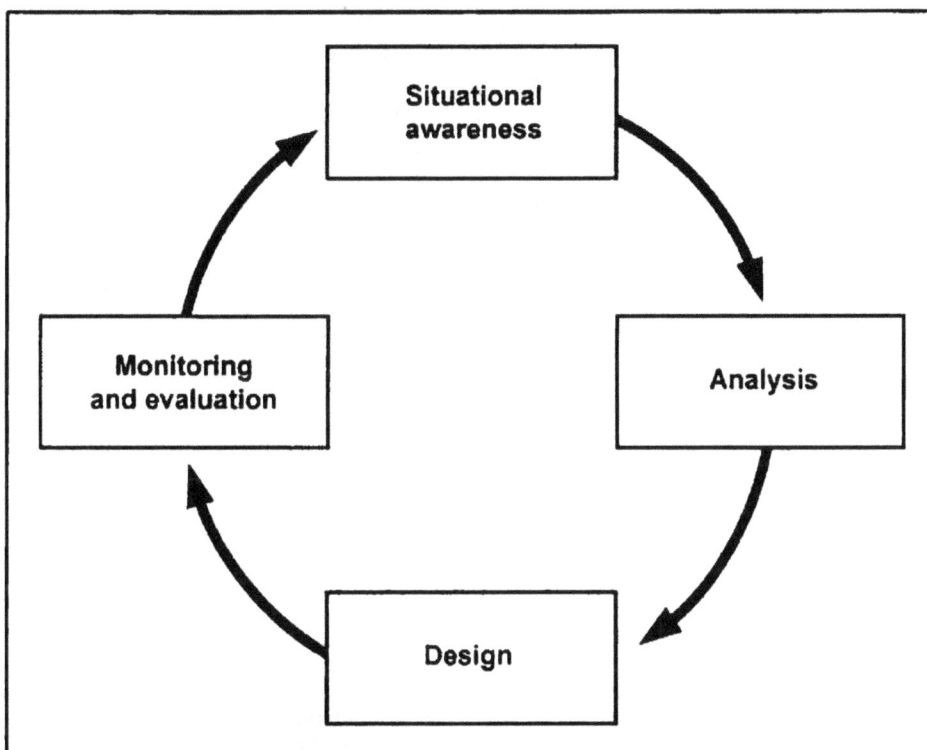

Figure C-16. District stability framework

## Situational Awareness

C-181. DSF does more than list facts about an operational environment. This framework identifies the relevance of the factors to the local population and an Army unit's mission. For example, the framework goes beyond identifying the fact that one tribal group dominates the local government. The DSF identifies that this inequity undermines the legitimacy and support for the government among other tribes. DSF emphasizes three subfactors to achieve a population-centric and stability-oriented understanding of an operational environment: culture, local perceptions, and stability and instability dynamics.

### *Culture*

C-182. The cultural matrix in figure C-17 identifies the factors about culture. It lists the tasks Army units achieve to determine the factors.

| Cultural Factors | Analysis Steps |
|---|---|
| Major cultural groups | Identify the major cultural and tribal groups in the area of operations. |
| Their interests | Identify the things these groups care about or consider to be valuable—both material and intangible. |
| Cultural codes, traditions, and values | Identify cultural codes, traditions, and values that the major cultural groups live by. |
| Traditional conflict resolution mechanisms | Identify the methods individuals and groups traditionally use to resolve conflicts. |
| Traditional authorities | Identify the traditional authorities to whom the locals respect and normally turn to for assistance |
| Disruptions to mechanisms and authorities | Describe new actors or conditions that potentially disrupt the traditional conflict resolution mechanisms and undermine the influence of traditional authorities. |
| How malign actors or stabilizing forces leverage cultural factors | Describe how actors exploit these cultural facts to their advantage. Identify how stabilizing forces do or do not leverage these facts. |

**Figure C-17. Cultural matrix**

### *Local Perceptions*

C-183. The population is frequently the center of gravity in stability. To be effective, Army units base stability efforts on a deep understanding of local conditions, local grievances, and local norms, rather than outsider assumptions. Units gain this understanding through several possible mechanisms, including population surveys, focus groups, key leader engagements, or polling conducted by external organizations.

C-184. Army units collect local perceptions using a tactical conflict survey. Army units on patrol, civilian agency implementing partners, and host-nation government and security forces use this simple, four-question survey. Personnel follow up each question by asking "why" to ensure they fully understand the interviewee's response and perspective. The four questions are:

- Has the number of people in the village changed in the last year? Why?
- What are the most important problems facing the village? Why?
- Who do you believe can solve your problems? Why?
- What should be done first to help the village? Why?

C-185. Mature individuals with good interpersonal skills, assisted with a good interpreter, conduct surveys. A poorly conducted survey program creates resentment, survey fatigue, unrealistic expectations among the population, and flawed data. Collectors document contextual information to facilitate further analysis, such as the location and the interviewee's characteristics (for example, ethnicity, tribe, age, and gender). Collectors aggregate and graphically represent the survey responses, allowing for a quick visual understanding of local perspectives—whether represented as a snapshot in time or as change over time. PSYOP personnel receive training and equipment to conduct surveys. These personnel analyze their results and provide the information to commanders and staffs, along with detailed studies of local populations and the factors influencing their behavior. Army units coordinate with the assigned PSYOP staff officer to conduct surveys. Figure C-18 shows a sample tactical conflict survey.

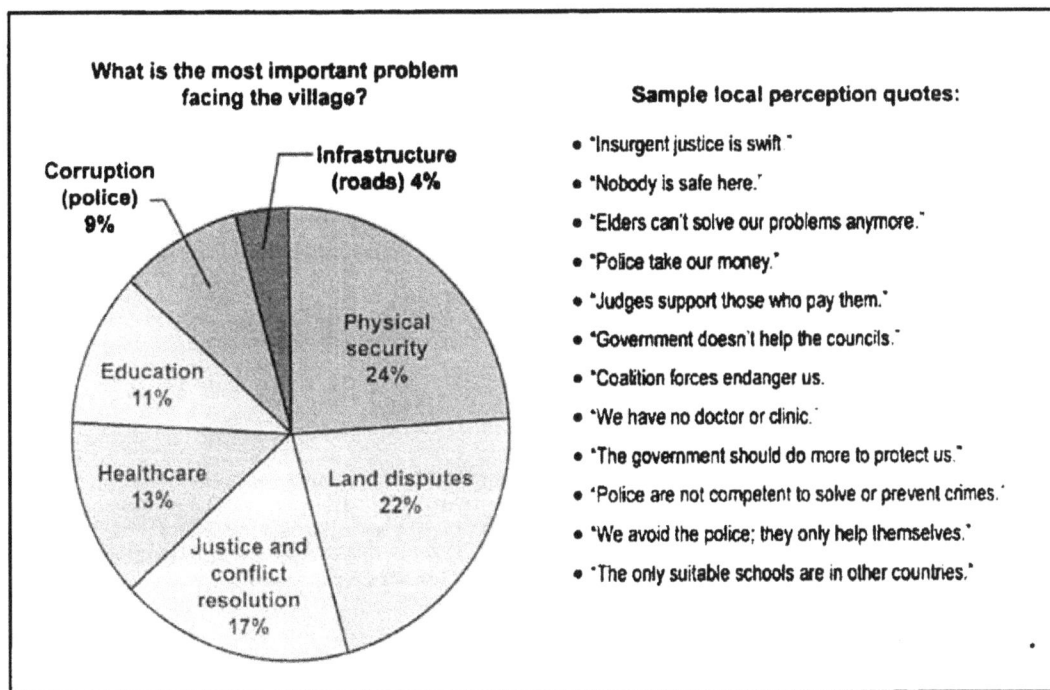

**What is the most important problem facing the village?**

- Infrastructure (roads) 4%
- Corruption (police) 9%
- Physical security 24%
- Education 11%
- Land disputes 22%
- Healthcare 13%
- Justice and conflict resolution 17%

**Sample local perception quotes:**

- "Insurgent justice is swift."
- "Nobody is safe here."
- "Elders can't solve our problems anymore."
- "Police take our money."
- "Judges support those who pay them."
- "Government doesn't help the councils."
- "Coalition forces endanger us."
- "We have no doctor or clinic."
- "The government should do more to protect us."
- "Police are not competent to solve or prevent crimes."
- "We avoid the police; they only help themselves."
- "The only suitable schools are in other countries."

**Figure C-18. Sample tactical conflict survey**

*Stability and Instability Dynamics*

C-186    Finally, DSF identifies potential stability and instability factors in the local environment. Factors of instability include grievances, events, and key actors (figure C-19, page C-58). Grievances of the local population come from various local perceptions and data sources. Events create a window of vulnerability in which threats undermine stability. Key actors—individuals or organizations—foment instability with their means, motives, and actions. Factors of stability include resiliencies, events, and key actors. Resiliencies in the society—institutions and mechanisms—help the society function peacefully. Events present a window of opportunity to enhance stability. Key actors—individuals or organizations—help enhance stability with their means, motives, and actions. Events are usually the same in both matrixes. Whether they end up reinforcing stability or instability depends mainly on how they play out.

## Analysis

C-187.   After gaining situational awareness, the DSF provides tools to analyze and identify potential sources of instability, their causes, the desired objectives, and the impact indicators that measure progress in addressing each source of instability. The second step of the DSF—analysis—consists of four steps: identify potential sources of instability, vet each issue against instability criteria, determine if the issue meets two of the three instability criteria, and prioritize sources of instability.

| Factors of Stability | | |
|---|---|---|
| **Resiliencies** | **Events** | **Key Actors** |
| What processes, relationships, or institutions enable the society to function normally and peacefully?<br><br>Are there any previous resiliencies that have been or are being undermined? | What potential or anticipated future situations could create an opening for key actors and their followers to further reinforce stability? | Which individuals or institutions in the society are attempting to preserve and strengthen stability?<br><br>What means do they possess, what are their motives, and what actions are they taking? |
| **Factors of Instability** | | |
| **Grievances** | **Events** | **Key Actors** |
| What issues or problems are the local populace concerned or upset about?<br><br>Whom do they blame for these conditions, and how severe are they? | What potential or anticipated future situations could create an opening for key actors and their followers to undermine stability? | Which individuals or institutions are leveraging popular grievances and events to create instability?<br><br>What means do they possess, what are their motives, and what actions are they taking? |

**Figure C-19. Sample stability and instability dynamics matrices**

C-188. Analysis typically results in a long list of issues, needs, and grievances that potentially drive instability. In the first step, analysis narrows this list down to fewer issues that actually cause instability and local people really care about. To begin narrowing down the list, the DSF groups closely related or logically connected issues in a symptom-cause relationship. For example, the DSF groups the issues of poor border control, police corruption, and violent crime under the heading of "physical insecurity." If physical insecurity causes instability, the DSF breaks apart these issues to address them in detail.

C-189. Grievances occur because people believe stability efforts do not meet their needs or defend their interests. Grievances do not necessarily result in instability unless events occur to translate these grievances into action. These events act as windows of vulnerability that other actors potentially exploit. Partners counterbalance instability with stability factors, such as resiliencies, windows of opportunity, and constructive actors.

C-190. In the second step, Army personnel enter each issue or group of related issues in the source of instability analysis matrix shown in figure C-20, page C-59. Personnel vet each issue against three instability criteria:

- Does this issue decrease support for the government and legitimate governance institutions?
- Does this issue increase support for malign actors or belligerents?
- Does this issue undermine the normal functioning of society?

C-191. The first instability criterion is: "Does this issue decrease support for the government and legitimate governance institutions?" Support for the government is based on what locals actually expect of their government rather than what outsiders might expect of their own government back home. Legitimate governance institutions refers to NGOs that help the society regulate itself, such as a tribal council.

C-192. The second instability criterion is: "Does this issue increase support for malign actors or belligerents?" Usually, support increases when the malign actors either directly help solve the problem or successfully leverage the issue in their propaganda. An example of helping to solve the problem can include providing security to a community that the police never visit. A sample excerpt for successfully leveraging the issue can include: "If we were in charge, we would reform and expand the police."

C-193. The third instability criterion is "Does this issue undermine the normal functioning of society?" The emphasis focuses on local norms. For example, if people have never had electricity, the continued lack of electricity can hardly be undermining the normal functioning of society.

C-194. In the third step, Army personnel determine if the issue meets two of the three instability criteria. The final step on the source of instability analysis matrix prioritizes identified sources of instability using local perceptions. Normally, practitioners focus first on sources of instability that are a priority grievance for the local population. Otherwise, locals perceive the stability efforts as disconnected from reality and focusing on problems that do not really matter to them.

| Potential Sources of Instability | Instability Criteria | | | Sources of Instability | Prioritization |
|---|---|---|---|---|---|
| | Does this issue decrease support for the government and legitimate governance? Explain. | Does this issue increase support for malign actors or belligerents? Explain. | Does this issue undermine the normal functioning of society? Explain. | Does the issue meet 2 of the 3 instability criteria? | Is the source of instability a priority grievance in the local population? If yes, priority? |
| Justice and conflict resolution (unclear land boundaries) | Yes<br>Formal and traditional mechanisms seen as ineffective | Yes<br>Spoilers increase their reputation by solving disputes. | Yes<br>Traditionally solved by tribal councils, now a source of violence. | Yes | Yes (#1) |
| Tribe A dominates | Yes<br>Undermines Tribes B and C support; increases resentment. | No<br>Spoilers not taking advantage. | No<br>Tribe A has dominated for several decades. | No | No |
| Lack of health care | Probably<br>Provincial government health care excludes North. | No<br>Spoilers do not provide health care. | No<br>North has never had health care. | No | Yes (#3) |
| Poor road infrastructure | No<br>Governor is working to build new road. | No<br>Spoilers are not building roads. | No<br>Road network has always been rudimentary at best. | No | No |
| Civilian government corruption | No<br>Corruption complaints directed solely at police. | No<br>No evidence that spoilers exploit this issue. | No<br>Locals not concerned; apparently within normal bounds. | No | No |
| Insecurity, ineffective police and civilian casualty | Yes<br>Civilian casualty, insecurity, police ineffectiveness reflect poorly on government. | No<br>Spoilers also blamed for civilian casualty and do not provide security. | Yes<br>Insecurity and police problems exceed local norms. | Yes | Yes (#2) |
| Lack of education | No<br>Despite limitations, people grateful for education improvements. | Yes<br>Spoilers promote radical ideology. | No<br>Despite low levels education has actually improved. | No | Yes (#4) |

**Figure C-20. Sample sources of instability analysis matrix**

C-195. After identifying and prioritizing, sources of instability practitioners fill out a tactical stability matrix for each source of instability. As shown in figure C-21, the tactical stability matrix helps further analyze and design activities to address each source of instability. It consists of nine columns. The first six support the analysis process while the final three support the design phase. Users fill out the columns in the tactical stability matrix by identifying the following six elements that support analysis:

- The targeted source of instability.
- Perceived causes—how locals perceive this situation and why they think it exists—usually presented as representative quotes from the local populace.
- Systemic causes—the root causes of the source of instability, such as the underlying conditions that led to the problem or allow it to continue.
- Objective—a succinct statement of the end state that addresses the source of instability—often simply the reverse of the source of instability, adding in the "who."
- Impact indicators—(also measures of effectiveness) changes in an operational environment that indicate progress toward reducing the systemic causes and achieving the objective.
- Impact indicator data sources—the source for the information to track the impact indicators.

| Analysis | | | | | | Design | | |
| Targeted sources of instability | Perceived causes | Systemic causes | Objective | Impact indicators | Impact indicators data sources | Activities | Output indicators | Output indicators data sources |
|---|---|---|---|---|---|---|---|---|
| Lack of gov't or traditional conflict resolution mechanism | "Spoilers provide swift justice"<br><br>"Judges support those who pay them the most"<br><br>"Elders can't solve our problems"<br><br>"Gov't doesn't help the councils" | Formal justice system is slow, inefficient, hard to access<br><br>Justice officials not paid in full or on time<br><br>Traditional conflict resolution structures are undermined | Foster conflict resolution mechanisms linked to gov't | Increased # disputes resolved by gov't-recognized entities<br><br>Decreased # disputes resolved by spoilers<br><br>Decreased violence linked to land disputes<br><br>Increase # land deeds registered | Tactical conflict survey<br><br>Gov't records<br><br>Public surveys<br><br>Patrol reports<br><br>Interviews<br><br>Assessments | Support training for justice officials<br><br>Facilitate judicial pay system reform<br><br>Establish mobile gov't dispute resolution unit<br><br>Facilitate councils<br><br>Link councils to gov't<br><br>IIA | # of justice officials trained<br><br>Pay reforms enacted<br><br>Mobile dispute unit established<br><br># of councils held<br><br># of councils with gov't involvement<br><br># of IIA radio spots | Gov't financial records<br><br>Interviews<br><br>Assessments<br><br>Tactical conflict surveys<br><br>Patrol reports<br><br>Radio |

Legend:
#      number
gov't      government
IIA      inform and influence activities

**Figure C-21. Sample tactical stability matrix**

C-196. The remaining three columns in the tactical stability matrix support design:

- Activities—things that will mitigate the systemic causes and achieve the objective (taken from the activity design worksheet in figure C-22, page C-61).

- Output indicators—(also measures of performance) metrics that indicate progress toward the completion of an activity. Ask: How can I confirm that the activity is progressing or has been completed?
- Output indicator data sources—the source for the information to track the output indicators.

| Design Process | | Design Outputs |
|---|---|---|
| **Possible brainstorming activities** | | Generate a list of potential activities that will address the systemic causes and contribute to achieving the objective for a given source of instability. |
| Stability criteria (must meet 2 of 3) | Explain how the activity increases support for government and legitimate governance. | Explain how the activity will increase support for the government and legitimate governance actors. |
| | Explain how the activity decreases support for malign actors. | Explain how the activity will decrease support for malign actors. |
| | Explain how the activity increases institutional and societal capacity and capability. | Explain how the activity will increase institutional and societal capacity and capability. |
| Design principles | • Sustainability<br>• Local ownership<br>• Short-term versus long-term results<br>• Leverage support from other organizations<br>• Culturally and politically appropriate<br>• Accountability and transparency<br>• Flexibility | For each potential activity that meets at least two of the three criteria, refine the proposed activity to make it meet as many of the seven design principles as possible. |
| Resources | • Money<br>• Personnel<br>• Expertise<br>• Time | Do you or your partners have the resources to complete the activity? If not, eliminate the proposed activity. |
| Select | Decide if activity is realistic. | Based on the stability criteria, design principles and resource availability, should the activity be implemented? |

**Figure C-22. Sample activity design worksheet**

## Design

C-197. In the third step of the DSF, practitioners design, prioritize, and synchronize stabilization activities. This process starts by brainstorming potential activities that address each systemic cause of the source of instability. Practitioners screen and refine these ideas using the three stability criteria, the seven design principles, and resource availability. The activity design worksheet (figure C-22) helps to guide this process. The results then feed into the activities column of the tactical stability matrix. The stability criteria essentially mirror the instability criteria in figure C-21, page C-60. Practitioners eliminate any proposed activity that does not meet at least two of these criteria.

C-198. Practitioners then refine proposed activities that meet the stability criteria using the seven design principles. To the extent possible, practitioners design or modify each activity such that it—
- Can be sustained by the local government or society.
- Maximizes local involvement to create local ownership

- Minimizes the trade-offs between short-term positive effects and any potentially negative long-term impacts (unintended consequences).
- Leverages or supports the programs of other government agencies, intergovernmental organizations, NGOs, and the host-nation government.
- Is appropriate to the local political and cultural context.
- Strengthens governmental accountability and transparency
- Includes the flexibility to adapt if circumstances change.

C-199. Next, practitioners screen each proposed activity against their available resources—money, personnel, appropriate expertise, and time. Those activities for which the necessary resources exist are entered into the activities column of the tactical stability matrix. Practitioners then complete the tactical stability matrix by identifying output indicators and output indicator data sources that enable practitioners to determine whether each activity is proceeding as planned and, ultimately, when it has been completed.

C-200. Last in the design process, practitioners prioritize and synchronize the selected activities. They prioritize activities based on their anticipated impact on the source of instability; practitioners implement activities with more anticipated "bang for the buck" first. Practitioners synchronize activities in time and space to build upon and reinforce other activities and operations conducted by the stability working group. A synchronization matrix (figure C-23) helps the stability working group.

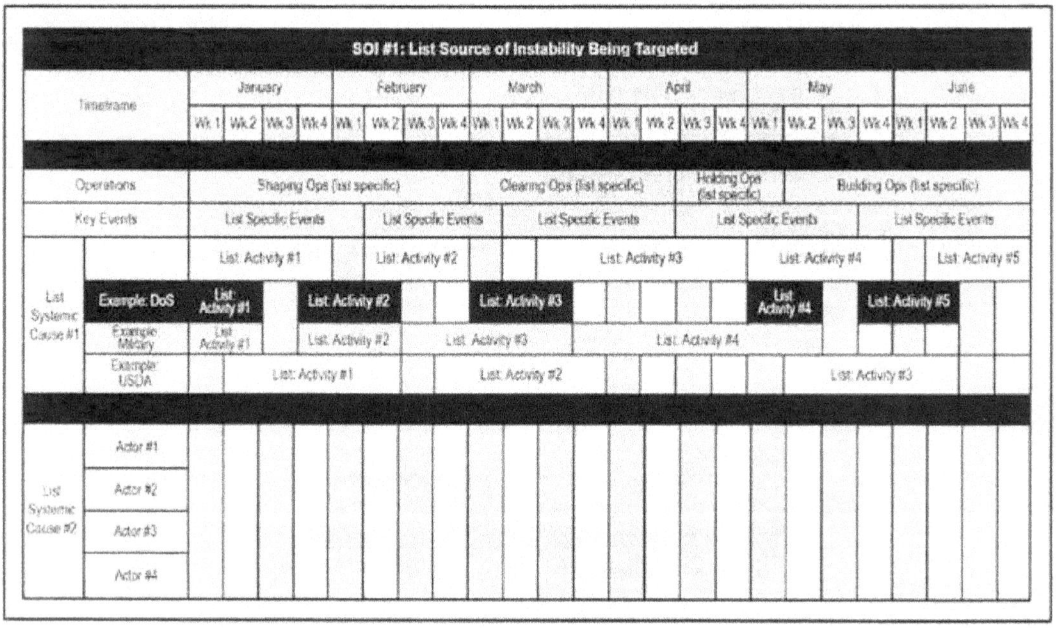

Figure C-23. Synchronization matrix

## Monitoring and Evaluation

C-201. The final step in DSF takes place during and after the implementation of stability activities. Army units conduct monitoring and evaluation on three levels: output indicators, impact indicators, and overall stability. The tactical stability matrix includes the indicators.

C-202. Output indicators (measures of performance) simply track implementation of an activity. They answer the question, "Is the activity progressing?" and in the long term, "Is the activity complete?" Examples might include the number of miles of road paved or number of police trained. Army units monitor output indicators during the implementation of an activity, until they complete the activity.

C-203. Impact indicators (measures of effectiveness) measure the effect that an activity achieved. Examples include decreased travel time (for a road project) or decreased criminal activity (for a police training activity). Generally, Army units evaluate input indicators only after they complete the activity.

C-204. Overall stability is the third level of monitoring and evaluation. Rather than measuring the impact of individual activities, it considers the stabilizing impact of all the activities conducted over a longer period, as well as the influence of external factors. It simply asks, "Is stability increasing or decreasing?" Effective measurements of overall stability establish good indicators and track the indicators at repeated intervals, starting as early as possible.

C-205. DSF uses the monitoring and evaluation matrix shown in figure C-24 as a program management tool to help track the output and impact of individual activities. Most information comes directly from the tactical stability matrix, with additional columns added to establish a baseline for the impact indicators and to measure change against this baseline.

| Source of instability | Activity | Measure of Performance | | Measure of Effectiveness | | | | |
|---|---|---|---|---|---|---|---|---|
| | | Output indicator data | Output indicator sources | Impact indicator | Baseline | Change | Impact data sources | Outcome |
| Taken from the tactical stability matrix | Taken from the tactical stability matrix | Data for output indicators identified on the tactical stability matrix | Taken from the tactical stability matrix | Taken from the tactical stability matrix | Baseline data for impact indicators identified on the tactical stability matrix | Change in baseline data | Taken from the tactical stability matrix | Taken from the tactical stability matrix |

Figure C-24. Sample monitoring and evaluation matrix

C-206. The best overall stability indicators reflect the local population's perceptions of stability, not perceptions or assumptions held by outsiders. They are based on the question, "What will local people do or say differently if they feel the environment is getting more stable?" The following are seven suggested indicators of overall stability illustrated in figure C-25, page C-64:

- Government recognition—the number of locals that take their problems to the government for resolution. This reflects trust and confidence in the government and its perceived legitimacy.
- Host nation on host-nation violence—a direct measure of insecurity.
- Host-nation security force presence—reflects security force confidence to range farther and more frequently into insecure areas.
- Freedom of movement—reflects security conditions.
- Security perceptions—a direct measure of perceived safety.
- Economic health—reflects freedom of movement and investor confidence.
- Governance perceptions—a direct measure of public confidence in the government's competence, transparency, and relevance.

C-207. Finally, as practitioners monitor and evaluate these three levels, they identify lessons about what worked, what did not work, and what partners can do to improve their stability efforts as they repeat the DSF process in the future.

C-208. The DSF can support effective monitoring, evaluating, and decision making. It focuses on the perceptions of the population and provides a common operational picture for Army units and their partnered civilian agencies. Furthermore, it helps inform and influence activities by identifying themes that resonate with the population.

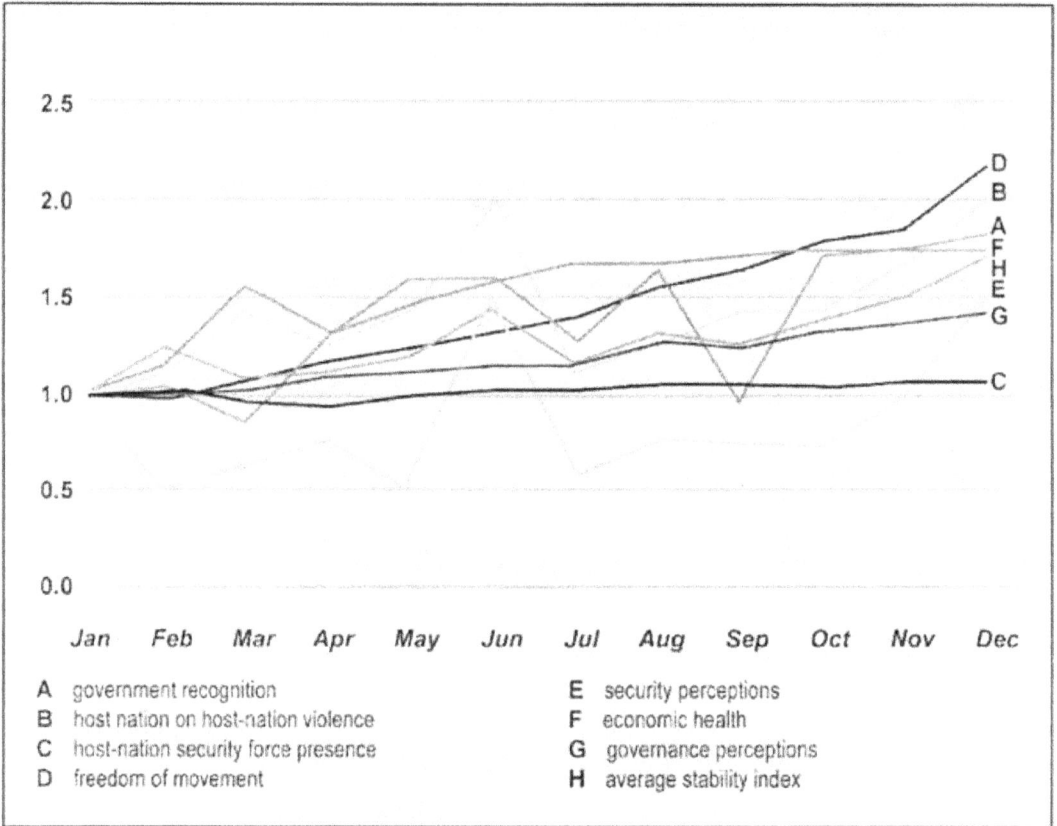

Figure C-25. Sample overall stability index

## Appendix D

# Varieties of Resistance

There are many types of resistance, and it is important to understand each specific example by its own unique characteristics, by its own objectives, and in its own historical circumstances. There is no one fixed model of resistance that applies to every case. However, below are some lists of famous examples and potential approaches to compare and contrast. Doing so should help sharpen the SF Soldier's understanding of the actual resistance reality before him.

## BASIC EXAMPLES OF RESISTANCE APPROACHES AND PHASING

D-1. Each insurgency or resistance movement has its own unique characteristics based upon its strategic objectives, operational environment, and available resources. Insurgencies normally seek to change the existing social order and reallocate power within the country. Typical insurgent goals include:
- Removal of the established indigenous regime or occupying military power.
- Establishment of an autonomous national territory within the borders of a state.
- Extraction of political concessions that the movement cannot attain through nonviolent means.

D-2. Strategic approaches include:
- Anarchist—destructively eliminate government entirely.
- Egalitarian—impose universal equality.
- Traditionalist—resist change and return to a perceived norm.
- Pluralist—break a monopoly on political discourse.
- Secessionist—break off some section from the polity.
- Reformist—modify the application of laws and mores.
- Preservationist—safeguard valued institutions from change.
- Globalist—provoke supranational reorganization.
- Apocalyptic—act as catalyst for an envisioned end times.
- Utopian—impose a theoretical vision of man's perfection.
- Commercialist—facilitate greed through violent illegalities.

D-3. Insurgents may use different approaches at different times, applying tactics that take advantage of circumstances. Insurgents may also combine a number of methods that include tactics drawn from any or all of the other approaches. In addition, different insurgent forces using different approaches may form loose coalitions when it serves their interests; however, these same coalitions may fight among themselves, even while engaging counterinsurgents. A single AO may have multiple, competing entities.

D-4. Typical insurgent approaches include:
- Urban.
- Military.
- Protracted.
- Subversive.

*Note:* FM 3-24 provides additional information on insurgent approaches.

D-5. Typical phasing and timing (Maoist Example) includes:
- Phase I—Strategic Defensive (Latent or Incipient).
- Phase II—Strategic Stalemate (Guerrilla Warfare).
- Phase III—Strategic Offensive (War of Movement).

*Note:* ATP 3-05.1 includes additional information on phasing and timing.

# SELECT EXAMPLES FROM THE MODERN ERA OF INSURGENT, RESISTANCE, OR REVOLUTIONARY WARFARE

D-6. The best-known era of using UW is the "classic era" during World War II. There are many sources of information on the exploits of the British Special Operations Executive, the American OSS, the French Resistance, Soviet Partisans throughout Eastern Europe, the American and Philippine Anti-Japanese Resistance, the Chinese Anti-Japanese Resistance, and OSS Detachment 101 in Burma. Following the World War II period, insurgency and revolutionary warfare theory and practice flourished during the Cold War and the anti-colonialist uprisings around the globe. Some of the most famous, influential, and enduring insurgent and resistance concepts that have influenced thinking on UW came from this period. Finally, a few of the current theorists of insurgency, resistance, revolutionary warfare, and jihad are noted. The user of this publication is encouraged to research these topics and the following selected list of significant personalities and events.

- **Ayman Al-Zawahiri (Egyptian).** Physician, member of Egyptian Muslim Brotherhood, founder of the Egyptian Islamic Jihad, and ideological and current political leader of al-Qaeda. Most important work is *Knights under the Prophet's Banner.* Key ideas include democracy must be destroyed because it says man rather than Allah is sovereign, jihad must be global, all Muslims everywhere must be held responsible for conducting jihad wherever possible, and that patience combined with mass killings through suicide bombings is required.

- **Abdullah Azzam (Palestinian).** Credentialed and authoritative Islamic scholar and theologian, teacher and mentor of Osama bin Laden, and most important figure in the creation of al-Qaeda. Most important work is *The Defence of the Muslim Lands.* His most important contribution is making the jihadist canon an internationally recognized concept; he was able to inspire and recruit fighters from across the Muslim world for Afghanistan. Azzam says jihad is obligatory and that no individual needs another's permission to wage it.

- **The Coming Insurrection (French).** Anarchist work by a French so-called "Invisible Committee" that proclaims the coming collapse of global capitalist society and political organizations will devolve around clandestine, underground identity groups, which will both seize local power and disrupt larger-scale mechanisms of industrial civilization.

- **Dau Tranh (Vietnamese).** Vietnamese variant of People's War. Central idea is that everything conceivable is an instrument of comprehensive struggle; there are no "noncombatants," everything is a political act. Its chief characteristic is its conceptual breadth, for it is of greater scope than ordinary warfare and requires the total mobilization of a society's resources and psychic energies.

- **Ernest "Che" Guevara (Argentinean).** Marxist theorist, revolutionary and guerrilla leader, physician, and author. Successful leader of the guerrilla campaign under Castro, which led to communist victory in Cuba. Unsuccessful in exporting his techniques to Congo, Bolivia, and elsewhere. Author of *Guerrilla Warfare: A Method.* Major variant of revolutionary warfare was his discredited (although popular) "foco" theory, which maintained that the armed struggle of rural guerrilla warfare vanguard elements could spark larger revolution.

- **T.E. Lawrence (British).** Successful British Army advisor to Arab forces resisting Turkey in World War I. Author of *The Seven Pillars of Wisdom.*

- **Ho Chi Minh (Vietnamese).** Marxist revolutionary, leader of the Viet Minh, and victor of Indochinese wars against France, the United States, and South Vietnam. President and founder of the Democratic Republic of Vietnam. Central figure of Vietnamese communism.

- **S. K. Malik (Pakistani).** As a serving Pakistani general, wrote *The Quranic Concept of Power*, a book that was endorsed and introduced by religious authority in Pakistan and the then President and Commander-in-Chief of Pakistan's armed forces. Serves as a strategic vision for jihad claiming that the center of gravity in the struggle is the enemy's heart, mind, and soul; that the object is to defeat that heart, mind, and soul; and that the primary means to that end is terrorism. Malik describes and advocates "holy war."

- **Mao Tse-Tung (Chinese).** Communist revolutionary, political philosopher, military strategist, founder and leader of the People's Republic of China from the 1920s to 1970s. Successful anti-Japanese resistance leader and Chinese Civil War victor. Notable for Marxist variant, which articulated revolutionary power of Chinese mass populace rather than the orthodox view of industrial proletariat of Marx or the revolutionary vanguard of Lenin. His selected military writings articulate arguably the most influential theory of insurgency, resistance, and revolutionary warfare ever advocated—protracted people's war. Mao emphasized the inseparability and primacy of political work in revolutionary warfare and identified three stages: the strategic defensive, the strategic stalemate, and the strategic offensive—otherwise known as the "latent and incipient," "guerrilla warfare," and "war of movement" phases, respectively.

- **Carlos Marighella (Brazilian).** Marxist revolutionary and writer. Most famous work was *Mini-Manual of the Urban Guerrilla*, which articulated his theories of urban guerrilla warfare in order to disrupt and overthrow an authoritarian regime.

- **Sayyed Qutb (Egyptian).** Islamist theorist, educator, leading member of the Egyptian Muslim Brotherhood, and influential ideologue for jihad. Key work is *Milestones*, which is a fundamental roadmap for subsequent jihadist thinking and action. His essential thesis is that the world is heretical, it is to be forcibly cleansed by Islam, and Islam is not to be understood as just a religion but as a revolutionary party. Furthermore, Islam is the last message given and is ordained for mankind. Anyone who hinders the spread of Islam must be fought until he is killed or declares his submission. Ironically, Qutb's theories were influenced by Lenin's concept of the revolutionary vanguard.

- **STUXNET.** Computer worm surreptitiously released into Iranian nuclear program networks by unknown actors to covertly sabotage the industrial control systems for gas centrifuges. This type of sophisticated emerging technology has enormous implications for future insurgency.

- **Josep Broz Tito (Yugoslavian).** Communist Commander-in-Chief of the Yugoslavian National Liberation Military Forces from 1941 to 1945. Successful Anti-Axis partisan leader.

- **Vo Nguyen Giap (Vietnamese).** Principal Commander of the Viet Minh and the People's Army of Vietnam. Victor of the Indochinese wars against France, the United States, and South Vietnam. Most famous work is *People's Army, People's War*.

This page intentionally left blank.

# Appendix E

# Varieties of Violent and Nonviolent Methods

## TERMS OF VIOLENCE

E-1.  There are many terms that can be used to convey the wide variety of violent methods a resistance could potentially use (figure E-1, pages E-1 and E-2). Just considering the nuanced differences in how such terms differ and how they might be employed should add breadth to the SF Soldier's operational art for UW.

| | | | | |
|---|---|---|---|---|
| Abuse | Accost Bellicosely | Act of Hostility | Act of Inhumanity | Altercation |
| Ambush | Annihilate | Arson | Assassinate | Assault |
| Assault with Intent to Murder | Atrocity | Attack from a Concealed Position | Attack Physically | Barbarity |
| Battery | Battle | Beat | Besiegement | Blind |
| Bludgeon | Bombard | Brawl | Break | Break to Pieces |
| Brigandage | Bruise | Brutality | Brutishness | Bully |
| Burn | Cause Injury | Charge | Choke | Clash |
| Collision | Combat | Combativeness | Come to Blows | Commit Hostilities |
| Contaminate | Create a Disturbance | Create a Riot | Create Havoc | Cripple |
| Cruel and Inhuman Treatment | Cruelty | Crush | Cut | Cut Out |
| Damage | Deal a Blow | Defacement | Defacement of Ballots | Defacement of Documents |
| Defacement of Labels | Defacement of Monuments | Defacement of Public Property | Defile | Degrade |
| Demolish | Deprive Of Life | Descend On | Desecration of Religious Structures | Desolate |
| Despoil | Destroy | Destruction of Property by Fire | Devastate | Disable |
| Disarrange | Disease | Disfigure | Dislocate | Dismantle |
| Disorder | Disorderly Conduct | Disorganize | Disrepair | Disrupt |
| Disruption | Dissolve | Do Violence to | Dominate | Drown |
| Duress | Eliminate | Endanger | Enforce | Engage in Hostilities |
| Eradicate | Erase | Eviscerate | Execute | Explosion |
| Expose to Danger | Expose to Loss | Exterminate | Fall Upon | Feud |
| Fight | Fire-Raising | Firing | Fisticuffs | Fly At |
| Force | Forcible Violation | Frighten | Frighten Away | Give Cause for Alarm |
| Grapple | Gut | Harass | Harm | Harry |
| Haze | Hit | Homicide | Horrify | Hostility |
| Hurt | Immolate | Impair | Impale | Incapacitate |

**Figure E-1. Terms of violence**

| Incite a Riot | Injure Fatally | Intentional Destruction of Public Utility Property | Interdict | Intimidate |
|---|---|---|---|---|
| Invade | Issue Threats | Jar | Jeopardize | Justifiable Killing |
| Kill | Lacerate | Lay Hands On | Lay Waste | Leave Defenseless |
| Leave Unprotected | Lie in Wait for | Liquidate | Loot | Lynch |
| Maim | Make a Shambles | Make Havoc | Make Insecure | Make Unsafe |
| Make Unsound | Make Vulnerable | Malicious Burning of Property | Malicious Killing | Mangle |
| Mar | Maraud | Massacre | Mêlée | Menace |
| Mistreat | Molest | Murder | Murder by Stealth | Mutilate |
| Obliterate | Open Fire Upon | Oppose | Oppress | Overthrow |
| Overturn | Perforate | Permanent Damage | Persecute | Physically Assault |
| Pillage | Piracy | Plague | Poison | Political Assassination |
| Pollute | Pounce Upon | Premeditated Killing | Prey Upon | Pulverize |
| Pummel | Push | Put Down | Put in Fear | Put in Hazard |
| Put in Jeopardy | Put to Death | Put Under Restraint | Raid | Rampage |
| Ransack | Rape | Ravage | Raze | Reave |
| Reduce to Nothing | Rend | Repress | Resort to Arms | Riot |
| Rub Out | Ruffianism | Ruin | Run At | Run Riot |
| Rush Upon | Sabotage | Sack | Savagery | Scar |
| Scare | Scratch | Scratch Out | Set a Trap for | Set Conflagration |
| Set Upon | Set Upon With Force | Set Upon With Violence | Setting of Fires | Sexual Assault |
| Shock with Sudden Fear | Shoot At | Sicken | Siege | Skirmish |
| Slaughter | Slay | Slaying | Smite | Snare |
| Spoil | Spring Upon | Squash | Start a Fight | State Of Siege |
| Storm | Strangle | Strike | Strike At | Strike With Overwhelming Fear |
| Strip | Subjugate | Subvert | Sudden Attack | Suppress |
| Tackle | Take Offensive Action | Taking Life | Terrify | Terrorize |
| The Destruction of Jail or Prison Property by Inmates | Threaten | Throw a Spanner in the Works | Throw Oneself Upon | Thrust At |
| Torture | Trample On | Traumatize | Treacherous Killing | Unprovoked Attack |
| Use Force Upon | Vandalize | Vanquish | Violence | Violent Death |
| Waylay | Willful Burning of Property | Wipe Out | Wound | Wreck |

**Figure E-1. Terms of violence (continued)**

# METHODS OF NONVIOLENT PROTEST AND PERSUASION

E-2. Gene Sharp of the Albert Einstein Institute developed the following 198 methods of nonviolent protest and persuasion (figure E-2, pages E-3 through E-8). Just considering the nuanced differences in how such terms differ and how they might be employed should add breadth to the SF Soldier's operational art for UW. Readers are encouraged to consult Sharp's original work, *The Politics of Nonviolent Action*, for a detailed explanation.

| THE METHODS OF NONVIOLENT PROTEST AND PERSUASION |
| --- |
| **Formal Statements** |
| 1. Public Speeches |
| 2. Letters of opposition or support |
| 3. Declarations by organizations and institutions |
| 4. Signed public statements |
| 5. Declarations of indictment and intention |
| 6. Group or mass petitions |
| **Communications with a Wider Audience** |
| 7. Slogans, caricatures, and symbols |
| 8. Banners, posters, and displayed communications |
| 9. Leaflets, pamphlets, and books |
| 10. Newspapers and journals |
| 11. Records, radio, and television |
| 12. Skywriting and earthwriting |
| **Group Representations** |
| 13. Deputations |
| 14. Mock awards |
| 15. Group lobbying |
| 16. Picketing |
| 17. Mock elections |
| **Symbolic Public Acts** |
| 18. Displays of flags and symbolic colors |
| 19. Wearing of symbols |
| 20. Prayer and worship |
| 21. Delivering symbolic objects |
| 22. Protest disrobings |
| 23. Destruction of own property |
| 24. Symbolic lights |
| 25. Displays of portraits |
| 26. Paint as protest |
| 27. New signs and names |
| 28. Symbolic sounds |
| 29. Symbolic reclamations |
| 30. Rude gestures |
| **Pressures on Individuals** |
| 31. "Haunting" officials |
| 32. Taunting officials |
| 33. Fraternization |
| 34. Vigils |

Figure E-2. Methods of nonviolent protest and persuasion

## THE METHODS OF NONVIOLENT PROTEST AND PERSUASION

### Drama and Music
35. Humorous skits and pranks
36. Performances of plays and music
37. Singing

### Processions
38. Marches
39. Parades
40. Religious processions
41. Pilgrimages
42. Motorcades

### Honoring the Dead
43. Political mourning
44. Mock funerals
45. Demonstrative funerals
46. Homage at burial places

### Public Assemblies
47. Assemblies of protest or support
48. Protest meetings
49. Camouflaged meetings of protest
50. Teach-ins

### Withdrawal and Renunciation
51. Walk-outs
52. Silence
53. Renouncing honors
54. Turning one's back

### Ostracism of Persons
55. Social boycott
56. Selective social boycott
57. Lysistratic nonaction
58. Excommunication
59. Interdict

### Noncooperation with Social Events, Customs, and Institutions
60. Suspension of social and sports activities
61. Boycott of social affairs
62. Student strike
63. Social disobedience
64. Withdrawal from social institutions

### Withdrawal from the Social System
65. Stay-at-home
66. Total personal noncooperation
67. "Flight" of workers
68. Sanctuary
69. Collective disappearance
70. Protest emigration (hijrat)

Figure E-2. Methods of nonviolent protest and persuasion (continued)

## THE METHODS OF ECONOMIC NONCOOPERATION: ECONOMIC BOYCOTTS

### Actions by Consumers
71. Consumers' boycott
72. Nonconsumption of boycotted goods
73. Policy of austerity
74. Rent withholding
75. Refusal to rent
76. National consumers' boycott
77. International consumers' boycott

### Action by Workers and Producers
78. Workmen's boycott
79. Producers' boycott

### Action by Middlemen
80. Suppliers' and handlers' boycott

### Action by Owners and Management
81. Traders' boycott
82. Refusal to let or sell property
83. Lockout
84. Refusal of industrial assistance
85. Merchants' "general strike"

### Action by Holders of Financial Resources
86. Withdrawal of bank deposits
87. Refusal to pay fees, dues, and assessments
88. Refusal to pay debts or interest
89. Severance of funds and credit
90. Revenue refusal
91. Refusal of a government's money

### Action by Governments
92. Domestic embargo
93. Blacklisting of traders
94. International sellers' embargo
95. International buyers' embargo
96. International trade embargo

### Symbolic Strikes
97. Protest strike
98. Quickie walkout (lightning strike)

### Agricultural Strikes
99. Peasant strike
100. Farm Workers' strike

### Strikes by Special Groups
101. Refusal of impressed labor
102. Prisoners' strike
103. Craft strike
104. Professional strike

Figure E-2. Methods of nonviolent protest and persuasion (continued)

**Ordinary Industrial Strikes**

105. Establishment strike
106. Industry strike
107. Sympathetic strike

**Restricted Strikes**

108. Detailed strike
109. Bumper strike
110. Slowdown strike
111. Working-to-rule strike
112. Reporting "sick" (sick-in)
113. Strike by resignation
114. Limited strike
115. Selective strike

**Multi-Industry Strikes**

116. Generalized strike
117. General strike

**Combination of Strikes and Economic Closures**

118. Hartal
119. Economic shutdown

THE METHODS OF POLITICAL NONCOOPERATION

**Rejection of Authority**

120. Withholding or withdrawal of allegiance
121. Refusal of public support
122. Literature and speeches advocating resistance

**Citizens' Noncooperation with Government**

123. Boycott of legislative bodies
124. Boycott of elections
125. Boycott of government employment and positions
126. Boycott of government depts., agencies, and other bodies
127. Withdrawal from government educational institutions
128. Boycott of government-supported organizations
129. Refusal of assistance to enforcement agents
130. Removal of own signs and placemarks
131. Refusal to accept appointed officials
132. Refusal to dissolve existing institutions

**Citizens' Alternatives to Obedience**

133. Reluctant and slow compliance
134. Nonobedience in absence of direct supervision
135. Popular nonobedience
136. Disguised disobedience
137. Refusal of an assemblage or meeting to disperse
138. Sitdown
139. Noncooperation with conscription and deportation
140. Hiding, escape, and false identities
141. Civil disobedience of "illegitimate" laws

**Figure E-2. Methods of nonviolent protest and persuasion (continued)**

## Action by Government Personnel

142. Selective refusal of assistance by government aides
143. Blocking of lines of command and information
144. Stalling and obstruction
145. General administrative noncooperation
146. Judicial noncooperation
147. Deliberate inefficiency and selective noncooperation by enforcement agents
148. Mutiny

## Domestic Governmental Action

149. Quasi-legal evasions and delays
150. Noncooperation by constituent governmental units

## International Governmental Action

151. Changes in diplomatic and other representations
152. Delay and cancellation of diplomatic events
153. Withholding of diplomatic recognition
154. Severance of diplomatic relations
155. Withdrawal from international organizations
156. Refusal of membership in international bodies
157. Expulsion from international organizations

## THE METHODS OF NONVIOLENT INTERVENTION

## Psychological Intervention

158. Self-exposure to the elements
159. The fast: (a) Fast of moral pressure; (b) Hunger strike; (c) Satyagrahic fast
160. Reverse trial
161. Nonviolent harassment

## Physical Intervention

162. Sit-in
163. Stand-in
164. Ride-in
165. Wade-in
166. Mill-in
167. Pray-in
168. Nonviolent raids
169. Nonviolent air raids
170. Nonviolent invasion
171. Nonviolent interjection
172. Nonviolent obstruction
173. Nonviolent occupation

## Social Intervention

174. Establishing new social patterns
175. Overloading of facilities
176. Stall-in
177. Speak-in
178. Guerrilla theater
179. Alternative social institutions
180. Alternative communications system

**Figure E-2. Methods of nonviolent protest and persuasion (continued)**

**Economic Intervention**

181. Reverse strike
182. Stay-in strike
183. Nonviolent land seizure
184. Defiance of blockades
185. Politically motivated counterfeiting
186. Preclusive purchasing
187. Seizure of assets
188. Dumping
189. Selective patronage
190. Alternative markets
191. Alternative transportation systems
192. Alternative economic institutions

**Political Intervention**

193. Overloading of administrative systems
194. Disclosing identities of secret agents
195. Seeking imprisonment
196. Civil disobedience of 'neutral' laws
197. Work-on without collaboration
198. Dual sovereignty and parallel government

Figure E-2. Methods of nonviolent protest and persuasion (continued)

## Appendix F

# Resistance Development Models

This appendix presents a collection of models from various sources useful to analyzing and understanding resistance at the basic cellular level and several that attempt to characterize the macro-structure and overall development of a resistance movement. The basic cellular structure is generic, has been in UW and related doctrine for a long time, and is widely considered unobjectionable across the resistance professional community. The macro-level models represent a common subject of interest with differing interpretations. All of these models have some merit in helping SF Soldiers analyze resistance, but none of them is the definitive template or sole analytical solution for the SF operational detachment.

## BASIC CELLULAR STRUCTURE MODELS

### OPERATIONS CELL

F-1. The operations cell is usually composed of a leader and a few cell members, operating directly as a unit (figure F-1). A cell can be as large as required to conduct a specific mission, but it is generally kept as small as possible for security reasons. The responsibilities of each cell member will be as defined by the cell leader and the cell's assigned functions. However, cell member responsibilities are also shaped by the cell member's societal status and skill sets. Operational cell members may or may not know each other or work together. Wherever possible, organizational security is enhanced by structuring intermediaries between operational members and higher-echelon leaders. This system of intermediaries permeates the entire insurgent structure all the way to the organization's strategic leadership.

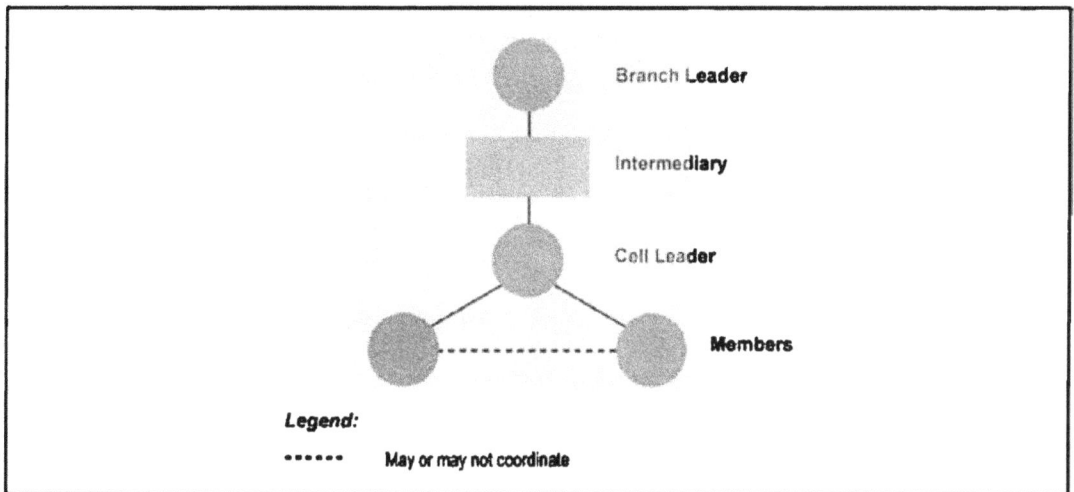

Figure F-1. Operations cell

## INTELLIGENCE CELL

F-2.   The intelligence cell exhibits similar organizational and security characteristics as operational cells. However, the need for cell security is of critical importance; the intelligence cell leader is seldom in direct contact with the members of the cell, and the members are rarely in contact with one another (figure F-2). Methods of assigning missions down and reporting information up these structures maximize third-party intermediaries or mechanical communications methods, which avoid human-to-human direct contact altogether.

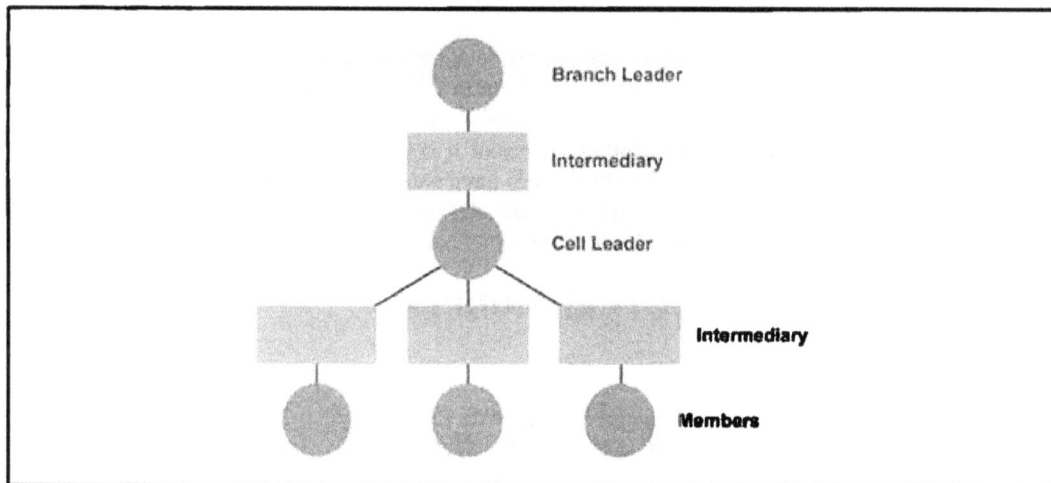

**Figure F-2. Intelligence cell**

## AUXILIARY CELL

F-3.   The auxiliary cell is commonly found in front groups or in sympathizers' organizations. It contains an underground cell leader, assistant cell leaders, and members (figure F-3).

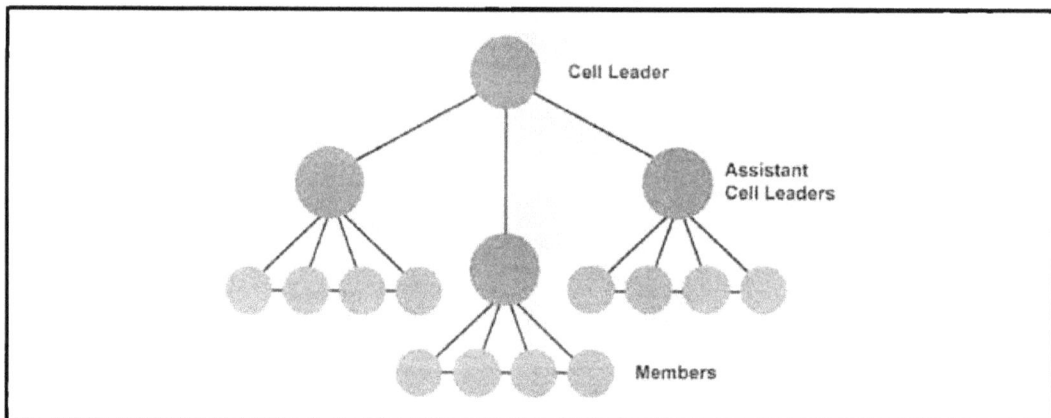

**Figure F-3. Auxiliary cell**

## PARALLEL CELLS

F-4.   The key idea of parallel cells is that many activities can be conducted that support a goal known only to a higher controlling echelon leader. For example, an underground planning to conduct an attack on a government facility may assign more than one intelligence cell to surveil the intended target. Although both cells have the same assignment, each cell is unaware of the membership, presence, or mission of the other cell. Using parallel cells provides the overall organization operational redundancy. Although unknown to the executing cells, redundant, secondary parallel cells are frequently set up to support a primary cell (figure F-4).

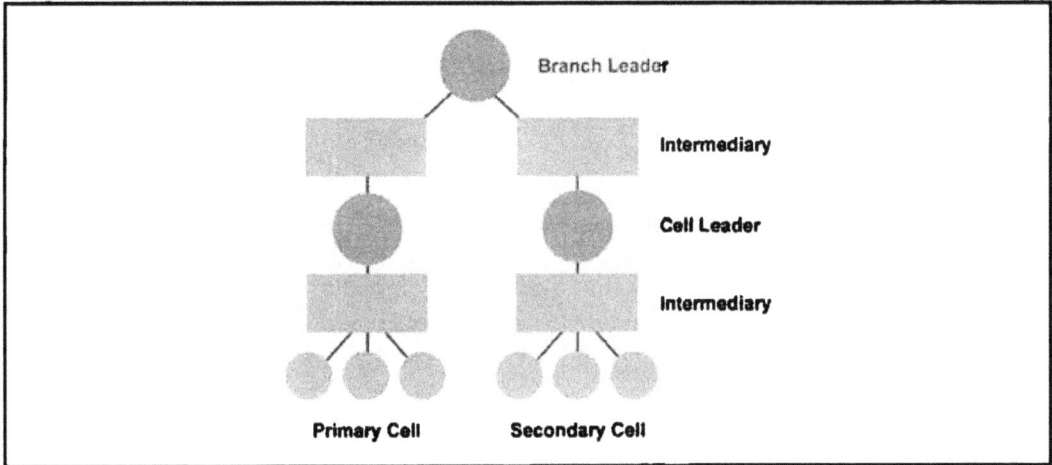

**Figure F-4. Parallel cells**

### CELLS IN SERIES

F-5.  The cells in series provide a division of labor in order to carry out underground and auxiliary functions, such as the manufacture of weapons, supply, escape and evasion, propaganda, and printing of newspapers. The task assigned to a particular cell must transition or carry over (depicted by arrows) to the next cell in order to accomplish the function in its entirety. For example, cell 1 purchases certain items, cell 2 assembles the items, and cell 3 distributes the assembled item (figure F-5). Based on the assigned mission, cell members do not communicate directly with one another. However, cell leaders will communicate indirectly through intermediaries.

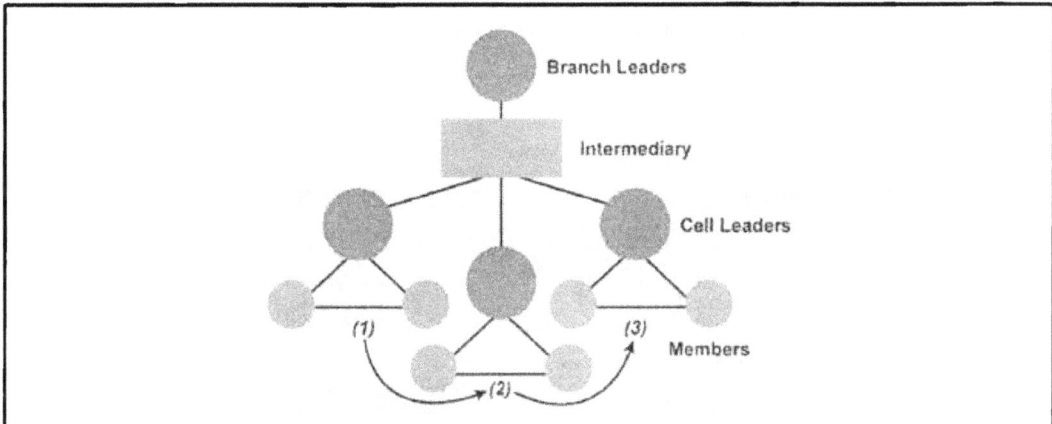

**Figure F-5. Cells in series**

## RESISTANCE MACRO-STRUCTURE MODELS

F-6.  The CIA's Guide to the Analysis of Insurgency (2012) shows insurgency primary and alternative trajectories and lists phases of insurgency as preinsurgency, and early, middle and end phases, also known as growth, mature, and resolution (figure F-6, page F-4).

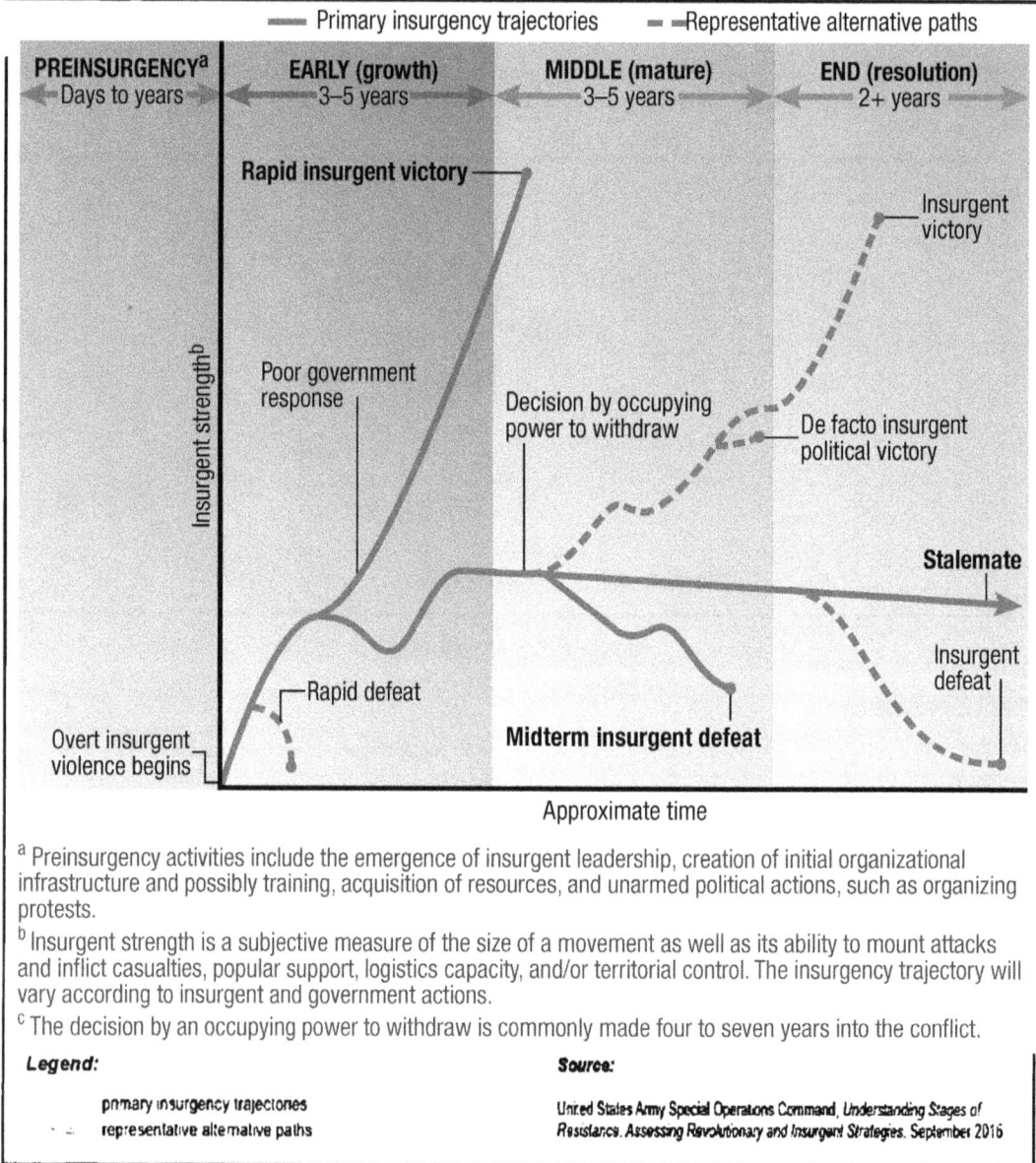

**Figure F-6. Central Intelligence Agency's life cycle of an insurgency**

F-7   Although the model (figure F-7, page F-5) graphically presents these states as linear, the construct allows for the flexibility of resistance movements to progress in a nonlinear fashion. The states include:

- **Preliminary State.** The most defining feature of this state is the manifestation of unorganized and unattributed unrest—unorganized because actors are unconnected, and unattributed because the unrest lacks a common narrative about the source of the problem

- **Incipient State.** Alternatively called "coalescence," the defining feature is the development of intentional organization and a common narrative. The movement has begun to take shape, with leaders and overt, strategic patterns of action as opposed to featuring leaderless short-term actions.

- **Crisis State.** A movement in this state features the essential characteristic of an escalated confrontation with opponents (most often the state authority) that marks a decisive moment, when the movement demonstrates itself to be a real and clear threat to the opponent's interests, such that the opponent must respond.

- **Institutional State.** At this point the resistance movement has survived the confrontation with its opponent and has established a role in society.
- **Resolution State.** A transition from any previous state through a variety of avenues. According to this study, examples include the varying types of success and failure below:
  - Dormancy.
  - Decline.
  - Radicalization.
  - Institutionalization.
  - Repression.
  - Facilitation.
  - Success (listed as a separate type, rather than a larger context).
  - Failure (listed as a separate type, rather than a larger context).
  - Co-optation.
  - Establishment with the mainstream.
  - Exhaustion.

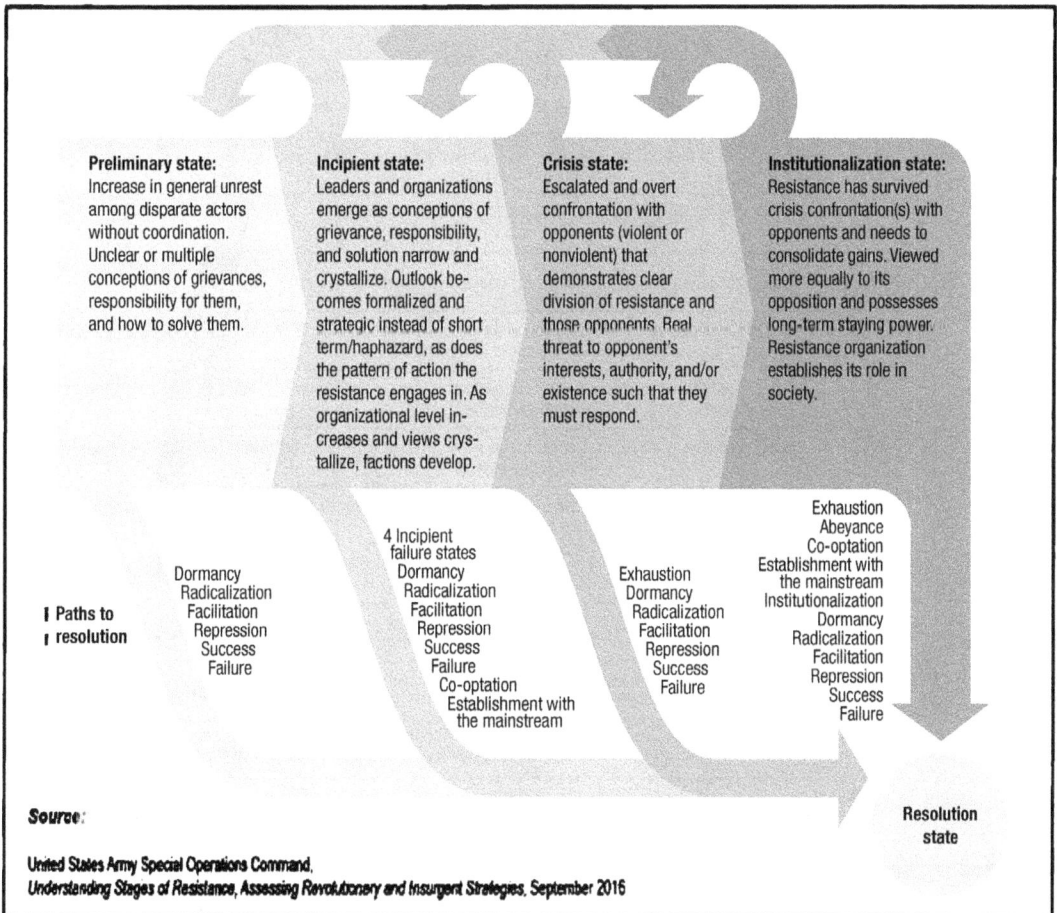

**Preliminary state:**
Increase in general unrest among disparate actors without coordination. Unclear or multiple conceptions of grievances, responsibility for them, and how to solve them.

**Incipient state:**
Leaders and organizations emerge as conceptions of grievance, responsibility, and solution narrow and crystallize. Outlook becomes formalized and strategic instead of short term/haphazard, as does the pattern of action the resistance engages in. As organizational level increases and views crystallize, factions develop.

**Crisis state:**
Escalated and overt confrontation with opponents (violent or nonviolent) that demonstrates clear division of resistance and those opponents. Real threat to opponent's interests, authority, and/or existence such that they must respond.

**Institutionalization state:**
Resistance has survived crisis confrontation(s) with opponents and needs to consolidate gains. Viewed more equally to its opposition and possesses long-term staying power. Resistance organization establishes its role in society.

| Paths to resolution

Dormancy
Radicalization
Facilitation
Repression
Success
Failure

4 Incipient failure states
Dormancy
Radicalization
Facilitation
Repression
Success
Failure
Co-optation
Establishment with the mainstream

Exhaustion
Dormancy
Radicalization
Facilitation
Repression
Success
Failure

Exhaustion
Abeyance
Co-optation
Establishment with the mainstream
Institutionalization
Dormancy
Radicalization
Facilitation
Repression
Success
Failure

Resolution state

Source:

United States Army Special Operations Command,
*Understanding Stages of Resistance, Assessing Revolutionary and Insurgent Strategies,* September 2016

**Figure F-7. Central Intelligence Agency model—life cycle of insurgency**

F-8. This Special Operations Research Office pyramid construct (figure F-8, page F-6) has informed ARSOF training and education in resistance and UW for decades, and it is still found in current UW doctrine. Despite its flaws, it still has some intuitive explanatory power.

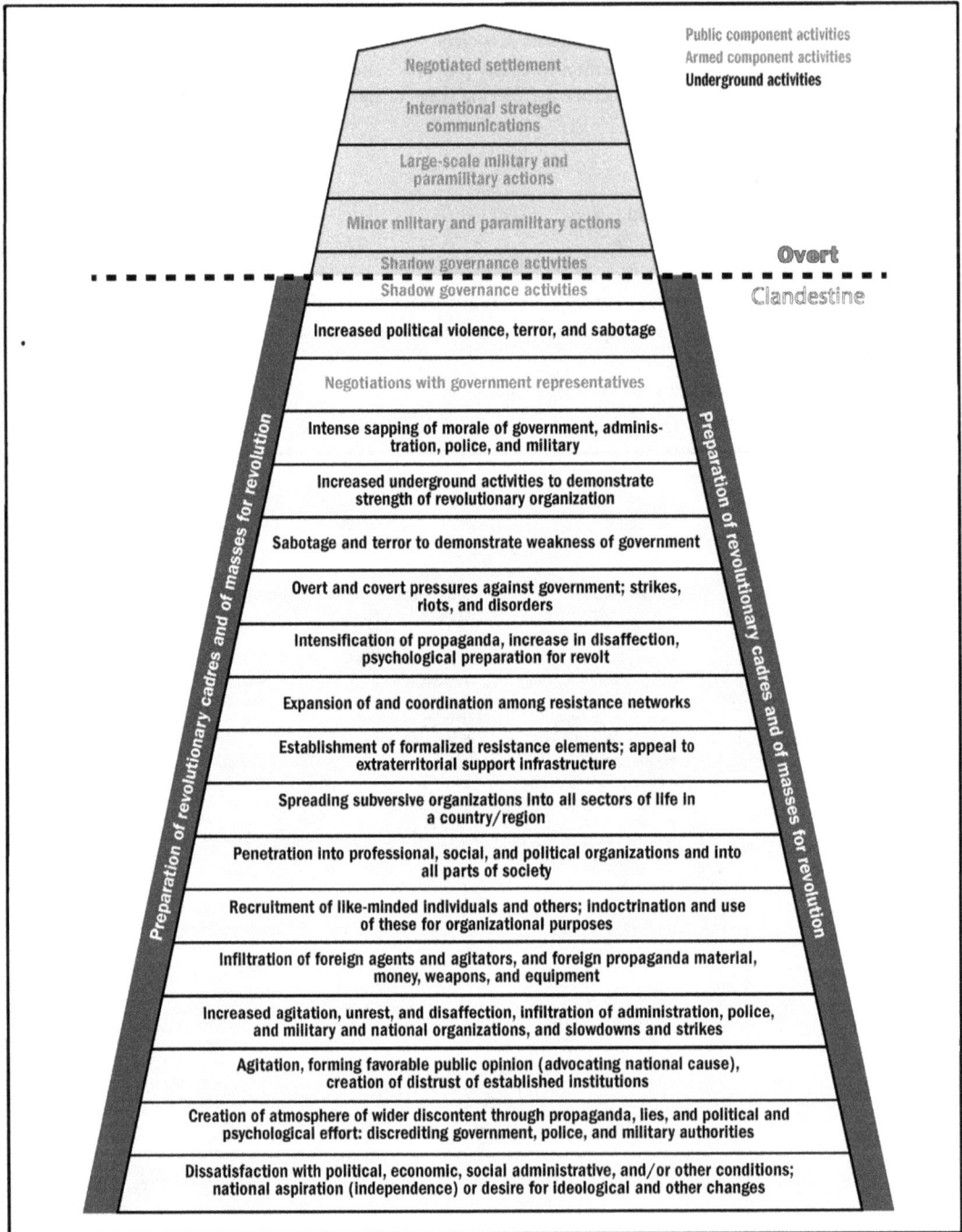

Figure F-8. Special Operations Research Office resistance pyramid

F-9. The following two models below were developed to suggest an alternate to the time-honored Special Operations Research Office pyramid, which does not distort the relative importance and proportion of the underground to the overall resistance effort. The first model (figure F-9) includes all of the activities in the Special Operations Research Office pyramid but not in a linear, choreographed order. It also emphasizes the primacy of political activities and organizational activities in overall resistance development.

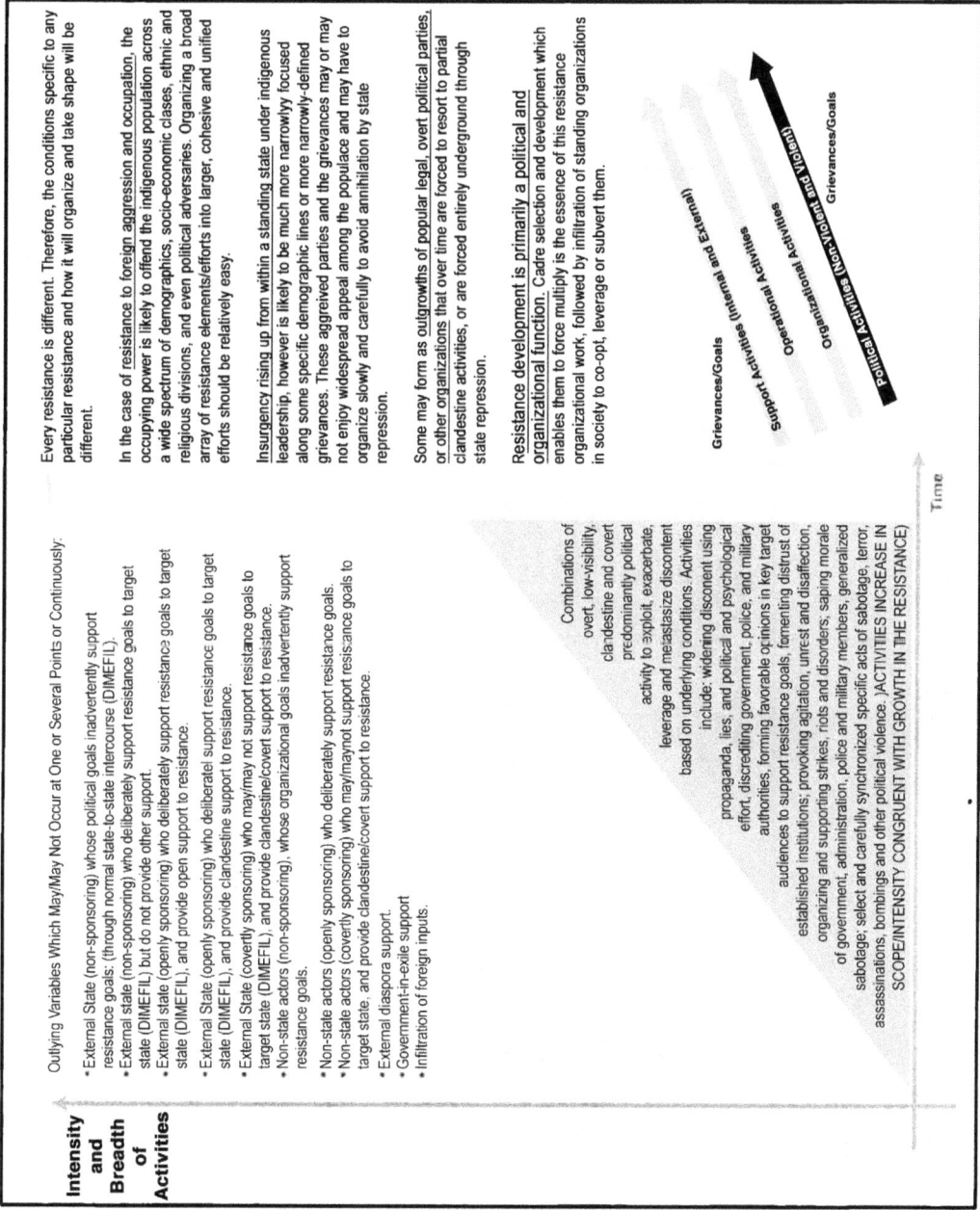

**Intensity and Breadth of Activities**

Outlying Variables Which May/May Not Occur at One or Several Points or Continuously:
* External State (non-sponsoring) whose political goals inadvertently support resistance goals: (through normal state-to-state intercourse (DIMEFIL).
* External state (non-sponsoring) who deliberately support resistance goals to target state (DIMEFIL) but do not provide other support.
* External state (openly sponsoring) who deliberately support resistance goals to target state (DIMEFIL), and provide open support to resistance.
* External State (openly sponsoring) who deliberate support resistance goals to target state (DIMEFIL), and provide clandestine support to resistance.
* External State (covertly sponsoring) who may/may not support resistance goals to target state (DIMEFIL), and provide clandestine/covert support to resistance.
* Non-state actors (non-sponsoring), whose organizational goals inadvertently support resistance goals.
* Non-state actors (openly sponsoring) who deliberately support resistance goals.
* Non-state actors (covertly sponsoring) who may/maynot support resistance goals to target state, and provide clandestine/covert support to resistance.
* External diaspora support.
* Government-in-exile support.
* Infiltration of foreign inputs.

Combinations of overt, low-visibility, clandestine and covert predominantly political activity to exploit, exacerbate, leverage and metastasize discontent based on underlying conditions. Activities include: widening disconent using propaganda, lies, and political and psychological effort, discrediting government, police, and military authorities, forming favorable opinions in key target audiences to support resistance goals, fomenting distrust of established institutions; provoking agitation, unrest and disaffection, organizing and supporting strikes, riots and disorders; saping morale of government, administration, police and military members, generalized sabotage; select and carefully synchronized specific acts of sabotage, terror, assassinations, bombings and other political violence. )ACTIVITIES INCREASE IN SCOPE/INTENSITY CONGRUENT WITH GROWTH IN THE RESISTANCE)

Every resistance is different. Therefore, the conditions specific to any particular resistance and how it will organize and take shape will be different.

In the case of resistance to foreign aggression and occupation, the occupying power is likely to offend the indigenous population across a wide spectrum of demographics, socio-economic clases, ethnic and religious divisions, and even political adversaries. Organizing a broad array of resistance elements/efforts into larger, cohesive and unified efforts should be relatively easy.

Insurgency rising up from within a standing state under indigenous leadership, however is likely to be much more narrowly focused along some specific demographic lines or more narrowly-defined grievances. These aggrieved parties and the grievances may or may not enjoy widespread appeal among the populace and may have to organize slowly and carefully to avoid annihilation by state repression.

Some may form as outgrowths of popular legal, overt political parties, or other organizations that over time are forced to resort to partial clandestine activities, or are forced entirely underground through state repression.

Resistance development is primarily a political and organizational function. Cadre selection and development which enables them to force multiply is the essence of this resistance organizational work, followed by infiltration of standing organizations in society to co-opt, leverage or subvert them.

Grievances/Goals (Internal and External)
Support Activities
Operational Activities
Organizational Activities (Non-Violent and Violent)
Political Activities (Non-Violent and Violent)
Grievances/Goals

Time

**Figure F-9. An alternate graphic model of resistance based on resistance-as-a-whole (overthrow)**

F-10. The second model (figure F-10) shows variations of up-and-down resistance tempo and varied culmination points, depending on the supporting sponsor's UW objective. These alternate examples are contrasted with the assumptions explicit and inherent in the classic Special Operations Research Office pyramid model.

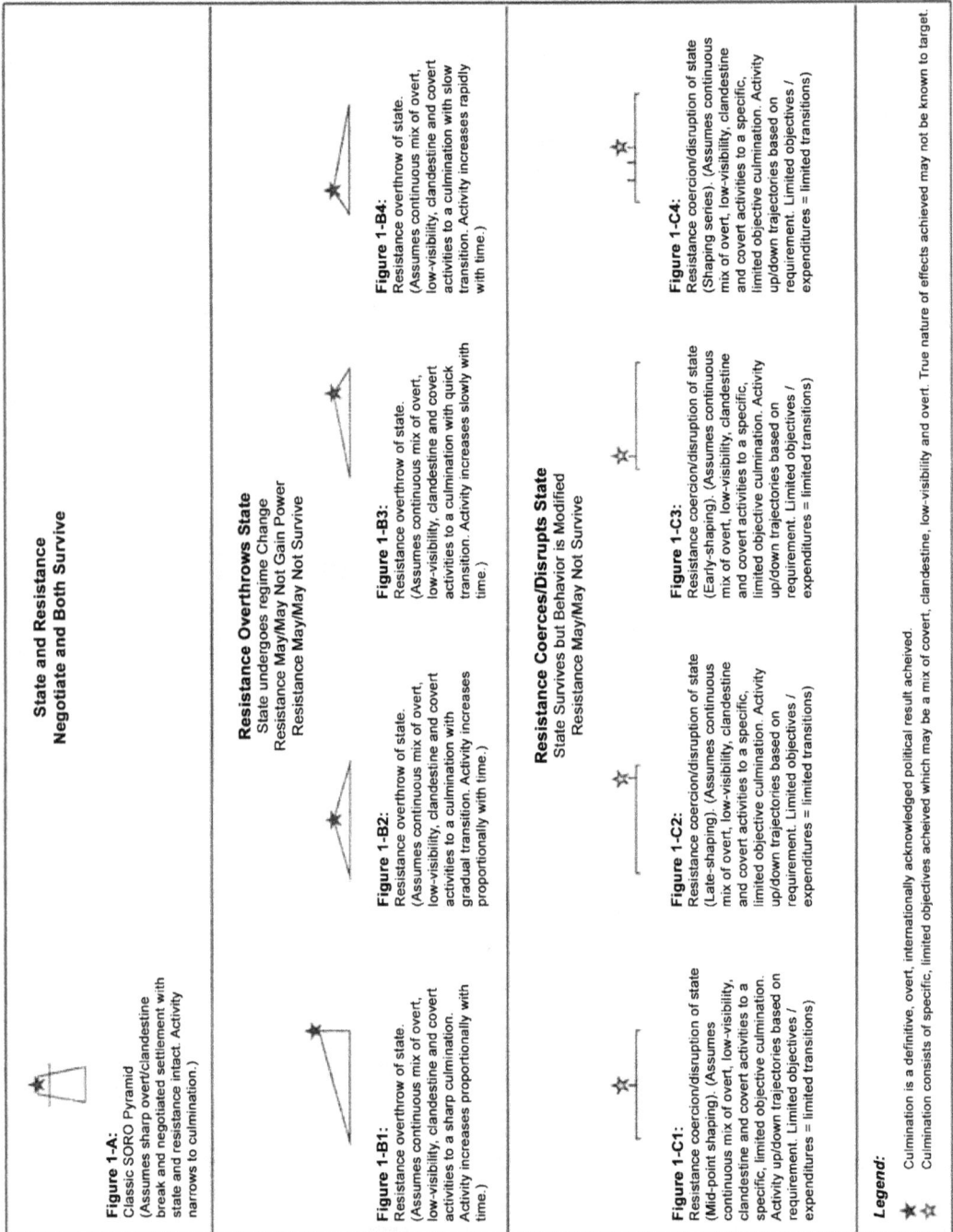

**State and Resistance Negotiate and Both Survive**

**Figure 1-A:**
Classic SORO Pyramid
(Assumes sharp overt/clandestine break and negotiated settlement with state and resistance intact. Activity narrows to culmination.)

**Resistance Overthrows State**
State undergoes regime Change
Resistance May/May Not Gain Power
Resistance May/May Not Survive

**Figure 1-B1:**
Resistance overthrow of state.
(Assumes continuous mix of overt, low-visibility, clandestine and covert activities to a sharp culmination. Activity increases proportionally with time.)

**Figure 1-B2:**
Resistance overthrow of state.
(Assumes continuous mix of overt, low-visibility, clandestine and covert activities to a culmination with gradual transition. Activity increases proportionally with time.)

**Figure 1-B3:**
Resistance overthrow of state.
(Assumes continuous mix of overt, low-visibility, clandestine and covert activities to a culmination with quick transition. Activity increases slowly with time.)

**Figure 1-B4:**
Resistance overthrow of state.
(Assumes continuous mix of overt, low-visibility, clandestine and covert activities to a culmination with slow transition. Activity increases rapidly with time.)

**Resistance Coerces/Disrupts State**
State Survives but Behavior is Modified
Resistance May/May Not Survive

**Figure 1-C1:**
Resistance coercion/disruption of state (Mid-point shaping). (Assumes continuous mix of overt, low-visibility, clandestine and covert activities to a specific, limited objective culmination. Activity up/down trajectories based on requirement. Limited objectives / expenditures = limited transitions)

**Figure 1-C2:**
Resistance coercion/disruption of state (Late-shaping). (Assumes continuous mix of overt, low-visibility, clandestine and covert activities to a specific, limited objective culmination. Activity up/down trajectories based on requirement. Limited objectives / expenditures = limited transitions)

**Figure 1-C3:**
Resistance coercion/disruption of state (Early-shaping). (Assumes continuous mix of overt, low-visibility, clandestine and covert activities to a specific, limited objective culmination. Activity up/down trajectories based on requirement. Limited objectives / expenditures = limited transitions)

**Figure 1-C4:**
Resistance coercion/disruption of state (Shaping series). (Assumes continuous mix of overt, low-visibility, clandestine and covert activities to a specific, limited objective culmination. Activity up/down trajectories based on requirement. Limited objectives / expenditures = limited transitions)

*Legend:*
★ ☆
★ Culmination is a definitive, overt, internationally acknowledged political result achieved
☆ Culmination consists of specific, limited objectives achieved which may be a mix of covert, clandestine, low-visibility and overt. True nature of effects achieved may not be known to target.

**Figure F-10. Multiple alternate graphic models of resistance—coerce, disrupt, and overthrow**

F-11. USASOC developed a model to help think about resistance from a legal perspective (figure F-11). This model, too, is useful for considering resistance from a different perspective than standard military mission analysis. The model is, nevertheless, flawed by an arbitrary temporal distinction between nonviolent and violent activities that does not exist in reality. Moreover, the categorical terms employed by this conceptual model also do not align with official doctrine, nor does doctrine accept the veracity of the linear trend.

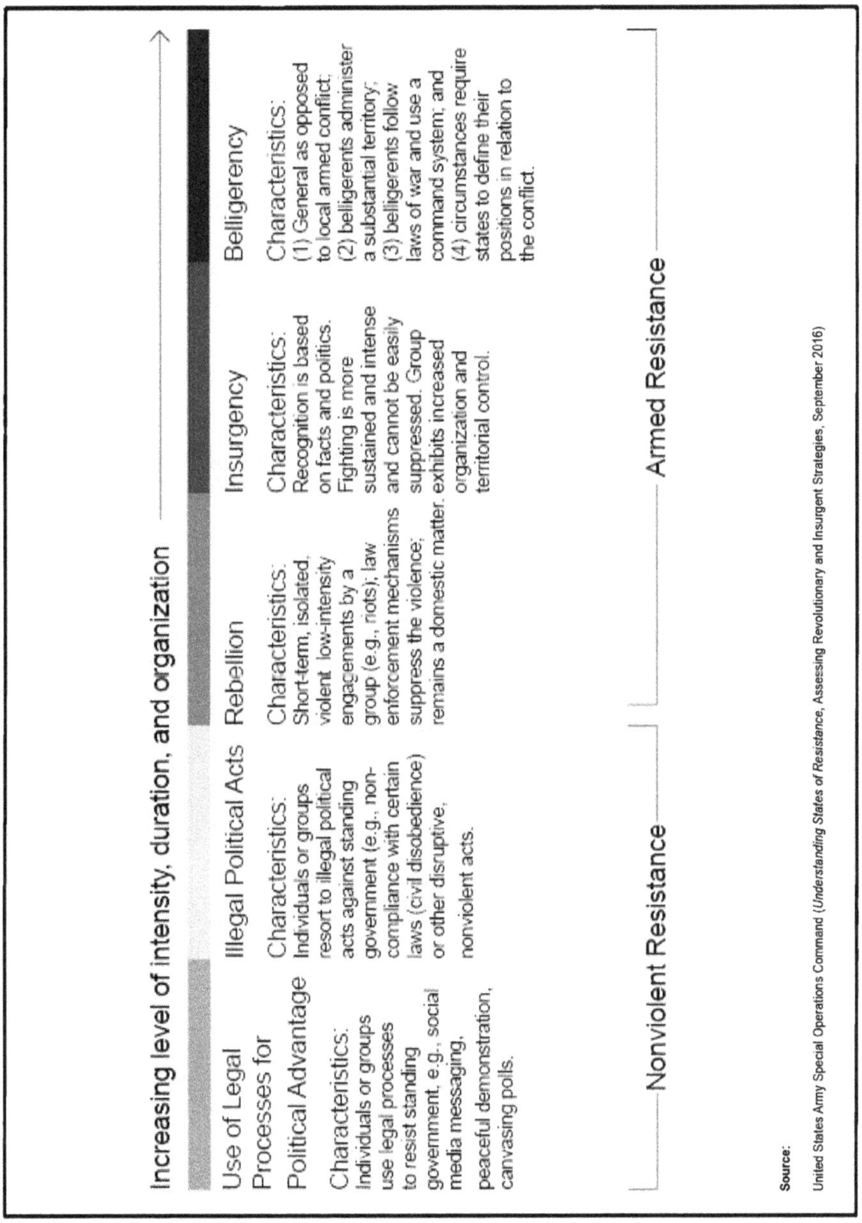

Figure F-11. Continuum from legal protests to insurgency and belligerency

This page intentionally left blank.

# Appendix G

# Theater Special Operations Command Mission Letter Examples

Mission letters can be of two types. Standing mission letters are based on GCC steady-state postures and TSOC analysis of GCC enduring and anticipated requirements for the future operating environment. Mission-specific mission letters are tailored to a particular campaign or operation that augment or are separate from standing mission letters. Used correctly, the former provides a wide range of mission and area specialization throughout the total SF regiment, which develops over time into deep and broad-based regimental expertise. By contrast, the latter provides guidance for immediate and short-term effect.

## STANDING MISSION LETTERS

G-1. Standing mission letters are based on GCC CONPLANs and anticipated enduring requirements inherent to assigned theaters. Standing mission letters provide a context and rationale for subordinate unit mission planning and related training development. Standing mission letters allow commanders and planners to efficiently provide for capabilities needed in anticipated future operations. Such mission letters provide a means for commanders to direct routine specialization among subordinate elements. Standing mission letters provide subordinate elements a measure of predictable continuity in long-range training development, and the letters provide a rationale for area, language, infiltration capability, and core-mission specialization and prioritization by SF element. In the absence of a mission-specific mission letter, units return to standing mission letters to orient their METL development, training plans, and personnel development (figure G-1, pages G-2 through G-4).

> *Note:* TC 18-01.1 (Appendix B, Sample Mission Letter) provides the SF battalion, TSOC, and SF Group (Airborne) standing mission letter examples. TC 18-01.2 and TC 18-01.3 also provide these examples.

## MISSION-SPECIFIC MISSION LETTERS

G-2. Mission-specific mission letters are based on current joint task force orders for specific current operations. Mission-specific mission letters are issued as needed to meet current operational requirements. Commanders may issue mission-specific mission letters that augment standing mission letters or which may have no relation to standing mission letter guidance whatsoever. The duration of the mission-specific mission letter will be driven by the duration of the specific operation. There is no prescribed format for a mission-specific mission letter.

(CLASSIFICATION)
DEPARTMENT OF THE ARMY
HEADQUARTERS, 1st BATTALION
9th SPECIAL FORCES GROUP (AIRBORNE)
1st SPECIAL FORCES
APO NEW YORK 99999

AANS-C-SC                                                                    06 July 2014

MEMORANDUM FOR: Commander Company A, 1st Battalion, 9th Special Forces Group
(Airborne), Fort Bragg NC 28307

SUBJECT: Company A Mission Planning Guidance (U)

1.  ( ) References:
    a.  ( ) COMSOCATL OPLAN 2640, SOCATL Subordinate Campaign Plan for Special Operations in
        USAATLCOM AOR (U)
    b.  ( ) SOCATL Memorandum Subject: 1st BN, 9th SFG (A) Mission Planning Guidance (U) 15 Mar 14.
    c.  ( ) 9th SFG (A) OPLAN 2640, Major Operations Plan for Special Operations in the USATLCOM AOR
        (U)
    d.  ( ) 9th SFG (A) Field Readiness SOP, 18 Dec 2012.
    e.  ( ) 9th SFG (A) Field SOP, 23 Sep 13.
    f.  ( ) 9th SFG (A) LOI Subject: Area Study and Mission Planning Program 01 Feb 2012
2.  ( ) General:
    a.  ( ) This memorandum provides specific guidance regarding the planning, training, and preparation
        of your ODAs for employment in the USAATLCOM AOR. It identifies detachment special skills and
        missions, area orientations, and high priority tasks.
    b.  ( ) This mission letter assumes that your battalion remains at authorized level of organization
        (ALO) 1 and that all assigned personnel and equipment are available for employment. You will
        coordinate with the battalion commander to modify mission priorities if this assumption becomes
        invalid.
3.  ( ) Mission. Continue to support special operations in the AOR. Prepare assigned elements for
    employment according to 9th SFG (A) OPLAN 2640. When directed, conduct special operations in your
    assigned sectors.
4.  ( ) Execution.
    a.  ( ) Concept of the Operation. See Ref c.
        (1) ( ) Primarily orient your company to conduct special operations in the SRN (see enclosure 1 for
            specific JSOA). Mission orientation is UW, SR, direct action, in that order. Maintain one ODA
            with MFF capability and one ODA with combat diving capability. Each ODA in the company will
            have at least four personnel with __/__ language proficiency in Portuguese. Remaining
            operational personnel will have __/__ language proficiency.
        (2) ( ) Prepare to execute preplanned special operations missions commencing N+4, and fully
            commit all operational elements NLT N+10.
    b.  ( ) Specified Tasks.
        (1) ( ) Provide one ODB and two ODAs to conduct a special activity in support of the UNITA
            resistance by operating a UW training facility center in Nuevo Basque.

(CLASSIFICATION)

Figure G-1. Sample Special Forces (standing) mission letter

(CLASSIFICATION)

      (2)  ( ) Prepare, according to Ref e, to conduct special operations in response to actual or threatened SRN aggression against Nuevo Basque. Give planning priorities to the following missions:

          (a)  ( ) Conduct SR in the SRN to provide early warning of hostile attack.

          (b)  ( ) Conduct direct action operations in the SRN to disrupt hostile preparations for attack and convey U.S. intent to successfully oppose aggression against Nuevo Basque.

          (c)  ( ) Conduct direct action and SR operations in support of conventional military operations. Give planning priority in support of UW operations, battlefield surveillance, target acquisition, and interdiction of hostile LOC from the SRN attack toward San Sebastian.

      (3)  Prepare for operations in (2) above in a chemical, biological, radiological, and nuclear (CBRN) environment.

  c.  ( ) Detachment Missions.

      (1)  ( ) Detachment 9110. Prepare to advise, assist and support the UNITA area command in central SRN and to execute mission command the UW activities of up to four ODAs and two special operations teams–alpha (SOT-As). Plan to conduct this mission as a clandestine operation.

      (2)  ( ) Detachment 9111. Prepare to conduct low-visibility or clandestine UW operations in northeastern SRN, either independently or under the OPCON of Detachment 9110.

      (3)  ( ) Detachment 9112. Prepare to conduct low-visibility or clandestine UW operations in southeastern SRN, either independently or under the OPCON of Detachment 9110.

      (4)  ( ) Detachment 9113. Prepare to conduct clandestine UW operations in western SRN, either independently or under the OPCON of Detachment 9110

      (5)  ( ) Detachment 9114. Prepare to conduct clandestine UW operations in the vicinity of the capital of Nagos, either independently or under the OPCON of Detachment 9110.

      (6)  ( ) Detachment 9115. Prepare to conduct clandestine SR and direct action operations in western and central SRN. Maintain combat diving capability.

      (7)  ( ) Detachment 9116. Prepare to conduct clandestine SR and direct action operations in western and central SRN. Maintain MFF capability.

  d.  ( ) Coordinating Instructions.

      (1)  ( ) Ensure each detachment has at least four current jumpmasters. ODA 9116 must have at least two current MFF jumpmasters. Detachment 9115 must have at least two diving supervisors and one diving medical technician.

      (2)  ( ) Maintain trained company air movement control teams according to Ref d.

      (3)  ( ) Establish and sustain a company area study and mission planning program according to Ref e.

      (4)  ( ) Specific ODA and SOT-A tasks and mission priorities can be found in the battalion METL and in applicable SOMPFs maintained by the battalion AST.

      (5)  ( ) Review this mission letter at least annually and recommend changes to the battalion S-3.

      (6)  ( ) Perform a company mission analysis at least annually and provide the battalion S-3 with an updated METL.

      (7)  ( ) Advise the CDR 1st BN, 9th SFG (A) of any shortfalls in your ability to meet your assigned tasks.

(CLASSIFICATION)

**Figure G-1. Sample Special Forces (standing) mission letter (continued)**

**(CLASSIFICATION)**

5.   ( ) Training Guidance. Focus your training to establish and maintain detachment readiness. Conduct consolidated company training only when necessary. Ensure detachment training is realistic and oriented on the priority missions identified above. Each detachment will participate in at least one live fire exercise and one field training exercise each quarter. Each detachment and SOT-A will perform at least one week of area study and mission planning semiannually.

Encl

King
Col, SF
Commander

Enclosure 1. Company Area of Operations Overlay

**(CLASSIFICATION)**

Figure G-1. Sample Special Forces (standing) mission letter (continued)

# Appendix H

# Sample Area Study Format

This appendix provides an outline format for an area study. This format provides a systematic means for compiling and retaining essential information to support SF operations.

H-1. Although the basic outline is general, it is flexible enough to permit detailed coverage of a given area As time is available for further study, the preparer should subdivide various subjects and assign them to detachment members to produce a more detailed analysis of specified areas of interest. Figure H-1, pages H-1 through H-5, provides a sample area study format.

---

1. **Purpose and limiting factors.**
   a. Purpose. Delineate the area being studied.
   b. Mission. State the mission the area study supports.
   c. Limiting factors. Identify factors that limit the completeness or accuracy of the area study.
2. **Geography, hydrography, and climate.** Divide the operational area into its various definable subdivisions and analyze each subdivision using the subdivisions shown below:
   a. Areas and dimensions.
   b. Strategic locations:
      (1) Neighboring countries and boundaries.
      (2) Natural defenses, including frontiers.
      (3) Points of entry and strategic routes.
   c. Climate. Note variations from the norm and the months in which they occur. Note any extremes in climate that would affect operations. Include data on the following:
      (1) Temperature.
      (2) Rainfall and snow.
      (3) Wind and visibility.
      (4) Light data. Include begin morning nautical twilight, end of evening nautical twilight, sunrise, sunset, moonrise, and moonset.
      (5) Seasonal effect of the weather on terrain and visibility.
   d. Relief:
      (1) General direction of mountain ranges or ridgelines and if hills and ridges are dissected.
      (2) General degree of slope.
      (3) Characteristics of valleys and plains.
      (4) Natural routes for and natural obstacles to cross-country movement.
      (5) Location of area suitable for guerrilla bases, units, and other installations.
      (6) Potential LZs, DZs, and other reception sites.
   e. Land use. Note any peculiarities especially in the following:
      (1) Former heavily forested land areas subjected to widespread cutting or dissected bypaths and roads. Also note the reverse—pastureland or wasteland that has been reforested.
      (2) Former wasteland or pastureland that has been resettled and cultivated and is now being farmed. In addition, note the reverse—former rural countryside that has been depopulated and allowed to return to wasteland.
      (3) Former swampland or marshland that has been drained, former desert or wasteland now irrigated and cultivated, and lakes created by dams.

**Figure H-1. Area study format**

---

f.  Drainage (general pattern):
    (1)  Main rivers, including direction of flow.
    (2)  Characteristics of rivers and streams, including widths, currents, banks, depths, kinds of bottoms, and obstacles. Note seasonal variations, such as dry beds and flash floods.
    (3)  Large lakes or areas with many ponds or swamps. Include potential landing zones for amphibious aircraft.
g.  Coast. Examine primarily for infiltration, exfiltration, and resupply points:
    (1)  Tides and waves. Include winds and current.
    (2)  Beach footing and covered exit routes.
    (3)  Quiet coves and shallow inlets or estuaries. General direction of mountain ranges or ridgelines, and if hills and ridges are dissected
h.  Geological basics. Identify types of soil and rock formations. Include areas for potential landing zones for light aircraft.
i.  Forests and other vegetation:
    (1)  Natural or cultivated.
    (2)  Types, characteristics, and significant variations from the norm at different elevations.
    (3)  Cover and concealment. Include density and seasonal variations.
j.  Water. Note ground, surface, seasonal, and potable.
k.  Subsistence:
    (1)  Seasonal or year-round
    (2)  Cultivated. Include vegetables, grains, fruits, and nuts.
    (3)  Natural. Include berries, fruits, nuts, and herbs.
    (4)  Wildlife. Include animals, fish, and fowl.
3.  **Political characteristics.** Identify friendly and hostile political powers and analyze their capabilities, intentions, and activities that influence mission execution:
    a.  Hostile power:
        (1)  Number and status of non-national personnel.
        (2)  Influence, organization, and mechanisms of control.
    b.  National government (indigenous):
        (1)  Government, international political orientation, and degree of popular support.
        (2)  Identifiable segments of the population, with varying attitudes and probable behavior toward the United States, its allies, and the hostile power.
        (3)  National historical background.
        (4)  Foreign dependence or allies.
        (5)  National capital and significant political, military, and economic concentrations.
    c.  Political parties:
        (1)  Leadership and organizational structure.
        (2)  Nationalistic origin and foreign ties (if single dominant party exists).
        (3)  Major legal parties with their policies and goals.
        (4)  Illegal or underground parties and their policies and goals.
        (5)  Violent opposition factions within major political organizations.
    d.  Control and restrictions:
        (1)  Documentation.
        (2)  Rationing.
        (3)  Travel and movement restrictions.
        (4)  Blackouts and curfews.
        (5)  Political restrictions.
        (6)  Religious restrictions.

**Figure H-1. Area study format (continued)**

4. **Economic characteristics.** Identify those economic factors that influence mission execution:
   a. Technological standards.
   b. Natural resources and degree of self-sufficiency.
   c. Financial structure and dependence on foreign aid.
   d. Monetary system:
      (1) Value of money and rate of inflation.
      (2) Wage scales.
      (3) Currency controls.
   e. Black market activities. Note the extent and effect of those activities.
   f. Agriculture and domestic food supply.
   g. Industry and level of production.
   h. Manufacture and demand for consumer goods.
   i. Foreign and domestic trade and facilities.
   j. Fuels and power.
   k. Telecommunications adequacy by U.S. standards.
   l. Transportation adequacy by U.S. standards:
      (1) Railroads.
      (2) Highways.
      (3) Waterways.
      (4) Commercial air installations.
   m. Industry, utilities, agriculture, and transportation. Note the control and operation of each.

5. **Civil populace.** Pay particular attention to those inhabitants in the area of operations who have peculiarities and who vary considerably from the normal national way of life:
   a. Total and density.
   b. Basic racial stock and physical characteristics:
      (1) Types, features, dress, and habits.
      (2) Significant variations from the norm.
   c. Ethnic and/or religious groups. Analyze these groups to determine if they are of sufficient size, cohesion, and power to constitute a dissident minority of some consequence:
      (1) Location or concentration.
      (2) Basis for discontent and motivation for change.
      (3) Opposition to the majority or the political regime
      (4) Any external or foreign ties of significance.
   d. Attitudes. Determine the attitudes of the populace toward the existing regime or hostile power, the resistance movement, and the United States and its allies.
   e. Division. Division between urban, rural, or nomadic groups:
      (1) Large cities and population centers. Rural settlement patterns.
      (2) Area and movement patterns of nomads.
   f. Standard of living and cultural (educational) levels:
      (1) Extremes away from the national average.
      (2) Class structure. Identify degree of established social stratification and percentage of populace in each class.
   g. Health and medical standards.
      (1) General health and well-being.
      (2) Common diseases.
      (3) Standard of public health.

**Figure H-1. Area study format (continued)**

(4) Medical facilities and personnel.

(5) Potable water supply

(6) Sufficiency of medical supplies and equipment.

h. Traditions and customs (particularly taboos). Note wherever traditions and customs are so strong and established that they may influence an individual's actions or attitude even during a war situation.

6. **Military and paramilitary forces.** Identify friendly and hostile conventional military forces (Army, Navy, and Air Force) and internal security forces (including border guards) that can influence mission execution. Analyze non-national (indigenous) forces using the subdivisions shown below.

a. Attitude. Morale, discipline, and political reliability.

b. Size. Personnel strength.

c. Structure. Organization and basic deployment.

d. Appearance. Uniforms and unit designations.

e. Identification. Ordinary and special insignia.

f. Control. Overall control mechanism.

g. Communication. Chain of command and communication.

h. Leadership. Note officer and noncommissioned officer corps.

i. External control. Non-national surveillance and control over indigenous security forces.

j. Practices. Training and doctrine.

k. Tactics. Note seasonal and terrain variations.

l. Mobility. Equipment, transportation, and degree of mobility.

m. Logistics.

n. Effectiveness. Note any unusual capabilities or weaknesses.

o. Internal security. Vulnerabilities in the internal security system.

p. Past and current reprisal actions.

q. Information network. Use and effectiveness of informers.

r. Populace. Influence on and relations with the local populace.

s. Mindset. Psychological vulnerabilities.

t. Activity. Recent and current unit activities.

u. Counterintelligence activities and capabilities. Pay particular attention to reconnaissance units, special troops (airborne, mountain, ranger), rotary-wing or vertical-lift aviation units, counterintelligence units, and units having a mass chemical, biological, radiological, and nuclear delivery capability.

v. Guard posts and wartime security coverage. Note the location of all known guard posts or expected wartime security coverage along the main lines of communications (railroads, highways, and telecommunications lines) and along electrical power and petroleum, oil, and lubricant lines.

w. Forced labor and detention camps. Note exact location and description of the physical arrangement (particularly the security arrangements).

x. Populace and resources control measures. Note locations, types, and effectiveness of internal security controls. Include checkpoints, identification cards, passports, and travel permits.

7. **Resistance organization.** Identify the organizational elements and key personalities of the resistance organization. Note each group's attitude toward the United States, the hostile power, various elements of the civil populace, and friendly political groups.

a. Guerrillas:

(1) Disposition, strength, and composition.

(2) Organization, armament, and equipment

(3) Status of training, morale, and combat effectiveness.

(4) Operations to date.

(5) Cooperation and coordination between various existing groups.

**Figure H-1. Area study format (continued)**

(6) Motivation of the various groups and their receptivity.

(7) Quality of senior and subordinate leadership.

(8) General health.

b. Auxiliaries and the underground:

(1) Disposition, strength, and degree of organization.

(2) General effectiveness and type of support.

(3) Responsiveness to guerrilla or resistance leaders

c. Logistics capability:

(1) Availability of food stocks and water. Include any restrictions for reasons of health.

(2) Agricultural capability.

(3) Type and availability of transportation of all categories.

(4) Types and location of civilian services available for manufacture and repair of equipment and clothing.

(5) Medical facilities, to include personnel, medical supplies, and equipment

(6) Enemy supply sources accessible to the resistance.

8. **Targets.** The objective in target selection is to inflict maximum damage on the hostile power with minimum expenditures of men and materiel. Initially, a guerrilla force may have limited operational capabilities to interdict or destroy hostile targets. Study the target areas. Identify and analyze points of attack. List targets in order of priority by system and according to mission requirements. As appropriate, address both fixed and mobile (generic) targets.

9. **Effects of characteristics.** State conclusions reached through analysis of the facts developed in the previous paragraphs:

a. Effects on hostile courses of action.

b. Effects on friendly courses of action.

**Figure H-1. Area study format (continued)**

This page intentionally left blank.

## Appendix I

# Area Assessment Formats–Initial and Principal

This appendix provides an outline format for an area assessment. This format provides a systematic means for compiling and retaining essential information to support SF operations. Although the basic outline is general, it is flexible enough to permit detailed coverage of a given JSOA.

## IMMEDIATE-INITIAL ASSESSMENT

I-1. The initial assessment includes items essential to the operational detachment immediately following infiltration. Detachments must satisfy these requirements as soon as possible after arriving in the JSOA. This initial information should include:

- Location and orientation.
- Physical condition of the detachment.
- Overall security, to include the—
  - Immediate area.
  - Attitude of the local populace.
  - Local enemy situation.
  - Status of the local resistance element.

## SUBSEQUENT-PRINCIPAL ASSESSMENT

I-2. The principal assessment is a continuous operation that includes collection efforts that support the continued planning and conduct of operations. The principal assessment forms the basis for all of the detachment's subsequent activities in the JSOA. Figure I-1, pages I-1 through I-4, shows the areas that the principal assessment should encompass.

---

### The Enemy

1. Disposition
2. Composition, identification, and strength.
3. Organization, armament, and equipment.
4. Degree of training, morale, and combat effectiveness
5. Operations such as—
   a. Recent and current activities of the unit
   b. Counterguerrilla activities and capabilities, with particular attention given to reconnaissance units, special troops (airborne, mountain, ranger), rotary-wing or vertical-lift aviation units, counterintelligence units, and units having a mass chemical, biological, radiological, and nuclear delivery capability.
6. Unit areas of responsibility
7. Daily routine of the units.
8. Logistics support, to include the following:
   a. Installations and facilities.
   b. Supply routes.
   c. Methods of troop movement.
   d. Past and current reprisal actions.

---

Figure I-1. Sample principal assessment

## Security and Police Units

1. Dependability to the existing regime or the occupying power
2. Disposition.
3. Composition, identification, and strength.
4. Organization, armament, and equipment.
5. Degree of training, morale, and efficiency
6. Use and effectiveness of informers.
7. Influence on and relations with the local populace.
8. Security measures over public utilities and government installations

## Civil Government

1. Control and restrictions, such as—
   a. Documentation.
   b. Rationing
   c. Travel and movement restrictions
   d. Blackouts and curfews
2. Current value of money and wage scales.
3. The extent and effect of the black market.
4. Political restrictions
5. Religious restrictions
6. Control and operation of industry, utilities, agriculture, and transportation

## Civilian Populace

1. Attitudes toward the existing regime or occupying power
2. Attitudes toward the resistance movement
3. Reaction to U.S. support of the resistance.
4. Reaction to enemy activities in the country, specifically that portion in the unconventional warfare operational area
5. General health and well-being.

## Potential Targets

1. Railroads
2. Telecommunication
3. Petroleum, oils, and lubricants.
4. Electric power.
5. Military storage and supply
6. Military headquarters and installations
7. Radar and electronic devices.
8. Highways
9. Inland waterways and canals.
10. Seaports.
11. Natural and synthetic gas lines.
12. Industrial plants.
13. Key personalities.

## Weather

1. Precipitation, cloud cover, temperature, visibility, and seasonal changes
2. Wind speed and direction.
3. Light data (begin morning nautical twilight, end of evening nautical twilight, sunrise, sunset, moonrise, and moonset).

**Figure I-1. Sample principal assessment (continued)**

**Terrain**

1. Location of areas suitable for guerrilla bases, units, and other installations.
2. Potential LZs, DZs, and other reception sites
3. Routes suitable for—
   a. Guerrillas.
   b. Enemy forces
4. Barriers to movement.
5. Seasonal effect of the weather on terrain and visibility.

**Resistance Movement**

1. Guerrillas:
   a. Disposition, strength, and composition.
   b. Organization, armament, and equipment.
   c. Status of training, morale, and combat effectiveness.
   d. Operations to date.
   e. Cooperation and coordination between various existing groups
   f. General attitude toward the United States, the enemy, and various elements of the civilian populace.
   g. Motivation of the various groups and their receptivity to U.S. presence
   h. Caliber of senior and subordinate leadership.
   i. Health of guerrillas
2. Auxiliaries and the underground:
   a. Disposition, strength, and degree of organization
   b. General effectiveness and type of support
   c. Motivation and reliability.
   d. Responsiveness to guerrilla or resistance leaders.
   e. General attitude toward the United States, the enemy, and various guerrilla groups

**Logistics Capability of the Area**

1. Availability of food stocks and water. Include any health-related water restrictions
2. Agricultural capability
3. Type and availability of all categories of transportation.
4. Types and location of civilian services available for the manufacture and repair of equipment and clothing.
5. Supplies locally available, including type and amount.
6. Medical facilities, including personnel, medical supplies, and equipment
7. Enemy supply sources accessible to the resistance.

**Preventive Medicine**

1. Weather:
   a. Area cold-weather impact on causes, treatment, and prevention of cold-weather injuries.
   b. Area hot-weather impact on causes, treatment, and prevention of hot-weather injuries
   c. Terrain impact on evacuation and medical resupply.
2. Indigenous personnel:
   a. Physical characteristics of local people, including endurance, ability to carry loads, and performance of other physical feats.
   b. Symbolism attached to various articles of clothing and jewelry, such as amulets.

**Figure I-1. Sample principal assessment (continued)**

## Preventive Medicine (continued)

3. Attitudes:
   a. Taboos and other psychological attributes present in the society.
   b. Rites and practices unconventional healers use during illness, including symbolic rites and Western medicine in use.
   c. Response of indigenous personnel to feelings, such as fear, happiness, anger, and sadness.
4. Housing:
   a. Analysis of the physical layout of the community.
   b. Infestation of ectoparasites and vermin.
5. Food:
   a. Food cultivation for consumption, including types of food.
   b. Influence of seasons in the area of operations on diet, including any migration.
   c. Types of foods provided by U.S. personnel, including preferences and rejections of foods.
   d. Types of crops raised.
6. Water:
   a. Urban water supply.
   b. Rural water supply, including numbers and types.
   c. Water treatment plants in use.
   d. Water treatment in rural areas. Attitudes of indigenous personnel toward standard U.S purification methods.
   e. Sewage disposal (when applicable):
   f. Types and locations of sewage treatment plants.
   g. System used in remote areas to dispose of human excrement, offal, and dead animals or humans.
   h. Attitudes of indigenous personnel to standard U.S hygiene methods, such as the use of latrines
7. Epidemiology:
   a. What specific diseases in the following two major categories are present among the guerrillas, their dependents, or their animals?
   b. Domestic animals:
      (1) Types of domestic animals present.
      (2) Typical forage.
      (3) Supplemental animal food supply, including food supplements
      (4) Animal housing (penned or free roaming).
      (5) Religious symbolism or taboos locals associate with animals (for example, an animal considered sacred).
      (6) Animal sacrifices for religious purposes.
      (7) Availability of local veterinarians for animal treatment and postmortem inspections of meats.
      (8) Training of local veterinarians.
   c. Local flora and fauna:
      (1) Species of birds, large and small mammals, reptiles, and arthropods present in the area
      (2) Describe unknown varieties for survival purposes. Keep a record.
      (3) Area plants known to be toxic through contact with the skin, inhalation of smoke from burning vegetation, or through ingestion.

**Figure I-1. Sample principal assessment (continued)**

# Appendix J

# Special Operations Mission Planning Folder Formats

The SOMPF consists of three parts: the mission tasking packet, the TIP, and the Plan of Execution. ARSOF units prepare SOMPFs when tasked by the GCC. The TSOC prepares the transmittal documents and assembles the mission tasking packets. The GCC mission guidance is the heart of the mission tasking packet. The GCC will task the Army component intelligence agencies to produce the TIP. The ARSOF unit assigned the mission will conduct a feasibility assessment, which will determine if the target is a valid ARSOF target. The feasibility assessment is reviewed by the TSOC and the theater targeting board. Following review and approval of the feasibility assessment, the theater targeting board tasks the executing unit (through the TSOC) to develop a plan of execution.

## MISSION TASKING PACKET

J-1.   The mission tasking packet consists of tasking and transmittal documents, target identification data, and GCC mission guidance. Figure J-1, page J-2, shows the format for the mission tasking packet. SF commanders conduct target feasibility analysis on nominated SOF targets. The feasibility assessment is used by the executing unit to analyze the viability of the target. The purpose of the feasibility assessment is to address the following questions:

- Is this a valid ARSOF mission?
- Is it within the command's capability, unilaterally or as a joint or combined operation to accomplish the mission within an acceptable level of risk?

J-2.   Figure J-2, pages J-2 and J-3, shows the format for an SF mission feasibility assessment. Detailed planning and the selection of a preferred COA is reserved for the development of a plan of execution should the feasibility assessment indicate that the mission is viable. All existing shortfalls that would preclude development of the feasibility assessment must be identified immediately. At a minimum, imagery, area study, maps, and order of battle are required to produce the feasibility assessment. The order of battle must be updated prior to deployment. The process outlined in this format may be compressed during adaptive targeting.

## TARGET INTELLIGENCE PACKAGE

J-3.   The TIP is tailored for each specific mission, and the UW TIP format will vary for foreign internal defense, direct action, special reconnaissance or other core SF missions. Figure J-3, pages J-4 through J-6, shows the UW TIP format for SF UW missions.

## PLAN OF EXECUTION FORMAT

J-4.   The plan of execution is a detailed plan of how the assigned SF will carry out the validated mission assigned to them. The plan of execution, in conjunction with mission rehearsals, is the end result of the mission planning process. Plans of execution are also developed by the unit responsible for infiltration or exfiltration of the SF unit to and from the target area. Figure J-4, pages J-6 through J-8, shows a plan of execution format for the executing unit. Figure J-5, pages J-8 through J-10, shows a plan of execution format for the infiltration and exfiltration platform unit. Figure J-6, page J-11, shows an air mission coordination meeting checklist.

---

**MISSION TASKING PACKET**

SECTION I – TASKING AND TRANSMITTAL DOCUMENTS

a. GCC tasking.

b. Subordinate tasking from TSOC.

c. Coordinating Instructions.

SECTION II – TARGET IDENTIFICATION DATA

a. Name.

b. Basic encyclopedia (BE) number.

c. Mission number (if applicable).

d. Mission tasks.

e. Functional classification code.

f. JSOA coordinates (latitude (LAT), longitude (LONG) and Universal Transverse Mercator (UTM).

g. Target area coordinates (LAT, LONG, and UTM).

h. General description and target significance

SECTION III – GCC MISSION GUIDANCE (GCC's mission statement and objectives).

a. Mission statement.

b. Specific objectives

c. Commander's guidance.

d. Mission command.

SECTION IV – RECORD OF CHANGES.

SECTION V – RECORD OF DISTRIBUTION

**Figure J-1. Mission tasking packet**

---

**SPECIAL FORCES MISSION FEASIBILITY ASSESSMENT**

SECTION I – MISSION

a. Target identification data.

b. Mission statement and commander's guidance.

SECTION II – COMMANDER'S ASSESSMENT

a. Feasibility as a target.

b. Probability of mission success.

c. Recommendation.

SECTION III – ASSUMPTIONS.

SECTION IV – FACTORS AFFECTING THE COURSE OF ACTION.

a. Characteristics of the JSOA.

    (1) Weather

    (2) Terrain.

    (3) Other pertinent factors.

b. Situation.

c. Enemy situation.

    (1) Composition

    (2) Disposition

**Figure J-2. Mission feasibility assessment**

---

(3) Strength.

    (a) Committed forces.

    (b) Location of reinforcements and estimated reaction times.

    (c) Nuclear, biological, and chemical capabilities.

(4) Significant enemy activity, intelligence, and counterintelligence capabilities.

(5) Peculiarities and weaknesses.

(6) Vulnerability to deception.

(7) Enemy capabilities.

    (a) Defensive.

    (b) Offensive.

    (c) Intelligence and counterintelligence.

(8) Reaction and Reinforcement.

(9) Security on target.

d. Mission command.

SECTION V – COURSE OF ACTION.

a. Identify COA.

b. Analyze COA.

c. Compare COA.

    (1) Advantages.

    (2) Disadvantages.

    (3) Risks.

d. Recommended COA.

SECTION VI – INTELLIGENCE REQUIREMENTS.

SECTION VII – SPECIAL REQUIREMENTS.

a. Personnel.

b. Logistics.

c. Other.

SECTION VIII – INDIGENOUS SUPPORT

a. Resistance partners.

b. Unilateral assets.

**Figure J-2. Mission feasibility assessment (continued)**

---

**TARGET INTELLIGENCE PACKET FORMAT (UW)**

1. The element analyzes the directed target or objective area and commander's planning guidance.

    a. Extracts intelligence preparation of the battlefield products related to the target/objective area.

    b. Complies intelligence reports and analytical products/tools developed relating to the target/objective area.

    c. Conducts mining of intelligence databases for current intelligence related to the target/objective area.

    d. Determines existing information collection tasks directed against the target/objective area.

2. The element develops the formatted target intelligence package (TIP) supporting an unconventional warfare mission.

    a. Defines the objective area.

        (1) Provides identification data concerning the objective area.

        (2) Describes the objective area.

        (3) Defines the significance of the objective area.

    b. Describes the natural environment of the objective area.

        (1) Provides geographical terrain data of the objective area.

            (a) Describes the military aspects of the terrain.

                <u>1</u> Observation and fields of fire.

                <u>2</u> Avenues of approach and mobility corridors.

                <u>3</u> Key terrain in the objective area.

                <u>4</u> Obstacles that are natural or man-made in the objective area.

                <u>5</u> Cover and concealment.

            (b) Describes hazards to movement.

        (2) Provides meteorological data.

            (a) Compiles climatological tables.

            (b) Visibility and illumination data.

            (c) Precipitation and cloud cover.

            (d) Temperature and humidity data.

            (e) Wind data.

            (f) 12-, 24-, 36-. and 72-hour forecasts.

        (3) Provides hydrographic data of the objective area.

            (a) Describes coastal areas.

            (b) Describes natural and man-made waterways.

            (c) Identifies lakes and ponds.

            (d) Identifies marshes and swamps.

        (4) Describes sources of water in the objective area.

        (5) Describes flora and fauna of tactical importance in the objective area.

    c. Describes threat forces in the objective area.

        (1) Threat order of battle in the objective country.

        (2) Irregular and paramilitary forces.

        (3) Opposition groups and resistance forces.

**Figure J-3. Target intelligence packet format**

d. Describes civil demographics, cultural, political, and social features in the objective area.

    (1) Area population characteristics.

    (2) Social conditions in the objective area.

    (3) Religious groups or factions.

    (4) Political characteristics of the area.

    (5) Available work force and skills.

    (6) Customs and social mores in the area.

    (7) Medical treatment capabilities in the area.

    (8) Health and sanitation conditions.

    (9) Economic conditions.

    (10) Local currency, holidays, and style of dress.

e. Describes lines of communication (LOC) in the objective area.

    (1) Airfields and air strips.

    (2) Railway systems and hubs.

    (3) Roadways.

    (4) Navigable waterways.

    (5) Ports.

    (6) Petroleum, oils, and lubricants (POL) storage and distribution facilities.

    (7) Media systems.

    (8) Telecommunications capabilities.

    (9) Civilian transportation modes.

    (10) Power grids.

    (11) U.S. provided equipment and technology.

    (12) War material or munitions manufacturing capabilities.

f. Describes routes or sites of infiltration and exfiltration in the objective area.

    (1) Identifies potential points of infiltration and extraction.

        (a) Landing zones (LZs).

        (b) Drop zones (DZs).

        (c) Helicopter LZs.

        (d) Beach landing sites.

    (2) Identifies chokepoints and engagement areas between point(s) of infiltration and the objective.

    (3) Identifies chokepoints and engagement areas between the objective and point(s) of extraction.

g. Describes survival, evasion, resistance, escape; and recovery data.

    (1) Provides escape and resistance data.

    (2) Identifies safe areas

    (3) Provides survival information.

h. Provides unique intelligence on mission specific requirements not covered elsewhere.

i. Describes intelligence shortfalls and gaps.

j. Develops appendixes to the TIP.

    (1) Bibliography.

    (2) Glossary of terms and acronyms.

    (3) Imagery products of the objective area.

    (4) Geospatial information and services products.

    (5) Sensitive compartmented information, if applicable.

**Figure J-3. Target intelligence packet format (continued)**

3. The element initiates actions to address information and intelligence gaps.

    a. Researches available intelligence databases and networks for relevant target/objective area information.

    b. Determines if collection can be accomplished with internal capabilities.

    c. Submits requests for information (RFIs) collection capabilities from external sources.

    d. Submits RFIs to higher headquarters.

4. The element presents the target information according to the unit standard operating procedures to the unit responsible for target engagement.

    a. Submits the formatted TIP.

    b. Briefs intelligence gaps related to the target.

    c. Presents ongoing information collection efforts and RFI submitted to address intelligence gaps.

**Figure J-3. Target intelligence packet format (continued)**

**SPECIAL FORCES MISSION PLAN OF EXECUTION**

SECURITY CLASSIFICATION

Headquarters

Issuing Location

Day, Month, Year, Hour, Zone.

COMMANDER'S ESTIMATE OF THE SITUATION

References:

    a. Maps and charts.

    b. Other pertinent documents.

SECTION I – MISSION

SECTION II – THE SITUATION AND COAs

1. Considerations affecting the possible COAs.

    a. Characteristics of the JSOA.

        (1) Military geography.

            (a) Topography.

            (b) Hydrography.

            (c) Climate, weather, illumination data, and so on.

        (2) Transportation.

        (3) Telecommunications.

        (4) Politics.

        (5) Economics.

        (6) Sociology.

        (7) Science and Technology.

    b. Relative combat power.

        (1) Enemy.

            (a) Strength.

            (b) Composition.

            (c) Location and disposition.

            (d) Reinforcements.

            (e) Logistics.

            (f) Time and space factors.

            (g) Combat efficiency.

        (2) Friendly.

            (a) Strength.

            (b) Composition.

**Figure J-4. Special Forces mission plan of execution**

   (c) Location and disposition.

   (d) Reinforcements.

   (e) Logistics.

   (f) Time and space factors.

   (g) Combat efficiency.

   (h) Friendly force assistance available.

 c. Assumptions.

2. Analysis of enemy capabilities.

3. Comparison of friendly COAs.

 a. Statement of COAs.

 b. Comparison of COAs.

4. Decision (recommended COA) mission profile.

 a. Method and location of infiltration.

 b. Movement to target area.

 c. Actions on objective.

 d. Movement to and method of exfiltration.

SECTION III – SUPPORTING PLANS.

1. Overall schedule.

 a. Characteristics of the JSOA.

  (1) Preparation.

  (2) Rehearsal.

  (3) Rendezvous.

  (4) Transit.

  (5) Execution

  (6) Recovery.

 b. Logistics.

 c. Communications-Electronics.

 d. Deception

 e. Indigenous force support.

  (1) Resistance partners.

  (2) Unilateral assets.

 f. Time and distance charts.

 g. Deployment

 h. Weapons employment.

 i. Target recoverability.

**Figure J-4. Special Forces mission plan of execution (continued)**

j. Resupply.

k. Exfiltration.

l. Survival, evasion, resistance, escape data.

m. Command relationships.

SECTION IV – LIMITING FACTORS.

1. Intelligence.

2. Weather.

3. Equipment.

4. Tactics.

5. Logistics.

6. Personnel.

7. Training.

8. Supporting forces.

9. Command relationships.

10. Law of war, rules of engagement, U.S. law.

11. Other factors.

(signature)

ANNEXES: (As required, by letter and title)

DISTRIBUTION: (According to policies and procedures of the issuing headquarters

**Figure J-4. Special Forces mission plan of execution (continued)**

**SPECIAL FORCES MISSION PLAN OF EXECUTION INFILTRATION AND EXFILTRATION**

SECTION I – MISSION

1. Target identification data.

2. Mission statement.

SECTION II – MISSION SUMMARY

1. Mission tasking.

2. Objective area.

3. General concept.

4. Summary of limiting factors.

5. Probability of mission success.

**Figure J-5. Special Forces mission plan of execution—infiltration and exfiltration**

SECTION III – ASSUMPTIONS

SECTION IV – THREAT ASSESSMENT

SECTION V – NAVIGATION AND OVERALL MISSION PORTRAYAL

1. Launch bases.

2. Intermediate staging bases.

*Note:* This will represent the entire infiltration and exfiltration route from launch to recovery on a suitable scale chart annotating all information deemed necessary by the planning cell for portrayal of the mission.

3. LZs, DZs, and beach landing sites.

4. Recovery bases.

5. Abort or emergency divert bases.

6. Air refueling tracks forward arming and refueling points (aircraft and boats).

7. Routes.

    a. Ingress.

    b. Egress.

    c. Orbiting and holding.

    d. Safe passage procedures.

    e. Strip charts, navigation logs, global positioning satellite receivers, and other aids (as required).

SECTION VI – SUPPORTING PLANS

1. Overall schedule of events.

2. Prelaunch requirements.

    a. Updates to order of battle.

    b. Essential elements of information

    c. Problem areas and key factors

3. Infiltration or exfiltration platform factors and logistics considerations.

4. Mission command.

    a. Security preparations.

    b. Departure procedures (overt or deception procedures).

    c. Communications equipment requirements.

       (1) Infiltration or exfiltration platforms.

       (2) Unit to platform.

    d. Specialized operational procedures and techniques.

    e. Radio silence areas.

    f. GO or NO-GO areas.

    g. Publish signal operating instructions (SOI).

    h. Deception.

5. Emergency procedures.

    a. Engine out capability.

    b. Weather.

    c. Faulty intelligence.

    d. Infiltration or exfiltration platform abort procedures.

       (1) Late departure procedures.

       (2) Maintenance problems.

       (3) Battle damage.

**Figure J-5. Special Forces mission plan of execution—infiltration and exfiltration (continued)**

(4) Platform destruction.

(5) Bump plan.

e. Drop or other fuel-related malfunctions.

f. Lost communications procedures.

g. Mission abort procedures.

6. Evasion plan of action.

a. Crew responsibilities.

b. Immediate actions upon sinking, ditching, or bailout.

c. Safe area intelligence descriptions (SAID).

d. Safe.

e. Evasion team communications.

f. Search and rescue contact procedures.

SECTION VII – LIMITING FACTORS

1. Intelligence.

2 Weather.

3. Equipment.

4. Munitions.

5. Tactics.

6. Logistics.

7. Personnel.

8. Training.

9. Supporting forces.

10. Rules of engagement and legal issues.

SECTION VIII – SPECIAL OPERATIONS AVIATION, SURFACE SHIP AND SUBMARINE REQUIREMENTS FROM ARSOF AND NAVSOF TO CONDUCT INITIAL ASSESSMENT.

1. Target coordinates.

2. Maximum and minimum distances the LZs and DZs can be from the target.

3. Time frame in OPLAN or CONPLAN scenario (Pre-D-Day or D+XX_.

4. Desired launch and recovery bases.

5 Type of delivery and recovery required (airborne, airmobile, fast rope, SEAL delivery vehicle. zodiac).

6. Number of personnel to be transferred and approximate combat weight per person.

7. Approximate size and weight of additional equipment.

8. Assumptions made during customer's feasibility assessment and plan of execution.

9. Desired time over target.

10. Resupply or exfiltration requirements.

*Note:* Some information may not be readily available. Information on hand will normally suffice to conduct the plan of execution. However, all efforts should be made to obtain the above information and incorporate it into the plan of execution.

**Figure J-5. Special Forces mission plan of execution—infiltration and exfiltration (continued)**

## AMCM (COA-D) CHECKLIST   Page 1 of 2

Roll Call:

Supported Unit: | Supporting Unit:

ENEMY SITUATION (MLCOA, MDCOA):

FRIENDLY SITUATION

Weather/Illum for INFIL | Weather Decision:
EXFIL | Go Or Abort Time/ Method

AATF MSN: | Key Tasks:
End-state:

Concept of Operation (GТР) | COA Sketch/Mission Graphics

H-Hour:
Total Force to Move:
- Minimum force to move (1st lift/ 2nd lift)
- Type of equipment to move
- # of Aircraft and Type
- # of seats required per aircraft
- Doors Open/Closed
- Internal/external max weight expected
- Identify NET, NLT, and delay time for MSN
- Any special equipment required in aircraft
- Priority equipment/personnel (bump plan)

PZ Operations: | (Attach Diagram/Image)
- PZ NAME
- Location/Frequency: (PACE )/Call sign:
- Description (Size, Surface)
- Markings Far/Near (Day/Night)
- Approach/Departure Direction
- Landing Direction
- Security (weapons control status)
- PAX and equipment location.
- Hot or Cold Load (right/left/both doors)
- Static Load Training (where and when)
- Hazards (Wires, Brownout, etc)

Route overview/Estimated Time En route:
- Areas to Avoid
- Time/Check point calls requested
- DOORS (Open/Closed)

LZ Operations: Primary HLZs | - Attach Imagery and TADPOLE
- Location | Frequency/Call sign | Diagrams
- Time/Event Driven | - Hot/Cold (conditions check/
- Land formation and heading (go around) | method/time)
- Suitability/Hazards | - Immediate/Emergency EXFIL plan/
- Markings Far/Near (Day/Night) | signal
- Weapon control status
- Troop Offload (Left/Right/Both)

ALTERNATE LZs:
- Location | - Frequency/Call sign
- Land formation and heading (go around)
- Description
- Suitability/Hazards
- Markings Far/Near (Day/Night)
- Weapon control status
- Troop Offload (Left/Right/Both)
- Trigger for use:

- BUMP Plan:

## AMCM (COA-D) CHECKLIST   Page 2 of 2

Laager Plan (Loc. REDCON, Comms)

Deception Plan: False Insertions

ATK/ RECON/ ORGANIC UAS: | - COA Sketch
- # Type aircraft | - RP/ABFs/BPs/NAIs/TAIs/TRPs/
- Task and purpose: | EAs/GRG
- Munitions requested | - Airspace Deconfliction Measures
- Call sign/ voice/ Data frequencies | - Ground target marking methods

ISR/UAS Requested/Approved | - Time on Station/ MSN window
Internal (CAB/BCT/DIV UAS Available) | - NAIs/ROZs/Corridors
- Task and Purpose | - Airspace Deconfliction Measures
- Call Sign/Voice/Data Freqs | - Ground target marking methods
- MIRC Address/ BFT URN
- Retrans Required (YES / NO)

Airspace Coordination/Control (CAS/
RW/UAS/ISR)

FIRE SUPPORT
- SEAD/ Pre-AASLT Fires
- Close Air Support Requested/
Approved

MEDEVAC/CASEVAC Procedures:

Service and support: FARP Plan (Lift/
Attack)

CONTINGENCIES | CONTINGENCIES
- Max mission delay time | - No EW/CAS/CAP/AWT
- Wx (Min. Max. Delay) | - En route Fallen Angel (Downed
- Key leader locations/BUMP info | Aircraft)
- Abort Criteria (INFIL) | - PZ Fallen Angel (Downed Air-
- Detection Compromise/Contact (En- | craft)- Ground Exfil Plan/ NLT
route, Objective) | - Emergency Resupply
- No COMM | - No Comms Extraction: (visual
| signal, NLT time)
| - Emergency Extraction (signal,
| brevity, location)

Command and Signal | ABN
AATFC | CAN
GFC | CMD
AMC | SER INT
FLT LEAD | PZ LZ
| SPARE
| COMSEC (current/change over
| time)

Proposed Timeline:
(ACB, rehearsals, static load, commex, update
briefs, weather call, H-Hour, EXFIL debrief)

**Figure J-6. Air mission coordination meeting checklist**

This page intentionally left blank.

# Appendix K

# Infiltration and Exfiltration Considerations

An SF pilot team may infiltrate into the JSOA before the SF operational detachment. The preferred method is for a resistance reception committee to receive the pilot team. Blind infiltrations may be considered. based on METT-TC. The number and type of pilot team personnel will be very dependent on the METT-TC factors. The pilot team may establish contact with the resistance area command element, assess resistance potential. and perform other missions as specified by higher HQ. Based on these factors, an SF operational detachment may be infiltrated later into the JSOA to conduct UW or other special operations. The pilot team may exfiltrate on order, according to the OPORD. or at the discretion of the team leader. Pilot team members may exfiltrate any time before the SF operational detachment's scheduled infiltration, or they may remain in the JSOA as the reception committee for a planned SF operational detachment infiltration. They could also remain and act as an advanced operations base for the infiltration of two or more SF operational detachments.

## ARRIVAL SITE

K-1. At the infiltration site, SF operational detachment members must put the sterilization plan into effect and ensure the site is clean and secure. The resistance may execute deception plans if required. After infiltration, all personnel must know the assembly and contact plans. The assembly area and the contact point should be close to the DZ or LZ (figure K-1, page K-2). The SF operational detachment's assembly time after infiltration is dependent on the threat reaction time. A minimum number of personnel make contact. The SF operational detachment must employ correct verbal, light, and visual signals. SF operational detachment members must enforce noise, light, and camouflage discipline and keep time spent in the area to a minimum. They must use stealth with proper movement techniques and know their RPs. If enemy contact is made, the SF operational detachment members use immediate action drills, based on METT-TC. A workable alternative CONPLAN, especially on the DZ or LZ, is imperative.

K-2. Movement from the contact site to a "safe area" should be toward the guerrilla base camp. Transportation, if available, would save valuable time; however, movement on foot is always an option. During initial assessment, security principles that must be adhered to are:

- Noise and light discipline during transportation or movement.
- Route and halt security. (Camouflage and immediate action drills are used while maintaining stealth.)
- Primary and alternate RPs.

K-3. During movement, SF operational detachment members must anticipate their reaction to orders from the detachment commander. The SF operational detachment commander conducts observation and security checks during all movement. He also keeps the SF operational detachment oriented to its location and aware of any immediate actions to take in case of enemy confrontation. The SF operational detachment commander must:

- Know the location of key weapons and heavy equipment in the column.
- Begin plans to re-establish contact with the resistance force if the enemy separates the column and the SF operational detachment.
- Check the physical condition of SF operational detachment members during halts.
- Establish initial communications with the JSOTF and submit the initial entry report (ANGUS) if possible.

**Note:**
Primary and alternate assembly areas are on same side of the drop zone. Primary and alternate contact points are on same side as the assembly area (same planning for the alternate drop zone).

**Legend:**

A    alternate
AA   assembly area (should be close to the drop zone)
CP   contact point (should be close to the assembly area)
P    primary

**Figure K-1. Assembly areas and contact points**

# METHODS OF INFILTRATION

K-4    For UW, mission success depends on secrecy. Speed is secondary. Enemy capabilities, disposition, reaction time, and security measures, as well as the ASCOPE factors, affect the selected infiltration method. A heavily guarded border may preclude land infiltration; a strongly defended and patrolled coastline may eliminate infiltration by water, and the enemy's air detection and defense systems may reduce air insertion potential. METT-TC is always an influencing factor.

K-5    SF operational detachments must consider geographic formations when selecting the infiltration method. Terrain affects the selection of an aircraft's altitude, approach and exit routes, landing areas, or parachute operations for any mission aircraft. Mountains could force aircraft to fly higher than desired altitudes, resulting in early aircraft acquisition by enemy radar and increasing its vulnerability.

K-6. Seasonal weather conditions also affect infiltrations. Factors to consider include temperature, precipitation, visibility, clouds, and wind. High-altitude or surface winds and their effects on surf conditions or periods of reduced visibility may prohibit the use of parachutes, inflatable boats, or surface and subsurface swimming as entry or recovery techniques. These same conditions generally favor a land infiltration. The adverse weather aerial delivery system reduces the impact of adverse weather as a limiting factor on air infiltrations. Hydrography is the science of describing the sea and marginal land areas and the effect on water operations. The hydrographic conditions on the far-shore and the near-shore sea approach influence water infiltration. These conditions include the offshore water depth, beach gradients, tide, surf, currents, sea bottom, and the location of reefs, sandbars, seaweed, or natural and man-made obstacles.

K-7. Other considerations for UW infiltrations include:
- Light data (periods of twilight, beginning morning nautical twilight, early evening nautical twilight, nautical sunrise, sunset, moon phase, moonrise, and moonset).
- The distance to the contact site, possible base camps, and enemy locations.
- Plans for fire support, deception, evasion and recovery, and countermeasures against fratricide. Commanders should use METT-TC as a planning guide.

K-8. The SF operational detachment uses one of three methods to infiltrate into the JSOA—air, land, or water. A combination of the three may also be used. A matrix can be used to determine the type and infiltration method best suited for the mission.

## AIR INFILTRATION

K-9. Commanders often consider using conventional rotary-wing/fixed-wing assets or Army Special Operations Aviation Command's resources for air infiltration. Parachute operations are one of the most rapid infiltration methods used by the SF operational detachments. It provides less exposure and risk to the aircraft. Usually, standard troop carrier fixed-wing aircraft are well equipped and satisfy airdrop requirements. Some circumstances may require nonstandard aircraft. These situations may require an aircraft capable of parachute delivery of equipment and personnel from high altitudes using military free-fall parachute techniques. Other techniques include static-line operations on small, unsurveyed DZs or hazardous, tree-covered terrain. Assault aircraft and amphibious or utility aircraft may be available. During infiltration, certain conditions may require that these aircraft conduct air-landing operations on relatively short, unprepared airstrips. Under other circumstances, longer-range tactical aircraft may be used.

---

*Note:* FM 3-04, *Army Aviation,* ATP 3-18.10, *Special Forces Air Operations,* ATP 3-18.11, *Special Forces Military Free-Fall Operations,* and the Combat Aviation Brigade *Army Aviation Handbook* provide detailed planning information on air infiltration.

---

## LAND INFILTRATION

K-10. During land infiltration, the SF operational detachment should move from the launch site clandestinely to an assembly area short of the JSOA border and establish local security. In this scenario, the SF operational detachment will move to the assembly area, conduct a reconnaissance of the crossover point, and conduct the infiltration, preferably during periods of low visibility. The assembly area should provide cover, concealment, and security. The reception committee, through prior coordination, may meet the SF operational detachment at the assembly area to conduct final coordination. In the event friendly forces occupy the border area, the SF operational detachment must establish liaison from higher HQ beforehand. The liaison element should remain near the crossover point until the JSOTF or SOTF has received the initial entry report.

## WATER INFILTRATION

K-11. Waterborne operations provide an expanded capability for all SF operational detachments. Through premission analysis, detailed planning, and a unit sustainment training program, the SF operational detachment can deploy for a successful waterborne infiltration. A thorough understanding of all factors affecting successful waterborne-related missions is essential due to the inherently high risk associated with even the most routine waterborne operations.

> *Note:* ATP 3-18.12, *Special Forces Waterborne Operations,* provides detailed planning information on waterborne operations.

# SPECIAL FORCES OPERATIONAL DETACHMENT EQUIPMENT AND SUPPLIES

K-12. The quantity and type of accompanying equipment and supplies carried by the SF operational detachment on initial infiltration are influenced by the:

- Infiltration and resupply airframes available. The aircraft used to infiltrate the SF operational detachment may be by Army Special Operations Aviation Command aircraft (MH-60M and the MH-47G), United States Air Force aircraft (MC-130, MH-60G, or CV-22), United States Army aircraft (UH-60M/V or the CH-47F), or other locally contracted assets.
- Political, psychological, and military threat situation in the JSOA.
- Size and training of the resistance force.
- Infiltration time and site location.
- Contingency operations in the JSOA.

# INITIAL ENTRY REPORT

K-13. Infiltration into the JSOA is not complete until the SF operational detachment transmits the initial entry report (ANGUS) by radio to the JSOTF or SOTF. The JSOTF or SOTF sends an acknowledgement back to the deployed SF operational detachment. If the SF operational detachment fails to receive the acknowledgment with the primary communications asset (tactical satellite radio), SF operational detachment personnel should use high frequency radio equipment as an alternate means of communicating. Upon receipt of the message, the JSOTF will send the SF operational detachment's acknowledgment at the next scheduled call-up, using the alternate high frequency radio equipment.

K-14. The SF operational detachment should send the initial entry report (ANGUS) within 24 hours after its infiltration time. If the JSOTF or SOTF does not receive the initial entry report within 72 hours after infiltration, another 72 hours will pass before the emergency resupply is dropped. This procedure gives the SF operational detachment the necessary flexibility to carry out CONPLANs precipitated by enemy action. If possible, it will attempt to contact the JSOTF. This flexibility gives the United States Air Force the necessary planning and reaction time. Another option may be to have a preplanned recovery operation on alert for a predesignated recovery site for any survivors. Launch and recovery depend on METT-TC.

# EMERGENCY RESUPPLY

K-15. If the deployed SF operational detachment fails to send its initial entry report (ANGUS) to the SF operational detachment or SOTF by radio within 72 hours after scheduled deployment, it may, according to the OPORD, start emergency resupply procedures. The SF operational detachment is now assumed to be in an evasion mode. The emergency resupply may be flown by any aircraft and airdropped or airlanded in the JSOA. Any waterborne vessel may transport an emergency resupply to a beach landing site. The resupply may also be pre-positioned in the evasion corridor by various U.S. agencies.

K-16. When the evasion corridor is long, more than one resupply mission may be needed. SF operational detachments must plan these resupply missions with realistic times and distances traveled. When wounded SF operational detachment members need to be transported, security is difficult but must be maintained. The resupply bundle is dropped primarily to re-equip the SF operational detachment for mission accomplishment. The resupply bundle may consist of just enough supplies for the SF operational detachment to establish communications with higher HQ and request further instructions. The resupply bundle might also contain a complete issue of all table of organization and equipment supplies and weapons for every SF operational detachment member.

# Appendix L
# Communications

*Organization of resistance movements and the operation of their component networks as well as guerrilla warfare, which in itself is a comprehensive area, including not only organization, tactics, and logistics, but specialized demolition; codes and radio communication; survival, the Fairbairn method of hand-to-hand combat, and instinctive firing.*

Colonel Aaron Bank
*NATO's Secret Armies: Operation GLADIO and Terrorism in Western Europe*

## INDIGENOUS RESISTANCE COMMUNICATIONS

L-1. A resistance will have communications equipment available, ranging from the most modern to the most primitive. Resistance forces can use high frequency, short-wave, and ham radio sets (to include citizen band-sets); cellular phones; the Internet; the mail; and couriers.

### CLANDESTINE COMMUNICATIONS NETWORKS

L-2. In order for a clandestine organization to function, it must be able to communicate. Clandestine operations may require a network of complex communications channels for transmitting instructions and material between echelons. Clandestine communications networks must provide for sustained transmissions to meet operational needs, ensuring a continuous flow of information and material. Clandestine communications network planning involves identifying requirements, such as the:
- Participants.
- AO.
- Type and volume of information or material to be passed.
- Desired speed and frequency.

### UNDERGROUND AND AUXILIARY COMMUNICATIONS

L-3 Underground and auxiliary elements rely heavily on concealing communications within their organizational structure and between similar or supporting organizations. They adapt appropriate means of communications that preserve security and compartmentation and are reliable. Face-to-face meetings, couriers, and cutouts are habitually combined into a communications system designed to serve the needs of these clandestine organizations. Techniques are changed frequently to avoid setting a pattern that can be easily detected by the enemy. Communications within or between these organizations facilitate·
- Countersurveillance.
- Cell and network sustainment when key leaders are detained or killed.
- Recruitment of new members to grow the organization or replace losses.
- Hiding of key individuals using safe areas or facilities.
- Security for operational locations for meetings or training.

### GUERRILLA COMMUNICATIONS

L-4. Guerrilla commanders may frequently visit subordinate units to give instructions, assess efficiency, and check the security of their commands (figure L-1, page L-2). Guerrilla units use all means of communications to serve their needs; however, they rely heavily on basic communications concealment, particularly when the widespread employment of electrical means is impractical. Under these conditions, guerrilla units rely heavily on face-to-face meetings, couriers, and relay points between echelons or units.

**Figure L-1. Guerrilla mission briefing**

## Guerrilla Communications—Outer and Inner Security

L-5   The outer perimeter usually consists of innocents—old men, women, and children—while the inner perimeter comprises members of the unit. When those in the outer perimeter spot an advancing government patrol or helicopter, they alert the inner perimeter by some predetermined signal. Those in the inner perimeter, in turn, alert the unit (figures L-2 and L-3, page L-3).

> "The Huks in the Philippines depended upon the villagers for intelligence information and improvised techniques to relay this information. If government troops approached a village and a man chopping wood observed them, he would increase the rate of his swing. A woman noticing his increased pace would place white and blue dresses side-by-side on the clothesline.     . At night, the Huks used light signals, such as opened windows on a certain side of the house."
>
> Luis Taruc, Born of the People, p. 121, COL Napoleon Valeriano:
> Speech at the Counterinsurgency Officer's Course,
> Special Warfare School, Fort Bragg, NC, 5 November 1963

## Face-to-Face Meetings

L-6.   Face-to-face meetings (figure L-4, page L-4) may occur between two or more members of an organization and serve many purposes. Meetings may be between a resistance leader and a courier, an informant, or a group assembled for a briefing prior to engaging in a clandestine mission, such as auxiliaries supporting an air resupply mission. Meeting locations should provide relative security to facilitate discussions or the exchange of material, allow for the placement of countersurveillance elements to provide early warning, and have suitable escape routes.

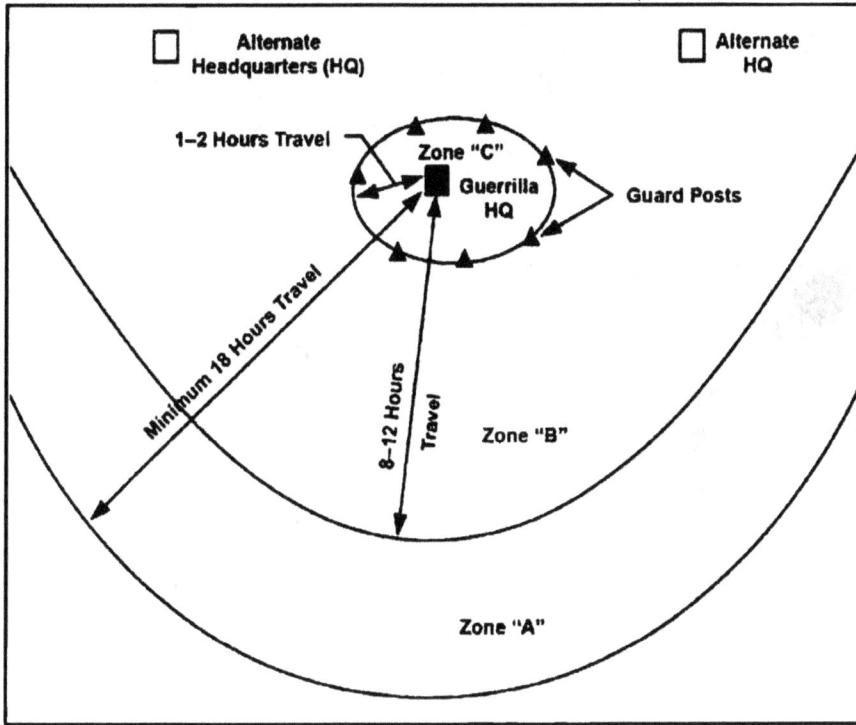

Figure L-2. Inner and outer security zones

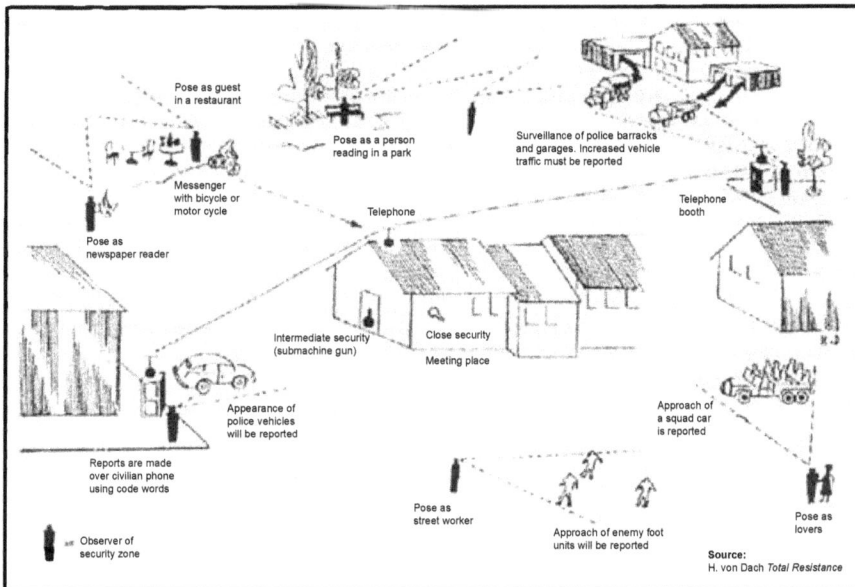

Figure L-3. Security zone participant examples

Figure L-4. Face-to-face meetings

### Belgian Underground

"The Belgian undergrounder was instructed     that it was important to have an alibi wherever he went, especially when meeting someone. The members prepared a subject of conversation in advance, such that if they were captured and questioned separately that they would agree upon the alibi. If they were to meet in public, they would agree in advance on a signal to indicate danger or that it was too dangerous to talk. For meetings in homes, prearranged signals were agreed upon to warn of danger. The undergrounder was instructed to examine the meeting place before any appointment. Appointments were not to be scheduled exactly on the hour and were to be kept punctually. If meetings were scheduled between contacts who met repeatedly, one individual should leave after the other so as not to be seen in public together too frequently. Meeting places were to be varied to be sure that the individual was not followed. If he was followed, he was to walk to some isolated spot so he could check to see if he was under surveillance. He was admonished to select clothes to fit the environment and his occupation.

Organizational contacts were to be limited to those within the cell or the leader or chief of the cell. All blackmail threats were to be reported immediately. He was also instructed to beware of telephone taps and postal censorship on any correspondence."

George K. Tanham, *The Belgian Underground Movement 1940–1944*, p. 226.
E. Bramstedt, *Dictatorship and Political Police*, pp. 213–224

## Courier Systems

L-7   Couriers are members of a clandestine network who carry information, material, or equipment to, or between, other members of an organization. Couriers facilitate a clandestine operation by bridging the association of resistance members that cannot meet due to social, economic, or security limitations. A successful courier is intelligent, motivated, dependable, flexible, and possess courage in the face of a hostile government or occupying power. Couriers may operate at a local, tactical level or traverse large expanses of terrain, to include crossing international borders at the operational level. Couriers must be absolutely loyal to the resistance organization and thoroughly familiar with the terrain and the indigenous people. Besides handling the flow of messages. courier routes may also serve as arteries over which personnel and supplies are moved.

> "The induction of a railway worker could significantly increase the underground's access to uncontrolled. uncensored transportation of communications and materials."
>
> Czelaw Stankiewicz, Polish Underground Veteran

L-8. Couriers are considered the most secure means of communications, however, the need for speed must be weighed against the need for security since couriers move relatively slow. Couriers may only be able to move on foot over long distances, in all weather conditions and. therefore, may require physical conditioning. They may need to be hardened to run long distances, swim across rivers in their clothes, or go into hiding for extended periods without food (figure L-5). Training for prospective couriers should include memory enhancement. as well as basic movement techniques common to infantry squads. Couriers must also be able to clandestinely communicate with and identify contacts.

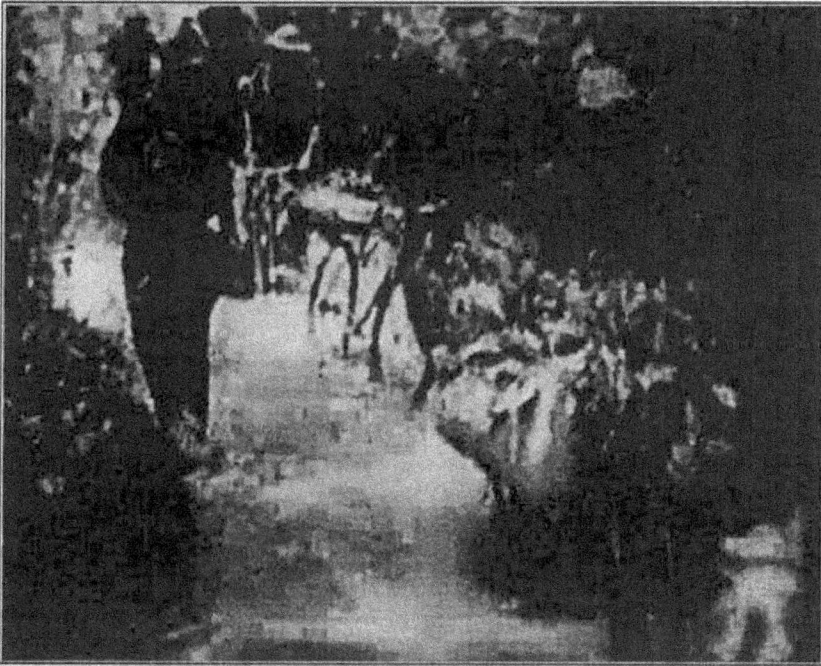

Figure L-5. Courier physical conditioning challenge

L-9.  Couriers can be employed in different configurations; each with its own advantages and risks. One courier used at all times is easier to control, but having a relatively large knowledge of the organization is a greater risk of compromising the entire effort (figure L-6). Using several couriers is a greater recruitment and management challenge. However, using several couriers provides built-in redundancy and, because each has a more limited knowledge of the entire organization, any compromised and interrogated courier presents a smaller risk to the entire effort (figure L-7). Figure L-8, page L-7, shows an unlikely courier.

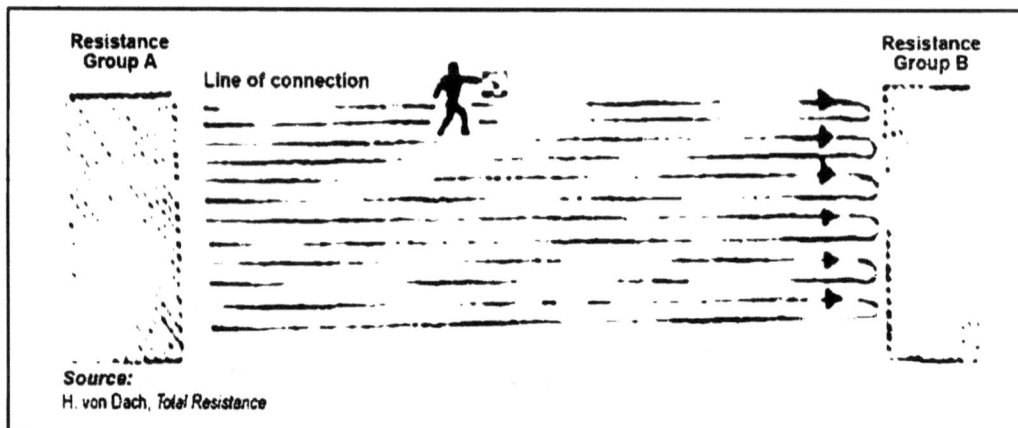

**Figure L-6. Single courier**

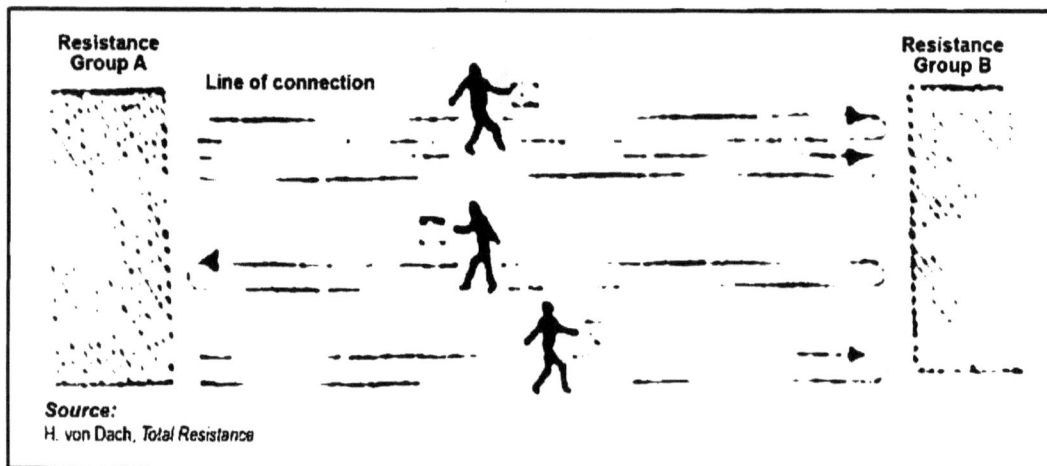

**Figure L-7. Multiple couriers**

**Figure L-8. Unlikely courier**

L-10. A simple courier system uses relay points (middle men), for compartmentation, usually located approximately halfway between resistance elements or guerrilla units. Couriers supporting guerrilla operations travel only between their parent unit and the relay point and, thus, do not know the location of other units or HQ. A more complex guerrilla courier system may involve multiple relay points between units and may also incorporate auxiliary support.

L-11. Relay points normally operate out of fixed locations, such as laundromats, barbershops, hotels, gas stations, general stores, farms, bakeries, meat markets, or taverns. Relay points can receive, hold, and transmit messages and reports; arms and ammunition; food, rations, and water; clothing and uniforms; petroleum products; medical supplies; repair parts; funds; and other supply or support materials. Some precautions to consider are:

- Material passed should balance with the nature of the facility used.
- The relay point must never be overloaded, overused, or visited unnecessarily.
- Material should remain at the relay point for the shortest amount of time.
- Safety and danger signals must be prearranged.

## Land-Line Telephone Systems

L-12. Communicators should limit the use of commercial telephone systems in the UW operational area, particularly during the initial phases of UW operations. The enemy can trace telephone calls quickly and easily, and even conversations consisting of a few code words are risky and should be avoided. Wires are very easy to find and can be followed directly to the points of termination, making field telephones also risky. Even if the points of termination are used sporadically, the enemy can surveil wire for indefinite periods, rendering them compromised and useless. Expedient ground return circuits and existing conductors, such as barbed wire fences or railroad rails can be used, provided both parties are hooked to the circuit at the proper time.

## Wireless Communications

L-13. Wireless communications are now widespread and regarded, even by poor indigenous populations, as cost-effective and convenient. However, these systems can be easily monitored by state and commercial threats (figure L-9 and figure L-10, page L-9).

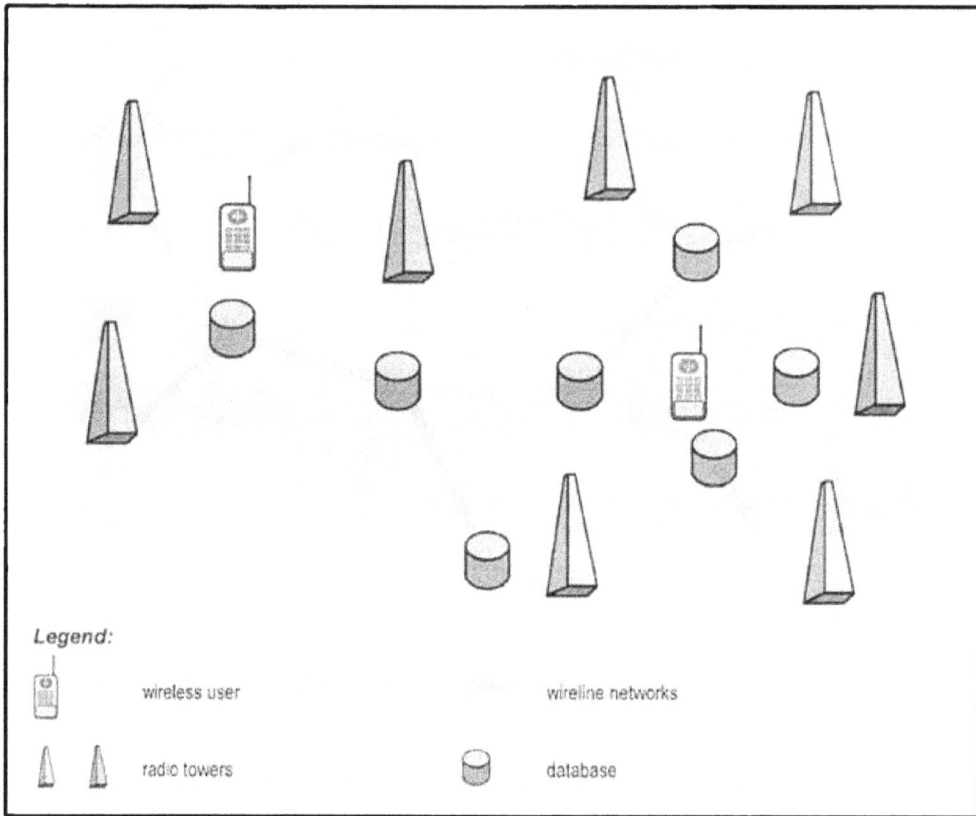

**Figure L-9. Wireless systems**

### Radio Security

L-14. During UW, the majority of wireless communications should be considered unsecured, presenting opportunities for enemy exploitation. Radio security measures are essential. Stations should be camouflaged and guarded, and messages should be sent in code.

### Secure Technical Measures

L-15. Communicators should limit the use of high frequency radios and directional super-high frequency transponders. They should maintain low noise emissions by operating at reduced power and turning off selected equipment when not in use. Communicators should also limit nonsecure computer email messages to nonresistance activities, prepare for cyberattack, and back up vital operational information on removable secure media. In addition, communicators should use encryption to protect voice, data, and video communications when available, as well as consider the use of forced-to-send and forced-to-receive signals or codes.

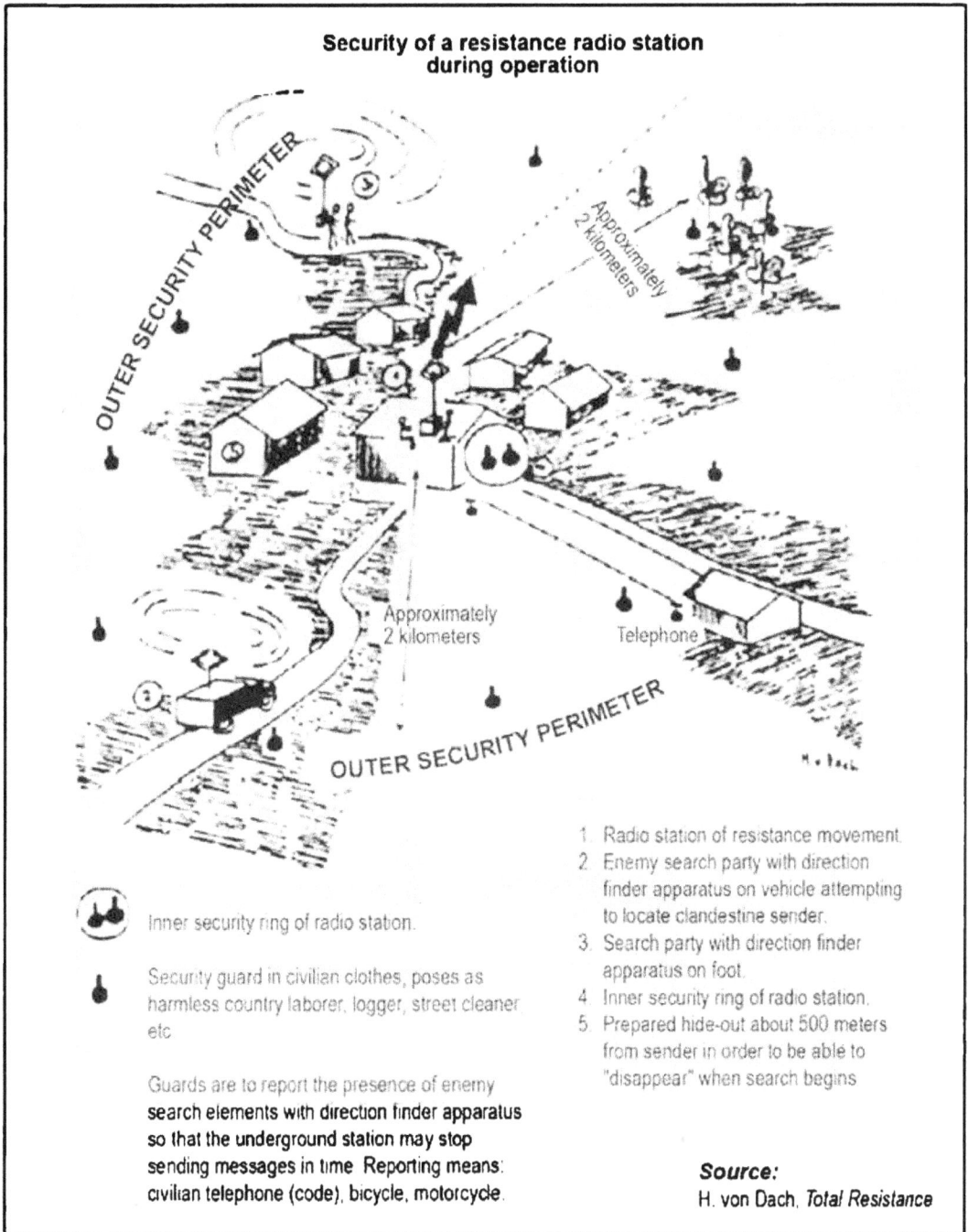

**Security of a resistance radio station during operation**

1. Radio station of resistance movement.
2. Enemy search party with direction finder apparatus on vehicle attempting to locate clandestine sender.
3. Search party with direction finder apparatus on foot.
4. Inner security ring of radio station.
5. Prepared hide-out about 500 meters from sender in order to be able to "disappear" when search begins

Inner security ring of radio station.

Security guard in civilian clothes, poses as harmless country laborer, logger, street cleaner etc

Guards are to report the presence of enemy search elements with direction finder apparatus so that the underground station may stop sending messages in time. Reporting means: civilian telephone (code), bicycle, motorcycle.

*Source:*
H. von Dach, *Total Resistance*

**Figure L-10. Radio security**

### Visual Signals Systems

L-16. Resistance members can use any device or object as a signal that can be displayed in a way to produce a specific meaning. Visual communication can relay information using any device, such as lights, mirrors, and flags that can be seen from a distance. Detailed planning and coordination are needed to ensure security.

Visual signals need an encryption system to ensure the meaning is discernible only to the intended party and to secure the information. Visual signals must have a unique way of being identified, a way that is unmistakable, yet not unusual for the area, and have a specific meaning known only to the receiver. The "Wig-Wag" system used one flag or one torch for operations. The position of the flags—left, right, front—represent the numerals 1, 2, 3, respectively, and combinations of these numerals are used to convey the message (figure L-11).

**Flag Positions of Myer's Original Two-Element Code**

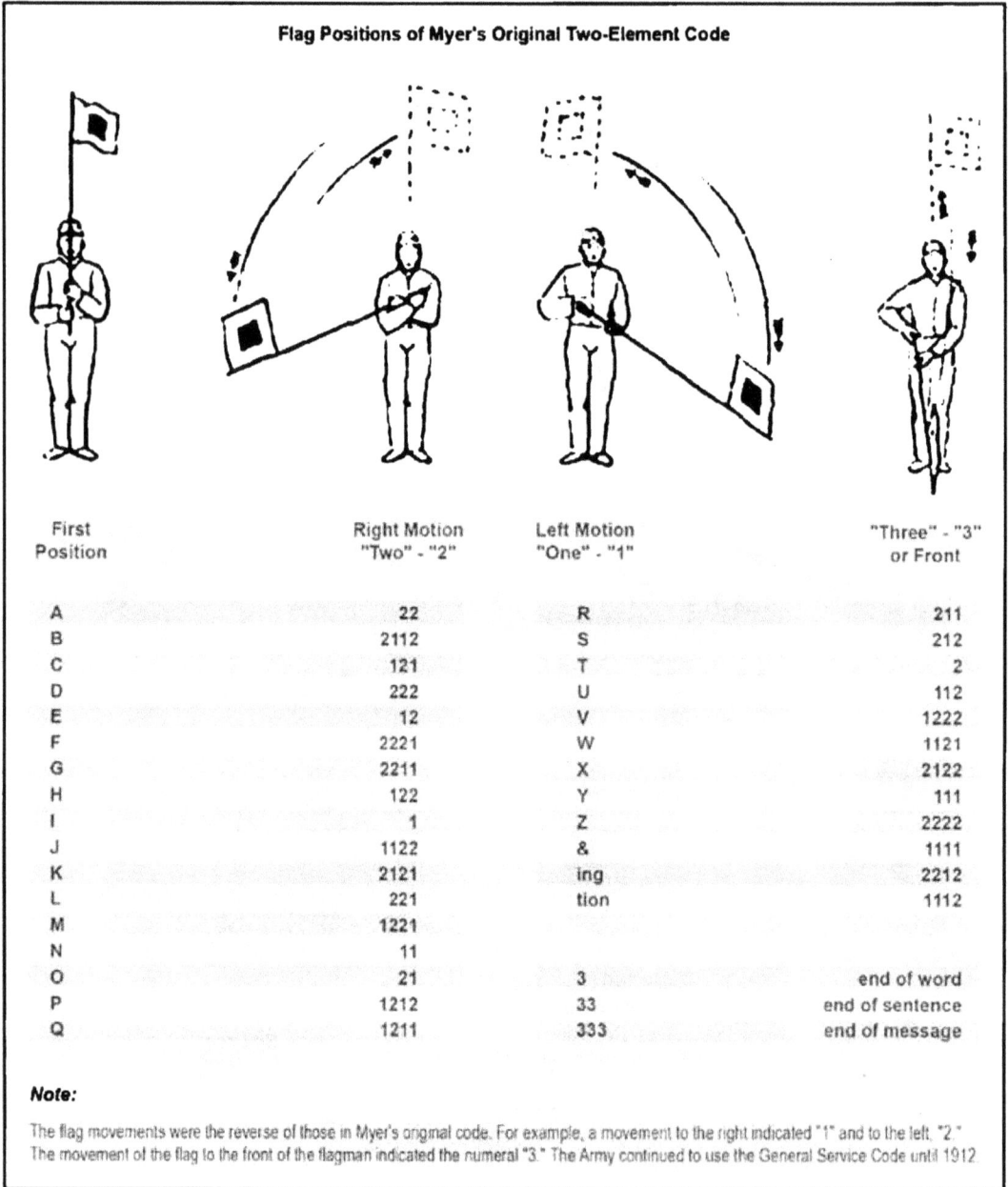

| First Position | Right Motion "Two" - "2" | Left Motion "One" - "1" | "Three" - "3" or Front |
|---|---|---|---|
| A | 22 | R | 211 |
| B | 2112 | S | 212 |
| C | 121 | T | 2 |
| D | 222 | U | 112 |
| E | 12 | V | 1222 |
| F | 2221 | W | 1121 |
| G | 2211 | X | 2122 |
| H | 122 | Y | 111 |
| I | 1 | Z | 2222 |
| J | 1122 | & | 1111 |
| K | 2121 | ing | 2212 |
| L | 221 | tion | 1112 |
| M | 1221 | | |
| N | 11 | | |
| O | 21 | 3 | end of word |
| P | 1212 | 33 | end of sentence |
| Q | 1211 | 333 | end of message |

**Note:**

The flag movements were the reverse of those in Myer's original code. For example, a movement to the right indicated "1" and to the left, "2." The movement of the flag to the front of the flagman indicated the numeral "3." The Army continued to use the General Service Code until 1912.

**Figure L-11. Wig-wag system**

# U.S. FORCES UNCONVENTIONAL WARFARE COMMUNICATIONS

L-17. In UW, SF detachments can be expected to deploy deep into areas that are hostile or denied. In these areas, communications requirements are more challenging and require a focus on operating in a clandestine, covert, or low-visibility manner. Under these conditions, planning for UW communications requirements is vital to mission success. UW communications requirements are usually long term, unique, or regulated, and they rely on redundancy.

## COMMUNICATIONS PLANNING

L-18. U.S. internal communications between a detachment in the UW operational area and the base station (advanced operations base, JSOTF, or SOTF) must be reliable and secure. To meet this demand, communicators should use a communications system that incorporates times and mode of contacts. Each segment of the communications plan should be thoroughly scrutinized for dependability and security, to include mechanisms used in every situation. A combination of systems (technical and nontechnical) in the plan may ensure that communications will continue if one system is compromised. Communicators' training and adherence to communications security procedures will always enhance the chances of communications success.

## COMMUNICATIONS PLAN: SCHEDULED, OPEN, AND GUARD

### Scheduled

L-19 Scheduled communications plans are routine communications between the SF detachment and the base station. Communications are sent using SAV SER SUP 7 message formats and brevity codes. Figure L-12, page L-12, shows a cover page of the SAV SER SUP 7. The first transmission from the base station to the infiltrated SF detachment is based on a set time. Message traffic will include instructions to change the frequency. The second transmission is from the SF detachment to the base station, and it is also based on a set time. The third transmission is from the base station acknowledging the receipt of the detachment's message.

### Open

L-20 The open net is established to allow either the base station or the SF detachment to send message traffic during an established time window if scheduled contact is missed or interrupted. This is usually a 30-minute time window. The open net is monitored 24 hours a day; therefore, a detachment may send messages as needed.

### Guard

L-21. The guard net is monitored 24 hours a day, 7 days a week for emergency communications from the SF detachment to the base station. This is likely to be used when the mission or force is compromised and the detachment is in danger. The guard net is usually unsecure and intended for short duration use.

## COMMUNICATIONS PLAN: PRIMARY, ALTERNATE, CONTINGENCY, AND EMERGENCY

### Primary

L-22. A UW primary communications system includes routine communications between the SF detachment and the base, but it will be mission-specific and will be driven by threat conditions. The primary usually consists of military issued radio equipment in high frequency or tactical satellite modes. A primary communications system should be capable of transmitting and receiving message traffic between the SF detachment and the base station in support of mission requirements. Communications should be employed using primary radios, frequencies, and times, and according to established SOI.

```
┌─────────────────────────────────────────────────────────────┐
│                                                               │
│                 FOR OFFICIAL USE ONLY                         │
│                                                               │
│                                                               │
│       SAV SER SUP 7                                           │
│                                                               │
│                                                               │
│                                                               │
│          Signal Audio Visual Service Supplement 7             │
│                                                               │
│                    Special Operations Forces                  │
│                  Signal Operating Instructions                │
│                  Supplemental Instructions (U)                │
│                                                               │
│                                                               │
│   DISTRIBUTION RESTRICTION: Distribution authorized to U.S. Government agencies and their contractors │
│   only to protect technical or operational information from automatic dissemination under the International │
│   Exchange Program or by other means. This determination was made on 16 January 2019. Other requests │
│   for this document must be referred to Commander, U.S. Army Special Operations Center of Excellence, │
│   USAJFKSWCS, ATTN: AOJK-SFD, 3004 Ardennes Street, Stop A, Fort Bragg, NC 28310-9610. │
│                                                               │
│   DESTRUCTION NOTICE: Destroy by any method that will prevent disclosure of contents or reconstruction │
│   of the document.                                            │
│                                                               │
│   FOREIGN DISCLOSURE RESTRICTION: This publication has been reviewed by the product developers │
│   in coordination with the U.S. Army Special Operations Center of Excellence, USAJFKSWCS, foreign │
│   disclosure authority. This product is releasable to students from foreign countries on a case-by-case │
│   basis only.                                                 │
│                                                               │
│                                                               │
│                 FOR OFFICIAL USE ONLY                         │
│                                                               │
└─────────────────────────────────────────────────────────────┘
```

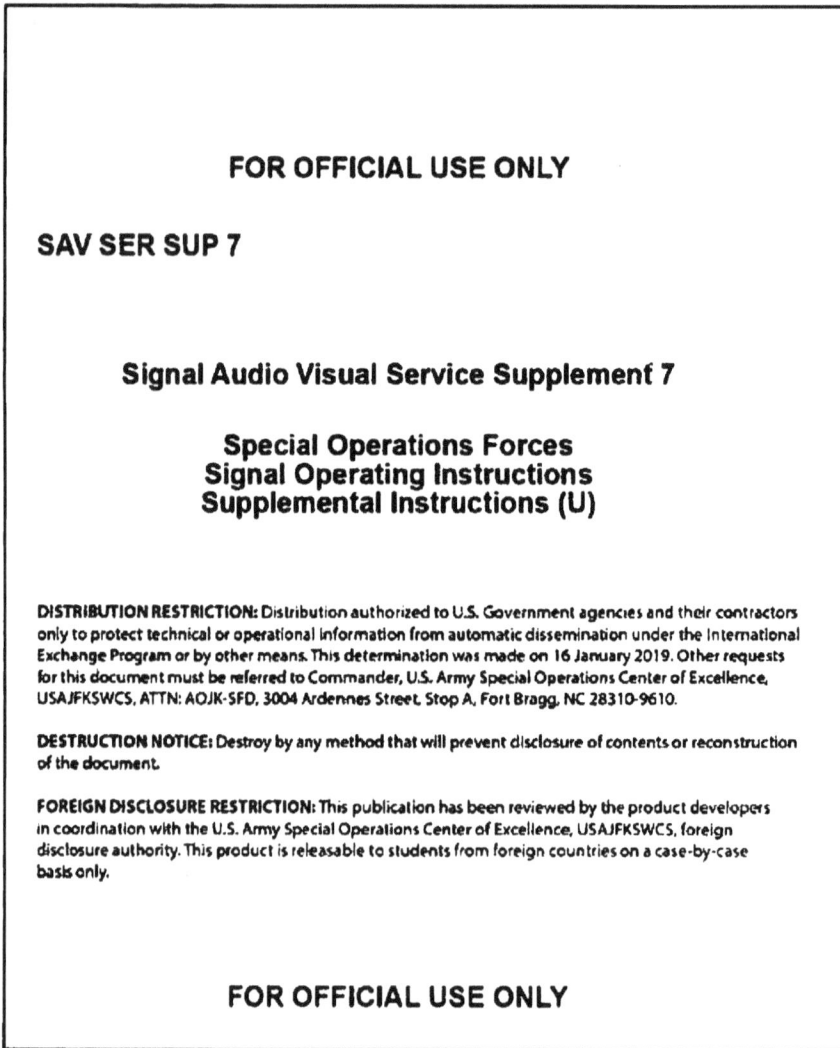

Figure L-12. SAV SER SUP 7 cover page

## Alternate

L-23. UW communicators plan and use an alternate communications system, which consists of alternate radios, frequencies, and times in open net contacts according to established SOI. The alternate communications system may or may not be of the same type, but it must have the capability to transmit and receive in the required mode of operation (high frequency or tactical satellite) to handle the same amount of traffic as the primary system. When initially used, it alerts the base station of a potential problem; however, alternate systems may be periodically used—even when the mission does not require it—to ensure they are still operational. Usually the same type of encryption system will be used; however, in the event of a primary encryption system compromise, UW communicators must plan, practice, and use an alternate encryption system.

## Contingency

L-24. The detachment plans and uses a contingency communications system in case of primary and alternate systems failure. While operating in this mode, the detachment is in a communications-critical stage; therefore, immediate steps must be taken to bring the primary and alternate systems on line as soon as possible. The contingency system should be able to provide the normal communications between the deployed detachment and base station, but only for short durations. Contingency communications may be executed through battlefield recovery of indigenous equipment. A guard net frequency, memorized by the detachment members while in isolation, may be used as a contingency when loss or compromise of the frequency encipher pad occurs.

## Emergency

L-25. An emergency communications system is a last resort medium for the detachment to inform the base station of its status, and it must be fail-safe, simple, and reliable. Emergency communications may not always be secure. The base station will maintain a continuous running "open net" to receive emergency and chance unscheduled contacts. Emergency communications may also be planned in a nontechnical medium, such as a one- or two-word message exfiltrated by resistance elements, or ground-to-air signals delivered from a predetermined location. If higher systems of communications are not available, the emergency communications system is used to inform the base station that:

- All detachment communications equipment is damaged or compromised.
- An emergency resupply is compromised.
- The detachment is in evasion mode.

## TECHNIQUES AND PROCEDURES

L-26. UW communications techniques and procedures cover the entire spectrum of communications available to the SF communicator. However, radio communications dominate between the detachment and the base station. Between elements within the UW operational area, UW communications will be predominantly nontechnical in nature, at least until the electronic warfare threat has been considerably lessened (figure L-13, page L-14). Security is the UW communicator's number one concern. Any communications within the UW operational area must be totally clandestine until a communications linkup is made with conventional forces.

L-27. Commanders must learn and accept a concept of communications, based on security and dependability rather than speed and ease of communication. An SF detachment normally operates under a maximum-security SOI, which does not provide for full-time, two-way communications between the SF detachments and the base communications station. The SOI provides for periodic, scheduled contacts from the base to the SF detachment and emergency (Guard Net) contacts from the detachment to the base at any time. The SF detachment must spend a minimum time on the air. The SOI is designed to confuse enemy intercept operators and make the intercept task as difficult as possible.

L-28. SF detachment to base station communications will be dictated by the mission. The deployed SF detachment will make contact with the base only as often as necessary to accomplish the mission. The detachment's ability to accomplish its mission could be drastically affected if the enemy had knowledge of the—

- Presence of the detachment in the UW operational area.
- Amount of time spent on message traffic being sent and received.
- Areas of concentration and message repetition.
- Extent of the sophistication and size of the communications net.

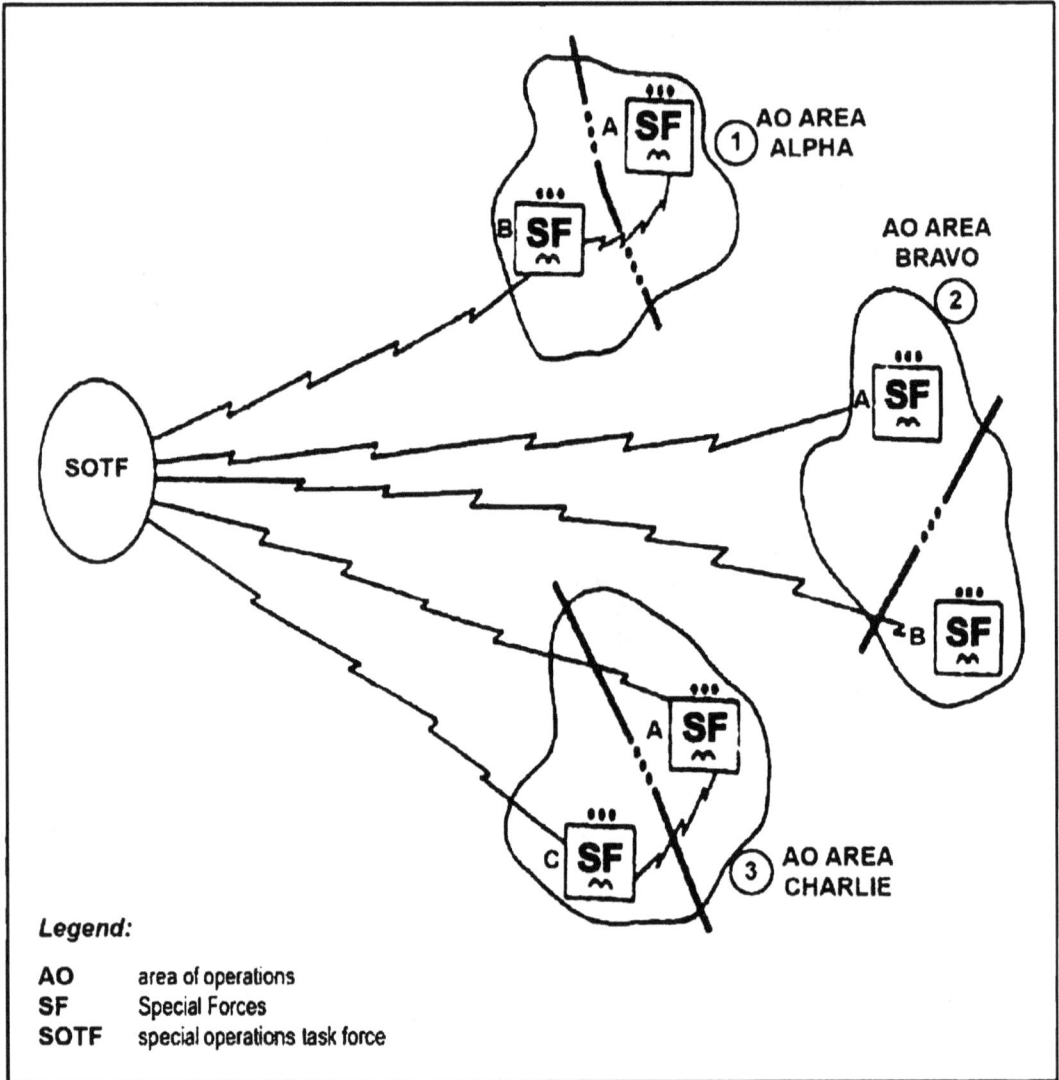

**Figure L-13. Base station to Special Forces detachment communications**

## UNCONVENTIONAL WARFARE COMMUNICATIONS SIGNAL SECURITY

L-29. Signal security, which includes physical, transmission, and cryptographic security, allows the SF detachment to operate longer without detection. The most effective means of ensuring signal security and avoiding detection is to minimize radio transmissions from and within the UW operational area. Signal security denies the enemy any usable intelligence from traffic analysis or radio direction finder data. It also keeps the enemy from using imitative deception against friendly use of the electromagnetic spectrum. For signal security efforts to be successful, all SF detachment and resistance elements must follow established communications security procedures.

### Physical Security

L-30. Physical security includes all measures taken to safeguard classified materials and equipment from unauthorized persons. These measures have a functional destruction plan for materials and equipment in anticipation of capture of classified materials and equipment by the enemy. The UW communicator must sterilize transmission sites after all transmissions. Only SF detachment members should have access into any area that has classified materials, communications equipment, or communications logs and other documentation. Indigenous personnel must be screened before being trained in communications assistance within the UW operational area. Under no circumstance will indigenous personnel be given access to the SF detachment's encryption systems or equipment used for communications with the base station.

## Transmission Security

L-31. Transmission security identifies measures taken to prevent interception, traffic analysis, and imitative deception. It also limits the effect of jamming friendly use of the electromagnetic spectrum. To reduce chances of detection, communicators use directional antennas and terrain masking from known radio direction finder stations. They also adhere to the transmission procedures according to ACP-125 (G), *Communications Instructions-Radiotelephone Procedures*, and SAV SER SUP 7 formats to lessen transmission time and the possibility of threat detection. A safe practice is to assume the threat has intercepted the signal and has located the transmission site. The communicator should keep transmissions short or risk the consequences.

## Cryptographic Security

L-32. Sufficient quantities and types of cryptographic systems should be infiltrated with the SF detachment to ensure all message traffic is encrypted and decrypted using one of these authorized systems. It is imperative that UW communicators use only authorized cryptographic systems for transmissions between the SF detachment and the base station. UW communicators will not use these systems for encrypting message traffic within the UW operational area. The SF detachment must infiltrate additional encryption techniques to ensure communications within the UW operational area are secure. Under no circumstances will U.S. cryptographic keys be used with indigenous forces.

L-33. Rendering plaintext unintelligible (cyphertext) and recovering the plaintext from the unintelligible text can be done electronically with cryptographic systems, code systems, and cypher systems. Code systems include brevity codes and secure codes. Brevity codes are used for the sole purpose of shortening a message. Secure codes are arbitrary groups of characters that represent units of plaintext (for example, KTC 600 tactical operations codes). Cypher systems include transposition and substitution (figure L-14, page L-16).

## ADDITIONAL CONSIDERATIONS

L-34. Underground, auxiliary, guerrillas, and SF detachments must focus on concealing communications employment to lessen the probability of detection and decrease the effect of a compromise. SF detachments should advise the resistance on techniques that will augment security, develop intelligence-gathering capabilities, and increase the effectiveness of communications between all elements of the resistance. UW intelligence activities should be highly compartmented, provide security to both personnel and operations, and conceal communications between elements from enemy observation or intercept. These SF roles are particularly critical during the early phases of UW, and the area assessment should include detailed considerations of communications capabilities and coverage indigenous to or capable of influencing the UW operational area.

L-35. Since UW, by definition, will involve U.S. STR personnel, innovative combinations of U.S. and indigenous communication methods are required. Upon infiltration into the UW operational area, SF detachments may find resistance communications systems to be of low quality, lacking reliability, and unsecured or nonexistent. SF detachments must take immediate steps to develop secure and reliable communications between various resistance groups and themselves. Due to the likelihood of a high electronic warfare threat, especially in the initial phases of the resistance movement, nontechnical communications should prevail (figure L-15, page L-17).

L-36. During UW SF, detachment communicators should anticipate the requirements for interoperability of U.S. communications with non-U.S., nonstandard systems or media (which may be seriously degraded). They should also anticipate the likelihood of rendering advice and support on rehabilitation of indigenous communication infrastructure in the UW operational area.

```
2 7 1 3 6 5 8 9 10 4
B L A C K H O R S E
P R O C E E D T O R
A L L Y P O I N T O
N E F O R N E X T M
I S S I O N D E T A
I L S A T T W O O N
E O N E E I G H T Z
E R O Z E R O Z U L
U A P R I L X X X X
```

**Transposition Cypher System**

```
OLFSS NOPPA
NIIEE UCYOI
AEZRR OMANZ
LXEON NTIRL
EPROT EEIRL
ESLOR ADIED
WGOXT NXEOH
ZXOTT TOTUX
```

**Substitution Cypher System**

*Key:*

| ABC | DEF | GHI |
|-----|-----|-----|
| JKL | MNO | PQR |
| STU | VWX | YZ |

*Message:*

**Destroy target one**

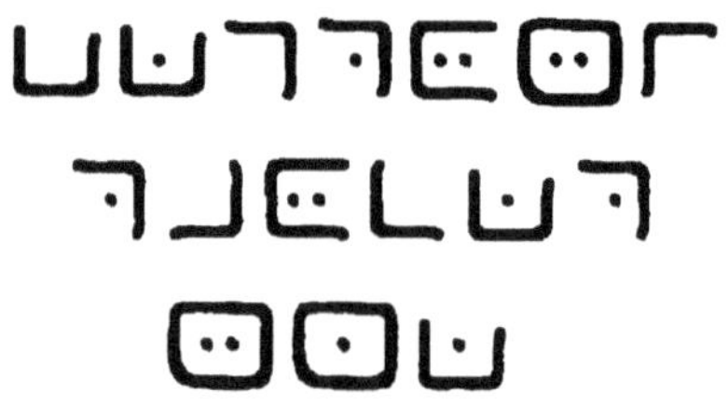

**Rosicrucian Cypher System**

**Figure L-14. Cypher systems**

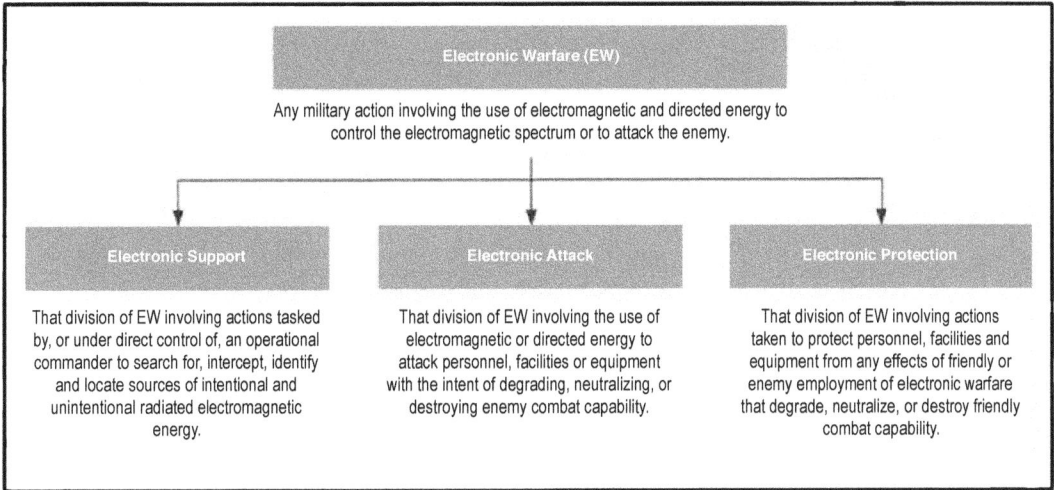

**Figure L-15. Electronic warfare**

This page intentionally left blank.

# Appendix M
# Logistics

The logistics support function needs to maintain proficiency in two specific areas. The first is in the processes used by conventional forces for their equipment acquisitions and maintenance to maximize commonality and minimize costs. The second is mastery of the process for nonstandard logistics needed for the unique sustainment of SF detachments in areas where it is necessary to avoid attribution. The training should support the planning, coordination, integration, and execution of all ARSOF activities, and the delivery of logistics in permissive, uncertain or hostile environments.

## INDIGENOUS RESISTANCE LOGISTICS

M-1.    There are three levels of logistics considerations for the SF UW planner. The base consideration (first level) is the resistance itself. How does the resistance sustain itself? What infrastructure already exists within the denied area? Is that infrastructure adequate to support current resistance operations and activities? What logistics expansion is required to sustain expanded resistance operations and activities? How is the infrastructure organized and dispersed? What external support, if any, might the resistance be receiving already from other sponsors or benefactors? What is the likelihood that the resistance will survive and progress with the level of logistics sustainment they already have? Living off the land and living off of the opposition is a standard technique of resistance sustainment over time with many examples, such as Washington's experts in guerrilla warfare Thomas Sumter, Andrew Pickens, and Francis Marion, and Mao's Military Principle to "Capture from the enemy so as to arm yourself," or the current mujahedeen.

---

"Marion's logistical support, more often than not, came from his opponents. The British often abandoned their arms and supplies to escape from this totally uncivilized warrior and his motley band. The Swamp Fox received food and shelter from the sympathetic colonists and his resourceful men forged swords from the saws they took from the sawmills and cast pewter mugs and spoons into badly needed bullets. Francis Marion, like all successful guerrilla leaders, never bit the hand that fed them—he was never cruel or uncharitable to the people upon whom he was dependent for support."

John Fiske, *The American Revolution, Volume II*, p. 188.

---

### The Mujahideen Fighter
### (circa 1979–1989)

"In practical terms the Mujahid can live off the land, or rather from the villages. . . .Even when he takes rations on the march all he needs is nan (flat bread) and tea to sustain him for days on end. The fatty bread is carried wrapped in a blanket, or cloth, and becomes rotten with age, making it the most revolting of meals. Nevertheless, it is eaten. The Mujahedeen can walk for days, even weeks, on the minimum of food; then, when the opportunity comes, they will stuff themselves with huge quantities, stocking themselves up like camels for the next journey.

**(continued on next page)**

---

---

**The Mujahideen Fighter (continued)**

. . . .After their weapon, the next most valued possession is their blanket. It is usually a grayish-brown colour, and is used day and night for a wide number of purposes. The Mujahedeen uses it as a coat, or cloak, for warmth in winter, or against the wind; they crouch under it to conceal themselves from enemy gunships, as it blends perfectly with the mud or rocks; they sleep on it; they use it as a sack; they spread it on the ground as a table cloth, or upon which to display their wares; often it becomes a makeshift stretcher and sometimes it is a rope; several times a day it becomes their prayer mat."

Mohammad Yousaf, *The Bear Trap (Afghanistan's Untold Story)*

---

M-2. The second level of logistics assessment regards the U.S.-unilateral sustainment requirements to infiltrate, assist the survival of, and exfiltrate U.S. detachments and personnel should the USG make the policy decision to support the resistance. These unilateral mechanisms do not rely on resistance assistance to the U.S. personnel. In fact, many of these mechanisms are deliberately set up to conceal their existence from the resistance and all others, without a need-to-know to maximize the mechanisms' security. Should the situation in the denied area deteriorate and the detachments must evade capture, the survival of the evaders may depend on this secrecy.

M-3. The third level is U.S. sponsor support to the resistance. The baseline means of resistance logistics will remain resistance exploitation of resources from within the AO itself, and as much as possible will be taken at the expense of the opposing regime and its security forces. However, some key items provided from external sources are likely to be critical to ultimate resistance success. The more severe the pressure of the state or occupying power on the resistance, the more difficult and dangerous the task will be of providing such external support. There are likely to be limited platforms and methods available for external resupply—each with limited cargo capacity. Being high risk operations, the number of external resupply missions will also be limited by opportunity and attrition. Therefore, prioritization of supplies, secure communications methods, and ease of handling will be key considerations.

## THE AREA COMMAND

M-4. The resistance will have its own conceptions and terminology to characterize how it is organized, led, and secured. U.S. UW doctrine uses the term *area command* to characterize the basic command organization structure (figure M-1, page M-3). The area command includes components and functions, such as—

- Centralized planning and decentralized execution.
- Intelligence, recruiting, shelter, and logistics support.
- Indigenous intelligence organization, to include:
  - Intelligence capability.
  - Counterintelligence capability
  - Agents and informers.
  - Civilian agents and informers.

## THE AREA COMPLEX

M-5. Instead of the term area complex, the resistance will have its own conceptions and terminology to characterize how it is sustained. U.S. UW doctrine uses the term *area complex* to characterize the basic resistance support structure. A majority of this structure will be clandestine to protect it from the state or occupying power. A large percentage of the area complex is the resistance logistics infrastructure and its supporting nodes and functions. The area complex (figure M-2, page M-4) includes—

- Security and intelligence systems.
- Logistics and supply installation.
- Communications systems.
- MSSs.

- Base camps.
- Reception sites.
- Training areas.
- DZs and LZs.
- Medical facilities.
- Supply caches.
- Evasion and recovery sites.

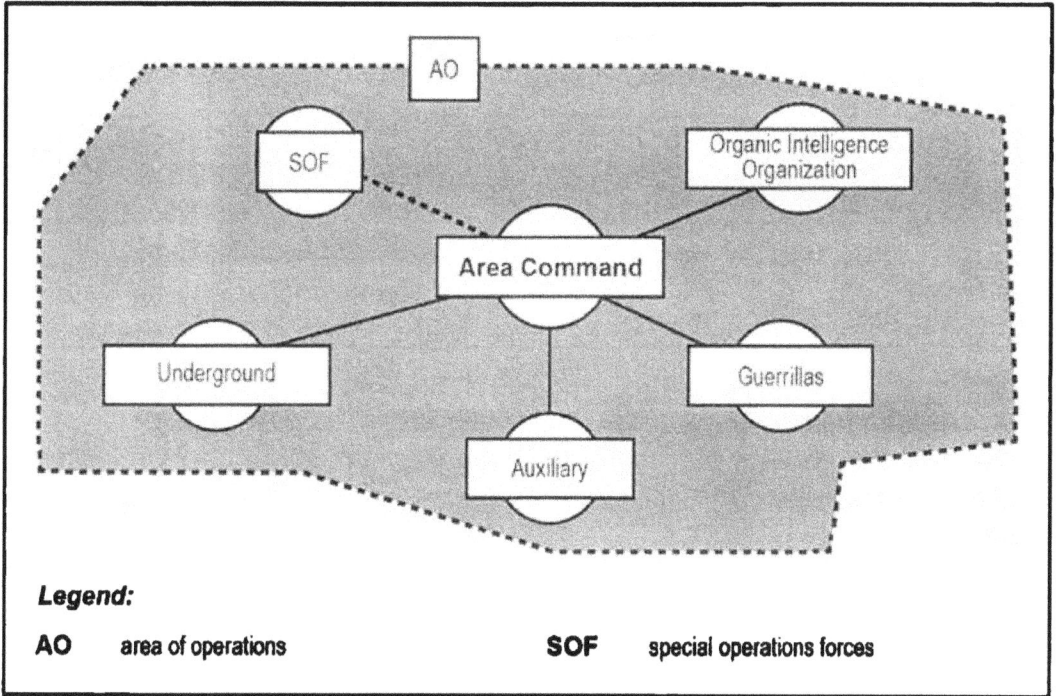

Figure M-1. Area command

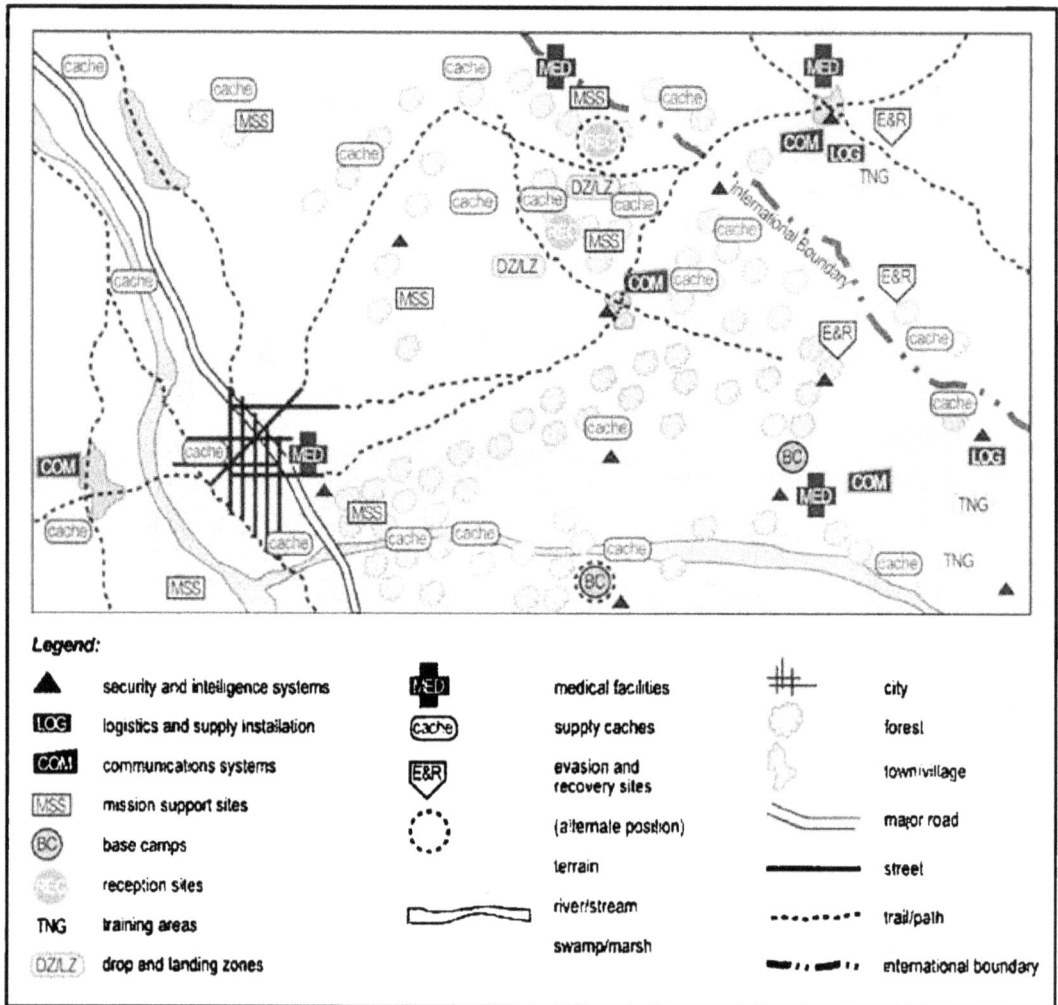

**Figure M-2. Area complex sketch**

## UNDERGROUND SUPPORT NETWORK

M-6.    The intensity and omnipresence of state or occupying power scrutiny and oppression will determine the degree of clandestine organization and activity and the corresponding requirement for compartmented security (figure M-3, page M-5). Underground support functions may include:

- Transportation (ground, water, or pack animal).
- Finance (procurement or distribution).
- Medical (intelligence, supply, or treatment).
- Safe sites, houses, and areas (rural or urban).
- Subsistence (procurement or packaging).
- Sanctuary (training or refit).
- Manufacturing (ammunition or explosives).
- Training (sabotage or courier).
- Recruiting (cadre or spotting).

M-7.

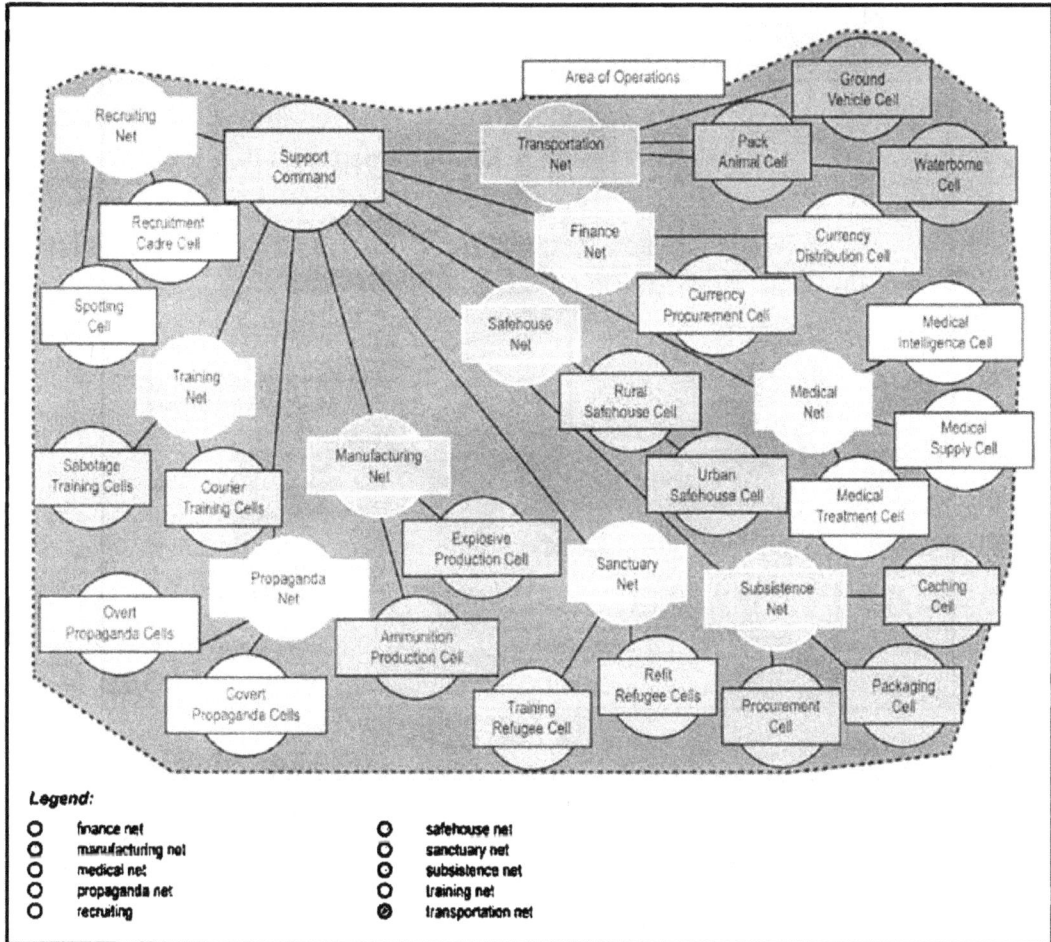

**Figure M-3. Underground support network**

M-8.    Undergrounds require a logistics staff section to plan and provide operational supplies and keep the commander advised on logistics matters within his area of responsibility. In the combat phase of resistance, this staff must ensure that subordinate elements are adequately supplied to carry out assigned tasks. At times, ad hoc supply units may be organized to support local groups in dealing with problems beyond their capability. Underground logistics operations are normally decentralized and are tailored and controlled by commanders to meet their specific needs.

M-9.    The organizational structure most commonly used is the cells-in-series in which each cell is assigned a specific supply task, commonly referred to as divisions of labor (figure M-4, page M-6). In some cases, logistics functions are conducted by special cells or activities supporting either the underground itself or to meet specific needs of the guerrilla force.

M-10.    A regime controlled economy usually gives rise to extensive black market activities. Exploiting corrupt government and military officials can also prove extremely useful. Undergrounds may also purchase supplies in what appears to be legal marketing through front organizations. Undergrounds may also be able to obtain supplies through theft by secretly removing them from plants and warehouses. Under extreme needs, raids can be conducted against warehouses, plants, moving convoys, and other installations by special underground action units. Undergrounds may also resort to manufacturing demand items due to shortages and difficulties involved in procuring sufficient quantities of war materiel. Manufacture of such items as mines, incendiaries, grenades, small arms, explosives, clothing, ammunition, and miscellaneous items is most common. In addition to manufacturing, undergrounds may also remake, repair, and adapt existing and captured weapons and equipment.

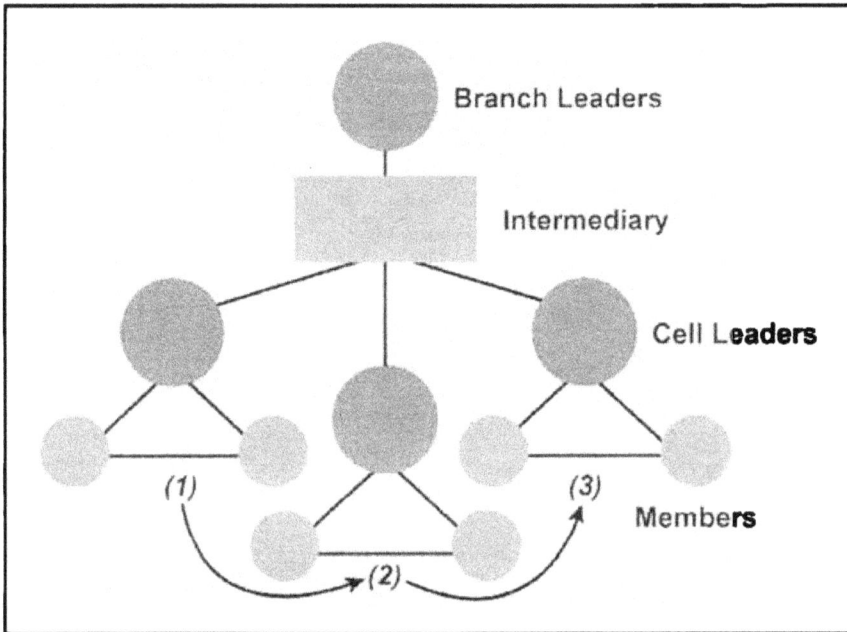

**Figure M-4. Cells in series**

M-11. During critical periods, it may be necessary to relocate manufacturing efforts from urban to rural areas, where intrusions from security forces are less frequent. To increase security, workshops are often small and mobile, employing less than a dozen workers. Production is conducted more or less in the open under the guise of legal work, thus avoiding the problem of minimizing shop noises. As a disguise, items may be fashioned in shops turning out similar products. For example, the production of flamethrowers may be constructed in a factory producing fire extinguishers.

---

**Indo-China
(circa 1941–1954)**

"As the targets of thievery became more wary, the Viet Minh began to establish their own primitive production capability. In fact local production became a primary source of arms and equipment. Of necessity these facilities were extremely small, mobile and completely non-mechanized operations consisting of 10–15 people engaged in a single endeavor e.g., the manufacture of crude, though effective, mines. These shops were operating on a direct support basis— mobility was essential so that they could move whenever the force being supported moved. Larger, less mobile facilities, employing up to 500 people were located in firmly held Viet Minh base areas to preclude capture by the French.

**(continued on next page)**

---

## Indo-China (continued)

To illustrate the effectiveness of these manufacturing operations, during the first six months of 1948, the Viet Minh reported that one sector had produced:

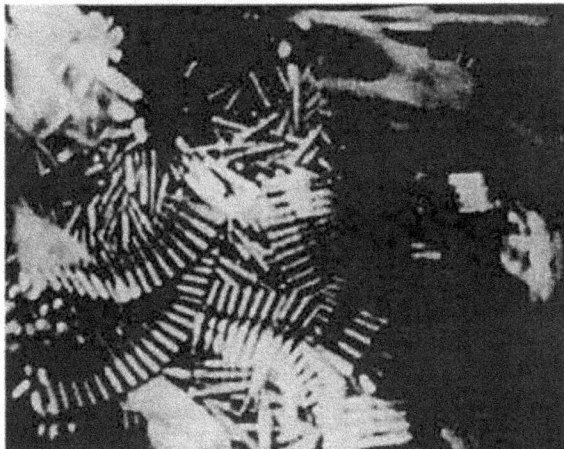

' 38,000 grenades, 30,000 rifle cartridges, 8000LMG cartridges, 60 rounds for bazookas, and pistols, and 00 mines Another sector during all of 1948 produced 61 light machine guns, 4 submachine guns, 20 pistols, and 7000 cartridges."

George K. Tanham, *Communist Revolutionary Warfare*, p. 68

(Passage from Caraccia. *Guerrilla Logistics*. 1966, p. 26)

## The Ayalon Institute—Palestine
### (circa 1945–1948)

The Ayalon Institute was a dangerous top secret operation that produced over 2.25 million bullets in a clandestine underground factory, built not far from the British who ruled the area. The 300 sq. yard factory was built under a kibbutz and its sole purpose was to hide the work of the young people underground who risked their lives daily. A bakery was built above ground, as well as a chicken coop, a laundry, a dining hall, a vegetable garden, workshops, and a barn—all to give the appearance of a normal kibbutz The factory required 2 entrances—one for the workers and one for the machinery. One entrance was hidden below a washing machine specially designed to swing away from it base The other entrance was under a huge 10 ton bakery oven, which was mounted on tracks To get the boxes out of the kibbutz every day, they were placed in a specially outfitted fuel truck, so as not to arouse suspicion, but highly dangerous. The raw materials were delivered to the kibbutz and the bullets were removed daily. The underground workers were not aware of how the materials got there in the morning and how the bullets were removed. The difficult work continued underground from 1945 to 1948.In that time, the Ayalon Institute produced over 2 25 million Sten bullets—10,000–14,000 bullets daily.

(continued on next page)

---

### The Ayalon Institute—Palestine (continued)

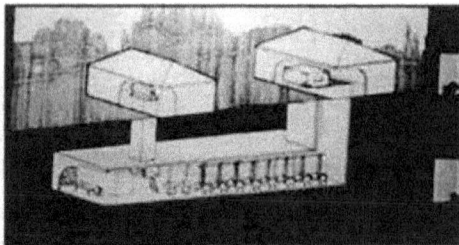

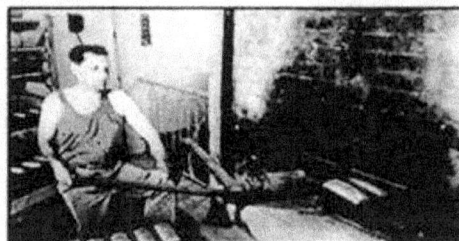

Risa Borsykowsky, The Ayalon Institute: Kibbutzim Hill Rehovot
https://www.jewishgiftplace.com/Ayalon-Institute.html

---

# UNILATERAL LOGISTICS (IN SUPPORT OF U.S. FORCES)

M-12. U.S.-unilateral sustainment requirements are designed to ensure the U.S. detachments are able to securely infiltrate and exfiltrate the denied area without relying on the resistance (figure M-5, page M-9). It is possible that indigenous surrogates in the employ of the United States and not members of the resistance will comprise some aspects of these mechanisms. These mechanisms may be created and exist prior to contact with, assessment of, or a decision to support a resistance. U.S. unilateral mechanisms will include:

- Intelligence networks and cells.
- Ratlines.
- Bugout lines.
- Escape and recovery mechanisms.

## ACCOMPANYING LOGISTICS

M-13. Accompanying logistics include supplies received, prepared, and rigged during isolation for delivery in conjunction with infiltration. Preparation includes packaging and load considerations to facilitate transportation subsequent to infiltration. Accompanying supplies may need to be cached following infiltration for later use.

> "Each team parachuted with the standard Jedburgh radio set, with which they were to contact SFHQ in London to arrange the delivery of additional weapons and supplies. Following a request for such supplies, it would take an estimated eight days for delivery."
>
> S.J. Lewis, *Jedburgh Team Operations in Support of the 12th Army Group,*
> August 1944

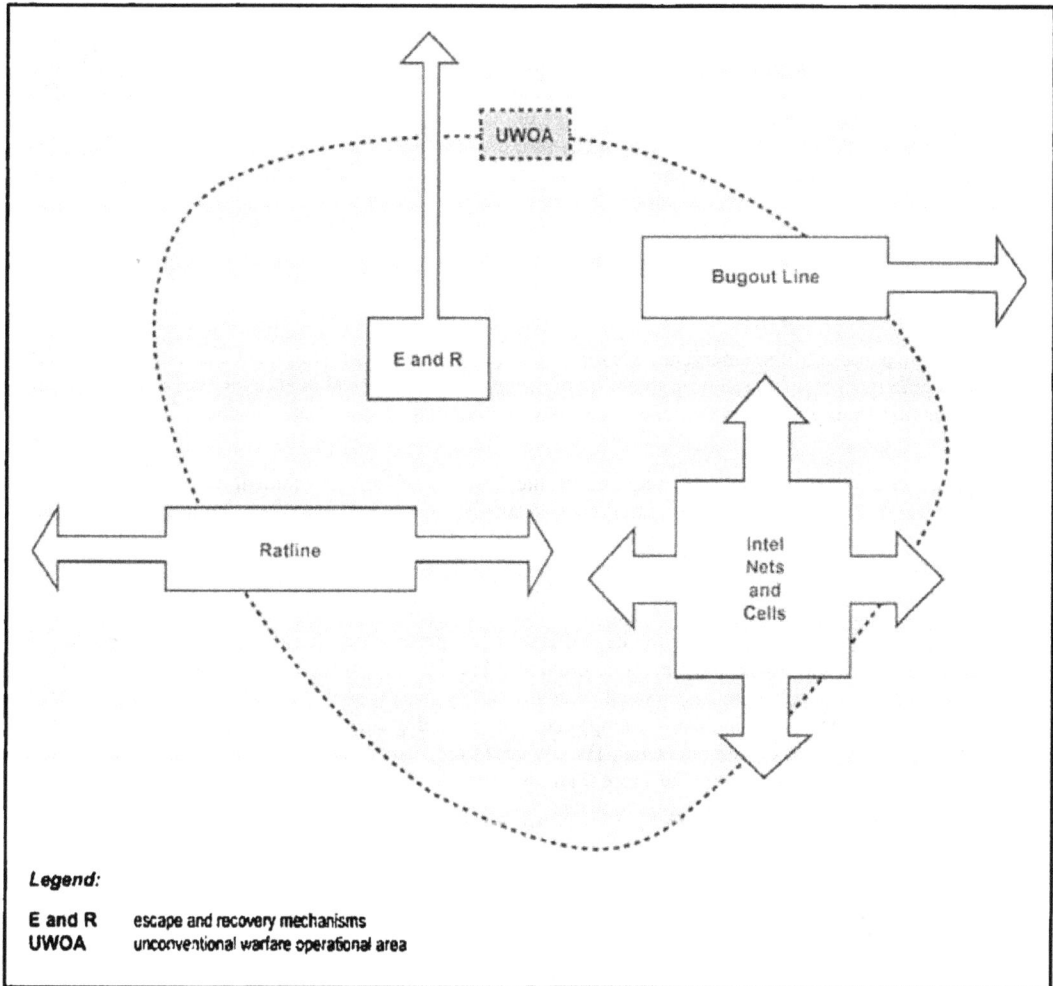

**Legend:**

**E and R**  escape and recovery mechanisms
**UWOA**  unconventional warfare operational area

**Figure M-5. Unilateral mechanisms**

## EXTERNAL LOGISTICS

M-14. External resupplies are procured and delivered to the UW operational area or JSOA by the sponsor (JSOTF), based on the needs of the detachment, as well as the resistance force or insurgents. Resupply is planned in isolation to be delivered after infiltration at a coordinated location and time automatically, as requested, or based upon a no-communications trigger. The detachment preselects resupply items and delivery merchandise during isolation to replenish or supplement supplies that its members consume or to fulfill other requirements. There are three types of external resupply: automatic resupply, on-call resupply, and emergency resupply.

### Automatic Resupply

M-15. The detachment plans for automatic resupply before infiltration, coordinating for the delivery time, location, contents, and the identification marking system or authentication. Automatic resupply is delivered after the detachment successfully infiltrates and establishes radio contact, unless delivery is cancelled, modified, or rescheduled. Automatic resupply may consist of items that are mission essential but could not feasibly accompany the detachment due to the nature or mode of infiltration.

## On-Call Resupply

M-16.  When the detachment establishes communications with the JSOTF or SOTF, external supply begins on call. Personnel use the abbreviated code of a catalog supply system contained in the SOI to request supplies based on operational need. These supplies consist of major equipment items that units do not consume at a predictable rate. Supplies are packed and rigged into man-portable loads and color coded according to type of supplies for immediate identification. If airdropped, supply personnel normally pack equipment and supplies in appropriate delivery containers that have a cargo capacity of 500 pounds or less to ease handling and transportation within the JSOA. To allow rapid clearance of the DZ, personnel ensure the contents of each container are in man-portable units of about 50 pounds each.

## Emergency Resupply

M-17.  The purpose of the emergency resupply is to provide essential equipment and supplies in order to restore operational capability and survivability of the detachment. Typical items contained in the bundle may be communications equipment, batteries, weapons, ammunition, money, and handheld Global Positioning Systems.

M-18.  A coded message, a radio request, or the absence of any detachment communication over a prearranged period can trigger an emergency resupply. As a minimum, resupply should consist of communications equipment and enough mission-essential supplies to establish base contact.

## MARGE BUNDLE CODES

M-19.  Communications security in the denied area puts a premium on limiting the size of messages. For resupply requests, the marge bundle codes provide a brevity code system by which the detachment can request prepacked supplies and equipment without lengthy messages. Figure M-6 depicts an example of marge bundle codes. Marge bundle (training) codes are a catalog supply system used for the requesting of supplies and equipment by the outstation. Bundle codes are issued during isolation consisting of either a number or a number and a letter. For more than one package, the number of packages requested is placed before the proword MARGE. Examples include:

- Code MARGE ONE requests 12 each meals, ready-to-eat, wt. 22 lbs., 1 piece.
- Code MARGE TEN BRAVO requests 1200 each cartridge, 5.56mm, ball.
- TEN MARGE TEN BRAVO requests 1200 each cartridge, 5.56mm, ball.

Example - Code MARGE FOUR requests Ration, Indigenous Personnel (30 men), 489 lbs., 10 pieces consisting of:

- 120 lbs. bacon
- 120 lbs. sugar
- 60 lbs. coffee
- 45 lbs. tobacco
- 15 lbs. salt
- 45 lbs. flour, grain
- 45 lbs. comfort items

Example - Code TWO MARGE TWENTYTHREE ALPHA requests 20 blocks composition C-4 (2 ½ lb. block) x two for a total of 40 blocks.

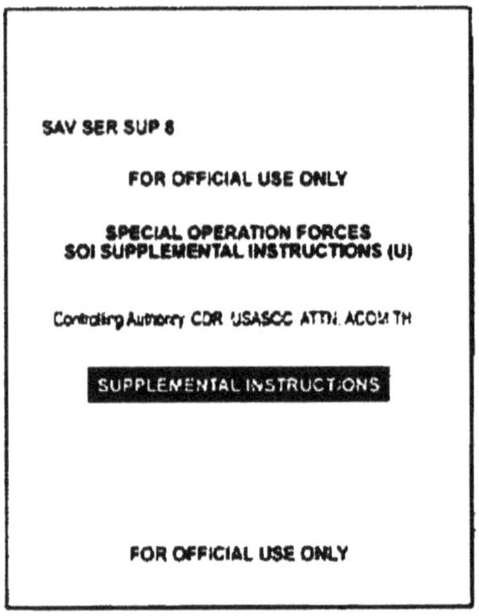

SAV SER SUP 8

FOR OFFICIAL USE ONLY

SPECIAL OPERATION FORCES
SOI SUPPLEMENTAL INSTRUCTIONS (U)

Controlling Authority CDR USASOC ATTN ACOM TH

SUPPLEMENTAL INSTRUCTIONS

FOR OFFICIAL USE ONLY

**Figure M-6. Example marge bundle codes**

# U.S. LOGISTICS SUPPORT TO RESISTANCE FORCES

M-20. The detachment will need to conduct an ongoing identification and assessment or resistance logistics infrastructure to identify gaps and priority requirements. These shortfalls will help determine what U.S. sponsor support will be delivered to the resistance.

## AREA ASSESSMENT

M-21. The ongoing assessment provides a basis for modifying operational and logistics plans that were drafted prior to infiltration, based on the situation on the ground. The pilot team conducts detailed area assessments—particularly human terrain analysis—prior to infiltration to expand its understanding of the operational environment. Assets exfiltrated from the denied area with in-depth knowledge of the area can aid this assessment. During and following infiltration, SF Soldiers continue the area assessment initiated by the pilot team to confirm or refute the information previously reported. Logistics capability of the area topics include:

- Availability of food stocks and water. Include any health-related water restrictions.
- Agricultural capability.
- Type and availability of all categories of transportation.
- Types and location of civilian services available for the manufacture and repair of equipment and clothing.
- Supplies locally available, including type and amount.
- Medical facilities, including personnel, medical supplies, and equipment.
- Enemy supply sources accessible to the resistance.

### Guerrilla Logistics

M-22. The U.S. concept of logistics support to guerrilla forces is dependent on the establishment of an SF operational base. From this base within the operational area, guerrilla forces are supported with arms, ammunition, communications equipment, and other critical supplies unavailable within the UW operational area. In general, guerrilla forces are expected to live off the land within the UW operational area, and with the support of the local inhabitants, they are expected to provide themselves with food, shelter, transportation, medical assistance and other services. However, the U.S. should be prepared to provide the guerrillas with total logistics support if necessary. The U.S. should normally be prepared to provide packages of armorer's tools, small arms repair kits, sewing kits, and weapons cleaning materials.

M-23. External procurement of logistics support is commonly through arrangements made by the government-in-exile or by the U.S. external sponsor. Depending on the geographical features of the terrain and operational security conditions, supplies are delivered by airdrop, amphibious, or overland methods.

### Overland Logistics

M-24. Smuggling operations may involve providing supplies overland and across borders of neutral and occupied countries by man-pack, animal, or by vehicle. Difficulties involved may be enormous; borders may be heavily guarded, patrolled, and covered by electronic ground surveillance and sensor devices, making movement dangerous. The amount and type of supplies that can be carried may be limited by the mode of transportation, route, terrain, security, and the situation. To conceal the illegal cargo, a number of methods can be used. Supplies may be hidden under logs, in moving vans, beneath a tarpaulin covered with a layer of reeking fresh manure, or oxygen cylinders ostensibly destined for hospitals (figure M-7, page M-12).

Figure M-7. Overland resupply

## Aerial Delivery Logistics

M-25.  Aerial delivery is the air transport of cargo, equipment, and/or personnel to a desired location on the ground by aircraft These aircraft may include military, contracted, and commercial as well as strategic and theater fixed-wing airlift. Aerial delivery also includes the use of rotary-wing as transportation platforms to move personnel, equipment, and sustainment supplies (See ATP 4-48, *Aerial Delivery*, for more detailed information )

M-26.  Aerial delivery can be a vital link in UW sustainment as surface means of logistics support may be impractical or impossible into and through the denied area. Aerial delivery eliminates the need for clearance of a ground route to deliver cargo or personnel. Aerial delivery also reduces time in transport for delivery of cargo or personnel and handling of supplies for limited quantities of material. However, aerial delivery will still be limited by adversary air defense capabilities, physical restrictions on air corridor survivability, and operational requirements for concealment.

M-27  Aerial delivery operations can be performed using three methods: airdrop, airland, and sling load. The practicality of using any of these techniques in UW logistics will be based on the characteristics of each unique situation. However, in the majority of cases, most aerial STR elements and/or advisors in denied areas will present conditions unfavorable for air landings or sling load operations. Airdrop will likely be the most feasible method of aerial logistics.

## AIRDROP

### Standard Airdrop Methods

M-28.  Aerial delivery by parachute drop to resistance reception committees on clandestine DZs is the most direct and fastest means of logistics supply. This method of delivery is usually conducted by the resistance with the assistance of special liaison agents from the allied sponsoring power or government-in-exile. DZs are primarily selected in areas devoid of enemy air defense systems. Where this is not possible, drops may be made in defended areas when it is felt enemy air defense systems can be successfully penetrated (figure M-8). Unlike the representative figure shown above, most resistance supply drops will be conducted during a period of darkness.

**Figure M-8. Airdrop logistics**

M-29.  Airdrop loads are classified as either standard or nonstandard loads:
- **Standard Loads.** Standard loads include all loads that are detailed in an associated technical publication.
- **Nonstandard Loads.** Nonstandard loads may include approved loads for U.S. Air Force aircraft without an associated technical publication. These loads can also be dropped in accordance with an associated technical publication but with a variance in the drop altitude or speed of the aircraft, and the loads may also include hazardous material.

M-30.  Other classifications of loads are determined by the weight of the load, the type of aircraft it is loaded on, by its class of supply and if it is explosive and/or hazardous material.

M-31.  Airdrop types are categorized based on the rate of descent of the load. The methods of airdrop are door-load, extraction, and gravity. Airdrop methods pertain to how loads exit the aircraft. These types and methods often use common components and systems. There are three standard types and methods of airdrop:
- **Free Drop.** Free drop is the delivery of certain non-fragile items of equipment or supply from a slow-flying aircraft at low altitude without the use of parachutes or other decelerators. The load descends at a rate of 130 to 150 feet per second. Energy-dissipating material (such as honeycomb) may be placed around the supplies to lessen the shock when the load impacts with the ground. Humanitarian daily rations, baled clothing, fortification, and barrier materials are other examples of non-fragile free drop items.
- **High-Velocity Airdrop.** High-velocity airdrop is used when threat conditions dictate that the aircraft remain at high altitudes to avoid hostile air defenses but, for accuracy, drift must be minimized. The load descends at a rate of 70 to 90 feet per second. The rapid rate of descent mitigates drift. A small parachute provides enough drag to hold the load in an upright position. Energy absorbing material is used to reduce the effect of the ground impact. Subsistence items, packaged petroleum products and ammunition are the most probable loads for this type of delivery.
- **Low-Velocity Airdrop.** Low-velocity drop is the preferred method to drop all supplies and equipment certified for airdrop. Low-velocity drop is a procedure in which the drop velocity is

less than 28 feet per second, and multiple parachutes can be used to achieve the desired rate of descent. Loads are specially prepared for airdrop either by packing the items in airdrop containers or by rigging the loads to airdrop platforms.

## Army Special Operations Forces Methods and Types of Aerial Delivery

M-32.  ARSOF aerial delivery involves all types and methods of air-to-ground delivery of equipment and supplies. Airdrop is one of the best and fastest means of resupply to ARSOF. In some cases, it may be the only means of resupply available to the SOTF commander because of the uniqueness of the UW environment and strategic strike objectives.

M-33.  The type and method of aerial delivery depends on the specific needs of the SOF mission. SOF uses a variety of special operations, conventional, foreign military, and nonstandard or contract aircraft for aerial delivery. The rigging and dispatching techniques are specific to each aircraft and vary during combat or training. Risks are mitigated based on the mission, situation, availability, and capability of the aircraft and aircrew.

*Note:* Further details can be found in ATP 3-18.10, *Special Forces Air Operations*; ATP 3-05.40, *Special Operations Sustainment*; ATP 3-18.11, *Special Forces Military Free-Fall Operations*; and TC 18-11, *Special Forces Military Free-Fall and Double Bag Static Line Operations*.

## Joint Precision Airdrop System

M-34.  The joint precision airdrop system (JPADS) is a high-altitude, precision-guided airdrop system that provides increased control of the canopy, allowing for increased accuracy from a high-altitude deployment. The system is available in a 2,200-pound (2K) container delivery system size capability and a 10,000-pound (10K) total rigged weight capability. JPADS provides additional protection for the aircraft by increasing the stand-off and altitude distance from potential enemy anti-aircraft weapons systems. JPADS can be dropped from a maximum altitude of 24,500 feet mean sea level. The 2K system requires a minimum 150-meter circular DZ, and the 10K system requires a minimum 250-meter circular DZ. Loads can be dropped from a single aerial release point and delivered to multiple or single DZs. The JPADS is controlled by the assistance of an airborne guidance unit and uses military global positioning satellite data to navigate to the DZ. Figure M-9, page M-15, depicts the JPADS methodology for enabling a single aircraft to deliver bundles to multiple DZs using steerable ram-air parachutes and Global Positioning System technology.

M-35.  The ultra-lightweight (also called ULW) JPADS is an aircraft-deployed steerable canopy system that is capable of delivering cargo loads of 250 to 699 pounds (rigged weight) safely and effectively from 4,500-feet to 24,500-feet mean sea level. Precision is accomplished through hardware, software altitude control, and glide management of system descent and landing. It satisfies high-priority air delivery requirements of low volume items for precision delivery locations and times. It is particularly valuable for troops in combat as an emergency resupply bundle of mission-essential equipment and supplies to restore or supplement the operational capability and survivability of the element. Ultra-light weight can also supplement essential items required for military freefall clandestine infiltrations by being dropped in combination (same time) with the military freefall parachutist at the on-set of the mission. In general an ultra-light weight bundle is similar in size to an A-21 bundle system. Ultra-light weight is a valuable aerial delivery resupply method for small DOD elements requiring precision and clandestine airdrop.

M-36.  The United States Air Force cargo aircraft mission planner software enables aircrews to plan and initiate load release at a precise computed air release point or within a launch acceptance region through the application of accurate JPADS component modeling. The mission planner provides the capability to model parameters of aircraft position, altitude, airspeed, heading, ground speed, course, onboard load position (station), roll-out and exit time, decelerator opening time, and trajectory to stabilization and descent rate. The mission planner software must be recovered and returned.

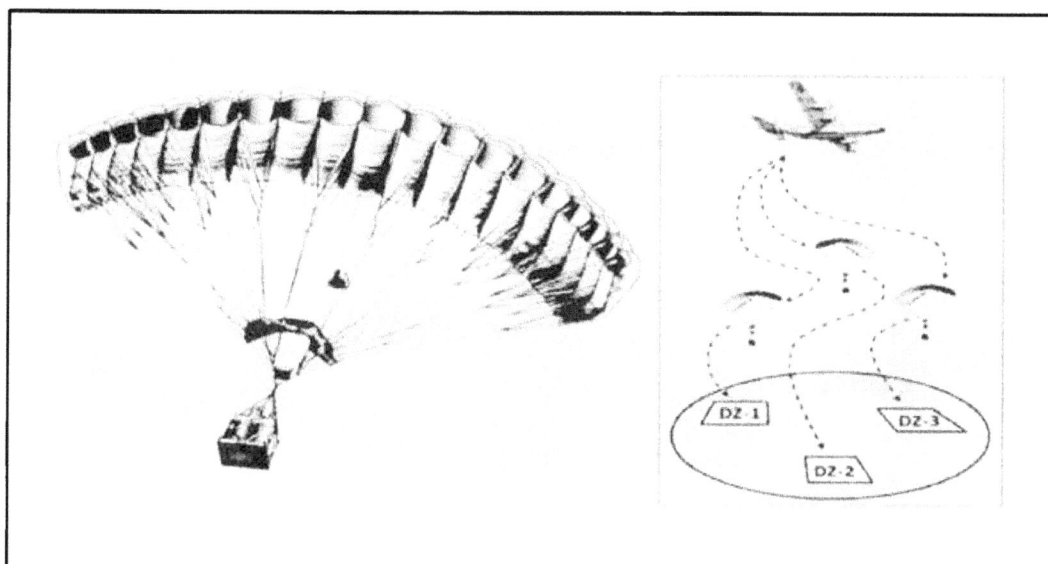

**Figure M-9. Joint precision aerial delivery system**

## Airdrop Containers

M-37. Aerial delivery containers used for supplying resistance forces in past eras, such as World War II, usually consisted of strong metal or plastic cylindrical cases under 6 feet in length and divided into one or more cells to separate types of supply. Because recovery parties would want to exit the recovery DZ quickly, the containers would either be designed to be man-portable or the supplies loaded inside the container would be divided into smaller, man-portable packages.

M-38. Today, bundles can be constructed in many different configurations as the mission requires. Small, discrete bundles can be specially made for small operations or multiple, large bundles can be dropped to larger resistance forces operating in more secure areas Even large bundles will likely be an assemblage of smaller packages that can be rapidly disassembled and quickly moved or hidden.

## AIRLAND

M-39. *Airland* is defined as move by air and disembark, or unload, after the aircraft has landed or while an aircraft is hovering (JP 3-17). Airland is the preferred method of aerial delivery because it is usually the most efficient and cost effective. It permits delivery of larger loads with less risk of cargo loss or damage than airdrop or sling load methods. It is also desirable because it makes the most efficient use of available cargo space.

M-40. Airland exists primarily as a rear-echelon capability. Although crews normally accomplish offloading from a stationary aircraft with engines shut down, procedures exist to load and offload with engines running when necessary to reduce ground time—or in the case of rotary-wing aircraft after it has entered a hover. In a higher threat environment or when sufficient materiel handling equipment is not available, procedures exist to use the combat offload method from a moving aircraft. This delivery method can be conducted at well-established airbases or it may involve tactical deliveries to unimproved dirt strip assault LZs.

### Airland-Capable Platforms

M-41. In UW, enemy area denial will often make normal airland delivery operations unusable. Resistance forces and advisors will often have to rely on remote, makeshift or austere landing strips. Opportunities to use theater medium-lift aircraft, such as MC-130s, will be relatively rare.

M-42. Figure M-10 shows possible methods of resupply into denied areas beyond standard fixed-wing transports and traditional helicopters. The M-22 Osprey vertical short takeoff and landing aircraft combines the benefits of both fixed- and rotary-wing flight. It is more conspicuous than the common, usually civilian, smaller fixed-wing short takeoff and landing aircraft, such as the M-28 Skytrain. Large and small rotary-wing unmanned aerial vehicles are steadily improving in range and payload and are capable of landing on and taking off again from unimproved sites. In addition, although fixed-wing drone unmanned aerial vehicles, such as the MQ-9 Reaper, are designed primarily for intelligence, surveillance, and reconnaissance or configured for combat, they are capable of being fitted with supply containers for extremely high-value cargo.

**MV-22 Osprey VSTOL**    **M-28 Skytrain STOL**    **Kaman K-Max Pilotless Cargo Helicopters**

**TRV-80 Rotary UAV**    **MQ-9 Reaper UAV**

**Figure M-10. Airland-capable platforms**

## Advantages and Disadvantages of Airland

M-43. There are several advantages and disadvantages of airland operations as it relates to airdrop and sling load operations. Careful consideration of the advantages and disadvantages must be taken when planning to incorporate the airland method as a means of aerial delivery.

### Airland Advantages

M-44. Airland delivery is usually the most efficient delivery method for moving equipment, personnel, and supplies for the following reasons:

- Allows equipment that is not airdroppable, such as tanks, some artillery, and rotary-wing aircraft, to be brought rapidly in-theater.
- Allows a greater degree of tactical integrity and the capability to rapidly employ units after landing.
- Reduces the risk of injury or damage to personnel and equipment.
- Permits the maximum use of aircraft loads by eliminating the volume and weight of preparing loads for airdrop deliveries.
- Allows aircraft to be used for backhaul or evacuation of personnel.
- Has a low cost per ton for cargo.
- Seldom requires special rigging materiel.
- Does not require specialized training for troops.

*Airland Disadvantages*

M-45. Airland delivery disadvantages revolve around landing area requirements, aircraft vulnerability, and lack of facilities—all of which are significant in support of UW operations in denied areas. Disadvantages include:

- May require landing strip improvements to ensure a moderately level, unobstructed LZ.
- More time required for delivery of an equivalent size force than when delivery is by airdrop, especially if using a small, restricted LZ.
- Greater number of support personnel and materials handling equipment required on the LZ.
- Exposure of aircraft to prolonged air and ground attack due to extended time on the ground at forward airfields.
- Engineer assets may be required to maintain the airfield based on the physical composition of the LZ and weight of the cargo aircraft.
- Specialty trained personnel are required to supervise, prepare, and inspect supply loads.
- Potential requirement for refueling at the LZ due to reduced range of cargo aircraft carrying heavy loads.

# SLING LOAD

M-46. A sling load is an external load carried beneath a utility or cargo rotary-wing aircraft. The load is held in place by one or more slings, cargo bags, or cargo nets. As in airdrop, weather conditions, mission requirements, threat environment, and equipment to be delivered determine the equipment and type of aircraft used for the delivery.

## Sling Load Rotary-Wing Aircraft Considerations

M-47. The flexibility of rotary-wing aircraft allows all three methods of aerial delivery: airland, airdrop, and sling load. Rotary-wing aircraft, however, are most commonly used for sling load missions. Several factors affect how much weight a rotary-wing aircraft can carry:

- **Altitude.** The rotary-wing aircraft rotor efficiency decreases at higher altitudes and requires more power to hover than at lower altitudes. This means less capability to lift cargo.
- **Temperature.** High air temperature has an adverse effect on the power output of rotary-wing aircraft engines. An increase in temperature decreases engine performance. This means decreased lift capability.
- **Humidity.** As the relative humidity increases, the rotary-wing aircraft's lift performance decreases.
- **Wind.** Wind affects rotary-wing aircraft performance by increasing rotor lift without an increase in engine power. Therefore, less power is required to hover into the wind than when no wind conditions exist. Also, with constant power, the rotary-wing aircraft can hover into the wind with higher payloads. This is why wind conditions and direction are important to the performance of the rotary-wing aircraft.
- **Fuel.** Fuel weighs approximately 7 pounds per gallon. The weight of the fuel required to fly and complete the mission and the distance the load must be flown may reduce the lift capability of the aircraft. Compared with fixed-wing aircraft, rotary-wing aircraft are usually slower, have less range, have less cargo capacity, and are more vulnerable to anti-air defenses.

## Methods of Sling Loading

M-48. All sling loads are configured in one of the following methods:

- **Single-Point.** Loads use one aircraft cargo hook on one rigged load during flight.
- **Dual-Point.** Loads use two aircraft cargo hooks on one rigged load during flight.
- **Tandem.** Loads use two aircraft cargo hooks on two rigged loads, one in front of the other, during flight.
- **Side-By-Side (Shotgun).** Loads use one or two aircraft cargo hooks on two rigged loads—one beside the other during flight.

M-49. Figure M-11 depicts the four methods of sling load using a CH-47.

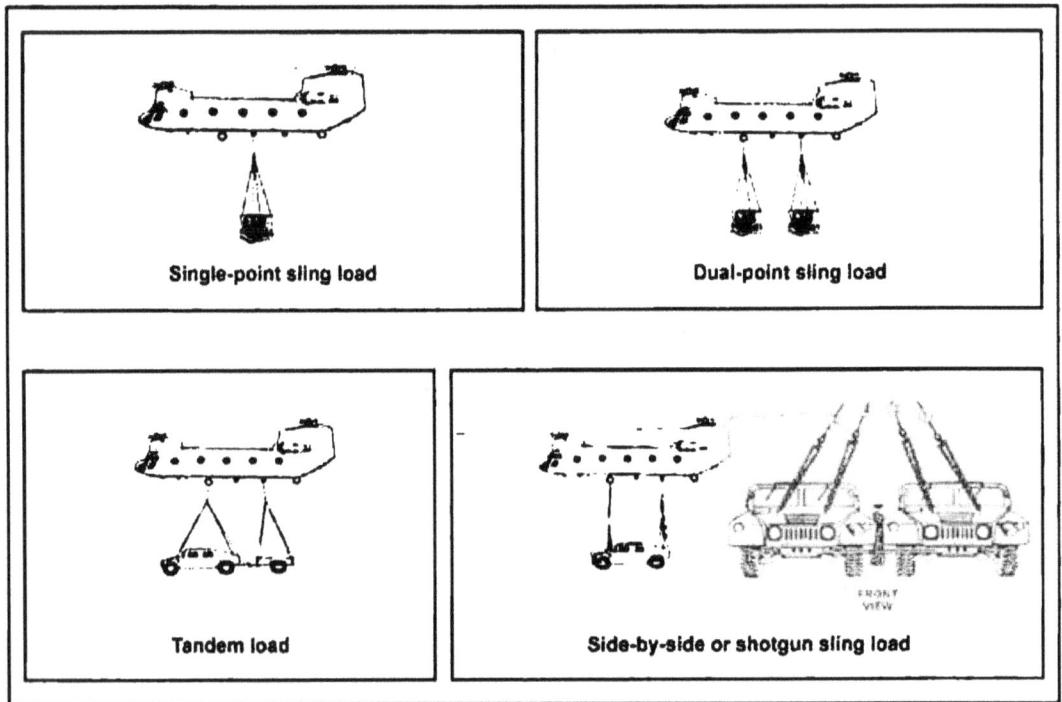

**Figure M-11. Sling load methods**

## Advantages and Disadvantages of Sling Load

M-50. There are several advantages and disadvantages of sling load:

- **Advantages.** Sling load advantages include the following:
  - Provides for rapid movement of heavy, outsized cargo directly to the user, bypassing surface obstacles.
  - Allows the use of multiple flight routes and landing sites, enhancing survivability of the aircraft and the flexibility afforded to the ground commander.
  - Reduces the planning cycle time, thus providing a far more flexible and responsive asset.
  - Is capable of positioning combat and support assets without materials handling equipment or the need for onward movement.
- **Disadvantages.** Sling load disadvantages include the following:
  - Rotary-wing aircraft weight capacities are generally less than that of fixed-wing aircraft.
  - Can cause the rotary-wing aircraft to be unstable during flight, which may restrict airspeed or maneuvering capabilities.
  - Sling load is more likely to be affected by adverse weather conditions. For example, rotary-wing aircraft lift capacity is affected by atmospheric pressure, altitude, temperature, humidity, and winds.
  - Limited number of rotary-wing aircraft currently available in the force may limit sling load operations.
  - LZ surface conditions (debris, dust, and snow) and the size of the LZ will impact the capability to conduct successful operations.
  - Sling load requires specialized training
  - Sling load increases aircraft detection because the aircraft has to fly above nap-of-the-earth.
  - Restricted flight time and distance due to increased fuel consumption rate.

## Amphibious Logistics

M-51. Resistance movements in countries bordered by coastal areas may be supplied by surface or subsurface means from the external sponsor. The biggest drawback to sea delivery is the difficulty of getting the reception committee and the supply vessel or submarine to the same place at the same time. One method may be to provide waterproof containers and cache them underwater for later retrieval. Containers may be equipped with self-surfacing floats and recovered by seamen working for the underground under the guise of fishermen.

M-52. Movement of logistics from offshore mother ships to onshore locations or recipients requires intermediate transport vessels (figure M-12). In some cases where enemy surveillance and coastal defenses are lacking, it may be possible to offload cargo directly from motherships to an unguarded shore. More often, the intermediate vessels will have to be discrete—either difficult to detect, such as submarines and semi-submersibles, or unobtrusive vessels, such as civilian trawlers or pleasure craft that can conceal their smuggled cargo even under direct threat observation.

**Figure M-12. Intermediate transport vessels**

## Preparation and Storage of Logistics

M-53 Equipment, not for immediate use, must be carefully preserved, packed, and stored as soon as possible to prevent its discovery and loss. To prevent crippling losses if a cache is discovered, like equipment items are stored throughout disbursed cache sites. The underground sometimes establishes small caches of equipment that can be betrayed by captured workers who feel that they can no longer withhold information from their interrogators.

M-54 Storage of supplies in urban areas may involve the use of homes where fewer persons will be subject to capture in the event of compromise.

M-55 Sites, such as cellars of factories, underground passages, sewers, derelict graves, mausoleums, abandoned structures, gardens, and even church steeples may be considered for use. Supplies hidden in remote rural areas ensure survival in the event of massive compromise of urban caches. Possible rural cache sites may include wells, caves, hollow trees, cisterns, unused culverts, underwater, to include sites along the banks below the water level of lakes, streams, or rivers.

# NONSTANDARD LOGISTICS

M-56. SF Soldiers will—to the maximum extent possible—sustain themselves on the local economy. This has two primary benefits. It simultaneously reduces the footprint of the operation, as well as lowers the force protection profile of the deployed forces. SF will continue to use the existing theater logistics systems, when appropriate, and nonstandard logistics, when necessary. Nonstandard logistics is composed of, or informed by, the following related disciplines and functions

## DEFINITION

M-57. The USSOCOM Directorate of Force Management and Development, Deputy Chief of Doctrine and Lessons Learned Division, is the proponent for nonstandard logistics. USSOCOM Publication 4-0, *Logistics Support to Special Operations Forces*, 6 February 2013—is USSOCOM's keystone publication for logistics. It contains the following draft definition for nonstandard logistics.

> "Nonstandard logistics (NSL) is the overt, covert clandestine and low-visibility provision of sustainment support, resources, supplies, and/or equipment to U.S. or foreign personnel across a range of missions, particularly in denied areas. NSL covers logistics support to unconventional warfare (UW) as well as to other special operations mission areas and activities. UW is distinctive for the challenge it presents to the NSL planner to provide clandestine logistics support on a potentially large scale. The scale of logistics requirements for a UW operation may outstrip organic SOF logistics capabilities, creating the need to access additional sources of support.
>
> The possible scale and diversity of support requirements, as well as the need to operate undetected in less than permissive environments, fundamentally separates NSL from conventional logistics. While conventional logistics operate in optimized supply chains, NSL transpire over multidimensional and multimodal NSL networks. For the nonstandard logistician, consideration of logistics support as an activity conducted by a network is fundamental, dictating the types, scale, and arrangement of material, as well as the collective and individual capabilities required to appropriately support UW. Establishment of effective NSL capability will likely require planning and execution of activities conducted prior to and distinct from doctrinal Phase 1 UW activities, including placement, infrastructure development, and pattern development. NSL networks are established to enable flows that generally cannot utilize conventional or overt means due to traceability, customs restrictions, or other compounding issues."
>
> USSOCOM Publication 4-0, *Logistics Support to Special Operations Forces*, 6 February 2013

M-58. Nonstandard logistics draws upon many different systems, stocks, functions, and offices. A list of nonstandard logistics support considerations across the DOD, other government agencies and departments, and commercial and multinational partners includes:

- Special mission funding authorities.
- Classified contracting.
- Acquisition capabilities.
- Chain of custody requirements.
- Attribution.
- SIGREDUX.
- Interagency operations.
- Embassy operations.
- Transportation and distribution systems.
- Concept of support and planning methods.

## CHENG'S NONSTANDARD LOGISTICS METHOD FOR SPECIAL OPERATIONS FORCES LOGISTICIANS

M-59. Colonel Steve Cheng was the logistician for Task Force Dagger involved in logistics support of the UW operation into Afghanistan in 2001. He outlined a method for SOF logisticians to consider nonstandard logistics problem solving (figure M-13, page M-21).

Below is one way for a SOF logistician to navigate the regulatory ambiguities and successfully support the mission:

1. Determine the mission requirement.
2. Assess the available support capabilities.
3. Determine whether those capabilities meet mission requirements and are appropriate to the mission's OPSEC parameters.
4. If the answer to 3 is "No", then a nonstandard logistics solution could be appropriate. The question, then, is what degree of nonstandard solution is necessary?
   a. Can a government-sponsored commercial solution work?
   b. Can an overt interagency support method work?
   c. Can a low-visibility, locally procured method work?
   d. Is there an indigenous network available to support the operation?
   e. Is there an ally or partner with the appropriate access and capability who is cleared and willing to provide support through a mutual support agreement?
   f. Is a higher degree of secrecy required than any of the options above?
5. What special authorities does the organization have that might allow appropriate support to the mission?
6. Are the requisite resources and appropriate funding sources available to meet the requirement?
7. If no special authority exists that meets mission requirements, what regulatory, policy, or statutory provisions are impeding mission success?
8. Who has the authority to either waive or approve a deviation from that policy?
9. Is the commander willing to request that approval in order to accomplish the mission?
10. Is the commander willing to accept the risks associated?
11. What accountability measures are in place to ensure proper oversight of nonstandard methods of support?
12. What policy changes might be appropriate to support future operations of this nature?

Of the steps outlined above, items 9 and 10 are critical. Staff officers do not make decisions of this magnitude. Commanders own the risk and therefore make the decisions. The fog of war is never an excuse to usurp that authority, particularly when seeking to execute in a way that is not in compliance with established regulation or higher policy. Items 11 and 12 are equally important. Surviving the scrutiny that follows a support operation that is not in compliance with established regulations or policies requires that:

(1) The unit can clearly demonstrate that the course of action was vital to mission success, and

(2) All efforts were made to properly document the approval process and maintain accountability of U.S. government funds and property.

Finally, a thorough, thoughtful after action review can help policy makers anticipate future mission requirements and implement new policies and/or authorities that will accelerate the speed of support for future operations.

COL Steve Cheng, *Navigating Ambiguity*, 2016

**Figure M-13. Cheng's special operations forces logistician's model for nonstandard logistics**

This page intentionally left blank.

# Appendix N

# Caching

Caching is the process of hiding equipment or materials in a secure storage place with the view to future recovery for operational use. The ultimate success of caching may well depend upon attention to detail—professional competence that may seem of minor importance to the untrained eye. Security factors, such as cover for the caching party, sterility of the items cached, and removal of even the slightest trace of the caching operations are vital. Highly important, too, are the technical factors that govern the preservation of the items in usable condition and the recording of data essential for recovery. Successful caching entails careful adherence to the basic principles of clandestine operations, as well as familiarity with the technicalities of caching.

## CACHING CONSIDERATIONS

N-1    Caching requirements—vital to the success of the caching operation—should be considered for a variety of operational situations (figure N-1, page N-2). For example, cached supplies can meet the emergency needs of personnel who may be barred from their normal supply sources by sudden developments or who may need travel documents and extra funds for a quick escape. Caching can help solve the supply problems of long-term operations conducted far from a secure base. Caching can also provide for anticipated needs of wartime operations in areas likely to be overrun by the enemy.

## PLANNING FOR A CACHING OPERATION

N-2.    Caching involves selecting items to be cached, procuring those items, and selecting a cache site. Selection of the items to be cached requires a close estimate of what will be needed by particular units for particular operations.

N-3.    Procurement of the items usually presents no special problems. In fact, the relative ease of procurement before an emergency arises is one of the prime considerations in favor of caching. When selecting a cache site, planners should always ensure that the site is accessible not only for emplacement but also for recovery. When planning a caching operation, the planner must consider the following six basic factors.

### PURPOSE AND CONTENTS OF THE CACHE

N-4.    Planners must determine the purpose and contents of each cache because these basic factors influence the location of the cache, as well as the method of hiding. For instance, small barter items can be cached at any accessible and secure site because they can be concealed easily on the person immediately after recovery from the cache. However, it would be difficult to conceal rifles for a guerrilla band immediately after recovery. Therefore, this site must be in an isolated area where the band can establish at least temporary control. Certain items, such as medical stock, have limited shelf life and require rotation periodically or special storage considerations, necessitating easy access to service these items. Sometimes it is impossible to locate a cache in the most convenient place for an intended user. Planners must compromise between logistics objectives and actual possibilities when selecting a cache site. Security is always the overriding consideration.

### ANTICIPATED ENEMY ACTION

N-5.    In planning the caching operation, planners must consider the capabilities of any intelligence or security services not participating in the operation. They should also consider the potential hazards the enemy and its witting or unwitting accomplices present. If caching is done for wartime operational purposes, its ultimate success will depend largely on whether the planners anticipate the various obstacles to recovery, which the enemy and its accomplices will create if the enemy occupies the area. What are the possibilities that the enemy will preempt an ideal site for one reason or another and deny access to it? A vacant field surrounded by brush may seem ideal for a particular cache because it is near several highways, but such a location may also invite the enemy to locate an ordnance depot where the cache is buried.

**Purpose**

Team E&E
Underground and guerrilla training
Equipment and operations

**Cache Contents**

- Money
- Weapons
- Ammunition
- Explosive ordnance
- Medical supplies
- Tools

- Radios
- Food
- Water
- Batteries
- Clothing
- POL

**Logistics**

1x weapon each man (12) per team + additional gear
= 500 + lb x 8 teams = over 4,000 lb
+ 1x weapon and gear each cell member = ???? lb

- Need for covert movement
- Large amount of material
- Many caches
- How to transport
- Lots of digging

- Security
- Availability of desirable cache placements
- Cache locations over large area
- Placement record maintenance
- Additional caches for contingencies

*Legend:*

E&E    escape and evasion
lb      pound
POL    petroleum, oils, and lubricants

**Figure N-1. Cache considerations**

## ACTIVITIES OF THE LOCAL POPULATION

N-6.  Deliberate, successful enemy action to find caches is probably less likely than inadvertent discovery by the populace. Normal activity, such as construction of a new building, may uncover the cache site or impede access to it. Unlucky circumstances cannot be anticipated, but it can probably be avoided by careful and imaginative observation of the prospective cache site and of the people who live near the site. If the cache is intended for wartime use, the planners must project how the residents will react to the pressures of war and conquest. For example, one of the more likely reactions is that many residents may resort to caching in order to avoid having their personal funds and valuables seized by the enemy. If caching becomes popular, any likely cache site will receive more than normal attention.

## INTENDED ACTIONS BY ALLIED FORCES

N-7. Using one cache site for several clandestine operations involves a risk of mutual compromise. Therefore, some planners should rule out otherwise suitable caching sites if they have been selected for other clandestine purposes, such as drops or safe houses. A site should not be located where it may be destroyed or rendered inaccessible by bombing or other allied military action, should the area be occupied by the enemy. For example, installations likely to be objects of special protective efforts by the occupying enemy are certain to be inaccessible to the ordinary citizen. Therefore, if the cache is intended for wartime use, the caching party should avoid areas, such as those near key bridges, railroad intersections, power plants, and munitions factories.

## PACKAGING AND TRANSPORTATION

N-8. Asset planners should assess the security needs and all of the potential obstacles and hazards that a prospective cache site can present. They should also consider whether the operational assets could be used for packaging and transporting the package to the site. The best results are attained when experts at a packaging center do the packaging. The first question, therefore, is to decide whether the package can be transported from the HQ or the field packaging center to the cache site securely and soon enough to meet the operational schedules. If not, the packaging must be done locally, perhaps in a safe house located within a few miles of the cache site. If such an arrangement is necessary, the choice of cache sites may be restricted by limited safe house possibilities.

## PERSONNEL ASSETS

N-9. All who participate directly in emplacement will know where the cache is located; therefore, it is best to use the fewest number of reliable personnel possible. Planners must consider the distance from the person's residence to the prospective cache site. Consideration must be given to the reason or story of why someone is involved in conducting this activity. Sometimes transportation and cover difficulties require the cache site to be located within a limited distance of the person's residence. The above considerations also apply to the recovery personnel.

# CACHING METHODS

N-10. Deciding which cache method to use depends on the situation. It is therefore unsound to lay down any general rules, with one exception—planners should always think in terms of suitability. For example, planners should determine the method most suitable for each cache, considering its specific purpose; the actual situation in the particular locality; and the changes that may occur if the enemy gains control.

## CONCEALMENT

N-11. Concealment requires the use of permanent man-made or natural features to hide or disguise the cache. Concealment has several advantages. Both employment and recovery can usually be done with minimum time and labor, and cached items concealed inside a building or dry cave are protected from the elements. Thus, they require less elaborate packaging. In some cases, a concealed cache can be readily inspected from time to time to ensure it is still usable. However, there is always the chance of accidental discovery in addition to all the hazards of wartime that may result in discovery or destruction of a concealed cache or denial of access to the site. The concealment method, therefore, is most suitable in cases where an exceptionally secure site is available or where a need for quick access to the cache justifies a calculated sacrifice in security. Concealment may range from securing small, gold coins under a tile in the floor to walling up artillery in caves.

## BURIAL

N-12. Adequate burial sites can be found almost anywhere (figures N-2 and N-3). Once in place, a properly buried cache is generally the best way of achieving lasting security. In contrast to concealment, however, burial in the ground is a laborious and time-consuming method of caching. The disadvantages of burial are that:

- Burial almost always requires a high-quality container or special wrapping to protect the cache from moisture, chemicals, and bacteria in the soil.
- Emplacement or recovery of a buried cache usually takes so long that the operation must be conducted after dark unless the site is exceptionally secluded.
- It is especially difficult to identify and locate a buried cache (figures N-2 and N-3).

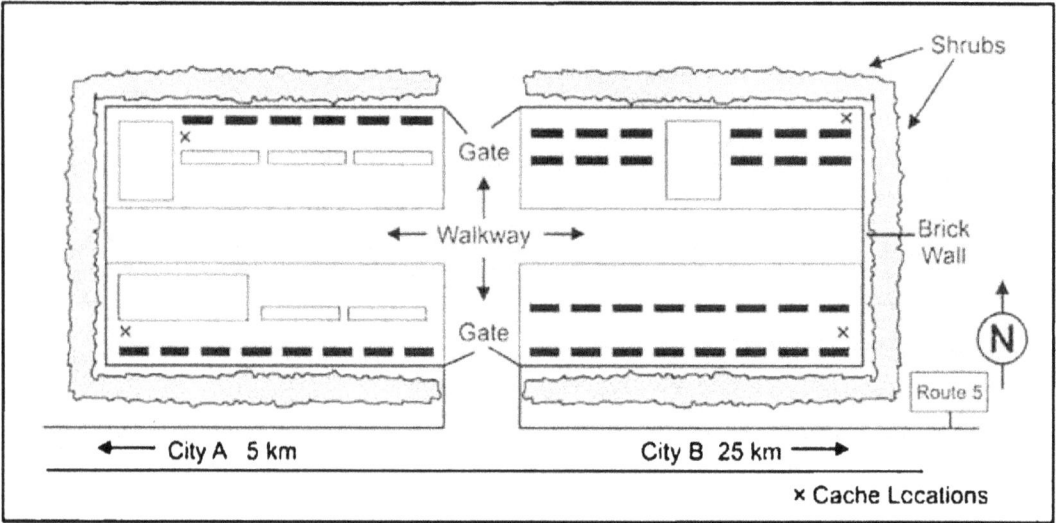

Figure N-2. Cache locations

Figure N-3. Cache location (adjacent to southwest corner of church on south side)

## SUBMERSION

N-13. Submersion sites that are suitable for secure concealment of a submerged cache are few and far between. The container of a submerged cache must meet such high standards for waterproofing and resistance to external pressure that the use of field expedients is seldom workable. To ensure that a submerged cache remains dry and in place, planners must determine not only the depth of the water but also the type of bottom, the currents, and other facts, that are relatively difficult for nonspecialists to obtain (figures N-4 and N-5).

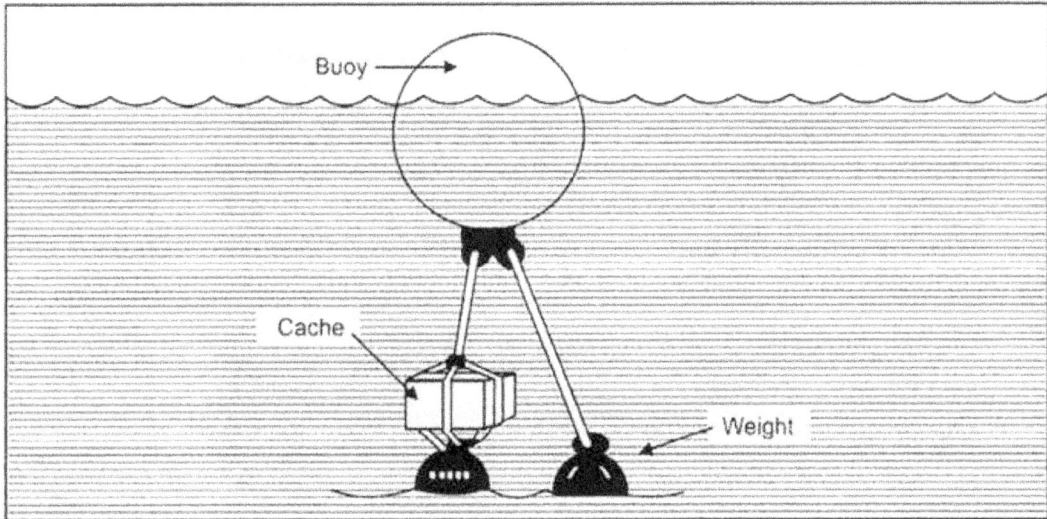

**Figure N-4. Submersible cache (example 1)**

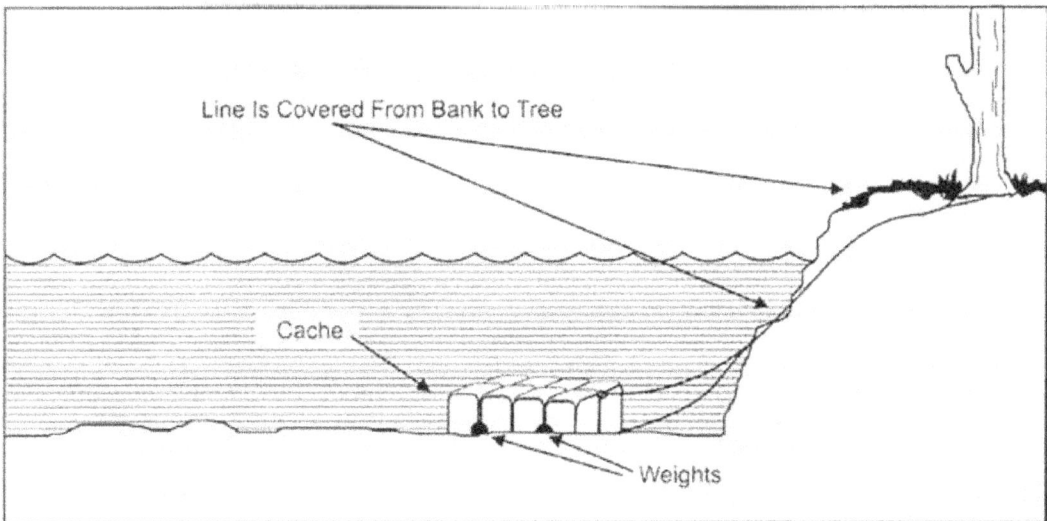

**Figure N-5. Submersible cache (example 2)**

## EMPLACEMENT

N-14. Emplacement, likewise, requires a high degree of skill. At least two persons are needed for emplacement, and it requires additional equipment. In view of the difficulties—especially the difficulty of recovery—the submersion method is suitable only on rare occasions. The most noteworthy usage is the relatively rare maritime resupply operation in which it is impossible to deliver supplies directly to a reception committee. Caching supplies offshore by submersion is often preferable to sending a landing party ashore to bury a cache.

# SELECTION OF THE SITE

N-15. The most careful estimates of future operational conditions cannot ensure that a cache will be accessible when it is needed. The following paragraphs address site selection considerations.

N-16. Criteria for a site selection can be met when the following three questions are answered: (1) Can the site be located by simple instructions that are unmistakably clear to someone who has never visited the location? A site may be ideal in every respect, but if it has no distinct, permanent landmarks within a readily measurable distance it must be ruled out. (2) Are there at least two secure routes to and from the site? Both primary and alternate routes should provide natural concealment so that the emplacement party and the recovery party can visit the site without being seen by anyone normally in the vicinity. (3) Can the cache be emplaced and recovered at the chosen site in all seasons of the year? Snow and frozen ground create special problems. Snow on the ground is a hazard because it is impossible to erase a trail in the snow. Planners must consider whether seasonal changes in the foliage will leave the site and the route dangerously exposed.

## MAP SURVEY

N-17. Finding a cache site is often difficult. Usually, a thorough systematic survey of the general area designated for the cache is required. The survey is best done with a large-scale map of the area, if available. By scrutinizing the map, the planners can determine whether a particular sector must be ruled out because of its nearness to factories, homes, busy thoroughfares, or probable military targets in wartime. A good military-type map will show the positive features in the topography, proximity to adequate roads or trails, natural concealment (for example, surrounding woods or groves), and adequate drainage. A map will also show the natural and man-made features in the landscape. A map will provide the indispensable reference points for locating a cache site, such as confluences of streams, dams and waterfalls, road junctures and distance markers, villages, bridges, churches, and cemeteries.

## PERSONAL RECONNAISSANCE

N-18. A map survey should normally show the location of several promising sites within the general area designated for the cache. To select and pinpoint the best site, however, a well-qualified observer must examine each site firsthand. If possible, whoever examines the site should carry adequate maps, a compass, a drawing pad or board for making sketch maps or tracings, and a metallic measuring line. (A wire knotted at regular intervals is adequate for measuring. Twine or cloth measuring tapes should not be used because stretching or shrinking will make them inaccurate if they get wet.) The observer should also carry a probe rod (if the rod can be carried securely) for probing prospective burial sites.

N-19. Since the observer seldom completes a field survey without being noticed by local residents, his story for his actions is of great importance. The observer's story must offer a natural explanation for his exploratory activity in the area. Ordinarily, this means that an observer who is not a known resident of the area can pose as a tourist or a newcomer with some reason for visiting the area. However, his story must be developed over an extended period before he undertakes the actual reconnaissance. If the observer is a known resident of the area, he cannot suddenly take up hunting, fishing, or wildlife photography without arousing interest and perhaps suspicion. The observer must build up a reputation for being a devotee of his sport or hobby.

# REFERENCE POINTS

N-20. When the observer finds a suitable cache site, he prepares simple and unmistakable instructions for locating the reference points (figure N-6, page N-7). These instructions must identify the general area (the names of general recognizable places, from the country down to the nearest village) and an immediate reference point. Any durable landmark that is identified by its title or simple description can be an immediate reference point (for example, the only Roman Catholic church in a certain village or the only bridge on a named road between two villages).

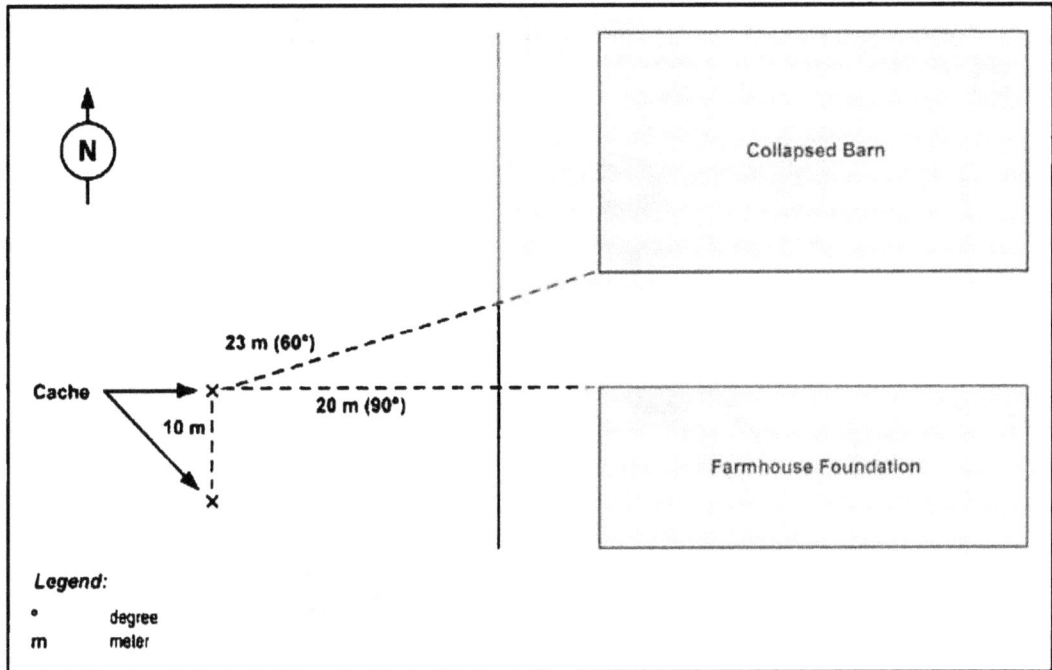

**Figure N-6. Reference points**

N-21. The instructions must also include a final reference point (FRP), which must meet the following four requirements. The FRP—

- Must be identifiable, including at least one feature that can be used as a precise reference point.
- Must be an object that will remain fixed as long as the cache may be used.
- Must be near enough to the cache to pinpoint the exact location of the cache by precise linear measurements from the FRP to the cache.
- Should be related to the immediate reference point by a simple route description, which proceeds from the immediate reference point to the FRP.

N-22. Since the route description should be reduced to the minimum essential, the ideal solution for locating the cache is to combine the immediate reference point and the FRP into one landmark—readily identifiable but sufficiently secluded.

N-23. The following objects, when available, are sometimes ideal reference points:

- Small, unfrequented bridges and dams.
- Boundary markers.
- Kilometer markers and culverts along unfrequented roads.
- A geodetic survey marker.
- Battle monuments and wayside shrines.

N-24. When such reference points are not available at an otherwise suitable cache site, natural or man-made objects may serve as FRPs, such as distinct rocks, posts for power or telephone lines, intersections in stone fences or hedgerows, and gravestones in isolated cemeteries.

## PINPOINTING TECHNIQUES

N-25. Recovery instructions must identify the exact location of the cache. These instructions must describe the point where the cache is placed in terms that relate it to the FRP. When the concealment method is used, the cache is ordinarily placed inside the FRP and pinpointed by a precise description of the FRP. A submerged cache is usually pinpointed by describing exactly how the moorings are attached to the FRP (figure N-7). With a buried cache, any of the following techniques may be used.

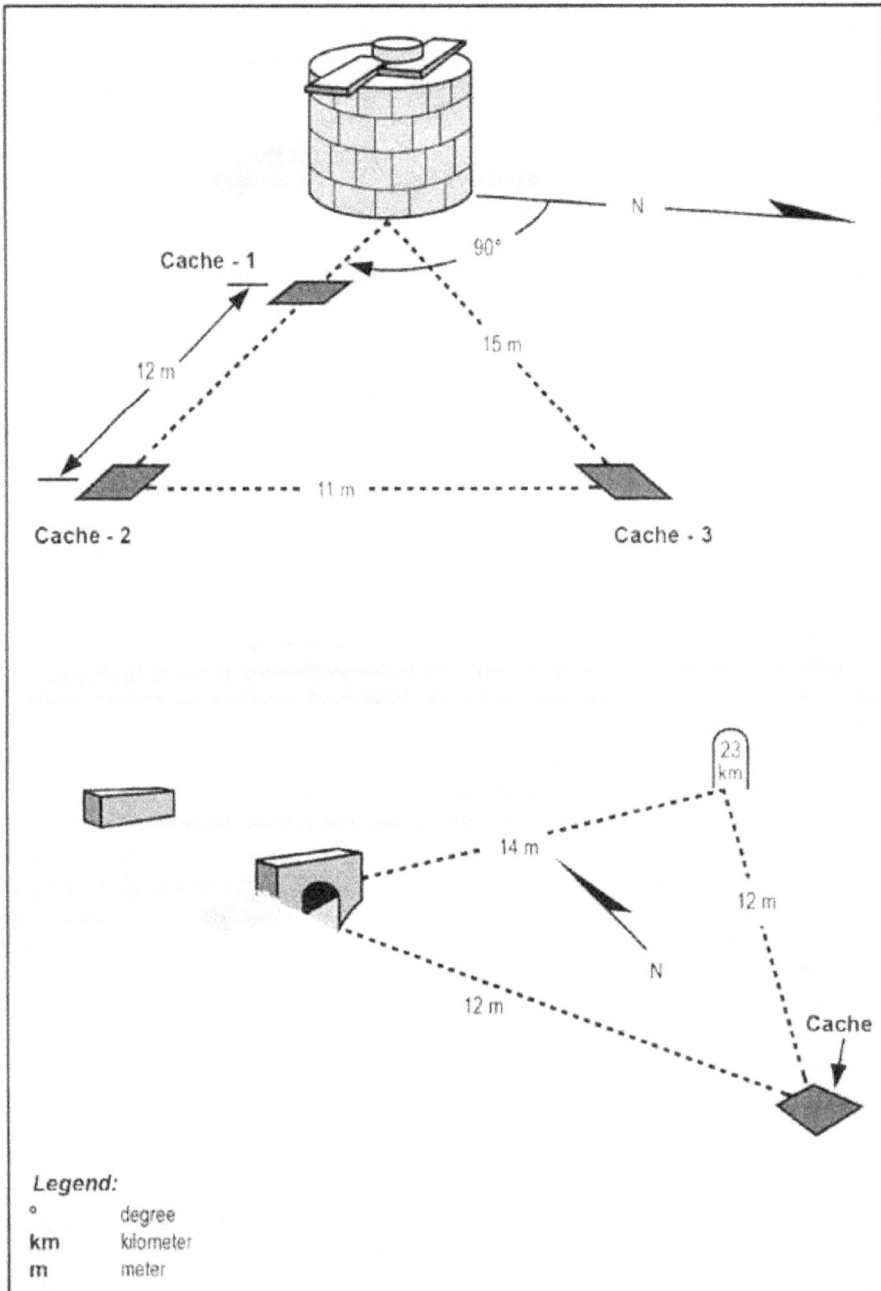

**Figure N-7. Pinpointing techniques**

### Placing the Cache Directly Beside the Final Reference Point

N-26. The simplest method is to place the cache directly beside the FRP. Then pinpointing is reduced to specifying the precise reference point of the FRP.

### Sighting the Cache by Projection

N-27. This method may be used if the FRP has one flat side long enough to permit precise sighting by projecting a line along the side of the object. The burial party places the cache a measured distance along the sighted line.

N-28. This method may also be used if two precise FRPs are available, by projecting a line sighted between the two objects. In either case, the instructions for finding the cache must state the approximate direction of the cache from the FRP. Since small errors in sighting are magnified as the sighted line is extended, the cache should be placed as close to the FRP as other factors permit. Ordinarily, this method becomes unreliable if the sighted line is extended beyond 50 meters.

### Placing the Cache at the Intersection of Measured Lines

N-29. If two FRPs are available within several paces, the cache can be one line projected from each of the FRPs. If this method is used, the approximate direction of the cache from each FRP must be stated. To ensure accuracy, neither of the projected lines (from the FRPs to the point of emplacement) should be more than twice as long as the base line (between the two FRPs). If this proportion is maintained, the only limitation upon the length of the projected lines is the length of the measuring line that the recovery party is expected to carry. The recovery party should carry two measuring lines when this method is used.

### Sighting the Cache by Compass Azimuth

N-30. If the above methods of sighting are not feasible, one measured line may be projected by taking a compass azimuth from the FRP to the point where the cache is placed. To avoid confusion, an azimuth to a cardinal point of the compass (north, east, south, or west) is used. Since compass sightings are likely to be inaccurate, a cache that is pinpointed by this method should not be placed more than 10 meters from the FRP.

## MEASURING DISTANCES

N-31. The observer should express all measured distances in a linear system that the recovery party is sure to understand—ordinarily, the standard system for the country where the cache is located. He should use whole numbers (6 meters, not 6.3 or 6.5) to keep his instructions as brief and as simple as possible. To get an exact location for the cache in whole numbers, the observer should take sightings and measurements first.

N-32. If the surface of the ground between the points to be measured is uneven, the linear distance should be measured on a direct line from point to point, rather than by following the contour of the ground. This method requires a measuring line long enough to reach the full distance, from point to point, and strong enough to be pulled taut without breaking.

## MARKING TECHNIQUES

N-33. The emplacement operation can be simplified, and critical time can be saved if the point where the cache is to be buried is marked during the reconnaissance. If a night burial is planned, the point of emplacement may have to be marked during a daylight reconnaissance. This method should be used whenever operational conditions permit. The marker must be an object that is easily recognizable but that is meaningless to an unwitting observer. For example, a small rock or a branch with its butt placed at the point selected for the emplacement may be used.

N-34. Since marking information is also essential to the recovery operation, it must be compiled after emplacement and included in the final cache report. Therefore, the observer should be thoroughly familiar with the Twelve-Point Cache Report before he starts a personal reconnaissance. This report is a checklist for the observer to record as much information as possible. Points 6 through 11 are particularly important. The personal reconnaissance also provides an excellent opportunity for a preliminary estimate of the time required for getting to the site.

## ALTERNATE SITE

N-35. As a general rule, planners should select an alternate site in case unforeseen difficulties prevent use of the best site. Unless the primary site is in a completely deserted area, there is always some danger that the emplacement party will find it occupied as they approach or that the party will be observed as they near the site. The alternate site should be far enough away to be screened from view from the primary site, but it should be near enough so that the party can reach it without making a second trip.

## CONCEALMENT SITE

N-36. A site that looks ideal for concealment may be revealed to the enemy for that very reason. Such a site may be equally attractive to a native of an occupied country to hide his valuables. The only real key to the ideal concealment site is careful casing of the area, combined with great familiarity with local residents and their customs. Some of the likely concealment sites may include:

- Walls (hidden behind loose bricks or stones or a plastered surface).
- Abandoned buildings.
- Infrequently used structures (stadiums and other recreational facilities, railroad facilities on spur lines).
- Memorial edifices (mausoleums, crypts, monuments).
- Public buildings (museums, churches, libraries).
- Ruins of historical interest.
- Culverts.
- Natural caves and caverns and abandoned mines and quarries.
- Sewers.
- Cable conduits.

N-37. The concealment site must be equally accessible to the person emplacing and to the person recovering. However, visits by both persons to certain interior sites may be incompatible with the story. For instance, a site in a house owned by a relative of the emplacer may be unsuitable because there is no adequate excuse for the recovery person to enter the house if he has no connection with the owner.

N-38. The site must remain accessible as long as the cache is needed. If access to a building depends upon a personal relationship with the owner, the death of the owner or the sale of the property might render it inaccessible.

N-39. Persons involved in the operation should not be compromised if the cache is discovered on the site. Even if a cache is completely sterile, as every cache should be, the mere fact that it has been placed in a particular site may compromise certain persons. If the police discovered the cache, they might suspect the emplacer because it was found in his relative's house.

N-40. The site must not be located where potentially hostile persons frequently visit. For instance, a site in a museum is not secure if police guards or curious visitors frequently enter the museum.

N-41. To preserve the cache material, the emplacer must ensure the site is physically secure for the preservation of the cached material. For example, most buildings involve a risk that the cache may be destroyed or damaged by fire, especially in wartime. The emplacer should consider all risks and weigh them against the advantages of an interior site.

N-42. A custodian may serve to ease access to a building or to guard a cache. However, the use of such a person is inadvisable, as a custodian poses an additional security risk. He may use the contents of the cache for personal profit or reveal its location.

## BURIAL SITE

N-43. In selecting a burial site, consider the following factors along with the basic considerations of suitability and accessibility.

## DRAINAGE

N-44. Drainage considerations include the elevation of the site and the type of soil. The importance of good drainage makes a site on high ground preferable unless other factors rule it out. Moisture is one of the greatest natural threats to the contents of a cache. Swamp muck is the most difficult soil to work in. If the site is near a stream or river, the emplacer should ensure that the cache is well above the all-year-high water mark so that the cache will not be uncovered if the soil is washed away.

## GROUND COVER

N-45. The types of vegetation at the site will influence the choice. Roots of deciduous trees make digging very difficult. Coniferous trees have less extensive root systems. The presence of coniferous trees usually means that the site is well drained. Does the vegetation show paths or other indications that the site is frequented too much for secure caching? Can the ground cover be easily restored to its normal appearance when burial is completed? Tall grass reveals that it has been trampled, but an overlay of leaves and humus can be replaced easily and will effectively conceal a freshly refilled hole.

## NATURAL CONCEALMENT

N-46. The vegetation or the surrounding terrain should offer natural concealment for the burial and recovery parties working at the site. Planners should carefully consider seasonal variations in the foliage.

## TYPES OF SOIL

N-47. Sandy loam is ideal because it is easy to dig and drains well. Clay soil should be avoided because it becomes quite sticky in wet weather, and in dry weather, it may become so hard that it is almost impossible to dig.

## SNOWFALL AND FREEZING

N-48. If the cache must be buried or recovered in winter, data on the normal snowfall, the depth to which the ground freezes in winter, and the usual dates of freezing and thawing will influence the choice of the site. Frozen ground impedes digging and requires additional time for burial and recovery. Snow on the ground is especially hazardous for the burial operation. It is practically impossible to restore the snow over the burial site to its normal appearance unless there is more snowfall or a brisk wind. It is very difficult to ensure that no traces of the operation are left after the snow has melted.

## ROCKS AND OTHER SUBSURFACE OBSTRUCTIONS

N-49. Large obstructions that might prevent use of a particular site can be located to some extent before digging by probing with a rod or stake at the exact spot selected for the cache.

# SUBMERSION SITE

N-50. To be suitable for a submerged cache, a body of water must have certain characteristics. The presence of these characteristics can be determined only by a thorough survey of the site. Their importance will be understood after familiarization with the technicalities of submersion.

N-51. Submersion usually requires a boat—first for reconnoitering, then for emplacement. Thus, the accessibility problems involved in submersion usually narrow down to the availability of a boat and the story for using it. If there is no fishing or pleasure boating at the site, the story for this peculiar type boating may be a real problem. In tropical areas, the course of streams or rivers is frequently changed by seasonal rainfall and can cause many problems. Planners should keep this fact in mind when choosing the site and when selecting reference points.

# CACHE RECOVERY

N-52. Since the method for recovering a cache is generally similar to that for emplacing a cache, it need not be described in full. However, several important considerations should be stressed in training for a recovery operation. Anyone who is expected to serve as a recovery person should have the experience of actually recovering dummy caches, if field exercises can be arranged securely. It is especially desirable for the recovery person to be able to master the pinpointing techniques. Mastery is best attained by practice in selecting points of emplacement and in drafting, as well as in following instructions.

## EQUIPMENT

N-53. Although the equipment used in recovery is generally the same as that used in emplacement, it is important to include any additional items that may be required in recovery in the cache report. A probe rod may not be essential for emplacement, but it is necessary to have some object roughly the same size as the cache container to fill the cavity left in the ground by removal of a buried cache. Some sort of container of wrapping material may be needed to conceal the recovered cache, while it is being carried from the cache site to a safe house. Recovery of a submerged cache may require grappling lines and hooks, especially if it is heavy.

## SKETCH OF THE SITE

N-54. If possible, the observer should provide the recovery person with sketches of the cache site and the route to the cache site. If the recovery person must rely exclusively on verbal instructions, as in the case when communications are limited to radio-telephone (RT) messages, he should draw a sketch of the site before starting on the recovery operation. He should use all the data in the verbal instructions to make the sketch as realistic as possible. Drawing a sketch will help to clarify any misunderstanding of the instructions. A sketch can be followed more easily than verbal instructions. It may also be helpful for the recovery person to draw a sketch of the route from the immediate reference point to the site. However, he should not carry this sketch on him because if he were apprehended, the sketch might direct the enemy to the cache.

## PRELIMINARY RECONNAISSANCE

N-55. Checking the instructions for locating the cache may be advisable, especially when the recovery operation must be performed under stringent enemy controls or when there is no extra time for searching. Careful analysis of the best available map can minimize reconnoitering activity in the vicinity of the cache and thus reduce the danger of arousing suspicion. If recovery must be conducted at night, the recovery person should find the cache by daylight and place an unnoticeable marker directly over it.

## PROBE ROD

N-56. The recovery person can avoid digging at the wrong spot by using a probe rod before starting to dig. He should push and turn the probe rod into the ground by hand, so that it will not puncture the cache's container. The recovery person should never pound the probe rod with a hammer. The recovery procedure is the same as for the burial, except—

- A pick should never be used for digging the hole because it might puncture the container and damage the cached items.
- It may be necessary to fill the hole with other objects in addition to soil after the cache is removed.

N-57. Sometimes it is possible to fill the hole with rocks, sticks, or other readily available objects at the site. If no such objects are found during the preliminary reconnaissance, the recovery person should carry an object roughly the same size as the cache container to the site.

## STERILIZATION OF THE SITE

N-58. As with emplacement, the recovery operation must be performed in such a way that no traces of the operation are left. Although sterilization is not as important for recovery as for emplacement, it should be done as thoroughly as time permits. Evidence that a cache has been recovered might alert the enemy to clandestine activity in the area and provoke countermeasures.

# PACKAGING

N-59. Packaging usually involves packing the items to be cached, as well as the additional processing in protecting these items from adverse storage conditions. Proper packaging is important because inadequate packaging very likely will render the items unusable. Since special equipment and skilled technicians are needed for best results, packaging should be done at HQ or a field packaging center whenever possible. However, to familiarize operational personnel with the fundamentals of packaging so that they can improvise field expedients for emergency use, this section discusses determining factors, steps in packaging, wrapping materials, and criteria for the container.

## DETERMINING FACTORS

N-60. The first rule of packaging is that all processing is tailored to fit the specific requirements of each cache. The method of packaging, as well as the size, shape, and weight of the package is determined by the items to be cached, by the method of caching, and especially, by the way the cache is recovered and used. For instance, if circumstances require one man to recover the cache by himself, the container should be no larger than a small suitcase, and the total weight of the container and contents should be no more than 30 pounds. Of course, these limits must be exceeded with some equipment, but the need for larger packages should be weighed against the difficulties and risks in handling them. Even if more than one person is available for recovery, the material should be divided, whenever possible, into separate packages of a size and weight readily portable by one man.

N-61. Another very important factor in packaging concerns adverse storage conditions. Any or all of the following conditions may be present: moisture, external pressure, freezing temperatures, and the bacteria and corrosive chemicals found in some soil and water. Animal life may present a hazard; insects and rodents may attack the package. If the cache is concealed in an exterior site, larger animals may also threaten it. Whether the packaging is adequate usually depends upon how carefully the conditions at the site were analyzed in designing the cache. Thus, the method of caching (burial, concealment, or submersion) should be determined before the packaging is done.

N-62. It is equally important to consider how long the cache is to be used. Since one seldom knows when a cache will be needed, a sound rule is to design the packaging to withstand adverse storage conditions for at least as long as the normal shelf life of the contents to be cached.

## STEPS IN PACKAGING

N-63. The exact procedure for packaging depends upon the specific requirements for the cache and upon the packaging equipment available. The following eight steps are almost always necessary in packaging.

### Inspecting

N-64. The items to be cached must be inspected immediately before packaging to ensure they are complete, in serviceable condition, and free of all corrosive or contaminated substances.

### Cleaning

N-65. All corrodible items must be cleaned thoroughly immediately before the final preservative coating is applied. All foreign matter, including any preservative applied before the item was shipped to the field, should be removed completely. Throughout the packaging operation, all contents of the cache should be handled with rubber or freshly cleaned cotton gloves. Special handling is important because even minute particles of human sweat will corrode metallic equipment. Any fingerprints on the contents of the cache may enable the enemy to identify those who packaged the item.

## Drying

N-66. When cleaning is completed, every trace of moisture must be removed from all corrodible items. Methods of drying include wiping with a highly absorbent cloth, heating, or applying desiccant. Usually, heating is best unless the item can be damaged by heat. To dry by heating, the item to be cached should be placed in an oven for at least 3 hours at a temperature of about 110 degrees Fahrenheit. An oven can be improvised from a large metal can or drum. In humid climates, it is especially important to dry the oven thoroughly before using it by preheating it to at least 212 degrees Fahrenheit. Then, the equipment to be cached is inserted as soon as the oven cools down to about 110 degrees Fahrenheit. If a desiccant is used, it should not touch any metallic surface. Silica gel is a satisfactory desiccant and it is commonly available.

## Coating With a Preservative

N-67. A light coat of oil may be applied to weapons, tools, and other items with unpainted metallic surfaces. A coat of paint may suffice for other metal items.

## Wrapping

N-68. When drying and coating are completed, the items to be cached are wrapped in a suitable material. The wrapping should be as nearly waterproof as possible. Each item should be wrapped separately, so that one perforation in the wrapping will not expose all items in the cache. The wrapping should fit tightly to each item to eliminate air pockets, and all folds should be sealed with a waterproof substance.

## Packing

N-69. The following rules must be observed when packing items in the container:

- All moisture must be removed from the interior of the container by heating or applying desiccant. A long-lasting desiccant should be packed inside the container to absorb any residual moisture. If silica gel is used, the required amount can be calculated by using the ratio of 15 kilograms of silica gel to 1 cubic meter of storage space within the container. (This ratio is based on two assumptions: the container is completely moisture proof and the contents are slightly moist when inserted.) Therefore, the ratio allows an ample margin for incomplete drying and can be reduced if the drying process is known to be highly effective.
- Air pockets should be eliminated as much as possible by tight packing. Thoroughly dried padding should be used liberally to fill air pockets and to protect the contents from shock. Clothing and other items, which will be useful to the recovery party, should be used for padding if possible. Items made of different metals should never touch, since continued contact may cause corrosion through electrolytic action.

## Enclosing Instructions for Using Cached Equipment

N-70. Written instructions and diagrams should be included if they facilitate assembly or use of the cached items. Instructions must be written in a language that recovery personnel can understand. The wording should be as simple as possible and unmistakably clear. Diagrams should be self-explanatory since the eventual user may not be able to comprehend written instructions because of language barriers.

## Sealing and Testing Seals by Submersion

N-71. When packing is completed, the lid of the container must be sealed to make it watertight. Testing can be conducted by entirely submerging the container in water and watching for escaping air bubbles. Hot water should be used if possible because hot water will bring out leaks that would not be revealed by a cold-water test.

## WRAPPING MATERIALS

N-72. The most important requirement for wrapping material is that it be moisture-proof. It should be self-sealing or adhesive to a sealing material; it should be pliable enough to fit closely, with tight folds; and it should be tough enough to resist tearing and puncturing. Pliability and toughness may be combined by using two wrappings: an inner one that is thin and pliable and an outer one of heavier material. A tough outer wrapping is essential unless the container and the padding are adequate to prevent items from scraping together inside the cache. Five wrapping materials are recommended for field expedient use because they often can be obtained locally and used effectively by unskilled personnel.

### Aluminum Foil for Use as an Inner Wrapping

N-73. Aluminum foil is the best of the widely available materials. It is moisture-proof as long as it does not become perforated and provided the folds are adequately sealed. The drawbacks to its use for caching are that the thin foils perforate easily, while the heavy ones (over 2 millimeters thick) tend to admit moisture through the folds. The heavy-duty grade of aluminum foil generally sold for kitchen use is adequate when used with an outer wrapping. Scrim-backed foil, which is heat-sealable, is widely used commercially to package articles for shipment or storage. Portable heat sealers that are easy to use are available commercially or sealing can be done with a standard household iron.

### Moisture-Resistant Papers

N-74. Several brands of commercial wrapping papers are resistant to water and grease. They do not provide lasting protection against moisture when used alone, but they are effective as an inner wrapping to prevent rubber, wax, and similar substances from sticking to the items in the cache.

### Rubber Repair Gum

N-75. Rubber repair gum is a self-sealing compound generally used for repairing tires; it makes an excellent outer wrapping. Standard commercial brands come in several thicknesses; 2 millimeters is the most satisfactory for caching. A watertight seal is produced easily by placing two rubber surfaces together and applying pressure manually. The seal should be at least 1/2 inch wide. Since rubber repair gum has a tendency to adhere to items, an inner wrapping of nonadhesive material must be used with it, and the backing should be left on the rubber material to keep it from sticking to other items in the cache.

### Grade C Barrier Material

N-76. Grade C barrier material is a cloth impregnated with microcrystalline wax that is used extensively when packing for storage of items for overseas shipment. Thus, it is generally available and it has the additional advantage of being self-sealing. Although it is not as effective as rubber repair gum, it may be used as an outer wrapping over aluminum foil to prevent perforation of the foil. Used without an inner wrapping, three layers of grade C barrier material may keep the contents dry for as long as three months, but it is highly vulnerable to insects and rodents. The wax wrapping has a low melting point, and it will adhere to many items, so it should not be used without an inner wrapping except in emergencies.

### Wax Coating

N-77. If no wrapping material is available, an outer coating of microcrystalline wax, paraffin, or a similar waxy substance can be used to protect the contents against moisture. However, a wax coating will not provide protection against insects and rodents. The package should be hot-dipped in the waxy substance, or the wax can be heated to molten form and applied with a brush.

## CONTAINER CRITERIA

N-78. The outer container serves to protect the contents from shock, moisture, and other natural hazards to which the cache may be exposed. The ideal container should be—
* Completely watertight and airtight after sealing.
* Noiseless when handled; its handles should not rattle against the body of the container.
* Resistant to shock and abrasion.
* Able to withstand crushing pressure.
* Lightweight in construction.
* Able to withstand rodents, insects, and bacteria.

- Equipped with a sealing device that can be closed and reopened easily and repeatedly.
- Capable of withstanding highly acidic or alkaline soil or water.

## STANDARD STAINLESS STEEL CONTAINER

N-79. The standard stainless steel container comes in several sizes. Since the stainless steel container is more satisfactory than any that could be improvised in the field, it should be used whenever possible. Ideally, it should be packed at HQ or at a field packaging center. If the items to be cached must be obtained locally, it is still advisable to use the stainless steel container because its high resistance to moisture eliminates the need for an outer wrapping. Packers should, however, use a single wrapping even with the stainless steel container to protect the contents from any residual moisture that may be present in the container when it is sealed.

## FIELD-EXPEDIENT CONTAINER

N-80. The ideal container cannot be improvised in the field, but the standard military and commercial containers discussed below can meet caching requirements if they are adapted with care and resourcefulness. First, a container must be sufficiently sturdy to remain unpunctured and retain its shape through whatever rough handling or crushing pressure it may encounter. (Even a slight warping may cause a joint around the lid to leak.) Second, if the lid is not already watertight and airtight, packers can make it so by improvising a sealing device. The most common type of sealing device includes a rubber-composition gasket or lining and a sharp, flat metal rim that is pressed against a threaded lid. Applying heavy grease to the threads can increase its effectiveness. (Metallic solder should not be used for sealing because it corrodes metal surfaces when exposed to moisture.) Whenever any nonstainless metal container is used, it is important to apply several coats of high-quality paint to all exterior surfaces.

### Instrument Containers

N-81. Ordinarily, aircraft and other precision instruments are shipped in steel containers with a waterproof sealing device. The standard instrument containers range from 1/2-gallon to 10-gallon sizes. If one of suitable size can be found, only minimum modifications may be needed. In the most common type of instrument container, the only weak point is the nut and bolt that tighten the locking band around the lid. These should be replaced with a stainless steel nut and bolt.

### Ammunition Boxes

N-82. Several types and sizes of steel ammunition boxes that have a rubber gasket closing device are satisfactory for buried caches. An advantage of using ammunition boxes as a cache container is that they are usually available at a military depot.

### Steel Drums

N-83. A caching container of suitable size may be found among the commercially used steel drums for shipping oil, grease, nails, soap, and other products. The most common types, however, lack an adequate sealing device, so a waterproof material should be used around the lid. Fully removable head drums with lock-ring closures generally give a satisfactory seal.

### Glass Jars

N-84. The advantage of using glass is that it is waterproof and does not allow chemicals, bacteria, and insects to pass through it. Although glass is highly vulnerable to shock, glass jars of a sturdy quality can withstand the crushing pressure normally encountered in caching. However, none of the available glass containers have an adequate sealing device for the joint around the lid. The standard commercial canning jar with a spring clamp and a rubber washer is watertight, but the metal clamp is vulnerable to corrosion. Therefore, a glass jar with a spring clamp and a rubber washer is an adequate expedient for short-term caching of small items, but it should not be relied upon to resist moisture for more than a year.

## Paint Cans

N-85. Standard cans with reusable lids require a waterproof adhesive around the lids. It is especially important to apply several coats of paint to the exterior of standard commercial cans because the metal in these cans is not as heavy as that in metal drums. Even when the exterior is thoroughly painted, paint cans will not resist moisture for more than a few months.

# METHODS OF EMPLACEMENT

N-86. Since burial is the most frequently used method of emplacement, this section describes first the complete procedure for burial, followed by a discussion of emplacement procedures peculiar to submersion and concealment. The last area discussed is the preparation of the cache report—a vital part of a caching operation.

N-87. When planners have designed a cache and selected the items for caching, they must carefully work out every step of the burial operation in advance.

## HORIZONTAL AND VERTICAL CACHES

N-88. Ordinarily, the hole for a buried cache is vertical—the hole is dug straight down from the surface. Sometimes a horizontal cache, with the hole dug into the side of a steep hill or bank, provides a workable solution when a suitable site on level or slightly sloping ground is not available. A horizontal cache may provide better drainage in areas of heavy rainfall, but it is more likely to be exposed by soil erosion and more difficult to refill and restore to normal appearance (figure N-8).

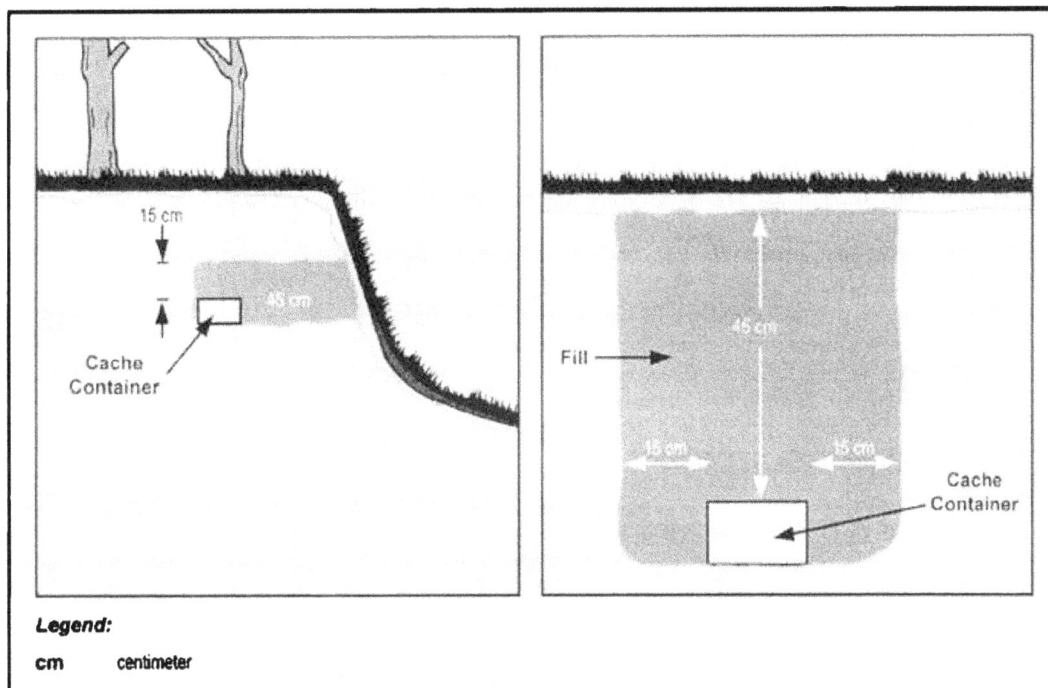

*Legend:*

cm     centimeter

**Figure N-8. Burial procedures for horizontal and vertical caches**

## DIMENSIONS OF THE HOLE

N-89. The exact dimensions of the hole, either vertical or horizontal, depend on the size and shape of the cache container. As a general rule, the hole should be large enough for the container to be inserted easily. The horizontal dimensions of the hole should be about 30 centimeters longer and wider than the container. Most importantly, it should be deep enough to permit covering the container with soil to about 45 centimeters.

This figure is recommended for normal usage because a more shallow burial risks exposure of the cache through soil erosion or inadvertent uncovering by normal indigenous activity. A deeper hole makes probing for recovery more difficult and unnecessarily prolongs the time required for burial and recovery.

## EXCAVATION SHORING

N-90. If there is a risk that the surrounding soil will cave in during excavation, boards or bags filled with subsoil may be used to shore the sides of the hole. Permanent shoring may be needed to protect an improvised container from pressure or shock.

## EQUIPMENT

N-91. Some items of equipment that may be helpful or indispensable in burying a cache, depending upon the conditions at the site, may include:
- Measuring instruments (a wire or metal tape and compass) for pinpointing the site.
- Paper and pencil for recording the measurements.
- A probe rod for locating rocks, large roots, or other obstacles in the subsoil.
- Two ground sheets on which to place sod and loose soil. An article of clothing may be used for small excavation if nothing else is available.
- Sacks (sandbags, flour sacks, trash bags) for holding subsoil.
- A spade or pickax if the ground is too hard for spading.
- A hatchet for cutting roots.
- A crowbar for prying rocks.
- A flashlight or lamp if burial is to be done at night.

# BURIAL PARTY

N-92. Aside from locating, digging, and refilling the hole, the most important factor in this part of the emplacement operation may be expressed with one word: personnel. Since it is almost impossible to prevent every member of the burial party from knowing the location of the cache, each member is a prime security problem as long as the cache remains intact. Thus, planners must keep the burial party as small as possible and select each member with utmost care. Once selected, each member must have an adequate story to explain his absence from home or work during the operation, his trip to and from the site, and his possession of whatever equipment cannot be concealed on the way. Transportation for the burial party may be a problem, depending on the number of persons, how far they must go, and what equipment they must take. When planners have worked out all details of the operation, they must brief each member of the burial party on exactly what he is to do from start to finish.

# OPERATIONAL SCHEDULE

N-93. The final step in planning the emplacement operation is to make a schedule to set the date, time, and place for every step of the operation that requires advance coordination. The schedule will depend mainly on the circumstances, but to be practical, it must include a realistic estimate of how long it will take to complete the burial. Here, generalizations are worthless, and the only sure guide is actual experience under similar conditions. Planners should consider the following with respect to scheduling.

N-94. A careful burial job probably will take longer than most novices will expect. Therefore, if circumstances require a tight schedule, a dry run or test exercise before taking the package to the site may be advisable. Unless the site is exceptionally well concealed or isolated, night burial will be required to avoid detection. Because of the difficulties of working in the dark, a nighttime practice exercise is especially advisable.

N-95. The schedule should permit waiting for advantageous weather conditions. The difficulties of snow have already been mentioned. Rainy weather increases the problems of digging and complicates the story. If the burial is to be done at night, a moonless or a heavy overcast night is desirable.

## SITE APPROACH

N-96. Regardless of how effective the individual's story is during the trip to the cache site, the immediate approach must be completely unobserved to avoid detection of the burial. To reduce the risk of the party being observed, planners must carefully select the point where the burial party disappears, perhaps by turning off a road into the woods. They should just as carefully select the reappearance point. In addition, the return trip should be by a different route. The burial party should strictly observe the rule for concealed movement. The party should proceed cautiously and silently along a route that makes the best use of natural concealment. Concealed movement requires foresight, with special attention to using natural concealment while reconnoitering the route and to preventing rattles when preparing the package and contents.

### Security Measures at the Site

N-97. The burial party must maintain maximum vigilance at the cache site since detection can be disastrous. The time spent at the site is the most critical. At least one lookout should be on guard constantly. If one man must do the burial by himself, he should pause frequently to look and listen. The burial party should use flashlights or lanterns as little as possible, and should take special care to mask the glare. Planning should include emergency actions in case the burial party is interrupted. The party should be so thoroughly briefed that it will respond instantly to any sign of danger. Planners should also consider the various escape routes and whether the party will attempt to retain the package or conceal it along the escape route.

### Steps in Digging and Refilling

N-98. Although procedures will vary slightly with the design of the cache, persons involved in caching operations must never overlook certain basic steps. The entire procedure is designed to restore the site to normal as far as possible.

### Site Sterilization

N-99. When the hole is refilled, a special effort should be made to ensure that the site is left sterile and restored to normal in every way, with no clues left to indicate burial or the burial party's visit to the vicinity. Since sterilization is most important for the security of the operation, the schedule should allow ample time to complete these final steps in an unhurried, thorough manner. These final steps are to:
- Dispose of any excess soil far enough away from the site to avoid attracting attention to the site. Flushing the excess soil into a stream is the ideal solution.
- Check all tools and equipment against a checklist to ensure that nothing is left behind. This checklist should include all personnel items that may drop from pockets. To keep this risk to a minimum, members of the burial party should carry nothing on their persons except the essentials for doing the job and disguising their actions.
- Make a final inspection of the site for any traces of the burial. Because this step is more difficult on a dark night, use of a carefully prepared checklist is essential. With a night burial, returning to the site in the daytime to inspect it for telltale evidence may be advisable if this can be done safely.

## SUBMERSION

N-100. Emplacing a submerged cache always involves two basic steps: weighting the container to keep it from floating to the surface and mooring it to keep it in place.

N-101. Ordinarily, container weights rest on the bottom of the lake or river and function as anchors, and the moorings connect the anchors to the container. The moorings must also serve a second function, which is to provide a handle for pulling the cache to the surface when it is recovered. If the moorings are not accessible for recovery, another line must extend from the cache to a fixed, accessible object in the water or on shore. There are four types of moorings.

### Spider Web Mooring

N-102. The container is attached to several mooring cables that radiate to anchors placed around it to form a web. The container must be buoyant so that it lifts the cables far enough off the bottom to be readily secured by grappling. The site must be located exactly at the time of emplacement by visual sightings to fixed landmarks in the water, or along the shore using several FRPs to establish a point where two sighted lines intersect. For recovery, the site is located by taking sightings on the reference points when a mooring cable is engaged, by dragging the bottom while diving. This method of mooring is most difficult for recovery. It can be used only where the bottom is smooth and firm enough for dragging or where the water is not too deep, cold, or murky for diving.

### Line-to-Shore Mooring

N-103. A line is run from the weighted container to an immovable object along the shore. The section of the line that extends from the shore to the container must be buried in the ground or otherwise well concealed.

### Buoy Mooring

N-104. A line is run from the weighted container to a buoy or other fixed, floating marker, and fastened well below the waterline. This method is secure only as long as the buoy is left in place. Buoys are generally inspected and repainted every 6 months or so. The inspection schedule should be determined before a buoy is used.

### Structural Mooring

N-105. A line for retrieving the weighted container is run to a bridge pier or other solid structure in the water. This line must be fastened well below the low-water mark.

### ESSENTIAL DATA FOR SUBMERSION

N-106. Whatever method of mooring is used, planners must carefully consider certain data before designing a submerged cache. The cache very likely will be lost if any of the following critical factors are overlooked.

### Buoyancy

N-107. Many containers are buoyant even when filled, so the container must be weighted sufficiently to submerge it and keep it in place. If the contents do not provide enough weight, emplacers must make up the balance by attaching a weight to the container. The approximate weight needed to attain zero buoyancy is shown in figure N-9.

| Container Dimensions | Empty Container Weight | Added Weight to Zero |
|---|---|---|
| 7 x 9 x 8 5 inches | 5 pounds | 15 pounds |
| 7 x 9 x 16.5 inches | 8 pounds | 31 pounds |
| 7 x 9 x 40 inches | 16 pounds | 77 pounds |
| 7 x 9 x 45 inches | 17.5 pounds | 88 pounds |
| 7 x 9 x 50 inches | 19 pounds | 97 pounds |
| *Note:* The third column is the approximate weight needed to attain a zero buoyancy. | | |

**Figure N-9. Buoyancy chart**

N-108. This figure applies to several sizes of stainless steel containers. The weighting required for any container can be calculated theoretically if the displacement of the container and the gross weight of the container plus its contents are known. This calculation may be useful for designing an anchor, but it should not be relied upon for actual emplacement. To avoid hurried improvisation during emplacement, emplacers should always test the buoyancy in advance by actually submerging the weighted container. This test determines only that a submerged cache will not float to the surface. Additional weighting may be required to keep it from drifting along the bottom. As a general rule, the additional weight should be at least one-tenth of the gross weight required to make the container sink; more weight is advisable if strong currents are present.

## Submersion Depth

N-109.  Planners must first determine the depth that the container is to be submerged to calculate the water pressure that the container must withstand. The greater the depth, the greater the danger that the container will be crushed by water pressure. For instance, the standard stainless steel burial container will buckle at a depth of approximately 4.3 meters. The difficulty of waterproofing also increases with depth. Thus, the container should not be submerged any deeper than necessary to avoid detection. As a general rule, 2.2 meters is the maximum advisable depth for caching. If seasonal or tidal variations in the water level require deeper submersion, the container should be tested by actual submersion to the maximum depth it must withstand.

## Depth of the Water

N-110.  Emplacers must measure accurately the depth of the water at the point where the cache is to be placed. This depth will be the submersion depth if the cache is designed so that the container rests on the bottom of the lake or river. The container may be suspended some distance above the bottom, but the depth of the water must be known to determine the length of moorings connecting the containers to the anchors.

## High- and Low-Water Marks

N-111.  Any tidal or seasonal changes in the depth of the water should be estimated as accurately as possible. Emplacers must consider the low-water mark to ensure that low water will not leave cache exposed. The high-water mark should also be considered to ensure that the increased depth will not crush the container or prevent recovery.

## Type of Bottom

N-112.  Emplacers should probe, as thoroughly as possible, the bed of the lake or river in the vicinity of the cache. If the bottom is soft and silty, the cache may sink into the muck, become covered with sediment, or drift out of place. If the bottom is rocky or covered with debris, the moorings may become snagged. Any of these conditions may make recovery very difficult.

## Water Motion

N-113.  Emplacers should consider tides, currents, and waves because any water motion will put additional strain on the moorings of the cache. Moorings must be strong enough to withstand the greatest possible strain. If the water motion tends to rock the cache, emplacers must take special care to prevent the moorings from rubbing and fraying.

## Clearness of the Water

N-114.  When deciding how deep to submerge the cache, emplacers must first determine how far the cache can be seen through the water. If the water is clear, the cache may need to be camouflaged by painting the container to match the bottom. (Shiny metallic fixtures should always be painted a dull color.) Very murky water makes recovery by divers more difficult.

## Water Temperature

N-115.  Planners must consider seasonal changes in the temperature of the water. Recovery may be impossible in the winter if the water freezes. The dates when the lake or river usually freezes and thaws should be determined as accurately as possible.

## Salt Water

N-116.  Since seawater is much more corrosive than fresh water, tidal estuaries and lagoons should not be used for caching. The only exception is the maritime resupply operation in which equipment may be submerged temporarily along the seacoast until it can be recovered by a shore party.

## CONCEALMENT

N-117. There are many different ways to conceal a cache in natural or readymade hiding places. For instance, if a caching party was hiding weapons and ammunition in a cave and was relying entirely on natural concealment, the emplacement operation would be reduced to simply locating the site. No tools would be needed except paper, a pencil, and a flashlight. On the other hand, if the party was sealing a packet of jewels in a brick wall, a skilled mason would be needed, along with his kit of tools and a supply of mortar expertly mixed to match the original brick wall.

N-118. When planning for concealment, planners must know the local residents and their customs. During the actual emplacement, the caching party must ensure the operation is not observed. The final sterilization of the site is especially important since a concealment site is usually open to frequent observation.

## CACHING COMMUNICATIONS EQUIPMENT

N-119. As a general rule, all equipment for a particular purpose (demolitions, survival) should be included in one container. Some equipment, however, is so sensitive from a security standpoint that it should be packed in several containers and cached in different locations to minimize the danger of discovery by the enemy. This is particularly true of communications equipment since, under some circumstances, anyone who acquires a whole radio transmitter set with a signal plan and cryptographic material would be able to play back the set. An especially dangerous type of penetration would result. In the face of this danger, the signal plan and the cryptographic material must never be placed in the same container. Ideally, a communications kit should be distributed among three containers and cached in different locations. If three containers are used, the distribution may be similar to this:
- Container 1: The radio transmitter set, including the crystals.
- Container 2: The signal plan and operational supplies for the radio transmitter operator, such as currency, barter, and small arms.
- Container 3: The cryptographic material.

N-120. When several containers are used for one set of equipment, they must be placed far enough apart so that if one is discovered, the others will not be detected in the immediate vicinity. On the other hand, they should be located close enough together so that they can be recovered conveniently in one operation. The distance between containers will depend on the particular situation, but ordinarily they should be at least 10 meters apart. One FRP ordinarily is used for a multiple cache. The caching party should be careful to avoid placing multiple caches in a repeated pattern. Discovery of one multiple cache would give the opposition a guide for probing others placed in a similar pattern.

## CACHING MEDICAL EQUIPMENT

N-121 A feasibility study must be performed to determine the need for the caching of medical supplies. The purpose of caches is to store excess medical supplies to maintain mobility and deny access to the enemy. In addition, caching large stockpiles of medical supplies allows pre-positioning vital supplies in anticipation of future-planned operations.

# TWELVE-POINT CACHE REPORT

N-122. The final step, which is vital in every emplacement operation, is the preparation of a cache report. This report records the essential data for recovery. The cache report must provide all of the information that someone unfamiliar with the locality needs to find his way to the site, recover the cache, and return safely. The report is intended merely to point out the minimum essential data. Whatever format is used, the importance of attention to detail cannot be overemphasized. A careless error or omission in the cache report may prevent recovery of the cache when it is needed.

N-123. The twelve-point cache report includes—
- Type of cache.
- Method of caching.
- Contents.

- Description of containers.
- General area.
- Immediate area.
- Cache location.
- Emplacement details.
- Operational data and remarks.
- Dates of emplacement and duration of the cache.
- Sketches and diagrams.
- Radio message for recovery.

## CONTENT

N-124. The most important parts of the cache report must include instructions for finding and recovering the cache. It should also include any other information that will ease planning the recovery operation. Since the details will depend upon the situation and the particular needs of each organization, the exact format of the report may vary slightly. A format for a cache report (figure N-10) and a sample message is included in the SAV SER SUP 7.

| ZERO FIVE UNDER | |
| --- | --- |
| AAA | ODA XXX ODA BUCKY |
| BBB | BURIAL |
| CCC | TWO CASES MRE XXX MRE XX TWO FIVE GALLON WATER CANS |
| DDD | ONE FOUR POUNDS EACH |
| EEE | CARNIS VILLAGE |
| FFF | SOUTHWEST CORNER OF CEMETERY XX TWO THREE METERS AT ONE NINE ZERO DEGREES |
| GGG | OAK TREE XX SIX METERS AT ONE ZERO ZERO DEGREES |
| HHH | THREE FEET XX HARD CLAY |
| III | SHOVEL XX PROBE XX NO SECURITY FORCES |
| JJJ | ZERO FOUR OCT ONE SEVEN XX ONE YEAR |

**Figure N-10. Cache report**

## PROCEDURES

N-125. The observer should collect as much data as possible during the personal reconnaissance to assist in selecting a site and planning emplacement and recovery operations. Drafting the cache report before emplacement is also advisable. Following these procedures will reveal the omissions, then the missing data can be obtained at the site. If the procedures are followed, the preparation of the final cache report will be reduced to an after-action check. This check ensures that the cache actually was placed precisely where planned and that all other descriptive details are accurate. Although this ideal may seldom be realized, two procedures always should be followed:

- The caching party should complete the final cache report as soon as possible after emplacement, as details are fresh in mind.
- Someone who has not visited the site should check the instructions by using them to lead the party to the site. When no such person is available, someone should visit the site shortly after emplacement, provided he can do so securely. If the cache has been emplaced at night, a visit to the site in daylight may also provide an opportunity to check on the sterilization of the site.

This page intentionally left blank.

# Appendix O

# Link-Up Operations

The JSOA commander plans and coordinates link-up operations if the JSOA is coming under the operational control of a conventional force. A physical juncture is necessary between the conventional and resistance forces. The mission of the conventional force may require SF and resistance force personnel to support conventional combat operations. All elements involved must conduct detailed centralized planning. The conventional force commander, in coordination with the JSOTF, prepares an OPLAN and coordinates a PSYOP program to simplify the link up of forces within the JSOA. Link-up planning involves the JSOA command. the conventional force or SOF commands, and any adjacent JSOA commands.

## LIAISON

O-1. The SF operational detachment and resistance leaders coordinate the plans for linkup. If linkup is with SOF, resistance liaison personnel coordinate with the senior SOF commander, his staff, and leaders of the link-up element. When the linkup is with a resistance force in an adjacent JSOA, both area commands provide liaison coordination.

O-2. The conventional and resistance forces coordinate communications between them. An obvious requirement is for all communications equipment to be compatible. The SF operational detachment will ensure that the radios and SOI used for linkup remain under its control.

O-3. A liaison party. consisting of SF operational detachment members and area commanders, may exfiltrate from the JSOA and assist in link-up planning. Conventional forces do not infiltrate into the JSOA to conduct their own liaison planning. SOF or designated U.S. agencies infiltrate the JSOA to conduct link-up planning and coordination. The JSOTF or SOTF monitors the linkup and provides administrative and logistics support to the resistance force until physical linkup is complete. After linkup, the operational control element assumes a support responsibility.

## PHYSICAL AND NONPHYSICAL LINKUPS

O-4 Normally. a joint or allied force uses a physical linkup when operating in the JSOA. A physical linkup is difficult to plan, conduct, and control. It requires detailed, centralized coordination and a planning conference between those involved. Commanders conduct physical linkups for the following reasons:
- Joint tactical operations.
- Resupply and logistics operations.
- Intelligence operations.
- Exfiltration of sick and wounded U.S. and indigenous personnel.
- Exfiltration of very important persons and EPWs.
- Infiltration of U.S. and indigenous personnel.
- Transfer of guides and liaison personnel to the conventional forces.

O-5. Forces must establish a nonphysical linkup when operations are conducted in a JSOA and a physical linkup is not required or desirable. The conventional force is the attacking force and the resistances are in support. A nonphysical linkup also requires coordination between the link-up forces. Commanders must state procedures before operations begin and when joint communications are established. Compatible communications equipment and current SOI must be available to the participants. Commanders use nonphysical linkups when the conventional force conducts a deep raid and resistances conduct security missions. Commanders also use linkups when the conventional force attacks and when the resistances—

- Serve as a blocking force.
- Screen flanks and block threats.
- Conduct deception operations.
- Conduct reconnaissance or surveillance.

# LINKUP WITH AIRBORNE OR AIR ASSAULT FORCES

O-6. Command link-up plans for airborne or air assault forces depend on the ground tactical plan. The airborne or air assault task force commander, the area command liaison personnel, and the SOCCE conduct a planning conference. Either an SF operational detachment or a SOCCE remains with the resistance force in the JSOA as an advisor for all postlinkup operations.

O-7. Precise timing of the airborne operation with the supporting resistance operation is desirable. Premature commitment of the resistance force may stop the surprise effect of the conventional force mission and lead to its defeat. If committed late, the desired effects from the resistance force may not occur. Resistance forces may have to secure DZs or LZs, seize objectives within the airhead, occupy reconnaissance and security positions, or delay or harass enemy movements toward the objective area.

O-8. The resistance force may act as a reception committee and mark the DZ and LZ. The reception committee links up with airborne or air assault force personnel at one of the designated LZ or DZ markers. The reception committee guides these personnel to the area commander. The linkup is complete when the airborne or air assault force commander establishes contact with the area commander.

O-9. Concurrent with the landing of the airborne or air assault elements, the resistance force furnishes current intelligence data, provides guides, and conducts reconnaissance and security missions. The resistance force interdicts approaches, controls areas between separate airheads, and attacks enemy reserve units and installations. It also helps control the civilian population within the objective area.

# LINKUP WITH AMPHIBIOUS FORCES

O-10. Timing is critical when the resistance force supports an amphibious operation. Premature commitment may alert the enemy and lead to the possible defeat of the resistance force. Late employment may not produce the desired effects. A SOCCE deploys early for the amphibious task force to ensure joint planning is complete and attempts to participate in the beach landing. The element establishes command relationships and support during the planning conference. During an amphibious assault or linkup, the task force commander assumes operational control of the resistance forces. When the amphibious force commander links with the area commander, the link-up operation is complete.

# MULTIPLE LINKUPS

O-11. Linkups can occur at several points and at separate times. Planning and conducting multiple linkups simultaneously is difficult. All resistance forces should not be involved in the linkup because deception and interdictions should continue during the linkup. A physical linkup with a large resistance force may not even be required or desired. A small reception committee can conduct the linkup and act as guides or liaisons with the link-up force. Conventional forces may conduct a relief in place after the resistances have conducted a raid or seized key terrain.

# CONDUCT OF LINKUP

O-12. When a linkup appears imminent, the SOCCE deploys to the conventional force HQ and begins detailed, centralized planning to involve all parties. The involvement of the SOCCE in the link-up planning process is especially critical, as it coordinates the actions between external forces and the area command in the JSOA. The SOCCE plans and coordinates at all levels, especially with the units conducting the linkup.

O-13. Units develop link-up planning and contingencies during premission training. The initial planning conference establishes command relationships. Planners consider the following link-up factors:

- The resistance force may continue to conduct UW operations under the umbrella of the unified commander.
- The resistance force may be under control of the joint task force unified commander or national command of the resistance force (shadow government or area commander).
- The conventional force commander will be advised of the capabilities and limitations of the resistance force, which may conduct combat operations and rear area security.
- The resistance force may be under tactical control to SOF in support of special operations.
- An adjacent JSOA resistance force may link with the internal resistance force and conduct joint UW operations. The resistance force is under the control of the unified commander, or both forces may be under tactical control to a conventional force or SOF.
- The resistance force may be under tactical control to SOF temporarily and then revert to the relationships described above.

O-14. The SF operational detachment coordinates operations, control measures, and the scheme of maneuver with the conventional and resistance forces. The SF operational detachment coordinates tactical control measures with the resistance force commanders to assist linkup. In addition, it establishes fire control measures to provide support for the linkup. Both forces must understand the plans for securing objectives and have access to dedicated assets to ensure mission success. Before physical linkup occurs, the resistance force first confirms the location of the link-up contact point and secures it. The SF operational detachment ensures security is emplaced and maintained at this location until the linkup is complete.

# CONTACT PROCEDURES

O-15. The participating forces establish simple, primary and alternate contact procedures and mutual control measures for physical and nonphysical linkups. At the initial planning conference, participating forces can modify the primary, alternate, and contingency plans developed during premission training, based on METT-TC.

O-16. At a coordinated date and time, the conventional and resistance forces move toward the contact point on a specific azimuth. The contact point is near a well-defined, easy-to-locate terrain feature. Both forces stop about 500 meters short of the contact point and send a small element forward. The elements display distinct, mutual, and simple recognition signals. They provide security for the person making the contact. The smallest possible contact element makes contact to preclude unnecessary personnel losses. The SF operational detachment and the resistance force begin the linkup by displaying prearranged recognition signals to identify themselves to the conventional force. The conventional force responds with its own prearranged signal. Radio contact before this action is highly advisable to decrease the chances for any mistaken identity and fratricide. These additional principles apply:

- SF may be included in both contact elements.
- Resistances dispatched during the planning conference may guide the conventional force to the contact point.
- Both forces will establish and maintain communications not later than 24 hours before linkup.
- Deception and feints may be used to cover link-up operations.

# POSTLINKUP OPERATIONS

O-17. After linkup, resistance forces convert to national control and reorganize into conventional forces or demobilize. The unified commander, in coordination with U.S. and allied officials, determines the further use of resistance forces following linkup. With this guidance, the tactical commander may employ these resistance forces. The SF operational detachment stays with and helps them become an effective combat unit operating under the joint task force or higher commanders.

# POSTLINKUP EMPLOYMENT CONSIDERATIONS

O-18. Conventional force commanders should be aware of several important factors when they employ resistance forces following linkup. Some of those factors include:

- Knowing the resistances, their organization, concepts of operation, capabilities, and limitations.
- Making sure the subordinate leaders understand the value of resistance forces and know how to use them.
- Anticipating the problems of providing administrative, logistics, and operational support to attached resistance forces.
- Anticipating possible language and political problems in establishing liaison.
- Knowing that the high value of resistance forces is limited to those operations conducted in areas familiar to them.
- Maintaining resistance force integrity as much as possible.
- Working through existing channels of the resistance command. (Imposing a new organizational structure may hamper their effectiveness.)
- Respecting resistance leaders. (Give them the same consideration as officers of the conventional forces.)
- Maintaining resistance morale by awarding decorations and letters of commendation. (Express appreciation whenever such action is justified and warranted.)
- Not making political commitments or promises to resistance forces unless authorized by higher HQ.
- Recognizing when the value of the resistance forces has ended and promptly returning them to the control of the unified commander.

# POSTLINKUP MISSIONS AND OPERATIONS

O-19. If the resistance force is to be employed as light infantry, it must undergo a period of retraining and re-equipping before its commitment in the new combat role. Commanders may reorganize, retrain, and re-equip SF advised resistance forces to conduct reconnaissance, airmobile, or other similar light infantry operations. Resistance, paramilitary, or irregular forces supporting conventional forces may conduct the following missions.

## RECONNAISSANCE

O-20. Familiarity with the terrain and people qualifies resistance forces for reconnaissance missions. Resistance forces provide the principal sources of intelligence on the enemy. They can patrol difficult terrain and gaps between units, establish roadblocks and observations posts, screen flanks, and provide guides.

## DEFENSIVE OPERATIONS

O-21. Control of terrain is rarely critical for the resistance. These resistance forces, with relatively light weapons and equipment, are normally inferior to the organized enemy forces in manpower, firepower, mobility, and communications. Resistance forces do not undertake defensive operations unless forced to prevent enemy penetration of resistance-controlled areas or gain time for their forces to accomplish a specific mission. Resistance forces may defend key terrain or installations for a limited time in support of conventional forces. When the resistance does defend an area, he modifies the principles of conventional, defensive combat to maximize their specific assets and minimize any deficiencies. The resistance demonstrates best-planned efforts using METT-TC.

## COUNTERGUERRILLA OPERATIONS

O-22. The experience and training of guerrilla forces make them very useful for counterguerrilla operations. They detect enemy sympathizers in villages and towns and implement control measures in unfriendly areas. Tactical commanders should exploit their knowledge of, and experience with, guerrilla techniques, language, terrain, and population. When properly supported, the guerrilla forces may be given complete responsibility for counterguerrilla operations in selected areas.

## REAR AREA SECURITY

O-23. Resistance forces may act as security forces within the theater Army area command. Tactical commanders assign these forces to a rear area security role based on their knowledge and experience. Whenever possible, resistance forces should be assigned on an area basis to guard lines of communications, supply depots, airfields, pipelines, railroad yards, or port facilities. They also patrol terrain that contains bypassed enemy units or stragglers, aids in recovering prisoners, helps control civilians and refugees, and polices towns and cities. When provided with appropriate transportation, resistance forces may act as a mobile security force in their JSOA.

## CIVIL SUPPORT

O-24 Because of their area knowledge and experience, resistance forces help restore an area to its "normal state." They perform dislocated civilian collection and control duties. They assist in PSYOP campaigns in rear areas, apprehend collaborators and spies, recruit labor, and guard key installations and public buildings.

## CONVENTIONAL FORCE OPERATIONS

O-25 Guerrilla forces conduct combat operations and augment, relieve, or replace conventional forces in the rear area. However, guerrilla forces usually cannot complete the same size mission as a conventional force. The operations or tactical commander considers the guerrillas' capabilities and takes advantage of their special expertise and area knowledge.

O-26. Shortages of adequate voice communications and transportation may severely limit the use of guerrilla forces for conventional force combat operations. The strength, organization, leadership, training, equipment, and extent of civilian support for guerrilla forces affect their combat capability.

O-27. After linkup with guerrilla forces, the SF operational detachment may have to retrain host-nation forces. The SF operational detachment can retrain and reconstitute host-nation forces that have suffered reversals in combat actions. The goal is to rapidly train the units' cadres in leadership, operations, and combat tactics, techniques, and procedures.

This page intentionally left blank.

# Appendix P

# Special Forces Resistance Advisor Considerations

The Merriam-Webster Online Dictionary defines *rapport* as a friendly, harmonious relationship: especially: a relationship characterized by agreement, mutual understanding, or empathy that makes communication possible or easy. When people discuss good rapport, they describe a relationship founded upon mutual trust, understanding, and respect. Relationships characterized by personal dislike, animosity, and other forms of friction are often described as no rapport. Rapport, for purposes of the SF advisor, is a term used to describe the degree of effectiveness in which the advisor can influence a counterpart toward the desired action for a long-term, attainable, and sustainable solution.

## RAPPORT

P-1. The requirement to establish rapport with resistance partners is the product of the advisor being in a unique military position. That position is one in which the advisor has no positional authority over the actions of the resistance partners. This lack of authority means that the traditional (doctrinal) view of military leadership must be modified to emphasize interpersonal relationships and de-emphasize authoritarian roles. Advisors use their interpersonal skills to influence the outcome of events.

P-2. SF advisors must be armed with certain knowledge before they can establish effective rapport. They must study ADRP 6-22, *Army Leadership*—the publication that provides the SF advisor with the basic leadership knowledge required to understand human nature and motivation. In addition to ADRP 6-22, advisors must incorporate information specific to the culture and society of their potential counterparts. This information may take the form of thorough area studies, operational area studies, and other research materials (shown, in part, in Appendixes H and I).

P-3. The SF advisor should strive to establish and maintain good rapport by conveying to his resistance partner that he is sincerely interested in him, his nation, and its cause; that he will not belittle him or his efforts; and that he has no intention of taking over from him. Advisors must demonstrate that they have come to help because they believe the resistance partner's goals are just, fair, and deserving of success. Genuine rapport is slow to develop and can be ruined in an instant.

P-4. The true measure of rapport is whether the advisors can motivate their counterpart to take a desired action. For SF, the basic techniques of motivation (in the absence of authority) are advising, setting the example, seeking compromise, and coercing.

### ADVISING

P-5. Advising the resistance partner to select a particular COA is only effective if the resistance partner perceives that the advisor is professionally competent to provide sound advice. If the resistance partner does not perceive the proposed solution to a problem as realistic, the advisor's competence may be questioned. The advisor must take the effort to explain to his counterpart that the recommended COA is realistic and will be effective.

### SETTING THE EXAMPLE

P-6. Setting the example for the resistance partner must be an ongoing effort to avoid the appearance of a do as I say, not as I do attitude. In setting the example, the advisor should make every effort to explain and demonstrate to the resistance partner that what he is doing is the most effective form of behavior for the situation. Thus, it is extremely important that SF advisors properly prepare their advisory mission and take their predeployment training seriously to hone the skill sets they need during their deployment. The ability to influence will be demonstrated when the SF advisors arrive in the resistance's AO—technically and tactically proficient, physically fit, culturally astute, and with a good plan that is properly resourced.

P-7.   The prospect that an advisor may—from time to time—have to share in the same risks of his resistance partners to properly set the example should be clear. This shared risk may especially be critical in a combat situation. By sharing the same risks as that of the resistance partners, the advisor will dispel any appearance of a privileged status that may potentially harm relations or lose rapport.

## SEEKING COMPROMISE

P-8.   When seeking compromise with the resistance partner in the desired COA or behavior, the advisor may create a situation in which the resistance partner has a personal interest in successful execution. In some cultures, seeking a compromise may allow the resistance partner to save face. Furthermore, in certain situations the resistance partner—because of practical experience—may have a better solution to the problem at hand.

P-9.   Advisors must also recognize that when they seek a compromise in certain cultures, their perceived competence may suffer. This may be mitigated somewhat by approaching the compromise as two professionals (the advisor and the resistance partner) in order to reach a mutual conclusion. To reach an effective compromise, SF advisors may have to conduct negotiations.

*Note:* There are two areas of concern that must never be compromised for the sake of maintaining rapport—force protection and human rights.

## EXERTING COERCION

P-10. Coercion is the least desirable method of motivation because it can cause irreparable damage to a relationship. Advisors should use coercion only in extreme circumstances (life or death). Although coercion may prompt surrogate "employee" forces to perform a specific action, it is inadvisable to try and coerce resistance partners. Moreover, trying to use coercion as a regular method can never lead to institutionalization of proper conduct and legitimacy. Lasting, long-term success may only be achieved through the patient application of influence. Exerting pressure is more desirable than outright coercion.

P-11. American advisors may enjoy a privileged status in the denied area. Their mere presence may garner personal benefits for the resistance partner, simply because the resistance partner has a one-on-one association with an American. Worried about losing these benefits, resistance partners may agree with advisors simply to avoid confrontation. SF advisors must be careful to avoid unintentionally forcing their resistance partners into action, and they must obtain buy-in from the resistance partner on a solution for a lasting and learned result.

P-12. Again, other means of influence are preferable; however, an SF advisor could also leverage U.S.-provided resources toward desired behaviors. Using U.S.-provided sustainment resources as a carrot and stick can help discipline a resistance partner that fails to provide resources through his own indigenous sources when they are available.

# POLICY AND RELATIONSHIPS

P-13. Being an SF operational detachment advisor is, in some ways, to be a diplomat. However, even career diplomats' primary duty is to advance the objectives of the USG.

## SUPPORTING UNITED STATES AND HOST-NATION POLICIES

P-14. Advisors must keep in mind that their primary aim is to forward U.S. policy. Advisors attempt to meet mutually agreed USG-resistance objectives, but they must ensure that activities are in keeping with U.S. objectives. Local customs, resistance organization, economy, customs, and education may dictate practices and procedures that appear inefficient or uneconomical. Advisors should avoid criticizing or condemning such practices and procedures until they are thoroughly understood. Recommendations that may be critical or contrary to local custom should be made in private.

## GOING NATIVE

P-15. Building and maintaining rapport has its limits. The SF advisor has to be close to his resistance partners to fully understand the challenges, to help meet resistance objectives, and to transmit ground truth and valuable insights to higher HQ. In doing so, the SF advisor may encounter situations that are not black and white, and he may face moral and ethical dilemmas on an almost daily basis. SF advisors are allowed greater autonomy and little to no supervision while operating in most denied areas; thus, they will have to be comfortable in their environment and present the moral courage and intellect to make sound decisions in assisting their resistance partner. However, if an SF advisor begins to pursue the agenda of the resistance to the detriment of the U.S. national security strategy, that SF advisor becomes ineffective and puts U.S. objectives at risk.

## UNDERSTANDING THE ENVIRONMENT

P-16. The SF advisor must have knowledge of the political, social, and military organizations in the AO and the manner in which they interrelate. In many countries, these relationships depend heavily on personal relationships between individuals. The SF advisor must understand the personalities, political movements, and social forces at work. Close observation of local practices and relationships is a key standard procedure.

P-17. Depending on the level of cooperation between varied resistance groups, an SF advisor may need to foster relations with and between all parties. Indigenous entities may have issues with trusting each other, or they may not see the benefits of working together. An SF advisor, as a welcomed USG representative, may enjoy the opportunity to interact with different factions and powerbrokers, who can provide him with the opportunity to network with those that he may not normally interact with. Networking and mediating between competing parties toward mutual goals may shorten time and lessen resources spent trying to accomplish similar goals by pooling efforts, thereby enabling all concerned.

## UNDERSTANDING COMMAND RELATIONSHIPS

P-18. Care must be taken to distinguish between the U.S. chain of command and the resistance chain of command. In particular, advisors must prevent the resistance partner from attempting to control subordinates through the U.S. chain of command. Advisors provide recommendations—not orders—to their resistance partners Only the resistance partner should issue orders to subordinates.

P-19. Effective communication is essential for the advisor. The use of proper channels should be stressed at all echelons. SF advisors must keep their resistance partners informed of advice given to their subordinates. Fellow U.S. personnel should be kept informed of advice offered to resistance partners.

P-20. Local customs and courtesies should be observed. Resistance partners that are senior in grade should be treated accordingly. If warranted by local military customs, senior-ranking personnel should be referred to by their rank (as customs permit) and shown respect. Although resistance officers may have no command authority over the advisor, effectiveness is greatly enhanced when the advisor displays respect for the resistance partner and the resistance chain of command.

P-21. Advising revolves as much around who to influence as it does influencing. Positive results in influencing require the advisor to identify the powerbrokers and stakeholders—the key decision makers. The advisor should keep in mind that the rank of a resistance partner can be meaningless in a foreign army. Gender, age, rank, social and political positions, and lineage are factors that can determine the decision maker in the sphere of influence. The advisor should not equate the level of rank with the power to make a decision and its execution.

P-22. An advisor's approach is important for both influencing and maintaining rapport. Advisors should refrain from overtly criticizing how resistance partners currently do business. Some cultures are much less direct in their suggestions, so the best way to get a resistance partner to accept advice can be as simple as changing the expression. For example, a direct approach would be expressed, "You should…" whereas a more palatable expression using an indirect approach would be expressed, "Wouldn't it be a good idea if someone did...?" Only when a comprehensive professional relationship has been established, based on trust in which the advisor has genuinely demonstrated the interest of his resistance partner, should the advisor seek to identify better processes more overtly. Figure P-1 provides additional recommendations of common dos and don'ts for advisors when interacting with their resistance partners.

| DO | DO NOT |
|---|---|
| Study the resistance partner's personality and attempt to influence from the background. | Attempt to command the resistance partner's organization. |
| Make recommendations in private. | Correct the partner in public, particularly in front of subordinates. |
| Respect the resistance partner's position and defend his position in disputes with others, as long as it is based on sound, reasoned judgment and is ultimately consistent with USG objectives. | Represent the partners or defend their misguided position out of blind loyalty or an attempt to win favor |
| Provide advice in a respectful manner | Present suggestions in a condescending manner or in a way that might be embarrassing. |
| Ensure the partner understands the recommendation. | Provide too little, too much, or disjointed guidance. |
| Ensure the partner has the authority and capability to follow recommendations | Attempt to force the partner into following guidance that he is unable or unauthorized to perform |
| Verify that the partner understands, agrees with, and intends to carry out the recommendation. | Accept "Yes" at face value. ("Yes" in his context may mean he understands but does not agree, or he does not understand but is being polite). |
| Present recommendations carefully and in detail, supported by sound reasoning and an explanation of the advantages being offered | Present recommendations that require an immediate decision. |
| Allow the partner to exercise their prerogative. | Make the partner overly-dependent on U.S guidance or allow such a perception to develop |
| Express U.S. customs with caution. | Assume behavior that is common in the United States is acceptable to the indigenous populace. |
| Praise the partner when he makes good choices. | Convey the impression that everything is wrong. |
| Select words carefully and watch the message being conveyed by tone and body language. | Use sarcasm, irony, or mockery. |
| Be willing to ask for the partner's advice, particularly in matters of the local situation, customs, and courtesies. | Present an attitude of intellectual superiority. |
| Present an accurate self-image and a realistic measure of capabilities. | Create unachievable expectations or promises that cannot be kept. |
| Share in the same risks as the resistance partner when necessary. | Rely on a privileged status to avoid hazardous duty. |
| Develop goals and milestones and encourage inspections. | Operate without a training plan or allow training to be unsupervised. |
| Demonstrate and encourage initiative and inventiveness. Encourage the resistance partner to clarify orders and to make recommendations when appropriate. Encourage partner to allow his subordinates to do the same. | Develop or encourage an environment of blind loyalty. |
| Remain alert to the larger indigenous populace outside of the resistance organization. | Become isolated from and blind to the status of the local populace. |

**Figure P-1. Special Forces resistance advisor considerations**

# Appendix Q

# Resistance Training Considerations

Conventional forces train to achieve the capability to conduct operations; training enables operations. While this is also true of UW, supported resistance movements are also expected to eventually conduct mutually beneficial operations—to a greater extent than conventional operations—the training (and advising and equipping) of the resistance is often the mission itself. The ability to be proficient trainers is critical to the SF Soldier.

## TRAINING THE RESISTANCE ORGANIZATION

Q-1. A major part of the SF operational detachment's mission is to plan, organize, conduct, and evaluate training of selected resistance cadre. This work prepares the SF operational detachment to be trainers and force multipliers for the resistance organization. During premission planning, the SF operational detachment develops a tentative training plan based on METT-TC. The SF operational detachment members prepare the training plan, program of instruction, and training aids in the indigenous language, if possible. They include sand tables with toy soldiers and vehicles as instruction aids. After commitment into the JSOA, SF operational detachment members evaluate the present level of the resistance forces' training and update the initial training plan, or they modify the training plan to ensure its effectiveness.

### TRAINING PLAN

Q-2. The training plan outlines how the SF operational detachment will best accomplish its training objective. SF operational detachment members develop the training plan in isolation based on an estimate of the training situation by the pilot team. Important factors in developing the training plan are mission, personnel, time, facilities, and the organization of the training. Other considerations are weather, climate, and the enemy situation.

### MISSION

Q-3. Based on operational directives from higher HQ, the SF operational detachment determines the specific tasks, conditions, and standards the resistance forces must accomplish. If the resistance force has to conduct multiple missions, the SF operational detachment must set up priorities for training. Because a guerrilla travels primarily on his feet, he must undergo rigorous physical training and numerous cross-country marches.

### TIME

Q-4. The available training time is a critical factor. The shorter the training time, the more time is taken to define the training objectives. SF operational detachment personnel will lose training time because of operational requirements. They must include makeup training as an integral part of the training program.

### FACILITIES

Q-5. The SF operational detachment will advise the area command and resistance force on selecting and establishing ranges, training areas, improvised classrooms, training aids, and other training facilities. Security is of prime importance. The SF operational detachment locates ranges and training areas away from the base camp. It can make a small valley with aerial concealment or a man-made tunnel into single-lane ranges. Air and BB rifles are excellent alternate, inexpensive, training aids in lieu of full-caliber marksmanship training.

# POLITICAL TRAINING

Q-6. Politically or religiously extreme insurgents use frequent criticism and self-criticism sessions as a form of catharsis (venting). These sessions allow members to voice fears and problems and to hear from other members. An individual who is disillusioned with the resistance movement will find it difficult to conceal his true feelings in these sessions. He will become influenced by what his friends and comrades think of him. Ideologically oriented unit leaders are more cohesive and effective in their training. Cadres in each unit must set the "politically correct" example during these daily training sessions. They must continually reinforce rank-and-file members, assuring them that what they are doing is needed to carry out the will of the people. Such confidence will elevate morale and fighting spirit.

> "Equal importance should be attached to the military and political aspects of the one-year consolidation and training program that has just begun, and the two aspects should be integrated. At the start, stress should be placed on the political aspect, on improving relations between officers and men, enhancing internal unity and arousing a high level of enthusiasm among the masses of cadres and fighters. Only thus will the military consolidation and training proceed smoothly and attain better results."
>
> Mao Tse-Tung, *The Tasks for 1945*, 15 December 1944

# TRAINING THE GUERRILLA FORCE

Q-7. SF operational detachment members evaluate how well the guerrillas are trained through personal observations and inspections and the results of limited (easy) combat missions. Characteristics of the guerrilla force that may present obstacles to training include:

- A wide range of education and military experience.
- Different personal motivation for joining the force.
- Possible language and dialect barriers requiring training through interpreters.

## TRAINING ORGANIZATION

Q-8. The requirements for physical security in the JSOA dictate that guerrilla forces be dispersed over a wide area. Consequently, the system and organization for training are decentralized, and hands-on training is emphasized. The SF operational detachment must train the guerrilla cadre and later help this cadre in additional training and combat employment. Throughout the organization, development, and training phases of guerrilla activities, the guerrillas conduct limited combat operations. These actions support training, instill confidence, and test the readiness of the force. The goals of these combat operations include:

- Attracting additional recruits to the guerrilla force.
- Assisting in gaining popular support from the civilian population.
- Allowing the area command and SF operational detachment an opportunity to evaluate the training.
- Increasing the morale and esprit de corps of the guerrilla forces with initial successful combat operations.

## TRAINING METHODS

Q-9. The SF operational detachment can use a three-tier concept for training. This concept is similar to the U.S. Army red, green, and amber training cycle. It allows for one-third of the guerrilla forces in training, one-third providing security for the training, and one-third providing support or doing small-scale operations. The three-tier concept accommodates the training of any size unit.

Q-10. Generic program of instruction development is the same anywhere. Trainers base all training (basic, advanced, and specialized) on METT-TC and the current combat skills of the guerrillas. They modify the training after contact with the guerrilla force. (Appendix U, Guerrilla Periods of Instruction, outlines a generic guerrilla program of instruction.)

> "As for the training courses, the main objective should still be to raise the level of technique in marksmanship, bayoneting grenade-throwing and the like and the secondary objective should be to raise the level of tactics, while special emphasis should be laid on night operations."
>
> Mao Tse-Tung, *Policy for the Work in the Liberated Areas for 1946,*
> Selected Works, Vol. IV, p. 76, 15 December 1945

## Basic Training

Q-11. METT-TC determines the need for base camps within the JSOA. All guerrillas, including the area command, may receive basic training. Each base camp may conduct its basic training in sectors independent of the other two, or it may conduct a part of basic training and rotate to another camp for more training. Either the SF operational detachment or guerrillas may rotate. The base camps may construct ranges and training areas for their use, or one base camp may construct a range and teach all marksmanship classes. Another base camp may construct training areas for raids, and the other may construct areas for ambushes and reconnaissance. The present combat abilities of the guerrillas and the threat determine how much basic training is needed.

Q-12. Basic training contains subjects on small arms, first aid, land navigation, and political or PSYOP classes. The time allotted for training needs to be flexible—between 21 and 31 days, depending on the knowledge and abilities of the force

## Advanced Training

Q-13. Each base camp conducts some advanced training or rotates to other base camps for additional training. This concept is similar to basic training, and the guerrillas or SF operational detachment may rotate. Trainers can conduct medical and demolitions training in one base camp; communications and intelligence in another; and tactics, operations, and heavy weapons in the third. At the end of the advanced training period, the guerrillas or SF operational detachment would rotate to another base camp for a new training cycle. For security reasons, all the engineer, communications, medical, and intelligence sergeants should not be in the same sector base camp at the same time.

Q-14. Advanced training may last from 40 to 80 days. This phase consumes the most time, but it also pays the most dividends. Some subjects covered in greater depth include combat orders, machine guns, tactics, survival, and specialty skill training in medical, communications, weapons, and demolitions.

## Specialized Training

Q-15. Specialized training is broadly subdivided into elementary and advanced specialized training. Personnel and units assigned to the theater SF may have received elementary specialized training. If not, such training becomes a responsibility of the theater SF commander. Specialized training, elementary and advanced, is conducted by a training center operated by the theater SF.

Q-16. Elementary specialized training consists of training individuals and teams to carry out their assigned functions. Intensive courses are conducted and include:
- Map reading and sketching.
- Patrolling.
- Close combat.
- Physical training.
- Fieldcraft.
- Tactics of both regular and guerrilla forces.
- Demolitions and techniques of sabotage.
- Use and care of weapons, including those of the enemy forces.
- First aid.
- Use of enemy and civilian motor transportation.

- Enemy organization and methods.
- Methods of organizing and training guerrillas.
- Security information.
- Methods of supply to guerrilla forces.

Q-17. Advanced specialized training prepares qualified individuals and teams for specific missions in enemy territory. During this training, individuals are organized into the teams in which they will be employed. Thereafter, the group trains, lives, and operates as a unit under simulated conditions of the area where they are to be employed. Special techniques, skills, and orientation are stressed to enable them to carry out their mission and to weld them into efficient, mobile, and self-sufficient teams. Parachute or amphibious training is given, depending on the contemplated means of entering enemy territory.

Q-18. Besides the training outlined above, technical training is given to radio operators, medical technicians, demolition experts, and other specialists.

# TRAINING THE STAFF AND AUXILIARY OR UNDERGROUND PERSONNEL

Q-19. While the SF operational detachment members are training the guerrilla force, the SF operational detachment commander, assistant detachment commander, and team sergeant are training the area command and staff, sector commander and staff, and selected members of the auxiliary or underground. The area commander and his staff may train in the same base camp with his guerrillas. Although the area command and auxiliary or underground may be trained anywhere in the JSOA, it should be compartmented for security reasons.

# TRAINING THROUGH TRANSLATORS

Q-20. Historically, SF operational detachment members in a UW environment have been faced with a social and political power struggle when they tried to select a translator to work with them. The struggle has not always been to their advantage. The next few paragraphs discuss some tips on how to select a translator and how to avoid some of the often-repeated pitfalls.

## NATIVE SPEAKER

Q-21. The ideal translator would be a native speaker from the AO who knows most of the area dialects. His speech, grammar, background, and social mannerisms should be understandable to the students. If he fits this ideal description, the students will listen to what he says—not how he says it. The students will understand the translator.

## SOCIAL STATUS

Q-22. A translator is often limited in his effectiveness with students if his social standing is considerably lower than that of his students. There may be significant differences among the students and the translator in military rank or memberships in ethnic or religious groups. When students are officers, it may be best to have an officer or civilian act as a translator. On the other hand, if students are enlisted personnel, an officer translator might intimidate the students and stifle class participation. An enlisted translator would be the best choice in that case. Most cultures recognize technical competence and international differences in military structure, so there should be no problem for the SF operational detachment. Despite personal feelings on social status, the instructor's job is to train all students equally—not act as an agent of social reform in a foreign land. The instructor must accept local customs as a way of life.

## ENGLISH FLUENCY

Q-23. If the SF operational detachment instructor and translator can communicate with each other in English, the translator's command of English is adequate. The instructor can check the translator's level of understanding by asking him to paraphrase in English what the instructor has just said. If the translator restates the comments correctly, both are "reading off the same sheet of music."

## INTELLIGENCE

Q-24. The translator should be quick, alert, and responsive to changing conditions and situations. He must be able to grasp complex concepts and discuss them without confusion in a logical sequence. Education does not equate to intelligence, but better educated translators will be more effective due to experience and maturity.

## TECHNICAL ABILITY

Q-25. If the translator has technical training or experience in the instructor's subject area, he will be more effective since he will translate meaning, as well as words. A doctor could interpret for a medic, and former military personnel could interpret best for the weapons and intelligence sergeants.

## RELIABILITY

Q-26. An instructor must be leery of any translator who arrives late for the class. Many cultures operate on a "flexible clock," where time is relatively unimportant. The translator must understand the concept of punctuality.

Q-27. It is safe to assume that any translator's first loyalty is to his country, not to the United States. The security implications are clear; the instructor must be very cautious in explaining concepts to give the translator a greater depth of understanding. Some translators, for political or personal reasons, may have a hidden agenda when they apply for the job.

## COMPATIBILITY

Q-28. The instructor must establish rapport with his translator early in their relationship and maintain compatibility throughout their joint effort. Mutual respect and understanding are essential to effective instruction. Some rapport-building subjects to discuss with the translator are history, geography, ethnic groups, political system, prominent political figures, monetary system, business, agriculture, exports, and hobbies. The SF operational detachment member is building a friendship on a daily basis.

Q-29. If several qualified translators are available, the instructor should select at least two. The exhausting nature of the job makes a half-day of active interpreting the maximum for peak efficiency. One translator, however skilled, will seldom be enough except for short-term courses conducted at a leisurely pace. If two or more translators are available, one of them can sit in the rear of the class and provide quality control of the instruction by cross-checking the active translator. Meanwhile, additional translators can conduct rehearsals, grade examinations, and evaluate the exercises. Mature judgment and a genuine concern that the students are learning important skills go a long way toward accomplishing the mission.

Q-30. Good instructors will tactfully ask about their translators' background. With genuine concern, they ask about the translator's family, aspirations, career, and education. They can start with his home life; it's very important to him and is neutral territory. Instructors can follow up with a discussion of cultural traditions to find out more about him and the land he lives in.

Q-31. The instructor should gain his translator's trust and confidence before embarking on sensitive issues, such as sex, politics, and religion. He must approach these areas carefully and tactfully. They may be useful and revealing in the professional relationship between instructor and translator. Once this stage is reached, the two are well on their way to a valuable friendship and a firm, professional working relationship.

## TRAINING THE TRANSLATORS

Q-32. Very early on, the instructor conveys to the translator that he will always direct the training. However, he must stress how important the translator is as a link to the students. He can appeal to the translator's professional pride by describing how the quality and quantity of learning are dependent on his interpreting skills and his ability to function as a conduit between instructor and students. The instructor must also stress patriotism—that the defense of his country is directly related to his ability to transfer the instructor's knowledge to the students.

Q-33. Because of cultural differences, translators may attempt to "save face" by concealing their lack of understanding. They may attempt to translate what they think the instructor meant without asking for a clarification, resulting in disinformation and confusion for the students. Ultimately, when the students realize they have been misled, they question the credibility of the instructor, not of the translator. If the instructor has established rapport with his translator, he is in a better position to appeal to the translator's sense of duty, honor, and country. A mutual understanding allows for clarification when needed, leads to more accurate interpretation, and keeps the instructor informed of any student difficulties

## CONDUCTING THE TRAINING

Q-34. To prepare for teaching, the instructor must have initial lesson plans available for basic, advanced, and specialized training. He must also have on hand the available supporting documentation, such as FMs. When the class begins, the instructor should—

- Express the training objective in measurable performance terms.
- Outline the course content with methods of instruction and the various training aids to be demonstrated and then used by the students.
- Supply and circulate all class handout material when needed.
- Modify the training schedule to allow more time to train foreign students due to language and translation constraints.

Q-35. A glossary of terms is a valuable aid for the instructor and the translator. Many English words and phrases do not translate literally into many foreign languages. Technical terms need to be clearly defined well ahead of class. A listing of the most common terms and their translated meaning will be a useful product.

Q-36. The instructor presents bite-sized information tailored to the student audience. He talks directly to the students in a relaxed and confident manner. The translator watches the instructor carefully and emulates his style and delivery as he interprets for the students. During the translation, the instructor observes the translator to detect any problems. The translator will do some editing as a function of the interpreting process, but it is imperative that he transmits the instructor's meaning without additions or deletions. A well-coordinated effort is the key to success.

Q-37. Although maximum improvisation must be used in all phases of operations, items accompanying deployed detachments that may prove useful in conducting training include:

- Grease pencils and colored chalk.
- Target cloth or ponchos (blackboard substitutes).
- Basic publications on weapons generally found in the area (in the language of the country, if possible).
- Graphic training aids improvised from parachutes or other such material.

## STUDENT QUESTIONS

Q-38. Whenever students have questions, the translator immediately relays them to the instructor for an answer. The students then realize the instructor, not the translator, is the subject-matter expert and is in charge of the class. When a problem occurs, neither the instructor nor the translator corrects each other in front of the students. They must settle all differences in a professional manner.

Q-39. Rapport is as important between student and instructor as it is between translator and instructor. When the translator and instructor treat the students as mature, valuable people capable of learning, rapport will build easily between the students and the instructor.

## COMMUNICATION

Q-40. An instructor learns by experience that a way to communicate is through a translator. Use of profanity, slang, colloquialisms, and military jargon with students is harmful. Often, these expressions cannot be translated and do not come out with the desired meaning. If he must use a technical term or expression, the instructor ensures the translator conveys the proper meaning in the indigenous language.

## TRANSITIONAL PHRASES

Q-41. Transitional phrases tend to confuse the learning process and waste valuable time. Expressions, such as "for example" and "in most cases" or qualifiers, such as "maybe" or "perhaps" are difficult to translate. Many native translators have learned much of their English from reading, rather than hearing English spoken. The instructor keeps the class presentation as simple as possible, using short words and phrases.

## TABOO GESTURES

Q-42. Social and cultural restrictions will manifest themselves during class. Gestures are learned behavior and vary from culture to culture. If the instructor does not know, he should ask the translator to relate the cultural taboos before class and avoid them. The instructor should know before class:

- When it is proper to stand, sit, or cross legs.
- If the index finger, chin, or eyes may be used for pointing.
- If nodding of the head means yes or no.

## MANNER OF SPEAKING

Q-43. The instructor should try to look at the students and talk directly to them, not the translator. He speaks slowly and clearly and repeats himself as needed. The instructor should not address the students in the third person through the translator. Instead, he should say something like, "I'm glad to be your instructor," and not "Tell them I'm glad to be their instructor."

Q-44. The instructor must speak to the students as if they will understand every word he says. He must convey enthusiasm and use all of the gestures, movements, voice intonations, and inflections he would use for an English-speaking audience. The students will reflect the same amount of energy, interest, and enthusiasm that the instructor conveys to them. The instructor must not let the translator "sabotage" training with a less than animated delivery and presentation of the material.

Q-45. When the translator is translating and the students are listening to get the full meaning of the translation, the instructor should do nothing that could be distracting. These distractions might be pacing the floor, writing on the blackboard, drinking water, or carrying on with other distracting activities.

Q-46. The translator should be checked periodically to make sure the students understand the instructor's meaning. A cadre member, qualified in the native language, may observe and comment on the translator's knowledge, skills, and abilities. When the instructor has been misunderstood, the point needs to be made clear immediately. If further clarification is needed, the instructor should phrase the instruction differently and illustrate the point as necessary.

This page intentionally left blank.

# Appendix R

# Shadow Government Operations Model

A *shadow government* is the governmental elements and activities performed by the irregular organization that will eventually take the place of the existing government. Members of the shadow government can be in any element of the irregular organization (underground, auxiliary, or guerrilla force) (ATP 3-05.1).

## SHADOW GOVERNMENT FUNCTIONS

R-1. The shadow government is an organization the underground forms in occupied territory. Ideally, the shadow government can perform normal governmental functions in a clandestine manner and synchronize those functions with the resistance movement. The shadow government is critical because it exercises a degree of control, supervision, and accountability over the population at all levels (district, village, city, province, and so on), and further discredits and delegitimizes the existing government. The Chinese and Vietnamese communists are classic examples of effective shadow government while Hezbollah and the Taliban are recent examples.

## SHADOW GOVERNMENT OPERATIONS

R-2. This appendix provides a simple four-frame progression showing the initiation, development, and expansion of a shadow government (figures R-1 through R-4, pages R-2 through R-5). Many casual observers assume that shadow government only refers to the political leadership of the indigenous resistance within the UW operational area. This is only partially true. The shadow government also refers to the leadership and systematic, organized, and continuous provision of essential services, such as food, water, sanitation, schools, medical clinics, community news, mail, local policing, emergency services, and so on. Who controls these services has immediate, visceral, and potentially decisive legitimate influence over population groups. Therefore, who controls these services and who can influence their effectiveness and reach—and the extent to which such services displace the presence and authority of the enemy government or occupying power—is not merely a CA concern, it is an operational concern fundamental to the success of a UW campaign. The underground (with or without sponsor advice) may take the initiative to form the nucleus of such shadow government entities, or it may coopt existing local population-driven efforts to band together.

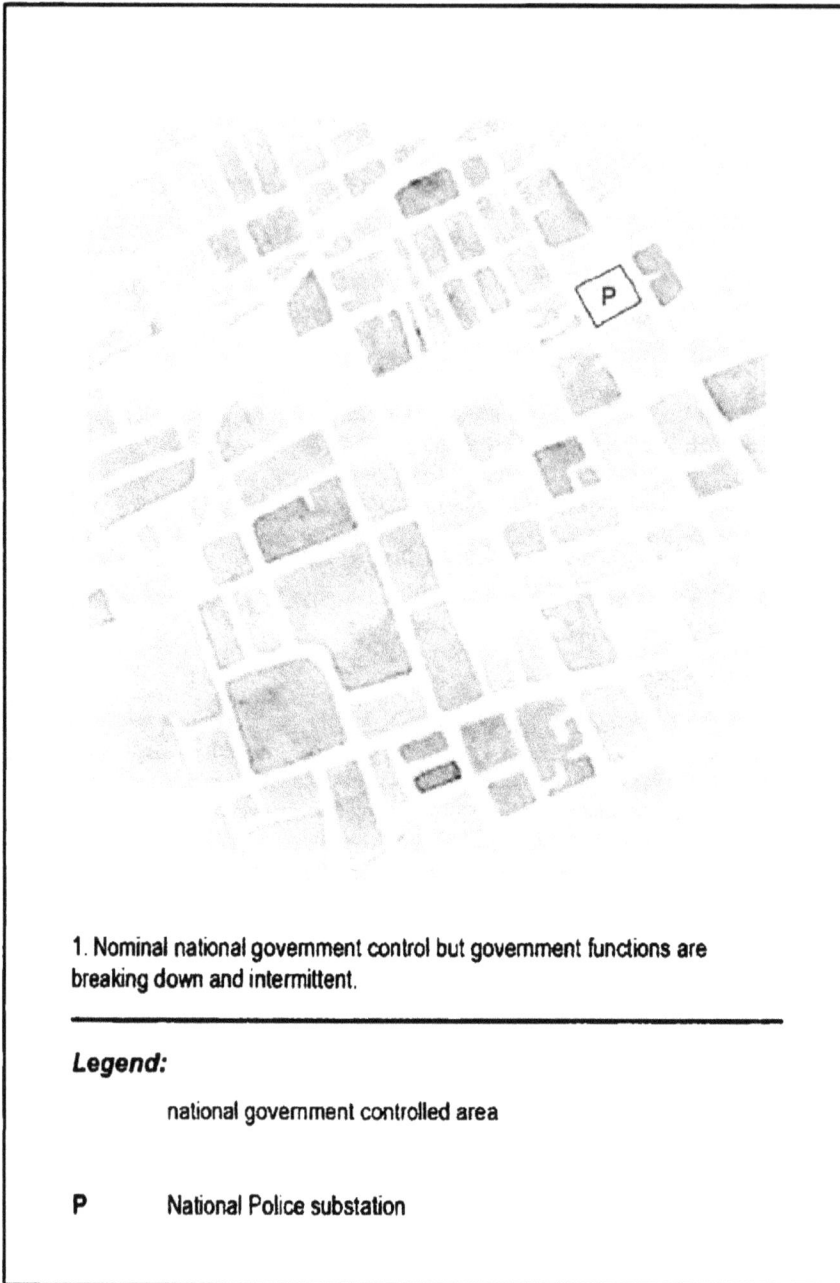

1. Nominal national government control but government functions are breaking down and intermittent.

---

*Legend:*

national government controlled area

P      National Police substation

**Figure R-1. Shadow government operations model—one**

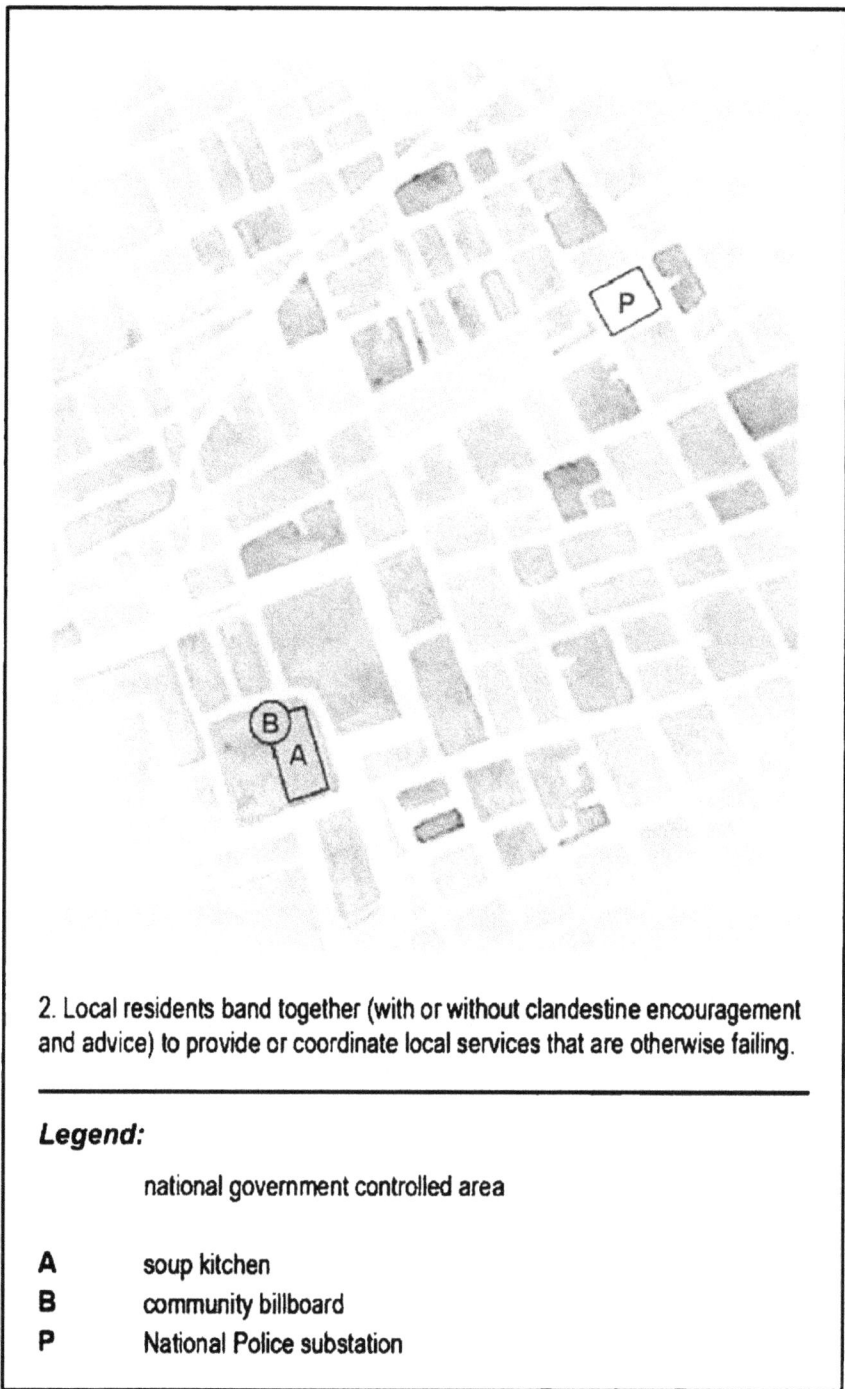

2. Local residents band together (with or without clandestine encouragement and advice) to provide or coordinate local services that are otherwise failing.

---

*Legend:*

national government controlled area

A       soup kitchen
B       community billboard
P       National Police substation

Figure R-2. Shadow government operations model—two

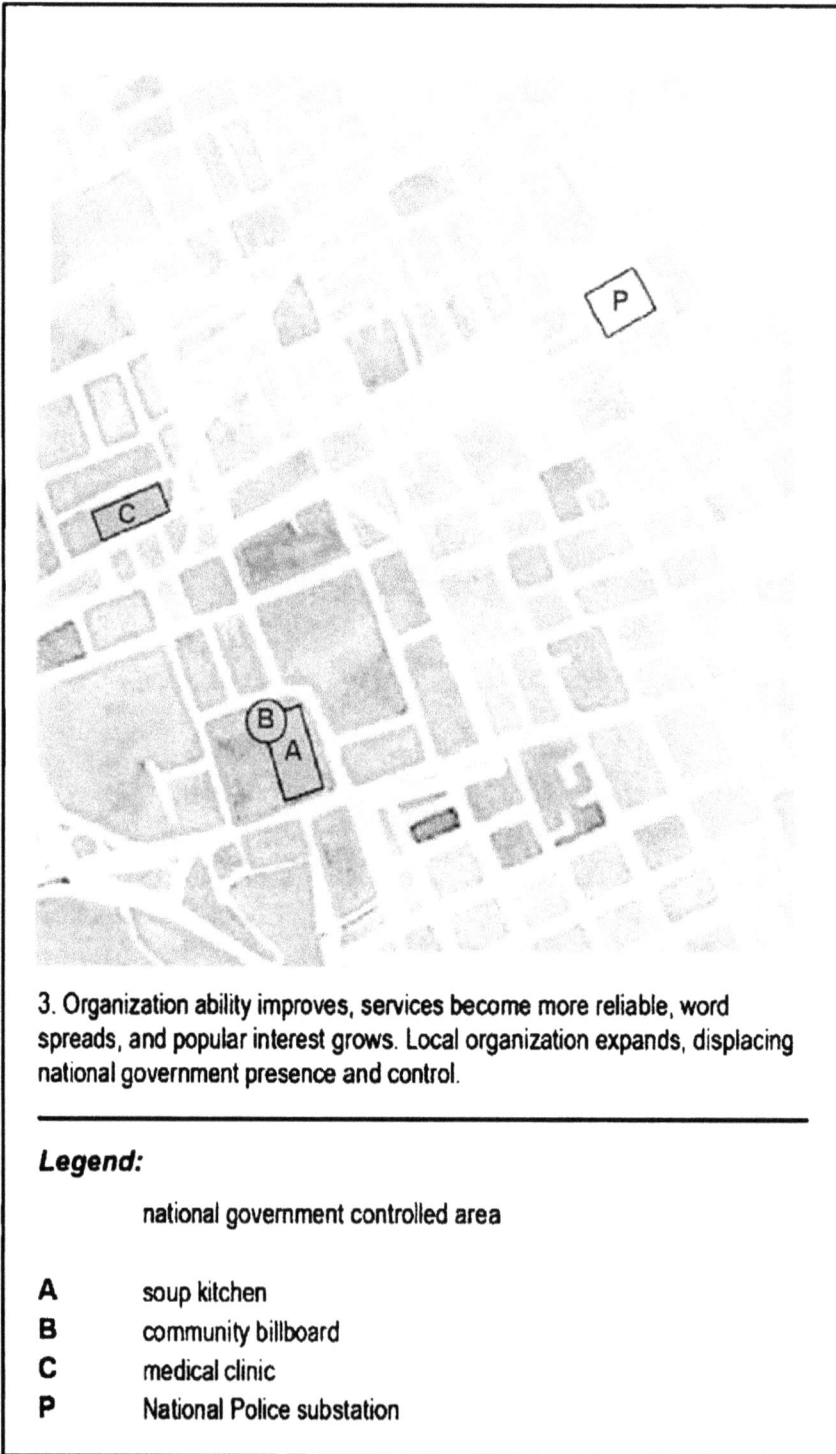

3. Organization ability improves, services become more reliable, word spreads, and popular interest grows. Local organization expands, displacing national government presence and control.

**Legend:**

national government controlled area

A     soup kitchen
B     community billboard
C     medical clinic
P     National Police substation

Figure R-3. Shadow government operations model—three

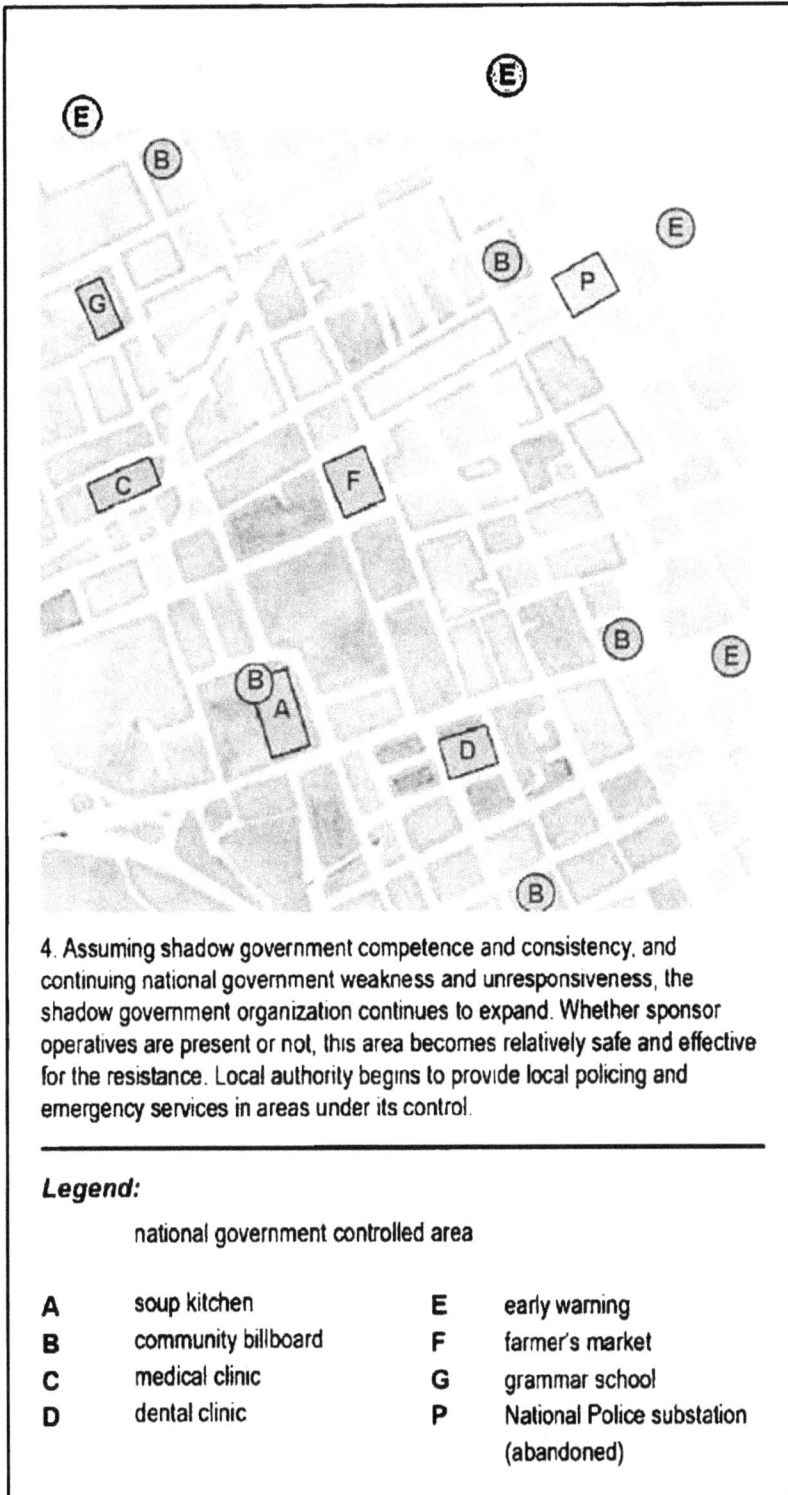

4. Assuming shadow government competence and consistency, and continuing national government weakness and unresponsiveness, the shadow government organization continues to expand. Whether sponsor operatives are present or not, this area becomes relatively safe and effective for the resistance. Local authority begins to provide local policing and emergency services in areas under its control.

*Legend:*

national government controlled area

| | | | |
|---|---|---|---|
| A | soup kitchen | E | early warning |
| B | community billboard | F | farmer's market |
| C | medical clinic | G | grammar school |
| D | dental clinic | P | National Police substation (abandoned) |

Figure R-4. Shadow government operations model—four

This page intentionally left blank.

# Appendix S

# Area Complex Development Model

An area complex is a clandestine, dispersed network of facilities to support resistance activities in a given area designed to achieve security, control, dispersion, and flexibility. The area complex is contested territory or an area that contains clandestine supporting infrastructure. It is not liberated territory. It represents the insurgent's AO. Insurgent forces can maintain their clandestine infrastructure in the area complex. The clandestine infrastructure provides insurgent forces with a measure of freedom of movement and support. These areas overlay areas under the control of the government or occupying military. These areas can eventually transform into liberated areas if the enemy's ability to challenge the insurgent forces degrades to a level of parity with the guerrilla forces. To support resistance activities, an area complex will include a security system, guerrilla bases, communications, logistics, medical facilities, and a series of networks capable of moving personnel and supplies. The area complex may consist of friendly villages, towns, or portions of urban areas under guerrilla military or political control (figure S-1, page S-2).

## SECURITY

S-1. Within the area complex, the resistance forces achieve security by:
- Establishing an effective intelligence net.
- Using the early warning with listening posts, observation posts, and security patrols.
- Practicing counterintelligence measures.
- Rehearsing withdrawals and CONPLANs.
- Employing mobility and flexibility.
- Using rapid dispersion techniques for personnel.
- Camouflaging and adhering to noise and light discipline.
- Organizing the active support of the civilian population.

S-2. The command and control base camp is in the heart of the area complex. Specially trained and equipped guerrilla forces control and defend this camp. A special guerrilla security detachment provides the internal protection for the area commander, his staff, and the SF operational detachment. Key personnel, critical equipment, and sensitive information are based from there and may include the following:
- The area commander and staff.
- The SF operational detachment and support personnel.
- Communications equipment.
- Controlled medical supplies and treatment facilities.
- Supply caches of weapons, ammunition, and explosives.

S-3. An area complex can be subdivided into three security zones: the enemy-controlled zone and the resistance outer and inner security zones (figure S-2, page S-3). These zones are sometimes referred to as Zones C, B, and A, respectively. There are no clear-cut boundaries between zones, and security responsibilities can overlap. Each zone is the responsibility of specific resistance elements whose mission is to provide for or support the security and defense of the zone in the area complex, to achieve total and overlapping security coverage.

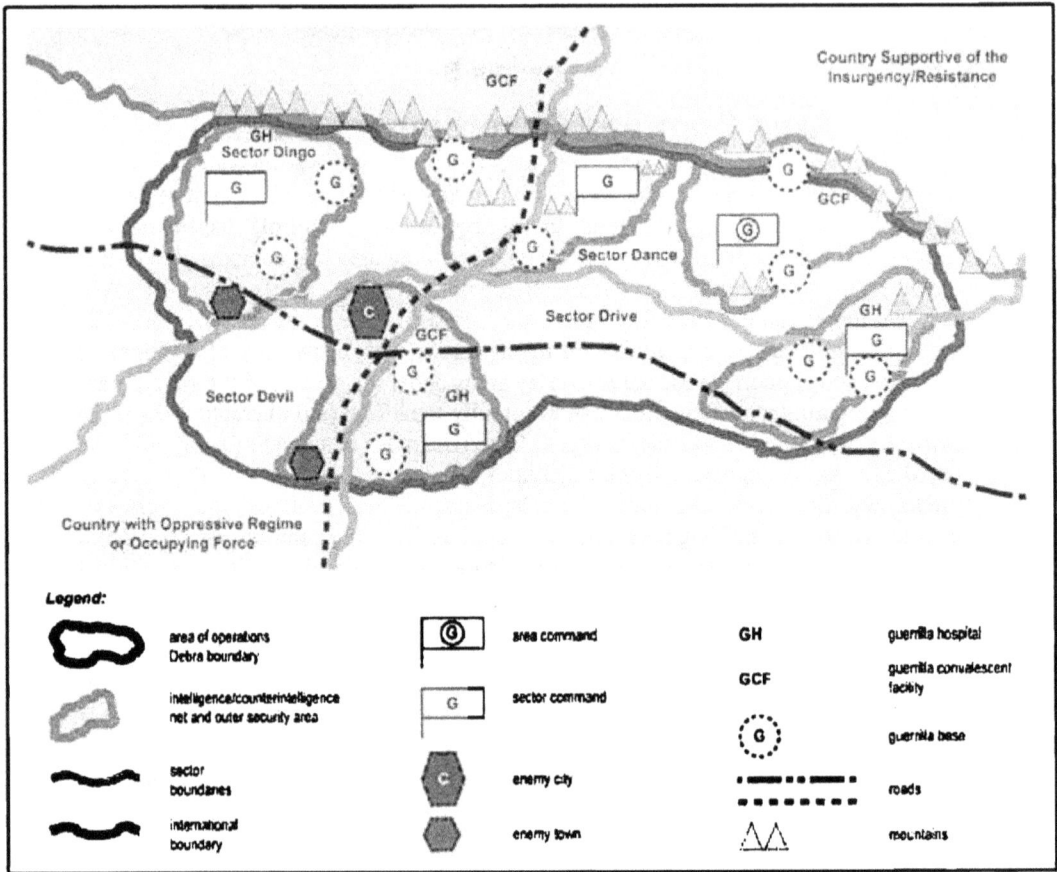

Figure S-1. Area complex sketch of joint special operations area "Debra"

## ENEMY-CONTROLLED ZONE (ZONE C)

S-4.    The enemy-controlled zone designates all areas in the operational area that are outside of routine resistance local control and/or observation (for example, outside of Zones B and A). It is where the enemy is the strongest, has the most repressive security, and has the most popular support. Resistance presence in this zone is relatively sporadic and higher risk. Resistance elements will attempt to gain early warning on these zones by clandestine information-gathering activities. Typically, the resistance will be required to go underground or use clandestine methods to remain in or pass through these areas. Those areas in which there is no resistance presence whatsoever and about which neither enemy nor resistance status are known (such as beyond observation) can be included with Zone C. Note that in Figure S-1 above, the sector commanders will likely have responsibility for the area inside Security Zone C.

## OUTER SECURITY ZONE (ZONE B)

S-5.    The outer security zone is vitally important to a guerrilla force. The local guerrilla forces and the civilian support infrastructure are organized and developed in the outer security zone. This area serves as the primary source of recruits for the guerrilla force. The outer zone also serves as the first line of in-depth security and defense for the area complex. Resistance elements in this area are responsible for providing the area command with timely and accurate information on enemy activities within the zone. Local guerrillas are the resistance element responsible for the zone's control and defense. This element is also responsible for the conduct of operations within the zone. The local guerrilla forces organize, employ, and serve as part of the civilian support element. The civilian support element gathers current information and intelligence and provides logistics, PSYOP, and operational support to the regular guerrilla forces and the area command. The initial screening, selection, and training of new resistance members takes place in this zone. The most promising and trusted recruits are then selected for membership in the regular or full-time guerrilla forces.

Legend:

   inner security zone

   outer security zone

   listening post or observation post

**Figure S-2. Area complex security zones**

## INNER SECURITY ZONE (ZONE A)

S-6.   The inner security zone encompasses the base camp of the regular or full-time guerrilla forces. These forces defend and control the zone, and they are constantly mobile within the area. The primary mission of the guerrilla forces operating in this zone is to temporarily delay any penetration made by the enemy. They watch trails and avenues of approach. They use observation posts or listening posts, fixed fighting positions, and pre-positioned obstacles, such as wire and minefields to delay enemy forces. They employ harassment, ambushes, sniping, and other interdiction tactics and practice rapid withdrawal procedures. These tactics increase the in-depth defense of the area complex. They should also use command-detonated anti-vehicular and antipersonnel mines as needed. Guerrilla forces may place mines along probable enemy vehicular and personnel avenues of approach, such as trails, creeks, and riverbeds. The guerrillas must avoid, at all costs, becoming decisively engaged while carrying out their delaying and defensive mission. Civilians do not normally occupy the inner security zone; therefore, it may serve as an area of food cultivation for the guerrilla population.

# GUERRILLA BASES

S-7.   A guerrilla base is HQ for any size guerrilla force. A base may be temporary or permanent, depending on the guerrilla's stage of development. Guerrilla command, control, communications, computers, and intelligence; support; facilities; and operational units are located within the base. LOCs connect the base and facilities within the area complex. The installations and facilities found within a guerrilla base are the command posts, training areas or classrooms, a communications facility, and medical services. The occupants and facilities must be capable of rapid displacement with little or no prior warning. There is usually more than one guerrilla base within a sector or JSOA. They are in remote, inaccessible areas and their locations are revealed only on a need-to-know basis. Personnel must use passive and active security measures to provide base security, employing overhead cover, concealment, and escape routes. A mandatory requirement for a guerrilla base camp is a source of water. Wells may be dug where permanent bases are established. Ideally, there will be an abundance of water sources to choose from in the area. All base camps should have an alternate location for contingency use. In case the enemy overruns the base, all personnel should plan for and rehearse rapid withdrawals.

---

"According to Mao Tse-Tung in *On Guerrilla Warfare*, a guerrilla base may be defined as an area, strategically located, in which the guerrillas can carry out their duties of training, self-preservation, and development. The ability to fight a war without a rear area is a fundamental characteristic of guerrilla action, but this does not mean that guerrillas can exist and function over a long period of time without the development of base areas. There is a difference between the terms base area and guerrilla base area. An area completely surrounded by territory occupied by the enemy is a 'base area.' On the other hand, a guerrilla base area includes those areas that can be controlled by guerrillas only while they physically occupy them."

FM 3-05.201, *Special Forces Unconventional Warfare Operations*,
30 April 2003, p. 3-8. para. 3-28

---

## TYPES OF BASES

S-8.   There are three types of guerrilla bases. In order of development, they are mobile, semipermanent, and permanent. Initially, all guerrilla base camps are mobile, and as the JSOA matures, semipermanent camps are constructed. When the JSOA matures enough to conduct battalion combat operations, the semipermanent camps become permanent. Normal occupation time is based on METT-TC.

### Mobile

S-9.   Full-time guerrillas and local guerrilla forces establish mobile bases. These bases are at the periphery of their zones of responsibility. Mobile bases are normally occupied for periods ranging from 1 to 7 days.

### Semipermanent

S-10. HQ elements or sector commands establish semipermanent bases in the inner security zones. These bases are in areas that provide a tactical advantage for the guerrilla. Semipermanent bases are normally occupied for periods ranging from 1 to 2 weeks.

## Permanent

S-11. This base is within the rear security zone of the area complex. The guerrilla command element, SF operational detachment, and key installations and facilities are located here. Adequate training areas are established to support all the training activities. The guerrilla force protects the training areas, and an SF operational detachment member, who is the subject-matter expert, monitors the training. When needed, personnel secure DZs and LZs to receive supplies and equipment. An SF operational detachment member accounts for supplies. These DZs and LZs must be accessible to the appropriate aircraft and be a safe distance from the guerrilla base camp. Permanent bases may normally be occupied for periods ranging from 1 to 2 months.

## BASE SECURITY MEASURES

S-12. The defense of any base includes strict adherence to camouflage, noise, and light discipline. Defense measures should also include inner security posts, listening posts and observation posts, security and tracking patrols, and other obstacles to concentrate, impede, or stop the enemy (figure S-3, page S-6). Personnel should plan contingencies for rapid withdrawal from the area before any enemy attack.

## Inner Security Posts

S-13. Inner security posts are normally established within 100 meters of the main body. The mission of the inner security posts is to delay a small reaction force that has penetrated the base perimeter and is closing in on the main body. This delay allows the main body to break out. During low visibility, inner security posts are closer, about 25 meters from the main body. A challenge and password system should be implemented.

## Listening and Observation Posts

S-14. Listening posts and observation posts are established in unit SOPs and based on observation and fields of fire, avenues of approach, key terrain, obstacles, and cover and concealment. At a minimum, listening posts and observation posts will be located on the most likely avenues of approach. They should be located on high and commanding ground surrounding the base, as per unit SOP. The mission of the posts is to detect and report in a timely manner enemy air and ground movement that threatens the guerrilla base. If the enemy is detected, post personnel may not fire on the enemy but radio a size, activity, location, unit, time, and equipment (SALUTE) report. This tactic saves giving away their position and possibly the position of the base. These posts are normally within 400 to 800 meters from the base. Posts located closer to the base may be able to use tactical landlines for a rapid, secure means of communications.

## Security and Tracking Patrols

S-15. Security and tracking patrols may be carried out at dawn and dusk to provide security and early warning for the base. Each patrol should carry a frequency modulated short-range radio for enabling the patrol to relay information to the base in a timely manner. Patrols must search all areas, but they should give priority of search to the high ground surrounding the base and to creek and riverbeds in the area. Patrols also search roads and trails for tracks or signs of enemy presence. If there are friendly civilians in the area, they may be questioned regarding enemy activity. Civilians unfamiliar to the patrol may be a threat or sympathizer. Information provided by the friendly civilians is critical to the security of the guerrilla base camp. The mission of the security patrols is to detect signs or other indicators of enemy presence or activity. These indicators include:

- Tobacco, candy, gum, and food wrappers.
- Human excrement or other waste products.
- Tracks made by bare feet or boots on recently used trails.
- Broken branches and bent twigs, suggesting direction of travel.
- Discarded rations, containers, and equipment.

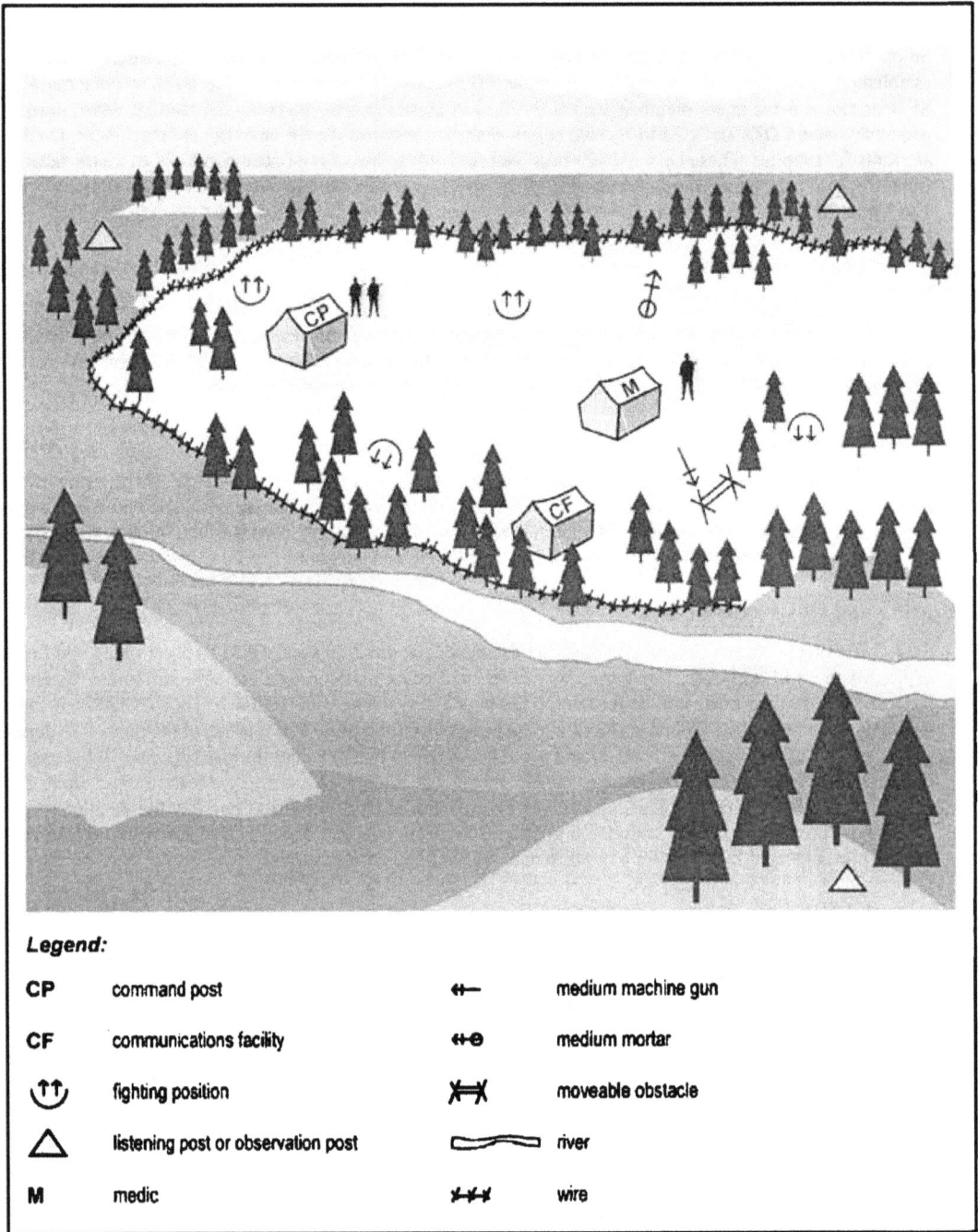

**Figure S-3. Permanent base security**

S-16. Passive security measures that can be taken include camouflaging dwellings and hutches with vegetation. Personnel should change vegetation daily. They can also camouflage trails and erase tracks. Personnel should avoid smoke from cooking fires, especially during daylight hours, and maintain noise and light discipline at all times.

### Obstacles

S-17. SF operational detachments may temporarily use obstacles, such as selectable lightweight attack munition, spider networked munitions, command detonated M18A1 claymores, concertina wire, and so on, along likely avenues of approach into the base. Creeks, riverbeds, and the surrounding elevations are good locations for placing obstacles. Personnel will warn the civilian population about the use of obstacles to preclude unnecessary civilian casualties. They may temporarily employ obstacles in the following areas not used by civilians:

- Near running water sources.
- Around fruit and shade trees.
- On little-used roads and trails.
- In and around abandoned fighting positions or around abandoned uninhabited dwellings.

S-18. Protective minefields and munition fields are recorded on DD Form 3007 (Hasty Protective Row Minefield Record) and should be submitted by the fastest and most secure means available. The form contains UNCLASSIFIED SENSITIVE INFORMATION for current operations at company level and below and CONFIDENTIAL for current operations above company level. Future operations are classified SECRET.

---

*Note:* Appendix C to ATP 3-34.20, *Countering Explosive Hazards*, provides more information on reporting and recording minefield information.

---

# MISSION SUPPORT SITES

S-19. An MSS is a temporary operational and logistics base for guerrillas who are away from their main base camp for more than a few days. It extends the range of guerrillas in the JSOA by permitting them to travel long distances without support from their base camps. The guerrillas should not occupy them for more than 24 hours. Guerrillas should always reconnoiter and surveil the MSS before occupying it.

S-20. Personnel establish an MSS to support a specific mission and should not use it more than once. Using the MSS only once protects the force from setting up repeated patterns of movement. However, it may be used before and after a mission, based on METT-TC. The MSS may contain food, shelter, medical support, ammunition, demolitions, and other operational items. To preclude unnecessary noise and movement in and out of the MSS, auxiliary personnel may establish supply caches in the surrounding vicinity before the combat force arrives.

S-21. METT-TC is very important when selecting the MSS. These sites must not be near LZs, DZs, or any other sites of heightened activity. When selecting the location for an MSS, personnel must consider the following:

- Proximity to the objective.
- Level of enemy activity.
- Cover and concealment.
- Preplanned routes of withdrawal.
- Tribal or factional and religious issues.

# INSURGENT SUPPORT NETWORKS

S-22. Just like other large organizations, insurgencies need support networks. These networks are usually run by either the auxiliary or underground or a combination of the two. Insurgent support networks include the following:

- Logistics support networks.
- Transportation networks.
- Medical support networks.
- Communications networks.
- Information and propaganda networks.
- Recruitment networks.

- Intelligence and counterintelligence networks
- Finance networks.
- Logistics support networks.

S-23. Each resistance organization must develop a logistics system to meet the specific requirements of their situation. In general, however, a resistance organization meets its logistics requirements through a combination of internal and external means.

S-24. The area complex must provide the bulk of an insurgent organization's logistics requirements. The area commander must balance his support requirements against his need for civilian cooperation. Imposing excessive demands on the population may adversely affect popular support. Logistics constraints may initially dictate the size of the resistance organization. (See Appendix M, Logistics, for more details on the logistics functions of an area complex.)

S-25. As the resistance organization expands, its logistics requirements may exceed the capability of the area complex to provide adequate support. When this situation occurs, an external sponsor provides supplemental logistics support, or the resistance organization reduces the scale of its activities. External support elements normally limit support to the necessities of life and the essential equipment and supplies the resistance needs to conduct combat operations. Internal sources of resistance supply include:

- Battlefield recovery.
- Purchase.
- Levy.
- Barter.
- Production.
- Confiscation.

S-26. Successful, offensive operations permit resistance forces to satisfy some of their logistics requirements through battlefield recovery. Capturing supplies from hostile forces also avoids alienating civilians. The resistance organization normally limits its purchases to critical items unavailable by other means. Excessive introduction of external currency may disrupt the local economy, which may not be in the interest of the resistance organization or the United States.

S-27. The resistance organization may organize a levy system to ensure an equitable system for obtaining supplies from the local population. Levy systems manifest in a variety of ways from tithes and collections, through sympathetic religious organizations, to cross-border fees and formal taxes on goods and income. Under a levy system, the resistance organization provides receipts and maintains records of levy transactions to facilitate reimbursement at the end of hostilities. Obstacles to a levy system may include:

- Chronic shortages among the local population.
- Hostile populace and resources control measures, including confiscation or destruction of local resources.
- Competition from the hostile power or rival resistance organizations.
- Chemical, biological, radiological, and nuclear or other contamination of local resources.

S-28. Barter may adversely affect the levy system. However, it is sometimes the only method of obtaining critical services or items, such as medical supplies.

S-29. Resistance forces often have to improvise their own field expedients. They may even have to plant and raise their own food, dig wells, and tend their own livestock. The area commander may consider establishing clandestine factories to produce unobtainable items.

S-30. Confiscation alienates the local population. The resistance organization should use confiscation only in emergencies or as punishment for individuals who refuse to cooperate or who actively collaborate with the hostile power. In all cases, resistance leaders must strictly control confiscation to ensure that it does not deteriorate into looting.

S-31. Guerrillas need the ability to acquire, store, and distribute large quantities of supplies without standard lines of supply and communications. They accomplish this by maintaining a decentralized network of widely distributed caches instead of large centralized stockpiles. This minimizes the loss of materiel if a guerrilla base moves quickly or faces destruction. This network allows the guerrillas to conduct operations across a wide area without a long logistics tail. The area commander caches extra supplies and equipment throughout the operational area. Caching is not a haphazard affair. Caches must support anticipated operational requirements or specified emergencies.

S-32. The logistics supply network also includes facilities for materiel fabrication, such as false documentation, improvised explosives and munitions, and medical aid. If the resistance is receiving external support, this network will extend to clandestine airstrips, DZs, seaports, and border-crossing sites.

S-33. The resistance organization obtains repair materiel from the local economy and through battlefield recovery to perform all maintenance and repairs within its capability. It may establish repair facilities within the area complex. The sponsor includes necessary maintenance and repair items with all equipment it provides the resistance. Introducing sophisticated equipment into the area complex may complicate the maintenance system.

## TRANSPORTATION NETWORKS

S-34. The resistance requires the capability to move personnel and logistics safely through enemy-controlled areas. Transportation networks include a compartmentalized series of safe houses or similar hiding locations. These locations allow the transport of personnel and materiel over long distances under the control of regional personnel who are familiar with the local enemy security measures. Security requires a complex series of recognition signals and communications that allow the individual segments to transfer the personnel and materiel safely, with minimum exposure of either compartment to the other. These networks can also facilitate the evacuation of wounded personnel or personnel evading the enemy, such as downed airmen.

S-35. The area commander normally obtains transportation support from the auxiliary on a mission basis. The guerrilla force may have its own organic transportation system to meet its immediate needs. In remote or undeveloped areas, the primary means may be human porters or pack animals.

## MEDICAL SUPPORT NETWORKS

S-36. The resistance initially confines clandestine medical treatment facilities to emergency and expedient care, with little preventive medicine. Once the area complex sufficiently develops, the clandestine facilities can expand and become a semipermanent medical organization, which serves to—
- Sustain and preserve combat power.
- Support the population.

S-37. If the area command has not established a degree of clandestine medical support, the result will be evident in the guerrilla force's morale. Historically, a lack of proper medical attention has led to serious illness and disability that reduced overall unit effectiveness.

S-38. Medical elements supporting the resistance forces must be mobile, responsive, and effective in preventing disease and restoring the sick and wounded to duty. It is unlikely the movement will have a safe rear area where it can take casualties for treatment. Medical personnel help during combat operations by operating casualty collection points, which allows the healthy guerrillas to keep fighting. Medical personnel evacuate casualties from these points to a guerrilla base or civilian care facility.

S-39. The resistance organization requires basic medicines and other medical supplies to treat its members. Preventive medicine is especially important to a resistance organization because it normally does not have adequate facilities to treat diseases.

S-40. Resistance personnel use existing logistics and transportation nets to gain supplies and move casualties. The movement of wounded personnel across enemy-controlled areas by auxiliary members is a clandestine operation, not a support function.

S-41. There are three levels of care in the medical support network beyond simple, self-aid and buddy-aid. Each represents a progressively higher quality of treatment and usually represents a greater distance traveled within the system and away from the area of fighting.

S-42. The three levels of care in the medical support network are as follows:

- Aid station.
- Guerrilla hospital.
- Convalescent facility

## Aid Station

S-43. Aid stations are locations where trained medical personnel provide emergency treatment. Evacuation of wounded personnel from the battle area begins at these stations. Because the condition of the wounded may prevent movement to the unit base, personnel hide them in secure locations and notify the auxiliary. The auxiliary cares for and hides the wounded or evacuates them to a treatment facility.

S-44. The evacuation of the dead is important for security reasons. If the enemy identifies the dead, the families of the guerrillas may be in danger. Personnel evacuate and cache the bodies of those killed in action until proper burial or disposal of the bodies according to the customs of the local population. Removal and burial of the dead denies the enemy valuable intelligence concerning indigenous casualties.

## Guerrilla Hospital

S-45. A guerrilla hospital is a medical treatment facility (or complex of smaller facilities) that provides inpatient medical support to the guerrilla force. The resistance movement establishes a guerrilla hospital during the organization and buildup phase of its development. The hospital must be ready for operation at the start of combat operations, and it must be able to continue providing medical support until the leadership directs otherwise.

S-46. A guerrilla hospital rarely, outwardly, resembles a conventional hospital. The requirement for strict security, flexibility, and rapid mobility prevents visible comparison with conventional military or civilian medical facilities. As the guerrilla force consolidates its hold on the area complex, all medical support functions tend to consolidate. Safe areas allow the resistance to establish a centralized system of medical care. Sophisticated hospitals provide more elaborate care because they provide a wider selection of trained personnel and specialized equipment. These hospitals can also render more extensive and prolonged treatment.

## Convalescent Facility

S-47. A convalescent facility is the area where guerrilla forces send patients to recuperate. A guerrilla convalescent facility may be a safe house in which one or two convalescents are recuperating with an appropriate alibi, or it could be in any base in guerrilla-controlled areas.

## COMMUNICATIONS NETWORKS

S-48. Guerrillas and underground leaders need to communicate with their subordinate elements in an area where enemy forces are always actively looking and listening for any indicators that would compromise the location of guerrilla forces or their supporting mechanism. Because of the likelihood of a high early-warning threat, especially in the initial phases of the resistance movement, nontechnical communications should prevail.

## INFORMATION AND PROPAGANDA NETWORKS

S-49. Special networks are responsible for providing information to the population, against the will of the controlling regime. This information will bolster the will of the population to support the insurgent cause, undermine the legitimacy of the regime or occupying power, and undermine the morale of enemy security forces. Guerrilla forces may produce and distribute bootleg radio broadcasts, underground newspapers, Internet sites, social media postings or chat rooms, and rumor campaigns. Guerrilla propaganda networks also draw new recruits to the movement. The networks may also coordinate with sympathetic elements outside the country to raise international favor and support. The resistance or insurgent leadership must have a degree of communication with the propaganda network to produce a coordinated effort.

## RECRUITMENT NETWORKS

S-50. The insurgency requires new recruits to join all aspects of the movement. The incorporation of these individuals requires special security measures to prevent the compromise of the components. The insurgency often sequesters recruits until it can check the recruit's validity and until the recruit can complete training and possibly participate in an operation to prove his loyalty.

## INTELLIGENCE AND COUNTERINTELLIGENCE NETWORKS

S-51. Aside from normal intelligence collection requirements, the resistance must recruit new members. The resistance screens new members to ensure they are not infiltrators. Further details on this topic are beyond the scope of this publication

## FINANCIAL NETWORKS

S-52. Financial networks have become a critical aspect of supporting, sustaining, and resourcing operations. UW planners must consider the protection of supporting financial networks, while targeting the funding sources of the targeted regime or government. Financial support is typically drawn from a combination of domestic and international fundraising. Typical domestic sources include voluntary donations by supporters, tithes conflating religious duty with support for political activities, and legitimate money-producing front enterprises. Examples of involuntary domestic sources include mandatory union dues and association fees that are then funneled into insurgent coffers, "revolutionary" taxes imposed directly on the populace, extortion from legitimate neutral businesses, and outright criminal activity, such as bank robbing or drug trafficking.

S-53. Insurgent financial networks can also enjoy a variety of international sources: direct donations from sponsoring or sympathetic states, sympathetic NGOs and purpose-built fundraising front groups, legitimate business enterprises, and diaspora communities. Whereas the primary emphasis of domestic financial networks is the clandestine accumulation, transfer, and expenditure of funds, the greater emphasis of international financial support is maintaining the covert nature of the sources through money laundering by banks and other legitimate businesses or series of front groups. In the information age, digital transference of enormous sums is often easily achieved. Meanwhile, ancient methods of illicit cross-border activity (such as smuggling) are as relevant today as ever.

This page intentionally left blank.

# Appendix T

# Administrative Considerations

Based on preinfiltration intelligence, the SF operational detachment makes tentative plans for formalizing administrative machinery to support a resistance force during UW operations in the JSOA. Before the SF operational detachment's infiltration and contact by the Secretary of Defense or DOD agencies with the government-in-exile or area command, the JSOTF or SOTF can guide the establishment of pay scales, rank structure, codes, and legal systems. This guidance will ensure uniformity in all AOs throughout the country and preclude inflated rank structures, unrealistic pay scales, and kangaroo courts. Necessary forms to support this administrative machinery may be drafted and printed during the SF operational detachment isolation phase. Final decisions on the administrative organization must be delayed until after infiltration. By then, the area commander, in concert with the SF operational detachment, will have resolved the majority of outstanding issues.

## PLANNING CONSIDERATIONS

T-1. Administrative systems can be established early in the planning stages of deployment and finalized in the JSOA. Administrative systems should be simple and effective. At a minimum, these systems should include:

- Accountability of sensitive items, to include weapons, radios, cryptographic material, and drugs.
- Accurate and updated personnel files on the guerrillas. Fingerprints and photographs can be used for identification. Identification cards can serve as a pay receipt.
- Records of the sick, wounded, and deceased.
- Records of awards, decorations, schools, and special skills.
- A daily staff journal. Written OPORDs and reports will be kept to a minimum, coded for security, and issued on a need-to-know basis.

### RECORDS

T-2. Each guerrilla base camp must have an administrative section to maintain essential records. As the area complex develops, personnel can centralize administration and duplicate information and records that could compromise the operation. These records should be forwarded to the administrative section of the area command for miniaturization, classification, and disposition.

T-3. Because of its ultimate historical importance, personnel should maintain an operational journal. Reports of combat engagements must include the following information concerning the guerrilla force: the designation and commander, type of action, approximate strength, and casualties. This data should be made available to CA and PSYOP personnel for their operations.

T-4. The command structure record should reflect the designation of the various units within the guerrilla force (similar to a modified table of organization and equipment) and the auxiliary. It should also include the names and designation of key personnel.

T-5. The personnel dossier should list members of the various organizations within the area command, and the record must be kept current. In the initial stages of individual unit development, duplicate copies of their dossier can be sent forward to the next-higher command when feasible.

T-6. Personnel records should be maintained on guerrillas and other key personnel in the resistance movement to prove or refute posthostility claims for wartime service. Opposition may be expected when someone recommends to the area commander that all personnel be photographed and fingerprinted. Such opposition can be overcome, however, with assurances that a viable, secure system can be established and that the records will be exfiltrated from the JSOA and maintained at the JSOTF or SOTF. Records may be microfilmed and placed in a secure cache in the JSOA as an alternate means to exfiltration. Photographing these documents and, subsequently, caching or exfiltrating the negatives provides a method of preserving and securing records not obtainable by other means. The SF operational detachment will find its organic photographic equipment useful in making identification photographs for population control. Photographs are also helpful in organizing and controlling resistance forces. Personnel should take photographs at frequent intervals to send forward and to avoid having a lot of sensitive material on hand. After processing the negatives and determining their acceptability, personnel may destroy the originals of unit records or send them to higher HQ.

T-7. The information placed on personnel records should include the full name of the individual and:
- His home village or city.
- The date he joined the resistance force.
- Whether an oath of enlistment was taken.
- The date he was discharged.
- Promotions and demotions.
- Acts of bravery.
- Awards and decorations.
- Rank or position attained in the resistance force.
- Any disciplinary action taken against him.

T-8. Posting to the initial records may be by serial number; thus, there is no reference to individuals by name and no incriminating data to associate them with the resistance organization. Code names and simulated records should be maintained to prevent any possibility of compromise if captured.

## OATH OF ENLISTMENT

T-9. Resistance leaders must be convinced of the need for a formal oath of loyalty to the resistance movement. This formality will solidify the union of U.S. and indigenous troops to a common goal—the freedom of their country.

T-10. At an appropriate ceremony, the local indigenous leader may administer an oath of enlistment to each new member of the resistance force. After hostilities, the local government can then recognize the jurisdictional authority commanded by guerrilla leaders over individual guerrillas.

T-11. Personnel must sign the oath. It will become a part of the individual's personnel record, which should be secured accordingly. Ideally, the oath refers to the guerrilla code and to punishment for violations.

## OTHER DOCUMENTS

T-12. Casualty records include the names of personnel killed, wounded, missing in action, or separated from the guerrilla force because of illness or for other reasons. Grave registration information, at a minimum, should include name, date, cause of death, and location of the remains as accurately plotted as possible

T-13. Medical records must include data on the type of prevalent diseases, preventive medicine actions taken, types of wounds, and general information on the organization of the medical structure for the area command.

T-14. The administrative section must maintain appropriate payroll records to support any commitment made to members of the resistance force for services rendered. The area command should maintain sufficient records to help settle claims after hostilities.

## AWARDS AND DECORATIONS

T-15. Guerrillas should formalize and establish valorous actions and meritorious acts and service within the force. Sometimes, U.S. awards may be recommended; however, the group S-1 is responsible for guidance in this area. He can provide such guidance in the isolation phase. A government-in-exile or the area command may wish to act as the final approving authority and can provide general guidelines for the establishment of an awards and decorations program. Once the S-1 approves an award or decoration, it is awarded at an appropriate ceremony consistent with security regulations.

# DISCIPLINE CONSIDERATIONS

T-16. Strict discipline is an integral part of command and control procedures instigated over any paramilitary force. Without discipline, the force cannot survive, let alone carry out effective operations against the enemy. Since guerrillas are usually not in similar uniforms and often appear dirty or bedraggled, an impression persists that discipline is loose in guerrilla units. Discipline must be hard but fair in guerrilla units. Orders should be executed without delay or question. Minor infractions of orders, especially during the conduct of operations, may have broad, negative consequences for guerrillas and supporting resistance elements.

T-17. More often than not, a guerrilla force will have a code, possibly not in writing, but there would certainly be an understanding of what is expected of all guerrillas. The resistance leader, with assistance from the SF operational detachment, can develop a written code. Each new recruit must know and understand its provisions and the penalties for treason, desertion, and dereliction of duty. Codes for guerrilla forces are usually simple but call for extreme punishment for what would be called minor infractions in conventional forces.

T-18. Any legal code for guerrilla forces should, if possible, be in line with the one that existed for the regular military forces of that country. If this code is impractical, the area commander and his staff may draft a new code.

T-19. Provisions must exist for punishments similar to those imposed under Article 15, *Uniform Code of Military Justice*, and for more severe punishments. The area commander will be advised to establish a court-martial or tribunal to try cases, rather than arbitrarily deciding the fate of the alleged perpetrator.

T-20. As soon as possible after infiltration, all parties must reach an understanding concerning the exercise of disciplinary and judicial authority over the SF operational detachment. The SF commander will not give up his disciplinary or judicial authority to the area commander or to any other resistance official, unless directed by U.S. higher HQ.

T-21. With decentralization of command and widely dispersed operations, individuals are habitually given mission orders with little guidance or supervision. They are expected to complete their missions—no excuses are accepted. Therefore, guerrillas must understand that their personal conduct has to be above reproach when interacting with the civilian population. Every act that loses civilian supporters is harmful to the resistance movement. PSYOP forces impress on the resistance organization's leaders that producing favorable reactions among the populace is vital. Such programs must stress proper individual and official conduct toward the populace. They must also point out the need for stringent disciplinary action against offenders.

T-22. The area commander ensures discipline is maintained. The SF commander normally provides advice that will ensure fair and consistent discipline. The ethnic culture of a group may indicate the most effective way to encourage self-discipline.

This page intentionally left blank.

## Appendix U

# Guerrilla Periods of Instruction

The master training program for the 10-day and 30-day leadership schools provides indigenous leaders and potential leaders with general knowledge of the subjects to teach subordinate personnel. Trainers place the primary emphasis on the role of the leader or commander in preparing leaders to supervise the activities of their subordinates. Trainers assume that most individuals in leadership positions have prior military service. Attendees should already possess a basic knowledge of the subjects that the trainers will cover.

## EXAMPLE OF A 30-DAY LEADERSHIP COURSE

U-1. Figure U-1, pages U-1 and U-2, shows an example of a 30-day master training program that leaders may use as a basis for preparing individual master training programs for each indigenous unit.

| Subject | Scope | Hours | | | |
|---------|-------|-------|-------|-------|-------|
| | | Day | Night | Total | PE |
| Map Reading with Compass (Day and Night) | Map reading; map orientation with compass; self-location; azimuth determination; and compass use. | 14 | 10 | 24 | (20) |
| First Aid, Field Sanitation and Hygiene | Basic wound treatment; infection prevention; simple bandaging; pressure points; shock prevention; splint placement; litter construction and use; field sanitation measures with water supply; waste disposal; and personal hygiene. | 6 | 4 | 10 | (7) |
| Individual Tactical Training (Day and Night) | Camouflage; cover; concealment; movement; observation; reporting; discipline; sounds; hand-to-hand combat; combat formations and night movement; night camouflage; preparation of equipment and clothing; night visions, sounds, and observation; night security and formations; message writing; immediate action drills; and security of operational bases. | 26 | 9 | 35 | (31) |
| Small-Unit Tactics, Raids, Patrols and Ambushes (Day and Night) | Planning; organization; preparation; formations; commands; control; security; communications; patrol reporting; objectives; target selection; raid force organization; reconnaissance and intelligence planning; raid preparation, movement, deployment, and conduct; raid force disengagement and withdrawal; ambush characteristics, definition, and objectives; ambush site selection; ambush force organization; ambush operation phases; planning, preparation, movement, deployment, execution, disengagement, and withdrawal of ambush forces. Trainers cover all subjects for day and night. | 26 | 44 | 70 | (60) |

**Figure U-1. Sample master training plan for 30-day leadership course**

| Subject | Scope | Hours | | | |
|---|---|---|---|---|---|
| | | Day | Night | Total | PE |
| U.S. and Foreign Weapons | Weapons include 5.56-mm and 7.62-mm automatic rifles; .38-caliber, .45-caliber, and 9-mm handguns; submachine guns; and machines guns. Training includes care and cleaning, loading, aiming, stoppages, and range firing. Trainers familiarize trainees with all weapons, as well as day and night firing. | 26 | 9 | 35 | (31) |
| Intelligence | Security measures; information gathering and reporting; captured documents and materiel; prisoner handling and interrogation; and counterintelligence procedures. | 26 | 44 | 70 | (60) |
| Air Operations | DZ establishment; marking and identification; security; and the transport and reception of supplies and equipment. | 28 | 10 | 38 | (32) |
| Demolitions | Nonelectric and electric firing systems; charge placement and calculation; rail and bridge destruction; booby traps; and expedient devices. | 21 | 8 | 29 | (24) |
| Squad Tests | Review and exercise covering all instructions. | 23 | 16 | 39 | (37) |
| Platoon Tests | Review and exercise covering all instructions. | 42 | 24 | 66 | (63) |
| Total Hours in Master Program | | 210 | 140 | 350 | (304) |

*Notes:*

1. Trainers will use the maximum number of trained, indigenous personnel to assist in training others. Trainers should identify potential indigenous cadre and leaders. In addition, trainers must identify personnel with substandard leadership ability, knowledge, skill, or desire.

2. Whenever possible, trainers will integrate intelligence collection, compass instruction, map familiarization, observation and reporting, individual tactical training, patrolling, weapons, demolitions, and field sanitation.

3. Trainers can break classes down to platoon-sized groups whenever possible.

4. Trainers should use practical work exercises, demonstrations, and conferences instead of lectures whenever possible.

5. Trainers must stress small-unit training (patrol, squad, and platoon) and develop teamwork and esprit de corps.

**Figure U-1. Sample master training plan for 30-day leadership course (continued)**

# EXAMPLE OF A 10-DAY LEADERSHIP COURSE

U-2. Upon completion of the 10-day leadership school, the leaders will return to work and train with their units, thus expanding their knowledge. Figure U-2, pages U-3 and U-4, is an example of 10-day master training program for a leadership school training program for select indigenous personnel.

| Subject | Scope | Day | Night | Total | PE |
|---|---|---|---|---|---|
| Map Reading and Compass | (Same general scope as in the 30-day program.) Include how to read scale and coordinates. | 4 | 2 | 6 | (4) |
| First Aid, Field Sanitation and Hygiene | (Same general scope as in the 30-day program.) Emphasis on field sanitation and responsibility of commanders. | 4 | | 4 | (1) |
| Individual Tactical Training (Day and Night) | (Same general scope as in the 30-day program.) Emphasis on security of operational bases, movements, formations, night control measures, and duties and responsibilities of commanders. | 10 | 9 | 19 | (16) |
| Mall-Unit Tactics: Raids, Patrols and Ambushes (Day and Night) | (Same general scope as in the 30-day program.) Emphasis on planning, organization, preparation, command, control, security, patrol execution, ambushes, and raids. | 10 | 29 | 39 | (25) |
| U.S. and Foreign Weapons | (Same general scope as in the 30-day program.) Familiarization firing. Primary emphasis on employment of weapons. | 8 | 2 | 10 | (7) |
| Intelligence | (Same general scope as in the 30-day program.) Emphasis on basic intelligence and counterintelligence, as well as night vision. | 6 | 4 | 10 | (8) |
| Air Operations | (Same general scope as in the 30-day program.) Primary emphasis on selection and reporting of DZs, organization of reception committee, and duties and responsibilities of commanders. | 6 | 8 | 14 | (11) |
| Demolition | Familiarization with demolition procedures, including demonstrating, planning, and safety. | 5 | | 5 | (3) |
| Communication | Communications includes available systems, communications security, and simple cryptographic systems. | 4 | | 4 | (2) |
| Leadership Principles and Techniques | Military leadership traits, principles, indications, actions, and orders; responsibilities and duties of the commander; human behavior problem areas and problem-solving process; selection of junior leaders; span of control and chain of command; combat leadership. | 6 | | 6 | (4) |
| Tactics and Operations | Characteristics of guerrilla warfare; guerrilla operations, principles, capabilities, and limitations; organization of operational bases; security; civilian support; logistics; counterintelligence; combat employment; missions; tactical control measures; target selection; mission support site and defensive measures; responsibilities and duties of indigenous leaders. | 7 | 5 | 12 | (9) |
| Total Hours in Master Program | | 70 | 59 | 129 | (80) |

Figure U-2. Sample master training plan for 10-day leadership course

Notes:

1. Identify personnel with substandard leadership ability, knowledge, skill, or desire.

2. Upon completion of leadership school, trainers may schedule one additional day for coordinating and planning future operations.

3. A suggested arrangement of scheduling is as follows:

   a. Preparation for training and selection of leaders: 29 April through 4 May.

   b. Leadership training: 5 May through 14 May.

   c. Troop training: 16 May through 14 June.

**Figure U-2. Sample master training plan for 10-day leadership course (continued)**

# EXAMPLE DATA CARD FORMAT—PERSONNEL AND TRAINING RECORD

U-3. Figure U-3 depicts an example of a personnel data card format that may also serve as a training record. Using this format for both purposes simplifies recordkeeping and minimizes the number of records that personnel must maintain in the AO. The type and amount of information recorded will vary by AO, as will the degree of security the U.S. element affords resistance personnel.

1. Personnel Data

    a.    Joint Special Operations Area
    b.    Full name
    c.    Serial number
    d.    Rank
    e.    Date of birth
    f.    Place of birth
    g.    Unit
    h.    Date of enlistment
    i.    Last civilian address
    j.    Civilian occupation
    k.    Languages
    l.    Special skills and aptitudes (civilian)
    m.    Financial data
        (1) Name, Date, Amount Paid
        (2) Name, Date, Amount Paid
        (3) Name, Date, Amount Paid
    n.    Left thumbprint
    o.    Right thumbprint
    p.    Photograph
    q.    Date of death or demobilization

2. Training Record

    a.    Basic Training (subjects/dates)
    b.    Advanced Training (subjects/dates)

**Figure U-3. Personnel data card format**

## Appendix V

# Special Forces Advisor's Guide to the Combat Employment of Guerrilla Forces

*We . . . strike to win, strike only when success is certain, if it is not, then don't strike.*
North Vietnamese General Vo Nguyen Giap, *Military Art of People's War*

The combat employment of guerrilla forces can no longer be viewed as simply operations conducted in rural areas supported by low-tech TTPs. In the 21st century, increased urbanization and the ease of access to information and media offer an altered environment. In addition to the classic notion of a rural fighter drifting between borders to find sanctuary confined by national or regional boundaries, the modern day guerrilla operates within a global human terrain of networks organized and motivated by cultural ties, religion, and ethnicity.

## ENVIRONMENTAL FACTORS

V-1. Major factors that commonly affect guerrilla warfare in the present-day environment include terrain, culture, and population control. In judging the potential for and effects of combat guerrilla operations, physical location is of great influence when considering organization, tactics, and the security of operations. Less industrialized countries characterized with rural mountains, swamps, large forests or jungles have historically nurtured overt or guerrilla type resistance. Flat plains areas and large towns or cities have typically proven more apt to lead to underground resistance activities; however, the possibility of organizing a guerrilla force in these areas should not be overlooked.

V-2. A population's cultural environment also has its effects on resistance movements. The success of a guerrilla force depends heavily on continuous moral and material support from a civilian population. The local community is usually under intense pressure from anti-resistance enemy units or collaborators. Punitive measures, such as reprisals, terrorism, deportation, restriction of movement, and seizure of goods and property are routinely conducted against resistance cohorts or supporters of guerrilla activity, making patronage dangerous and difficult.

V-3. The urge to bear arms, escape, and fight the enemy is largely dependent on the courage and cultural background of the people. Individuals from rural or peasant environments, not subjected to tight governmental control, may have more opportunity to display their hatred of the enemy by overt and violent means, such as guerrilla warfare. People from an industrialized and highly urbanized culture may be constrained to resist with such activities as sabotage, propaganda, passive acts, or espionage. If the local populace has courage and a strong will to resist, enemy retaliations usually result in an increase in underground activities. The civilian community assists the guerrilla force by furnishing information or early warnings, providing safe houses, securing supplies, offering recruits, supporting evasion and recovery, assisting with subversion or sabotage, and supporting other activities.

## MOTIVATIONAL FACTORS

V-4. Along with environmental factors influencing guerrilla forces, the sociological climate produces many motivating factors that have a profound effect upon a resistance movement. Strong, individual motivations, conviction, and resolve are essential to the development of a resistance movement. In guerrilla units, some individuals develop strong ideological motives for taking up arms. These ideologies typically take root in two broad areas politics and religion. The individual guerrilla tends to subordinate his own personality to these ideologies—working constantly and solely for the "cause". The ideology of some guerrilla fighters is extremely strong and often considered the most reliable motive. Many individuals foster hate and revenge due to negative experiences with the enemy. People who have lost loved ones or witnessed atrocities may willingly fight as a result of agitation, incitement, or provocation. However, uncontrolled hatred can pose problems, as it is sometimes difficult to restrain or regulate passion and properly direct an individual's efforts. Others join resistance movements to keep from losing their livelihood. This may lead some to volunteer for personal gain or materialism, manifesting as a desire for money, material plunders, or to better himself/herself for personal rewards. Total materialism can be the worst motivational factor. An individual so motivated may be inclined to change sides or sell his services to the highest bidder if he believes he can gain more by fighting for an opposing force.

V-5. If a resistance movement is strong or gives the impression of being powerful, many individuals may join out of a feeling of personal safety, stemming from fear of enemy reprisals against themselves or their families. Some may even join in order to escape recruitment into the service of the enemy. Usually, these situations occur only after the resistance movement is well organized and the enemy has been weakened to some degree. In some instances, personal association with a resistance organization may satisfy power, pride, or the eagerness for adventure. Depending upon the moral fiber of the individual, these motives can prove potentially dangerous or may sustain him in times of great stress or adversity.

V-6. In addition to environmental and motivational factors, active participation in any resistance movement is highly influenced by its chance for success. A population must feel that they ultimately have a stake in victory or there can be no effective resistance development or undertaking. Resistance movements stand or fall on the caliber of leaders and committed individuals within the organization. An analysis of these factors plays an important part in evaluating any potential resistance organization, particularly guerrilla forces.

# POLITICAL AND LEGAL FACTORS

V-7. Guerrilla warfare may function more politically in nature than militarily. Though guerrilla operations are tactical in nature, guerrilla warfare is usually rooted in local political factions and power struggles. Guerrilla leaders may recognize a common enemy, while at the same time be politically opposed to each other. This may lead to dissipated efforts resulting in "in-fighting" or withholding of cooperation with sponsors until promises of political significance are resolved. The political imprint on guerrilla warfare is another aspect that must be closely studied.

V-8. Guerrilla warfare is bound by the rules of the Geneva Conventions as much as conventional warfare. As outlined in appropriate international agreements and FM 27-10, four important factors give a guerrilla legal status:
- Is commanded by a person responsible for the actions of his subordinates.
- Wears a fixed and distinctive insignia or sign recognizable at a distance.
- Conducts operations in accordance with the laws and customs of war.
- Carries arms openly.

V-9. If these four factors are present, the guerrilla is entitled to the same treatment from his captors as a conventional military soldier. During World War II, General Eisenhower sent a proclamation to Nazis and Frenchmen alike, formally recognizing the French Resistance Maquis as members of the Allied Forces, and warned the Germans that all guerrillas were to be given the same honorable treatment as the regular soldiers under him in the Allied Expeditionary Force.

# ORGANIZATION AND DEVELOPMENT

V-10. The organization of a JSOA during UW involves initial development of the area and buildup of resistance forces. Initial organization includes establishing the required command and administrative structure, taking necessary security precautions, and training and advising a nucleus of leadership and cadre. Buildup is the expansion of the original nucleus into an operational unit capable of accomplishing assigned missions. Detachments infiltrate the operational area either before or after initial organization has taken place. The plans for the organization of a guerrilla force start when the theater commander designates certain areas within the JSOA as UW operational areas. The UW operational area is the name given to a geographic area where SOF is responsible for the conduct of all UW activities, to include guerrilla warfare. Initially, one detachment may be given responsibility for an entire UW operational area. With the development of the UW operational area and an increase in detachments, subdivision into sectors may be necessary. Each sector has the same characteristics as an operational area but remains a subdivision of the UW operational area. Typically, an operational detachment becomes responsible for each sector.

V-11. Upon linkup with resistance elements, the detachment continues to develop a thorough area assessment, verifying information acquired through previous area studies and briefings, and revises plans as necessary to reflect the local situation. The area assessment serves as the detachment's estimate of the situation, and it is the basis for plans to carry out UW missions. It considers all the major factors involved, including the enemy situation and security measures, the political background of the resistance movement, and the attitude of the civilian population. In addition to the area assessment, the detachment immediately focuses on the development of the resistance into an effective force. To facilitate this development, several tasks must be considered, to include:

- Establishment of a SOF and resistance forces command relationship.
- Establishment of security, intelligence collection, and communications systems.
- Organization of a logistics system.
- Provision for administrative services.
- Establishment of a training program.
- Planning and execution of tactical operations.
- Expansion of forces to support theater objectives.
- Civilian support.

# COMMAND AND CONTROL

V-12. During UW inclusive of guerrilla operations, the detachment's primary concern is the development and control of the guerrilla forces. However, because the guerrilla unit is only one part of the forces generated by a resistance movement, other resistance forces—auxiliary and underground—must be considered. The other aspects of the total resistance movement are brought in only as they bear upon the UW mission.

V-13. In its early stages, a guerrilla element often is highly unorganized. In general, the people who become guerrillas have suffered a reduction of their living standards. Their main concern is grouping together for food, shelter, and mutual protection. Oftentimes, several groups begin independent operations in rural or urban areas with very little concern for coordination among them. Detachments may find that guerrillas are not cooperating with each other and may even be working for conflicting objectives. The detachment must obtain control of the guerrilla groups and advise, train, and coordinate their actions to ensure that missions are accomplished in accordance with operational and strategic end-states. The degree of control varies in different parts of the world and with the specific personalities involved. As the scope of operations increases, closer coordination between guerrilla units is required.

V-14. Although the military advantages of close cooperation between guerrilla units are obvious, a detachment may find that guerrillas resist efforts to unify. This opposition may be based on personal antagonisms or political or ethnic differences. The detachment can control supplies as a means to persuade the guerrillas to form a united force and cooperate. The detachment should not openly threaten to use this power except as a last resort. In conferences with the resistance leadership, detachment leadership should be careful not to become involved in political differences. The detachment would certainly lose the respect and cooperation of the guerrillas by taking sides in their internal disputes.

V-15. Once guerrillas have been convinced of the advantages of close cooperation, the detachment must decide on a command structure. While local conditions must be adapted to, certain factors should be considered in any situation. There must be sufficient control over the guerrillas to ensure they carry out assigned missions. At the same time, it should be understood that the nature of guerrilla operations requires individual units to be given a large measure of freedom in carrying out their missions. Perhaps the most delicate part of a detachment's leadership mission is ensuring that competent leaders occupy command positions. If leaders of the original groups are not capable of filling the positions they hold, the detachment should arrange for their removal without creating dissension, which could endanger the success of the mission.

# AREA COMMAND DEVELOPMENT

V-16. The area command is the formal organization integrating the detachment(s) and the resistance forces within a UW operational area. It is established as soon as the development process requires it. There can be no rigid pattern for the organization of an area command. It must carry out the basic functions for which it is responsible and be tailored in strength and composition to fit the situation and mission. When a UW operational area is subdivided, the subdivisions are called sector command, district command, subsector command, and so on. The basic doctrinal composition of the area command consists of a command group and the three recognized resistance forces (underground, auxiliaries, and guerrillas); however, all three resistance forces may not be established in a UW operational area. The command group and staff is made up of the detachment, the local resistance leader, and representatives from the resistance forces in the area, and it organizes a staff as necessary. During guerrilla operations, the command group may be located with the guerrilla force; however, it is always located where it can best control the resistance movement.

V-17. Detachments may infiltrate a UW operational area in different sequences to establish an area command. The order and composition of detachment infiltration may depend upon many factors. Some of the more important factors include characteristics of the resistance movement, capabilities of the detachment, and needs of the theater commander. One detachment may be infiltrated when the situation is not well known, the resistance movement is not extensive, or the guerrilla force is so well organized that minimum coordination is needed. Another possible solution is for two or more detachments to infiltrate concurrently—each setting up a separate sector command This solution is adopted when the topography, enemy situation, or problems peculiar to the resistance movement prohibit the initial establishment of an effective area command (figure V-1).

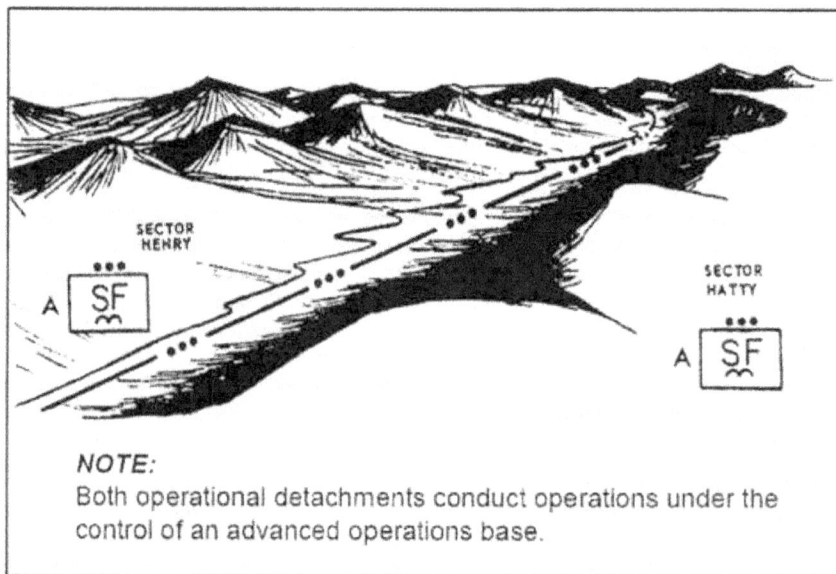

Figure V-1. Independent sector commands

V-18. After an area command has been established, other detachments can be infiltrated to set up sector commands within the area (figure V-2, page V-5). Either a B or C detachment is infiltrated or an initial A detachment is redesignated a B detachment. With a B or C detachment initially in the area, A detachments may be infiltrated to establish the sector commands. Subsequent infiltration of other operational detachments takes place with the expansion of the resistance forces, an increase in operations, or for political reasons.

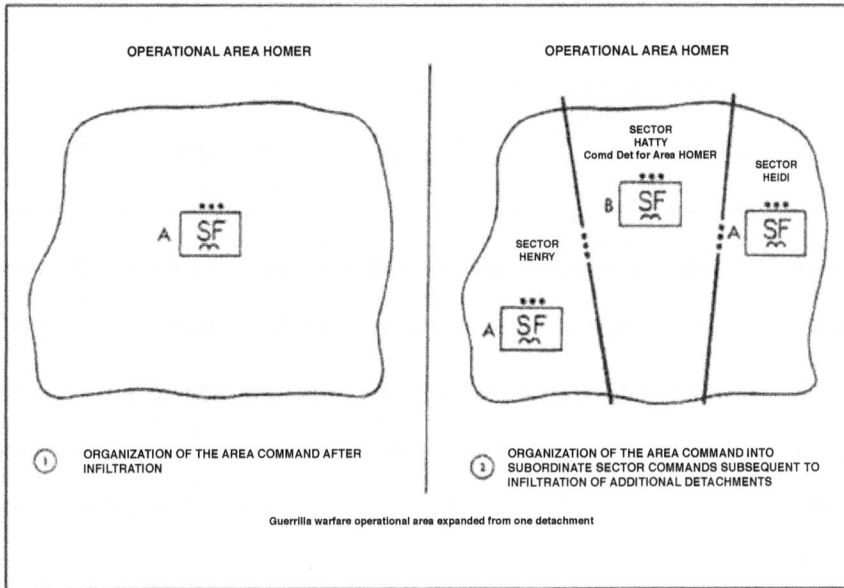

**Figure V-2. Area command divided into sectors**

V-19. After separate commands have been established, a detachment B or C may be infiltrated to establish an area command for the same reasons as previously mentioned (figure V-3).

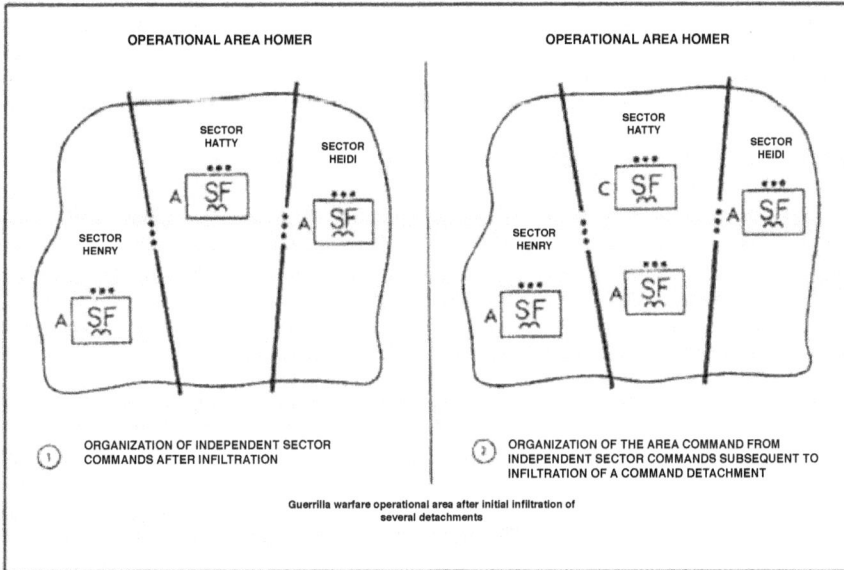

**Figure V-3. Area command established over sectors**

V-20. Because of the nature of operations and the distances involved, control measures are not as effective within an area command as they are in a conventional military organization. Thus, certain criteria are established to increase effective control. Sufficient guidance to subordinate units is outlined in the OPORD to cover extended periods of time. This is especially true when operations preclude frequent and regular contact. OPORDs include long-term guidance on such matters as PSYOP, intelligence, target attack, air support, external logistics support, evasion and recovery, and political and military relationships through the resistance. Another technique used to maintain control is the use of SOPs. SOPs standardize recurring procedures and allow the detachment and SOTF to anticipate prescribed actions when communications have been interrupted.

# PHYSICAL ORGANIZATION

V-21. The physical organization of the area, together with the command structure, is another priority task of the detachment following infiltration. In some situations, the organization of the area is well established, but in others, organization is lacking or incomplete. In all cases, however, some improvement in the physical dispositions will most likely be necessary. Organization is dictated by a number of requirements and depends more on local conditions than upon any fixed set of rules. Among the factors considered are degree of guerrilla unit organization, extent of cooperation among resistance forces, amount of civilian support, enemy activity, and topography. In practice, the detachment can expect to make compromises in organization because it is difficult to bring together in one area an ideal set of circumstances.

V-22. The basic guerrilla establishment within the UW operational area is the guerrilla base as discussed in Appendix S. A guerrilla base is a temporary site where the installations, HQ, and units are located. There is usually more than one guerrilla base within a UW operational area. Guerrilla bases add to the infrastructure of the overall area complex. From a base, lines of communication stretch out and connect other bases and various elements of the area. Installations normally found at a guerrilla base are command posts, training and bivouac areas, supply caches, communications sites, and medical facilities. In spite of the impression of permanence of the installations, a guerrilla base is considered temporary, and tenant guerrilla units must be able to rapidly abandon the base when required.

V-23. By virtue of their knowledge of the terrain, guerrillas should be able to recommend the best areas for locating guerrilla bases and supporting installations. Whereas inaccessible areas are best for the physical location of guerrilla bases, the lack of these remote areas does not preclude guerrilla operations. For instance, there may be times when guerrillas are able to fight effectively in towns and on the plains. Approaches to the base are well guarded and concealed. The locations of guerrilla bases and supporting installations are disseminated on a need-to-know basis. Since guerrilla forces seldom defend fixed positions for extended periods of time, alternate areas are established to which the guerrillas withdraw if their primary area is threatened or occupied by the enemy.

# ORGANIZATIONAL GOAL

V-24. The ultimate organizational goal is to integrate all elements of the resistance and the detachment into a unified force. The degree of unification depends upon many factors. The organization that combines the detachment and the resistance, regardless of the degree of cohesion, is called the area or sector command. The guerrilla force is the overt, militarily organized element of the area command. If a guerrilla force does not exist within a resistance movement, one is established when there is an operational requirement and the area commander agrees to accept U.S. sponsorship. Once the guerrilla force is officially recognized, it is the detachments responsibility to unite and control it to the best of its ability.

V-25. Active support from some of the civilian population and passive support from most of the remainder is essential to extended guerrilla operations. To ensure that both active and passive support is responsive to the area command, some form of organization and control is required. Control of civilian support is accomplished primarily through the auxiliaries. Auxiliary forces compose that element of the area command established to provide for and organize civilian support of the resistance movement. Auxiliaries may be organized in groups or operate as individuals. Auxiliary forces normally organize to coincide with or parallel the existing political administrative divisions of the country. This method of organization ensures that each community and the surrounding countryside is the responsibility of an auxiliary unit. It is relatively simple to initiate since auxiliary commands may be established at each administrative level, for example, regional, county, and district or local (communities and villages). This organization varies from country to country depending upon the existing political structure. Organization of auxiliary units can commence at any level or at several levels simultaneously and is either centralized.

V-26. The basic organization at each level is the command committee. This committee controls and coordinates auxiliary activities within its area of responsibility. In this respect, it resembles the command group and staff of a military unit. Members of the command committee are assigned specific duties, such as logistics, recruitment, transportation, communications, security, intelligence, and operations. At the lowest level, one individual may perform two or three of these duties. The command committee may organize civilian sympathizers into subordinate elements or employ them individually. When possible, these subordinate elements are organized functionally into a compartmented structure. However, because of a shortage of loyal personnel, it is often necessary for each subordinate auxiliary element to perform several functions.

V-27. The home guard is the paramilitary arm of the auxiliary force. Home guards are controlled by the various command committees. All auxiliary elements do not necessarily organize home guards. Home guards perform many missions for the local auxiliary force, such as conducting tactical missions, guarding caches, and training recruits. Their degree of organization and training depends upon the extent of effective enemy control in the area. Auxiliary units derive their protection in two principal ways, a compartmented structure and operating under cover. While enemy counterguerrilla activities often force the guerrillas to move temporarily away from given areas, the auxiliaries survive by remaining in place and conducting their activities so as to avoid detection. Individual auxiliary members carry on their normal, day-to-day routine, while secretly carrying out the many facets of resistance activity. Auxiliary units frequently use the passive or neutral elements of the population to provide active support to the common cause. Usually this is done on a one-time basis because of the security risks involved in repeated use of such people. The ability of auxiliary forces to manipulate large segments of the neutral population is further enhanced by the demonstrated success of friendly forces.

V-28. Enemy security measures and/or the antipathy of certain segments of the population often deny selected portions of an operational area to the guerrilla force or the auxiliaries. Since these areas are usually essential to the support of enemy operations, the resistance force attempts to extend its influence into them. The element used to conduct operations in such areas is the underground. The underground, then, is that element of the resistance force established to reach targets not vulnerable to other elements. The underground is employed to achieve objectives that would otherwise be unattainable. In many respects, the underground closely resembles the auxiliary force. They conduct operations in a similar manner and perform many of the same functions. The major differences are twofold:

- The underground is tailored to conduct operations in areas that are normally denied to the auxiliary force and guerrillas.
- The underground is not as dependent upon control or influence over the civilian population for its success. The degree to which the underground achieves this objective is a byproduct of other operations. Control of the population is not the primary underground objective.

# COMBAT OPERATIONS

V-29. The guerrilla force should carefully select, plan, and execute UW combat operations to ensure success with a minimum number of casualties. A combat defeat in the early stages of training demoralizes the guerrilla force. Combat operations should be commensurate with the status of training and equipment available to the resistance force. As training is completed and units are organized, guerrilla forces with SF assistance can plan and execute small-scale combat operations against soft (easy) targets—an important confidence builder. Later, they progress to larger and more complex targets. The simple, straightforward idea for conducting combat operations provided by General Giap above is the key concept of an SF operational detachment advising guerrillas should never disregard.

V-30. No word describes guerrilla combat operations better than "fluid". Both enemy and guerrilla units move and change their relative positions as the result of tactical maneuvers. The area of guerrilla activity is rarely, if ever, static; the situation changes constantly as the enemy reacts to guerrilla actions. The enemy holds localities where he occupies in force, and the guerrillas conduct their operations in those regions where the enemy is weakest. These factors reduce the enemy's capability to mount coordinated operations quickly against the guerrillas, allowing sufficient time for guerrilla units to avoid becoming involved in static defensive combat. These opposing courses of action create an operational environment that is fluid.

V-31. Guerrilla combat operations are rarely focused with seizing and holding terrain. However, the guerrilla force is committed to establishing area control in order to expedite operations. Temporary control of a specific area is attained through maximum use of the principles of surprise, mass, and maneuver. Complete area control is attained whenever the enemy is incapable of effective interference with guerrilla operations. Maximum effective results are attained through offensive guerrilla operations. Normally, the guerrilla force is primarily interested in the interdiction of lines of communications and the destruction of critical enemy installations. Conversely, the enemy force must provide security for its critical installations and seek to contact and destroy the guerrilla force. Because the guerrilla force can initiate offensive operations while employing a variety of methods of attack, against widespread target systems, and at a time and place of their choosing, complete security by the enemy is virtually impossible.

V-32. In urban areas, the guerrilla employs aggressive tactics of offensive character following the basic principles of attack and survive. Urban tactics focus on attacking and withdrawing swiftly to preserve forces and wear out, demoralize, and distract enemy forces. Urban guerrilla attacks are typically launched away from enemy logistics centers, making it difficult for the enemy to support a counterattack. Surprise is essential to the success of any guerrilla operation in an urban area. The urban guerrilla must fully blend in with the local population and urban lifestyle. He often makes his living in a normal manner when not forced to go underground. The urban guerrilla relies heavily on his knowledge of the terrain for survival. Knowing where to maneuver, where to rendezvous, where to escape to, and where to hide are critical to the preservation of the unit, and the individual guerrilla fighter.

## PRINCIPLES FOR SUCCESS

V-33. Guerrillas have proven themselves effective during all stages of conflict from the outbreak of hostilities until the end of fighting. However, in the early stages of guerrilla development, when the enemy is still strong, resistance operations normally tend to be conducted less openly. During this period, security is a prime concern. If the resistance movement is to survive and develop while surrounded by strong enemy forces, security precautions must be extensive and effective. Primary security measures ensure that they are well hidden and guarded and that there are secure methods in place to prevent enemy security forces from infiltrating their organization or discovering critical infrastructure. Activity is generally limited to information gathering, recruiting, training, organization, and small-scale operations. In time and when the situation changes to favor the organization of guerrilla forces, either due to enemy weakness or resistance-created favorable circumstances, operations become more overt, making large-scale actions possible.

V-34. Successful UW combat operations depend on five principles. Those principles are speed of movement, surprise, exploit low enemy morale, security, and collaboration with the local population:

- **Speed of Movement.** To achieve speed, guerrillas practice rapid force concentration and rapid deployment from march formations. They also practice movements and attacks during periods of limited visibility, pursuit of disorganized enemy units with little time wasted on reorganization after an engagement, and fast withdrawals. Guerrillas use an MSS and travel light to increase their element of speed and surprise.
- **Surprise.** To surprise the enemy, guerrillas plan to conduct and integrate deception operations into every UW mission.
- **Exploit Low Enemy Morale.** Guerrillas take advantage of every opportunity to undermine enemy morale by including PSYOP.
- **Security.** Guerrillas prepare the battlefield as far in advance as possible. Reconnaissance elements gather all available information on the terrain, installations, enemy units, and civilian activities. They also reconnoiter escape and withdrawal routes well in advance.
- **Collaboration with the Local Population.** The auxiliary in the area provides the guerrillas information, transportation, supplies, hideouts, and guides familiar with the objective.

## SECURITY WITHIN THE AREA OF OPERATIONS

V-35. The AO requires special security measures that apply particularly to guerrilla forces. The survival of the resistance movement depends upon constant vigilance on the part of every member of the organization, plus the ability to transmit warnings. Effective counterintelligence is also essential. Security measures must prevent losses by enemy action, ensure freedom of action, and minimize interruption of guerrilla activities. Dependable security can be achieved by intensive training in security discipline, establishment of warning systems, and extensive counterintelligence.

## Responsibility

V-36. The area commander is responsible for the overall security of the resistance forces, although commanders of subordinate units must take individual measures for their own local protection. The chief of the security section of the area command controls all security operations, except counterintelligence. He prescribes necessary measures and coordinates those adopted by subordinate commanders. Counterintelligence is the responsibility of the chief of the intelligence section of the area command. Again, subordinate commanders must establish local counterintelligence for their own security.

## Factors Affecting Security

V-37. Security measures developed by the chief of the security section of the area command are affected by the following factors:

- Mission.
- Local situation of individual units.
- Physical characteristics of the AO.
- The enemy situation.
- Capabilities and limitations of the resistance forces.
- Considerations affecting the civilian population.
- Operations of conventional and coalition forces.

V-38. During the early phases of UW, the mission of resistance forces will necessitate organization of a counterintelligence system alongside the intelligence system, development of a communications system that will facilitate warnings, and establishment of physical security for installations. Particular attention should be directed toward the enemy's state of internal security formations and their intelligence and communications systems.

V-39. Military actions against the enemy, initiated during the early phases of operations, should be planned and executed in such a way that will not lead to wholesale enemy anti-resistance activity, reprisals against the civilian population, or compromise of external logistics support, in the latter stages of insurgent warfare. Operations are not curtailed for security reasons because the established security system provides greater protection for the resistance. Also, resistance control over the area may rival the enemy's own influence.

V-40. Security consciousness is impressed upon guerrilla forces from inception and continues throughout operations. Commanders at all levels constantly strive to improve security measures. Particular attention is devoted to those units and elements that have recently been inactive or are located in relatively safe areas. Security measures enforced include:

- Camouflage discipline.
- Isolation of units from each other.
- Proper selection and rigid supervision of courier routes between the HQ and units.
- Police of camp sites and installations.
- Movement control within and between guerrilla bases.
- Isolation of guerrilla units from the civilian population at large. (Contact with civilians is accomplished through auxiliary elements.)
- Thorough indoctrination of all units in resistance to interrogation.

## Principles of Security

V-41. Guerrilla forces avoid a large concentration of troops. Even though logistics conditions may permit large troop concentrations, commands should be broken down into smaller units and widely dispersed. The dispersion of forces facilitates concealment, mobility, and secrecy. Large forces may be concentrated to perform a specific operation, but on completion of the operation, they should again be quickly dispersed. The principle of dispersion is also applied to command, service, and technical installations. For example, a large area command element is divided into several echelons and scattered over the area.

V-42. In the event of a well-conducted, large-scale enemy operation against the guerrilla force, the area commander may find it necessary to order the division of units into smaller groups to achieve greater dispersion and facilitate escape from encirclement. This action should be taken only when all other means of evasive action are exhausted because such dispersion renders the force inoperative for a considerable period of time, lowers the morale of guerrilla forces, and weakens the will of the civilians to resist. To assure successful reassembly of dispersed units, emergency plans must include alternate assembly areas.

V-43. All guerrilla installations and forces must have a high degree of mobility. Their evacuation plans must ensure that all traces of guerrilla activity are eliminated before abandonment of the area. Forces maintain evacuation mobility by ensuring that equipment that must be moved can be disassembled into one-man loads. The area commander ensures suitable caches for equipment that would reduce mobility are provided, materiel that could provide intelligence for the enemy is destroyed, the area is policed, and signs of the route of withdrawal are eliminated.

## Security of Information

V-44. Information concerning guerrilla operations is limited to those who need to know it. Only necessary copies are made or maintained. Each person is given only that information that is needed to accomplish his mission. Special efforts are made to restrict the amount of information given to individuals who are exposed to capture, which is why installations are not marked on maps or papers that could be taken off the base. Guerrilla force members habitually memorize the location of installations and areas to which they have access. Administrative records are kept to a minimum and cached, and the location is made known only to a required few. Whenever possible, references to names and places are coded, and the key to the code is given on a need-to know basis. Records that are no longer of value to operations or for future reports must be destroyed.

V-45. Strict security discipline is necessary and all security measures must be rigidly enforced. Security instruction of personnel must be extensive. They must be impressed with the importance of not divulging information concerning guerrilla activities to persons not requiring it. Individuals seeking such information must be reported to proper authorities. Security violations are extremely serious and demand severe punishment. All cases involving a possible breach of security must be reported immediately. The key to successful security of information, however, is the individual guerrilla who must always be security-conscious. One careless individual can destroy the best security system devised. During the training phase, security consciousness must be stressed. Special emphasis should be placed on safeguarding documents, security of information, and resistance against interrogation.

## Security of Movement

V-46. Security of movement can be provided only by an accurate knowledge of the enemy's location and strength. Intelligence regarding enemy disposition and activities is essential. The intelligence section of the area command, informed through its various nets, must provide this vital information for security of movement. After the routes have been selected, the units must be briefed on enemy activity, primary and alternate routes, dispersal and reassembly areas along the way, and security measures to be observed en route. If the route leads through areas outside resistance influence, auxiliary civilian organizations must provide security during movement.

V-47. While on the move, the guerrilla forces employ march security techniques, such as advance, rear, and flank guards. Pre-selected bivouacs are thoroughly screened by patrols prior to their occupation by guerrilla units. Contact is established with local auxiliary units designated to support the guerrilla movement. The auxiliaries are thus able to furnish the latest enemy information to guerrilla commanders

## Security of Installations

V-48. Most installations are located in isolated regions known as MSSs or guerrilla base. They are mobile and secured by guards and warning systems. Alternate locations are prepared in advance so that any installation threatened by enemy action can be evacuated from the endangered base area to a more secure area. Location of these alternate areas is given to personnel only on a need-to-know basis.

V-49. Physical security of installations will include terrain counterintelligence. This may vary from simple deceptive measures, such as camouflage or destruction and reversal of road signs and mileposts, to the creation of physical barriers, such as roadblocks and demolition of roadbeds and bridges. The use of civilian guides to misdirect enemy troops (for example, into ambush) can also be effective.

## Tri-Zonal Security System

V-50. A typical means of providing adequate security for the guerrilla base area is a tri-zonal security system. This system provides the following series of warning nets:

- Zone A is the guerrilla base area itself. It is secured by a regular guard system. Outpost, patrols, sentinel systems, warning devices, and methods of planned deception are designed to conceal the location, intent, or strength of the guerrilla force. Zone A largely depends on its safety from advance warnings received by clandestine agents in Zone C, or posted observers in Zone B. If enemy action threatens, the guerrillas move to another location before the arrival of enemy forces.

- Zone B, lying beyond the populated Zone C, is territory not well controlled by the enemy in which the guerrilla forces can operate overtly. It is usually open, rugged terrain, and the warning system depends upon stationed observers watching for enemy movements in the area.

- Zone C, the farthest from the guerrilla base area, is usually well populated and located inside enemy-controlled territory. Enemy security forces, police, and military units exercise relatively effective control, and the populace may be predominately hostile to the resistance or guerrilla forces. At the same time, there are excellent and rapid LOCs, whereby clandestine agents are able to warn the guerrillas quickly of enemy activity. This area is known as the clandestine zone and the functions of the warning system are the responsibility of both the auxiliary and the underground. Incidental to everyday operations, the resistance operating in this zone uncovers enemy activity or indications which, when evaluated, discloses potential danger to the guerrilla force. Resistance observers establish specific systems designed to provide a warning of the approach of enemy units. Resistance operatives in this zone intimidate any collaborators and attempt to elicit information from enemy personnel, local officials, and the police.

## Security of Communications

V-51. The enemy's capability to intercept electronic or nonelectronic communications and the operational situation are the two primary factors to be considered when planning communications within an operational area. During the early phases of guerrilla development, messengers are typically the chief means of communication. Security is enhanced by cellular structure of the messenger organization, use of secure cryptographic systems, and proper authentication.

V-52. Radio communications within an area or a sector depend on the operational situation, the physical location of the command echelons, terrain barriers, the training of the guerrilla force, the enemy capability in electronic interception, the security of the area, and the communications equipment available. Until the area is secure, the use of radio equipment should be restricted to those operational missions from which little intelligence data will accrue to enemy interceptors.

V-53. Resistance radio communications facilities are rigidly regulated by the SOI. These measures include restrictions on what may be transmitted; the use of codes and ciphers; and means of concealment, deception, and authentication. Particular emphasis is placed on restricting the time and number of radio transmissions to the absolute minimum. The modern ubiquity of cell phones—even among otherwise impoverished populations—significantly multiplies serious communications security concerns the resistance leadership and SF advisors will have to mitigate. The resistance can attempt to prohibit or restrict personal electronic devices entirely or can attempt to establish disciplined protocols for where, when, and by whom such devices may be used. Unrestricted personal use of electronic devices will almost surely result in compromise by the enemy.

## Outlaw Bands

V-54. Outlaw bands, operating as guerrillas, also endanger security by alienating the civilian population through their depredation. The area commander cannot tolerate outlaw bands, which are not willing to join the organized resistance effort. Every effort must be made to persuade these bands to join forces. If all other methods fail, it may be necessary to conduct operations against these groups.

## Reaction to Enemy Operations

V-55. Inexperienced guerrilla commanders and troops are often inclined to move too soon and too frequently to escape enemy troops conducting counterguerrilla operations. Unnecessary movement caused by the presence of the enemy may expose guerrillas to greater risks than remaining calm and concealed. Such moves disrupt operations and reduce security by dislodging previously established nets and exposing guerrillas to enemy agents, informants, and collaborators.

## Intelligence

V-56. A sound intelligence system is vital to successful guerrilla force planning and operations. Guerrilla tactics stress striking the enemy where he is weak and where he least expects to be hit. Guerrilla intelligence typically focuses on a detailed study of the terrain and of enemy strength, movements, dispositions, armament, and habits. Immediate dissemination of information, especially up-to-date changes, is a necessity. Besides combat intelligence, guerrilla forces are interested in strategic considerations, such as the enemy's political, economic, administrative, and propaganda policies and measures.

V-57. The organization of guerrilla force intelligence staffs and nets normally parallels that of regular forces. To collect information, the guerrilla force uses extensively organized nets of spies, informers, and agents. Civilians living near or working in enemy installations report on industrial operations, equipment, weapons, dispositions, habits, and morale. Government officials, sympathetic to the resistance movement, supply valuable information on the enemy's political, economic, administrative, and propaganda measures. Agents and observers located at strategic points report on ground, air, and naval movements.

V-58. Guerrilla activity readily lends itself to supplying valuable information to allied regular forces. During certain phases of an active war campaign, the primary mission assigned to guerrilla forces may be intelligence. Regular forces allied to a resistance movement may be expected to infiltrate personnel and equipment to aid and direct the guerrilla forces in the collection and evaluation of information and the production and dissemination of intelligence. Unless supervised by trained personnel, guerrilla forces may lack the ability to correctly evaluate reports and observations, and they may often pass on exaggerated information as true.

## Counterintelligence

V-59. Guerrilla security depends not only on security measures taken to safeguard information, installations, and communications, but also on an active counterintelligence program to neutralize the enemy's intelligence system and especially to prevent the penetration of the resistance organization and guerrilla forces by enemy agents. The intelligence section of the area command implements the counterintelligence program. Specially selected and trained counterintelligence personnel carefully screen all members of the resistance organization and protect guerrilla forces from enemy infiltration. Counterintelligence personnel can implement active campaigns of deception, disseminating false information to mislead the enemy. Counterintelligence active measures also include penetration of enemy intelligence and counterintelligence organizations by selected personnel and the manipulation of defectors and double agents.

V-60. Counterintelligence personnel must keep a constant check on the civilian population of the area through clandestine sources to ensure there is no presence of enemy agents within their midst. Civilians upon whom the guerrilla force depend heavily for support may compromise the resistance warfare effort as easily as a disloyal guerrilla fighter. False rumors and false information concerning guerrilla strength, location, operations, training, and equipment can be disseminated by counterintelligence through clandestine nets. Facts may be distorted intentionally to minimize or exaggerate guerrilla force capabilities at any given time. Although such activities are handled within the intelligence section, they must be coordinated with the security section in order to prevent inadvertent violations of security.

V-61. Counterintelligence efforts in support of guerrilla force operations is as important to security as intelligence is to the actual operations. Counterintelligence agents are used extensively to cover all fields of activity and zones of guerrilla force operations. Guerrilla force agents and loyal civilian supporters methodically seek out enemy spies and informers. Personnel joining the guerrilla force ranks are screened thoroughly to ensure they are not collaborating with the enemy. Normally, they are required to undergo a period of minimal responsibility, during which time they are closely observed before they are accepted as bona fide members of the organization.

## LOGISTICS OPERATIONS

V-62. UW missions cannot be accomplished without adequate support and sustainment. All units need food, clothing, water, medical, and personnel services sustainment. The types, quantity, and phasing of supplies influence the guerrillas, their capabilities and limitations, and the type of missions they undertake. Supplies and equipment made available to the guerrillas may influence their morale since each shipment represents encouragement and assurance of support from the outside world. Once a channel of supply is established, the guerrillas will continue to rely on that source for support.

V-63. Inherently, guerrilla units depend upon their own personnel for basic maintenance and repair of equipment and take overt action to satisfy logistics requirements. When difficulties in procuring supplies exist, rigid supply discipline is dictated. All personnel are expected to perform first-echelon maintenance. Guerrilla logistics plans typically provide for the maximum use of available supplies and the establishment of local repair facilities to prolong the life of equipment.

V-64. In addition, each guerrilla unit is assigned a portion of the operational area for logistics support through the sector or area command. Within this operational area, guerrilla units serve as a satellite to an auxiliary region and receive direct logistics support from local auxiliary units. In dealing with the auxiliaries, guerrillas must balance their requirement for supplies against the need to maintain the cooperation of their civilian supporters. A procurement program designed without regard to the needs of the population may impose such heavy commitments on the civilians that they refuse to cooperate. Non-cooperation of the civilian population limits auxiliary capabilities, impedes guerrilla operations, and increases requirements for external supply.

V-65. External supplies from a sponsor are most often delivered directly to a guerrilla unit. If this is not desirable or possible supplies can be delivered to a designated location and distributed to various users. Although this system takes much time and effort, it permits centralized control over sponsor-provided supplies, and it is the preferred method when the situation requires close supervision of subordinate elements.

V-66. Guerrilla units do not maintain excess stocks of supplies since large quantities of equipment limits mobility, without increasing combat effectiveness. Supplies in excess of current requirements are cached in a number of isolated locations to minimize the risk of discovery by the enemy. Caches permit a high degree of operational flexibility for the guerrilla force, and they are established and secured by both guerrilla and auxiliary units in support of the guerrilla force. Guerrilla caches are established for items used on a regular basis, materials required for future operations, and as a reserve for emergency use throughout the area. Items for cache are carefully packaged so that damage from weather and exposure is minimized. Specialized packaging of supplies can also be accomplished by the external sponsor.

## MEDICAL OPERATIONS

V-67. Due to the nature of guerrilla operations, battle casualties are normally lower in guerrilla units when compared to conventional units. However, the incidences of disease and sickness is often higher in guerrilla forces. Medical services are typically organized into two distinct functional areas: the area medical support system and the medical system within the operational area. The area medical support system is based primarily upon local facilities supplemented by sponsor-provided medical supplies. The medical system in the operational area features both organized guerrilla medical units and auxiliary medical facilities for individuals and small groups. The former are located in guerrilla base areas and staffed by guerrilla medical detachments The auxiliary facility is a location in which one or a small number of patients are held in a convalescent status.

V-68. Regardless of the varying size of guerrilla units, the medical detachments retain essentially the same structure and functions. Their duties are to maintain a high state of health in the command, to render efficient treatment and evacuation of casualties, and to ensure the earliest possible return to duty of those who are sick or injured. The detachment may also provide treatment and drugs to auxiliary and underground elements.

V-69. The organization of the medical detachment consists essentially of three sections, to include the—

- Aid station, which is charged with the immediate care and evacuation of casualties.
- Hospital, which performs defensive treatments of casualties and coordinates medical resupply and training.
- Convalescent section, which cares for patients who require rest and a minimum of active medical attention before their return to duty. The convalescent section is not located near the hospital area as this increases the size of the installation and thus the security risk. Instead, the patients are placed in homes of local sympathizers or in isolated convalescent camps.

V-70. During the early stages of development, the medical organization is small and probably combines the aid station and the hospital into one installation. The use of auxiliary convalescent facilities is found at all stages of development. Every effort is made to evacuate wounded personnel from the scene of action. The condition of wounded guerrillas may preclude movement with the unit to the base. In this event, the wounded are hidden in a covered location and the local auxiliary unit is notified. The local auxiliaries then care for and hide the wounded until they can be returned to their own organizations. The evacuation of dead from the scene of action is most important for security reasons. The identification of the dead by the enemy may jeopardize the safety of their families as well as that of their units. The bodies of those killed in action are evacuated, cached until they can be recovered for proper burial, or disposed of by whatever means is consistent with the customs of the local population.

V-71. As the operational area develops and the overall situation favors the sponsor, evacuation of the sick and wounded to friendly areas may be feasible. This lightens the burden upon the meager facilities available to the sector and area command and provides a higher standard of medical care for the patient. As the area command expands it is more efficient, from a medical standpoint, to establish a centralized system to provide advanced medical care. Field hospitals permit more flexibility because of the wider selection of trained personnel, equipment available to provide special treatment, and capability to relieve the aid stations of the responsibility for the prolonged treatment of patients. Since this type of installation may be fairly large and may have sizeable amounts of equipment, its mobility will suffer. For that reason, it is located in a relatively isolated area away from troop units, the HQ, and other sensitive areas to receive the maximum protection from guerrilla units.

V-72. Certain actions are taken to prevent the hospital from becoming so large that it attracts undue attention. First, as trained personnel, supplies, and equipment become available, additional hospitals are established. Second, patients are transferred to a convalescent home as soon as possible to complete their recovery. If the patient is placed in a civilian home unhidden, he should be provided the proper documents to make his presence there plausible. In some cases, the local population may not be able to support the sector or area command with qualified medical personnel. As the requirement for doctors and specialized personnel increases, the SOTF may have to provide additional medical personnel over and above the detachment's organic medical personnel.

# COMBAT EMPLOYMENT

V-73. The combat employment of guerrilla forces requires detachments to direct the efforts of indigenous resistance elements in combat operations. Integrated with these combat operations are PSYOP, evasion and recovery, and intelligence activities. Raids and ambushes are the principal offensive techniques of the guerrilla force. Raids and ambushes may be combined with other actions, such as mining and sniping, or these latter actions may be conducted independently. When raids, ambushes, mining and sniping are directed against enemy lines of communications, key areas, military installations, and industrial facilities, the total result is interdiction.

## CHARACTERISTICS

V-74. Detailed intelligence of enemy dispositions, movements, and tactics; thorough planning and preparation; and knowledge of the terrain are prerequisites of effective guerrilla combat operations. Combat operations of guerrilla forces take on certain characteristics that must be understood by SF personnel who direct and coordinate the resistance effort. These characteristics include planning, intelligence, decentralized execution, surprise, short-duration action, and multiple attacks.

## Planning

V-75. Careful and detailed planning is a prerequisite for guerrilla combat operations. Plans provide for the attack of selected targets and subsequent operations designed to exploit the advantage gained. In addition, alternate targets are designated to allow subordinate units a degree of flexibility in taking advantage of sudden changes in the tactical situation. Plans must be thorough and flexible enough to allow commanders who are responsible for an operation or series of operations to adopt alternate, predetermined courses of action when contingencies arise.

## Intelligence

V-76. The basis of planning is accurate and up-to-date intelligence. Prior to initiating combat operations, a detailed intelligence collection effort is made in the projected objective area. This effort supplements the regular flow of intelligence. Provisions are made for keeping the target or objective area under surveillance up to the time of attack.

## Decentralized Execution

V-77. Guerrilla combat operations feature centralized planning and decentralized execution. Actions of all resistance elements are directed and coordinated by the area command. However, within the guidance furnished by the area commander, subordinate units are allowed the widest possible latitude in the conduct of operations.

## Surprise

V-78. Guerrilla combat operations stress surprise. Attacks are executed at unexpected times and places, and set patterns of action are avoided. Maximum advantage is gained by attacking the enemy's weaknesses. Low visibility and adverse weather are exploited by guerrilla forces. Surprise may also be enhanced by the conduct of concurrent diversionary activities.

## Short-Duration Action

V-79. Usually, combat operations of guerrilla forces are marked by a short-duration action against the target, followed by a rapid withdrawal of the attacking force. Prolonged combat action from fixed positions is avoided.

## Multiple Attacks

V-80. Another characteristic of guerrilla combat operations is the employment of multiple attacks over a wide area by small units tailored to an individual mission. This is not piecemeal commitment of units against single targets but a number of attacks directed against several targets or portions of the target system. Such action tends to deceive the enemy as to the actual location of guerrilla bases, causes him to overestimate guerrilla strength, and forces him to disperse his rear area security and counterguerrilla efforts (figure V-4, page V-16).

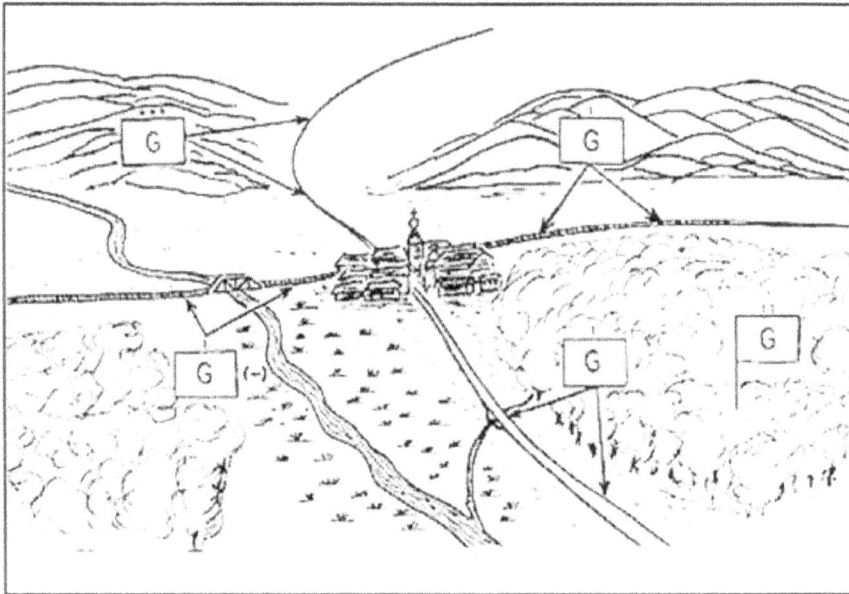

**Figure V-4. Multiple attacks by guerrilla units**

## TACTICAL CONTROL MEASURES

V-81. The area commander uses tactical control measures to aid him in directing and coordinating guerrilla combat operations. Common tactical control measures are:

- Targets (objectives).
- Zones of action.
- Axis of advance.
- MSSs.

### Targets (Objectives)

V-82. The area commander designates targets or objectives for attack by subordinate units. These targets are usually lines of communications, military installations and units, or industrial facilities. Normally, targets or objectives for guerrilla forces are not held for any length of time nor are they cleared of determined enemy fighters.

### Zones of Action

V-83. Zones of action are used to designate areas of responsibility for operations of subordinate units. Within the zone of action, the subordinate commander exercises considerable freedom in the conduct of operations. Movement of other guerrilla units, through an adjacent zone of action, is coordinated by the sector or area command. The auxiliary forces within a zone of action provide support to the guerrilla unit responsible for the area. Boundaries of zones of action are changed by the commander who established them as required (figure V-5, page V-17).

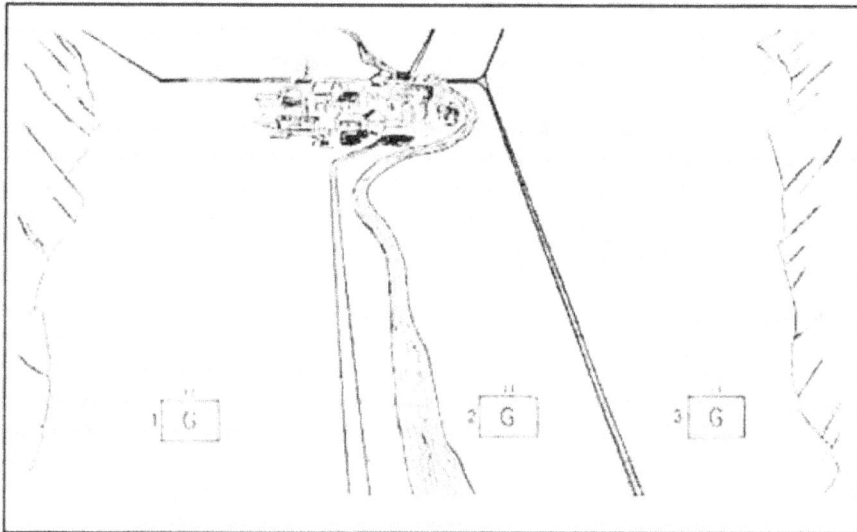

Figure V-5. Tactical control measure—zones of action

## Axis of Advance

V-84. Guerrilla commanders may prescribe axis of advance for their unit or subordinate units in order to control movement to targets. Guerrilla units move to the objective area either by single or multiple routes.

## Mission Support Sites

V-85. MSSs are used by guerrilla units to add reach to their operations and enable them to remain away from guerrilla bases for longer periods of time. The MSS is a pre-selected area used as a temporary stopover point, and it is located in areas not controlled by the guerrilla force. MSSs are used prior to and/or after an operation. The sites are occupied for short periods of time—seldom longer than a day. As in an assembly area, the using unit prepares for further operations and may be provided with supplies and intelligence by auxiliary forces.

## Additional Tactical Control Measures

V-86. Additional control measures may be employed by smaller guerrilla units, such as RPs, direction of attack, assault positions, and lines of departure. These control measures are employed in a manner similar to their use by conventional military units.

## DEFENSIVE OPERATIONS

V-87. Defensive operations are exceptional forms of combat for guerrilla forces. The guerrilla force may engage in defensive operations to—

- Prevent enemy penetration of guerrilla-controlled areas.
- Gain time for their forces to accomplish a specific mission.
- Assemble their main forces for counterattacks.

V-88. Guerrillas normally lack supporting fire (artillery, antitank weapons, and other weapons) to face conventional forces. Historically, guerrillas have avoided a prolonged position type of defense. When committed, they modify the principles of defensive operations to best meet their needs and offset the difficulties. They are aware of their limitations. The guerrillas choose the terrain that gives them every possible advantage. They seek terrain that denies or restricts the enemy's use of armor and complicates his logistics support. In the guerrilla-position defense, they raid, ambush, and attack the enemy's LOCs, flanks, reserve units, and supporting arms and installations. The guerrillas provide camouflaged sniper fire on officers, radio operators, and other high-value targets. They mine or booby-trap approach and departure routes.

V-89. Guerrillas may resort to defensive operations to contain enemy forces in a position favorable for attacking their flanks or rear. They often begin or intensify diversionary actions in adjacent areas to distract the enemy. Guerrillas use skillful ruses to lure the attacking forces into dividing their troops or hold objectives pending the arrival of conventional or allied coalition forces.

V-90. Defensive guerrilla operations may also focus on the concept of preservation of an area supportive of resistance activities and guerrilla forces against an enemy engaged in counterguerrilla operations. Under these circumstances, the goal is to undermine the enemy's counterguerrilla efforts and drive them out of the area through drawn-out, protected low-intensity conflict combined with PSYOP. Typical of guerrilla defensive counterguerrilla tactics is the use of a raid to conduct lightning fast attacks to destroy enemy outposts or logistics nodes or divert their efforts to other areas.

## Offensive Operations

V-91. Raids and ambushes are the principal techniques of the guerrilla force during offensive operations. Raids and ambushes may be combined with other actions, such as mining and sniping, or these latter actions may be conducted independently. When raids, ambushes, mining and sniping, are directed against enemy lines of communications, key areas, military installations, and industrial facilities, the total result is referred to as "interdiction."

V-92. Offensive guerrilla operations often extend the range of operations into highly contested areas where the enemy is in control of the terrain and, to some degree, the civilian population. Operations into these remote, enemy controlled areas may be conducted in unfamiliar terrain with logistics capabilities stretched to their limits. The civilian population may or may not be accustomed to resistance activities; they may be unfriendly or uncooperative toward resistance members and their organization. The success of a guerrilla offensive often relies on the penetration of underground organizations into denied areas to establish an initial resistance presence, collect intelligence on the enemy, develop clandestine infrastructure, and conduct subversion or sabotage operations.

V-93. Capitalizing on the unpopularity of the enemy, the underground works to gain traction with the civilian population to establish a base of auxiliaries. From this base of support, guerrilla forces can build out their tactical capabilities to conduct offensive operations. The basis of successful guerrilla offensive combat is aggressive action combined with surprise. During periods of low visibility, guerrillas attack, try to gain a momentary advantage of firepower, execute the mission to capture or destroy enemy personnel or equipment, and leave the scene of action as rapidly as possible. Normally, the guerrilla does not consistently operate in one area but varies operations so that no pattern is evident. If possible, two or three targets are struck simultaneously to divide the enemy pursuit and reinforcement effort.

---

**World War II Broadcast to Russian Soldiers and Civilians Caught Behind German Enemy Lines**

"In areas occupied by the enemy, guerrilla units. . .must be formed, diversionist groups must be organized to combat the enemy troops, to foment guerrilla warfare everywhere, to blow up bridges and roads, damage telephone and telegraph lines, set fire to forests, stores, and transports. In the occupied regions conditions must be made unbearable for the enemy and all his accomplices. They must be hounded and annihilated at every step, and all their measures frustrated."

Joseph V. Stalin, Radio Broadcast, July 1941

Bert "Yank" Levy, *Guerrilla Warfare*, 1964, p. 72.

---

V-94. The degree to which the offensive operations of guerrilla forces can be sustained depends, in the long run, on the MSS or base camp support available to them. When operating remotely from, or not with, conventional forces, the guerrilla forces establish and hold bases of their own. They locate their bases, if available, with a view to isolation and difficulty of approach by the opposing forces. They also consider strong defensive characteristics and closeness to neighboring, supporting states. The bases should be organized for defense and tenaciously defended by trained, motivated forces.

# RAIDS

V-95. Conventional U.S. doctrine describes a *raid* as an operation to temporarily seize an area in order to secure information, confuse an enemy, capture personnel or equipment, or to destroy a capability culminating with a planned withdrawal (JP 3-0). A guerrilla raid follows the same principles, with additional emphasis on ensuring surprise and survivability of the raiding force in both rural and urban areas.

V-96. Rural raids are characterized by secret movement to the objective area; brief, violent combat; rapid disengagement from action; and a swift deceptive withdrawal. Rural raids are conducted by guerrilla units to destroy or damage supplies, equipment or installations such as command posts, communication facilities, depots, radar sites, etc.; capture supplies, equipment or key personnel; or cause casualties among the enemy and his supporters. Other effects of raids are to draw attention away from other operations; keep the enemy off balance and force him to deploy additional units to protect his rear areas.

V-97. Urban guerrilla raids are characterized by rapid attacks on infrastructure located in densely populated areas, such as boroughs, municipalities, or even in city centers. The urban guerrilla operates in a three-dimensional environment; they can act on targets at ground level, above the ground, or subsurface. Targets may include police stations or substations; military support facilities; food production, storage, or distribution sites; inter-city road and rail transportation networks; or large clinics or hospitals.

V-98. The purpose of urban guerrilla raids is to foster anxiety among the enemy; capture arms, ammunition, or equipment; rescue detainees, prisoners, or combatants hospitalized and under enemy guard; destroy vehicles and equipment; or damage motor pools, hangers, or warehouses. In addition, guerrilla raids may be planned only to harass or demoralize the enemy.

V-99. Specific targets may require a small-scale raid operation to penetrate homes of government officials or military commanders; administrative or legal offices; or archives or public records centers. The purpose of targeting such locations may be to capture compromising documents or media for bribery or exposure of corruption or criminal activities.

## PLANNING CONSIDERATIONS

V-100. SF and guerrilla or area sector commanders must consider the nature of the terrain (METT-TC) and the combat efficiency of the raid force. Commanders base target selection on a decision matrix using CARVER. The SF operational detachment assesses the criticality and recuperability of various targets during the area study. Accessibility and vulnerability are situation-dependent, and these assessments must be supported by the most current area intelligence. CARVER factors are discussed in the following paragraphs.

V-101. **Criticality** is the importance of a system, subsystem, complex, or component. A target is critical when its destruction or damage has a significant impact on the output of the targeted system, subsystem, or complex, and, at the highest level, on the threat's ability to make or sustain war. Criticality depends on several factors:

- How rapidly will the impact of target destruction affect enemy operations?
- What percentage of output is curtailed by target damage?
- Is there an existence of substitutes for the output product or service?
- What is the number of targets and their position in the system or complex flow diagram?

V-102. **Accessibility** is the ease with which a target can be reached, either physically or by fire. A target is accessible when an action element can physically infiltrate the target or if the target can be hit by direct or indirect fire. Accessibility varies with the infiltration and exfiltration, the survival and evasion and security situation en route to and at the target, and the need for barrier penetration, climbing, and so on, at the target. The use of standoff weapons should always be considered when evaluating accessibility. Survivability of the attacker is usually most closely correlated to a target's accessibility.

V-103. **Recuperability** is a measure of the time required to replace, repair, or bypass the destruction or damage inflicted on the target. Recuperability varies with the sources and ages of targeted components and with the availability of spare parts. The existence of economic embargoes and the technical resources of the enemy nation will influence recuperability.

V-104. **Vulnerability** is a measure of the ability of the action element to damage the target using available assets (both men and material). A target is vulnerable if the unit has the capability and expertise to successfully attack it. Vulnerability depends on the—

- Nature and construction of the target.
- Amount of damage required.
- Assets available (manpower, transportation, weapons, explosives, and equipment).

V-105. **Effect** is the positive or negative influence on the population as a result of the action taken. Effect considers public reaction in the vicinity of the target, but it also considers the domestic and international reaction as well. Effects to consider include the following:

- Will reprisals against friendlies result?
- Will national PSYOP themes be reinforced or contradicted?
- Will exfiltration or evasion be helped or hurt? What will be the allied and domestic reaction?
- Will the enemy population be alienated from its government or will it become more supportive of the government?

---

*Note:* Effect is often neutral at the tactical level.

---

V-106. **Recognizability** is the degree to which a target can be recognized under varying weather, light, and seasonal conditions without confusion with other targets or components. Factors that influence recognizability include the size and complexity of the target, the existence of distinctive target signatures, and the technical sophistication and training of the attackers.

V-107. Target selection factors may be used to construct a CARVER matrix. The matrix is a decision tool for rating the relative desirability of potential targets and for wisely allocating attack resources (figure V-6). To construct the matrix, analysts list the potential targets in the left column. For strategic-level analysis, analysts list the enemy's systems or subsystems (electric, power, rail). For tactical-level analysis, analysts list the complexes or components of the subsystems selected for attack by their higher HQ.

| Potential Targets | C | A | R | V | E | R | TOTAL (Higher is Better) |
|---|---|---|---|---|---|---|---|
| Fuel Tanks | 2 | 5 | 3 | 5 | 5 | 5 | 25 |
| Fuel Pumps | 3 | 4 | 3 | 5 | 5 | 4 | 24 |
| Boilers | 4 | 2 | 5 | 4 | 3 | 3 | 21 |
| Turbines | 4 | 2 | 5 | 4 | 3 | 3 | 21 |
| Generators | 2 | 3 | 3 | 4 | 4 | 5 | 21 |
| Condensers | 4 | 2 | 4 | 4 | 3 | 3 | 20 |
| Feed Pumps | 3 | 4 | 3 | 4 | 4 | 3 | 21 |
| Circular Water Pumps | 3 | 4 | 4 | 4 | 3 | 3 | 21 |

Figure V-6. Sample CARVER matrix

V-108. Next, analysts develop concrete criteria for evaluating each CARVER factor. For instance, time may be used to evaluate criticality. If loss of a component results in an immediate halt of output, then that component is very critical. If loss of the component results in a halt of output, but only after several days or weeks, then that component is less critical. Similarly, percentage of output curtailed might be used as the evaluation criterion.

V-109. Once the evaluation criteria have been established, analysts use a numerical rating system (for example, 1-to-5 or 1-to-10) to rank the CARVER factors for each potential target. In a 1-to-10 numbering system, a score of 10 would indicate a very desirable rating (from the attacker's point of view), and a score of 1 would reflect an undesirable rating. The evaluation criteria and numerical rating scheme shown in the matrix are only included as examples. The analyst must tailor the criteria and rating scheme to suit the particular strategic or tactical situation and the particular targets being analyzed.

V-110. The area commander considers the possible adverse effects target destruction will have on future operations and the civilian population. Targets that will hinder or hurt the civilian population may be attacked only as a last resort. The goal is to diminish the enemy's military potential, not destroy the only footbridge in the area for civilians to go to work. However, an improperly timed operation may provoke enemy counteraction for which resistance units and the civilian population are unprepared. An unsuccessful guerrilla attack often may have disastrous effects on troop morale. Successful operations raise morale and increase prestige in the eyes of the civilians, making them more willing to provide support. PSYOP exploit the impact of successful raids. If a raid is unsuccessful, PSYOP personnel need to diminish the adverse effects on the friendly local indigenous force.

V-111. Although detailed, the plan for a raid must be practical and simple. The raid force commander plans activities so that the target is not alerted. He carefully considers time available, allowing enough time for assembly and movement. The best hours for the operation are between midnight and dawn when limited visibility ensures surprise. Personnel favor early dusk when knowledge of the installation is limited or other factors require tight control of the operation. A successful guerrilla withdrawal late in the day or at night makes close, coordinated pursuit by the enemy much more difficult.

V-112. The commander must strictly enforce OPSEC measures during planning. Only those personnel directly involved with the operation must be informed. Civilian sympathizers should never be informed of upcoming operations unless they provide support to the guerrilla forces. Personnel should carefully rehearse all raids and contingencies using real-time and full size mock-ups. They must also select and rehearse an alternate plan and escape route for use in case of emergencies.

V-113. The raid unit must also plan for medical support. Reactive planning in the medical arena is predictably unsuccessful, resulting unnecessarily in loss of life or limb. Adequate and visible medical planning has considerable positive psychological effects on the morale of the raid force. Personnel should plan to handle anticipated casualties with aid and litter teams at the objective, at planned RPs, and in the base area. Considerations should include evacuation routes at all levels and priorities for evacuation, nonevacuation, and hospitalization. Personnel should coordinate with treatment facilities before a raid but not divulge the target or timing of the mission.

## ORGANIZATION

V-114. The size of the raid force depends on METT-TC. The raid force may vary from a few personnel attacking a checkpoint to a battalion attacking a large supply depot. Regardless of size, the raid force consists of four basic elements: command, assault, security, and support with strategic placement of medical personnel within all elements.

## Command Element

V-115. The raid force commander and key personnel normally make up this element. They provide general support to the raid, such as medical aidmen, radio operators and, if a fire support element is part of the raid, a forward observer. The command element is not normally assigned specific duties with any element. Personnel may work with any of the major elements of the raid force. The raid force commander locates himself where he may best control and influence the action.

## Assault Element

V-116. Applying METT-TC, the assault element is specifically task-organized by what is needed to accomplish the objective. If the raid objective is to attack and render unusable critical elements of a target system, such as a bridge or tunnel, the raid force assaults and demolishes the bridge or tunnel. If the target is enemy personnel, the raid force conducts its attack with a high proportion of automatic assault weapons, covered by mortar fire from the support element. In some instances, the assault element may be required to physically move on or into the target—this method is the least preferred. A more preferred method is for the assault element to complete its task from a standoff distance. The assault element can attack from a standoff using lasers, antitank weapons, and other heavy weapons.

V-117. A guerrilla raid assault element is primarily organized and trained to accomplish the objectives of the raid. Due to the unconventional, isolated, and unsupported nature of guerrilla raids, the assault element typically consists of a "main action group" to execute the raid mission and personnel detailed to execute "special tasks," when required. The main action group executes the major task—the accomplishment of which ensures the success of the raid. The efforts of other elements of the raid force are designed to allow the main action group access to the target for the time required to accomplish the raid mission.

V-118. A special task detail may be required to assist the main action group in reaching the principle target. They can be tasked to execute such complementary missions as conducting a surreptitious entry; deactivating alarm systems; breaching obstacles with quickie tools, saws, or cutting torches; or clearing and securing off target withdrawal or escape routes. A special task detail may precede, act concurrently with, or follow the main action group.

## Security Element

V-119. The security element supports the raid by securing withdrawal routes, providing early warning of enemy approach, blocking avenues of approach into the objective area, preventing enemy escape from the objective area, and acting as the rear guard for the withdrawing raid force. The size of the security element depends on the enemy's capability to intervene and disrupt the mission. If the threat has armor, then the element needs anti-armor weapons. Where the enemy is known to have aircraft, the security element employs antiaircraft weapons. As the assault element moves into position, the security element keeps the command group informed of all enemy activities, firing only if detected and on order from the command group. Once the assault begins, the security element prevents enemy entry into or escape from the objective area. As the raid force withdraws, the security element, enhanced by sniper teams, conducts a rearguard action to disrupt and ambush any enemy counterattacks or pursuits.

V-120. In urban areas, the security element also focuses on interdicting police, paramilitary, or military reinforcements arriving on foot, horseback, or in vehicles. Those arriving on foot can be delayed by well-placed fires and hand grenades to pin them down or force their withdrawal. Those arriving on horseback can be defeated with ropes, ball bearings, or marbles in channelized areas. Those arriving in vehicles can be delayed or halted by forming road blocks with other vehicles, laying spike strips or similar devices, attacking with Molotov cocktails, or using gun fire aimed at critical areas, such as tires or engine blocks.

## Support Element

V-121. The support element of the raid force conducts diversionary or coordinated attacks at several points adjacent to or on the periphery of the main target to help the assault element gain access to the target. It uses ambushes, roadblocks, and mortar fire on the threat. Support personnel also execute complementary tasks in eliminating guards, breaching and removing obstacles, and conducting diversionary or holding actions. They assist by providing fire support and acting as demolition teams to set charges and neutralize, destroy, or render secondary parts of the target unusable. Historically, the support element has covered the withdrawal of the assault element from the immediate area of the objective and then withdrawn on order or prearranged signal.

## INTELLIGENCE AND RECONNAISSANCE

V-122. The raid force commander must have maximum information on the target site, enemy reaction forces (including routes, strength, and avenues of approach), and the routine activities and attitudes of the indigenous population in the area. Guerrilla force intelligence and reconnaissance personnel, scouts, or guides conduct a premission survey of the routes to the target, locations for friendly support weapons, enemy defenses (to include key weapons, minefields, and weak points), critical nodes to be destroyed within the target site, and withdrawal routes. The raid force gains access to the target site itself. Civilian auxiliary supporters may help in these attempts if they have a good cover for action status and are able to conduct plausible activity in or around the target area. If tactically feasible, personnel may conduct surveillance of the target to learn last minute requirements.

V-123. Intelligence and reconnaissance personnel conduct detailed intelligence gathering and leader reconnaissance before beginning the raid. Personnel conducting detailed reconnaissance construct a basic SALUTE report to compile information gathered, to include the following:

- Strength and location of the threat and its combat effectiveness.
- The threat's armaments and its location.
- Reaction time, security, and protection.
- Positions of key and automatic weapons.

V-124. Information gathering should also include answering the following questions:

- Are reserve threat troops in the vicinity?
- Are they waiting with armor or aircraft?
- What are their strengths, time to reinforce, and communications abilities?
- Is the terrain accessible?
- Can the terrain be blockaded or defended?
- What are the locations and capabilities of local inhabitants?
- What routes to and from the raid site provide cover, concealment, and security and simplify movement?
- Does the threat have armor or air support?
- Where should key support weapons—antitank, antiair, sniper teams, and machine guns—be placed?

V-125. In addition, the SF operational detachment—and when available, intelligence and reconnaissance personnel—should consult with supporting CA team members to consider the nonmilitary threats to the planned raid. They analyze the civilian component of the target area using ASCOPE. Typical questions are as follows:

- What civilian areas exist between the line of departure and the objective? What activities are employed in these areas?
- What civilian structures (permanent, semipermanent, or temporary) may be encountered along the route? What protection status is assigned to these structures?
- What civilian capabilities exist that could intercede or support the raid as part of a contingency? Is there a credible police capability?
- What organizations (host nation, United Nations, NGO, multinational corporation, criminal, or terrorist) exist in and around the objective area? What activities are they engaged in? What assistance might we obtain from them?
- What types of civilians might we encounter in and around the objective area? What general activities are they engaged in? What might be their reaction to contact with raid forces? What might be their reaction to combat operations?
- Are there any civilian events that may affect the conduct of the military operation, such as call to prayer or church services, festival celebrations, rush hour traffic, and planting or harvest season activities?

## PARTICIPANT REHEARSALS

V-126. Raid participants conduct realistic, timely rehearsals for the operation using terrain similar to the target area whenever possible. Participants use sand tables, full size mock-ups, sketches, and photographs, to assist in briefings. They practice immediate action drills along with contingency and emergency actions. When feasible, guerrillas hold full scale, final rehearsals under conditions and visibility realistically expected in the objective area at the time of attack. At a minimum, actions on the objective are always rehearsed in some capacity.

## NIGHT RAIDS

V-127. The best time for a raid is during limited visibility. Darkness allows units to maneuver even closer to the enemy. Enemy reinforcements will have difficulty in moving to assist their troops under attack, and air assets will be at a disadvantage. However, maneuvering at night is more difficult to accomplish, and command, control, and communications are more difficult to maintain.

## DAY RAIDS

V-128. Units conduct raids during daylight when enemy troops at the target location are lacking in security, morale, or discipline. A key daytime raid planning factor is whether they will get help from adjacent units, especially under adverse weather conditions of sandstorms, rain, or snowstorms.

## MOVEMENT

V-129. The raid force commander plans and conducts movement to the objective area so that the raid force's approach to the target is undetected (figure V-7, page V-25). Movement may be by single or multiple routes. The preselected route or routes may end in assembly areas, one or more patrol bases, or MSSs, which enhance mission success. The raid force makes every effort to avoid contact with the enemy during movement. Upon reaching the objective RP, security and leader reconnaissance parties deploy and make final coordination before the assault force moves to the attack position.

## FINAL INSPECTIONS

V-130. Final inspections prior to movement to the objective occur at one or multiple designated MSS(s) or objective RP(s). The raid force commander or key leader conducts or delegates to subordinate leaders a final inspection of personnel, weapons, and equipment before moving to the objective area. He ensures—

- Each member of the raid force is equipped and configured correctly with uniform and equipment common to all.
- Individual and crew-served weapons are assembled correctly, operational, and accompanied with the correct basic load of ammunition.
- Demolitions are constructed and configured correctly with redundant initiation systems.
- Communications equipment is operational.
- There is no disparity in the physical or mental condition of each raider.

V-131. He also checks personal belongings to ensure that no incriminating documents are carried during the operation. Prior to movement, the raid commander ensures each key leader of the raid force understands all primary, alternate, contingency, or emergency plans of the mission, to include actions on compromise prior to, during, and after the mission. This inspection assures the raid force commander that his unit is equipped and ready for a successful mission.

## ACTIONS ON THE OBJECTIVE AREA

V-132. Security and leader reconnaissance parties deploy to set eyes on the target and secure designated areas prior to the departure of the support and assault elements from the MSS or objective RP. On order, the support elements move to their positions and prepare to execute other assigned tasks to facilitate the movement of the assault elements to the target. If assigned, those members of the assault element with special task details move to their positions and prepare to execute specific tasks to clear the final approach to the target. The main action group of the assault element quickly follows into the target area and pre-positions for the assault. On initiation of the raid, each element conducts its designated actions on and around the objective area. Once the objective of the raid has been accomplished, the entire assault element withdraws and is covered by fires from the support element. If the attack is unsuccessful, the raid force ends the action to prevent undue loss of personnel, and the support elements withdraw according to plan. The assault and support elements assemble at one or more RPs, while the security elements further covers the withdrawal of the entire raid element according to plan.

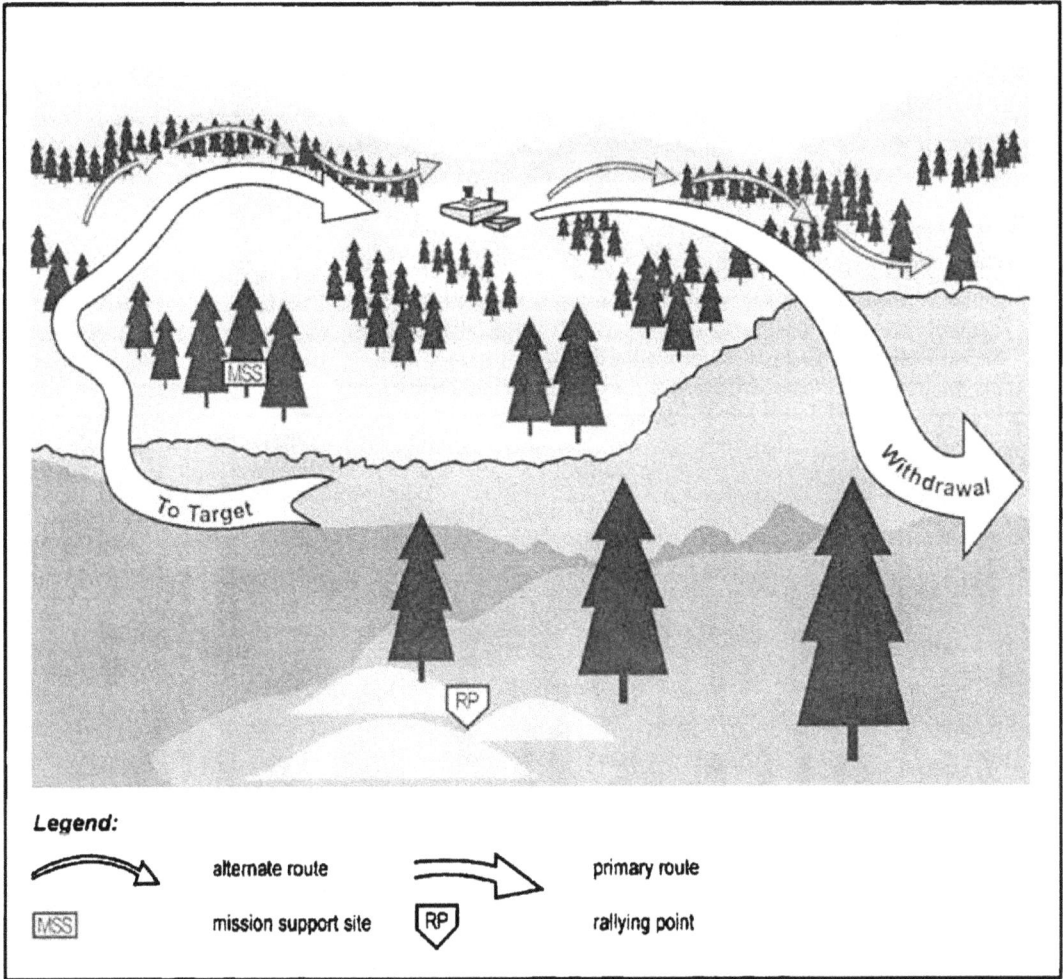

**Figure V-7. Movement to and withdrawal from the objective area**

## WITHDRAWAL

V-133. The raid force commander designs withdrawal to achieve maximum deception against the enemy and minimum danger to the raid force. The various elements of the raid force withdraw on order, or at a prearranged time, but never the same way twice. The movement uses many doglegs over the previously reconnoitered routes to the base camp through a series of RPs. Should the enemy organize a close pursuit of the assault element, the security element (covering force) assists by fire and movement, harassing the enemy and slowing it down. Other elements of the raid force do not attempt to reach the initial RP but, on their own initiative, separate into small groups or even individuals and lead the enemy away, attempting to lose them by evasive action in difficult terrain.

V-134. If follow-on missions are planned, the raid force may disperse into smaller units, withdraw in different directions, and reassemble at a later time at a predesignated location to conduct additional operations. For example, specific elements of the raid force may be further tasked to conduct operations such as ambushes of pursuing enemy forces during withdrawal. When compromised, and time permits, the raid force commander issues specific instructions concerning contingencies. The commander decides which COA to follow based on time and distance to be traveled, firepower or fire support, and the physical condition of the raid force. Predesignated contingencies plans can be issued through the use of codes for ease and speed of instruction to subordinate elements. Displaced or detached raid force elements then attempt to reestablish contact with the main force at other RPs or continue to the assembly area. MSS, patrol base, or base camp as separate groups in accordance with METT-TC.

## BATTALION (LARGE) RAIDS

V-135. When a target is large and well-guarded, a much larger raid force conducts the mission to ensure a successful attack. Large raids involve the use of a battalion-sized unit. Conduct is similar to that of smaller raids, but command and control becomes more difficult as the force increases in size.

### Movement to the Objective Area

V-136. Surprise is a priority in all raids, but it is more difficult to achieve during battalion operations. The number of troops to assemble and deploy requires additional MSSs farther from the target to preserve secrecy. The force requires a longer route to the attack position A large raid force usually moves by small components over multiple routes to an MSS, then to the objective (figure V-8).

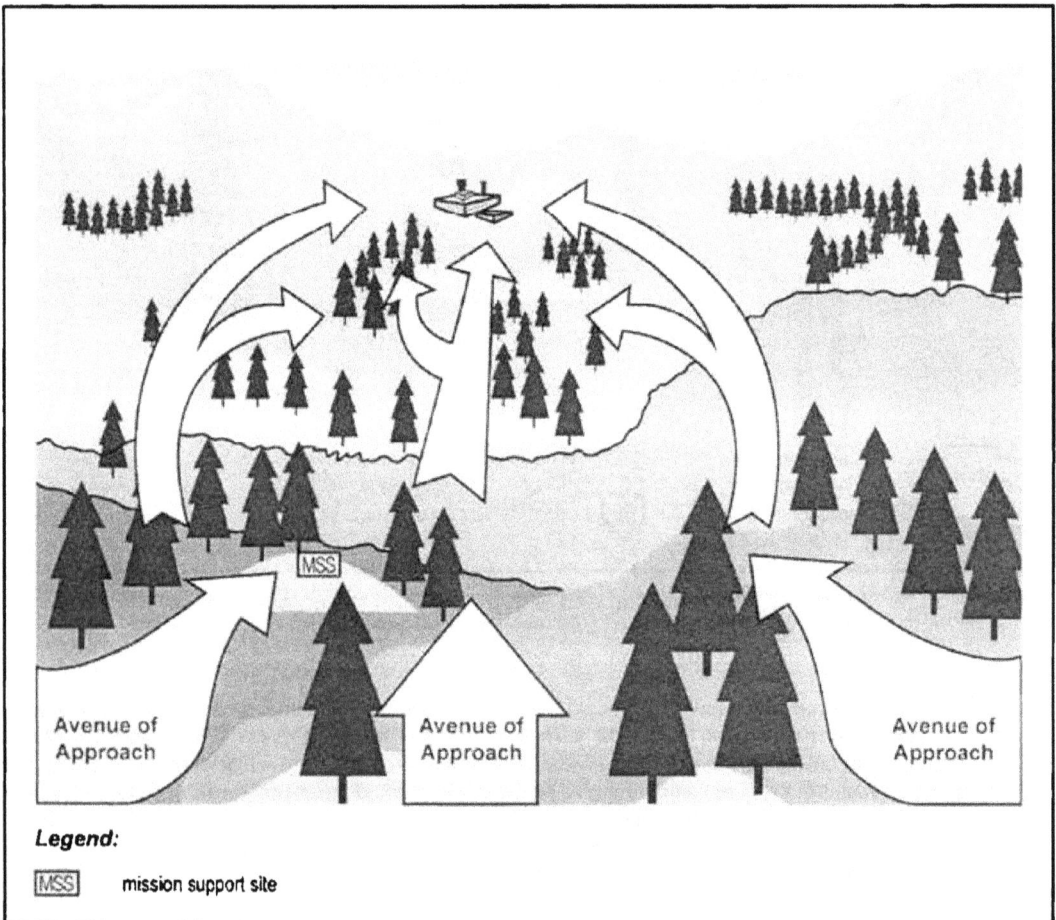

**Figure V-8. Movement to the objective area for a battalion raid**

### Control

V-137. Command and control is required for large raids. However, in an active electronic warfare environment, freedom of communications by radio is difficult to achieve. Recognizing this limitation, large raid planners typically revert to the use of immediately visible pyrotechnics (such as flares, star clusters, and so on), audible signals, or runners to coordinate actions at designated times. Even under optimum conditions, massing of the raid force at the objective is extremely difficult to control. Lights, armbands, or scarves also enhance control measures. During planning, the raid force commander considers the complexity of the plan and the possibility of overall failure if subordinate elements do not arrive on time. He plans for these possible contingencies to ensure mission success.

### Training

V-138.   Executing a large raid requires a high degree of training and discipline. Extensive rehearsals help prepare the force for the mission. In particular, commanders and subordinate key leadership must be prepared to command and control large numbers of troops as a cohesive and coordinated fighting force. At a minimum, actions on the objective is the primary rehearsal focus.

### Fire Support

V-139.   Raids usually require additional fire support. In the JSOA, such support may mean secretly caching ammunition in assembly areas, MSSs, or patrol bases over a period of time before the raid. In addition, during movement to the objective area, each member of the raiding force can be instructed to carry an extra mortar round, recoilless rocket round, or can of machine-gun ammunition to drop off at an MSS or support by fire position prior to initiation of the raid.

### Timing

V-140.   Timing is both crucial and much more difficult for a large raid. The time of the raid takes on increased importance because of the large number of personnel involved. More time is required to coordinate and move units, and the main action element usually needs more time to do its mission. As a result, larger raids typically require larger security elements to isolate the objective for longer time periods. Movement to the objective area is usually accomplished during limited visibility, but because of fire support coordination requirements and the large number of personnel, actions on the objective may not begin until early daylight hours.

### Withdrawal

V-141.   Elements usually best withdraw from a large raid in small groups, over multiple routes, to deceive the enemy and discourage enemy pursuit. Dispersed withdrawal also denies a priority target to enemy air and fire support elements. The raid force commander considers the possibility of an alert and aggressive enemy counterattacking the dispersed elements of the force. He carefully weighs all METT-TC factors before deciding how, when, and where he will conduct his withdrawal.

# AMBUSHES

V-142.   The ambush is a surprise attack from a concealed position upon a moving or temporarily halted target. It is one of the oldest and most effective types of guerrilla tactics. An ambush is executed to reduce the enemy's overall combat effectiveness by destroying or harassing his soldiers and their will to fight. An ambush may include an assault to close with and decisively engage the target, or the attack may be by direct or indirect fire to harass the enemy. In an ambush, the enemy sets the time and the attacker sets the place.

V-143.   Guerrilla ambushes are conducted to destroy or capture personnel and supplies; harass and demoralize the enemy; delay or block movement of personnel and supplies; and channelize enemy movement by making certain routes useless for traffic. The result usually is concentration of the majority of movements to principal roads and railroads where targets are more vulnerable to attack.

V-144.   In rural areas, passes and ravines in mountainous or wooded areas are good sites for guerrilla ambushes. Enemy forces can be lured into these areas by creating a detour or road block or altering road signs. The guerrilla ambush force typically employs sufficiently strong forces to enable them to completely smother and destroy the enemy by quick, violent action in a very short period of time. Commanding ground, concealment, and camouflage are fully exploited. Close proximity ambushes compensate for poor marksmanship to gain maximum effect. The guerrilla ambush element uses extensive security measures to cover its movement to the ambush position, conceal the action, and expedite a speedy withdrawal.

V-145. Silence and immobility are scrupulously observed in the area of the ambush. Guerrilla ambushes on enemy columns are typified by allowing small enemy advance guards to pass through the ambush position to be dealt with by separate elements. A prearranged signal is given to launch the ambush—followed by violent action, usually short in duration. If the enemy is incapable of counteraction and there is no threat of local enemy reinforcements, the guerrilla force quickly salvages usable supplies and equipment, destroys what cannot be recovered, and withdraws rapidly.

V-146. When the guerrilla ambush force is not strong enough to destroy the enemy completely, action is terminated by a prearranged signal as the enemy's counteraction begins to form. A planned withdrawal covered by security details is executed. Often, guerrillas will withdraw by detachments in several directions to frustrate and complicate enemy pursuit. Prearranged details are used to maintain visual contact and to report on enemy pursuing units. Subsequent ambushes will often be conducted to delay or destroy a pursuing enemy force.

V-147. In favorable terrain and during periods of low visibility, guerrillas may simultaneously ambush enemy columns moving toward each other. After deceiving the enemy and inducing a fire fight, the guerrilla forces clandestinely withdraw, leaving the two enemy forces to fight each other. Another tactic may be to act as a collaborator offering to guide enemy forces to attack a resistance element while leading them into a guerrilla ambush.

V-148. In urban areas, guerrillas take advantage of available three-dimensional terrain to channelize, confine, and fix enemy forces. In urban terrain, guerrillas can create a ruse or false alarm to bait enemy personnel or forces to a specific ambush site. As a result, enemy forces may by trapped in traffic or lured into surrounding a house or building suspected of harboring resistance members where they are then ambushed. The urban guerrilla ambush may incorporate snipers in elevated rooms or located at ground level, shooting through subterranean basement windows. Such urban snipers are often hard to spot when they are set up deep into a room away from open doors, windows, and loopholes. In certain cases, brief work interruptions, strikes, or protest marches can present opportunities to ambush state police or paramilitary forces. Tunnels between buildings and hidden sub-basements allow guerrilla fighters to move quickly between positions or hide out and wait until enemy forces move past them, surfacing later to their rear and ambushing from areas considered cleared.

V-149. Like the raid force, the ambush force is organized into assault, support, and security elements. The assault and support elements conduct the main attack against the ambush target, which includes halting a column, killing or capturing personnel, recovering supplies and equipment, and destroying unwanted vehicles or supplies which cannot be moved. The security force isolates the ambush site using roadblocks, other ambushes, and outposts. Security elements are also responsible for covering the withdrawal of the entire ambush element.

*Note:* The following article was originally serialized in Red Thrust Star, dated July and October 1995 and October 1996.

"Afghanistan is not Europe, yet the Soviet Army that occupied Afghanistan in late December 1979 was trained to fight NATO on the northern European plain. Consequently, the Soviet Army had to reequip, reform and retrain on-site to fight the insurgent mujahidin [holy warrior] guerrillas. The Soviets were forced to revise their tactics and tactical methodologies in order to meet the demands of this very different war. One of the tactical areas which the Soviets thoroughly revised was the conduct of ambushes. The Soviets planned to use ambushes in the European theater, but they were primarily ambushes against attacking or withdrawing NATO armored columns. The Soviets constructed most of their ambushes around tanks and tank units. They planned to employ concealed individual tanks, tank platoons and tank companies along high-speed

**(continued on next page)**

---

**(continued)**

avenues of approach or withdrawal to engage the enemy from the flank and then to depart. Such ambushes were part of security zone defensive planning as well as planning for the deep battle and pursuit. The Soviets also trained their squad and platoon-sized reconnaissance elements to conduct dismounted ambushes to capture prisoners and documents. They employed a command element, a snatch group and a fire support group in these small-scale ambushes.

In Afghanistan, the mujahidin seldom used armored vehicles and seldom advanced along high-speed avenues of approach. Instead, they infiltrated light-infantry forces through some of the most inhospitable terrain on the planet to mass for an attack or ambush. The Soviets soon discovered that they had difficulty maintaining control of the limited road network, which constituted the Soviet lines of communication. The guerrillas constantly cut the roads and ambushed convoys carrying material from the Soviet Union to the base camps and cities in Afghanistan. The Soviet ability to maintain its presence in the country depended on its ability to keep the roads open and much of the Soviet combat was a fight for control of the road network. During the war, the guerrillas destroyed over 11,000 Soviet trucks (and reportedly even more Afghan trucks) through ambush. The Soviets learned from mujahidin ambushes and used the ambush to interdict the guerrilla supplies coming from Pakistan and Iran. The Soviets conducted ambushes mainly with reconnaissance and other special troops (airborne, air assault, spetsnaz and elements from the two separate motorized rifle brigades which were designed as counter-guerrilla forces). The composition and employment of ambush forces differed with the units involved and the part of Afghanistan in which they were employed."

*Red Thrust Star*, July and October 1995 and October 1996

---

## OPERATIONAL CONSIDERATIONS

V-150.   In preparing the ambush plan, consideration is given to the:

- **Mission.** This may be a single ambush against one route or a series of ambushes against one or more routes.
- **Size of the Enemy Force.** Probable sizes, strength, and composition of the enemy force that is to be ambushed; formations likely to be used; and his ability to reinforce.
- **Terrain.** Terrain along the route favorable for an ambush, including unobserved avenues of approach and withdrawal.
- **Timing of the Ambush.** Ambushes conducted during periods of low visibility offer a wider choice of positions and better opportunities to surprise and confuse the enemy than daylight ambushes. However, control and movement to and during the night ambush is more difficult. Night ambushes are more suitable when the mission can be accomplished during or immediately following the initial burst of fire. Night ambushes also require a maximum number of automatic weapons to be used at close range. Night ambushes are effective in hindering the enemy's use of routes by night. Daylight ambushes facilitate control and permit offensive action for a longer period of time. A day ambush also provides the opportunity for more effective aimed fire of such weapons as rocket launchers and recoilless rifles.

## Destruction

V-151.   Destruction is the primary purpose of an ambush. The number of men killed, wounded, or captured and loss of equipment and supplies critically affect the enemy. Guerrillas benefit from the capture of equipment and supplies through battlefield recovery.

## Harassment

V-152.   Frequent ambushes harass the enemy and force him to divert men from counterguerrilla patrol operations to guard convoys, troop movements, and installations. When enemy patrols fail to accomplish their missions because they are ambushed, the enemy is deprived of the valuable contributions these patrols make to its combat effort. A series of successful guerrilla ambushes cause the enemy to be less aggressive and more defensive-minded. The enemy becomes apprehensive and overly cautious and reluctant to go on patrols, to move in convoys, or to move in small groups. The enemy wants to avoid night operations, is more subject to confusion and panic if ambushed, and eventually becomes mentally defeated.

## Element of Surprise

V-153. Surprise allows the ambush force to seize control of any situation. The force achieves surprise by carefully planning, preparing, and executing the ambush. Guerrillas attack the targets when, where, and in a manner for which the enemy is least prepared.

## Coordinated Fires

V-154. The ambush force commander positions and coordinates the use of all weapons, mines, and demolitions. He coordinates all fires, including artillery and mortars when available. Coordinated fire support ensures isolation of the kill zone. This isolation prevents enemy escape or reinforcement as a result of the large volume of accurate, concentrated fire.

## Control Measures

V-155. The ambush force commander maintains close control measures during the ambush operation. These control measures include provisions for—
- Early warning signals of target approach.
- Withholding fire until the target has moved into the killing zone.
- Opening, shifting, and halting fire at the proper time.
- Initiating proper actions if the ambush is prematurely detected.
- Timely and orderly withdrawal to a recognized RP.

## INTELLIGENCE

V-156. In preparation for an ambush, the guerrilla force needs vital information about the plans and movements of the enemy, to include where they are, how they move, their means of communication, and the secret activities they carry out both day and night. Particularly in rural areas, it is seldom possible to learn in advance the exact composition, strength, and time of movement of convoys; therefore, intelligence efforts are directed toward determining the convoy patterns of the enemy. Using this information, guerrilla commanders are able to decide on the type of targets to be attacked by ambush. In addition, intelligence considerations for a raid are equally applicable to an ambush.

## SITE SELECTION

V-157. In selecting the ambush site, the basic consideration of the guerrilla force is favorable terrain. However, limitations that may exist, such as deficiencies or proficiencies in the firepower of guerrillas and lack of resupply during actions, will govern the choice of the actual ambush site. The guerrilla ambush site ideally funnels the enemy into an engagement area, which offers friendly concealment and favorable fields of fire. Whenever possible, firing is done through a screen or foliage.

V-158. The entire killing zone is covered by fire to avoid dead space that would allow the enemy to organize counteraction. In rural areas, the guerrilla force takes advantage of natural obstacles, such as gorges, swamps, and cliffs, which will restrict enemy maneuver against the ambush force or away from the ambush site. When natural obstacles do not exist, guerrilla support elements emplace mines and demolitions prior to the ambush to channelize and contain the enemy.

V-159. In urban areas, enemy communications, reconnaissance, maneuver, and organized pursuit are all degraded. Dense infrastructure, congested throughways, crowded locations, and limited maneuver space provide the urban guerrilla with ripe terrain for ambush missions. Buildings above the street can be used for observation, security, sniping, anti-armor firing positions, placement of heavy machine guns, or launching of grenades or Molotov cocktails into a kill zone. Underground passages, sewers, and sub-basements provide concealed areas to maneuver, mass, and withdraw elements or store weapons, ammunition, and explosives in proximity of the ambush site. Within and around the kill zone, ground level entrances and doorways can be sealed off to prevent the enemy from fleeing into covered and concealed locations.

V-160. Security elements are placed on roads and trails leading to the ambush site to warn the assault element of any enemy approach. The proximity of security to assault elements is dictated by the terrain. In many instances, it may be necessary to organize secondary ambushes and roadblocks to intercept or delay enemy reinforcements.

## MOVEMENT

V-161. The guerrilla ambush force moves over a preselected route or routes to the ambush site. One or more MSSs are usually necessary en route to the ambush site. Last minute intelligence is provided by reconnaissance elements and final coordination for the ambush is made at the MSS.

## ACTION AT THE AMBUSH SITE

V-162. Guerrilla forces are moved to an assembly area near the ambush site. Security elements take their positions first, followed by support elements, and then the main ambush body (assault element). As the approaching enemy is detected, or at a predesignated time, the ambush commander decides whether to execute the ambush. This decision depends on the size of the element, guard and security measures, and estimated worth of the target in light of the mission. When a decision is made to execute the ambush, initiation is signaled when the most damage can be done to the largest amount of the enemy element within the kill zone. If ambushing an enemy column, advance guards can be allowed to pass through the main position. When the head of the main column reaches a predetermined point, it is halted by fire, demolitions, or obstacles.

V-163. At the initiation signal, the assault element and support element(s), when employed, open fire along predetermined sectors of fire within the kill zone. Designated details engage the advance and rear guards to prevent reinforcement of the main column. The volume of fire is rapid and directed at enemy personnel manning automatic weapons, vehicle engine and driver compartments, and riflemen. Anti-tank grenades, rocket launchers, and recoilless rifles fired from multiple locations are used against armored vehicles targeting fuel cells, engines compartments, and road wheels or tracks. As a secondary field of fire, machine guns lay bands of fixed fire across escape routes. Mortar shells and hand and rifle grenades are fired into the kill zone, likely areas of escape, or dead zones when employed. If the commander decides to assault, it is launched under covering fire on a prearranged signal.

V-164. After enemy resistance has been nullified, special parties can be signaled into the kill zone to recover supplies, equipment, and ammunition. When the commander desires to terminate the action because either the mission has been accomplished or superior enemy reinforcements are arriving, he withdraws the assault and support element(s) first. Security elements cover the withdrawal of the entire ambush element and withdraw last.

V-165. If the purpose of the ambush is to harass and demoralize the enemy, a different approach may be adopted. The advance guard is selected as the target of the ambush and the fire of the assault element is directed against them. Repeated attacks against enemy advance guards have the following effects:

- They cause the use of disproportionately strong forces in advance guard duties. This may leave other portions of the column vulnerable or require the enemy to divert additional troops to convoy duty.
- They have an adverse psychological effect upon enemy troops. Continued casualties incurred by the advance guard make such duty unpopular.

## WITHDRAWAL

V-166. Withdrawal from the ambush site is covered by the security elements in a manner similar to the withdrawal from a raid. To facilitate a good head count off the ambush site, all elements can be funneled through one or more sites at the beginning of the withdrawal route(s).

## AMBUSH PATROLS

V-167. An ambush patrol is a combat patrol whose mission is to—

- Harass a target.
- Destroy a target.
- Capture personnel or equipment.
- Execute any combination of these.

V-168. An ambush patrol is planned and prepared in the same general manner as other patrols-that is, by using patrol steps (troop leading procedures). Each step is explained below.

## Planning and Preparation

V-169. Planners must first consider whether the ambush is to be a deliberate ambush or an ambush of opportunity. In a deliberate ambush, the greater amount of target intelligence available permits planning for every COA at the target. Plans for an ambush of opportunity must include consideration of the types of targets that may be ambushed, as well as varying situations. In both, plans must be flexible enough to allow modification, as appropriate, at the ambush site.

V-170. All plans must be rehearsed in detail. Considerations include:

- Every person must thoroughly understand what he/she is to do at every stage of the operation. In ambush, more so than in other operations, failure of even one person to perform exactly as planned can cause failure.
- Many factors affect the types of ambush (point, area, and so on). These factors include organization, the number of men required, the equipment and communications required, and all other aspects of the patrol.
- Each possible formation must be considered for its advantages and disadvantages.
- An attack may be by fire only (harassing only) or may include an assault of the target (destruction ambush).
- The patrol is tailored for its mission. Two men may be adequate for a harassing ambush. A destruction ambush may require the entire unit (squad, platoon, company).

V-171. An ambush patrol is organized in the same manner as other combat patrols, to include a patrol HQ, an assault element, a support element, and a security element (figure V-9, page V-33). The assault and support elements are the attack force; the security element is the security force. When appropriate, the attack force is further organized to provide a reserve force. When an ambush site is to be occupied for an extended period, double ambush patrols may be organized. One ambush patrol occupies the site while the other rests, eats, and tends to personal needs at the objective RP or other concealed location. They alternate each 8 hours. If the waiting period is more than 24 hours, three ambush patrols are organized (figure V-10, page V-33).

V-172. The selection of accompanying equipment and supplies is based on the—

- Mission.
- Enemy threat.
- Size of the guerilla force.
- Means of transportation.
- Distance and terrain.
- Weight and bulk of equipment.

V-173. A primary route is planned that will allow the patrol to enter the ambush site from the rear. The killing zone is not entered if entry can be avoided. If the killing zone must be entered to place mines or explosives, great care must be taken to remove any tracks and signs that might alert the target and compromise the ambush. If mines, mantraps, or explosives are to be placed on the far side, or if the appearance of the site from the target's viewpoint is to be checked, a wide detour around the killing zone is made. Here, too, great care must be taken to remove any traces that might reveal the ambush. An alternate route from the ambush site is planned, as in other patrols.

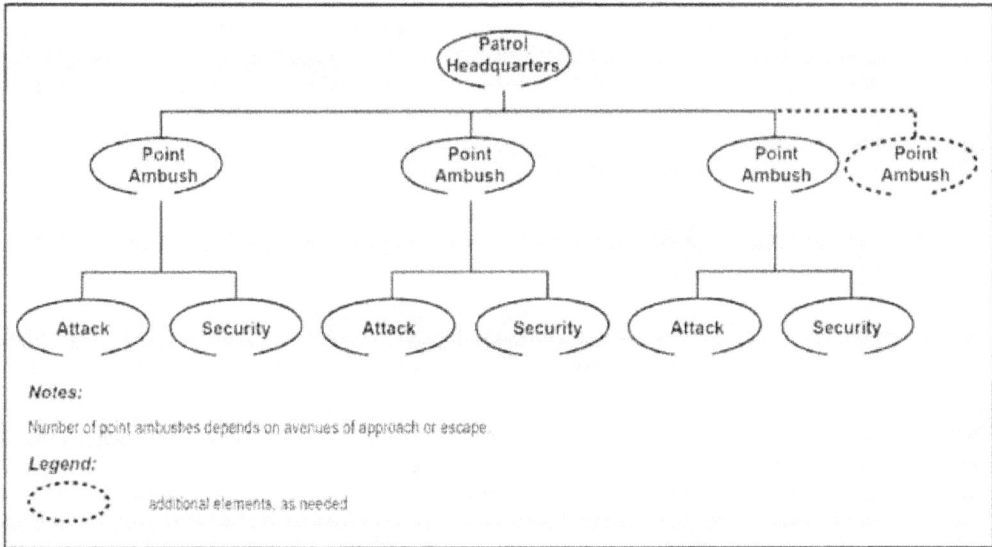

Figure V-9. Organization of ambush patrols—example 1

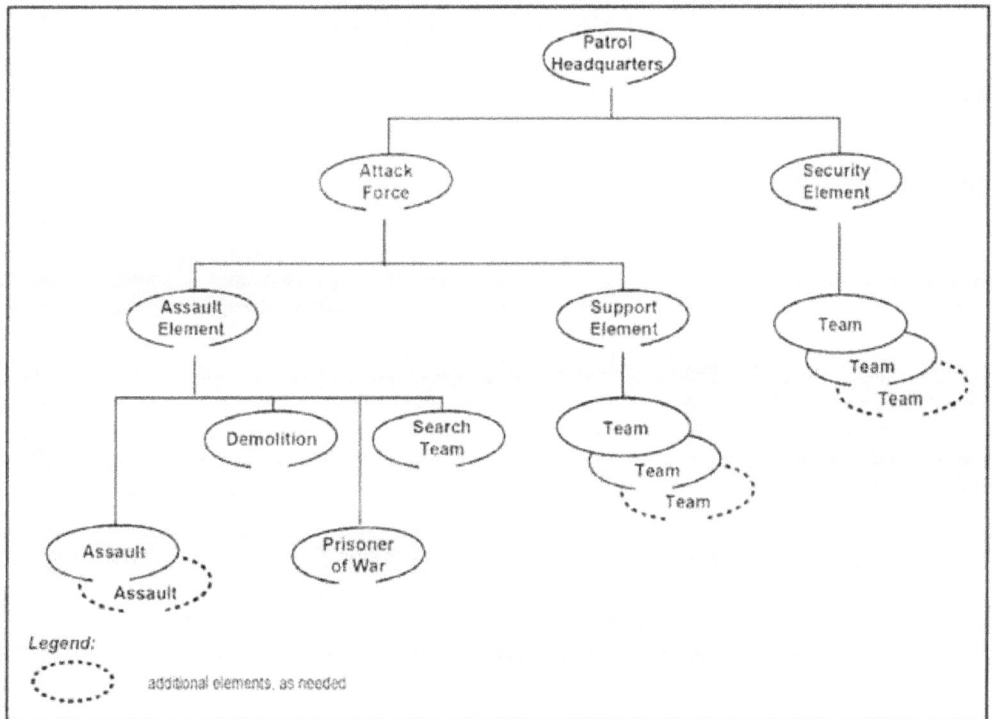

Figure V-10. Organization of ambush patrols—example 2

V-174. Maps and aerial photographs are used to analyze the terrain. When possible, the patrol makes an on-the-ground reconnaissance. Against an experienced enemy, so-called ideal ambush sites should be avoided. An alert enemy is suspicious of these areas, avoids them if possible, and increases vigilance and security when they must be entered. Surprise is even more difficult to achieve in these areas. Instead, unlikely sites are chosen that offer—

- Favorable fields of fire.
- Occupation and preparation of concealed positions.
- Channelization of the target into the killing zone.
- Covered routes of withdrawal to enable the ambush patrol to break contact and avoid pursuit by effective fire.

V-175. As a general rule, the ambush patrol occupies the ambush site at the latest possible time permitted by the tactical situation and by the amount of time required to perform site preparation required. This not only reduces the risk of discovery, but it also reduces the time men must remain still and quiet in position.

V-176. The patrol moves into the ambush site from the rear, as discussed earlier. Security elements are positioned first to prevent surprise while the ambush is being established. Automatic weapons are then positioned so that each can fire along the entire killing zone. If this is not possible, they are given overlapping sectors of fire so that the entire killing zone is covered. The patrol leader then selects his position-located where he can tell when to begin the ambush. Riflemen and grenadiers are then placed to cover any dead space left by automatic weapons. All weapons are assigned sectors of fire to provide mutual support. The patrol leader sets the position preparation time. The degree of preparation depends on the time allowed. All men work at top speed during the allotted time.

V-177. Camouflage is of utmost importance. Each man must be hidden from the target. During preparation for the patrol, each man camouflages himself and his equipment and secures his equipment to prevent noise. At the ambush site, positions are prepared with minimum change in the natural appearance of the site. All debris resulting from preparation of positions is concealed.

## Execution

V-178. Effective command and control is essential to mission success. The patrol leader establishes the communications plan and control measures for execution. As the patrol leader makes contact, communications stand a good chance of breaking down. He must plan using the primary, alternate, contingency, and emergency method. Rehearsals are conducted to ensure everyone knows and understands the following crucial points.

V-179. Three signals, often four, are needed to execute the ambush. Audible and visual signals, such as whistles and pyrotechnics, must be changed often to avoid establishing patterns. Too frequently, use of the same signals may result in their becoming known to the enemy. A target might recognize a signal and be able to react in time to avoid the full effects of the ambush. For example, if a white star cluster is habitually used to signal withdrawal in a night ambush, an alert enemy may fire one and cause premature withdrawal.

V-180. A signal by the security force to alert the patrol leader to the target's approach may be given by:

- Arm and hand signals.
- Radio (as a quiet voice message), by transmitting a prearranged number of taps, or by signaling with the push-to-talk switch.
- Field telephone, when there is no danger that wire between positions will compromise the ambush.

V-181. A signal to begin the ambush, given by the patrol leader or a designated individual, may be a shot or the detonation of mines or explosives. A signal for lifting or shifting fires, if the target is to be assaulted, may be given by voice command, whistles, or pyrotechnics. All fire must stop immediately so that the assault can be made before the target can react. A signal for withdrawal may also be by voice command, whistles, or pyrotechnics.

V-182. This is a key part of the ambush. Fire must be withheld until the signal is given, then immediately delivered in the heaviest, most accurate volume possible. Properly timed and delivered fires achieve surprise as well as destruction of the target. When the target is to be assaulted, the lifting or shifting of fires must be equally precise. Otherwise, the assault is delayed and the target has the opportunity to recover and react.

V-183. The objective RP is located far enough from the ambush site that it will not be overrun if the target attacks the ambush. Routes of withdrawal to the objective RP are reconnoitered. Situation permitting, each person walks the route he is to use and picks out checkpoints. When the ambush is executed at night, each person must be able to follow his route in the dark.

V-184. On signal, the patrol quickly, but quietly, withdraws to the objective RP, reorganizes, and begins its return march. If the ambush was not successful and the patrol is pursued, withdrawal may be by bounds. The last group may arm mines, previously placed along the withdrawal route, to further delay pursuit.

V-185. Contingency plans should include removal of the wounded, both friendly and hostile, under pursuit or at a more measured pace. Treatment location and moves from the target site to a rearward position must be flexible. Plans should also include insertion of medical assets within the assault element, as well as within the HQ. Security and support elements should be considered, depending on the mission.

## CATEGORIES OF AMBUSHES

V-186. Ambushes have two general categories: point and area. A point ambush, whether independent or part of an area ambush, positions itself along the target's expected route of approach. It attacks a single kill zone. When there is not sufficient intelligence for a point ambush, the commander establishes an area ambush. An area ambush uses multiple point ambushes around a central kill zone.

V-187. These two variations succeed best in situations where routes of approach by relieving or reinforcing units are limited to those favorable for ambush by the guerrillas. Both variations were used extensively by the North Vietnamese guerrilla forces in Vietnam against U.S. forces in the Republic of Vietnam.

## Point Ambush

V-188. A point ambush, whether independent or part of an area ambush, is positioned along the target's expected route of approach. Formation is important because, to a great extent, it determines whether a point ambush can deliver the heavy volume of highly concentrated fire necessary to isolate, trap, and destroy the target.

V-189. The formation to be used is determined by carefully considering possible formations and the advantages and disadvantages of each in relation to terrain, conditions of visibility, forces, weapons and equipment, ease or difficulty of control, target to be attacked, and overall combat situation.

V-190. The following paragraphs discuss a few formations that have been developed for the deployment of point ambushes. Those discussed are named according to the general pattern formed on the ground by the deployment of the attack element.

### Line Formation

V-191. The attack element is deployed generally parallel to the target's route of movement (road, trail, or stream). This deployment positions the attack element parallel to the long axis of the killing zone and subjects the target to heavy flanking fire. The size of the target, which can be trapped in the killing zone, is limited by the area the attack element can effectively cover with a heavy volume of highly concentrated fire. The target is trapped in the killing zone by natural obstacles, mines (claymore, anti-vehicular, antipersonnel), demolitions, and direct and indirect fires (figure V-11, page V-36). A disadvantage of the line formation is the chance that lateral dispersion of the target may be too great for effective coverage. Line formation is appropriate in close terrain that restricts target maneuver and in open terrain, where one flank is restricted by mines, demolitions, mantraps, or sharpened stakes. Similar obstacles can be placed between the attack element and the killing zone to provide protection from the target's counterambush measures. When a destruction ambush is deployed in this manner, access lanes are left so that the target can be assaulted (figure V-12, page V-36). The line formation can be effectively used by a rise from the ground ambush in terrain seemingly unsuitable for ambush. An advantage of the line formation is its relative ease of control under all conditions of visibility.

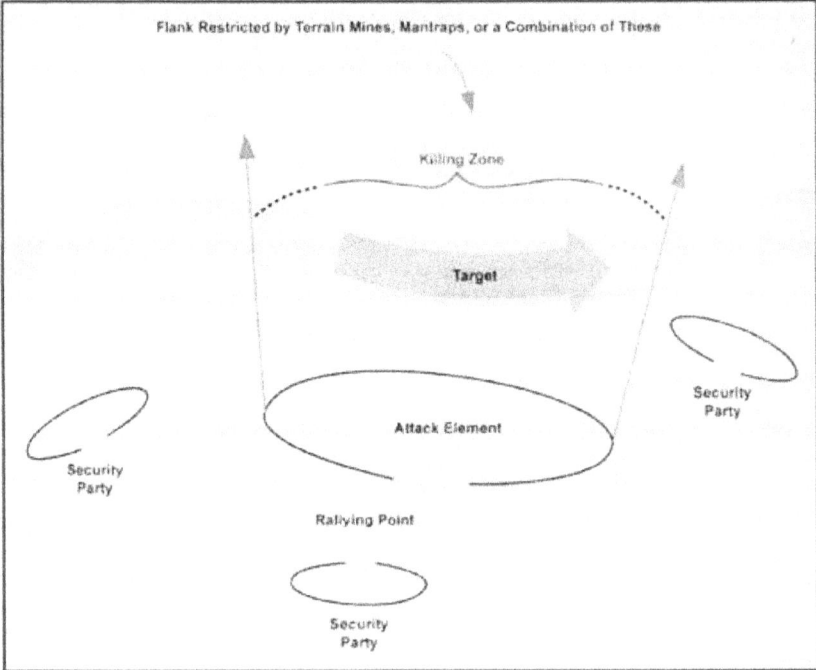

**Figure V-11. Line formation for harassing or destruction ambush**

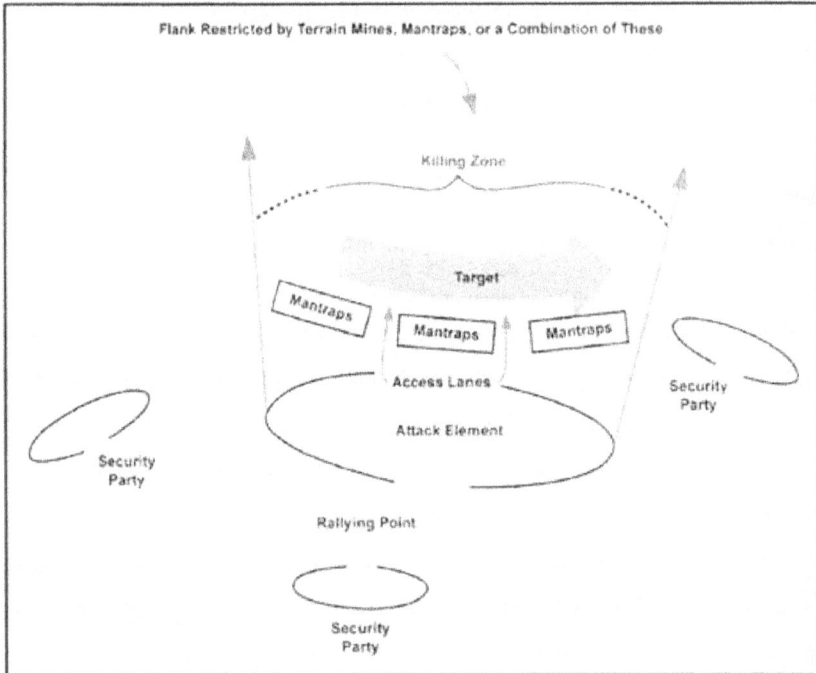

**Figure V-12. Line formation for destruction ambush**

*L Formation*

V-192. The L-shaped formation is a variation of the line formation. The long side of the attack element is parallel to the killing zone and delivers flanking fire. The short side is at the end of and at right angles to the killing zone and delivers enfilading fire that links with fire from the other leg. This formation is very flexible. It can be established on a straight stretch of a trail or stream (figure V-13) or a sharp bend in a trail or stream (figure V-14) When appropriate, fire from the short leg can be shifted to parallel the long leg if the target tries to assault or escape in the opposite direction. In addition, the short leg prevents escape in the direction of the attack element and reinforcement from its direction (figure V-15, page V-38).

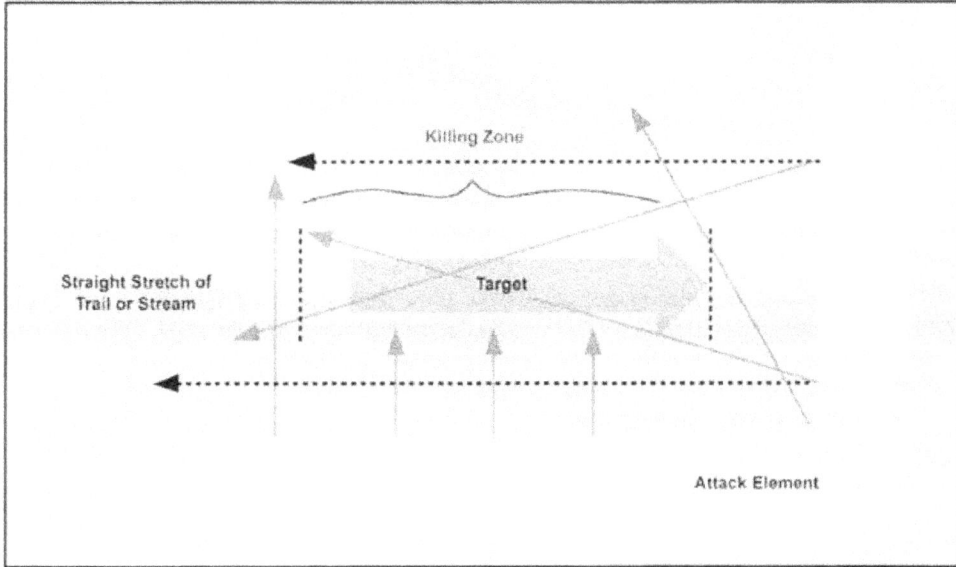

**Figure V-13. L formation for destruction ambush**

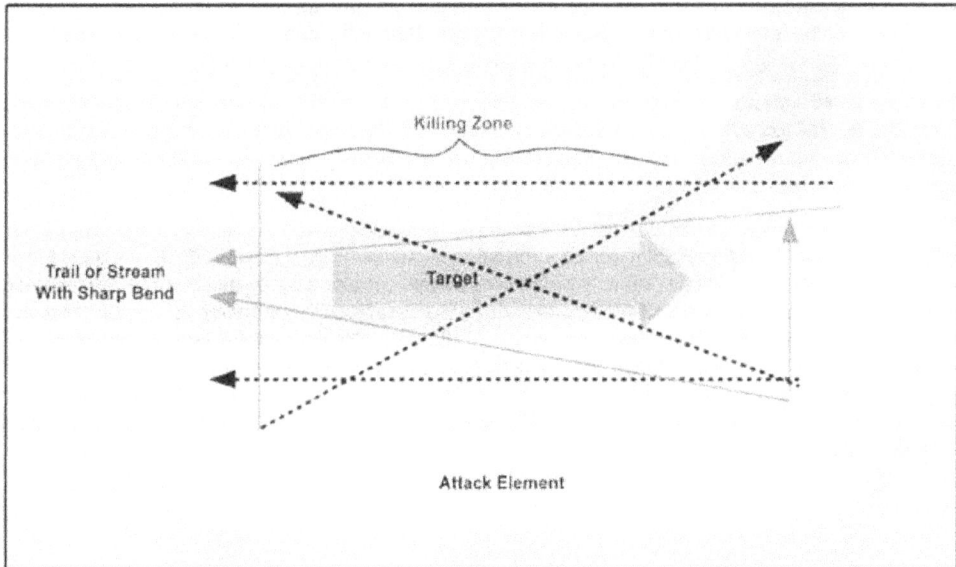

**Figure V-14. L formation for destruction ambush on bend of trail or stream**

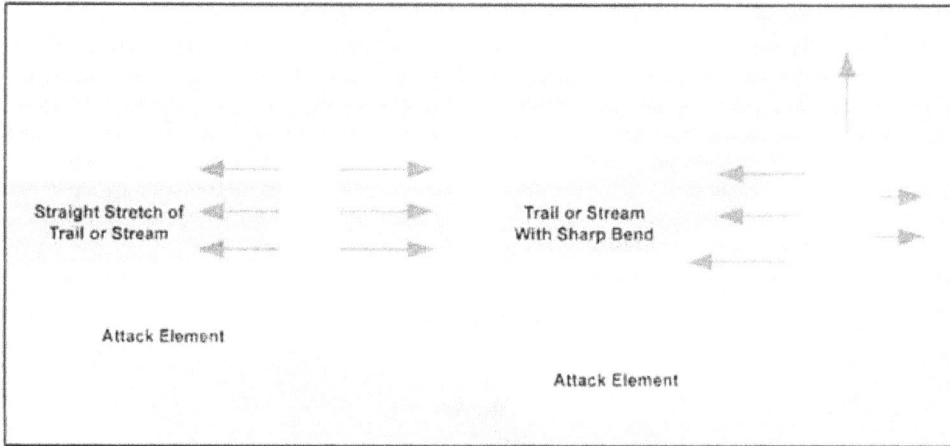

Figure V-15. L formation where short leg of attack element prevents escape or reinforcement

## Z Formation

V-193. The Z-shaped formation is another variation of the line formation. The attack force is deployed as in the L formation but with an additional side so that the formation resembles the letter Z. The additional side as depicted in figure V-16, page V-39, may serve to—

- Engage a force attempting to relieve or reinforce the target.
- Seal the end of the killing zone.
- Restrict a flank.
- Prevent envelopment.

## T Formation

V-194. In the T-shaped formation, the attack element is deployed across and at right angles to the target's route of movement so that it and the target form the letter T. This formation can be used day or night to establish a purely harassing ambush (figure V-17, page V-39 and figure V-18, page V-40) and at night to establish an ambush to interdict movement through open, hard-to-seal areas (such as rice paddies).

V-195. A small group of persons can use the T formation to harass, slow, and disorganize a larger force. When the lead elements of the target are engaged, they will normally attempt to maneuver right or left to close with the ambush. Mines, mantraps, and other obstacles placed to the flanks of the killing zone slow the enemy's movements and permit the ambush patrol to deliver heavy fire and withdraw without becoming decisively engaged.

V-196. The attack element can also use the T formation to interdict small groups attempting night movement across open areas. For example, the attack element is deployed along a rice paddy dike with every second person facing in the opposite direction. The attack of a target approaching from either direction requires only that every second person shift to the opposite side of the dike. Each person fires only to his front and only when the target is at very close range. Attack is by fire only and each person keeps the target under fire as long as it remains on his front. If the target attempts to escape in either direction along the dike, each man takes it under fire as it comes to his vicinity. The T formation is very effective at halting infiltration, but it has one chief disadvantage: while spread out, the ambush may engage a superior force. Use of this formation must, therefore, fit the local enemy situation.

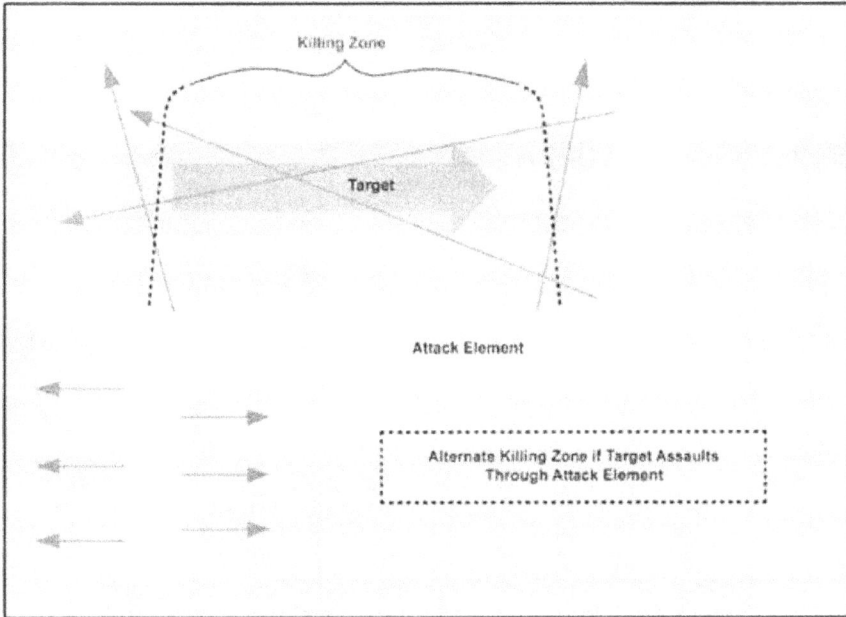

**Figure V-16. Z formation for destruction ambush**

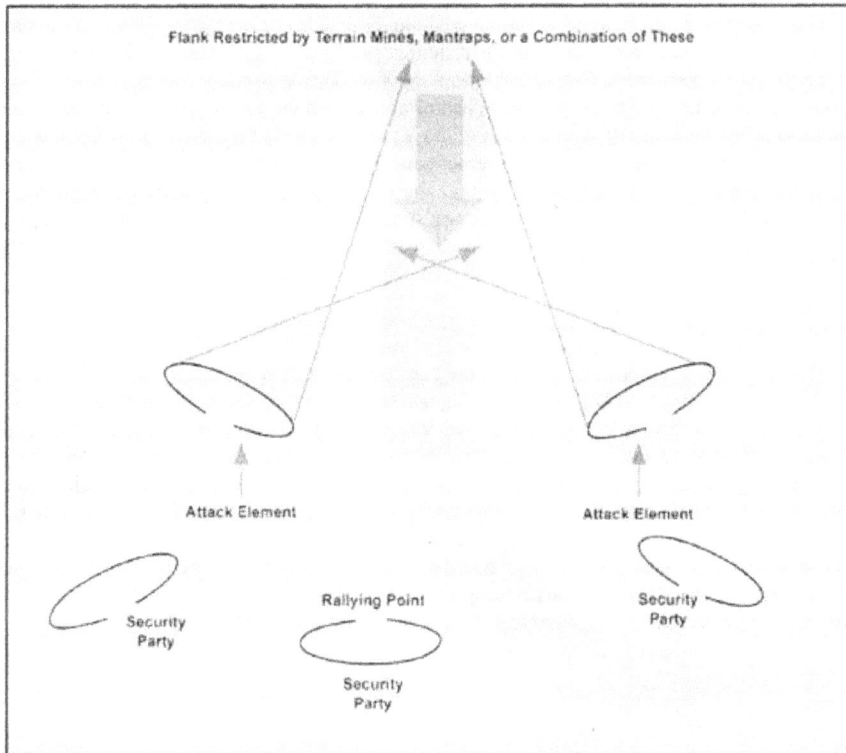

**Figure V-17. T formation for harassing ambush**

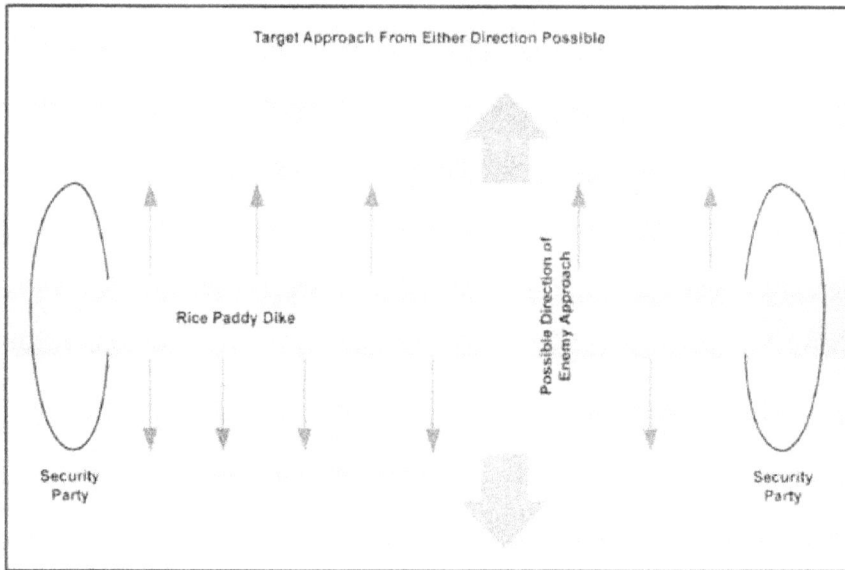

**Figure V-18. T formation for harassing ambush in rice paddy**

### V Formation

V-197. The V-shaped attack element is deployed along both sides of the target's route of movement so that it forms the letter V; care is taken to ensure that neither group (nor leg) fires into the other. This formation subjects the target to both enfilading and interlocking fire. The V formation is best suited for fairly open terrain, but it can also be used in the jungle. When established in the jungle, the legs of the V close in as the head elements of the target approach the apex of the V; the attack element then opens fire from close range. Here, even more than in open terrain, all movement and fire must be carefully coordinated and controlled to ensure that the fire of one leg does not endanger the other. The wider separation of elements makes this formation difficult to control, and there are fewer sites that favor its use. Its main advantage is that it is difficult for the target to detect the ambush until it has moved well into the killing zone (figures V-19 and V-20, page V-41).

### Triangle Formation (Closed V)

V-198. This formation is a variation of the V and can be used in three different ways. One way is the closed triangle (figure V-21, page V-42), in which the attack element is deployed in three groups or parties, positioned so that they form a triangle (or closed V). An automatic weapon is placed at each point of the triangle and positioned so that it can be shifted quickly to interlock with either of the others. Men are positioned so that their fields of fire overlap. Mortars may be positioned inside the triangle. When deployed in this manner, the triangle ambush becomes a small unit strongpoint. It is used to interdict night movement through rice paddies and other open areas when target approach is likely to be from any direction. The formation provides all-around security, and security parties are deployed only when they can be positioned so that if detected by an approaching target, they will not compromise the ambush. Attack is by fire only, and the target is allowed to approach within close range before fire is opened.

In mountain terrain where plunging fire is obtained, legs may be closed in more (nearly parallel to the killing zone).

Security Element    Security Element

Attack Element    Attack Element

Killing Zone    Target

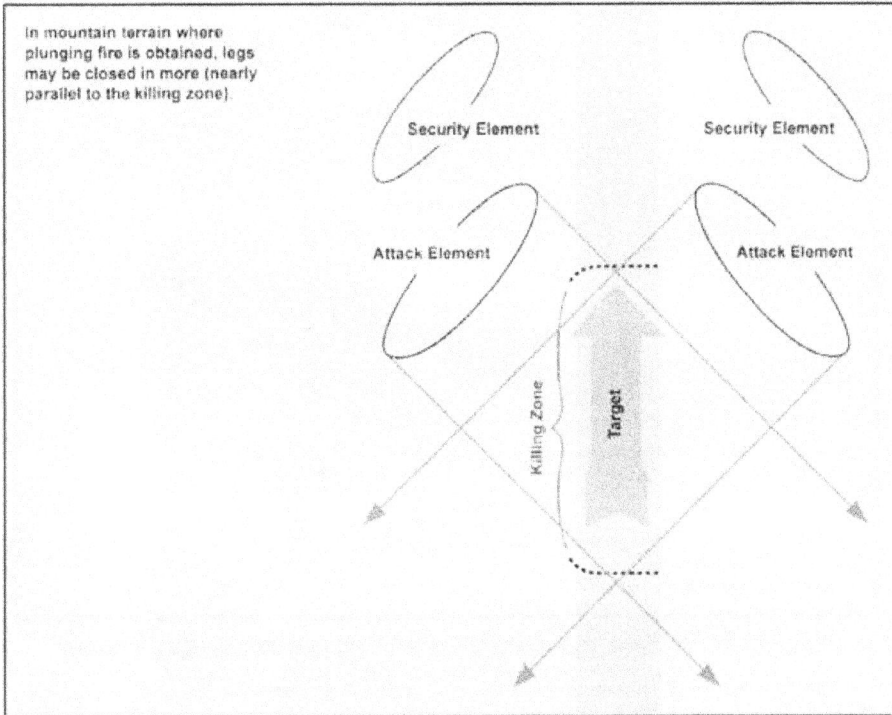

**Figure V-19. V formation for open mountain terrain triangle formation**

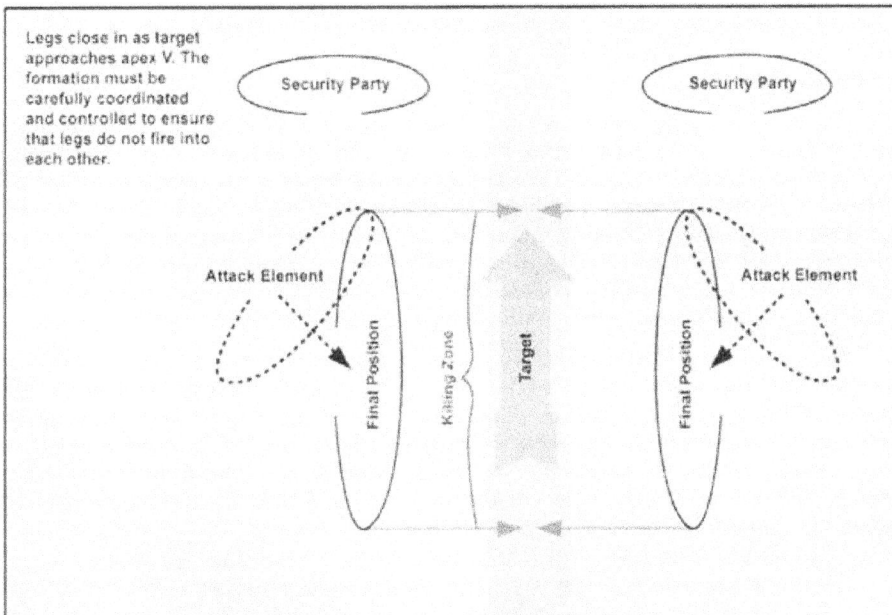

Legs close in as target approaches apex V. The formation must be carefully coordinated and controlled to ensure that legs do not fire into each other.

Security Party    Security Party

Attack Element    Attack Element

Final Position    Killing Zone    Target    Final Position

**Figure V-20. V formation for jungle terrain**

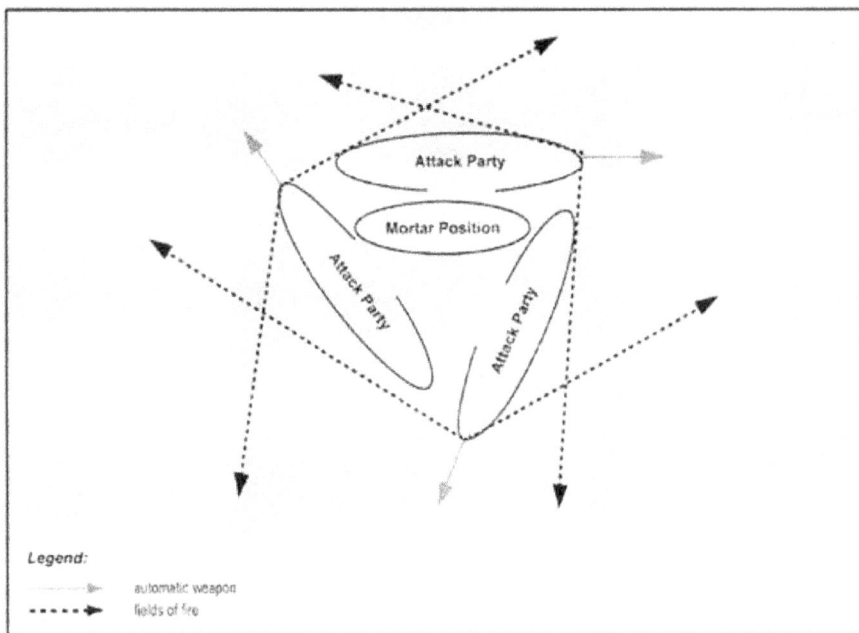

**Figure V-21. Triangle formation (closed V) for night harassing ambush**

V-199. Some advantages of the closed V triangle formation are the ease of control and all-around security. In addition, a target approaching from any direction can be brought under fire of at least two automatic weapons. There are several disadvantages. For example, an ambush patrol-sized or larger is required to reduce the danger of being overrun by an unexpectedly large target. One or more legs of the triangle may come under enfilade fire. Lack of dispersion, particularly at the points, increases danger from enemy mortar fire.

## Triangle Formation (Open V)

V-200. This formation is the second variation of the V formation. This open triangle (during a harassing ambush) is designed to enable a small force to harass, slow, and inflict heavy casualties upon a larger force without itself being decisively engaged. The attack element is deployed in three parties, positioned so that each party becomes a corner of a triangle containing the killing zone. When the target enters the killing zone, the party to the target's front opens fire on the leading element. When the target counterattacks, the group withdraws and an assault party to the flank opens fire. When this party is attacked, the party opposite flank opens fire. This process is repeated until the target is pulled apart. Each party reoccupies its position, if possible, and continues to inflict the maximum damage possible without becoming decisively engaged (figure V-22, page V-43).

V-201. In an open triangle (or open V) during a destruction ambush, the attack element is again deployed in three parties, positioned so that each party is a point of the triangle, 200 to 300 meters apart. The killing zone is the area within the triangle. The target is allowed to enter the killing zone; the nearest party attacks by fire. As the target attempts to maneuver or withdraw, the other groups open fire. One or more assault parties, as directed, assault or maneuver to envelop or destroy the target. As a destruction ambush, this formation is suitable for platoon-sized or larger forces. A unit smaller than a platoon would be in too great a danger of being overrun. More disadvantages of the triangle may be—

- In assaulting or maneuvering, control is very difficult. Very close coordination and control are necessary to ensure that assaulting or maneuvering assault parties are not fired on by another party.
- The ambush site must be a fairly level, open area that provides (around its border) concealment for the ambush patrol (unless it is a rise from the ground ambush).

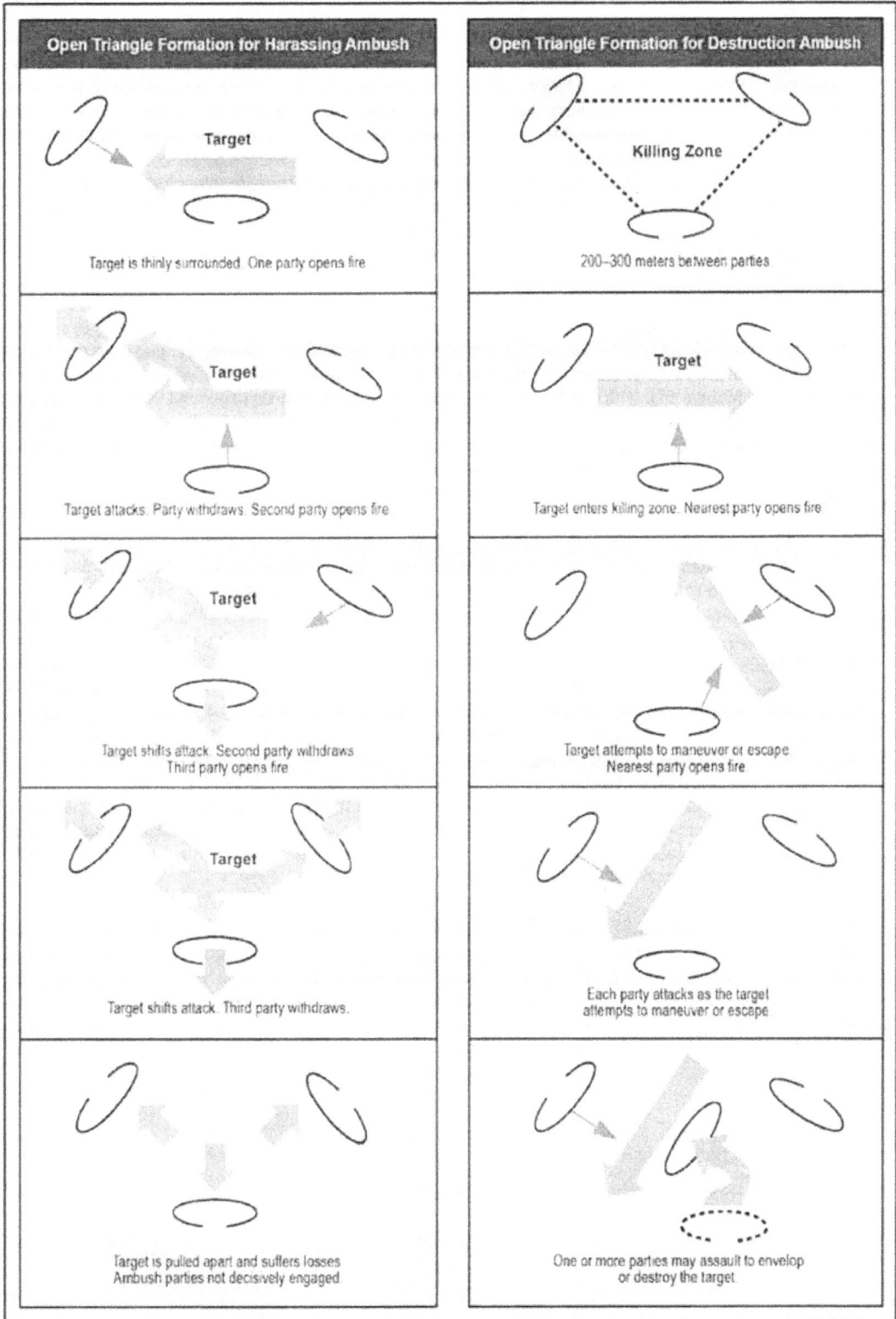

**Open Triangle Formation for Harassing Ambush**

Target is thinly surrounded. One party opens fire

Target attacks. Party withdraws. Second party opens fire

Target shifts attack. Second party withdraws.
Third party opens fire

Target shifts attack. Third party withdraws.

Target is pulled apart and suffers losses.
Ambush parties not decisively engaged.

**Open Triangle Formation for Destruction Ambush**

Killing Zone

200–300 meters between parties.

Target enters killing zone. Nearest party opens fire

Target attempts to maneuver or escape.
Nearest party opens fire.

Each party attacks as the target
attempts to maneuver or escape

One or more parties may assault to envelop
or destroy the target.

**Figure V-22. Open triangle formation (open V)**

*Box Formation*

V-202. This formation is the third variation of the V formation. This formation is similar in purpose to the open triangle ambush. The attack element is deployed in four parties, positioned so that each party becomes a corner of a square or rectangle containing the killing zone (figure V-23, page V-45). The box formation can be used as a harassing or destruction ambush in the same manner as the two variations of the open triangle ambush.

*Note:* The left column of figure V-22, page V-43, depicts the open triangle formation for harassing ambush and the right column depicts the open triangle formation for destruction ambush. The left column of V-23, page V-45, depicts the box formation of harassing ambush and the right column depicts the box formation for the destruction ambush.

## AREA AMBUSH

V-203. The origin of the type of ambush now called area ambush is not known. Hannibal used the area ambush against the Romans in the second century B.C. More recently, it was modified and perfected by the British Army in Malaya and, with several variations, used in Vietnam. The British found that point ambushes often failed to produce heavy casualties. When ambushed, the Communist guerrillas would immediately break contact and disperse along escape routes leading away from the killing zone. The British counteracted this tactic by blocking escape routes leading away from the killing zone with point ambushes. They called these multiple-related point ambushes the area ambush.

### British Version

V-204. The British Army version of the area ambush involves a point ambush that is established at a site having several trails or other escape routes leading away from it. The site may be a water hole, an enemy campsite, a known rendezvous point, or along a frequently traveled trail. This site is the central killing zone. Point ambushes are established along the trails or other escape routes leading away from the central killing zone. The target, whether a single group or several groups approaching from different directions, is permitted to move to the central killing zone. Outlying ambushes do not attack unless discovered. The ambush is initiated when the target moves into the central killing zone. When the target breaks contact and attempts to disperse, escaping portions are intercepted and destroyed by the outlying ambushes. The multiple contacts achieve increased casualties, harassment, and confusion (figure V-24, page V-46). The British Army version of the area ambush is best suited to counterguerrilla operations in terrain where movement is largely restricted to trails. It produces the best results when it is established as a deliberate ambush. When there is not sufficient intelligence for a deliberate ambush, an area ambush of opportunity may be established. The outlying ambushes are permitted to attack targets approaching the central killing zone, if within their capability. If too large for the particular outlying ambush, the target is allowed to continue and is attacked in the central killing zone.

## Baited Trap Version

V-205. A variation of the area ambush is the baited trap version (figure V-25, page V-46) in which a central killing zone is established along the target's route of approach. Point ambushes are established along the routes over which relieving or reinforcing units will have to approach. The target in the central killing zone serves as bait to lure relieving or reinforcing units into the killing zones of the outlying ambushes. The outlying point ambushes need not be strong enough to destroy their targets. They may be small, harassing ambushes that delay, disorganize, and eat away the target by successive contacts.

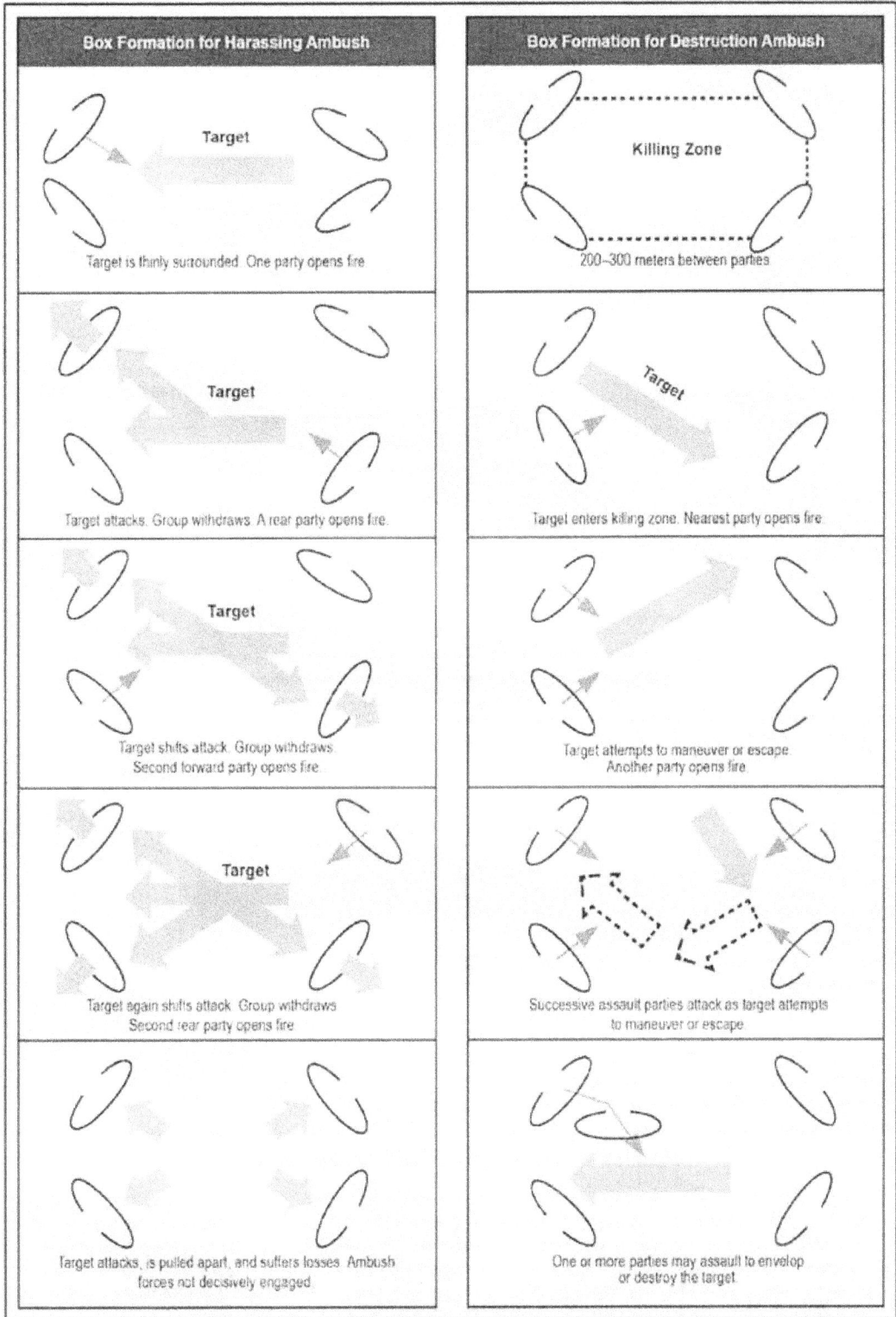

**Figure V-23. Box formation**

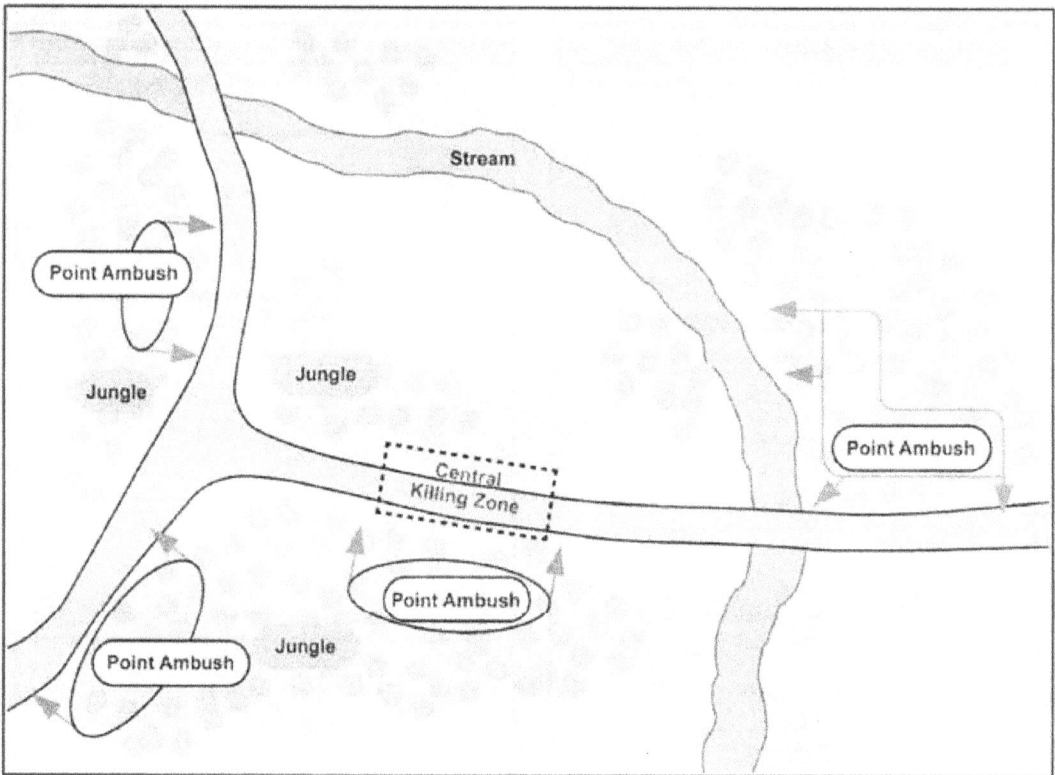

**Figure V-24. Area ambush, British version**

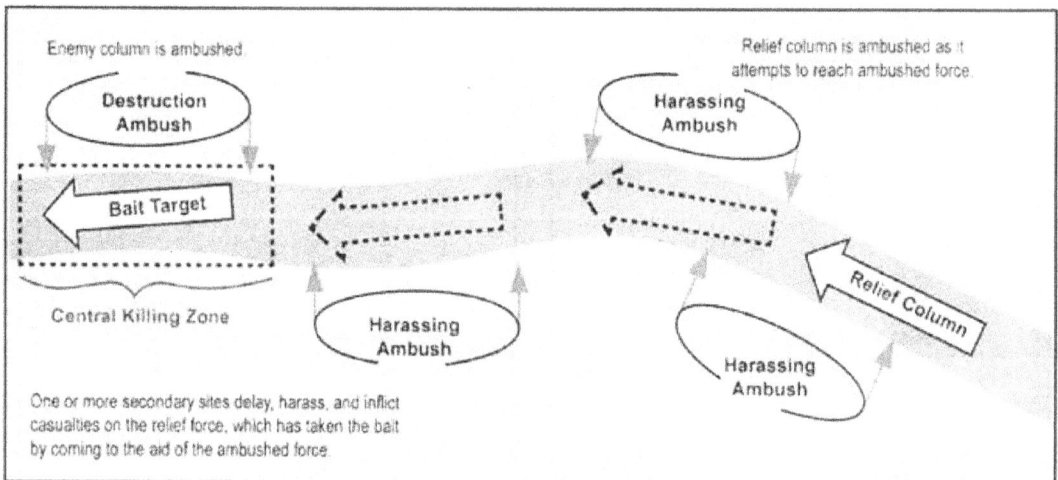

**Figure V-25. Area ambush, baited trap version**

V-206. This version can be varied by using a fixed installation as bait to lure relieving or reinforcing units into the killing zone of one or more of the outlying ambushes. The installation replaces the central killing zone and is attacked. The attack may intend to overcome the installation or may be only a ruse. These two variations are best suited for situations in which routes of approach for relieving or reinforcing units are limited to those favorable for ambush. They are also best suited for use by guerrilla forces, rather than counterguerrilla forces. Communist guerrilla forces in Vietnam used both variations extensively

## Rise From the Ground Ambush

V-207. The attack element uses this type of ambush (figure V-26) in open areas that lack the good cover and concealment and other features normally desirable in a good ambush site. The attack element is deployed in the formation best suited to the overall situation. It is completely concealed in the spider-hole type of covered foxhole. Soil is carefully removed and positions expertly camouflaged

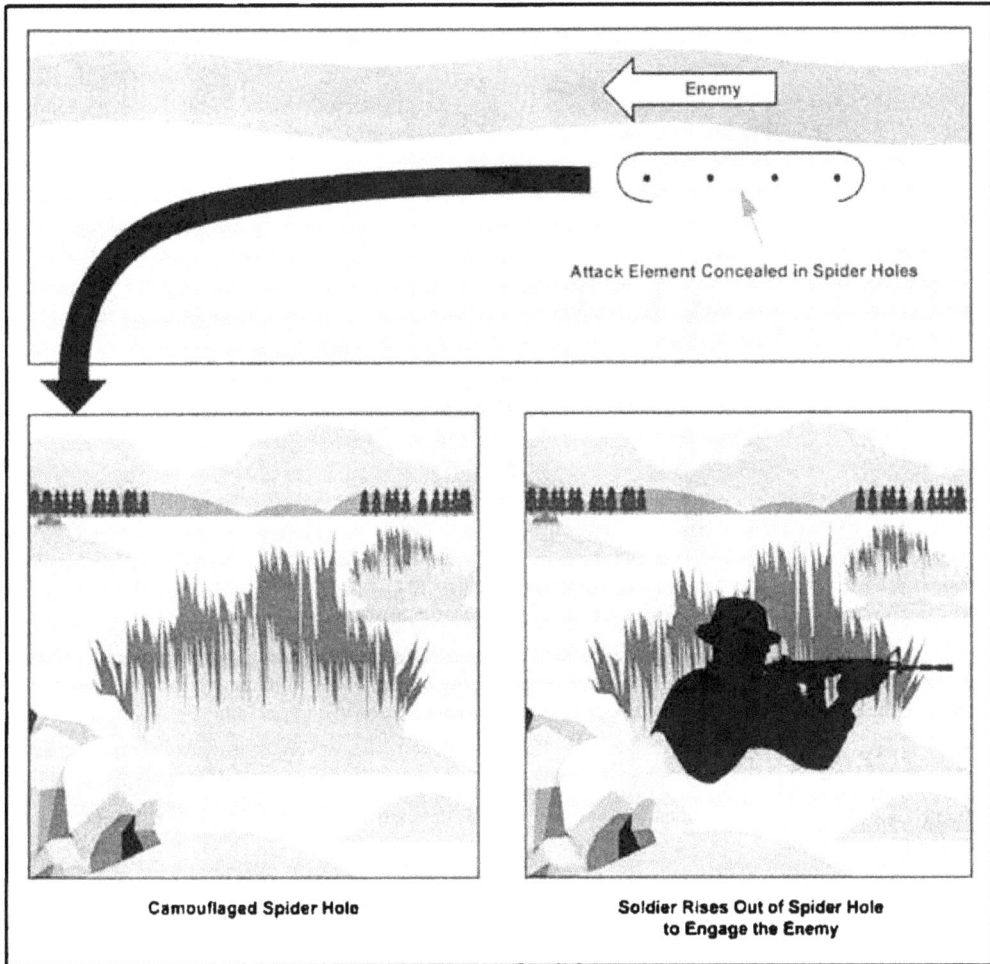

**Figure V-26. Rise from the ground ambush**

V-208. When the ambush begins, the attack element throws back the covers and literally rises from the ground to attack. This ambush takes advantage of the tendency of patrols and other units to relax in areas that do not appear to favor ambush. The chief disadvantage is that the ambush patrol is very vulnerable if prematurely detected.

## Demolition Ambush

V-209. Electrically detonated mines or demolition charges, or both, are positioned in an area (figure V-27) over which a target is expected to pass. This area may be a portion of a road or a trail, an open field, or any location that can be observed from a distance. Activating wires are run to a concealed observation point, which is sufficiently distant to ensure safety of the ambushers.

**Figure V-27. Demolition ambush**

V-210. As large a force as desired or necessary can be used to mine the area. Two men remain to begin the ambush; others return to the unit. When a target enters the mined area (killing zone), the two men remaining detonate the explosives and withdraw immediately to avoid detection and pursuit.

## SPECIAL AMBUSH SITUATIONS

V-211. The following techniques are not considered standard ambush scenarios and, therefore, require special considerations.

## Columns Protected by Armor

V-212. Attacks against columns protected by armored vehicles depend on the type and location of armored vehicles in a column and the weapons of the ambush patrol. If possible, armored vehicles are destroyed or disabled by fire of antitank weapons, land mines, Molotov cocktails, or by throwing hand grenades into open hatches. An effort is made to immobilize armored vehicles at a point in which they are unable to give protection to the rest of the convoy and block the route of other supporting vehicles.

## Ambush of Trains

V-213. Moving trains may be subjected to harassing fire, but the most effective ambush is derailment. Train derailment is desirable because the wreckage remains on the tracks and delays traffic for long periods of time. Derailment on a grade, at a sharp curve, or on a high bridge will cause most of the cars to overturn and result in extensive casualties among the passengers. Fire is directed on the exits of overturned coaches, and designated parties, armed with automatic weapons, rush forward to assault coaches or cars still standing. Other parties take supplies from freight yards and then set fire to the train. Rails are removed from the track at some distance from the ambush site in each direction to delay the arrival of reinforcements by train. In planning the ambush of a train, Soldiers must remember that the enemy may include armored railroad cars in the train for its protection, and important trains may be preceded by advance guard locomotives or inspection cars to check the track.

## Ambush of Waterway Traffic

V-214. Waterway traffic, such as barges or ships, may be ambushed similar to a vehicular column. The ambush patrol may be able to mine the waterway and thus stop traffic. If mining is not feasible, fire delivered by recoilless weapons can damage or sink the craft. Fire should be directed at engine room spaces, the waterline, and the bridge. Recovery of supplies may be possible if the craft is beached on the banks of the waterway or grounded in shallow water.

## MINES AND BOOBY TRAPS

V-215. Guerrilla forces may employ both mines and booby traps to enhance their combat operations. Mines can be antitank, anti-vehicular or antipersonnel. Some destroy by blast effect while others rely on fragments spread by the explosive charge. They can be pressure, trip wire, or command detonated. Some are buried in the ground, some bounce out of the ground and explode about waist level, and others are placed above ground level.

V-216. Mines are cheap, easy to manufacture, easy to deploy, relatively low risk, and provide an effective countermeasure to modern mechanized forces. Countering of mines requires increased efforts by the enemy to detect and clear their presence. Mines that wound are considered more effective—as they also tie up enemy support and medical resources. Mining also forces the enemy to spend time and resources developing countermeasures, to include issuing individual protective gear, increasing vehicle armor, deploying mine detecting, or clearing personnel, equipment, or vehicles. To defeat these countermeasures, guerrillas often manufacture smaller devices, reduce the number of metal components, or bury the mines deeper.

V-217. Mines allow the user to move away from the mined site before they are activated by the enemy. Mines may be employed in conjunction with other operations or used alone. When used alone, they are emplaced along main routes or known enemy approaches within an area at a time when traffic is light. This allows personnel emplacing the mines to complete the task without undue interference and then escape undetected.

V-218. Guerrillas use mines both offensively and defensively. The use of mines to cover the withdrawal of a raiding or ambush force slows enemy pursuit. The use of mines in roadbeds of highways and railroads interferes with movement. Mines may be emplaced around enemy installations, causing casualties to sentinels and patrols resulting in limited movement outside enemy installations. Guerrillas use mines in ambushes to kill or maim enemy personnel and damage military vehicles. Pressure-detonated mines are employed around guerrilla bases to prevent or impede enemy infiltration or attack.

V-219. Guerrillas frequently use military issued or improvised antipersonnel mines. Improvised mines are typically manufactured using duds, discarded ammunition, and materials thrown away by the enemy. Materials discarded as trash, such as improperly destroyed rations, ammunition, beer and soda cans, batteries, waterproof packing materials, and ammunition bandoleers, provide the guerrilla force a valuable source of supplies for mining and booby trap operations. Locally available ingredients, such as chemical fertilizer and plastic or steel pipe for the casing and shrapnel, are also used.

V-220. In areas occupied and protected by the enemy, guerrillas employ mines to impede, delay, and disrupt traffic using roads and trails. These actions cause the enemy to divert valuable forces to guard and clear those routes. The personnel and equipment patrolling the roads to detect and remove mines are prime targets for guerrilla mines and snipers. In congested areas where the enemy conducts offensive operations or patrol activities, the guerrillas could employ mines and mechanical booby traps. The mines and booby traps will inflict casualties, delay and channelize movement, and damage or destroy equipment. Mines should be deployed to reduce accidental injury of noncombatants.

---

*Note:* ATP 3-90.8, *Combined Arms Countermobility Operations*, contains more information on countermobility. Per U.S. Landmine Policy, DOD components will not assist, encourage, or induce anyone outside the Korean Peninsula to engage in activity prohibited by the Ottawa Convention.

---

## SNIPER OPERATIONS

V-221. Sniping as an interdiction technique has a very demoralizing effect on the enemy. Well-trained and properly used snipers can inflict many casualties. They can hinder or temporarily deny the use of certain routes or areas. Snipers also cause the enemy to use a disproportionate number of troops to clear and secure the area. They must have mission orders outlining priority targets to include key threat personnel. Snipers may cover an area that has been mined to prevent removal or breaching of the minefield. Snipers may be part of a raid or ambush to stop threat personnel from escaping the area under attack. They may also prevent or impede the enemy from reinforcing the objective. Besides their sniping mission, they may collect information for the area command or sector commands. All tactical plans can incorporate sniper missions. Provisions must be made for the sniper's rest and recuperation after continuous operations to prevent fatigue.

---

*Note:* TC 18-32, *Special Forces Sniper Training and Employment*, provides more information on sniper operations.

---

## MAN-PORTABLE AIR DEFENSE SYSTEMS

V-222. The most recent and large-scale UW operation occurred in Afghanistan between the Soviet Union and the Afghanistan freedom fighter (Mujahidin). They were a formidable guerrilla force against Soviet airborne, air assault, Spetsnaz, and ground forces. Initially, the U.S. Army supplied the Mujahidin with Redeye missiles in the early 1980s but soon followed with an improved man-portable air defense system (MANPADS), the Stinger. With the new system, the warhead did not have to get a direct hit; hitting close would cause an explosion.

V-223. The premier Soviet helicopter (Hind-D) in Afghanistan had a dual-role capability as an air assault vehicle and a gunship platform. This helicopter was quickly rendered out of action with a well-placed hit on the transmission. The Stinger team easily found and exploited this weakness by aiming at and hitting the large red star behind the cockpit. The Soviets had to alter some of their basic tactical doctrine—use of vehicle-equipped ground forces in conjunction with either a helicopter assault or gun run on suspected Mujahidin targets.

V-224. Using MANPADS in a UW role can have a significant tactical and operational impact. MANPADS are relatively new U.S. weapons—light and very mobile. They can be concealed easily for movement or cached for future operations. Most are relatively simple to operate. Guerrillas can quickly learn how to use them, as demonstrated very effectively in Afghanistan. They require little maintenance because the missile is self-contained. Personnel can use these systems in various ways—from the traditional defensive coverage to offensive tactics. Included are aerial ambushes, direct action or attacks on specific targets, and harassment attacks meant either to produce a psychological impact or to change enemy tactics. The degradation of the enemy's close air support pays great dividends, both tactically and psychologically, for the guerrilla.

### Considerations

V-225. There are four employment considerations for MANPADS: mass, mix, mobility, and integration. Each of these considerations is discussed in the following paragraphs.

V-226. Units achieve mass employment by allocating enough MANPADS to defend an asset. Soldiers move all the available MANPADS to the key assets or operations that need them.

V-227. Air defense operations are more effective when the guerrillas use a mix of weapons. This mix of weapons prevents almost any aircraft from countering the weakness of a solitary system with overlapping and concentrated fire. Although a guerrilla force is not likely to use ZSU-23-4s, Vulcans, and Hawks to any large extent, it will more than likely have some of the following or similar weapons systems available:
- RPG-7 launchers.
- DShK machine guns.
- M2HB caliber .50 machine guns.
- Redeyes.
- Stingers.
- Light machine guns and assault rifles.

V-228.  The guerrilla force must be able to move on short notice. Air defense assets must also be able to displace quickly in a UW environment.

V-229.  Massing all air defense weapons in a common, coordinated effort provides integration. Units can integrate MANPADS with other weapons for the best effect, based on terrain, enemy aerial tactics, and desired effect of the air defense operation, using METT-TC. Air defense personnel may use Stingers to force enemy aircraft to fly at lower altitudes. At lower altitudes, personnel can shoot down the enemy using massed heavy machine guns, rocket-propelled grenades, and Stingers.

## Employment

V-230.  Defensive and offensive uses of MANPADS provide for a balanced defense, overlapping fires, weighted coverage, mutual support, and early engagement. Each use of MANPADS is described below.

V-231.  Critical guerrilla assets may be subject to enemy attack as targets of opportunity. Since the attack can come from any direction, it is desirable to have equal firepower in all directions. The best COA is a balanced defense because the terrain may not favor a most probable avenue of approach by the enemy.

V-232.  Teams should position MANPADS at a distance of 2 to 3 kilometers apart, and one team should overlap another. Other types of weapons should be mixed in to complement the MANPADS. This overlapping prevents the MANPADS team from being overwhelmed by multiple aircraft and increases the chances of their successful air defense against any enemy aircraft.

V-233.  Teams can weight a defense in circumstances in which the terrain restricts low-level attacks to only particular avenues of approach. They can also weight a defense when intelligence has established that air attacks will come from a particular direction. Balance may be sacrificed with a weighted defense since most air defense weapons would be positioned to cover the probable direction of approach. The weighted defense then becomes the best COA.

V-234.  Support from another MANPADS team allows one to fire into the dead space of the other. If the terrain or situation will not allow covering each other's dead space, teams should make use of similar weapons to cover these areas using Stingers or Redeye missiles.

V-235.  Teams should position MANPADS and other similar systems well forward of the guerrilla force's main body or key facilities. This early engagement provides the best opportunity to identify and fix the enemy aircraft before they can attack the guerrillas.

## Technical and Tactical Requirements

V-236.  In addition to the principles and guidelines previously discussed, there are certain technical and tactical requirements that need to be considered before employing MANPADS. Among the questions are:
- What type of aircraft, ordnance, and electronic countermeasures has the threat been using in the area?
- What aerial tactics have the enemy pilots been using in the area?

## Terrain and Weather

V-237.  Mountains and hills may present terrain-masking problems for MANPADS. Whenever possible, MANPADS teams should position along the commanding heights to detect and engage enemy aircraft effectively. Weather can also adversely affect MANPADS that need an infrared source to lock on. In addition, poor weather conditions, such as snow, fog, or rain, can obscure the gunner's vision.

## Routes of Approach

V-238.  There are two general categories of routes of approach: probable and forced. A probable route of approach is the one the enemy is most likely to use but to which he is not restricted. A pilot of an aircraft traveling at 500 knots and 150 meters above the ground can see little detail on the ground. He can, however, see large objects (highways, rivers, and buildings) and use them as aids to navigation. If these landmarks lead to key assets, they may be considered a sign of probable approach. A forced route of approach is the one an aircraft will be forced to use and with no options. The forced route will be to the advantage of the guerrilla because he knows the terrain and where he can hide to best engage the aircraft.

## Map Analysis and Planning

V-239. Terrain analysis is necessary to find good observation points, fields of fire, routes of approach, and any terrain that may inhibit the full capabilities of MANPADS. The ideal planning range for MANPADS is 3 to 5 kilometers from the target. This positioning greatly enhances their survivability by optimizing the lock-on range to enemy aircraft.

## Position Selection

V-240. When selecting positions for MANPADS, personnel should consider observation and fields of fire, communications position, physical security, cover and concealment, alternate positions, and safety considerations.

## Offensive Operations

V-241. Aerial ambushes are similar to the ground "baited trap" ambush (figure V-28, page V-53). If the enemy is known to reinforce outposts or ground units with air support, personnel select a target for a ground attack just to draw an aerial response from the enemy. Personnel also select a target that causes a probable or forced avenue of approach for the reinforcing aircraft. The guerrilla MANPADS teams, together with other air defense assets, are positioned at key points along the aircraft's probable approach route. This pattern is very effective in mountainous terrain where valleys are the prime flight routes. An early warning post radios a timely, forewarning alert to prepare an ambush for the aircraft.

## Direct Action

V-242. Units may use MANPADS in a direct action role to take out a specific type of aircraft or aircraft with key personnel. This operation is most effective when employed around airfields. An aircraft is very vulnerable when taking off and, to a lesser extent, when landing. The concept is to use at least two MANPADS against the target. Personnel locate firing positions on a curve, 3 to 5 kilometers from the runway, within range and observation of the probable flight path of the aircraft. Personnel must study carefully the flight patterns to confirm this critical information. If the distance to the airfield is kept to 3 kilometers, a centralized positioning can cover flight routes either approaching or departing the airfield. With longer ranges and longer airfields, the MANPADS team must confirm the aircraft approach and takeoff direction and position the MANPADS toward that end of the airfield. The actual employment will depend heavily on the type of MANPADS available and the terrain around the target. When in doubt, personnel should use METT-TC.

## Harassment

V-243. The harassment campaign focuses on disrupting the operational procedures of the airport and aircrews. The intended results are to force the pilots to lower their flight altitudes, making them more vulnerable to guerrilla ground fire. Harassment also forces the enemy to decrease its air reconnaissance and support effort.

## Defensive Operations

V-244. In addition to the principles and guidelines of MANPADS employment discussed previously, defense planners must take other considerations into account. Personnel must establish air defense priorities first. Developing a priority list is a matter of assessing each asset to be defended. Air defense priorities include criticality, vulnerability, and recuperability. Despite the type of defense used, the same principles, guidelines, and air defense priorities still apply. Among the types are stationary point, moving point, integrated, and pre-positioned defenses.

V-245. The key to a stationary point defense is early engagement so that the enemy force cannot destroy the target. If the target is large, such as a series of facilities or units concentrated in a relatively small area, personnel should use a star-type defense. This type of defense makes use of interlocking fields of fire, bunkers, trenches, and concertina and tanglefoot wire along with mines and machine guns. Each leg of the star has central and alternate control capability to defend the base camp.

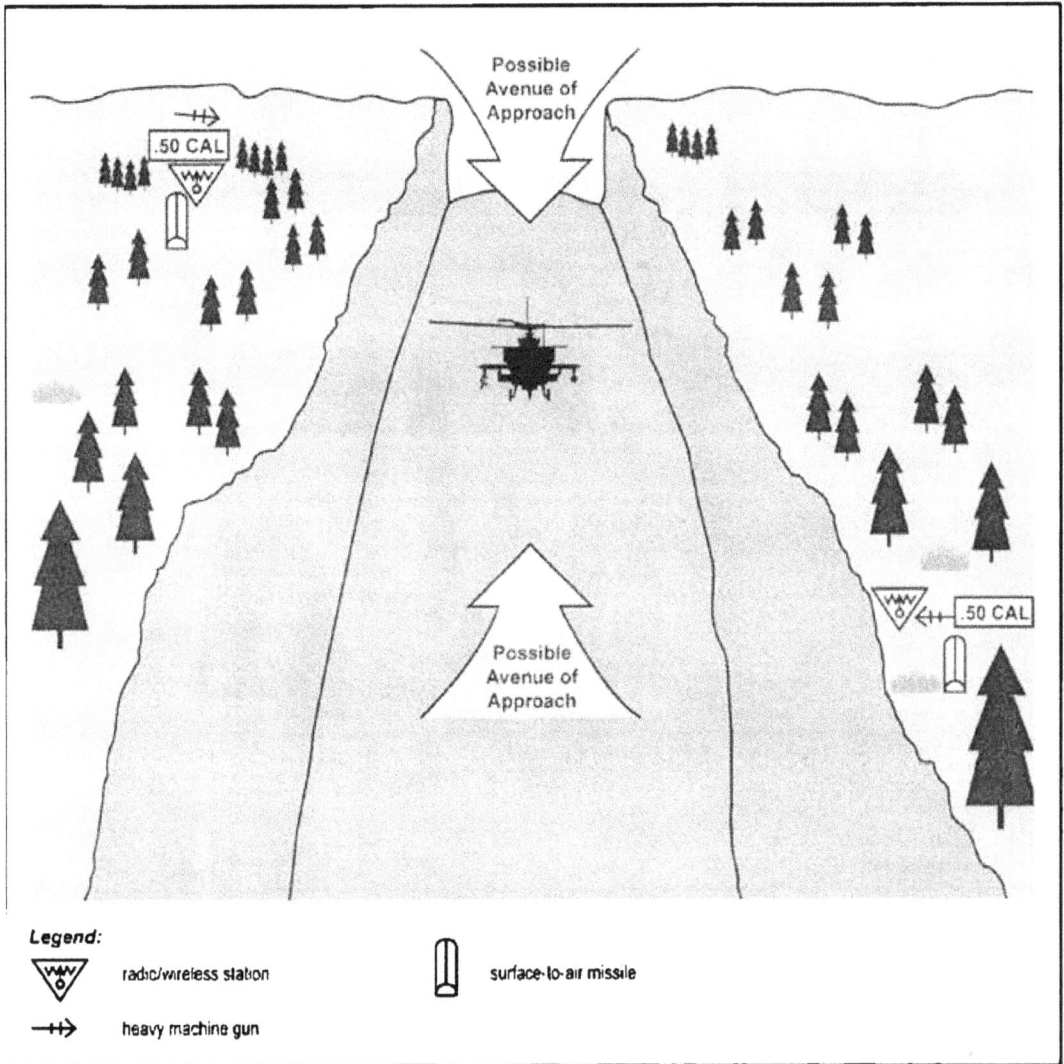

**Figure V-28. MANPADS in offensive operation, aerial ambush**

V-246. In the past, units have used this defense to defend march columns. In a UW environment, personnel and supplies may have to move in march columns. These columns consist of vehicles, carts, pack animals, bicycles, and personnel traveling on foot. Personnel use MANPADS to defend the columns by integration or pre-positioning. If personnel decide to integrate MANPADS into the march column, they should deploy them evenly along the length of the column. This pattern ensures other weapon systems are tied in to complement the overall air defense plan. When only one MANPADS team of two men is available, both men should only be gunners. A single MANPADS should be placed in the column where it can provide the best air cover.

V-247. Personnel pre-position MANPADS to defend a march column as it passes a critical point along the route. This method is preferred for defending a march column. Personnel use it when the distance to be traveled by the march column is relatively short. They also use it when air defense is required at only a few locations along the route. The MANPADS teams may join and integrate with the column after it passes the critical point (figure V-29, page V-54). The MANPADS teams may each receive orders, positioning themselves at a given location. They are then given engagement instructions for a specific window of time.

This plan allows for maintaining OPSEC and receiving air defense coverage Pre-positioned teams should be used only if the route to be used is relatively secure from enemy patrols (METT-TC) or if current guerrilla intelligence reports reflect minimal enemy patrols

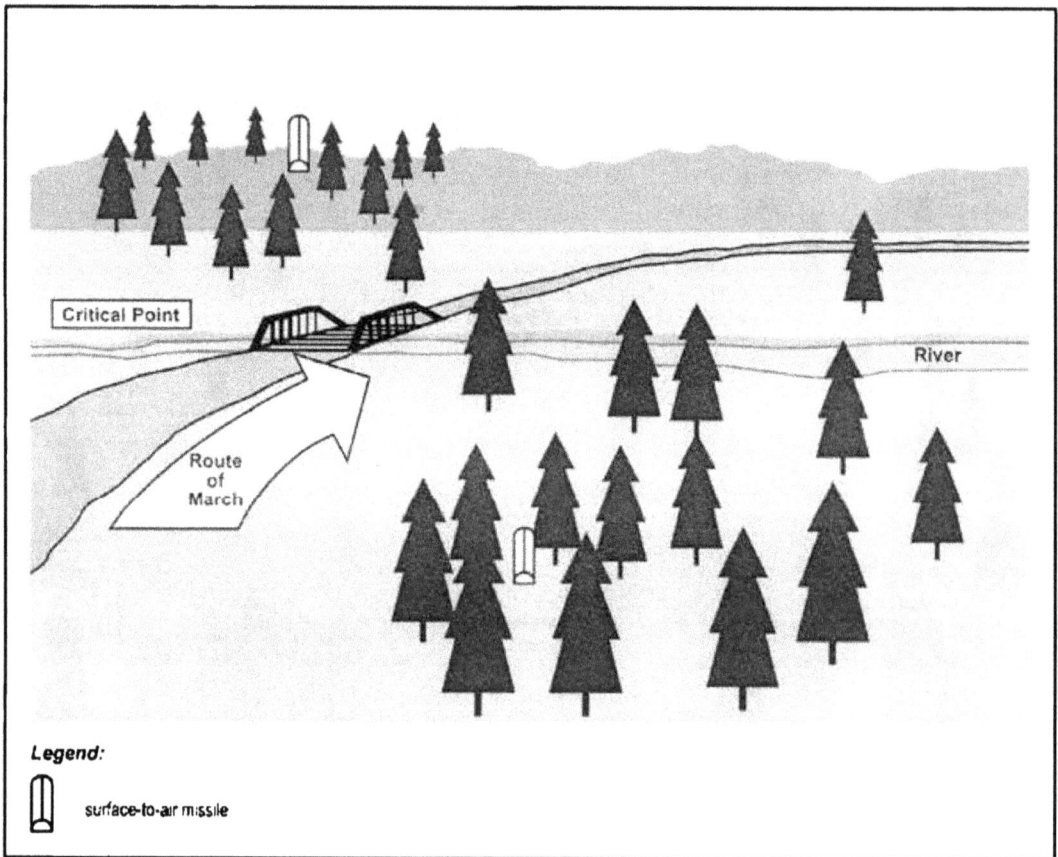

**Figure V-29. MANPADS pre-positioning at critical point defending a march column**

## UNMANNED AERIAL SYSTEM DEFENSE

V-248. Low, slow, and small (LSS) unmanned aircraft systems (UASs) are slow, small, tactical-level UASs operating at relatively low altitudes. The capability of operating within such low altitudes decreases the likelihood of friendly forces detecting the threat in a timely manner. LSS UASs provide a cost effective, high payoff means of surveillance and reconnaissance. Resistance forces and U S advisors may be able to exploit intelligence networks with visibility of known large enemy UAS launch and recovery airfields for early warning. However, resistance forces will have difficulty tracking, identifying, and defeating LSS UASs because they can easily be launched and recovered without significant launch and recovery infrastructure.

---

*Note:* ATP 3-01.81 provides detailed information on UASs.

---

V-249. Advancements in unmanned technologies allow asymmetrical approaches to conduct attacks, collect information, or trigger other threatening events. This problem can escalate as UAS technologies become less expensive and more capable, accessible, and adaptable.

V-250. Small units operating in and around combat areas should assume they can potentially be observed by the enemy While not all hostile air threats require engagement using active air defense measures, there is still a requirement to find, detect, identify and be prepared to defeat UASs.

V-251. Not all encounters with unknown LSS UASs means your unit is at risk or under attack. However, spotting unidentified LSS UAS operating near the location of the units or forces may be a precursor to an imminent attack. Combined arms units must react quickly and appropriately (respond and report) when recognizing signs of possible enemy observation or attack. Whether a counter response is available or not, units must implement passive air defense measures, to include camouflage, cover, concealment, and hardening, in order to protect lives and equipment.

V-252. Counter-UAS operations, at a minimum, should include techniques to detect, identify, respond to, and report threat UAS. Considerations for executing these techniques include:
- Integrating and networking sensors to develop the enemy threat UAS situation.
- Maneuvering to provide positions of optimal observation.
- Developing and sharing the common operational picture.
- Conducting observation (air guard) actions to detect and report threat UAS platforms.
- Ensuring communications with airspace management and aviation personnel and fires elements supporting airspace clearance and identification when these assets are available, within range, and their engagement would not risk exposure of the resistance.
- Coordinating with higher HQ for nonorganic support, such as early warning sensors or electronic warfare capabilities.
- Exploiting indigenous resources for early warning, pattern identification, and nonstandard air defense tactics and techniques.
- Establishing an immediate counter-UAS engagement area (fratricide prevention) and SOPs for immediate action drills.

## COUNTERGUERRILLA OPERATIONS

V-253. Tactical counterguerrilla operations are conducted to reduce the guerrilla threat or activity in the area. To effectively combat the enemy's counterguerrilla operations, Soldiers must be familiar with the indicators of counterguerrilla operations, effective offensive and defensive tactics, and countertracking methods.

V-254. Security of the UW JSOA requires guerrilla intelligence measures to identify indications of impending counterguerrilla action, population control measures, and guerrilla reaction to enemy counterguerrilla actions. Some activities and conditions that may indicate impending enemy counterguerrilla actions may be:
- Suitable weather.
- New enemy commander.
- Changes in battle situation elsewhere.
- Arrival of new enemy units with special training.
- Extension of enemy outposts, increased patrolling, and aerial reconnaissance.
- Increased enemy intelligence effort.
- Civilian pacification or control measures.
- Increased PSYOP against guerrillas.

V-255. Some measures that may be used to control the population of an area are:
- Mass registration.
- Curfews.
- Intensive propaganda.
- Compartmentation with cleared buffer zones.
- Informant nets.
- Party membership drives.
- Land and housing reform.
- Relocation of individuals, groups, and towns.
- Rationing of food and goods.

## DEFENSIVE TACTICS

V-256. The existence or indication of counterguerrilla operations requires the SF and guerrilla force commanders to plan and use defensive tactics. Some of the defensive tactics applicable against counterguerrilla operations are discussed below.

### Diversion Activities

V-257. A sudden increase in guerrilla activities or a shift of such activities to other areas assists in diverting enemy attention. For example, intensified operations against enemy LOCs and installations require the enemy to divert troops from counterguerrilla operations to security roles. Full use of underground and auxiliary capabilities assists in creating diversions.

### Defense of Fixed Positions

V-258. The rules for a guerrilla defense of fixed positions are the same as those for conventional forces, except there are few supporting fires, and counterattacks are generally not practicable. In conjunction with their position defense, elements of the guerrilla force conduct raids, ambushes, and attacks against the enemy's LOCs, flanks, reserve units, supporting arms, and installations. Routes of approach are mined, and camouflaged snipers engage appropriate enemy targets. Diversionary actions by all elements of the resistance movement are increased in adjacent areas.

### Delay and Harassment Activities

V-259. The objective of delay and harassment tactics is to make the attack so costly that the enemy eventually ends its operations. Defensive characteristics of the terrain are used to the maximum. Mines and snipers are employed to harass the enemy, and ambushes are positioned to inflict maximum casualties and delay.

V-260. As the enemy overruns various strong points, the guerrilla force withdraws to successive defensive positions to again delay and harass. When the situation permits, the guerrilla force attacks the enemy's flanks, rear, and LOCs. If the enemy continues its offensive, the guerrilla forces should withdraw and leave the area. Under no circumstances should the guerrilla force become so engaged that it loses its freedom of action and permits enemy forces to encircle and destroy it.

### Withdrawal

V-261. In preparing to meet enemy offensive action, the SF and guerrilla force commanders may decide to withdraw to another area not likely to be included in the enemy offensive. Key installations within a guerrilla base are moved to alternate bases, and essential records and supplies may be transferred to new locations. Less essential items will be destroyed or cached in dispersed locations. If the commander receives positive intelligence about the enemy's plans for a major counterguerrilla operation, he may decide to withdraw and leave his main base without delay.

V-262. When faced with an enemy offensive of overwhelming strength, the commander may disperse his force in either small units or as individuals to avoid destruction. This COA, however, renders the guerrilla force ineffective for an undetermined period of time and, therefore, should not be taken unless absolutely necessary.

### Counterambush

V-263 The very nature of ambush—a surprise attack from a concealed position—places the ambushed unit at a disadvantage. Obviously, the best defense is to avoid being ambushed, but this is not always possible. A guerrilla unit must, therefore, reduce its vulnerability to ambush and reduce the damage it will sustain if ambushed. These measures must be supplemented by measures to destroy or escape from an ambush.

---

*Note:* FM 3-90-1 contains more detailed information on ambush and counterambush procedures.

---

*Reduction of Vulnerability to Ambush*

V-264. No single defensive measure or combination of measures can prevent or effectively counter all ambushes in all situations. The effectiveness of counterambush measures is directly related to the state of training of the guerrilla unit and the leadership ability of its leader.

V-265. In avoiding ambush, dismounted units have an advantage over mounted units. They are less bound to the more obvious routes of movement, such as roads and trails (as in armored units). However, dismounted units are at a disadvantage when—

- Terrain, such as heavy jungle, restricts or prohibits cross-country movement.
- The need for speed requires movement on roads, trails, or waterways.

*Preparation for Movement*

V-266. In preparing for movement, the leader must use METT-TC and observation and fields of fire, avenues of approach, key terrain, obstacles, and cover concealment. In doing so, he studies maps of the area and, if possible, makes an aerial reconnaissance.

*Map Reconnaissance*

V-267. In studying maps of the terrain over which the leader will move his unit, the leader first checks the map's marginal data to determine reliability at the time the map was made. If reliability is not good, or if the map is old, he evaluates its reliability in light of all other information he can obtain. For example, a 20-year-old map may not show several nearby roads and trails, or more recent building development in the area will not be shown. The leader considers the terrain in relation to all available information of known or suspected enemy positions and previous ambush sites. His map study includes evaluation of the terrain from the enemy's viewpoint: How would the enemy use this terrain? Where could the enemy position troops, installations, and ambushes?

*Aerial Reconnaissance*

V-268. If possible, the leader makes an aerial reconnaissance. The information gained from the aerial reconnaissance enables him to compare the map and terrain. He also obtains current and more complete information on roads, trails, man-made objects, type and density of vegetation, and seasonal condition of streams. An aerial reconnaissance may reveal any—

- Movement or lack of movement in an area (friendly, enemy, civilian).
- Indications of enemy activity. Smoke may indicate locations of campsites, patrols, or patrol bases. Freshly dug soil may indicate positions or ambush sites. Shadows may aid in identifying objects. Unusual shapes, sizes, shadows, shades, or colors may indicate faulty camouflage.

V-269. Despite its many advantages, aerial reconnaissance has limitations. Some examples include the following:

- Strength of bridges cannot be determined.
- Terrain surface may be misinterpreted.
- Mines and booby traps cannot be seen.
- Presence of aircraft may warn enemy.

*Route Selection*

V-270. Cover and concealment are desirable, but a route with these features may obstruct movement. Terrain that provides a moving unit with cover and concealment also provides the enemy increased opportunities for ambush. Identification of areas where ambushes may be concealed allows the leader to develop plans for clearing these areas. How the terrain affects observation and fields of fire available to the unit and to the enemy will influence the selection of and movement over a route, formations, rates of movement, and methods of control.

V-271. Key terrain is an earth feature that has a controlling effect on the surrounding terrain. It must be identified and actions planned accordingly. If, for example, a hill provides observation and fields of fire on any part of a route, the leader must plan for taking the hill from the enemy or avoiding it altogether. Obstacles may impede movement, limit maneuver along a route, or limit enemy action.

### Current Intelligence

V-272. All available information is considered, including:
- Known, suspected, and previous ambush sites.
- Weather and light data.
- Reports of units or patrols that have recently operated in the area.
- Size, location, activity, and capabilities of guerrilla forces in the area.
- Attitude of the civilian population and the extent to which they can be expected to cooperate or interfere.

V-273. The unit must provide its own intelligence support. Members must be alert to report information, and leaders must be able to evaluate the significance of this information in relation to the situation. Obvious items from which intelligence may be gained are:
- Signs of passage of groups, such as crushed grass, broken branches, footprints, cigarette butts, or trash. These may reveal identity, size, direction of travel, and time of passage.
- Workers in fields, which may indicate absence of the enemy
- Apparently normal activities in villages, which may indicate absence of the enemy.

V-274. Less obvious items from which negative information can be gained are the absence of:
- Workers in fields, which may indicate presence of the enemy.
- Children in the village, which may indicate they are being protected from impending action.
- Young men in the village, which may indicate the enemy controls the village.

V-275. Knowledge of enemy signaling devices is very helpful. Those listed below are some that were used by communist guerrillas in Vietnam:
- A farm cart moving at night shows one lantern to indicate that no government troops are on the road or trail behind. Two lanterns mean that government troops are close behind.
- A worker in the field stops to put on or take off his shirt. Either act can signal the approach of government troops. This is relayed to the guerrilla unit.
- A villager (fishing) holds his pole out straight to signal all clear and up at an angle to signal that troops are approaching.

### Counterintelligence

V-276. In counterguerrilla operations, in particular, a key feature of preparing for movement is denying the enemy information. A unit is especially vulnerable to ambush if the enemy knows the unit is to move, what time it is to move, where it is to go, the route it is to follow, and the weapons and equipment it is to carry. The efforts made to deny or delay enemy acquisition of this information comprise the counterintelligence plan. As a minimum, the plan restricts dissemination of information.

V-277. The leader gives out mission information only on a need-to-know basis. This procedure is especially important when the native personnel operating with the unit might possibly be planted informers. Once critical information is given, personnel are isolated so that nothing can be passed out. If it is likely that the enemy or enemy informers will observe the departure of a unit, deception plans should be used.

### Communications

V-278. The leader plans how he will communicate with elements of his unit; with air, artillery, or other supporting units; and with higher HQ. On an extended move, a radio relay or a field-expedient antenna may be necessary. An aircraft might be used to help communicate with air, artillery support, or other units on the ground.

## *Fire Support*

V-279. The leader plans artillery and mortar fires to deceive, harass, or destroy the enemy. They may be planned as scheduled or on-call fires. Fires are planned—

- On key terrain features along the route. These can serve as navigational aids or to deceive, harass, or destroy the enemy.
- On known enemy positions.
- On known or suspected ambush sites.
- On the flanks of identified danger areas.
- Wherever a diversion appears desirable. For example, if the unit must pass near an identified enemy position, artillery or mortar fires on the position may distract the enemy and permit the unit to pass undetected.
- At intervals along the route, every 500 to 1,000 meters for example. With fires so planned, the unit is never far from a plotted concentration from which a shift can be quickly made.

V-280. Coordination with the supporting unit includes:

- Route to be followed.
- Scheduled and on-call fires.
- Call signs and frequencies.
- Checkpoints, phase lines, and other control measures.
- Times of departure and return.

## *Security*

V-281. Security is obtained through organization for movement, manner of movement, and by every man keeping alert at all times. Some examples of these security measures include:

- A two-man patrol can maintain security by organizing into a security team with sectors of responsibility.
- A larger unit can use any standard formation (file, column, V) and establish a reaction force. This reaction force can be positioned to the front, rear, or flanks of the main body, so that it does not come under direct contact. Any unit of squad or larger, regardless of the formation used, should have security forces to the front, flanks, and rear.
- A dismounted unit moves by the same methods as a motorized patrol. These methods include continuous movement and traveling, traveling overwatch, and bounding overwatch formations.

## COUNTERTRACKING

V-282. To be more effective in combatting counterguerrilla operations, Soldiers should be familiar with countertracking techniques. If the person tracking the Soldier is not an experienced tracker, some of the following techniques may throw him off.

## Moving From a Thick Area to an Open Area

V-283. While moving in any given direction from a thick area to a more open area, Soldiers walk past a large (10-inch diameter or larger) tree toward the open vegetation for five paces and then walk backward to the front of the tree and change direction 90 degrees. Soldiers must step carefully and leave as little sign as possible. If this is not the direction the Soldiers want to travel, they must change direction again at another large tree in the same manner. The purpose is to draw the trackers into the open area where it is harder to track. This technique may lead the trackers to search in the wrong area before realizing they have lost the track.

V-284. When Soldiers are being tracked by trained, persistent enemy trackers (those the Soldiers are unable to lose because the trackers keep hearing or seeing them), the Soldiers' best COA is to outrun or outdistance the trackers or double back and ambush them. This depends, however, on the Soldiers' strength compared to that of the trackers.

## Crossing a Road

V-285. Soldiers approach a trail from an angle and enter the trail in the direction they want to be followed, leaving considerable signs of their presence. After about 30 meters, Soldiers walk backward to the point they entered the trail and exit in another direction leaving no sign. Soldiers move off on an angle opposite the one they entered the trail on for about 100 meters and change direction to their desired line of march.

## Leaving Footprints

V-286. Soldiers walk backward over soft ground to leave reasonably clear footprints. They try not to leave every footprint clear and do not leave an impression of more than 1/4 inch deep. Soldiers continue this deception until they are on hard ground. They select the ground carefully to ensure that they have at least 20 to 30 meters of this deception. This technique should always be used when exiting a river or stream and can be used in conjunction with all other techniques as well. To add even further confusion to the tracking party, this tactic can be used several times to lay false trails before actually leaving the stream.

## Crossing a Stream

V-287. When approaching a stream, Soldiers approach at an angle in the same manner as a road. They move downstream for about 30 meters, backtrack, and move off into the intended direction. To delay the trackers, Soldiers set up false tracks and leave footprints as described above.

V-288. Some additional tactics that Soldiers can use to aid in eluding a tracking party may include:
- Staying in the stream for 100 to 200 meters.
- Keeping in the center of the stream and in deep water.
- Watching (near the banks) for rocks or roots that are not covered with moss or vegetation and leaving the stream at that point.
- Walking out backward on soft ground.
- Walking up small, vegetation-covered tributaries and replacing the vegetation to its natural position.
- Walking downstream until coming to the main river and then departing on a log or pre-positioned boat.

*Note:* Using a stream as a deception technique is one of the best ways to slow down and lose a tracking party. The deception starts 100 meters from the stream and the successful completion of the tactic is to ensure that the tracking party does not know where to exit from the stream.

## Camouflage Techniques

V-289. Walking backward to leave confusing footprints, brushing out trails, and moving over rocky ground or through streams are examples of camouflage techniques that may be used to confuse the tracker. Moving on hard surfaces or frequently traveled trails may also aid in eluding the tracker. Soldiers should avoid walking on moss-covered rocks as they can be easily displaced.

## Techniques Used to Confuse Dogs

V-290. Enemy tracking teams may use dogs to aid in tracking the Soldiers. Soldiers may confuse or delay dogs by:
- Scattering black or red pepper or, if authorized, a riot control agent (such as CS powder) along the route.
- Using silence-suppressed weapons against animals.

## ENCIRCLEMENT

V-291. An encirclement is the greatest danger to guerrilla forces because it prevents them from maneuvering. Once the enemy has succeeded in encircling a guerrilla force, he may adopt one of several possible courses of action (figure V-30, page V-61). The simplest is to have his troops close in from all sides, forcing the guerrillas back until they are trapped in a small area which is then assaulted. Differences in terrain make it almost impossible for his troops to advance at an equal rate all around the perimeter, thus creating the possibility of gaps between individuals and units.

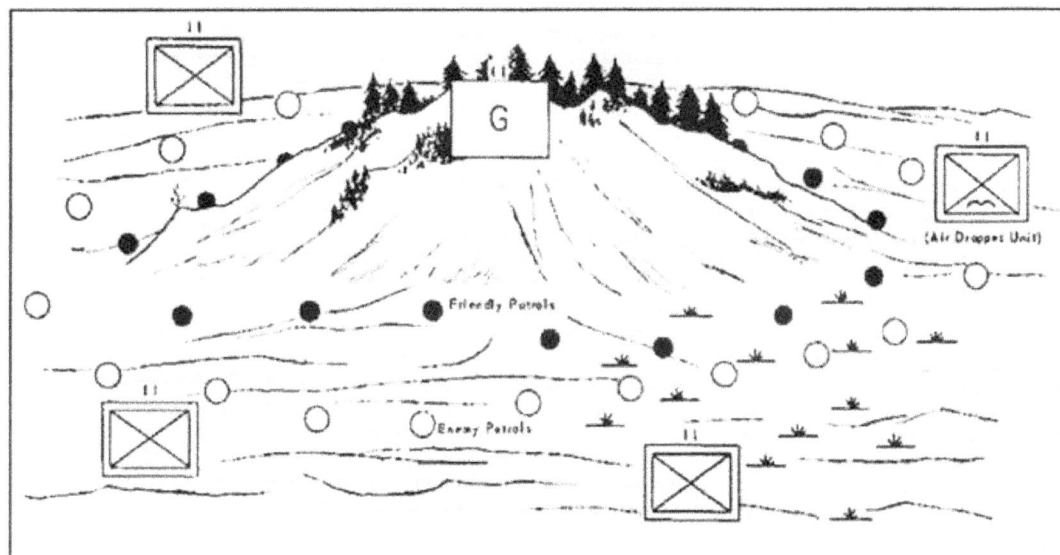

**Figure V-30. Encirclement**

V-292. In other cases, the enemy may decide to break down the original circle into a number of pockets, which will be cleared one by one. The creation of these pockets is a repetition of the original encirclement. In this situation, the guerrillas must either break out or escape through gaps, which may appear as enemy forces are maneuvering into new positions.

V-293. During a counterguerrilla move by a superior enemy in which they become encircled, the guerrillas do not engage in a showdown—instead they withdraw, disperse, or attempt a breakout. Perhaps the most difficult situation for guerrillas to counter with is an assault after encirclement has been accomplished. In this maneuver, enemy forces on one side of the encircled area either dig in or use natural obstacles to block all possible escape routes, while the forces on the opposite side advance—driving the guerrillas against the fixed positions. As the advance continues, enemy forces, which were on the remaining two sides, are formed into mobile reserves to deal with any breakouts (figure V-31, page V-62).

## Defense Against Encirclement

V-294. A guerrilla commander must be constantly on the alert for indications of an encirclement. When he receives indications that an encircling movement is in progress, such as the appearance of enemy forces from two or three directions, the guerrilla commander immediately maneuvers his forces to escape—while enemy lines are still thin and spread out and coordination between advancing units is not yet well established. Records and surplus equipment are either cached or destroyed. Thus, the guerrilla force either escapes the encirclement or places itself in a more favorable position to meet it.

V-295. If for some reason, escape is not initially accomplished, movement to a ridge line is recommended. The ridge line affords observation and commanding ground, which allows movement in several directions. The guerrillas wait on this high ground until periods of low visibility or another favorable opportunity for a breakout attempt.

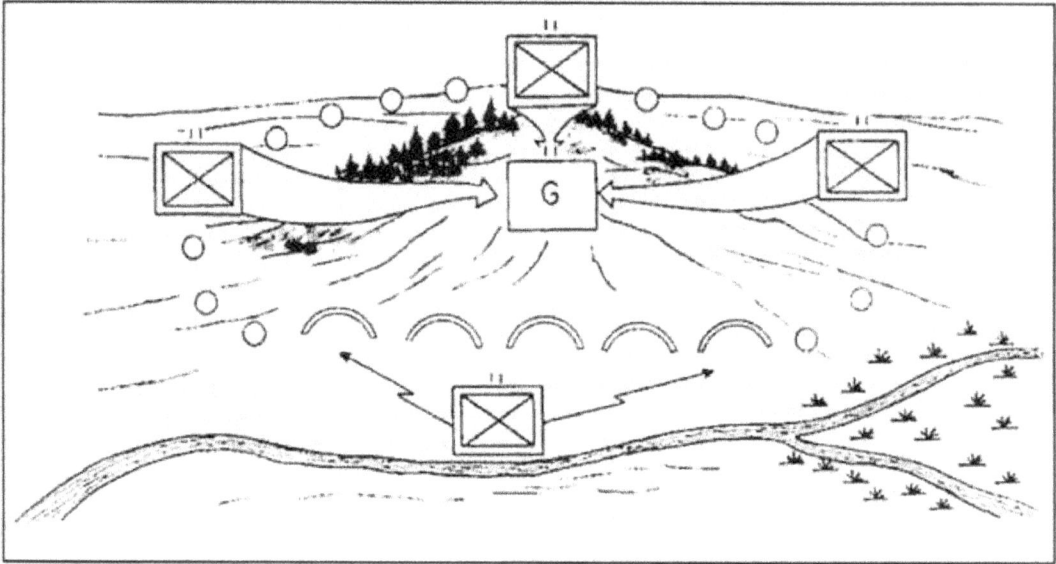

Figure V-31. Encirclement and assault

## Breakout

V-296. During a breakout, two strong combat detachments precede the main body, which is covered by flank and rear guards. If gaps between the enemy units exist, the combat detachments seize and hold the flanks of the escape route. When there are no gaps in the enemy lines, these detachments attack to create and protect an escape channel. The breakout is timed to occur during periods of poor visibility, reducing the potential for enemy observation and accurate fire. During the attempt, guerrilla units—not included in the enemy circle—make attacks against their rear to lure forces away from the main breakout attempt and help to create gaps. After successfully breaking through, the guerrilla force should increase the tempo of its operations whenever possible, thus raising guerrilla morale and making the enemy cautious in the future about leaving his bases to attack the guerrilla areas (figure V-32, page V-63).

### Action if Breakout Fails

V-297. If the breakout attempt is unsuccessful, the commander divides his force into small groups and instructs them to infiltrate through the enemy lines at night or hide in the area until the enemy leaves. This action should be taken only as a last resort, as it means the force will be inoperative for a period of time and the morale of the unit may be adversely affected. Reassembly instructions are announced before the groups disperse (figure V-33, page V-63).

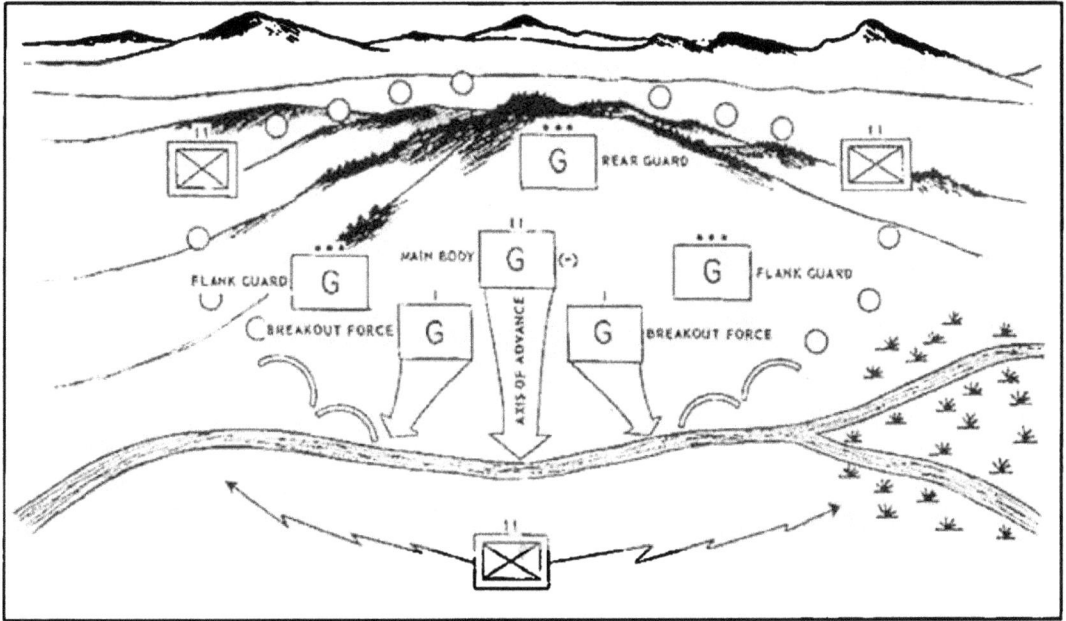

**Figure V-32. Guerrilla breakout from encirclement**

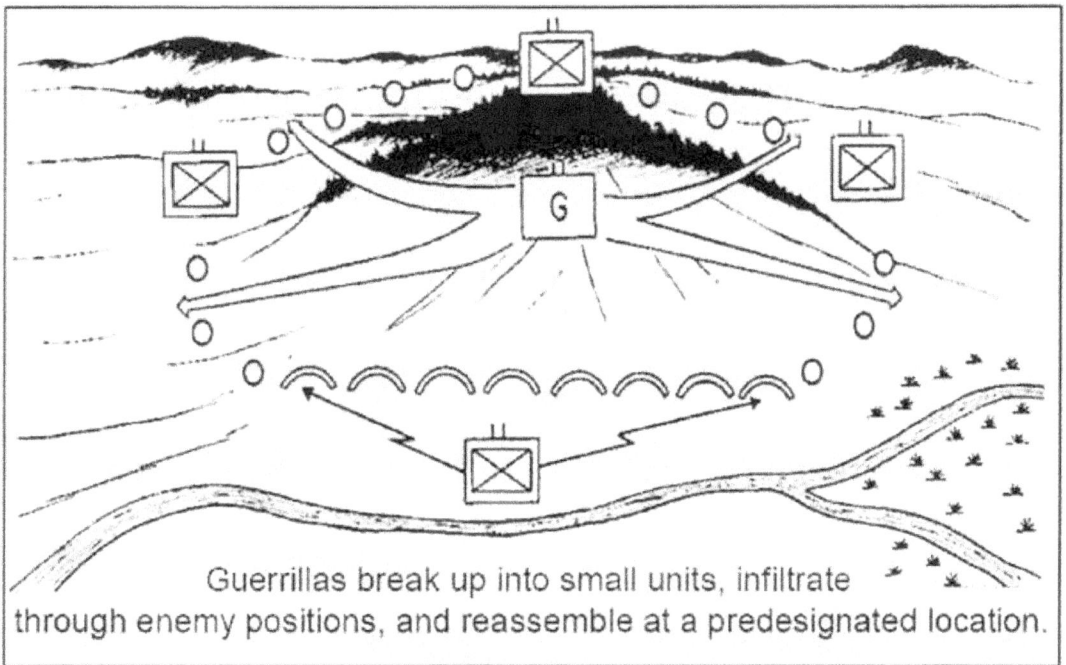

Guerrillas break up into small units, infiltrate through enemy positions, and reassemble at a predesignated location.

Figure V-33. Actions if breakout fails

This page intentionally left blank.

## Appendix W

# Unconventional Warfare in an Urban Environment

*The urban guerrilla must know how to live among the people, and he must be careful not to appear strange and different from ordinary city life. He should not wear clothes that are different from those that other people wear. . . .The same care has to be taken if the urban guerrilla must move from the South of the country to the North, and vice versa.*

*The urban guerrilla must make his living through his job or his professional activity. If he is known and sought by the police, he must go underground, and sometimes must live hidden. Under such circumstances, the urban guerrilla cannot reveal his activity to anyone, since this information is always and only the responsibility of the revolutionary organization in which he is participating.*

*The urban guerrilla must have a great ability for observation. He must be well-informed about everything, particularly about the enemy's movements, and he must be very inquisitive and knowledgeable about the area in which he lives, operates, or travels through.*

*But the fundamental characteristic of the urban guerrilla is that he is a man who fights with weapons; given these circumstances, there is very little likelihood that he will be able to follow his normal profession for long without being identified by the police. The role of expropriation thus looms as clear as high noon. It is impossible for the urban guerrilla to exist and survive without fighting to expropriate.*

Carlos Marighella, *Mini-Manual of the Urban Guerrilla*

The phenomenon of worldwide urbanization continues to increase. In virtually every country of the world, urban areas are rapidly expanding, both in area and population. The U.S. northeastern seaboard has become a vast, open megalopolis in only a few short years. In the developing nations, the rate of urbanization is even more pronounced. Their cities' rate of population growth is much greater than the general rate. As resistance is a human political activity, and the percentage of global urban population increases relative to total population, it is to be expected that resistance—and STR—will similarly increase in urban areas. The SF Soldier must therefore expect to conduct and prepare for UW in urban areas.

## OVERVIEW

W-1. Some of the most famous practitioners and theorists of modern resistance have advocated for an essentially rural-first approach to overthrowing their state or occupying power opponents. Anti-fascist partisans in World War II harassed enemy conventional supply lines from one end of Eurasia to the other. Mao Tse-Tung and the Viet Cong are classic examples of organizing in the countryside to eventually choke off the cities. Che Guevara postulated that heroic armed action in the countryside would serve as focal points of resistance uprising nationwide. Each of these famous examples was appropriate to its own era and unique situation with varying results.

W-2. However, a rural approach is not appropriate for every case of resistance. Karl Marx argued that the vanguard of the revolution would be the industrial proletariat—the properly organized, skilled, industrial urban worker. The Chinese communists attempted to put Marx's theories into practice in the urban cities of the Chinese south in the 1920's and were nearly massacred by superior government forces. This necessitated the Long March escape into the north China hinterland and a change of strategy to a rural-based approach. The Bolsheviks maintained that a relatively small, elite party cadre could function as the vanguard by swift, strategic strikes at the nodes of largely centralized and urban-based power. The Tupamaros of Bolivia began in the rural areas but eventually moved into the cities. As stated by one Tupamaro, "We do not have unassailable strongholds in our country where a lasting guerrilla nucleus could be installed. On the other hand, we have a large city with buildings covering more than 300 square kilometers and that allows the development of an urban struggle." Marighella stated that the Brazilian resistance should focus its revolutionary expropriations from the big businesses, industrial enterprises, and state political and military institutions—all of which were largely concentrated in the urban areas. In some cases, mere survival in rural areas is a challenge due to lack of resources or inadequate distance and sheltering from government operations. Sparsely populated wastelands are sparsely populated for a reason. In other cases, effective government rural pacification operations can make it impossible to survive in the countryside, forcing the resistance to hide amongst larger populations.

W-3. Cities have always been centers of human activity. The result is a greater centralization and consolidation of a country's entire range of social, political, cultural, and economic activities in the urban centers. In many countries, these activities are centered in only one or two key cities.

W-4. In most cases, the capital city of the nation is the primary area of conflict. The capital city of practically every national political entity is the nerve center of the nation. Even in the democratic nations where decentralization is more common, the capital city occupies the key position in the political control of a country. This is even truer in the case of the developing nations.

W-5. All present-day dissident groups are well aware of this situation and realize they must center the insurgent and terrorist activities on the political center of the country to be successful. The political institutions of a state are centered in its capital, with controls emanating from this hub of sovereign power. Communications media are located in a capital and other urban areas to cover the area of densest population. National police and military forces have their control HQ in the capital with subordinate elements in other urban centers.

W-6. In addition to the political power, most capital cities have symbolic value in terms of custom, culture, traditions, and religion. An insurgent force that can paralyze the capital can effectively paralyze the entire nation, and one that can take over the city with its institutions intact can use these institutions to exercise control over the remainder of the nation.

# CHARACTERISTICS OF THE CITY

W-7. Each city is unique. It may be built on hillsides in a mountainous area or lie on a plain. It can be located along a river or the seashore or in the middle of a jungle or desert. Wherever it is located, the site wholly or partially limits or restricts its development and in some respects determines the external form and dimensions of the city. No city is uniform in its internal composition. Rather, just as no two cities are alike, each city varies widely within itself. A city is composed of multiple nuclei—marked, distinct zones, which differ greatly. These zones, quite disparate although the boundaries between them may be blurred, can be residential, ranging from the ghetto neighborhood to the comfortable, affluent suburbs. There are business, service, and industrial areas and sometimes a combination of these. There are often parks and large unbroken expanses of woods. Like activities tend to cluster together while those that are incompatible, such as high-pollutant industries and nurseries, are widely separated.

## A RURAL AREA IS COMPLETELY SELF-SUSTAINING; A LARGE URBAN AREA IS NOT

W-8. The material base on which the city is dependent is largely external to it. The resources cities need and the goods and services it provides determine the functions of the city. Some of the standard functions are: economic, political, religious, educational, residential, or any combination of these.

## MOST CITIES ARE EITHER PREINDUSTRIAL OR INDUSTRIAL

W-9. In preindustrial cities, the upper class lives at or near the city center and the lower class lives on the outskirts. The reverse is usually true in the industrial city. The preindustrial city tends to be polynuclear, since it is based on a bazaar-type economy, whereas the industrial city is more likely to have a central business district and a distinct industrial zone. Internally, distinctive street patterns place constraints on the type of construction within the city. Preindustrial cities tend to be tightly knit with irregular street patterns and buildings of a fairly uniform height. This height is usually three to six stories, resulting from the limits imposed at the time of construction by lack of technology, building materials available, or weak foundations. A low, nearly uniform skyline characterizes this type of city with buildings very densely concentrated.

## THE INDUSTRIAL CITY PRESENTS A QUITE DIFFERENT FORM

W-10. The business district is usually concentrated near the point of peak accessibility. In this core, buildings are as high as possible to minimize the use of ground space. The result is a cluster of high-rise buildings with the skyline falling away in all directions. In larger cities, smaller versions of the downtown cluster appear at other points throughout the city.

# CHARACTERISTICS OF URBAN TACTICAL OPERATIONS

W-11. Despite their differences, urban areas have certain similarities that provide us with some general characteristics of urban tactical operations. These have special significance regardless of the nature or level of conflict.

## CITIES ARE COMPLEX AND HETEROGENEOUS

W-12. The nature of the cities presents difficulty to the planner as well as the operator. Planning must be in much greater detail than is normally required. The use of large-scale city plans and street maps is imperative. The texture of the city varies greatly and often changes abruptly. Operators must function in the city core, open suburbs, industrial areas, transportation centers, parks, woods, and waterfronts.

## THE NATURE OF THE CITY FAVORS THE INHABITANT AND DEFENDER

W-13. The city dweller is the man on the ground with an inherent knowledge of his daily surroundings. These surroundings appear to the outsider as a hopelessly tangled web of buildings, streets, and alleyways, but they are as familiar to the inhabitant as his own living room. Buildings and street complexes restrict movement and reduce the attacker's mobility but provide fields of fire to the defender. Obstacles are fairly easy to construct, as is the establishment of population control measures, checkpoints, and traffic control points.

## SMALL UNIT ENGAGEMENTS WILL PREDOMINATE

W-14. The normal building and street patterns will reduce any operation to a series of small unit engagements. The advantage of the principle of mass is greatly reduced. Flexibility is decreased for larger units and increased for smaller ones. Cells or teams operating in the city must be limited to no more than five or six members.

## THE UNIQUE VERTICAL DIMENSION MUST BE CONSIDERED

W-15. The elevation of the buildings provides a vertical dimension not normally encountered in tactical operations. This vertical dimension is not only above the surface but extends to the subterranean level in the form of subways, sewers, tunnels, and basements. Elevated positions offer good observation and fields of fire plus cover and concealment, particularly effective for snipers. Subterranean areas offer protected and concealed areas for storage, as well as effective routes of movement and communication.

W-16. Combat occurs at extremely short ranges, which limits or nullifies the effectiveness of long-range weapons and increases the usefulness of individual weapons. Operators must be proficient in the quick, accurate firing of their personal weapons systems; the emphasis is likely to be on pistols, submachine guns, carbines, shotguns, grenades and small charges, and pyrotechnics like smoke and flares.

## LARGE NUMBER OF NONCOMBATANTS WILL BE PRESENT

W-17. A large number of noncombatants is to be expected to restrict the use of firepower, munitions, and explosives on the part of the operator. The urban resistor will operate amongst hundreds or thousands of potential witnesses and informers. At the same time, the presence of large numbers of noncombatants can be used to the enemy's advantage, such as the creation of confusion. Large numbers of people also increase the operator's blending-in capability.

## THERE WILL BE COLLATERAL DAMAGE CONSTRAINTS

W-18.   These constraints may affect the selection of targets or the means by which the targets are attacked. The limitations may be imposed by a higher HQ or could take the form of moral constraints, as in the case of critical life-support services. Another form of constraint is when an assigned task would have a significantly adverse effect on the mission—for example, alienating a segment of the population on which the operator's existence depended.

## INTELLIGENCE COLLECTION IS DIFFICULT FOR THE OUTSIDER

W-19.   Accurate and timely intelligence is of extreme value. For the inhabitant, day-to-day activities and occurrences are routine and taken in stride, but these same activities could provide a situation for which the operator is totally unprepared. In the final analysis, there is no substitute for on-the-ground experience, training, and conditioning.

W-20.   UW conducted mostly in urban areas is likely to take on even more significance as the world population continues to coalesce into ever-larger groupings In most cases, the capital city of the nation is the primary area of conflict. The capital city of practically every national, political entity is the nerve center of the nation. Even in the western democratic nations where decentralization is more common, the capital city occupies the key position in the political control of a country

## TRUER IN THE CASE OF THE DEVELOPING NATIONS

W-21.   The political institutions of a state are centered in its capital, with controls emanating from this hub of sovereign power. Communications media are located in a capital and other urban areas to cover the area of densest population. National police and military forces have their control HQ in the capital, with subordinate elements in other urban centers. In addition to the political power, most capital cities have symbolic value in terms of custom, culture, traditions and religion An insurgent force that can paralyze the capital has a better chance of effectively paralyzing the entire nation, while one which can take over the city with its institutions intact can use these institutions to exercise control over the remainder of the nation. Present and future resistance groups, therefore, are likely to concentrate their activities on the political center of the country in order to be successful.

# ANALYZING THE URBAN ENVIRONMENT

W-22.   When conceptualizing urban operations, commanders must understand two important terms: urban area and urban environment. The first is a subset of the second. An urban area is a topographical complex where man-made construction or high population density is the dominant feature. Focusing on urban areas means concentrating on the physical aspects of the area and their effects on weapons; equipment; line-of-sight; and tactics, techniques, and procedures. The urban environment includes the physical aspects of the urban area, as well as the complex and dynamic interaction and relationships between its key components—the terrain (natural and man-made), the society, and the supporting infrastructure—as an overlapping and interdependent system of systems.

W-23.   Importantly, commanders must also understand and consider that critical elements of the infrastructure may lie far beyond the area's physical confines. For example, the generating source providing power to the urban energy system is part of that system, but it may be located well outside of the urban area. Similarly, effects of the interaction between components of the infrastructure, located both inside and outside the urban area, extend well into smaller, neighboring urban areas and surrounding rural areas and often form their political, economic, and cultural focus. Understanding the total urban environment is essential to planning and conducting full spectrum urban operations.

W-24.   The urban environment is too multifaceted for a single-service, single-agency, or single-dimensional solution. Generating desired effects and avoiding unintended negative consequences in this complex environment requires careful integration of joint (and often multinational) forces and interagency capabilities, throughout all phases of the operation. Effective interagency collaboration will help plan effects, supporting actions, and measures of effectiveness to ensure that military actions complement diplomatic, economic, and informational activities.

## URBAN TERRAIN

W-25. Although complex and difficult to penetrate with many intelligence, surveillance, and reconnaissance assets, the terrain is the most recognizable aspect of an urban area. Truly understanding it, however, requires comprehending its multidimensional nature. The terrain consists of natural and man-made features, with man-made features dominating; an analysis considers both. Buildings, streets, and other infrastructure have varied patterns, forms, and sizes. The infinite ways in which these factors can intertwine make it difficult to describe a "typical" urban area. However, these various factors provide a framework for understanding the complex terrain in an urban area. Furthermore, man-made features significantly affect military systems and Soldiers and, thus, tactics and operations.

## MULTIDIMENSIONAL BATTLEFIELD

W-26. Urban areas present an extraordinary blend of horizontal, vertical, interior, exterior, and subterranean forms superimposed on the natural relief, drainage, and vegetation. An urban area may appear dwarfed on a map by the surrounding countryside. In fact, the size and extent of the urban AO is many times that of a similarly sized portion of undeveloped natural terrain. A multistoried building may take up the same surface area as a small field, but each story or floor contains approximately an equal area as the ground upon which it sits. In effect, a ten-story building can have eleven times more defensible area than "bare" ground—ten floors and the roof. It is the sheer volume and density created by this urban geometry that makes urban operations resource intensive in time, manpower, and materiel.

W-27. Like natural disasters, urban operations can radically alter the physical characteristics of the urban terrain in ways not experienced in other environments. These disasters may cause (either intentionally or not) uncontrollable fires or the loss of electricity. A power outage can cause flooding (especially in subsurface areas) by shutting down pumping stations. Entire buildings may be destroyed, eliminating reference points, leaving large piles of rubble, altering fields of fire, and making movement and transportation extremely difficult. In addition, buildings and other urban structures, damaged but not destroyed, can become (or remain) effective obstacles and possible booby traps. Even without enemy exploitation, their weakened construction and unstable structure increase the risk of injury to Soldiers and civilians moving within them. (Engineer expertise will often be needed to determine whether the buildings can support occupation by Army forces or civilians.) Yet, even the total collapse of a building may not eliminate its defenders. Of additional concern, the likely presence of toxic industrial material can create additional obstacles and health hazards.

W-28. Commanders in other environments normally address the depth, breadth, and height of their AO in terms of two areas: airspace and surface. In an urban environment, they broaden their scope to include supersurface and subsurface areas (figure W-1, page W-6) that voluminously extend the commander's AO. Although spatially separated, each area may be used as an avenue of approach or mobility corridor, LOC, or engagement area.

W-29. Supersurface and subsurface areas magnify the complexity of the urban physical environment. Commanders must consider activities that occur outside of buildings and subterranean areas (the external space), as well as the activities happening unseen in buildings and subterranean systems (the internal space). This internal volume further challenges command, control, and intelligence collection activities and increases the combat power required to conduct urban operations. Commanders must develop methods and techniques to help themselves, their staffs, and their subordinate commanders and staffs to represent, visualize, and reference these multiple dimensions. Increasing the difficulty, such dimensions can change rapidly simply because of continued urban growth or, as described earlier, the effects of nature and urban operations themselves.

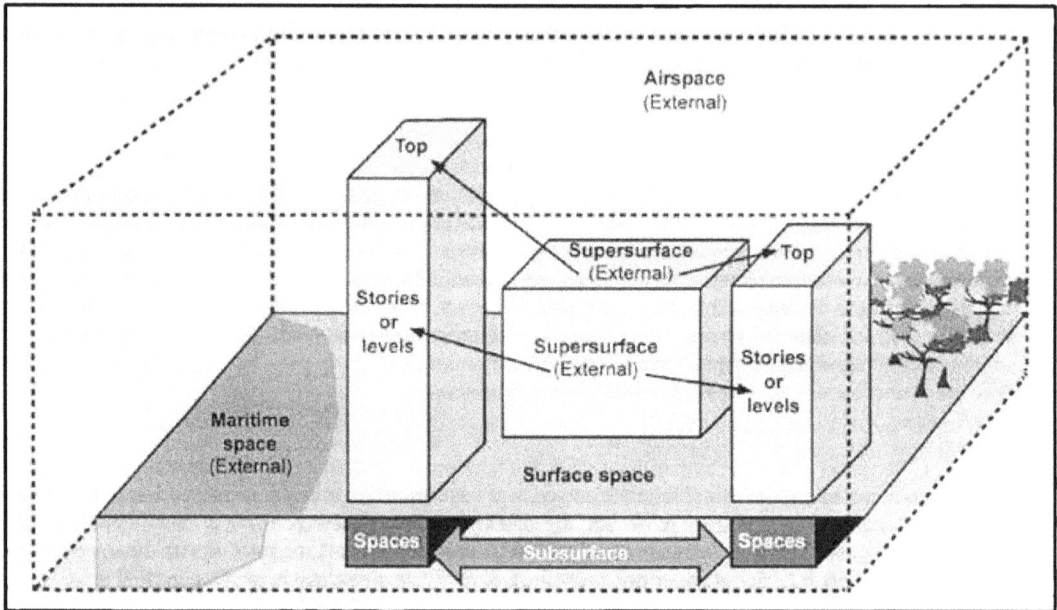

**Figure W-1. Multidimensional urban battlefield**

## Airspace

W-30. As in all other environments, aircraft and aerial munitions use the airspace as rapid avenues of approach in urbanized areas. Forces can use aviation assets for observation and reconnaissance, aerial attack, or high-speed insertion and extraction of Soldiers, supplies, and equipment. Some surface obstacles in an urban area, such as rubble, do not affect flight (though they may prevent the take-off and landing of aircraft). Buildings of varying height and the increased density of towers, signs, power lines, and other urban constructions, however, create obstacles to flight and the trajectory of many munitions (masking). Similarly, these obstacles can restrict a pilot's line of sight, as well as physically limit low-altitude maneuverability in the urban airspace. Excellent cover and concealment afforded enemy gunners in an urban area increases aviation vulnerability to small arms direct fire weapons and MANPADS, particularly when supporting ground forces. The potential for a high volume of air traffic (military and civilian) over and within urban airspace (including fixed-wing, rotary-wing, and UASs) may become another significant hazard and necessitate increased airspace command and control measures.

## Surface

W-31. Surface areas apply to exterior ground-level areas, such as parking lots, airfields, highways, streets, sidewalks, fields, and parks. These areas often provide primary avenues of approach and the means for rapid advance. However, buildings and other structures often canalize forces moving along them. As such, obstacles on urban surface areas usually have more effect than those in open terrain since bypass often requires entering and transiting buildings or making radical changes to selected routes. Where urban areas border the ocean or sea, large lakes, and major rivers, the surface of these bodies of water may provide key, friendly and threat avenues of approach or essential LOCs—a significant consideration for Army commanders. As such, amphibious, river-crossing, and river operations may be integral parts of the overall urban operation.

W-32. Larger open areas—such as stadiums, sports fields, school playgrounds, and parking lots—are often critical areas during urban operations. These areas can provide locations for displaced civilians, interrogation centers, and prisoner of war holding facilities. These areas also can afford suitable aircraft landing and pickup zones and artillery firing locations. They can provide logistics support areas and aerial resupply possibilities because they are often centrally located. Finally, large open areas (and immense or unusually shaped structures) within urban areas are often easier to see—especially from the air—and can serve as excellent target reference points from which to shift or control fires.

## Supersurface

W-33.   These areas include the internal floors or levels (intrasurface areas) and external roofs or tops of buildings, stadiums, towers, or other vertical structures. They can provide cover and concealment; limit or enhance observation and fields of fire; and restrict, canalize, or block movement. However, forces can move within and between supersurface areas, creating additional, though normally secondary, avenues of approach. Rooftops may offer ideal locations for landing helicopters for small-scale air assaults and aerial resupply. First, however, engineers must analyze buildings for their structural integrity and obstacles. Such obstacles include electrical wires, antennas, and enemy-emplaced mines (although personnel may be inserted by jumping, rappelling, or fast roping from a hovering helicopter and extracted by hoist mechanisms). Some rooftops are designed as helipads. Roofs and other supersurface areas may also provide excellent locations for snipers; lightweight, handheld antitank weapons; MANPADS; and communications retransmission sites. They enable top-down attacks against the weakest points of armored vehicles and unsuspecting aircraft. Overall, elevated firing positions reduce the value of any cover in surrounding open areas and permit engagement at close range with less risk of immediate close assault. This area (and the subsurface area) requires commanders to think, plan, and execute ground operations vertically as well as horizontally. In this latter regard, urban operations share strong similarities with mountain operations.

## Subsurface

W-34.   Subsurface areas are below the surface level. They may serve as secondary and, in fewer instances, primary avenues of approach at lower tactical levels. When thoroughly reconnoitered and controlled, they offer excellent covered and concealed LOCs for moving supplies and evacuating casualties. They may also provide sites for caching and stockpiling supplies. Subsurface areas include subterranean areas, such as subways, mines, tunnels, sewers, drainage systems, cellars, civil defense shelters, and other various underground utility systems. In older cities, they may include ancient hand-dug tunnels and catacombs. Both attacker and defender can use subsurface areas to gain surprise and maneuver against the rear and flanks of a threat and to conduct ambushes. However, these areas are often the most restrictive and easiest to defend or block. Their effectiveness depends on superior knowledge of their existence and overall design. Army commanders may also need to consider potential avenues of approach afforded by the subsurface areas of rivers and major bodies of water that border urban areas. This particularly applies when operating as part of a joint task force, task-organized with SOF, or when opposing a threat with similar capabilities.

## MAJOR URBAN PATTERNS

W-35.   Four major urban patterns (satellite, network, linear, and segment) can influence urban operations (figure W-2, page W-8). Central to two of the patterns (satellite and network) is the hub or dominant urban area or pattern around which outlying urban areas or patterns radiate. (A segmented urban area, because it tends to be a larger urban area, can often be a hub.) In offensive and defensive operations, the hub serves as a pivot or strong point; as such, it can become a major obstacle to an attacker. If the attacker chooses to bypass the urban area (hub), located along his axis of advance without first isolating the area, he may expose his flank or LOC to attack from the hub, as well as dependent urban areas or subordinate satellite patterns. Because the focus of stability and civil support operations is normally on people, commanders should understand the value and influence of the hub to the economic, political, or cultural well-being of the surrounding area. In general, the larger the hub, the greater influence it has on satellite urban areas and surrounding rural areas. Commanders must remember that urban areas are not islands; all are connected to the surrounding rural (and other urban) areas through fluid and permeable boundaries and LOCs.

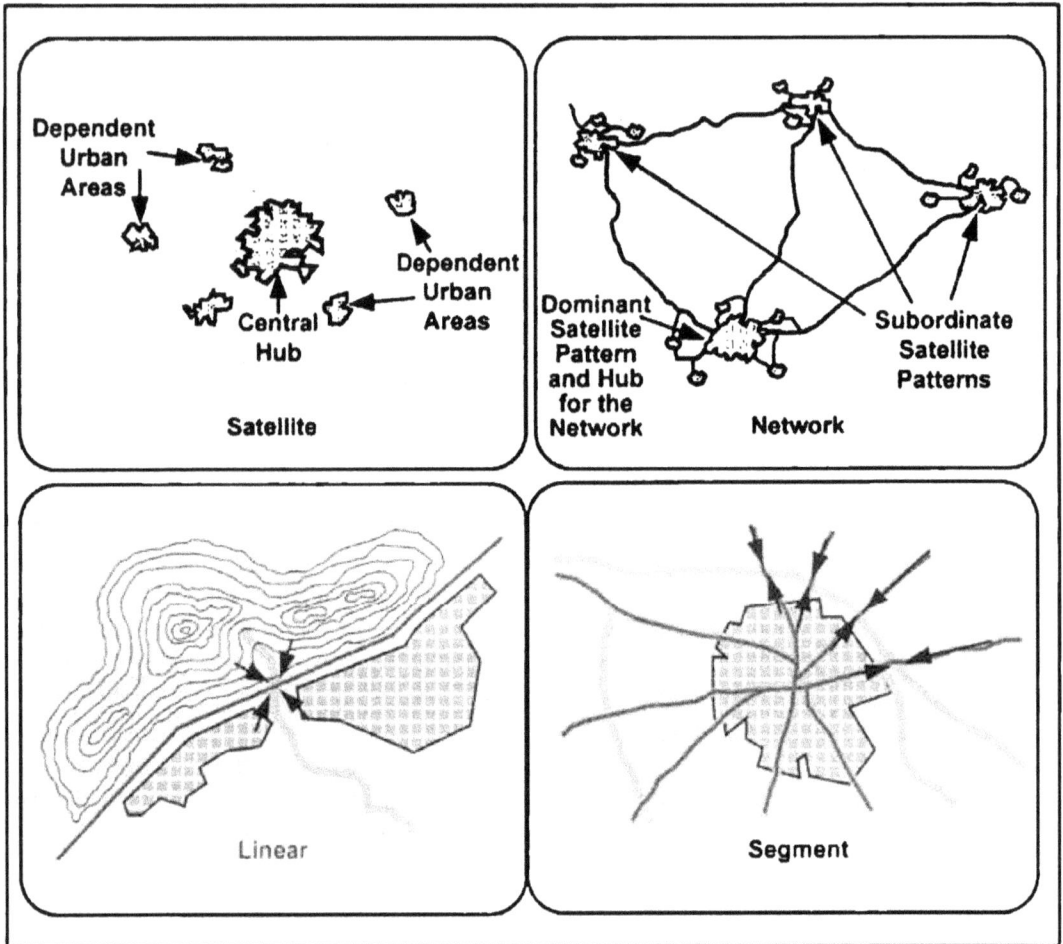

**Figure W-2. Major urban patterns**

**Satellite Pattern**

W-36. This common pattern consists of a central hub surrounded by smaller, dependent urban areas. LOCs tend to converge on the hub. Outlying areas often support the principal urban area at the hub with means of reinforcement, resupply, and evacuation. In some instances, the areas may serve as mutually supporting battle positions. Commanders should consider the effects of the outlying urban areas on operations within the hub, and, conversely, the effects of operations within the hub on outlying urban areas. Information operations (IO), for example, targeted primarily at key leaders and other civilians located within the hub of a satellite pattern may subsequently influence civilians in outlying urban areas. This dependence, therefore, may achieve necessary effects without having to commit specific resources to these outlying areas.

**Network Pattern**

W-37. The network pattern represents the interlocking of the primary hubs of subordinate satellite patterns. Although a dominant hub may exist, its elements are more self-sufficient and less supportive of each other. Major LOCs in a network extend more than in a satellite pattern and take more of a rectangular rather than a convergent form. Its natural terrain may vary more than in a single satellite array. Operations in one area may or may not easily influence, or be influenced by, other urban areas in the pattern.

## Linear Pattern

W-38.   Potentially a subelement of the previous two patterns, the linear pattern may form one ray of the satellite pattern or be found along connecting links between the hubs of a network. Most frequently, this pattern results from the stringing of minor urban areas along a confined natural terrain corridor, such as an elongated valley, a body of water, or a man-made communications route. In offensive and defensive operations, this latter form of the linear pattern facilitates developing a series of strong defensive positions in-depth, effectively blocking or delaying an attacking force moving along the canalized terrain.

## Segment Pattern

W-39.   When dominant natural terrain, such as a river or man-made features (canals, major highways, or railways), divides an urban area, it creates a segmented pattern. This pattern often makes it easier for commanders to assign areas of operations to subordinate commanders. However, this pattern may fragment operations and increase risks to an operation requiring mutual support between subordinate units. Still, the segmented urban areas may allow commanders to isolate threats more easily in these areas and focus operations within segments that contain their decisive points. Although an integral part of the whole (the urban area), each segment may develop distinct social, economic, cultural, and political characteristics. This social segmenting may benefit commanders faced with limited assets to influence or control the urban populace. After thoroughly analyzing the society, they may be able to focus information operations and populace and resources control measures against only specific segments that affect decisive operations. Commanders may need only to isolate other segments or may need to just monitor for any significant changes in the attitudes, beliefs, or actions of the civilians located there.

## LESSER STREET PATTERNS

W-40.   Lesser patterns in the urban area result from the layout of the streets, roads, highways, and other thoroughfares. They evolve from influences of natural terrain, the original designer's personal prejudices, and the changing needs of the inhabitants. Street patterns (and widths) influence all warfighting functions; however, they greatly affect movement and maneuver, command and control, and sustainment. (In some portions of older Middle Eastern urban areas, the labyrinths of streets were designed only to allow two loaded donkeys to pass each other; tanks are too wide.) Urban areas can display any of three basic patterns and their combinations: radial, grid, and irregular (figure W-3).

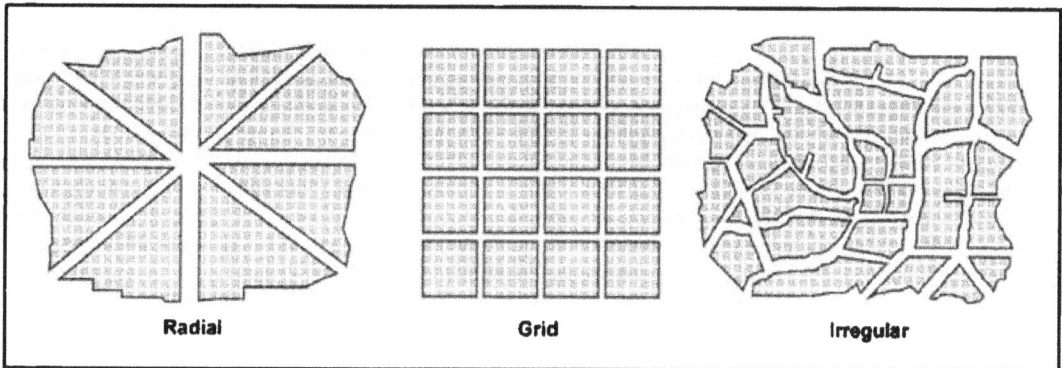

Radial | Grid | Irregular

**Figure W-3. Basic internal street patterns**

## Radial

W-41.   Societies of highly concentrated religious or secular power often construct urban areas with a radial design: all primary thoroughfares radiating out from the center of power. Urban areas with this design may signal an important historical aspect in the overall analysis of the urban society. Terrain permitting, these streets may extend outward in a complete circle or may form a semicircle or arc when a focal point abuts a natural barrier, such as a coastline or mountain. To increase mobility and traffic flow, societies often add concentric loops or rings to larger radial patterns. Unless commanders carefully plan boundaries, routes, and axes of advance, their subordinate units' movement or maneuver may be inadvertently funneled toward the center of urban areas, with this pattern resulting in congestion, loss of momentum, and an increased potential for ambush or fratricide.

## Grid

W-42.   The most adaptable and universal form for urban areas is the grid or rectangular pattern: lines of streets at right angles to one another forming blocks similar to the pattern of a chessboard. A grid pattern can fill in and eventually take over an original radial pattern. Grid patterns often appear to ease the assignment of boundaries for subordinates units. However, commanders also consider how the natural terrain influences operations and the establishment of graphic control measures. They also consider the influence of the buildings and other structures lining these streets, such as their height and construction, before assigning boundaries and developing other control measures. Commanders should also consider the following when developing urban graphic control measures:

- Describing boundaries, phase lines, checkpoints, and other graphic control measures by easily recognizable features. This is as important in urban areas as elsewhere. While easily identifiable urban structures, such as unusually tall or oddly shaped buildings, cemeteries, stadiums, or prominent rail or highway interchanges can be useful references, available natural features are a better descriptor than man-made features that may be altered or unrecognizable. As an aid to air-to-ground coordination, commanders should select features that can be identified by both ground and air forces. Those that help in controlling ground forces may not be easily visible from the air and vice versa.
- Determining whether a boundary along an easily recognizable terrain feature, such as a river, will also be easy to identify by the threat, who may seek to "find the seam" and exploit the likely control and coordination difficulties associated with boundaries (especially between higher-level units). This often requires commanders to carefully position a control measure away from the key feature to provide a designated subordinate force with the terrain and space necessary to control the feature. On the other hand, commanders working closely with local authorities during stability and civil support operations may not need to thoroughly understand the physical effect of street patterns on the assignment of boundaries as they might for combat urban operations. Instead, commanders may choose to assign boundaries overlaid on existing geopolitical boundaries, which are used by local agencies to increase interoperability and aid in unity of effort.

## Irregular

W-43.   In most urban areas, regardless of the original intent, plan, or vision, existing street patterns emerge from successive plans overlaid one on another. Some are well planned to fit with previous plans, while others are a haphazard response to explosive urban growth. The result may mix patterns. Urban engineers and planners may specifically design irregular patterns for aesthetic reasons (as in many suburban housing developments) or to conform to marked terrain relief. Irregular street patterns may alert commanders and analysts that the underlying natural terrain may exert greater influence over operations than in other portions of the urban area. Finally, irregular street patterns make the movement and maneuver of forces less predictable. However, a labyrinth of irregular or "twisting" street patterns may increase the possibility of fratricide, particularly for units that are trained or accustomed only to grid patterns.

## AN URBAN MODEL

W-44.   Throughout the world, urban areas have similar form and function. In form, urban areas contain like characteristics, readily divisible into distinct sections or areas. Functionally, they tend to be the centers of population, finance, politics, transportation, industry, and culture. While urban areas may be modeled by several different means, figure W-4, page W-11, illustrates the general forms and internal functions. Some forms and functions may overlap. For example, high-rise buildings are located in core areas, as well as in outlying areas, and may be used for residential purposes. With the rapid urbanization associated with developing nations, the areas displayed in this urban model often manifest themselves less clearly there than in developed nations.

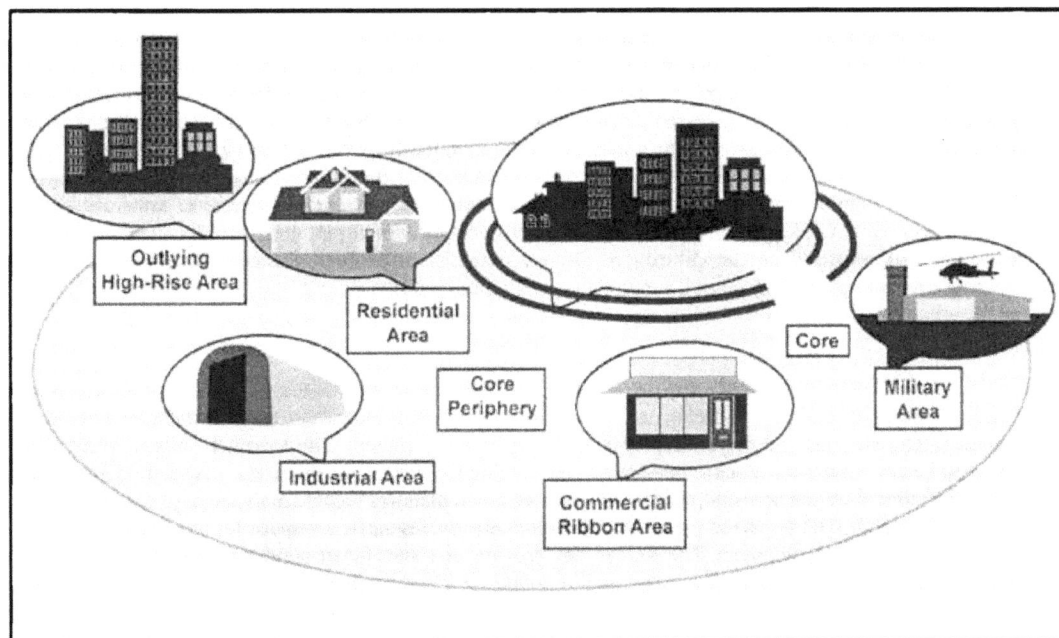

**Figure W-4. Urban functional zones**

W-45. This analysis helps to determine, in general terms, potential advantages and disadvantages each portion of the urban area may have toward accomplishing the urban operation. However, construction materials and methods can vary drastically. Commanders and their staff will often need to identify specific building types and construction and understand weapons effects on them. If a commander desires precise effects, the chosen munitions or weapons system must be sufficiently accurate, capable of penetrating the target structure (without exiting the other side), and create effects within. Noncombatants, critical infrastructure, or protected targets are often in the vicinity. Commanders may need to determine if the surrounding walls or structures will sufficiently absorb or negate the blast or thermal effects of the weapon. Regardless, understanding the structure and composition of buildings and other structures in the urban AO may be necessary to allow commanders to determine the best means to accomplish the mission.

## Core

W-46. The core is the heart of the urban area—the downtown or central business district. Relatively small and compact, it contains a large percentage of the urban area's shops, offices, and public institutions. Often, it houses the HQ for commercial and financial activities and contains important cultural, historical, and governmental buildings. These activities prefer the core because of its accessibility. As the focal point of the transportation network, residents find the core the easiest part of the urban area to reach. It normally has the densest concentration of multistory buildings and subterranean features (underground parking garages, underground shopping centers, and basements).

W-47. High-rise buildings, varying greatly in height (possibly 50 stories above ground and 4 stories below ground), make up the cores of many of today's urban areas. Buildings routinely abut one another, with little or no setback from the sidewalks. Building height and density (except in outlying high-rise areas) often decreases from the core to the edge of the residential areas, while the amount of open areas frequently increases. Modern urban planning allows for more open spaces between buildings than found in the cores of older urban areas. Most core areas have undergone constant redevelopment, resulting in various types of construction. Most commonly, brick building abound in the oldest part of the core; framed, heavy-clad structures exist in the next oldest part; and a concentration of framed, light-clad buildings abound in the newest part. The outer edge of the core, the core periphery, has ordinarily undergone less change than the core resulting in buildings of uniform height (commonly two to three stories in towns and five to ten stories in larger urban areas).

W-48.    In general, offensive operations focused in core areas (even when effectively isolated) will require greater resources—particularly manpower, time, and information—than in many other parts of the urban area. Mounted maneuver often proves more difficult in core areas because of fewer open areas, buildings closer to the streets, and an increased density of civilian vehicles. Razed buildings in central core areas (especially high-rise buildings) become greater obstacles to mobility as they can collapse on and easily block thoroughfares. Rubble piles provide excellent covered and concealed positions for dismounted threat forces. Consequently, commanders often use more dismounted forces as part of their combined arms operations. Conversely, the core may be critical to urban defensive operations, particularly older areas of heavier construction that afford greater protection. Despite potential difficulties, the core area may be key to accomplishing many stability or civil support missions, since it houses much of the human activity that occurs in the urban area.

## Industrial Area

W-49.    Industrial areas often develop on the outskirts of the urban areas where commercial transportation is easiest (along airfields and major sea, river, rail, and highway routes). The road networks in and around industrial areas are generally more developed and suitable for transportation assets. These areas will likely displace farther from the core and residential areas, as urban planners recognize the potential threat of toxic industrial material. The dispersed pattern of the buildings provides sufficient space for large cargoes, trucks, and material handling equipment. These areas may provide ideal sites for sustainment bases and maintenance sites. While older heavier-clad structures may be found, new construction generally consists of low, large, flat-roofed factory and warehouse buildings, with large parking areas and work yards. These structures generally have steel frame and lightweight exterior walls. Multistory structures usually have reinforced concrete floors and ceilings.

W-50.    Toxic industrial material may be transported through an urban area (by rail, barge, truck, or pipeline) or found stored throughout. However, larger concentrations will exist in industrial areas, and their presence should concern Army forces operating near them. Some toxic industrial material may be heavier than air and tend to settle in low-lying and subsurface areas.

W-51.    Each year, over 70,000 different chemicals are produced, processed, or consumed globally. An estimated 25,000 commercial facilities around the world produce, process, or store chemicals that have a legitimate industrial use, yet are also classified as chemical warfare agents. Many other chemicals (not classified as weapons) may still be sufficiently hazardous to pose a considerable threat to Army forces and civilians in urban areas as choking agents or asphyxiates, flammables or incendiaries, water contaminants, low-grade blister or nerve agents, or debilitating irritants. These chemicals can be released either accidentally or deliberately. For example, on 2 December 1984, nearly 40 tons of methylisocyanate used to produce pesticides leaked from a storage tank at Bhopal, India, killing thousands and injuring hundreds of thousands. The most common chemicals that may pose an immediate risk to Army forces are highly toxic irritant gases, such as ammonia, chlorine, hydrogen chloride, and sulfur dioxide.

W-52.    Standard chemical defense equipment may not protect against—and chemical detection devices may fail to detect—many toxic industrial chemicals. Therefore, the risk to Soldiers operating near the chemicals may increase. Commanders must vigilantly identify these potential hazards, carefully consider them as part of their overall vulnerability analysis, factor the analysis into their risk assessment, and execute necessary contamination avoidance measures. (Local urban firefighters may be a critical source of information for determining the likely locations of toxic industrial material.) Any assessment includes the chance that toxic industrial chemicals may be deliberately released by a threat to gain advantage or accidentally released by collateral damage.

## Outlying High-Rise Area

W-53.    High-rise areas consist of multistoried apartments, commercial offices, and businesses separated by large open areas, such as parking lots, parks, and individual one-story buildings. High-rise buildings are generally of framed, light-clad construction with thin walls of brick, lightweight concrete, or glass. The automobile, mass transit systems, and improved road networks encourage these areas to grow and function further from the urban core.

W-54. Similar to the urban core, units given the mission to clear these areas, or even portions therein, will need more resources—most notably personnel and time—to accomplish their mission. Commanders should consider courses of action that isolate these entire areas, multiple sections within these areas, or even individual buildings, before assigning tasks. Without careful consideration and analysis, some tasks in these areas could unintentionally—but rapidly—drain a unit's resources or unhinge other portions of the major operation. When defending, commanders who can integrate these areas in the defense will present the attacker with similar resource problems and may be appropriate in a defense to delay. However, defending commanders must ensure that the defense is arranged so that this portion cannot be easily isolated and bypassed. Defensive positions in structures may require extensive reinforcement because of light-clad construction.

## Residential Area

W-55. Residential areas can be found dispersed throughout the urban area; however, large suburban areas (or sprawl) normally form on the outskirts. Residential areas often consist of row houses or single-family dwellings set in a grid or ringed pattern in a planned development project. Yards, gardens, trees, and fences may separate the buildings in a residential area. In some areas of the world, residential areas may be located in high-walled compounds with houses built right up to the edge of the street. Modern residential construction is more often of light-clad, framed wood construction, or brick; however, residential homes formed by poured or precast concrete can be found throughout many parts of the world. The combined population of surrounding suburban areas often far outnumbers that of the urban area. Specific suburbs typically trend toward homogeneity based on ethnicity, religion, economics, or some other social aspect. Commanders must locate and analyze these areas to determine their impact on operations—often the most critical importance is the people located there. (See the subsequent discussion in this appendix on the urban society.)

W-56. In offensive and defensive operations, commanders should determine whether operations pose an unacceptable physical risk to civilians. If so, they may have to relocate civilians to a safer area, perhaps another residential area. If not, commanders may implement a "stay-put" policy for that area and attempt to isolate the effects of the operation from them. During civil support and stability operations, residential locations may be the initial focal point for operations since most of the permanent population is located there.

W-57. This area also contains an urban phenomenon known as shantytowns. These areas are commonly on unoccupied, low-value land in and around many urban areas in underdeveloped countries. Shantytowns may contain over 50-percent of the total urban population. They usually lack streets and public utilities. The lean-to structures tend to be irregularly laid out, connected by walking paths, and made of any scrap material available: lumber, brick, sheet metal, cardboard, cloth, or vegetation. The random arrangement of structures, the absence of formal street naming and numbering, and often the lack of easily identifiable buildings and terrain create challenges. These challenges include navigation, coordination, and the transmission of accurate information and intelligence.

W-58. Depending on the operation, the temporary nature of the structures can also mean that mobility can be either more or less restricted than other sections of the urban area. A military force may easily knock down and traverse structures without affecting mobility at all. However, their destruction may cause unacceptable civilian casualties, in which case mobility becomes more restrictive as the narrow paths often do not accommodate vehicular traffic. Similarly, the makeshift materials inhibit weapons effects less than many other parts of the urban area built more solidly. A tank round, for example, may go much farther and injure many more noncombatants than in an area where the primary building material is stone. Regardless, commanders should consider the effects of their operations in this area, to include vehicles and weapons, as the weak structures increase the risk of fratricide, civilian casualties, and large, rapidly spreading fires.

## Commercial Ribbon Area

W-59. Commercial ribbon areas are rows of stores, shops, and restaurants built along both sides of major streets that run through and between urban areas. These same types of areas often develop along the roads that connect one urban area to another (strip areas). The buildings uniformly stand two to three stories tall (about one story taller than the dwellings on the streets behind them).

## Military Area

W-60.   Fortifications and military installations may be found in or near urban areas throughout the world. Historically, they may have been the "seed" responsible for initiating the growth of the present-day urban area. Many countries possess long coastlines and borders with potentially hostile neighbors. To meet their defensive needs, they developed coastal and border defense works that include extensive subsurface facilities, many contiguous to urban areas. North Korea, for example, has built numerous hardened artillery, missile, and command and control facilities along both its coasts and the demilitarized zone that separates them from South Korea. While some fortifications may be inactive (particularly ancient fortifications), they can be rapidly activated and modified by threats (or Army forces) to accomplish their original role. Even if not activated, they may still serve as choke points and major obstacles to movement and maneuver.

W-61.   Permanent-type fortifications can be made of earth, wood, rock, brick, concrete, steel-reinforced concrete, or any combination of the above. Some contemporary constructions have been built subsurface and employ heavy armor, major caliber weapons, internal communications, service facilities, and chemical, biological, radiological, and nuclear overpressure systems. Because they have been built specifically for military purposes, commanders and planners should carefully consider the effects of these military constructions on the conduct of urban operations.

## URBAN SOCIETY

W-62.   Although intricate, understanding the urban terrain is relatively straightforward in comparison to comprehending the multifaceted nature of urban society. Unlike conventional operations in which operations are kept separate from the civilian populace wherever possible, UW is deliberately and unavoidably conducted through and with elements of the indigenous population. This population's attitudes, actions, communications with the media, political goals, and so on, will be factors in the conduct of operations. Analysis and understanding of these societal factors is critical to a successful UW operation in urban areas.

## GENERAL POPULATION SIZE

W-63.   Urban areas are commonly classified according to the general size of their population instead of landmass (figure W-5). These categories are useful to establish commonality and standardize terms that shape ideas, discussion, and concepts. Smaller populations usually suggest homogeneity among the inhabitants. Homogeneity can make consensus or compromise easier to achieve because fewer opposing viewpoints exist. Given this homogeneity, effects of change are more certain and often easier to determine. However, homogenous does not mean identical. If major social divisions exist (either physical or ideological), commanders can more easily determine those divisions and their fundamental causes with smaller populations. Treating an urban population as a completely homogenous entity can lead to false assumptions, cultural misunderstandings, and poor situational understanding. Moreover, in most cases, homogeneity decreases as population size increases.

| Category | Population |
|---|---|
| Village | 3,000 or less |
| Town | Over 3,000 to 100,000 |
| City | Over 100,000 to 1 million |
| Metropolis | Over 1 million to 10 million |
| Megalopolis | Over 10 million |

Figure W-5. Urban areas by population

W-64.   As urban areas expand, the urban patterns begin to blur and the social complexity increases. For example, as satellite patterns continue to grow, the LOCs between a central hub and outlying urban areas may develop and begin to assume a linear urban pattern. Simultaneously, a hub and outlying urban areas may continue to expand until they merge into a single, large metropolis. On a larger scale, a network pattern can grow and unite as a single megalopolis. This growth physically unites smaller urban areas but cannot force conformity of needs and beliefs. It also increases the physical and social complexity of an urban area.

## GROUP SIZE, LOCATION, AND COMPOSITION

W-65.   Understanding how specific elements of the urban society affect operations (and vice versa), begins with analyzing their size, location, and composition. In this analysis, planners should consider that urban areas, on many levels, are in constant motion. The densities of circulating people and other traffic often vary according to cultural events or the time of day, such as religious holidays or sporting events and rush hours or market times. Therefore in planning urban operations, planners must consider the timing, rhythms, or patterns of the population and their vehicular movements in the urban area. Identifying and understanding trends and patterns of activity (and disruptions to them) may provide critical information.

W-66.   Commanders normally determine the composition of, or the identifiable groups or organizations within, the civilian urban population. Groups may be categorized by race, religion, national origin, tribe, clan, economic or social class, party affiliation, education level, union memberships, age, gender, occupation, or any other significant social demographic. Physical and ideological overlaps (and divisions) often exist between groups. Overlaps may provide early focus for analysis and suggest ways to affect more than one group simultaneously. In some cases, groups may have radically different ideologies but are (or can be) united by a single characteristic. SF Soldiers must understand the intricacies of "who does what to whom." Such understanding furthers identifying the urban society's sources of power, influence (both formal and informal), and decisive points that hold the keys to shaping the human environment to meet mission objectives. Expert, detailed, and current knowledge and information is necessary to inform the ongoing development of the area study and area assessment.

## LEADERSHIP AND ORGANIZATION

W-67.   Commanders must also understand how authority and responsibility are held or shared within and between each of the identified urban groups; they must understand leadership and the social hierarchy. For groups to exert meaningful influence, leadership provides vision, direction, and organized coherence. This leadership can be a function of personality as well as organization. Some groups depend on a charismatic leader to provide cohesion; although in some cultures, the spokesman is not the leader. Others de-emphasize individual leadership and provide redundancy and replacement in decision making. Others combine elements of both these types of leadership and organization. Based solely on personality, a leader may centralize power or, while still being in ultimate control, decentralize decision making and execution to subordinates. In contrast, a single person may head a group while a ruling council actually makes and executes policy. Groups centered on one leader (which may or may not be the officially designated leader) can often produce decisions and initiate actions rapidly, but they are vulnerable to disruptions if key personalities are removed or co-opted. Groups with shared or redundant leadership take longer to make decisions, yet they are more resistant to change and outside influence.

W-68.   During UW in an urban environment, SF Soldiers will devote considerable effort to identifying and cultivating relationships with civilian leaders in their AO. This civilian leadership will include political, religious, tribal or clan, ethnic, and economic leaders. Attention by the sponsor may increase (or decrease) the targeted leaders' prestige and power. While this may be intentional, SF Soldiers must often ensure that the leaders they chose to deal with are the legitimate and accepted in the eyes of the urban population. Otherwise, they may further imbalance an already weak power structure and exacerbate the situation. In some circumstances, SF Soldiers may need to identify and interact with the leadership of criminal organizations.

## INTERESTS AND ACTIONS

W-69.   Identifying and analyzing groups also helps commanders focus on specific segments of the urban society to determine their beliefs, needs, and agendas. It also helps commanders determine how those interests motivate groups to future action (or inaction)—previous patterns of activity are critical in this regard. This analysis seeks to determine why groups (and their leaders) act as they do. SF Soldiers should consider political, economic, cultural, and religious factors in this analysis. These factors affect all groups to some extent and often provide the basis for their beliefs, needs (actual or perceived), and subsequent behavior. Size and location considerations also apply to each group to help determine to what extent its beliefs or ideologies, needs, and actions may impact the urban operation. However, size and proximity may not accurately indicate actual or potential capabilities. Individuals, small groups, and groups located some distance from the actual conduct of the urban operation may be able to influence large portions of the population. These individuals or groups may have a capability disproportionate to their size and proximity.

## INTERACTION, INFLUENCE, OR CONTROL

W-70.   SF Soldiers must cultivate an understanding of a group's—

- Size, location (and proximity to operations), and composition (to include leadership and organization).
- Interests.
- Capabilities.
- Potential actions (intent) and their operational effectiveness using their own assets and resources.

W-71.   SF Soldiers can then develop or modify courses of action with and through the indigenous partner leadership as appropriate. Certain courses of action may be needed to improve the urban resistance's ability to survive government security force scrutiny and repression. Other courses of action will contribute to sustaining the resistance elements and their constituent supporters through shadow government structures designed to displace or fill the void of state control. Ultimately, these cooperative measures between the resistance and the sponsor will lead to the fulfillment of common objectives. Some representative courses of action may include:

- Establishing buffer zones and restricted areas between pro-state and pro-resistance segments of the population and, where possible, between resistance "safe areas" and state security patrols.
- Setting up checkpoints and roadblocks with other travel restrictions on people and goods—both for internal resistance movement discipline and for counterintelligence security against penetration of the organization.
- Screening civilians for political attitudes and allegiances and potential value or threat to the resistance organization.
- Conducting negotiations between indigenous factions to achieve short-term requirements and for long-term growth of the resistance base.
- Providing or protecting rations, water, and other critical resources—both to the resistance operatives and to the constituent population supporting the resistance movement.
- Restoring or improving specific, key infrastructure to build legitimacy with the mass base of the population.
- Enforcing curfews as needed to strengthen internal security.
- Inspecting facilities to maintain a constant assessment of exploitable resources to sustain resistance.
- Directing amnesty programs when possible to aid recruitment and delegitimize government security forces.
- Conducting internment and resettlement operations to constantly maintain the image as the legitimate governance of resistance liberated areas.
- Continually monitoring the population by resistance cadres and internal security agents.

# URBAN INFRASTRUCTURE

W-72.   Urban infrastructures are those systems that support urban inhabitants and their economy. They form the essential link between the physical terrain and the urban society. When the objective of UW is overthrow of the state, efforts to build legitimacy of the resistance, while simultaneously subverting the legitimacy of the state, will include the effective control of such infrastructure. The more the resistance values legitimacy and aspires to rule after the removal of the current regime, and the more the allegiance of the population is required to achieve and sustain such control, the more important will be control of this infrastructure to meet the needs of the constituent population. During transition to resistance rule of the abandoned government structure, continued control of, safeguarding, restoration or repair of such urban infrastructure may prove decisive to population support of the new order. When the objective of UW is more limited to coerce or disrupt, careful assessment, selection and control—even if temporary—of key infrastructure may be a component of the operational plan. In such cases, destruction of infrastructure that results in calculated costs, degradations, and popular anger at the government may be an attractive option.

W-73.   During urban combat operations, destroying, controlling, or protecting vital parts of the infrastructure may be a necessary shaping operation that can isolate state security forces from potential sources of support. State security forces operating in an urban area may rely on the area's water, electricity, and sources of bulk fuel to support threat operations. This is true, particularly, when his bases or facilities are physically located in or near the urban area. Isolating threat forces from these routine sources may require them to generate their own electricity and transport their own water and fuel from outside the urban area. To transport supplies, the threat may rely on roads, airfields, sea or river lanes, and rail lines. Controlling these critical transportation nodes may prevent the threat from resupplying its forces. The control of key radio, television, and newspaper facilities may isolate the threat from the urban populace.

## INTERDEPENDENCE

W-74.   Destroying or disrupting any portion of the urban infrastructure can have a cascading effect (either intentional or unintentional) on the other elements of the infrastructure. So, the gaining of an operational advantage, while minimizing the risk of unwanted and unintended effects, has to be a planning consideration. Resistance elements can sabotage, control, seize, or secure an essential facility or structure by using small unit tactics, explosives, accidents, popular unrest, electronic disruption of communications and other means. To gain this advantage, resistance planners will rely on the expertise of local urban engineers, planners, public works employees, and others with infrastructure-specific expertise. SF advisors may be able to provide technical analysis by reachback through the SOTF to outside experts. After understanding the technical aspects of the area's systems and subsystems and the effects desired on the operational area by taking calculated actions, the resistance commanders can then develop the best COA.

*Note:* TC 18-02, *Special Forces Advisor Guide*, provides additional information.

## A SYSTEM OF SYSTEMS

W-75.   Hundreds of systems may exist. Each system has a critical role in the smooth functioning of the urban area. Simple or complex, all systems fit into six broad categories: communications and information; transportation and distribution; energy; economics and commerce; administration and human services; and cultural. Resistance planners and SF advisors should analyze key facilities in each category and determine their role and importance throughout all phases of the UW operation in an urban environment. As there is much overlap between infrastructure systems, this analysis considers each system individually and in relation to others.

## A COMBINATION OF STRUCTURES AND PEOPLE

W-76.   Urban infrastructure often consists of both a physical (terrain) and human component. For example, the physical component of the electrical segment of the energy infrastructure consists of power stations, substations, a distribution network of lines and wires, and necessary vehicles and repair supplies and equipment. The human component of this same segment consists of the supervisors, engineers, linemen, electricians, and others who operate the system. Planners must understand and recognize the physical and human components in their assessments.

## RESOURCE INTENSIVE

W-77.   Requirements to protect, restore, or maintain critical infrastructure may divert substantial amounts of resources and manpower needed elsewhere and place additional constraints on subordinate commanders. Civilian infrastructure is often more difficult to secure and defend than military infrastructure. The potentially large and sprawling nature of many systems (such as water, power, transportation, communications, and government) make their protection a challenge.

## COMMUNICATIONS AND INFORMATION

W-78.   This system is comprised of the facilities and the formal and informal means to transmit information and data from place to place. Understanding the communications and information infrastructure of an urban area is important because it ultimately controls the flow of information to the population and the opposition. It includes:

- Telecommunications, such as telephone (to include wireless), telegraph, radio, television, and computer systems.
- Police, fire, and rescue communications systems.
- Public address, loudspeaker, and emergency alert systems.
- The postal system.
- Newspapers, magazines, billboards and posters, banners, graffiti, and other forms of print media.
- The informal human interaction that conveys information, such as messengers, open-air speeches and protests, and everyday conversations.
- Other inventive informal means, such as burning tires and honking horns.

W-79.   Perhaps more than any other element of the infrastructure, communications and information link all the other elements in an interdependent system of systems. It is a critical enabler that helps coordinate, organize, and manage urban activities and influence and control the urban society. Loss or degradation in communications can have dramatic effects on government security force and popular effectiveness and morale. Urban governments and administrations may find it difficult to repair such losses in an unsecure, dangerous environment.

W-80.   Functioning urban communications and information systems may be leveraged by the resistance that has recruited or infiltrated members into the system. This co-optation of the existing indigenous system may be preferable to interdicting or destroying it. Messages on commercial systems, tailored to specific subaudiences within the larger urban population, may be intertwined with legitimate civilian users, making it unpalatable for the state to prevent use of these assets. The resistance may also use these systems to influence public opinion, gain information and intelligence, support deception efforts, or otherwise support the resistance narratives.

### Increasing Impact of Computers

W-81.   In many urban areas, computers link other elements of the urban infrastructure, and the greatest density and node centrality can usually be expected to be located in urban areas. They link functions and systems in the urban area and connect the area to other parts of the world. This latter aspect creates important implications for commanders of UW operations and activities. Operations involving this cyber function may produce undesirable effects on a greater scale than initially intended. For example, the resistance may be able to close or obstruct an urban area's banking system; however, this system may impact the international monetary exchange with unwanted or even unknown effects that is more than the sponsor is willing to accept or support. The authority to conduct these types of cyberspace operations will often be retained at the strategic level.

W-82.   However, cyberspace operations can have clear advantages. For example, if an entire power system can be taken off-line by interdicting its controlling electronics through a virtual attack, it may obviate the need for more costly, dangerous and irreversible physical destruction. Considerations of how to engage networks include: the differences between physical and virtual components; destruction versus disruption considerations; considerations of scale; ability to co-opt or "ride along" on the enemy's networks; the need in some cases for operations to remain clandestine; the degree to which access of an indigenous networks is done in the AO itself or done from remote locations, and whether such access is done by indigenous operatives, infiltrated U.S. or other operatives or remotely; and many, many more considerations.

### Pervasive Media

W-83.   The media is central to the communications and information infrastructure, and the extent it influences the indigenous population is a critical operational concern in UW. Compared to other operational environments (jungles, deserts, mountains, and cold weather areas), the media has more access to urban operations and more potential impact because of the concentrated population. Some of the same considerations for cyberspace operations apply to media. Sometimes co-optation or "riding along" will be preferable to interdiction or destruction of media systems. Much of the importance of media is its ability to carry the desired messages—whether government-crafted or supportive of the resistance. Planners should also consider nonobvious, traditional, and low-tech media as alternates to mass media. He who controls the message often controls the people. Therefore, carefully considered and applied content can be critical. This is a natural role for PSYOP to contribute to the UW plan.

## TRANSPORTATION AND DISTRIBUTION

W-84.   This element of the infrastructure consists of:

- Networked highways and railways, to include bridges, subways and tunnels, underpasses and overpasses, ferries, and fords.
- Ports, harbors, and inland waterways.
- Airports, seaplane stations, and heliports.
- Mass transit.
- Cableways and tramways.
- Transport companies and delivery services that facilitate the movement of supplies, equipment, and people. Similar to communications and information, this facet provides the physical link to all other elements of the infrastructure.

W-85.   Normal security operations of government security forces depend on routine access to ports, airfields, roadways, railways, and so on. Transportation and distribution systems in the urban area can contribute greatly to the movement of state forces, maneuver, and logistics operations throughout the entire AO. State control of decisive points in this infrastructure may be important to the military operation and to the normal functioning of the urban area and surrounding rural areas. Supplies traveling through the transportation and distribution system may be military-specific supplies (such as ammunition and repair parts) and supplies for both the military and urban population (such as food, medicine, oil, and gas). The system may also support the movement of military forces and the urban area's population (for which it was designed). Therefore, resistance seizure or disruption of these critical transportation nodes may impact the state's ability to conduct law enforcement, counterinsurgency, or other operations. Resistance leaders will want to consider developing innovative methods that limit the transit of state supplies and reinforcements, while facilitating the movement of their own resistance resources and those of supportive civilians.

W-86.   Most urban areas (particularly in developing countries) have two forms of transportation and distribution systems that exist simultaneously: a formal system and an informal or "paratransit" system. Large organizations, bureaucracy, imported technology, scheduled services, and fixed fares or rates characterize formal systems. Low barriers to entry; family and individual entrepreneur organizations; adapted technology; flexible routes, destinations, and times of service; and negotiated prices characterize the informal system. The informal system is more decentralized and covers a much greater portion of the urban area than the formal system. The informal transportation and distribution system often includes a waterborne element, is more likely to function through turbulence and conflict, and can extend hundreds of kilometers beyond the urban area. Accordingly, SF Soldiers should understand both systems to establish effective movement control.

## ENERGY

W-87.   The energy system provides the power to run the urban area. It consists of the industries and facilities that produce, store, and distribute electricity, coal, oil, wood, and natural gas. This area also encompasses alternate energy sources, such as nuclear, solar, hydroelectric, and geothermal power. Energy is needed for industrial production and is therefore vital to economics and commerce. Among many other things, this system also provides the fuels to heat, cool, and light homes and hospitals, cook and preserve food, power communications, and run the transportation necessary to move people and their supplies throughout the urban area. Loss of an important energy source, such as electricity or gasoline, especially for those accustomed to having it, is likely to become an immense area of discontent. The urban resistance leader may be able to deliberately use such interdictions as part of a planned series of effects to be exploited to erode support for the state and government forces. Alternatively, the resistance needs to be prepared to mitigate the popular discontent that may result from resistance disruptions or by a state regime shutting off power to certain city sectors or communities in reprisal.

W-88. Sources of energy may be tens or hundreds of miles away from the urban area itself. Therefore, resistance leaders may be able to exert control without applying combat power directly to the urban area itself, by controlling or destroying the source (power generation or refinement plant) or the method of distribution (pipelines or power lines). With electrical energy that cannot be stored in any sizable amount, the latter may be the best means as most major urban areas receive this energy from more than one source in a network of power grids. Lengthy pipelines and power lines are a security challenge for the owner and should be vulnerable to interdiction somewhere. However, assuming the resistance has the necessary access, control may be as simple as securing a power station or plant and turning off switches or removing a vital component that could later be restored.

W-89. The number of nations that have invested in nuclear power and nuclear research is increasing. With this increase, the potential for damage to one of these facilities and potential radiation hazards will present special challenges to resistance leaders. The destruction of such facilities and the potential catastrophic effects on a large, condensed urban population are likely to be too dangerous to consider for direct, destructive targeting. However, use of indigenous expertise with a stake in avoiding mass casualties in their own homeland—possibly fortified by sponsor-facilitated external expert support—may be able to temporarily or benignly interdict power from such facilities to achieve a specific, limited effect.

## ECONOMICS AND COMMERCE

W-90. This system encompasses:
- Business and financial centers, to include stores, shops, restaurants, hotels, marketplaces, banks, trading centers, and business offices.
- Recreational facilities, such as amusement parks, golf courses, and stadiums.
- Outlying industrial, mineral, and agricultural features, to include strip malls, farms, food processing and storage centers, manufacturing plants, mines, and mills.

W-91. Resistance leaders will look to engage select economic and commercial targets for specific effects to achieve clearly outlined objectives. Some of these targets will be inside the urban area itself, while others may be located outside of the urban area. Such targeting may be intended to have direct effects at undermining the support and resources of the government and its security forces. Some may be less about degrading economic power but important as a symbolic attack to make a political statement against domestic—and possibly international—supporters of the regime. In some cases, a spontaneous disgruntled civilian population attack on commercial targets can be leveraged for resistance propaganda. In other cases, the cumulative effects of economic scarcity, resulting from a resistance subversion campaign, can inflame that same civilian dissatisfaction and unrest. Resistance campaigns against the economy must also consider the mid- and long-term needs of the supporting population. Excessive economic disruption and resultant hardship on the population risks the population turning against the resistance rather than the regime.

W-92. This element of the infrastructure also consists of the production and storage of toxic industrial chemicals used in agriculture (insecticides, herbicides, and fertilizers), manufacturing, cleaning, and research (to include biological agents). Fertilizer plants may be of special interest as they contribute to providing a key material in resistance bomb-making activities. Like the potential unwanted, uncontrolled effects on nuclear facilities, the resistance leadership must give careful thought to avoiding unintended consequences from, and negative public reaction to, mass contamination and casualties from toxic industrial material release.

## ADMINISTRATION AND HUMAN SERVICES

W-93. This wide-ranging system covers urban administrative organizations and service functions concerned with an urban area's public governance, health, safety, and welfare. These services comprise much of what the resistance shadow government will seek to control, infiltrate, and exploit on behalf of the supportive resistance population. In cases of resistance in which gaining and marshalling the allegiance of the population is key to success, dedicated resistance cadres will be identified and organized to expropriate these services from state control or, alternatively, to replicate them from resistance-controlled resources and efforts, thereby displacing the state's role in providing for the population. Examples of these services and functions include:

- Governmental services that include embassies and diplomatic organizations.
- Activities that manage vital records, such as birth certificates and deeds.
- The judicial system.
- Hospitals and other medical services and facilities.
- Public housing and shelter.
- Water supply systems.
- Waste and hazardous material storage and processing facilities.
- Emergency and first-responder services, such as police, fire, and rescue.
- Prisons.
- Welfare and social service systems.

W-94.   The extent to which these services and functions are underground or in the shadows is a function of the ubiquity and effectiveness—or lack thereof—of state control. In areas that are undergoverned or completely ungoverned, most of these routine, civil activities could be out in the open. Sometimes the resistance may openly demonstrate their control of this infrastructure, and sometimes it will do so discreetly or even clandestinely through public components or front organizations. In situations where government presence and oppressive capabilities remain a concern, resistance leaders should consider the techniques used by criminal organizations or elements to conduct their activities and influence subgroups within the population. There is also a vast record in political history about the community organization of parallel structures to inform the development of shadow governance infrastructure.

## CULTURAL

W-95.   This system encompasses many organizations and structures that provide the urban populace with its social identity and reflect its culture. This infrastructure system overlaps with many recreational facilities included under the economics and commerce infrastructure. For example, an urban society may radically follow soccer matches and teams. Hence, soccer stadiums relate to the society's cultural infrastructure. Some of these facilities, particularly religious structures and cultural or historical icons have huge potential emotional appeal for the populace. Therefore, the symbolic importance of such venues can be leveraged by the resistance. In most cases, the resistance leadership will want to protect such venues from being targeted by the state, the resistance organization itself, or from any third-party actor looting and pilferage. Resistance deference to, and protection of, such cultural symbols can contribute to direct popular support for the resistance, just as shrewd exploitation and messaging of any state damage, destruction, or denial of them can breed popular anti-regime anger. Conversely, certain cultural venues may have negative connotations of state oppression, exploitation, and injustice. Resistance leaders can potentially reap powerful operational and strategic effects from the deliberately choreographed destruction or occupation and concurrent, resistance media exploitation of hated symbols of state tyranny. Some examples of cultural infrastructure include:

- Religious organizations, places of worship, and shrines.
- Schools and universities.
- Museums and archeological sites.
- Historic monuments.
- Libraries.
- Theaters.

## MEGACITIES

W-96.   At the extreme end of the urban environment scale is the megacity; an entity so large that potentially an entire campaign could be carried out within its boundaries. The megacity challenge is a matter of scale so enormous that it merits a brief, separate consideration. Context, scale, density, connectedness, and flow are megacity characteristics the resistance leader, the UW planner, and the SF advisor need to consider (figure W-6, page W-22).

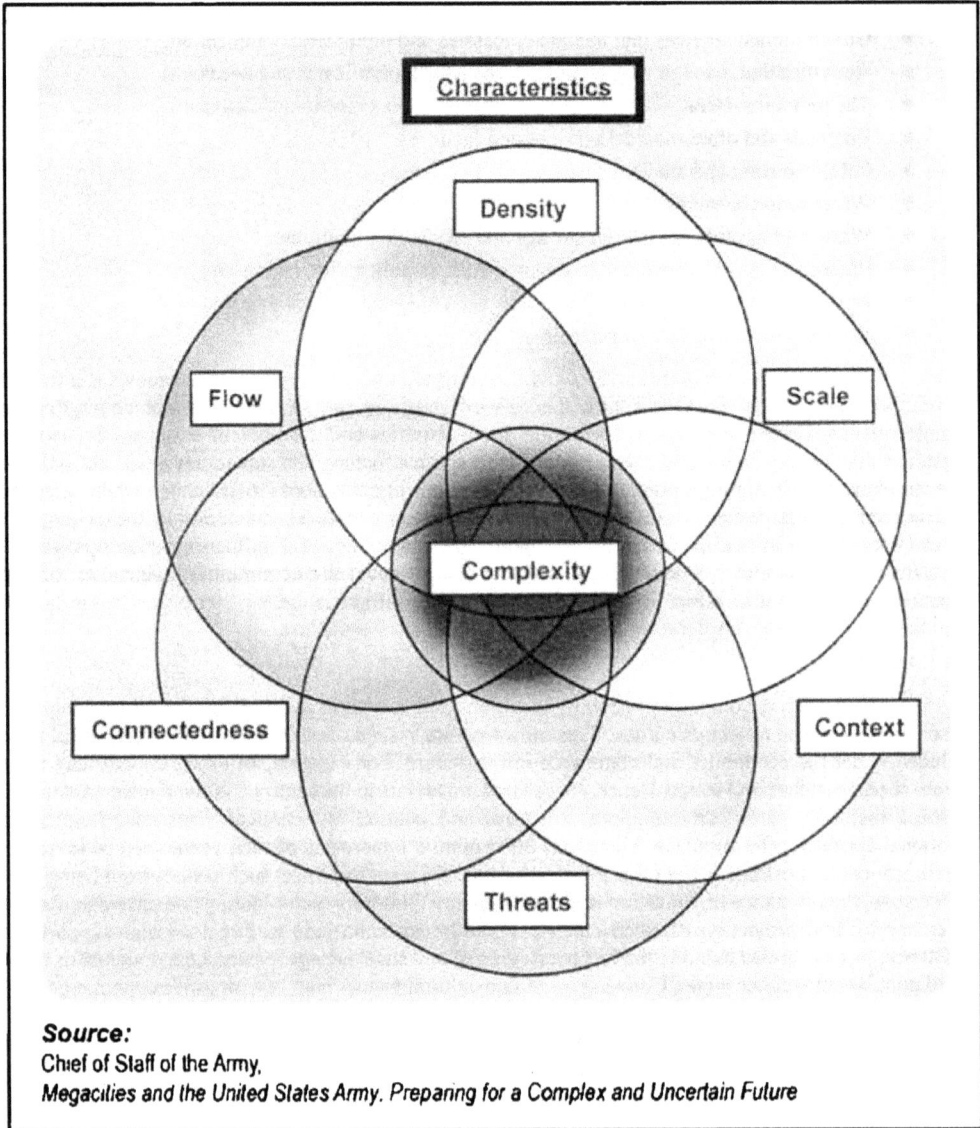

**Figure W-6. Megacity characteristics**

## Context

W-97. Every megacity is unique and must be understood within its own historical, cultural, local, regional and international context. Knowledge of the rate and characteristics of a megacities' growth may enrich contextual understanding, as will knowledge about certain drivers of instability. Some megacities have ancient roots and are inseparable from the national populations' self-identity. In other cases, the steady agglomeration of megacity sprawl is the relatively recent product of massive, rural poverty and the pursuit of whatever limited opportunity can be found in the urban hive. Some megacities are rich and powerful while others—despite being the locus of centralized national power—are predominantly occupied by a permanent underclass of the underemployed and undereducated. Megacities differ by whether they are growing steadily out of consistent, opportunity-based processes or explosively out of processes based on dependency, exploitation, and lack of alternatives.

## Scale

W-98.   The relative size of megacities differentiates them from other urban environments and presents a fundamental challenge to the feasibility of UW. Density, connectedness, flow, and context can be studied in any environment, but understanding these elements at scale is what makes megacities a different problem set. Can a small, elite, professional revolutionary cadre-driven, Bolshevik-style approach work in a city of 20 million people? If a Maoist, mass-based approach is sought, how are problems of organization and time expanded? How does a resistance movement and its sponsors influence tens of millions? Resistance planners will need to rethink traditional force size assumptions in a megacity.

## Density

W-99.   Population, infrastructure, and signals all pose significant challenges with regard to density. Population density can, intentionally or unintentionally, disrupt flows on fixed capacity lines of transportation and communications in and around the urban environment. Structural density limits maneuverability and places limitations on a formation's ability to mass, which disaggregates combat power. Electronic signal density presents problems with bandwidth congestion and signal-based targeting. What operational opportunities exist in megacity subareas with varying densities? Is there more potential security or threat if operating in an extremely densely populated area? Do the challenges of density affect the state's ability to control more than it affects the ability of the resistance to survive and conduct operations? What scale of traditional military activity would be required to isolate a megacity the size of New York, Shanghai, Tokyo, or Lagos? What scale of resistance activity would be required? These questions can be useful in focusing on the essential role played by leveraging the indigenous populations to affect change themselves, rather than assuming a unilateral approach might be effective.

## Connectedness

W-100.   It is obvious today that cities do not exist in isolation. Attempting to isolate one, as has been taught as a classic maneuver in traditional and current doctrine, would be difficult, resource-intensive (if not prohibitive), and would likely lead to unforeseen consequences. Instantaneous information transfer, robust international surface and air shipping, and mass migration (legal and illegal) connect the cities around the world in ways undreamed of only a decade ago. Robust and redundant external connectedness makes isolating a modern city nearly impossible. The larger the city, the more likely this is to be true. The very size of megacities makes notions of surrounding, containing, or isolating such entities improbable. However, this robust connectedness is not necessarily an advantage to the central government. The breadth and depth of modern connectedness multiplies the opportunity for resistance entities to circumvent central scrutiny and restrictions.

## Flow

W-101.   Flow is the movement of people, resources, or things into or out of a megacity. Just as a living organism relies on flows in (food, air and water), and flows out (waste) to stay alive, a city also requires flows. Vast amounts of energy and other vital goods must flow into the megacity. These goods must circulate throughout the urban space, and waste must flow out if the megacity is to remain healthy. As other systems and the human population rely on such flows, and control of the same flows plays a role in population influence efforts, the analysis of such "organic city flows" becomes an operational consideration for resistance planners. Moreover, flow-in and flow-out connote access and egress as well as critical concerns for sustainment of and potential, external sponsor support to the resistance.

## Threats

W-102.   Megacities are constantly challenged by threats to their stability. The nature of these environments manifest multiple dynamics of observable friction, which operates against the city or emanates from within it. These man-made and natural threats contribute significantly to the complexity of the megacity. For the resistance, such threats may present opportunities that may be exploited. Disruptions of state provisions of the necessities of life, breakdowns in security or health conditions, rampant criminality, social-economic and demographic frictions, and so on, occur normally. These disruptions may be co-opted, exploited for anti-government agitation purposes, or exploited for recruitment purposes. The disruptions also may serve as camouflage for resistance activities, as a distraction which consumes state resources and attention, and many other potential imaginative purposes.

## DYNAMICS OF INSTABILITY AND CAPACITY

W-103. While each megacity is unique and must be understood in its own context, megacities share some fundamental characteristics which contribute to their complexity. Tipping points and triggers are analogies that describe the cumulative effect of various inputs or stressors, on or within a megacity, whose massing precipitates a dramatic shift in a system, or systems, from a state of equilibrium to a state of relative imbalance. There is no set, universal measure for these tipping points: each city has its own equilibrium, which must be understood in its own unique context. In some mega cities, this equilibrium is a delicate condition (fragility); in others, order, security, and solvency are robustly maintained with heavy investment in redundancy and contingency management (resilience).

W-104. Regardless of the fragility or resilience of the city, its stable functioning is dependent on systems of finite capacity. When these systems—formal or informal, real or virtual—experience demand that surpasses their capacity, the load on the city's systems erodes its support mechanisms, increasing system fragility. These systems are then more vulnerable to triggers, which can push the city past its tipping point and render it incapable of meeting the needs of its population.

W-105. Some dynamics of friction are observable in all megacities to varying degrees. Population growth and migration, separation and gentrification, environmental vulnerability and resource competition, and hostile actors are all present in some fashion within every megacity.

### Population Growth and Migration

W-106. One of the hallmarks of megacities is rapid hetero and homogeneous population growth that outstrips city governance capability. Many emerging megacities are ill-prepared to accommodate the kind of explosive growth they are experiencing. Resistance movements are sometimes natural outgrowths of these demographic frictions; other times, they can hide within or exploit them.

### Separation and Gentrification

W-107. Radical income disparity, and racial, ethnic, and subcultural separation are major drivers of instability in megacities. As these divisions become more pronounced, they create delicate tensions, which if allowed to fester, may build over time, mobilize segments of the population, and erupt as triggers of instability. These social-economic, cultural, and demographic fracture lines are the tectonic stresses, which create the raw material for revolution and resistance. Resistance leaders and SF advisors need to be well-studied in these population stress lines. In a megacity, the larger the population, the larger the potential to find disaffected and mobilized subgroups to organize.

### Environmental Vulnerability and Resource Competition

W-108. Unanticipated weather events and natural disasters can be powerful catalysts, which can devastate city systems, interrupting governance and service delivery. While natural cataclysms occur across the globe and have throughout human history, these events will affect larger populations, densely packed into urban centers in ways and on a scale never before seen. Environmental disasters and resource scarcity (real or perceived) can produce relative resource disparity, competition, and instability that can rapidly exceed the capability of local authorities to address. Whether from natural or man-made causes, the alert resistance leader will notice and be prepared to exploit the popular grievances resulting from deprivation.

### Hostile Actors

W-109. If internal or foreign actors conducted offensive operations or repressive actions, which exceeded a city's capacity to accept and endure them, external intervention could be required to return the megacity to something resembling its previous state. In the case of major war, this has multiple posthostility governance challenges. For a resistance movement, perhaps awaiting outside liberation, the presence of "hostile actors" is a natural narrative, which sells itself and should make recruitment and motivating to action easier.

## Capacity

W-110. Every city has a unique way of organizing, equipping, and connecting the resources required to maintain its systems. Understanding the systems that keep the city functioning is an essential component of understanding the nature and logic of the city itself. Similarly, understanding the surge capacity of the city (the emergency response capability, the extent of planning and exercising emergency procedures, the material resources and reserve or mobilization capability, and so on) is essential to forecasting the city's ability to return to steady-state and determining how much external assistance may be necessary to help it do so. Resistance leaders and SF advisors need to be well-studied in the megacity's capacity at rest and when stressed. They should look for operationally and strategically significant nodes where more than one system capacity chokepoints connect. Limits to capacity and capacity bottlenecks can be exploitable.

## ANTI-FRAGILITY AND RESILIENCE

W-111. Cities differ widely on their ability to adapt to volatility and stress. Some cities respond poorly to adversity, making bad situations worse. Others return quickly to a normal state, expending resources to minimizing the impact of the adversity. These characteristics are fragility and resilience, respectively. Many cities learn and grow from adversity—a characteristic that is coming to be known as anti-fragility.

W-112. There is a strong correlation between highly integrated systems and anti-fragility. In large urban environments, a highly integrated city like New York exhibits anti-fragile characteristics when it learns from setbacks and then designs systems that prevent future disruptions from similar events. Loosely integrated cities, on the other hand, show little or no improvement in the aftermath of adverse events. Unless a city learns and evolves from adverse events, it is not anti-fragile and future events can overwhelm it. Even the most resilient cities will eventually wear down under constant and increasing pressure.

W-113. This model is a gross oversimplification of an extraordinarily complex dynamic. There are no absolute measures to quantify stress or capacity. Stressors and capacity must be understood in context of the specific city. The same stressor may exist in two different cities, but the scope of the stressor may be completely different, as might the capacity of the city or approach to responding to the stressor. The model is not intended to be all inclusive; rather, it is meant to provide a framework for analysis, feasibility assessments to provide national leaders strategic options and, subsequently, an aid to mission planning.

## COMPARING MEGACITIES

W-114. The ability of megacities to withstand and recover from expected or unexpected stressors is largely based on the city's degree of integration. Some megacities are more integrated than others and, when compared to one another, this degree of integration can be used to categorize megacities into a basic typology. This typology ranges from cities that are highly integrated, such as New York City or Tokyo, with hierarchical governance and security systems, to cities that are loosely integrated, such as Lagos, Nigeria, or Dhaka, Bangladesh, with alternatively governed spaces and security systems. Some cities exhibit a combination of the two. Bangkok, Rio de Janeiro, and Sao Paulo show moderate degrees of integration (figure W-7, page W-26).

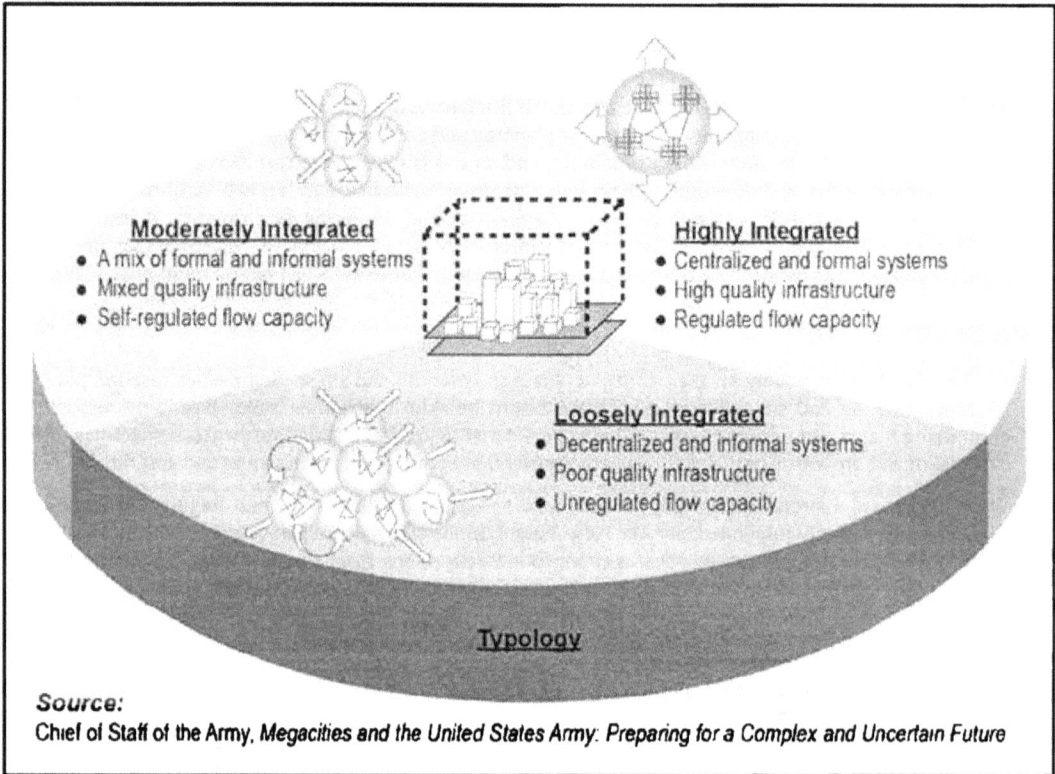

Source:
Chief of Staff of the Army, *Megacities and the United States Army: Preparing for a Complex and Uncertain Future*

**Figure W-7. Megacity integration levels**

W-115. Highly integrated systems are characterized by strong, formal and informal relationships among its component parts. These relationships manifest as highly ordered hierarchical structures with formalized procedures and norms and open communication among its various parts. Highly integrated systems are inherently stable, show high degrees of resilience (ability to absorb stress), and manage growth in a relatively controlled manner. Highly integrated cities are likely to be the richest and best resourced on earth, with professional security and intelligence forces and significant ability to monitor and respond to unlawful activity. Opportunities for lawful change and economic advancement are relatively abundant. The lower economic class and criminal communities are likely to comprise a smaller proportion of the overall population. These characteristics combined are likely to present a greater challenge to resistance, survival, and growth.

W-116. Loosely integrated cities, on the other hand, lack many of the formal relationships that keep highly integrated cities stable. Weak control and communications systems and lack of consistent rules for interaction among component parts lead to low resilience and unregulated growth. This growth, in turn, contributes more component parts that are not formally integrated into the system, creating a downward spiral of instability. Loosely integrated cities are largely incapable of dealing with the challenges presented them today and there should be little expectation of their ability to meet the growing challenges of tomorrow. Loosely integrated cities are likely to be the poorest and worst resourced of the megacities, with security and intelligence forces challenged to monitor significant subsegments of the population, and with significant limitations on response to unlawful activity. Opportunities for lawful change and economic advancement are relatively scarce and meaningfully open to only a tiny percentage of the city's overall population. The lower economic class is likely to comprise a majority of the overall population and criminal communities are numerous. These characteristics are likely to present a greater challenge to survival of the state under serious threat by organized resistance, and the resistance is like to enjoy advantages in surviving state repression and organizational growth. However, if the ultimate objective of the resistance is to overthrow the current regime and assume power itself in such a situation, the postseizure challenges are likely to remain intractable for the new rulers, presenting transition challenges to them and to any external sponsors.

W-117. Moderately integrated cities show some characteristics of both highly and loosely integrated megacities. In these environments, formal governments may be able to control portions of the city and episodically control other less integrated parts of the city. These conditions are brought about by rapid, unplanned growth, which is compounded by separation. Each city and megacity is different and must be analyzed for its unique attributes.

# FORMS AND TYPES OF URBAN OFFENSE

W-118. Traditional forms of offensive maneuver include envelopment, turning movement, infiltration, penetration, and frontal attack. Traditional types of offensive operations are movement to contact, attack, exploitation, and pursuit. These traditional forms and types listed apply to urban combat. Some have greater application to an urban environment than others do. Moreover, success will belong to commanders who imaginatively combine and sequence these forms and types throughout the depth, breadth, and height of the urban battlefield. This is true at both the lowest tactical level and in major operations.

W-119. Those who would conduct or advise the prosecution of UW in urban environments must be completely comfortable with these basic concepts. The illustrations and descriptions below are drawn from Army doctrine for urban operations and assume a conventional approach. However, indigenous resistance partners often live and operate in cities. There are, therefore, at least three reasons why SF Soldiers and advisors to resistance need to master these basic principles:

- First, a resistance may be called upon to defend against or survive state or occupying power major operations against the resistance and supporting population's cities.
- Second, not all instances of resistance and UW are the same scale. During World War II, the Chinese Civil War, Afghanistan 2011–2012, and so on, it has been demonstrated that major operations involving sizeable resistance forces have maneuvered on enemy-controlled cities.
- Thirdly, the examples given below in the figures often show large-scale efforts involving forces up to multidivision size. This is just to illustrate doctrinal principles to senior Army leaders; the basic maneuvers are classic and scalable down to any size. The principles of envelopment, infiltration or frontal assault, for example, do not change and would be useful against a threat village. Remember too, that SF advisors to resistance have and are likely again, in some future cases to cooperate with large-scale resistance partner forces capable of marshalling large maneuver units.

*Note:* ATP 3-06, *Urban Operations*, provides additional information.

## ENVELOPMENT

W-120. The envelopment is the ideal maneuver for isolating threat elements in the urban area or isolating the area itself. A deep envelopment effectively isolates the defending forces and sets the conditions for attacking the urban area from the flank or rear. Yet, enveloping an objective or threat force in the urban area is often harder since achieving speed of maneuver in the environment can be difficult (figure W-8, page W-28). Given the previous discussion of city fragility and dependence on outside systems that flow in and out of the urban area itself, cities remain vulnerable to envelopments, which disrupt one or more critical support system as part of a well-coordinated resistance plan. This is one of many useful lessons demonstrated by the Chinese communist example—that which cannot be taken directly by resistance forces battling state major forces is nonetheless vulnerable to being surrounded and cut off.

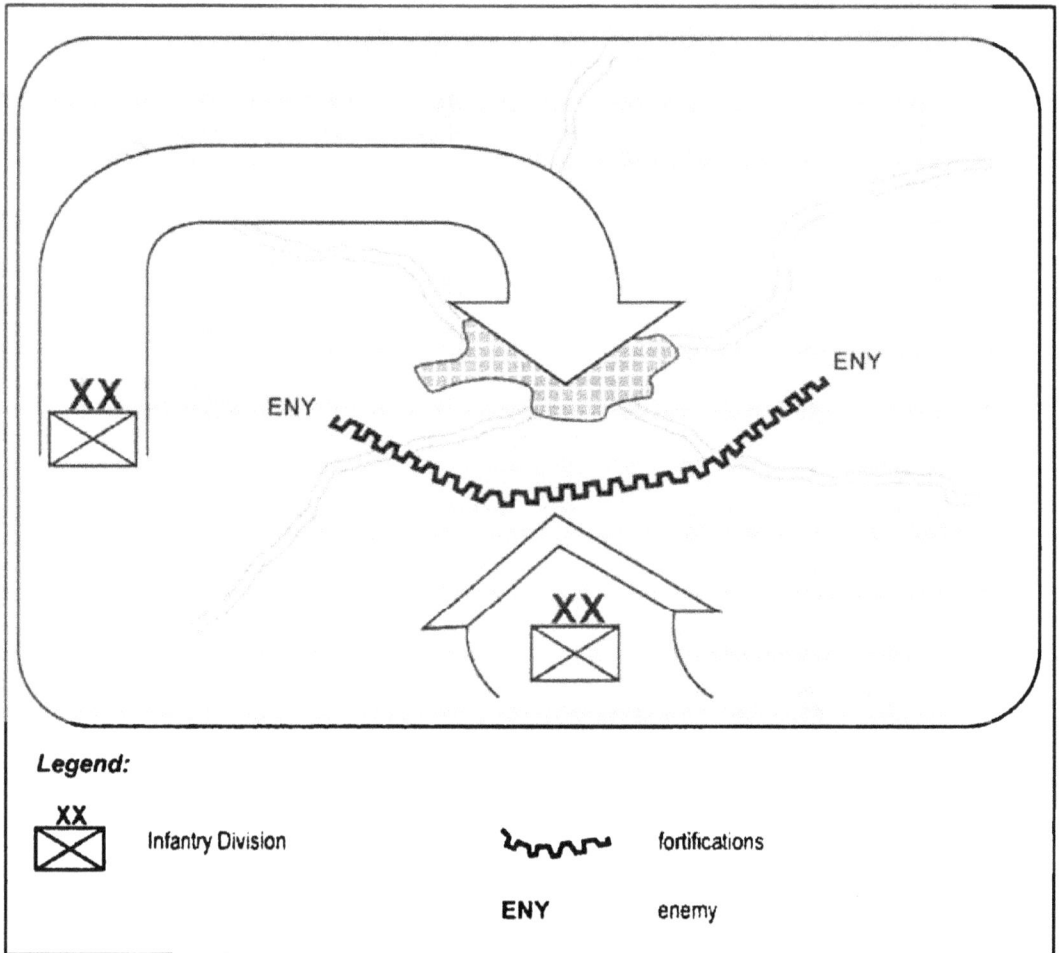

**Figure W-8. Envelopment isolates an urban area**

## TURNING MOVEMENT

W-121. Turning movements can also be extremely effective in major operations (figure W-9, page W-29). By controlling key LOCs into the urban area, Army forces can force the threat to abandon the urban area entirely. These movements may also force the threat to fight in the open to regain control of LOCs. The resistance should understand the opponent well enough to predict whether he can be forced to withdraw rather than stand and fight. A record of resistance (tactical and operational) successes in series, combined with an effective resistance PSYOP campaign, may convince state forces to relinquish controlled territory once it is perceived that resistance forces appear to be threatening the enemy's rear area. Successful use of this maneuver has the added virtue of reducing friendly casualties and preserving the force.

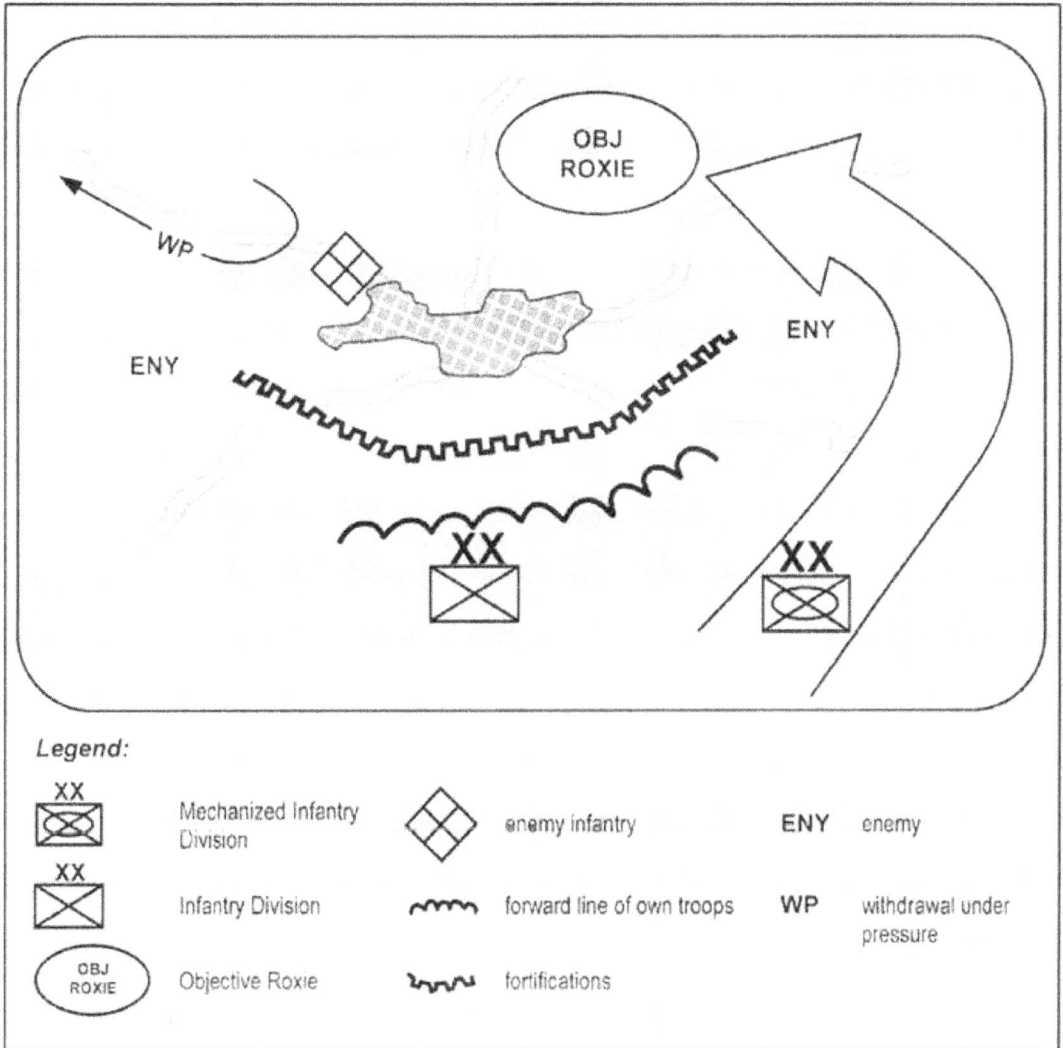

**Figure W-9. Turning movement**

## INFILTRATION

W-122. Infiltration secures key objectives in the urban area, while avoiding unnecessary combat with threat defensive forces on conditions favorable to them (figure W-10, page W-30). This technique seeks to avoid the threat's defense using stealthy, clandestine movement through all dimensions of the urban area to occupy positions of advantage in the threat's rear (or elsewhere). Infiltration depends on the careful selection of objectives that threaten the integrity of the threat's defense and a superior common operational picture. Well-planned and resourced deception operations may potentially play a critical role in masking the movement of infiltrating forces. The difficulty of infiltration attacks increases with the size and number of units involved. It is also more difficult when Army forces face a hostile civilian population. Under such circumstances, infiltration by conventional forces may be impossible. Armored forces are generally inappropriate for infiltration operations; however, they may infiltrate large urban areas if the threat is not established in strength and has not had sufficient time to prepare defenses.

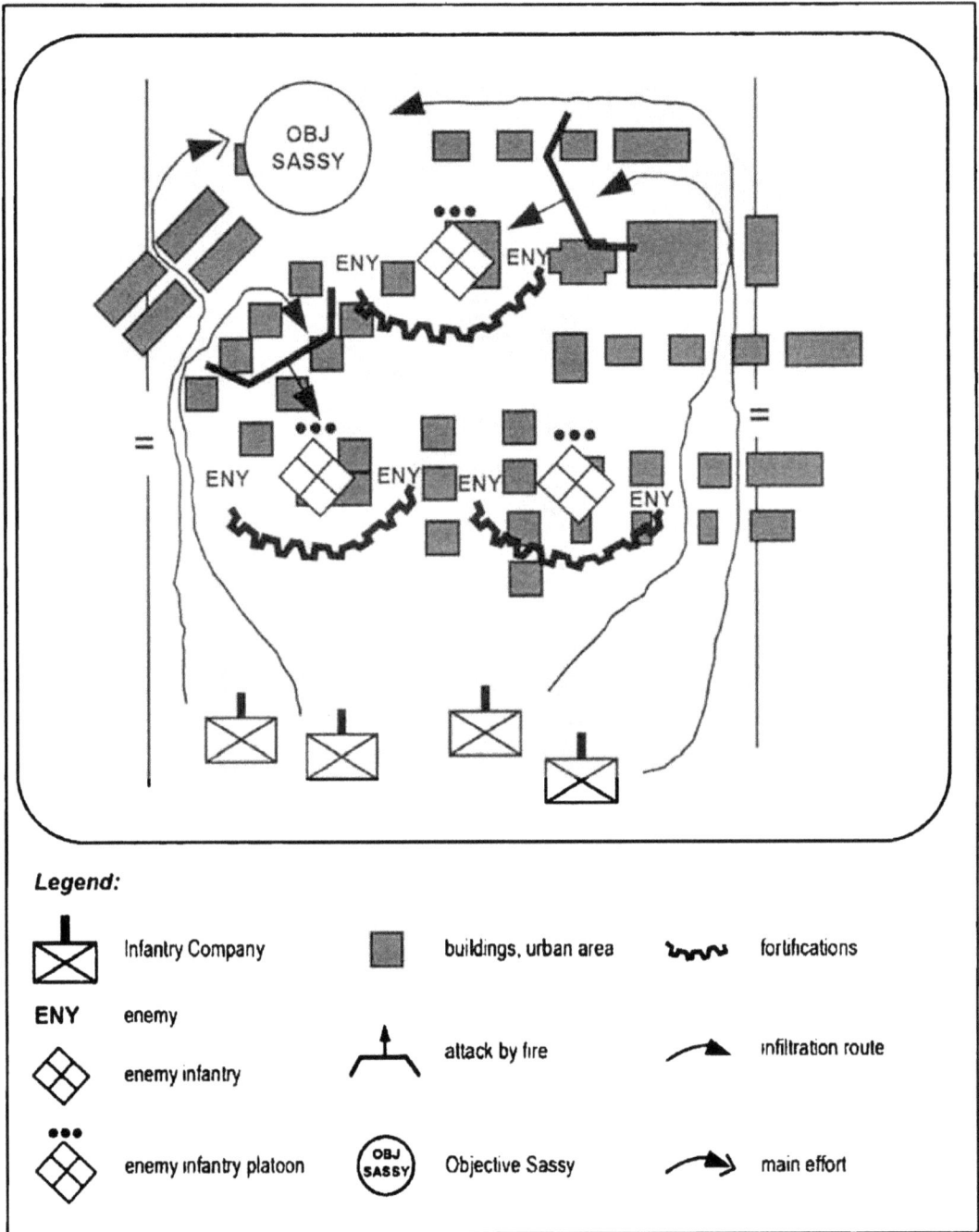

**Figure W-10. Infiltration**

## PENETRATION

W-123. Penetration is often the most useful form of attack against a prepared and comprehensive urban defense. Figure W-11, page W-31, focuses on successfully attacking a decisive point or on segmenting or fragmenting the defense—thereby weakening it and allowing for piecemeal destruction. The decisive point may be a relatively weak or undefended area that allows Army forces to establish a foothold for attacks on the remainder of the urban area.

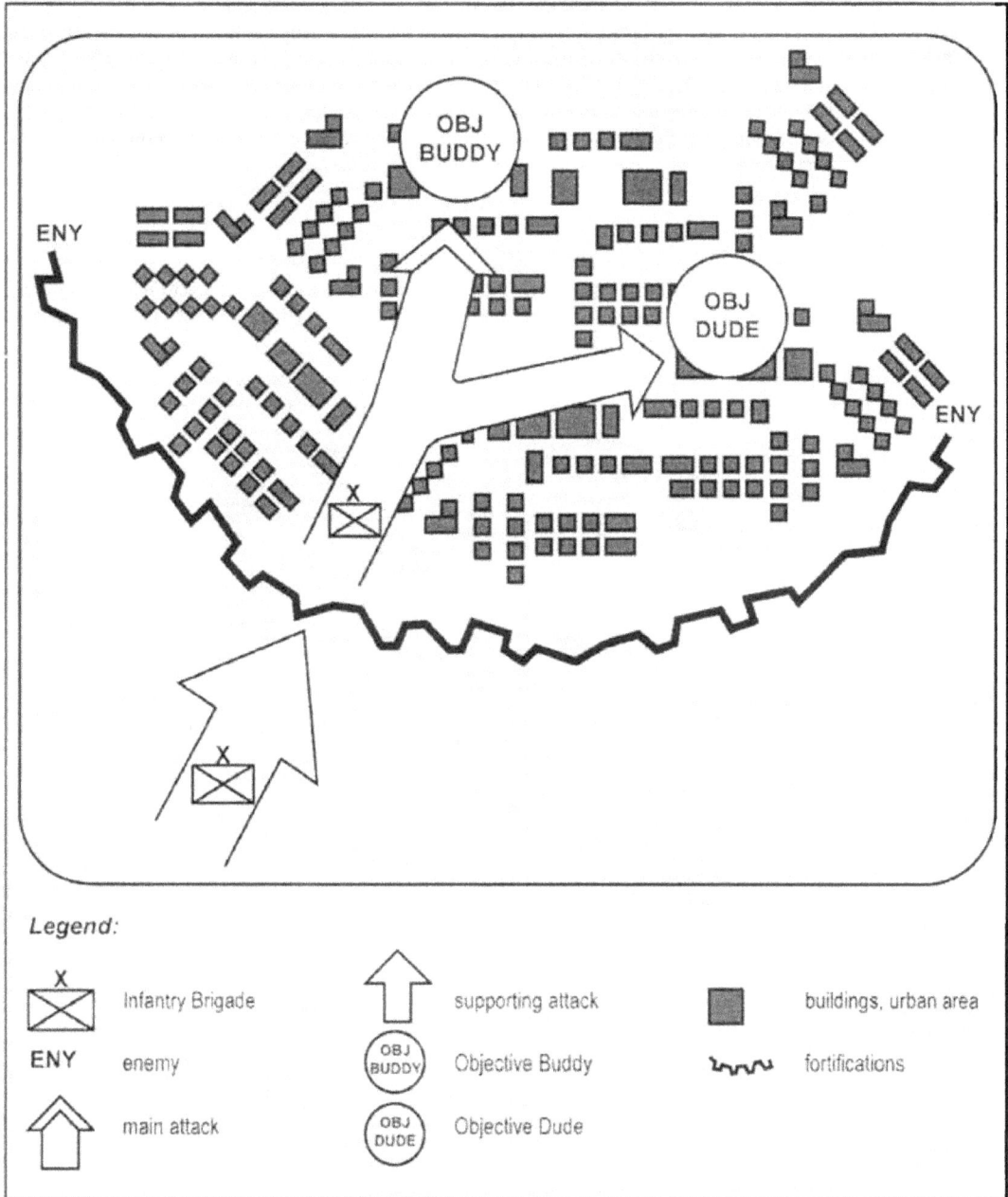

**Figure W-11. Penetration**

W-124. Ideally in urban combat, multiple penetrations in all dimensions are focused at the same decisive point or on several decisive points simultaneously. In urban combat, the flanks of a penetration attack are secure, and resources are positioned to exploit the penetration once achieved. Although always a combined arms team, rapid penetrations are enhanced by the potential speed, firepower, and shock action of armored and mechanized forces.

W-125. Importantly, commanders must consider required actions and resources that must be applied following a successful (or unsuccessful) penetration. A penetration may result in the rapid collapse and defeat of the threat defense and complete capitulation. On the other hand, success may cause threat forces to withdraw but leave significant stay-behind forces (or disperse into the urban population as an insurgent-type force), which may necessitate methodical room-to-room clearance operations by significant dismounted forces. In addition, securing portions (or all) of an urban area requires occupation by Army forces to prevent re-infiltration of threat forces, thereby further increasing manpower requirements. Based on the factors of the situation, commanders may conclude that, over time, methodical clearance operations conducted from the outset, while frontloading time and resource requirements, will be less costly than a penetration followed by systematic clearing operations.

## FRONTAL ATTACK

W-126. For the commander of a major operation, the frontal attack is generally the least favorable form of maneuver against an urban area, unless the threat is at an obvious disadvantage in organization, numbers, training, weapons capabilities, and overall combat power (figure W-12).

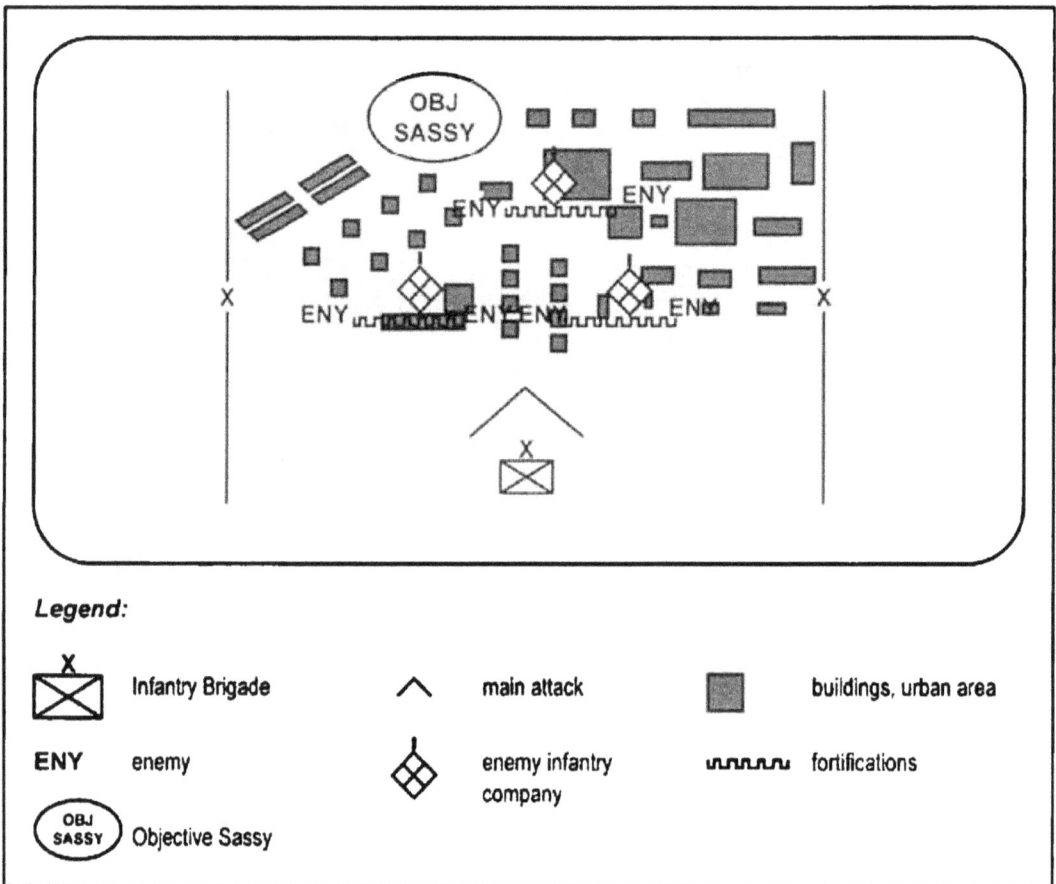

**Figure W-12. Frontal attack**

W-127. Frontal attacks require many resources to execute properly, risk dispersing combat power into nonessential portions of the area, and risk exposing more of the force than necessary to threat fires. In urban offensive combat, forces most effectively use the frontal attack at the lowest tactical level once they set conditions to ensure that they have achieved overwhelming combat power. Then the force of the frontal attack overwhelms the threat with speed and coordinated and synchronized combat power at the point of attack. The assigned frontage for units conducting an attack on an urban area depends upon the size and type of the buildings and the anticipated threat disposition. In general, a company attacks on a one- to two-block front and a battalion on a two- to four-block front (based on city blocks averaging 175 meters in width).

## FLANK ATTACK

W-128. For the commander of a major urban operation, the flank attack is a movement of an armed force around a flank to achieve an advantageous position over an enemy (figure W-13). Flanking is useful because a force's offensive power is most often concentrated in its front. Therefore, to circumvent a force's front and attack a flank is to concentrate offense in the area when the enemy is least able to concentrate on offense. A flank may be created by using fires or by a successful penetration. Usually, a supporting effort engages the enemy's front by fire and maneuver while the main effort maneuvers to attack the enemy's flank. This supporting effort diverts the enemy's attention from the threatened flank.

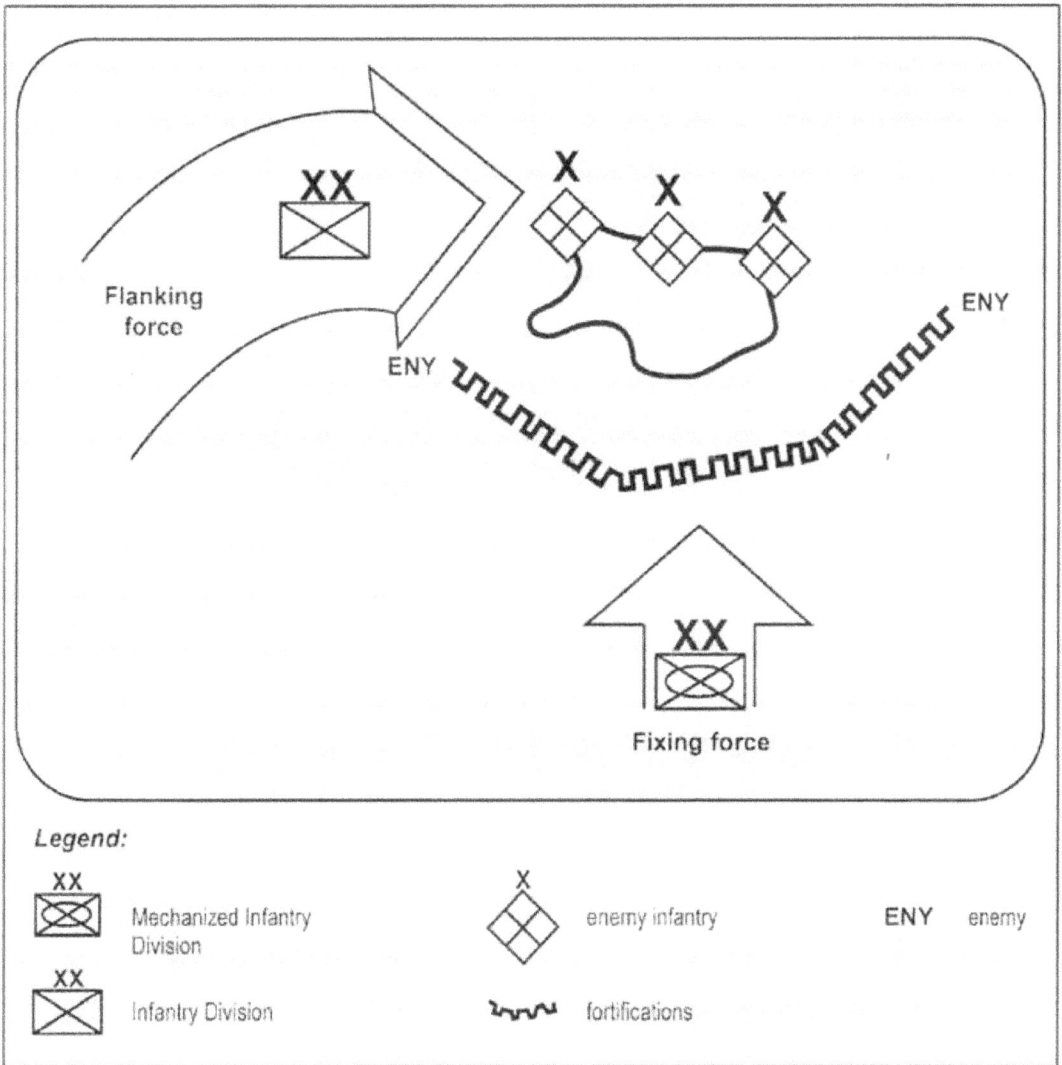

Figure W-13. Flank attack

# SHAPE

W-129. Commanders of major operations have a primary contribution to urban operations—the planning and conduct of effective shaping operations that set the conditions for subordinate tactical success. In urban operations, isolation will be a critical condition. Effective isolation will require persistent, continuous surveillance and reconnaissance, innovative use of fires and maneuver (including effective force allocation decisions), and well-established sensor-to-shooter links. These efforts—combined and synchronized with SOF direct actions, information operations that minimize noncombatant influences, and necessary shaping attacks (particularly the seizure of a foothold)—establish the conditions necessary for the subsequent offensive domination of the area.

## Isolation is Essential

W-130. One key to success in the history of urban operations has been the effective isolation of the threat force (figure W-14, page W-35). This applies today and equally well to major urban offensive operations as it does to smaller-unit attacks. This isolation not only denies access to the urban area from outside but also contains threat forces within. In a modern metropolis or megalopolis, this can appear a daunting task. Operational isolation requires dominating all physical and electronic contact between the threat in the urban area and supporting threat forces outside the urban area. This does not necessarily require physically encircling the urban area, but it does require that Army forces be able to exert control over the area's entire perimeter, as well as decisive points within. For a sprawling urban area, successful isolation may require the commitment of a large amount of resources.

W-131. Successful isolation of the urban area depends as much on the nature of the threat as it does on any other factor. A conventional threat in a large urban area may be much easier to isolate than an insurgent threat in a much smaller urban area. The forces needed in the former situation may be less than those needed in the latter. The more the characteristics of the threat are conventional in nature, generally the easier it will be to isolate him using standard combat methods and equipment. Isolating a more unconventional force requires many of the same techniques as used against conventional forces. However, it also requires a much greater ability to simultaneously conduct offensive information operations, to integrate CA units and CMO , and to work with allies, NGOs, and local authorities. Fundamentally, isolating a less conventional threat puts increased emphasis on separating combatants from noncombatants.

### Offensive Isolation Objectives

W-132. Isolating the threat in the urban area from external support, as well as isolating him from sources of support within the urban area, weakens his overall defense. The defense is weakened through a combination of attrition (the threat cannot replace his losses) and the diversion of his combat power from the defense to operations to counter the isolation effort. Isolation can also prevent the threat from shifting his forces to reinforce decisive points in the urban area or to conduct counterattacks. Isolation seeks to achieve two primary objectives with respect to defeating a threat's urban defense:

- Weaken the overall coherence of his defense.
- Limit or manipulate his maneuver options.

W-133. Commanders may choose not to isolate the urban area completely—or at least make it appear so to threat forces. Instead, they may afford the threat an apparent means of escape, create the conditions for its use through effective fire and maneuver against the defenders, and then destroy the threat through various ambush methods. While friendly forces may be able to move undetected to appropriate ambush sites, it is more likely that this technique will necessitate rapidly mobile air and ground forces moving along carefully chosen routes through the urban area. Commanders must consider maintaining the ability to complete the isolation of the urban area to prevent reinforcement and escape of urban threat forces—particularly if the ambush attack does not achieve desired effects.

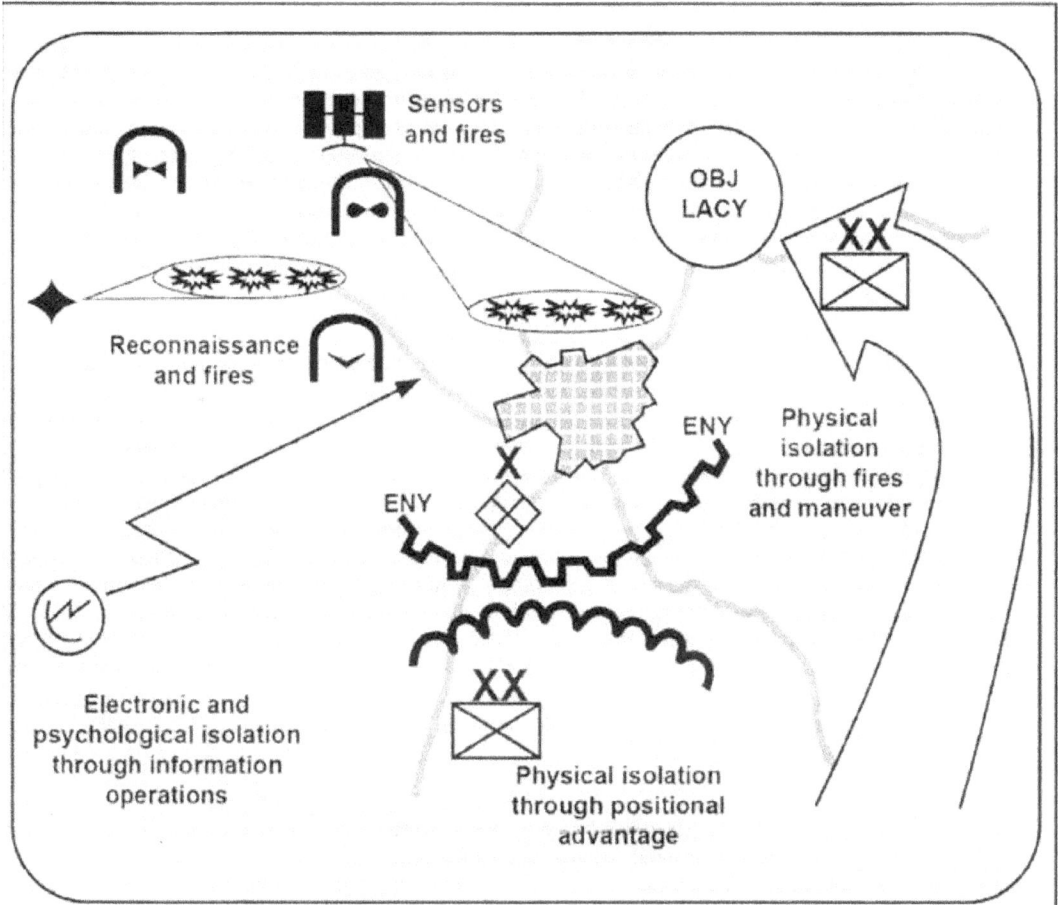

Figure W-14. Shaping through isolation

*Persistent Surveillance*

W-134. Persistent surveillance of the urban area is essential to all types of actions used to isolate an urban area and as complete as resources will allow. Surveillance of the urban area relies on either reconnaissance forces or sensors continuously observing or monitoring urban avenues of approach. This network of intelligence, surveillance, and reconnaissance assets updates the commander's situational understanding and provides the means to quickly identify and, if necessary, attack threat elements as they move. However, particularly with sensors, commanders must consider that each detection is not necessarily an enemy to be attacked. Noncombatant activity clutters the environment, making it easier for threats to disguise themselves. It also increases the burden (and the number of resources required) on Army forces to distinguish friend from foe.

*Fires and Maneuver*

W-135. Fires and maneuver may be used to achieve isolation, either singly or in combination. (As always, effective obstacles, monitored by sensors or observation, are integral to any isolation technique.) First, attacking forces can pre-position themselves along avenues of approach to deny entry and exit through positional advantage. Relying primarily on this method of isolation, particularly around a large urban area with multiple avenues of approach, can be overly resource intensive. Instead, the pairing of fires and maneuver provides attacking commanders more flexibility and allows them to isolate several avenues of approach with fewer resources. Highly mobile attack helicopters, operating outside threat-controlled portions of the urban area, are ideal for this purpose. Inside threat-controlled areas, it is more difficult to identify, eliminate, or effectively suppress the air defense threat. The threats may have numerous MANPADS, direct fire weapons, and enhanced effects of small arms used for air defense. However, mobile ground units—such as an air assault (subject to the same air defense threat considerations as attack aviation), armored, or mechanized forces—can also rapidly move to attack and destroy a threat moving in or out of an urban area. Potential disadvantages of the combined, fires and maneuver, option are that:

- Critical assets, on standby and dedicated to isolation efforts, may be unavailable for other missions.
- The attacking force may not locate the threat in time to complete its mission (an inherent risk to any attack).

W-136. Another alternative relies on indirect or joint fires alone to destroy the threat force. Its disadvantage is that fires alone often cannot completely destroy or stop a determined force from moving into or out of an urban area. Although targets and avenues of approach will require continual surveillance, it is usually a less resource-intensive option than those that include maneuver. It also does not normally require fires assets to remain on standby to accomplish the mission. However, fires must be able to reliably and quickly respond. For Army field artillery units and naval gunfire, the units must also be in range, which requires careful positioning. A skilled threat can avoid interdiction fires by using the geometry of the area to identify gaps as a result of obstructing terrain or the firing unit's range limitations. It can also use concealment and weather to avoid observation. However, effective sensor-to-shooter links throughout the urban battlefield will reduce the threat's ability to hide. A resolute threat may risk significant losses to fires to prevent isolation or may attempt to use noncombatants as a shield. Ultimately, commanders should use innovative combinations of all techniques discussed. Some units will physically block key avenues of approach. Surveillance will monitor less important routes and avenues. Artillery fires, joint fires, and maneuver units will then respond to the results of surveillance depending on the circumstances.

*Threat Reactions*

W-137. The reaction of the threat to the effects of isolation will depend on his mission, morale, force structure, and overall campaign plan. The threat may recognize isolation actions early and withdraw from the urban area before isolation is completed instead of risking destruction. On the other hand, the threat, based on a different or flawed assessment (perhaps a perception shaped by the Army force commander), may choose to:

- Continue to defend (or hide) and conduct local ambushes and counterattacks.
- Attack to break into the urban area or infiltrate forces and supplies in.
- Attack to break out of the urban area or infiltrate forces out.
- Execute any combination of the above (figure W-15, page W-37).

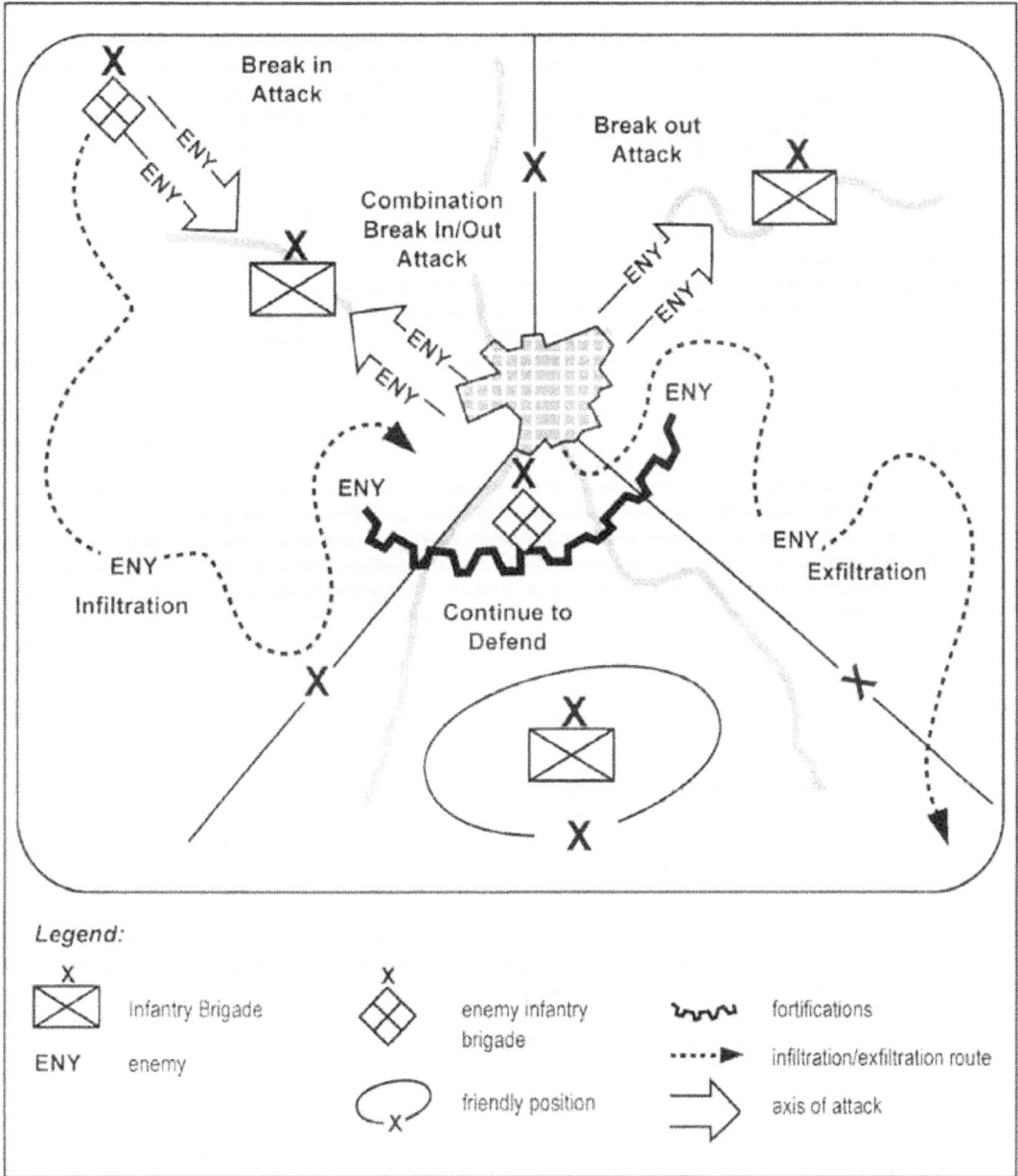

**Figure W-15. Reactions to isolation**

W-138. Attacking commanders must consider how the threat leadership's subsequent actions will affect the continuance of overall offensive operations. They deliberate many considerations, to include the allocation of more—

- Forces to the shaping operations to isolate the urban area.
- Combat power to achieve rapid penetration and seizure of objectives to take advantage of developing threat dispositions in the urban area.

*Civilian Reactions*

W-139. Commanders must also consider the potential effects on and reactions and perceptions of the population living in the urban areas that they choose to isolate and bypass—either as a direct effect or as a response of the threat force being isolated. Isolation to reduce the threat's ability to sustain itself will likely have similar (and worse) effects on the civilian population remaining in the isolated area. (If food and water are in short supply, threat forces may take from noncombatants to satisfy their needs, leaving civilians to starve.) Isolation may also create a collapse of civilian authority within an urban area as it becomes apparent that the military arm of their government is suffering defeat. Because of their isolation, elements of the population may completely usurp the governmental and administrative functions of the former regime and establish their own local control, or the population may lapse into lawlessness. Returning later, Army commanders may find that these self-governing residents are proud of their accomplishments and, in some instances, less willing to allow Army forces to assume control since they may be perceived as having done nothing to earn that privilege. Alternatively, as witnessed in some urban areas during Operation IRAQI FREEDOM in 2003, a power vacuum may lead to intra-urban conflicts among rival factions, coupled with general public disorder, looting, and destruction of the infrastructure.

## Direct Action by Special Operations Forces

W-140. Although SOF in urban offensive operations will likely conduct essential reconnaissance, they also have a direct action capability to shape the offensive operation. SF and Rangers can use direct action capabilities to attack targets to help isolate the urban area or to directly support decisive actions subsequently or simultaneously executed by conventional forces. Successful attacks against urban infrastructure, such as transportation or communications centers, further the area's physical and electronic isolation. Direct action against command centers, logistics bases, and air defense assets can contribute to the success of conventional attacks by destroying or disrupting key threat capabilities. Direct action can also secure key targets, such as airports, power stations, and television stations necessary for subsequent operations. Direct action by SF and Rangers in these operations can help achieve precision and reduce potential damage to the target or noncombatant casualties.

## INTELLIGENCE FOR URBAN OPERATIONS

W-141. The intelligence function facilitates understanding of the threat and the environment. The urban environment affects this critical function in many ways. Impacts of the environment on the intelligence function include degraded reconnaissance capability, more difficult IPB process, an increased importance of credible HUMINT (including the contribution of local civilian liaisons), and an established intelligence reach capability. The Army forces' response to these effects can result in timely, accurate, and actionable intelligence that permits the effective application of other warfighting functions to the mission within the urban environment.

## Degraded Reconnaissance and Surveillance Capability

W-142. The physical environment creates a major challenge to the intelligence function. The man-made construction in the urban areas provides nearly complete cover and concealment for threats. Although improving many sensor capabilities cannot penetrate the subsurface facilities and much of the space within supersurface areas. The mass of buildings can also defuse electronic signatures. Tall buildings shield movement within urban canyons from aerial observation except from directly overhead. Urban threats may be less technology dependent and may thwart some signals intelligence efforts simply by turning off their radios and using messengers. Threat forces will likely use elements of the civilian telecommunications infrastructure for command and control. These systems may include traditional landline phones, cellular telephones, and computer-to-computer or Internet data communications. Most urban telecommunications systems use buried fiber or cables or employ modern digital signaling technology. Such systems are difficult to intercept and exploit at the tactical level.

W-143. From the above, it is evident that these characteristics make it more difficult for the intelligence function to use electronic means to determine threat dispositions and, in offensive and defensive urban operations, identify decisive points leading to centers of gravity. While the environment limits some typical collection methods, all enemy electronic and human activity creates some form of observable signature and exposes the enemy to potential collection. Seeking ways to take advantage of these vulnerabilities will provide the commander an information advantage over his opponent.

## Challenging Intelligence Preparation of the Battlefield Process

W-144. The sheer complexity of the environment also challenges the intelligence function. The intelligence function applies the IPB process to the urban environment in accordance with Army doctrine. With more data points for the IPB process to identify, evaluate, and monitor, this application becomes more demanding. The human and societal aspects of the environment and the physical complexity primarily cause this difference. Relationships between aspects of the environment, built on an immense infrastructure of formal and informal systems connecting the population to the urban area, are usually less familiar to analysts. Thus, the urban environment often requires more specifically focused intelligence resources to plan, prepare for, execute, and assess operations than in other environments.

W-145. Compounding the challenges is the relative incongruity of all urban environments. No two urban areas are alike physically, in population, or in infrastructure. Thus, experience in one urban area with a particular population and pattern of infrastructure does not readily transfer to another urban area. Any experience in urban operations is valuable and normally serves as a starting point for analysis, but the intelligence function cannot assume (and treat as fact) that patterns of behavior and the relationships in one urban area mirror another urban area. The opposite is as likely to hold true. The intelligence function will have to study each urban area individually to determine how it works and understand its complex relationships.

W-146. Each characteristic of the urban environment—terrain, society, and infrastructure—is dynamic and can change radically in response to urban operations or external influences. Civilian populations pose a special challenge to commanders conducting urban operations. Civilians react to, interact with, and influence—to varying degrees—Army forces. Commanders must know and account for the potential influence these populations may have on their operations. Intelligence analysts must revisit or continuously monitor the critical points, looking for changes, relationships, and patterns.

W-147. The actions of Army forces will affect, positively or negatively, their relationship with the urban population and, hence, mission success. NGOs may deliberately or inadvertently influence civilians. The intelligence function can monitor and predict the reactions of the civil population. However, accurate predictive analysis of a large population requires specific training and extensive cultural and regional expertise.

## Increased Importance of Human Intelligence

W-148. HUMINT is the collection of foreign information—by a trained HUMINT collector—from people and multimedia to identify elements, intentions, composition, strength, dispositions, tactics, equipment, personnel, and capabilities. It uses human sources as a tool and a variety of collection methods, both passively and actively, to collect information.

W-149. The intelligence function adjusts to the degradation of its technical intelligence-gathering systems by increasing emphasis on HUMINT in urban operations. HUMINT operations may be the primary and most productive intelligence source in urban operations. In urban offensive and defensive operations, HUMINT gathers information from refugees, immigrants, and former citizens (especially previous civil administrators), civilian contractors, and military personnel who have operated in the area. Credible intelligence of this type can help meet requirements, provide more detail, and alleviate some of the need to physically penetrate the urban area with reconnaissance forces. In many urban operations where HUMINT is the primary source of intelligence, acting on single-source reporting is a constant pitfall. Yet, situations may arise where commanders must weigh the consequences of inaction against any potential negative consequences resulting from acting on uncorroborated, single-source information.

W-150. In urban stability operations, HUMINT identifies threats and monitors the intentions and attitudes of the population. A chief source of information contributing to the development of accurate HUMINT, particularly at the tactical level, is reconnaissance forces—especially small-unit dismounted patrols. Urban reconnaissance forces and patrols should be thoroughly and routinely debriefed by unit intelligence personnel to obtain information that aids in developing a clearer picture of the threat and the urban environment.

Reliable and trustworthy HUMINT is particularly important in foreign internal defense, counterterrorism, and support to counterdrug operations. Leaders must organize intelligence resources appropriately and learn and apply valuable techniques, such as pattern and link analysis (FM 34-3). In addition, Soldiers, as part of reconnaissance and patrolling training, should be taught to handle captured documents, weapons, material, and equipment as legal evidence much like military and civilian police. Proper evidence handling is often a critical intelligence concern in counterterrorism and counterinsurgency operations.

## Developing Local Liaisons

W-151. Whenever Soldiers encounter the urban populace, the resulting interaction may become an important source of information the commander can use to answer questions about the threat and the urban environment. While military intelligence units are the primary collectors and processors of HUMINT, commanders are unlikely to have enough trained HUMINT Soldiers to satisfy their requirements—particularly in a larger urban environment and during longer-term stability operations. Commanders, therefore, may need to cultivate and establish local civilian associations to provide relevant information for decision making and to support the overall HUMINT effort.

W-152. Urban liaisons can be developed through positive civil-military interaction with the urban populace. Critical information may be acquired through interface with the urban leadership (both formal and informal), administration officials, business owners, host-nation support workers, inhabitants along a unit's patrol route, pedestrians at a checkpoint, civilian detainees, or any other human source willing to volunteer information to Army forces or who respond positively to tactical questioning. (Noncombatants are never coerced to provide information.) Commanders may also direct unit leaders to conduct liaisons with specific local leaders and key members of the community to obtain command-directed information. Critical information may also come from other U.S. and coalition forces and intelligence organizations operating near or within the commander's AO. To this end, commanders should ensure that collectors operating in an urban area coordinate and deconflict activities and, if possible, outbrief subordinate, geographically responsible commanders with any relevant information that may affect their current operations. Any relevant information obtained as a result of civilian liaison activities should be routinely provided to intelligence staffs, not only to gain assistance in verifying the credibility of the information but also to share the information with all affected echelons and units.

W-153. NGOs operating in urban areas can also be especially beneficial resources for credible and relevant information about the urban environment. (However, they are generally not a good source for information about the threat, since providing such information can violate their neutrality—thereby making it difficult for them to achieve their humanitarian aid objectives.) During the 1999 fighting in Kosovo, for example, the Red Cross provided the most accurate figures regarding the number of Kosovar refugees, helping U.S. and other coalition forces to estimate the appropriate level of support required to handle their needs. In addition to a developed understanding of the current needs of the local urban populace, NGOs may also have:

- A network of influential associations.
- Historical archives.
- Extensive understanding of the urban infrastructure.
- Key knowledge of political and economic influences.
- A keen awareness of significant changes in the urban environment.
- Insight into the current security situation.
- Up-to-date websites and maps.

W-154. While productive, civilian associations may become long term, they should not be confused with HUMINT source operations. Only trained HUMINT personnel can recruit and task sources to seek out threat information. Information obtained from these societal connections is normally incidental to other civil-military relationships. For example, as part of infrastructure repair in an urban stability operation, a commander may be instrumental in obtaining a generator for a local hospital. Within the context of this relationship, the commander may develop a rapport with one or more of the hospital's administrators or health practitioners. These civilians may be inclined to provide valuable information about the threat and the urban environment—often on a continuing basis. In any civil-military relationship, however, commanders ensure that the information provided is not tied to promises of assistance or that such assistance is in any way perceived as a means to purchase civilian loyalty.

W-155. Commanders also understand that repeated interaction with any one individual may put that individual and his family in danger from threat forces. Before this potential danger becomes a reality, they should refer their civilian connections to trained HUMINT personnel for protection and continued exploitation. In addition to civilian protection considerations, commanders may also deem it necessary to turn their civilian associations over to trained HUMINT collectors anytime during the relationship if they consider the information that the contact is providing (or may provide) is credible, relevant, and—

- Provides essential threat information on a repetitive basis.
- Helps answer higher-level commander's critical information requirements.
- Affects operations in another AO.
- Requires interrogation or monetary compensation to obtain.

W-156. In developing these civilian liaisons essential to understanding the urban environment, commanders must avoid the distinct possibility of conducting unofficial source operations by non-HUMINT Soldiers. While prohibited by regulatory guidance, such actions also run the additional risks of—

- Obtaining unevaluated information that cannot be crosschecked with and verified by other sources of information.
- Creating inequities that result from illegally rewarding contracts, which can undermine HUMINT Soldiers who are constrained by intelligence contingency fund regulations.
- Disrupting ongoing HUMINT operations when different sources are seen to be treated differently by non-HUMINT Soldiers vice HUMINT Soldiers.
- Providing non-HUMINT and HUMINT Soldiers with the same information, potentially leading to a false confirmation of information.
- Increasing the likelihood that untrained Soldiers may fall victim to a threat deception and misinformation.

## Established Intelligence Reach

W-157. Understanding the complex urban environment, particularly the infrastructure and the society, will require more sources of information beyond a unit's organic intelligence capabilities. Therefore, commanders will have to make extensive use of intelligence reach to access information and conduct collaboration and information sharing with other units, organizations, and individual subject-matter experts. Before deployment (and throughout the operation), units should establish a comprehensive directory of intelligence reach resources. These resources may include national, joint, Army, foreign, commercial, and university research programs. (Prior to deployment for Operation IRAQI FREEDOM, some units established contacts within the local community outside their bases, such as police, fire department, and government officials that expanded their reach once in theater—particularly for information regarding civilian infrastructure and urban administration.) Once deployed, intelligence reach includes effective information sharing and collaboration among adjacent units, sister Services, coalition partners, and other governmental and nongovernmental agencies operating in the area. Effective information sharing and collaboration requires common network analysis software and databases to be used among all Army forces and, if possible, other governmental agencies.

## URBAN INTELLIGENCE TOOLS AND PRODUCTS

W-158. Developing tools and products to assist with situational understanding is not the task of intelligence sections alone. Commanders and all elements of their staffs must develop and adapt products and tools suited to their particular requirements. Listed below are some of the analytic tools and products that may help meet those requirements in a complex environment (figure W-16, page W-42). Standard tools and products include: modified combined obstacle overlays and doctrinal, situation, event, and decision support templates or matrices. In addition to these standard aids, staffs and analysts may develop or produce other innovative tools to assist commanders in their situational understanding of the complex urban environment. These tools may contain overlapping information as different types of information are compared to determine patterns among them. Staffs and analysts may also initiate requests for products (or information) from their higher HQ or other agencies, with the technical means or control over assets, when the capability lies outside the Army force's means. The tools that may be developed or requested include:

- Imagery.
- Three-dimensional representations.
- Infrastructure blueprints.
- Hydrographic surveys.
- Psychological profiles.
- Matrices, diagrams, or charts.
- Various urban overlays.

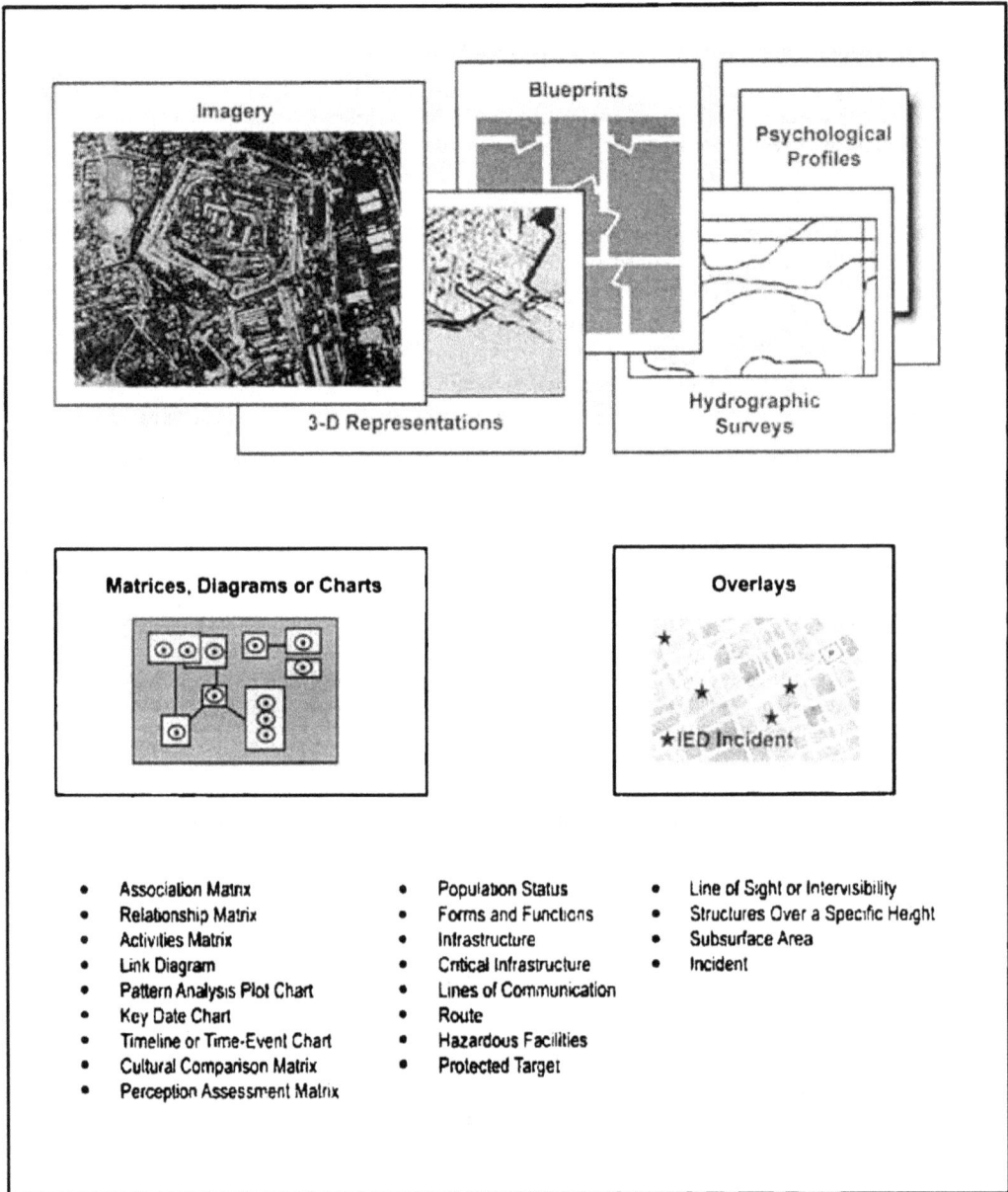

Figure W-16. Urban intelligence tools and products

W-159. Many software applications are available to the Army that can be used to conduct intelligence analysis and create many of the relevant products described above. These applications currently range from such programs as the Analyst Notebook and Crimelink—which have link analysis, association matrix, and pattern analysis software tools—to the Urban Tactical Planner developed by the Topographic Engineering Center as an operational planning tool. The focus of the following information, however, is the various types of tools that can be developed and used to understand the urban environment, rather than the constantly evolving and changing software or hardware.

# IMAGERY

W-160. Recent satellite imagery or aerial photography will be required for most types of urban operations. Such images clarify vague and inaccurate maps and other graphic representations. Satellite assets provide responsive data input into the geographic information systems. (The National Geospatial-Intelligence Agency is one important source of imagery and map data.) The geographic information system will often form the basis for creating the three-dimensional representations and the various overlays described below. Frequently updated (or continuous real-time) satellite or aerial imagery may be required for preparing a detailed pattern analysis and maintaining an accurate situational understanding. For example, imagery taken during an area's rainy season may appear significantly altered during the summer months. Digital, hand-held still and video cameras, particularly at the tactical level, will also be critical in developing and attaining situational understanding in a complex urban environment.

## THREE-DIMENSIONAL REPRESENTATIONS

W-161. Often, physical or computer-generated (virtual) three-dimensional representations may be required to achieve situational understanding. These models or computer representations include specific sections of the urban area or specific buildings or structures. Such detail is particularly important for SOF and tactical-level units. These units require detail to achieve precision, increase the speed of the operation, and lessen friendly casualties and collateral damage.

## INFRASTRUCTURE BLUEPRINTS

W-162. Urban police, fire, health, public utilities, city engineer, realty, and tourist agencies, and other urban organizations often maintain current blueprints and detailed maps. Such documents may prove useful to update or supplement military maps or to clarify the intricacies of a specific infrastructure. They may prove critical in operations that require detailed information to achieve the speed and precision required for success. Without such detail, analysts determine interior configurations based on a building's outward appearance. That task generally becomes more difficult as the building size increases.

## HYDROGRAPHIC SURVEYS

W-163. Many urban areas are located along the world's littoral regions and major rivers. Commanders, therefore, may need current hydrographic surveys to support amphibious, river crossing, riverine, and sustainment operations.

## PSYCHOLOGICAL PROFILES

W-164. Psychological profiles analyze how key groups, leaders, or decision makers think or act—their attitudes, opinions, and views. They include an analysis of doctrine and strategy, culture, and historical patterns of behavior. The degree to which the attitudes, beliefs, and backgrounds of the military either reflect or conflict with the urban populace's (or civilian leadership's) core values is extremely important in this analysis. Psychological profiles help to assess the relative probability of a threat (or noncombatant group) adopting various COAs, as well as evaluating a threat's vulnerability to deception. These profiles are derived from open-source intelligence and signals and HUMINT.

## MATRICES, DIAGRAMS, AND CHARTS

W-165. Matrices, diagrams, and charts help to identify and understand key relationships among friendly and threat forces and other significant elements of the urban environment. While similar, each looks at information in a different way to uncover hidden patterns and connections useful in understanding the complex urban environment, particularly the threat and the urban society. These tools and products might include:

- **Association Matrix.** The association matrix helps identify the nature and relationship between individuals. Association matrices also help to identify those personalities and associations needing a more in-depth analysis necessary to establish the degree of relationship, contacts, or knowledge between individuals. Threat organizational structure and functions are uncovered as connections between personalities are made.

- **Relationship Matrix.** Relationship matrices are an extension of association matrices described above. They are used to examine the relationship between groups and organizations—threat forces, friendly forces, NGOs, the media, and the various elements of the urban population. Urban terrain and infrastructure may also be included as analytical factors to help expose relationships. These matrices graphically depict how human elements of the urban area interact with the physical elements to promote or degrade mission success. A significant relationship matrix may be a comparison of cultural perspectives—ideology, politics, religion, acceptable standards of living, and mores—between urban population groups and Army (and multinational) forces to help understand and accurately predict a civilian element's actions. The relationships discovered may suggest ways that commanders can shape the environment. For example, conflicting ideals or issues between two disparate threat forces may be emphasized to cause threats to focus their resources against each other rather than Army forces. Conversely, stressing common ideals identified between opposing noncombatant organizations may cause these groups to unite and work harmoniously toward a goal reflecting their commonalities.

- **Activities Matrix.** An activities matrix connects individuals to organizations, professions, events, activities, or addresses. Information from this matrix (combined with information from association and relationship matrices) also helps to link personalities, uncover the structure of an organization (threat or otherwise), and recognize differences, similarities, and dependencies for possible exploitation.

- **Link Diagram.** This tool graphically depicts many of the relationships and associations described above—people, groups, events, locations, or other factors deemed significant in any given situation. It helps commanders and analysts visualize how people and other factors are interrelated to determine key links.

- **Pattern Analysis Plot Chart.** This chart depicts the times and dates of a selected activity (such as ambushes, sniper and mortar attacks, bombings, and demonstrations) to search for patterns of activity for predictive purposes, as well as to discern intent. This analysis can be conducted using a time-event wheel, as well as by plotting events on maps using multiple historical overlays (analog or digital). (Compare with Incidents Overlay covered in the next paragraph.) During pattern analysis, commanders and analysts consider not only what is occurring but also what is not occurring (or ceases to occur).

- **Key Event Chart.** In many urban operations, particularly stability operations, key holidays, historic events, and other significant cultural or political events can be extremely important for commanders to understand. More than just dates, these charts depict what can be expected to happen on each particular event. In Bosnia, for example, weddings are often held on Fridays and celebratory fire is a common occurrence. Understanding this cultural phenomenon could reduce collateral damage and accompanying civilian deaths.

- **Timeline or Time-Event Chart.** Timelines are a list of significant historical dates with relevant information and analysis that provide the commander with a record of past activities necessary to understand current operational conditions. A timeline may highlight a specific feature of the present situation, such as population movements or political shifts, or outline the general chronological record of the urban area, perhaps highlighting the activities of a certain population sector. Understanding past events also helps commanders predict future reactions to proposed COAs. Similarly to link diagrams, time-event charts display large amounts of information in a small space. These charts may help commanders discover larger-scale patterns of activity and relationships.

- **Cultural Comparison Matrix.** Commanders must avoid ethnocentricity—assuming that only one cultural perspective exists (often their own) and using that as the single lens with which to view the situation. To avoid this common obstacle, commanders use this matrix to compare local ideology, attitudes, beliefs, and acceptable standards of living with associated U.S. and Army norms and values. This matrix not only highlights differences but also similarities.

- **Perception Assessment Matrix.** All of the above tools contribute to developing the perception assessment matrix, allowing commanders to see the urban environment from the perspective of its inhabitants. The perception assessment matrix is another tool that uses this refined perspective and the framework outlined in the matrix to predict the urban population's perception of a proposed COA. Although perceptions are not actions, they—more than reality— drive decision making and resultant civilian activity. (As an example, legitimacy is a critical factor in developing many COAs that affect the civilian population, particularly during stability operations. This legitimacy must be viewed from the perspective of the inhabitants and not based on the Army commander's own culturally shaped perceptions. Moreover, the legitimacy of an operation may need to be considered from other perspectives as well. These might include those of the American public, coalition partners and their publics, and those of neighboring states in the region.) This same matrix can later be used as a foundation to track and gauge the effectiveness of the chosen COA with favorable perception as one criterion of success. Commanders must be wary that perceptions are not fixed. They may change based on factors that the commander may not be able to control or influence; therefore, commanders must continually monitor perceptions for deviation.

## VARIOUS URBAN OVERLAYS

W-166. Staffs can produce various map overlays. These overlays depict physical locations of some aspect critical to the planning and conduct of the urban operation. Given adequate lead time, the National Geospatial-Intelligence Agency can produce many overlays as an integrated map product (including satellite imagery). These overlays can include the:

- **Population Status Overlay.** This tool depicts the physical location of various groups identified by any significant social category, such as religion, tribe, or language. During offensive and defensive operations, it may simply be where significant numbers of people are "huddled" or located throughout the battlefield. Population dispersal can vary significantly through the day, particularly at night, and must be considered as part of the overall analysis leading to the development of this tool.

- **Forms and Functions Overlay.** Based on the urban model, this overlay depicts the urban core or central business district, industrial areas, outlying high-rise areas, commercial ribbon areas, and residential areas, to include shantytowns.

- **Infrastructure Overlay.** This overlay is actually a series of overlays. It depicts identifiable subsystems in each form of urban infrastructure: communications and information, transportation and distribution, energy, economics and commerce, administration and human services, and cultural. Each subsystem can be broken down into more detail. Infrastructure data may be used to develop three other overlays:

  - Critical Infrastructure Overlay. This tool displays specific elements of the urban infrastructure that, if harmed, will adversely affect the living conditions of the urban society to the detriment of the mission. These elements may include power generation plants, water purification plants and pumping stations, and sewage treatment plants. This information could be coded as part of the overall infrastructure overlay.

  - Lines of Communications Overlay. The LOCs overlay highlights transportation systems and nodes, such as railways, road, trails, navigable waterways, airfields, and open areas for DZs and LZs. It also includes subsurface areas and routes, such as sewage, drainage, and tunnels and considers movement between supersurface areas. The LOCs overlay and the route overlay (below) consider traffic conditions, times, and locations, to include potential points where significant portions of the urban population may congregate.

- Route Overlay. This overlay emphasizes mobility information to assist commanders and planners in determining what forces and equipment can move along the urban area's mobility corridors. Pertinent data includes street names, patterns, and widths; bridge, underpass, and overpass locations; load capacities; potential sniper and ambush locations (which may be its own overlay); and key navigational landmarks, such as major roads and highways, rivers and canals, cemeteries, bridges, stadiums, and churches. The structures over a specific height overlay and subsurface overlay may assist in its development. As with the LOCs overlay, commanders, planners, and analysts think in all dimensions.

- **Line-of-Sight or Inter-Visibility Overlay.** This product creates a profile view (optical or electronic) of the terrain from the observer's location to other locations or targets. It can show trajectory or flight-line masking, as well as obstructed or unobstructed signal pathways.

- **Structures Over a Specific Height Overlay.** This level of detail may also be critical to communications, fires, and Army airspace command and control (air mobility corridors especially low-level flight profiles). Incorporated as part of this overlay, it may include floors or elevations above limitations for particular weapon systems at various distances from the structure.

- **Subsurface Area Overlay.** As an alternate to the building or structure height overlay, this product provides the locations of basements, underground parking garages, sewers, tunnels, subways, naturally occurring subterranean formations, catacombs, and other subsurface areas. Similar to elevation "dead spaces," this overlay may show areas that exceed depression capabilities of weapon systems and potential threat ambush locations—again, affecting maneuver options.

- **Urban Logistics Resources Overlay.** This product identifies the locations of urban logistics resources that may contribute to mission accomplishment. It may contain specific warehouse sites, hospitals and medical supply locations, viable food stores, building material locations, fuel storage areas, car or truck lots, maintenance garages, and appliance warehouses. (NGO locations, taken from an NGO relationship matrix, may be an essential, overlapping element of this overlay.)

- **Hazardous Facilities Overlay.** This overlay identifies urban structures with known or suspected chemical, biological, or radiological features, such as nuclear power plants, fertilizer plants, oil refineries, pharmaceutical plants, and covert locations for producing weapons of mass destruction. These locations are critical to maneuver and fires planning.

- **Protected Target Overlay.** This overlay depicts terrain that should not be destroyed or attacked based on restrictions due to international, host-nation, or U.S. law and subsequent rules of engagement. These may include schools, hospitals, historical or other culturally significant monuments, and religious sites. This overlay may incorporate no-fire areas, such as SOF locations, critical infrastructure, logistics sources, and hazardous sites that must be protected as part of the commander's concept of the operation.

- **Incident Overlay.** Similar to the pattern analysis plot chart, this product depicts the location of different threat actions and types of tactics employed to uncover recurring routines, schemes, methods, tactics, or techniques and overall threat interests, objectives, or the desired end state.

W-167. The above IPB tools and products constitute a small sampling of what staffs and analysts can produce. They are limited only by their imaginations and mission needs; not all tools presented above may be relevant or necessary to every operation. Many products can be combined into a single product or each can generate further products of increasing level of detail. This is similar to transparent overlays positioned one atop another on a map. As discussed earlier, technology—software and hardware—will allow for more urban data to be combined, compared, analyzed, displayed, and shared—ultimately allowing scalable products that can expand and contract to meet the IRs of any echelon of command down to the individual Soldier. The challenge will remain, however, to provide timely, accurate, complete, and relevant information in an understandable, usable, and share-capable form—share-capable among different levels of command, as well as among multinational partners and various governmental and, as required, NGOs—without overloading the commander.

# Appendix X

# Medical Aspects

Before a UW deployment, the SF operational detachment medic should acquire as much medical information as possible at the home station to ensure he has current medical intelligence about the JSOA. An SF operational detachment medic prepares a medical area study of the JSOA as part of the overall area study. Some of the sources he will use to prepare this study include CA area studies (available through the CA database), reports by United Nations agencies, and information from NGOs operating in the area. Because the study is continuous, he updates it whenever he receives new medical intelligence.

## MEDICAL AREA STUDY

X-1. The medical area study format is an adjunct to the basic detachment area study format (Appendix H, Sample Area Study Format) to organize systematic inquiries into medically specific information. The recommended base format and the areas of focus that follow are suggestions; they are flexible to allow the medic to tailor them to the needs of the mission (figure X-1).

| Medical Area Study | |
|---|---|
| General | Provides a brief summary of the nation's health status. |
| Environmental Health Factors | Discusses the country's topography and climate, to include effects on health, medical evacuation, and logistics. |
| Demographics | Includes population, ethnic groups, life expectancy, and infant mortality. |
| Nutrition | Discusses nutrition and facilities for refrigeration and food inspection programs. |
| Water Supply | Discusses the method of supply, location, treatment, and health hazards as they apply to drinking, bathing, and swimming. |
| Fauna of Medical Importance | Focuses on disease vectors, hosts, reservoirs, poisonous mammals, reptiles, and spiders. |
| Flora of Medical Importance | Covers poisonous plants, plants with medical value, and edible plants used for survival. |
| Epidemiology | Discusses prevalent diseases and their contributing factors. It focuses on diseases of military importance, including communicable diseases and susceptibility to cold and heat injury. It discusses the concerns of indigenous personnel, such as physical characteristics, their unique attitudes, dress, religious taboos, and psychological attributes. It also discusses preventive veterinary medicine programs that deal with prevalent animal diseases and their diseases that can be transmitted to man. |
| Public Health and Military Medical Services | Focuses on public health and military medical services. |
| Village Organization | Covers such important village concerns as social, physical, and family organization; housing, diet, water and waste disposal, and local medical practices; and rapport with neighboring tribes. |
| Domestic Animals | Discusses the types and uses of domesticated animals and any possible religious symbolism or taboos associated with these animals. |

Figure X-1. Medical area study format

# MEDICAL AREA ASSESSMENT

X-2. The initial medical area assessment begins immediately upon infiltration. The medic establishes rapport (using his medical skills) with the local population and guerrillas. He gains an opportunity to receive information not available elsewhere. This information may include captured medical order of battle intelligence from medical supplies and documents through battlefield recovery and from indigenous sources. It assesses the actual extent of medical training for the resistance force, availability of medical supplies and facilities, and the state of the sanitation and health within the JSOA. Medics should consider the following when conducting the initial medical area assessment:

- Physical condition and morale of the SF operational detachment.
- Medical status of the guerrilla forces.
- Identification of any immediate threats to the health of the command, to include epidemics and environmental conditions, weather, terrain, lack of sanitation, food, and water problems.

X-3. The medical area assessment is a continuous process based on observations and firsthand factual reports by the deployed SF operational detachment medic. The medical area assessment is a continuation of the initial medical area assessment. It confirms, refutes, or clarifies previously researched information in the medical area study. The medical area assessment provides information to other units, supplements and supports area studies, and forms the basis of the after action review. Results of, and information on, the medical area assessment should not be transmitted out of the JSOA unless significant differences exist between previous intelligence reports and impact current or planned operations. The following are suggestions and questions to guide SF operational detachment medics in making a medical area assessment.

## INDIGENOUS PERSONNEL

X-4. SF operational detachment medics indicate the name of the tribe or native group and describe them in terms of height, build, and color and texture of skin and hair. Medics describe the native group's endurance and ability to carry heavy loads, while performing physical labor, and determine if they need more physical training.

X-5. SF operational detachment medics describe clothing and adornments. Are shoes or other footgear worn? What symbolism is attached to various articles of clothing and jewelry, if any? Are amulets worn? What do they symbolize? Medics should furnish photographs if possible.

X-6. SF operational detachment medics describe attitudes toward birth, puberty, marriage (monogamy or polygamy), old age, sickness, death, and so on. They consider any rituals associated with these events. They describe principal taboos, especially those about food, animals, and water. They determine what the attitudes are toward doctors and western medicine. Medics describe rites or practices by witch doctors during illnesses. What do the practices symbolize? Do the natives respond to events in the same manner as the medics would? Do they show appropriate feelings of sadness, happiness, anger, fear, and love from the medic's point of view?

## VILLAGE ORGANIZATION

X-7. SF operational detachment medics determine the status of priests, witch doctors, and chiefs. What is their relationship with the people? How do other community specialists, such as carpenters, weavers, and hunters, fit in socially? Medics describe the physical layout of the community.

## FAMILY ORGANIZATION

X-8. SF operational detachment medics determine whether intermarriage occurs outside the clan or between neighboring villages or tribes. Do males or females leave home when they marry? What are the attitudes toward old men and women, children, and deceased ancestors?

## RELIGION

X-9. SF operational detachment medics describe the religious beliefs of the tribe or native group. Are they Buddhist, Hindu, Muslim, Christian, or animist? Do they worship the sun, moon, or inanimate objects? What role do good and evil spirits play? Are sicknesses, deaths, or births ascribed to evil spirits? Does a priesthood exist, and what influence does it have? Are witch doctors and priests synonymous or different?

## HOUSING

X-10. SF operational detachment medics describe construction and materials used. They describe infestations, ectoparasites, and vermin. How many persons inhabit a dwelling? Are there community houses? Are animals housed in the same dwelling? Medics provide photos and diagrams if possible.

## FOOD

X-11. SF operational detachment medics describe the native diet of the indigenous population. They describe agricultural practices, such as slash-and-burn or permanent farms. Is human excrement used for fertilizer? What domestic crops contribute significantly to the diet? What wild vegetables are consumed? How is food prepared? What foods are cooked, pickled, smoked, or eaten raw? How are foods preserved? At what age are children weaned? What is fed to weaned infants? Which family member is given preference at the table? Are there food taboos and why? Does migration in search of food occur? Which foods, provided by U.S. personnel, do indigenous personnel prefer or reject?

## WATER

X-12. SF operational detachment medics determine the primary sources of potable water (river, spring, and well). Is the water used for bathing, washing, drinking, and cooking? How far is the water source from the village? Is water plentiful or scarce? What is the seasonal relationship to water availability? Is water boiled, filtered, or subjected to other purification processes before consumption? What are native attitudes toward standard U.S. purification methods (iodine and bleach)?

## WASTE DISPOSAL

X-13. SF operational detachment medics describe the system used for disposal of human excrement, offal, dead animals, and human bodies. Is any excrement collected for fertilizer? What scavengers assist in the process of disposal? What is the relationship of disposal sites to primary watering sites? What are the attitudes of indigenous personnel to standard U.S. methods, such as sit-down latrines?

## SOURCE OF INCOME

X-14. SF operational detachment medics determine if there is a monetary system or exchange in the form of barter or if property and service are communal. What items have monetary value or are equivalent to money (food, jewelry, pelts, and drugs)?

## LOCAL MEDICAL PRACTICES

X-15. SF operational detachment medics describe the types of local medical practitioners. Are there conventional doctors available, to include missionaries and other nonindigenous personnel? What is their level of formal training and specialization? Do their techniques conform to American practices? Do they have government support or affiliation? Who is the local medical leader? Is he a licensed doctor, witch doctor, or herbalist? What is his status in the community? How do his practices relate to local religious practices or taboos? What drugs or herbs does he use? Does he use surgical techniques? What is the attitude of the local medical leader toward SF operational detachment medics or aidmen? What status does he accord the SF operational detachment medics?

X-16. SF operational detachment medics determine the location of the nearest dispensary, hospital, or laboratory. They describe capabilities, if known, for surgery, radiology, and pharmacy. Do local hospitals accept indigenous personnel from the AO? What pharmaceuticals or biologicals are produced and used locally? What is the overall level of medical supplies? Where are the medical supplies manufactured? What is the quality of the medical supplies? Who provides them? How are they delivered?

X-17. SF operational detachment medics describe practices associated with childbirth. Do midwives attend the patient? Is there a period of confinement or banishment before or after delivery? What is the attitude of the father during pregnancy, at delivery, and after birth? How are the placenta and cord treated? What rituals accompany the birth process? Will natives request the assistance of SF operational detachment medics?

## LANGUAGE

X-18. SF operational detachment medics describe language, dialect, and speech variants of the basic tongue from village to village. What other tongue does the tribe understand?

## RELATIONSHIPS

X-19. SF operational detachment medics describe the relationships with neighboring tribes. Does intermarriage occur? Are the general characteristics of language, dress, and physique the same? What other forms of cooperation occur and for what purpose?

## FAUNA

X-20. SF operational detachment medics record species of wild animals, birds, reptiles, and arthropods found on the march, around campsites, and in villages. If their names are unknown, medics describe them. Medics note relationships among these species, including burrows and nesting sites to human habitation, food supplies, and watering sites.

X-21. SF operational detachment medics note the occurrence of dead or dying animals, especially if they involve large numbers of a given species. They note the relationship of any die-off to the occurrence of human diseases.

X-22. SF operational detachment medics report any methods used by indigenous personnel to defend against local ectoparasites (leeches). How effective are these methods compared to standard U.S. protective measures (insect repellent)?

# ORGANIZATION OF MEDICAL ELEMENTS WITHIN THE JOINT SPECIAL OPERATIONS AREA

X-23. The goals of medical operations in support of UW are to conserve the fighting strength of guerrilla forces and to help secure local population support for U.S. and guerrilla forces operating within the JSOA. Medical requirements within the JSOA differ from those posed by a conventional force conducting operations for two reasons: guerrilla forces normally suffer fewer battle casualties and their incidence of diseases and malnutrition is often higher.

X-24. In UW, commanders tailor the organization of medical elements to fit the particular situation. Depending on the skills required, organizers might be able to bring personnel from other medical units into the JSOA. The basic medical organization can expand by using guerrilla force members and recruiting professional medical personnel to establish and operate guerrilla hospitals.

X-25. Clandestine facilities are, at first, confined to emergency and expedient care with minimum preventive medicine. Once the area command develops sufficiently, the clandestine facilities can expand and become part of the unit's medical organization. A wounded guerrilla allowed to fall into enemy hands can be forced to reveal what he knows and may compromise the mission. Patients with appropriate cover stories can be infiltrated into civilian or enemy military hospitals to receive the care not otherwise available.

## MEDICAL REQUIREMENTS

X-26. Because of their high motivation and adaptation to frequent hardship and discomfort, guerrilla forces frequently ignore minor injuries and illness. Historically, the lack of proper medical attention has led to serious illness and disability, causing reduced unit combat effectiveness.

X-27. Health standards in many areas will be below those of the United States. Indigenous personnel may not accept treatment that is desirable for U.S. personnel because of religious beliefs or superstitions. Natives may have an acquired immunity to certain diseases of the area.

X-28. A broad range of medical support may be available in the JSOA—although at first, treatment may be limited to rudimentary medical procedures, such as first aid and personal hygiene. Some guerrillas in the past have developed highly organized and effective medical support units and installations. Their organizations paralleled those of conventional forces, to include field hospitals in inaccessible areas.

X-29. Medical elements supporting the resistance forces must be mobile, responsive, and effective in preventing disease and restoring the sick and wounded to duty. There may be no safe rear area where the guerrilla can take casualties for treatment. Wounded and sick personnel become a tactical problem rather than a logistics problem; medical support is a major tactical consideration in all operations. The civilian infrastructure of the guerrillas contributes to medical support by setting up and operating medical facilities. Medical personnel help during combat operations by starting casualty collection points, permitting the remaining members of the guerrillas to keep fighting. Casualties at these points are evacuated to a guerrilla base or civilian care facility.

X-30. In UW, the attitude of the sick and wounded is extremely important. The emotional importance the individual Soldier attributes to the medical service goes a long way in his care and treatment. Because the sick and wounded can find themselves in difficult conditions, they have simple wants for shelter, food, and medical treatment. Standards of care are not lowered; these Soldiers may just be unaware of any shortcomings in the medical care they receive. Experience has shown that a Soldier may have major surgery under extreme hardship conditions and yet demonstrate remarkable recuperative power.

## MEDICAL NET

X-31. The medical net in the JSOA is kept as simple as possible—just enough to provide security and fit the estimated needs of future expansion of the JSOA. Medical personnel refine and modify the net after it is functioning and secure. When setting up a medical net, personnel must consider the following.
- Scale of activities already in existence and those planned.
- Potential increase in strength, activities, and operations.
- Physical factors, including topography, climate, and geography, plus transportation and communications.
- The number, availability, and dependability of medically qualified and semiqualified personnel in the JSOA.
- The attitude of the population, government, and guerrillas toward medical problems and the medical standards accepted in the area.
- Existing nonmedical operational facilities of the area command.

X-32. Medical personnel may use the existing intelligence and security nets to start a separate medical net for collecting medical intelligence. They can also use the existing logistics net to transport medical supplies.

## AID STATION

X-33. Overall, mission planning includes locating and operating an aid station. Medical personnel provide emergency medical treatment at this location. Evacuation of wounded personnel from the battle area begins here. Because the condition of the wounded may preclude movement to the unit base, they are hidden in secure locations and the auxiliary is notified. The auxiliary will care for and hide the wounded or evacuate them to a treatment facility.

X-34. The evacuation of the dead is important for security reasons. If the enemy identifies the dead, the safety of the guerrilla families may be jeopardized. Personnel evacuate and cache the bodies of those killed in action until they can be properly buried or disposed of, according to the customs of the local population. Removal and burial of the dead will deny the enemy valuable intelligence concerning indigenous casualties.

X-35. As the operational area develops and the situation favors the sponsor, evacuation of the more seriously injured or diseased personnel to friendly areas may become possible. This action will lighten the burden on local facilities and provide a higher standard of medical care for the remaining patients. Air evacuation is the most logical evacuation means but the disadvantage is its inherent threat to security. Landing sites must be located well away from sensitive areas, and guerrilla forces must secure and control the surrounding area until the aircraft leaves.

## CONVALESCENT FACILITY

X-36. The area where patients are sent to recuperate is called a convalescent facility. These patients are discharged as soon as possible. A convalescent facility may be a safe house in which one or two convalescents are recuperating with their necessary cover stories, or it could be in any base camp in guerrilla-controlled areas.

## GUERRILLA HOSPITAL

X-37. The guerrilla hospital is a medical treatment facility or complex of smaller facilities providing inpatient medical support to the guerrilla force. A guerrilla hospital is established during the organization and buildup phase of a resistance organization. The hospital must be ready for operation at the start of combat operations, and it must continue to provide medical support until directed otherwise. The hospital is generally in the JSOA it supports but considerations of METT-TC may dictate otherwise. An indigenous medical officer with advice and assistance of the U.S. Army SF group or battalion surgeon will usually command the guerrilla hospital. However, depending on circumstances within the JSOA, the group or battalion surgeon may also be the commander.

X-38. The guerrilla hospital rarely, if ever, outwardly resembles a conventional hospital. The requirement for strict security, flexibility, and rapid mobility precludes visible comparison with conventional military or any civilian medical facilities. As the guerrilla force consolidates its hold on the JSOA, all medical support functions will tend to consolidate. Safe areas allow the establishment of a centralized system of medical care. Sophisticated hospitals permit care that is more elaborate because they provide a wider selection of trained personnel, specialized equipment, and the capability of more extensive and prolonged treatment. Hospital considerations will depend on the following factors.

### Location

X-39. The guerrilla hospital staff conducts a reconnaissance for possible hospital sites and coordinates the training of guerrilla members who will support hospital operations. The hospital should be in a secure area but accessible to casualties. Site planners must consider security, topography, distance, mobility, and enemy counterguerrilla activities. A sanctuary across an international border is ideal for a guerrilla hospital.

### Security

X-40. There must be strict security measures to protect the covert nature of the guerrilla hospital operations. Security compromises can lead to the capture of the hospital staff, patients, and supplies, which may compromise members of the auxiliary or underground or jeopardize the entire operation.

### Communications

X-41. Rapid communications are essential between the hospital command and the area or sector commander to maintain adequate medical support and ensure survival. Coordinating hospital movement, receiving casualties and supplies, requesting support, and disseminating intelligence all depend on rapid and secure communications.

### Medical Supplies

X-42. After infiltration, the SF operational detachment requests adequate medical supplies for initial hospital operations. Plans must provide for automatic and on-call medical resupply, although the staff should make maximum use of locally available supplies. The staff coordinates with the guerrilla force to acquire food and rations for the hospital patients.

## Sections

X-43. A guerrilla hospital has several sections. Some sections are collocated; others should be dispersed for security reasons. Staffing a guerrilla hospital depends on the mission and availability of trained medical personnel. Personnel must be attached to the hospital to provide security, communications, and logistics support. There are seven recommended sections of a guerrilla hospital.

### Command and Control

X-44. The command and control section provides command over hospital personnel and supervision of hospital functions. It maintains communications with the area or sector commander of the hospital, and it coordinates security for the hospital.

### Security

X-45. Security is primarily a function of location, early warning, and movement. The guerrilla hospital should be located in an area where the local populace is friendly or sympathetic to U.S. personnel and the guerrillas. Security should rely heavily on early warning, diversionary tactics, and movement to alternate locations.

### Logistics

X-46. The logistics section provides logistics support, to include supply, transportation, and graves registration.

### Sorting

X-47. Personnel in this section establish and maintain one or more sites that serve as staging areas for limited medical care and movement of patients and supplies to the treatment section. For security reasons, these sites are the only contact the guerrilla force has with the hospital.

### Treatment

X-48. Medical, surgical, and immediate postoperative care comes from this section. It is the central activity of the hospital, and all other sections support it. Individuals staffing triage sites should not be told the location of the treatment center. Only treatment center staff members will pick up patients at triage sites to take them to the central treatment center.

### Convalescent

X-49. This convalescent section establishes facilities to care for patients who no longer requiring the intensive support provided by the treatment section. Such facilities increase the dispersion of patients. Depending on the condition of the patients placed in convalescent facilities, hospital personnel may not be required to continually staff this area.

### Outstaging

X-50. The outstaging section establishes sites where patients may be transported once they have received maximum benefits from hospitalization. These sites are basically unmanned geographic points used as drop-off locations where patients may be returned to their units ready for combat.

## MEDICAL TRAINING OF GUERRILLA FORCES

X-51. The SF operational detachment medic begins medical training of guerrilla forces at the earliest possible opportunity. He selects the personnel to be trained and screens them for their abilities. Those with potential are trained as company medics or nurses. He then develops the training program for each of the different functions needed within the guerrilla medical system. He teaches the principles of self-aid and buddy-aid to the entire guerrilla force. Other training includes preventive medicine procedures, basic sanitation, personal hygiene, and individual protective measures. He also ensures immunizations of the guerrilla force and their families.

X-52. The newly trained medical personnel should make maximum use of all medical facilities. They also help the medic train additional medical personnel. The amount and quality of the training depends on the situation, facilities, and instructors. However, the knowledge and ability of the SF operational detachment medic will ultimately decide the success of the program.

## PERIODIC MEDICAL REPORTS

X-53. Depending on the tactical and medical situations in the JSOA, the JSOTF or SOTF will determine the frequency and contents of periodic medical reports. Before infiltration, the JSOTF or SOTF briefs the SF operational detachment on the health of the guerrilla forces, casualty rate (wounds or diseases), and overall guerrilla medical condition.

X-54. The training status report will include:
- New training programs since last report.
- Change in number of trainees.
- Change in number of graduates from training programs.
- Adequacy of training aids.

X-55. The medical program report will include:
- Medical supply status, to include—
  - Supplies reduced to critical level.
  - Success rate with supply procurement within the AO.
  - General condition of supplies and equipment following air resupply operations.
- New projects and facilities since the last report and any losses due to enemy activity.
- Any serious medical problems that may exist and the level of care patients are receiving in facilities of friendly forces.
- Whether other medical personnel are aiding friendly forces more than expected.
- Number of new guerrilla personnel (both full- and part-time) and their skills and combat effectiveness.
- Any special medical problems unique to the area.
- Any knowledge of enemy medical situation, such as—
  - Standards of medical care.
  - Attitudes or actions toward wounded guerrillas and U.S. personnel.
  - Anti-guerrilla activities toward medical facilities.

# Appendix Y

# Resistance Transition and Demobilization Considerations

Transition will differ depending on the ultimate policy goal of the UW campaign. When the objective is overthrow, the resistance may transition to a new, victorious regime resulting in the resistance being demobilized, or alternatively incorporated into the new regime's security service, or a combination of both. When the objective of UW is more limited, such as for coerce or disrupt objectives, transition may mean to disband, go dormant, change methods, change sponsors, proceed to a simple termination of activities, and so on.

## TRANSITION

Y-1. Transition refers to turning over an activity or task to a new indigenous government, allied or coalition force or government, or to private sector agencies when the objective is overthrow and means limited changes when the objectives are more limited. Examples include—

- Transfer of civil authority from military (guerrilla) forces to civil government (overthrow).
- Establishment of indigenous police or security forces (overthrow).
- Privatization or return of facilities, such as public works and utilities, airports, and seaports, to indigenous control (overthrow).
- Privatization of humanitarian demining operations (overthrow).
- Continue but change support to other clandestine indigenous operations other than resistance (coerce/disrupt).
- Hand off support to other USG departments or agencies or foreign partners (coerce/disrupt).
- Support is terminated or goes dormant (coerce/disrupt).

Y-2. Transition is a critical consideration in UW. From the beginning of planning, leaders and planners should consider the second- and third-order effects of desired strategic outcomes and how the mission will transition. This is particularly true for overthrow, which, if improperly planned for, could result in dangerous instability or counterproductive regimes, which the USG may then have to establish relationships with. Planners must also anticipate outcomes in which the supported indigenous partner is unsuccessful in, unable, or unwilling to assume the responsibility of, taking political control after an overthrow. The sustainment of a successfully terminated insurgency depends, in large part, on how quickly and efficiently the posthostility's government can take control and provide the sustainable stability and functioning infrastructure required to support a population. Transition may begin in areas where conflict has subsided, while combat operations continue elsewhere in the JSOA. The emphasis in this appendix is on transition following overthrow and demobilization of the resistance force.

Y-3. When planning to transfer an activity or task to indigenous populations or institutions, it is important to:

- Know the capabilities and limitations of the elements of the existing infrastructure, to include:
  - Host governments.
  - Bilateral donors.
  - United Nations agencies.
  - International organizations, especially the International Crisis Group and the International Committee of the Red Cross.
  - NGOs by type (for example, assistance, advocacy, and indigenous organizations).
- Define the desired end state (for example, continuity of current operations or modification of current operations to some other format).
- Identify the organizational structure required to perform the activity or task.

- Identify competent, trustworthy individuals to fill positions within the incoming organizational structure.
- Determine how to conduct demilitarization of indigenous forces and incorporation of former belligerents into the private sector.
- Attempt to fully understand implications of withdrawal, as well as tribal or factional breakdown, to include historical boundaries and differences not yet resolved.
- Identify equipment and facilities required to perform the activity or task and who will provide them. Prepare the appropriate property control paperwork if transferring equipment or facilities to the relieving organization.
- Create timelines that provide sufficient overlap between the outgoing and incoming organizations.
- Determine the criteria that will dictate when the incoming organization will assume control of the activity or task (for example, a target date, task standard, or level of understanding).
- Orient the incoming organization to the activity or task, to include providing procedures, routine and recurring events, and other information critical to the conduct of the activity or task. Demonstrate the activity or task, if possible.
- Supervise the incoming organization in performing the activity or task. The departing organization retains control of the activity or task during this process, providing critiques and guidance as needed.
- Transfer the task according to the transition plan.
- Provide continued support to the incoming organization, as required.
- Bolster the image and prestige of the incoming organization in the eyes of the populace and the international community.

# DEMOBILIZATION

Y-4.  Demobilization is a major activity of transition. It involves demilitarizing the combatants of a resistance movement and reorienting them (or, in some cases, orienting them for the first time) from life as a guerrilla to life as a peaceful, productive member of society. Demobilization planning must be closely coordinated and synchronized with transition planning. As the need for resistance fighters diminishes, the incoming government decides whether to place guerrilla forces in a supporting role or to begin demobilizing them.

Y-5.  How a guerrilla force is demobilized will affect the postwar attitudes of the civilian population toward the "new" government. To ensure the guerrilla force is aligned with the new government, the combatant commanders, through the TSOC, coordinate between SOF, the area commander, and other military organizations. When demobilizing the guerrilla force, the following COAs could be considered:

- Retain the force with weapons and equipment as a new military or police force.
- Integrate personnel, weapons, and equipment into new local, state, or country public services and retrain personnel into new civil sector occupations.
- Turn in all weapons and equipment and return personnel to their former occupations. (The collection of weapons is a time-consuming and sensitive issue. An "arms for cash" reward system may be effective.)
- Combine any of the above methods.

Y-6.  After the war is over, the guerrilla force's hospitals will be kept operational until patients can be transferred to military or civilian hospitals. Permanently disabled patients should be granted pensions by the new government. Rehabilitation assistance must be available to the guerrillas to assure their place in the civilian community.

# CIVIL AFFAIRS SUPPORT

Y-7.  During combat operations, the SF operational detachment and the guerrilla force conduct postmission activities, such as after action reviews, reconstitution, or prepare for the next mission. CA forces participate by conducting a CA battle damage assessment. This evaluation assesses the results of the mission on the factors of ASCOPE, validates the CA and CMO concept of operations, and determines whether the established measures of effectiveness for the operation have been met. It also helps the SF operational detachment and area commander decide when and how to adjust the plan, when to develop new plans to address unforeseen consequences of operations, and when to terminate or transition an operation.

Y-8. In areas where conflict has subsided and conditions indicate a successfully terminated insurgency, transition and demobilization activities may begin. Transition and demobilization require the collaborative efforts of many military and civilian, U.S. and international, and government and nongovernment agencies. CA participation is more prevalent in transition and demobilization than in most other aspects of the UW campaign. CA specialists provide the preponderance of CA support, although CA generalists will still have an important role.

Y-9. CA forces support transition by—

- Addressing transition issues early in the planning stages of the UW campaign, including establishment of measures of effectiveness and other indicators that determine points at which to transition to different phases of the operation.
- Identifying individuals among the insurgency movement who would provide the experience, knowledge, and leadership required to create and sustain a post-UW infrastructure. Individuals should be identified who can fill roles in government, economics and commerce, public facilities, and special functions at the national, provincial, and local levels.
- Separating those individuals from the main body of guerrilla forces and, with the assistance of other government and nonmilitary agencies, orienting and training the individuals to their future roles in the post-UW infrastructure. Depending on METT-TC, this may occur inside or outside the JSOA.
- Refining the transition plan based on periodic assessments of the situation within the JSOA, as well as among the international community.
- Observing and validating the measures of effectiveness and other indicators that help determine when conditions are right to begin the transition phase.
- Supporting transition operations by providing advisory SCA in a friendly territory.
- Maintaining the civil-military operations center to facilitate collaborative interagency planning, coordination, and synchronization throughout all phases of the UW movement and contracting the civil-military operations center according to the transition plan.
- Providing postredeployment oversight and support mechanisms for the new indigenous government and infrastructure (for example, "reach back" points of contact, periodic visits, and combatant commanders' theater engagement programs).

Y-10. CA forces support demobilization by—

- Considering demobilization issues in transition plans, to include interpreting the combatant commander's policy on demobilization, demilitarization, and disarming of the former warring factions.
- Supporting SF operational detachment foreign internal defense missions.
- Identifying civil sector manpower requirements in government, economics and commerce, public facilities, and special functions at the national, provincial, and local levels.
- Establishing programs to screen and identify candidates among the demobilizing guerrilla force to fill required positions in the post-UW infrastructure.
- Establishing and overseeing training programs for former guerrillas to gain new skills and knowledge and to make them peaceful, productive members of the new indigenous society.

# PSYCHOLOGICAL OPERATIONS SUPPORT

Y-11. The SF operational detachment establishes liaison with the PSYOP element before demobilization. When supporting a UW mission, the supporting PSYOP element advises and assists the indigenous personnel in the conduct of effective PSYOP, assists in gaining converts and recruits for the resistance as they develop the necessary governmental infrastructure, and conducts operations to create popular support for the resistance movement.

Y-12. The senior PSYOP officer at the joint special operations force HQ plans PSYOP to support the linkup and demobilization of the guerrillas. The PSYOP commander prepares and coordinates link-up plans with PSYOP personnel supporting the resistance movement. Because of their extreme sensitivity and importance, demobilization plans must begin in the early phases of operations and have continuous support.

Y-13. PSYOP elements in the JSOA prepare the civilian population to cooperate fully with the conventional tactical forces. PSYOP elements urge the civilian population to remain in place and not hinder operations. PSYOP elements brief the resistance organization's leaders on the importance of cooperating with the tactical force commanders and accepting the conventional force leadership. PSYOP elements psychologically prepare the resistance organization to assume whatever roles the legitimate government wants it to play. These roles include, but are not limited to, their incorporation into the national army, paramilitary organizations, national police, or demobilization. PSYOP programs explain the demobilization process. They promote the insurgents' orderly transition to peaceful civilian life. PSYOP personnel attempt to prevent the formation of quasi-military or political groups opposing the recognized government. Loyalty to the legitimate government is the major concern.

# DANGERS OF DEMOBILIZATION

Y-14. The primary concern in any demobilization program is the guerrilla. His personal and political motives vary. The resistance organization can include peasants, laborers, bandits, criminals, merchants, and a few social and intellectual leaders. During the conflict, some guerrillas may have achieved status and leadership positions that they now are reluctant to relinquish. Others may have found adventure in combat that they would not now trade for peace or prosperity. Hostile groups may have clandestinely infiltrated the guerrilla force to continue their own personal or political agenda. They may take advantage of the demobilization program to organize paramilitary or political groups that will be in conflict with the new provisional government or U.S. authorities.

Y-15. It is imperative that demobilization programs and procedures be executed quickly and with major political support. The programs and procedures begun are a direct result of decisions made by high-level civilian and military authorities. Their successful implementation requires maximum effort and coordination among CA, PSYOP, and SF.

Y-16. Because of their knowledge and history of the guerrilla force, SF operational detachments initially remain in their operational areas to assist in demobilization. The SF operational detachment commanders and their supporting CA and PSYOP elements ensure transfer of U.S. responsibility without loss of control, influence, or property accountability. The key to long-term strategic success in UW is planning and executing postmission responsibilities. Appendix M, Logistics, provides detailed guidance on the SF operational detachment's responsibilities.

# DETAILED RESPONSIBILITIES FOR DEMOBILIZATION OF GUERRILLA FORCES

Y-17. During the demobilization phase of UW, all members of the SF operational detachment perform duties and assume responsibilities as outlined in subsequent paragraphs. In addition, detachment members not mentioned perform duties as directed by their immediate supervisor, the operations sergeant, or the detachment commander. Users may deviate from the guidance provided in this appendix as required by special mission considerations or peculiarities of a UW JSOA.

## DETACHMENT COMMANDER

Y-18. The SF operational detachment commander is responsible for command of the detachment members. He directs, advises, and gives guidance to all staff and special staff sections and resistance force leaders in accomplishing the following:

- Conducting inventory, inspection, and check of:
  - Weapons and other serial-numbered items.
  - Rosters of tactical and logistics units.

- Rosters of auxiliary units and all other supporters.
- Rosters of leaders, staff, and special staff sections.
- Administrative records (for example, war records for decorations and awards pay and personnel records).
- All equipment, ammunition, supplies, and material in storage, cache sites, or in the possession of resistance force units and personnel.
- Minefield, obstacle, and hazardous material data and records.
- Compiling, posting, correcting, and making available:
    - The principal area assessment.
    - Supply records and inventory documents.
    - Personnel and pay records.
    - Intelligence and security files.
    - Evasion and recovery files.
    - OPLANs, OPORDs, and estimates.
    - Communication records and cryptographic material.
    - Medical records.
    - Rosters of potential, political assistance personnel.

Y-19. In addition, the detachment commander directs and coordinates detailed briefings for resistance force personnel to ensure that everyone understands:

- The need for demobilization.
- Demobilization requirements on the part of commanders and individuals.
- The importance of remaining in preselected assembly areas.
- Police and sanitation functions.
- Collection of arms, ammunition, and equipment.
- Immediate care for and processing of sick and wounded.
- The CA team's requirements, to include:
    - Necessary records and rosters:
    - Roster of legal survivors of killed in action or missing in action Soldiers as a direct result of hostile action.
    - Roster of resistance force personnel.
    - Other supporters who are auxiliary, bartering assets, doctors, and persons holding levy receipts.
- Combat records for preparation of decorations and awards.
- Pay records.
- Medical records.
- Personnel records for preparation of and discharge.
- Preparation and conduct of ceremonies.
- Implementation of rehabilitation procedures as planned within stipulations by higher HQ.
- Preparation for exfiltration of SF operational detachment members.

## ASSISTANT DETACHMENT COMMANDER

Y-20. The assistant detachment commander supervises and assists in the preparation of the following:

- Rosters of all resistance force personnel, guerrillas, and auxiliary.
- Rosters of other known supporters, such as bartering and levy assets or doctors.
- Rosters of persons who possess potential assistance qualifications, including—
    - Outstanding political attitudes.
    - Leadership ability (political).
    - Administrators.
    - Police and security qualifications.

- Inventory documents and records of the following supplies and equipment reflecting serial number, amount, and location:
  - Weapons.
  - Other serial-numbered items and sensitive items.
  - Ammunition and munitions.
  - Clothing.
  - Field equipment.
  - Rations.

Y-21. The assistant detachment commander also makes available to CA personnel all records and rosters necessary to accomplish their mission. He assists the CA team in the administration of resistance force personnel. The assistant detachment commander prepares SF operational detachment members for briefing and debriefing by CA officers. He provides guidance for and assists in the—

- Preparation and completion of principal area assessment.
- Preparation of intelligence and security files.
- Collection and accountability of cryptographic and other sensitive communication material.

Y-22. The assistant detachment commander assists CA personnel in execution of demobilization plans. He provides guidance and assistance to SF operational detachment members in preparation for exfiltration.

## INTELLIGENCE SERGEANT

Y-23. The intelligence sergeant prepares all classified documents and material for destruction or exfiltration as directed by the detachment commander. The intelligence sergeant compiles and prepares intelligence and security files for the assistant detachment commander or for use in briefing the CA team and subsequent exfiltration. The intelligence sergeant should be prepared to present the intelligence and security briefing to the CA team and the security briefing to operational detachment members.

Y-24. The intelligence sergeant also prepares and presents any assistant detachment commander files on and rosters of potential political assistance personnel to the operational detachment, including—

- Persons with outstanding political attitudes and qualifications.
- Political leadership ability and experience.
- Potential administrators.
- Police and security assets.
- Other favorable assets.

Y-25. The intelligence sergeant also assists SF operational detachment members in:

- Police and sanitation functions.
- Demobilization procedures.
- Collecting arms, ammunition, munitions, supplies, and equipment.
- Preparing for exfiltration.

## OPERATIONS SERGEANT

Y-26. The operations sergeant selects assembly areas for resistance force command posts or units. He assists the SF operational detachment commander in directing, supervising, and coordinating the collection of—

- Personnel.
- Resistance force units.
- Weapons.
- Ammunition and munitions.
- Clothing and equipment.
- Records and rosters.
- Equipment and supplies.
- Classified material.

Y-27. The operations sergeant directs and advises conduct of assembly, briefings, and control of resistance force personnel. He completes and compiles the—

- Principal area assessment.
- OPORDs and OPLANs.
- Estimates.
- Damage assessments.

Y-28. The operations sergeant plans and conducts briefings to the CA team. He assists CA personnel in the conduct of demobilization procedures and advises on a rehabilitation program. The operations sergeant also plans operational detachment exfiltration, including—

- Selects and secures landing site.
- Reports and confirms landing site.
- Directs and supervises preparation of detachment members, documents, and supplies for exfiltration.
- Organizes brief and rehearses reception committee.
- Plans and conducts demobilization ceremony.

## MEDICAL SUPERVISOR

Y-29. The medical supervisor prepares, posts, and makes information available to the assistant detachment commander, such as—

- All medical records.
- Immunization records.
- Plans for continued treatment of patients.
- Medical rehabilitation programs.
- Recommendations for awards and decorations for both detachment and guerrilla medical personnel.

Y-30. The medical supervisor prepares, posts, and makes information available to the K-4 sergeant, including—

- Plans for collection, use, storage, issue, exfiltration, or destruction of medical supplies and equipment.
- Inventory documents and records reflecting serial number, location, type, and amount of:
  - Medical equipment.
  - Instruments.
  - Installations and facilities.
  - Medicine.
  - Other equipment.

Y-31. The medical supervisor executes plans for medical treatment, hospitalization, and evacuation during demobilization. Guerrilla hospitals are kept in operation until the patients can be taken over by military hospitals or civilian institutions. He briefs and assists CA personnel. The medical supervisor executes plans for the disposition of medical supplies, equipment, facilities, and instruments, and for the handover of medical installations. He also assists SF operational detachment members in preparation of personnel and equipment for exfiltration. The medical supervisor coordinates the—

- Final contribution of medical intelligence to K-2.
- Latest contribution to the medical portion of the principal area assessment (K-3).
- Assistance in sanitation, police, and sterilization functions.

Y-32. The medical supervisor also supervises the final discharge physical examinations.

## SENIOR WEAPONS LEADER

Y-33. The senior weapons leader assists the assistant detachment commander in the preparation of administrative matters. The senior weapons leader advises the detachment commander on the future disposition of weapons, ammunition, and explosives, including—
- Use.
- Issue.
- Storage.
- Exfiltration.
- Destruction.
- Methods of collecting and accounting.

Y-34. The senior weapons leader advises and assists the operations sergeant in the—
- Selection, reconnaissance, and occupation of unit assembly areas.
- Selection and reconnaissance of exfiltration landing sites.
- Organization and rehearsals of exfiltration committee.

Y-35. The senior weapons leader plans, organizes, and conducts briefings of the CA team as directed by the operations sergeant. The senior weapons leader assists the CA team in the conduct of demobilization procedures. He also assists detachment members in the plans and preparation of equipment for:
- Use.
- Issue.
- Storage.
- Infiltration.
- Exfiltration.
- Destruction.

Y-36. The senior weapons leader makes final contribution to the principal area assessment. He also continues to assist in, direct, and supervise—
- Police functions.
- Sanitation measures.
- Sterilization requirements.

## COMMUNICATIONS SUPERVISOR

Y-37. The communications supervisor provides necessary communications for—
- Administrative contact with the:
  - JSOTF.
  - Conventional link-up command.
  - Resistance and auxiliary force units during demobilization.
  - Supporting naval and air units.
  - Exfiltration activities.
- Population control with:
  - Police units.
  - Other security elements.
  - CA teams.

Y-38. The communications supervisor advises the detachment commander on the amount and type of equipment necessary to accomplish the demobilization mission, the number of personnel required for operation of equipment, and the number and location of sites to be occupied.

Y-39. The communications supervisor collects, inventories, and accounts for all classified documents and material and makes the documents available to the intelligence sergeant (K-2) for future use, exfiltration, and destruction.

Y-40. The communications supervisor plans for the collection and disposition of communications equipment and supplies. He inventories, documents, and records up-to-date information reflecting serial number, amount, and location of all communications equipment and supplies. He also executes plans for collection, issue, use, storage, exfiltration, and destruction.

## ENGINEER SERGEANT

Y-41. The engineer sergeant collects all demolitions and engineer equipment and prepares it for turn-in. He inventories mine, obstacle, and hazardous material data and ensures it is properly recorded. He assists the assistant detachment commander in preparing inventories and supply documents and in readying equipment and weapons for exfiltration or turnover to civil authorities. The engineer sergeant presents engineer portions of briefings as directed. He also assists the operations sergeant in preparing assigned portions of the area assessment.

This page intentionally left blank.

# Appendix Z

# Stay-Behind Unconventional Warfare Considerations

Resistance can be an effective way to coerce, disrupt, and contribute to the defeat of occupying powers. However, resistance to—and oftentimes mere survival of—powerful, determined and ruthless occupiers is dangerous and difficult. The challenges of establishing such resistance after the invasion or occupation has already occurred is multiplied several fold. To the extent that such resistance can be preplanned and resourced prior to the actual invasion or occupation, the dangerous challenges of such resistance can be partially mitigated. This is the purpose of stay-behind resistance and stay-behind UW.

## OVERVIEW

A *stay-behind operation* is an operation in which the commander leaves a unit in position to conduct a specified mission while the remainder of the forces withdraw or retire from an area (FM 3-90-1).

A *stay-behind resistance operation* is an operation in which indigenous authorities leave personnel and resources in position before, during, and after a foreign occupation to conduct anticipated resistance activities against the occupying power (ATP 3-18.1, proposed for inclusion in ADP 1-02)

*Stay-behind unconventional warfare* is unconventional warfare operations and activities conducted by a sponsoring government in support of indigenous authorities' stay-behind resistance operations (ATP 3-18.1, proposed for inclusion in ADP 1-02)

## STAY-BEHIND OPERATIONS

Z-1. A *stay-behind operation* is an operation in which the commander leaves a unit in position to conduct a specified mission while the remainder of the forces withdraw or retire from an area (FM 3-90-1). The force should consist of enough combat, functional and multifunctional support, and sustainment elements to protect and sustain its fighting capability for the duration of the mission. A stay-behind force may also result from enemy actions that bypass friendly forces.

Z-2. The main purpose of a stay-behind force is to destroy, disrupt, and deceive the enemy. This force has a high-risk mission because of the danger that it will be located, encircled, and destroyed by the enemy. Resupply and casualty evacuation are also extremely difficult. A commander considers assigning this mission only after a thorough analysis of the mission variables of METT-TC. The stay-behind force attacks enemy combat forces and command nodes, functional and multifunctional support, and sustainment elements from unexpected directions. These attacks may cause enemy follow-on forces to be more cautious and to slow down to clear possible attack and ambush sites. The commander may require the stay-behind force to conduct a breakout from encirclement and link-up operations after it completes its mission.

*Note:* FM 3-90-2 discusses the conduct of a breakout from an encirclement.

Z-3. A light infantry, stay-behind force surprises an enemy by conducting a series of raids and ambushes. The light infantry force can be inserted via infiltration, air assault, or parachute; it can also be a bypassed force. Attacks in the enemy support area by friendly armored forces can cover a larger area than attacks by light infantry forces.

Z-4.  Stay-behind operations eventually require the force to reenter friendly lines or link up with other elements, often in more than one location. The commander must carefully coordinate this reentry to prevent friendly fire incidents. The return routes for the stay-behind force are the best-covered and concealed routes available. Commanders place guarded gaps or lanes near obstacles along these routes that cannot be bypassed.

Z-5.  A stay-behind operation is not a suicide mission. The commander conducts this operation only when there is confidence that the stay-behind force will rejoin the main body, extract itself in alternative ways, or the main body will fight its way forward to link up with the stay-behind force.

# STAY-BEHIND RESISTANCE OPERATIONS

Z-6.  Stay-behind resistance is a form of stay-behind operation, but it differs significantly in who executes the resistance and the expectation of control. Unlike stay-behind operations, which assume the executing elements are detachments from military organizations, stay-behind resistance organizations are likely to be a mix of military, paramilitary, civil government, and society organizations and civilians. There will be times when the civilian individuals and organizations can be expected to lead the stay-behind resistance effort. Unlike stay-behind operations, which assume the designated units will eventually reintegrate with the larger parent organizations within a relatively short period of time, stay-behind resistance elements may never leave the operational area. Elements engaged in stay-behind resistance are, therefore, often subject to unrelenting scrutiny, risk of discovery, interrogation, and elimination. In addition, while stay-behind resistance operations are likely to conduct some destroy, disrupt, and deceive the enemy activities, it is equally likely that its main purpose is to survive enemy occupation so that its infrastructure may serve as a platform from which other operations and activities may occur, such as intelligence gathering or personnel recovery. Finally, indigenous stay-behind resistance organizations may or may not receive external support.

## STAY-BEHIND RESISTANCE ORGANIZATION

Z-7.  If the stay-behind function in a given area is to be primarily intelligence gathering, the organizational requirement will differ from those set up primarily for harassment of the enemy. The size of the organization will be governed also by the dimensions of the tasks assigned to it. Thus, an unconventional assisted recovery organization charged with moving personnel over an extensive land route will need more spotters, safe house keepers, and couriers than one having a relatively small area to cover.

Z-8.  As in any clandestine organization, security factors in the stay-behind organization are vitally important. Maximum security may be achieved through compartmentation and the use of cover or concealment. By keeping individual units sealed off from each other, damage to the organization through compromise, penetration, or accident can be limited to one unit. Good cover will protect members of the unit in their activities, relationships to each other, and means of communication. Concealment will provide a means of removing persons or things from places of danger.

Z-9.  A decentralized or compartmented stay-behind organization has the advantage of providing a high degree of security, since for each mission, there is a separate operational team responsible through independent channels of communications and support to a base in a secure area. Problems of control and direction are thus simplified, cover problems are reduced, and the risk of one team's compromise endangering others is eliminated.

Z-10.  One danger inherent in separate operational teams is the security hazard created by a number of individual teams operating in an area containing a concentration of targets. Another adverse factor is the number of independent communications links required to service the teams. Sometimes this is an equipment or expense challenge, and sometimes it is a lack of trained radio operators and base-station capability—all of which are critical assets. Support of independent teams likewise magnifies the effort required to mount independent support operations, such as air or maritime reception. The compartmentation, originally established by creating independent teams, could be forfeited if a need for substantial supply or resupply surpasses security concerns.

Z-11.  A further consideration is the fact that combining certain tasks is inconsistent with security. For example, it would be unwise to link surveillance activity with a sabotage operation, although sabotage units might find it necessary to provide a certain amount of operational intelligence—that as a by-product would contribute to the overall intelligence picture. As another example, a clandestine movement organization cannot accept the risks of military escapers and evaders without sacrificing security unless it is designed specifically for that purpose.

Z-12. The choice of a stay-behind organization must be made on the basis of many factors, including those mentioned above, in conjunction with the best possible appraisal of facts about the area and indigenous population and estimates of probable enemy and friendly operational circumstances in the future. However, any stay-behind system should include the following basic operational elements:

- Target (task and location).
- Trained personnel in place.
- Communications links to serve them.
- Base in secure area (for support and direction).

## STAY-BEHIND RESISTANCE PERSONNEL REQUIREMENTS

Z-13. Personnel requirements for a stay-behind organization are basically the same as those required for the conduct of all clandestine operations; however, they differ from conventional forces in that they should not be linked together into complete operational units, and they do not become operational until events for which they have been created occur. The responsibility for activating stay-behind personnel rests with the HQ that will ultimately control and support operations. Activation will be contingent upon the occurrence of events within the specific area. These units will be operating in areas where security measures against the population will probably be most severe and restrictive. They must, therefore, possess resilience in order to survive under security controls imposed by hostile forces. In addition to surviving during the initial stages of occupation, stay-behind personnel must be capable of finding or making opportunities to function operationally in an environment that can only be roughly estimated in advance. They must be able to function with a minimum of close support and without continuous supervision and direction.

Z-14. The above factors point clearly to the realization that stay-behind personnel must possess qualities over and above those of normal clandestine personnel. These qualities can be obtained only by careful selection of personnel in terms of stability, high standards of operational capability, unquestioned motivation, and ability to function independently. The standards applied to the stay-behind member in terms of ability to function under severe, operational conditions must be very high. The standards of technical and tactical proficiency in setting up drops, safe houses, and other clandestine communications links must be equally high. The training given the member must provide him with all the necessary tools of his trade, and it must be absorbed so thoroughly as to enable him to train other personnel effectively.

Z-15. The normal criteria for selection of clandestine personnel—that is motivation, qualifications accessibility and security—apply to stay-behind personnel, with the added factor that they must be applied more strictly than in a less demanding situation. Selection of stay-behind personnel is further complicated by the need for resilience and by the two different types of tasks to be performed: the short range or tactical and the long-range or strategic.

## CRITERIA FOR STAY-BEHIND RESISTANCE PERSONNEL: SELECTION FOR SHORT-RANGE TASKS

Z-16. For short-range tasks, low-level personnel can observe and report a mass of information on enemy order of battle, targets of opportunity, control techniques, personality data, and all other information of an operational nature. With the short-range tasks, the candidate's motivation and qualifications for the particular task may be the primary criteria for his selection, since it is anticipated that this mission will be carried out prior to consolidation of enemy controls. Experience with Soviet control methods indicates that during the initial or occupation phase, disturbed conditions exist that permit circumvention of controls more easily than will be the case later. The short-range member's resilience, therefore, will not be subjected to the severe tests that will prevail later. If the candidate has the motivation and qualifications needed to perform the task and a reasonable chance of surviving to perform it, he will be acceptable for a short-range, tactical mission.

## CRITERIA FOR STAY-BEHIND RESISTANCE PERSONNEL: SELECTION FOR LONG-RANGE TASKS

Z-17. For the long-range strategic tasks, however, high-level personnel are needed who can report on political, economics, and industrial data, as well as organize operations that impair the enemy's ability to wage war by extensive destruction of his basic resources and by inciting widespread popular resistance to his control. Selection of personnel for the long-range stay-behind program, however, poses more serious problems. The nature of the long-range tasks and the need for continuity of operation call for the highest degree of motivation, superior qualifications, and assurance of resilience. Qualifications required for a long-range stay-behind member must include considering—

- Final responsibility for the selection and training of the members of his team. Spotting of potential assets for his team may be carried on by others prior to the stay-behind situation; in fact, whenever this is possible it should be done.
- Responsibility for details of organization of his team (for example, proper division of work, delegation of authority, communications to be established and relationships between members of his team which these communications channels create).
- Responsibility for procurement of operational data that the team must have to plan operations and accomplish its tasks.
- Responsibility for detailed operational planning prior to undertaking the tasks assigned to his team.
- Responsibility for directing operations of his team in accordance with the operational plan agreed upon or directed by the higher echelons to which he is subordinate. Team leaders should possess the judgment and objectivity to select subordinates, knowledge and ability to train the team the qualities of leadership to command loyalty, and the capability to organize and direct the team in carefully planned and managed operations

## SECURITY

Z-18. Security of a clandestine organization must be one of its prime considerations. Each team member is responsible for the security of the team. Three functions are especially important:

- Security factors must be constantly reviewed and reevaluated which affect the team and its operations.
- Proper evaluation of the operational risks must be undertaken.
- A means for testing and checking security factors of the team's operations must be established.

Z-19. Security factors of stay-behind operations include all those that are necessary in any clandestine operation, but again, there are added considerations imposed by the nature of the stay-behind situation itself. The resilience of the personnel will, in many cases, depend upon the manner in which they are selected and trained. If compromised by lack of cover, the person or persons will almost certainly be lost as an asset when the area comes under hostile control. Stay-behind personnel must be willing and able to contribute to building up their own security and to demonstrating an appreciation for the security problems connected with operational tasks. Only when all security requirements have been met satisfactorily can they be permitted to undertake any activity.

Z-20. In view of the above, it is imperative that investigation and evaluation of each member be as extensive as possible before acceptance. After selection and employment of stay-behind personnel, there must be a systematic and extensive training program conducted under secure conditions. The training program should not only provide him with the knowledge and skills required for the performance of his tasks but also afford an opportunity for detailed assessment and evaluation of the member. Unlike other types of operations, the nature of the stay-behind operation does not permit the gradual acquisition of knowledge about personnel during the course of the operation nor does it permit the gradual development of the person's capabilities under operating conditions. Intensive assessment and evaluation during the training period must therefore replace these normal processes. This should be done whenever possible through constant observation.

Z-21. One critical problem that must be faced is that all personnel, who are known members of the Service concerned with stay-behind operations and who have intimate knowledge of stay-behind plans and personnel, must be evacuated prior to enemy occupation. This category includes planners, organizers, and trainers.

Obviously, the number of such officers should be held to an absolute minimum. The HQ that organizes, trains, and controls a stay-behind unit will not remain behind when the area is occupied.

## LEADERSHIP AND TRAINING FOR STAY-BEHIND RESISTANCE

Z-22. To attract operatives of high quality for stay-behind cadre or organizations demands personnel of a high caliber. The commander of a stay-behind unit must possess all the qualities he seeks in stay-behind personnel, and he must also possess them to the highest degree. In addition, he must be able to command respect and inspire confidence and loyalty in the personnel with whom he deals. His age and experience are not as important as his ability to advise personnel at all levels and his unquestioned motivation. If the commander is to select people for serious work under severe, operating conditions, he must reflect in his person a serious, professional, and competent organization. He must demonstrate a thorough knowledge and understanding of the problems and spirit of resistance, professional skill in the techniques of intelligence, and real respect and humility in dealing with people.

Z-23. Stay-behind operations embrace all fields of clandestine operations, but no single individual can be a specialist in all fields. However, all must have a working proficiency in the following subjects:

- Security methods, both personal and those affecting the clandestine organization.
- Means of establishing and maintaining secure communications links.
- Techniques of cover and concealment for his activity.
- Probable enemy control methods and evasive measures that he probably will need to employ to avoid detection.

Z-24. The type and amount of training to be given to each member will depend on operational requirements, to the extent that they can be determined in advance, and on the individual's aptitudes and ability to absorb training. As the operational responsibilities of the individual increase, his need to be trained in special techniques also increases. For example, an individual earmarked to organize and lead a sabotage operation will require more specialized training than one selected for information reporting only. Each stay-behind member must have thorough training in the principles of clandestine activity previously mentioned. In addition, he must also receive intensive instruction in communications techniques, including operation and maintenance of radio equipment, ciphers and codes, danger signals, and so on.

Z-25. The list of subjects to be taught in a stay-behind training program may be expanded to include almost any subject pertaining to the various fields of clandestine activity. The individual's training needs are determined by the degree of knowledge and skill required for the specific activity in which he is to take part. Providing the necessary training for stay-behind personnel usually entails several administrative problems for the clandestine organization. It must provide competent instructors and suitable training facilities, establish cover for the training and for the trainee, and maintain the security of the training program from start to finish. In order for training to be effective, it must be led by competent, professional instructors. The number of personnel to be trained in a given area and their training requirements will determine what form the program will take. In some areas it may be desirable and feasible to conduct formal courses for a number of trainees. Thorough training and daily practice in clandestine principles and techniques by all concerned, in advance of the stay-behind situation, are the best insurance that stay-behind personnel will survive and be able to operate and train to operate under grueling conditions.

# STAY-BEHIND UNCONVENTIONAL WARFARE

Z-26. External sponsors who provide support to stay-behind resistance are conducting stay-behind UW. SF Soldiers and operational detachments may be pre-positioned in the friendly state prior to a foreign invasion or may infiltrate the postinvasion, occupied denied area afterwards. Likewise, SF detachments may be infiltrated either before or after the indigenous stay-behind resistance organization has formed and become effective. SF Soldiers are likely to advise and assist the stay-behind resistance organization in a manner similar to other types of UW; however, they will do so with the advantage of preplanned organization, personnel, resources, and targeting. SF stay-behind UW expertise in regional orientation and familiarity, language ability, tactical and operational proficiency and strategic acumen, and advisor skills is gained and substantially developed prior to the onset of open hostilities.

---

### Stay-Behind Resistance Operations/Stay-Behind
### Unconventional Warfare

#### World War II British "Auxiliary Units of GHQ:"

In March 1938, shortly after Hitler's annexation of Austria, a new department was created in MI6, labelled Section D, with the task to develop subversive operations in Europe. Section D began to establish 'stay-behind' sabotage parties in countries threatened by German invasion. When in 1940 the German invasion of southern England seemed imminent, . . . . the first British Gladio units received special training and were instructed to 'stay-behind' enemy lines in case of a German invasion of the island. Operating from secret hideouts and arms caches, they would be able to carry out sabotage and guerrilla warfare against the German invaders.

#### Cold War era "Operation Gladio:"

Conceived as a network of clandestine resistance within NATO countries to confront an eventual Soviet invasion. Composed of secret armies equipped by the CIA and the MI6 with machine guns, explosives, munitions and high-tech communication equipment hidden in arms caches in forests, meadows and underground bunkers across Western Europe. Leading officers of the secret network trained together with the U.S. Green Berets Special Forces in the United States of America and the British SAS Special Forces in England. Equipment was buried in caches in forests, meadows and even churches and cemeteries. Caches included portable arms, ammunition, explosives, hand grenades, knives and daggers, 60mm mortars, several 57mm recoilless rifles, sniper rifles, radios, binoculars and various tools.

# Source Notes

This section lists sources by page number.

2-1      "Man is born free, and everywhere...": Jean Jacques Rousseau, *The Social Contract*, 1762

2-7      **The Three Stages of the Protracted War** "...[I]t can reasonably be assumed...": Mao Tse-Tung, *On Protracted War*, 1938. https://archive.org/details/MaoTse-tungOnGuerrillaWarfare

2-7      "Mao conceived this type of war..." General Samuel B. Griffith, *Strategy, Tactics, and Logistics in Revolutionary War*, Praeger Publishers, 1961. Translation of *Introduction to On Guerrilla War* (Mao Tse-Tung), 1937, pp. 20–21.

2-8      "In the next phase, direct action assumes...": General Samuel B. Griffith, *Strategy, Tactics, and Logistics in Revolutionary War*, Praeger Publishers, 1961. Translation of *Introduction to On Guerrilla War* (Mao Tse-Tung), 1937, p. 21.

2-9      "Following Phase I (organization, consolidation...": General Samuel B. Griffith, *Strategy, Tactics, and Logistics in Revolutionary War*, Praeger Publishers, 1961. Translation of *Introduction to On Guerrilla War* (Mao Tse-Tung), 1937, pp. 20–21.

2-29      Figure 2-10. Organization of the French resistance. Will Irwin, *Historical Case Study: The Jedburghs and Unconventional Warfare in World War II*, Unconventional Warfare Symposium, Joint Base Lewis, McCord, WA, 25–26 September 2014.

2-33      "The Chinese Communist Party is a party leading...": Mao Tse-Tung, *The Role of the Chinese Communist Party in the National War*, Selected Works, Vol. II, October 1938, p. 202.

3-1      "He that would make his own...": Thomas Paine, *Dissertation on First Principles of Government*, 23 December 1791.

5-1      "The shifting from political struggle...": Vo Nguyen Giap, *Peoples' War, Peoples' Army*, Praeger, 1962.

5-4      "There is no doubt that most partisan action...": Otto Heilbrunn, *Partisan Warfare*, 1962.

5-13      "Arranging for and organizing the speedy...": Vladimir Ilyich Lenin, *What is to be Done?* Collected Works, Vol 5, 1961.

5-26      "The strategy is one to ten...": Li Tso-Peng, Commentary on Mao's *On Protracted War*.

6-1      "The whole is greater than the...": Aristotle, *Aristotle's Metaphysics*. Oxford [England]: Clarendon Press, 1981.

6-1      "In guerrilla warfare, more than...": CIA Tayacan, *Psychological Operations in Guerrilla Warfare*, 18 October 1984, pp. 7–8.

E-3      Figure E-2. Methods of nonviolent protest and persuasion. Gene Sharp, *The Politics of Nonviolent Action* (Boston: Porter Sargent, 1973), reproduced by The Albert Einstein Institution. https://www.aeinstein.org/nonviolentaction/198-methods-of-nonviolent-action/.

F-5      Figure F-7. Central Intelligence Agency model—life cycle of insurgency. USASOC, *Understanding Stages of Resistance, Assessing Revolutionary and Insurgent Strategies*, September 2016.

F-9      Figure F-11. Continuum from legal protests to insurgency and belligerency. USASOC, *Understanding Stages of Resistance, Assessing Revolutionary and Insurgent Strategies*, September 2016.

L-1      "Organization of resistance movements...": Colonel Aaron Bank, *NATO's Secret Armies: Operation GLADIO and Terrorism in Western Europe* (Daniele Ganser), 2005.

| | |
|---|---|
| L-2 | "The Huks in the Philippines depended...": Luis Taruc, *Born of the People*, p. 121 Colonel Napoleon Valeriano; Speech at the Counterinsurgency Officer's Course, Special Warfare School, Fort Bragg, NC, 5 November 1963. |
| L-3 | Figure L-3. Security zone participant examples. H. von Dach, *Total Resistance*, 1965. |
| L-4 | **Belgian Underground** "The Belgian undergrounder was instructed...": George K. Tanham, *The Belgian Underground Movement, 1940-1944*, p. 226, E. Bramstedt, Dictatorship and Political Police, pp. 213–24 |
| L-5 | "The induction of a railway worker...": Czeslaw Stankiewicz, *Polish Underground Veteran*, Interview with the Author, 4–5 December 1993. |
| M-1 | "Marion's logistical support, more often...": John Fiske, *The American Revolution, Volume II*, p. 188. |
| M-1 | **The Mujahideen Fighter** "In practical terms the Mujahedeen can...": Mohammed Yousaf, *The Bear Trap: Afghanistan's Untold Story*, September, 1992. |
| M-7 | **Indo-China** ".... 38,000 grenades, 30,000 rifle cartridges...": George K. Tanham, *Communist Revolutionary Warfare*, p. 68, Published in *Guerrilla Logistics*, U.S. Army War College (Caraccia, Marco J.), 8 April 1966. |
| M-7 | **The Ayalon Institute—Palestine** "The Ayalon Institute was a dangerous...": Risa Borsykowsky, The Ayalon Institute: Kibbutzim Hill Rehovot, https://www.jewishgiftplace.com/Ayalon-Institute.html |
| M-8 | "Each team parachuted with the standard...": S.J. Lewis, *Jedburgh Team Operations in Support of the 12th Army Group*, August 1944. |
| M-21 | Figure M-13. Cheng's special operations forces logistician's model for nonstandard logistics. COL Steve Cheng, *Navigating Ambiguity* (unpublished essay), October 2016. |
| Q-2 | "Equal importance should be attached...": Mao Tse-Tung, *The Tasks for 1945*, 15 December 1944. |
| Q-3 | "As for the training courses, the main...": Mao Tse-Tung, *Policy for the Work in the Liberated Areas for 1946*, Selected Works, Vol. IV, p. 76, 15 December 1945. |
| V-1 | "We ....strike to win, strike only...": Vo Nguyen Giap, *The Military Art of People's War*, Monthly Review Press, 1 August 1970. |
| V-18 | **World War II Broadcast to Russian Soldiers and Civilian Caught Behind German Enemy Lines** "In areas supported by the enemy...": Joseph V. Stalin, Radio Broadcast, July 1941; Bert "Yank" Levy, Guerrilla Warfare, 1964, p. 72. |
| V-28 | "Afghanistan is not Europe, yet the...": *Red Thrust Star*, July and October 1995 and October 1996. |
| W-1 | "The urban guerrilla must know how...": Carlos Marighella, *Mini-Manual of the Urban Guerrilla*, 1969. |
| W-22 | Figure W-6. Megacity characteristics. Chief of Staff of the Army, Strategic Studies Group, *Megacities and the United States Army: Preparing for a Complex and Uncertain Future*, June 2014. |
| W-26 | Figure W-7. Megacity integration levels. Chief of Staff of the Army, Strategic Studies Group, *Megacities and the United States Army: Preparing for a Complex and Uncertain Future*, June 2014. |

# Glossary

| | |
|---|---|
| ADP | Army doctrine publication |
| ADRP | Army doctrine reference publication |
| AO | area of operations |
| AR | Army regulation |
| ARSOF | Army special operations forces |
| ASCOPE | areas, structures, capabilities, organizations, people, and events |
| ATP | Army techniques publication |
| CA | Civil Affairs |
| CARVER | criticality, accessibility, recuperability, vulnerability, effect, and recognizability |
| CMO | civil-military operations |
| COA | course of action |
| CONPLAN | concept plan |
| DA | Department of the Army |
| DIMEFIL | diplomatic, information, military, economic, financial, intelligence, and law enforcement |
| DOD | Department of Defense |
| DOS | Department of State |
| DSF | District Stability Framework (USAID) |
| DZ | drop zone |
| EPW | enemy prisoner of war |
| FM | field manual |
| FRP | final reference point |
| GCC | geographic combatant commander |
| GTA | graphic training aid |
| HQ | headquarters |
| HUMINT | human intelligence |
| IPB | intelligence preparation of the battlefield |
| IR | intelligence requirement |
| JFC | joint force commander |
| JIATF | joint interagency task force |
| JP | joint publication |
| JPADS | joint precision airdrop system |
| JSOA | joint special operations area |
| JSOTF | joint special operations task force |
| LOC | line of communications |
| LSS | low, slow, and small |
| LZ | landing zone |

| | |
|---|---|
| MANPADS | man-portable air defense system |
| METL | mission-essential task list |
| METT-TC | mission, enemy, terrain and weather, troops and support available–time available and civil considerations |
| MISO | military information support operations |
| MSS | mission support site |
| NGO | nongovernmental organization |
| ODA | operational detachment–alpha |
| ODB | operational detachment–bravo |
| OPLAN | operation plan |
| OPORD | operation order |
| OPSEC | operations security |
| OSS | Office of Strategic Services |
| PE | preparation of the environment |
| PIR | priority intelligence requirement |
| PMESII-PT | political, military, economic, social, information, infrastructure, physical environment, time |
| PSYOP | Psychological Operations |
| RFI | request for information |
| RP | rallying point |
| RT | radio-telephone |
| SCA | support to civil administration |
| SF | Special Forces |
| SOCCE | special operations command and control element |
| SOF | special operations forces |
| SOI | signal operating instructions |
| SOMPF | special operations mission planning folder |
| SOP | standard operating procedure |
| SOTF | special operations task force |
| ST | special text |
| STR | support to resistance |
| TASKORD | tasking order |
| TC | training circular |
| TIP | target intelligence packages |
| TSOC | theater special operations command |
| UAS | unmanned aircraft system |
| USAJFKSWCS | United States Army John F. Kennedy Special Warfare Center and School |
| USASOC | United States Army Special Operations Command |
| USC | United States Code |
| USG | United States Government |

| USSOCOM | United States Special Operations Command |
|---------|------------------------------------------|
| UW | unconventional warfare |

## SECTION II – TERMS

\* - indicates proponent term and definition

**airland**

Move by air and disembark, or unload, after the aircraft has landed or while an aircraft is hovering. (JP 3-17)

**area command**

In unconventional warfare, the irregular organizational structure established within an unconventional warfare operational area to command and control irregular forces advised by Army Special Forces. (ATP 3-05.1)

**\*area complex**

A clandestine, dispersed network of facilities to support resistance activities in a given area designed to achieve security, control, dispersion, and flexibility (ATP 3-18.1, proposed for inclusion in ADP 1-02)

**auxiliary**

For the purpose of unconventional warfare, the support element of the irregular organization whose organization and operations are clandestine in nature and whose members do not openly indicate their sympathy or involvement with the irregular movement (ADRP 3-05)

**guerrilla**

An irregular, predominantly indigenous member of a guerrilla force organized similar to military concepts and structure in order to conduct military and paramilitary operations in enemy-held, hostile, or denied territory. Although a guerrilla and guerrilla forces can exist independent of an insurgency, guerrillas normally operate in covert and overt resistance operations of an insurgency. (ATP 3-05.1)

**guerrilla force**

A group of irregular, predominantly indigenous personnel organized along military lines to conduct military and paramilitary operations in enemy-held, hostile, or denied territory (JP 3-05)

**\*guerrilla warfare**

Military and paramilitary operations conducted in enemy-held, hostile, or denied territory by irregular, predominantly indigenous, guerrilla forces to reduce the effectiveness, industrial capacity, and morale of the enemy. (ATP 3-18.1, proposed for inclusion in ADP 1-02)

**intelligence operations**

The tasks undertaken by military intelligence units through the intelligence disciplines to obtain information to satisfy validated requirements. (ADP 2-0)

**intelligence preparation of the battlefield**

The systematic process of analyzing the mission variables of enemy, terrain, weather, and civil considerations in an area of interest to determine their effect on operations. Also called **IPB**. (ATP 2-01.3)

**pilot team**

A deliberately structured composite organization comprised of Special Forces operational detachment members, with likely augmentation by interagency or other skilled personnel, designed to infiltrate a designated area to conduct sensitive preparation of the environment activities and assess the potential to conduct unconventional warfare in support of U.S. objectives (ATP 3-05.1)

**preparation of the environment**

An umbrella term for operations and activities conducted by selectively trained special operations forces to develop an environment for potential future special operations. Also called **PE** (JP 3-05)

**psychological action**

Lethal and nonlethal actions planned, coordinated, and conducted to produce a psychological effect in a foreign individual, group, or population. (FM 3-53)

**\*public component**

An overt political manifestation of a resistance. (ATP 3-18.1, proposed for inclusion in ADP 1-02)

**raid**

An operation to temporarily seize an area in order to secure information, confuse an enemy, capture personnel or equipment, or to destroy a capability culminating with a planned withdrawal. (JP 3-0)

**\*resistance area command**

The largest territorial resistance organization commanded by a senior resistance leader inside a defined resistance area of operations. (ATP 3-18.1, proposed for inclusion in ADP 1-02)

**\*resistance partner**

A partner conducting resistance with whom the United States Government mutually establishes agreements to cooperate for some specified time in pursuit of mutually supporting specific objectives. A resistance partner is influenced; he is not an employee or subordinate to be commanded and controlled. (ATP 3-18.1, proposed for inclusion in ADP 1-02)

**\*safe area**

A designated area in hostile or denied territory that offers a reasonable chance for a resistance organization to conduct clandestine activities without compromise. (ATP 3-18.1, proposed for inclusion in ADP 1-02)

**shadow government**

Governmental elements and activities performed by the irregular organization that will eventually take the place of the existing government. Members of the shadow government can be in any element of the irregular organization (underground, auxiliary, or guerrilla force). (ATP 3-05.1)

**stay-behind operation**

An operation in which the commander leaves a unit in position to conduct a specified mission while the remainder of the forces withdraw or retire from an area. (FM 3-90-1)

**\*stay-behind resistance operation**

An operation in which indigenous authorities leave personnel and resources in position before, during, and after a foreign occupation to conduct anticipated resistance activities against the occupying power. (ATP 3-18.1, proposed for inclusion in ADP 1-02)

**\*stay-behind unconventional warfare**

Unconventional warfare operations and activities conducted by a sponsoring government in support of indigenous authorities' stay-behind resistance operations. (ATP 3-18.1, proposed for inclusion in ADP 1-02)

**subversion**

Actions designed to undermine the military, economic, psychological, or political strength or morale of a governing authority. (JP 3-24)

**\*subversive political action**

A planned series of activities designed to accomplish political objectives by influencing, dominating, or displacing individuals or groups who are so placed as to affect the decisions and actions of another government. (ATP 3-18.1, proposed for inclusion in ADP 1-02)

**\*support to resistance**

A United States Government policy option to support foreign resistance actors that offers an alternative to a direct U.S. military intervention or formal political engagement in a conflict. Also called **STR**. (ATP 3-18.1, proposed for inclusion in ADP 1-02)

**\*surrogate**

Someone who acts on behalf of another. A surrogate is an employee or subordinate that an employer commands and controls and for whose actions the employer bears some legal and moral responsibility. (ATP 3-18.1, proposed for inclusion in ADP 1-02)

**unconventional warfare**

Activities conducted to enable a resistance movement or insurgency to coerce, disrupt, or overthrow a government or occupying power by operating through or with an underground, auxiliary, and guerrilla force in a denied area. Also called UW. (JP 3-05 1)

**underground**

A cellular covert element within unconventional warfare that is compartmentalized and conducts covert or clandestine activities in areas normally denied to the auxiliary and the guerrilla force. (ADRP 3-05)

This page intentionally left blank.

# References

All URLs accessed on 20 February 2019.

## REQUIRED PUBLICATIONS

These documents must be available to intended users of this publication.

*DOD Dictionary of Military and Associated Terms*, as of February 2019.

ADP 1-02, *Terms and Military Symbols*, 14 August 2018.

## RELATED PUBLICATIONS

These documents contain relevant supplemental information.

### JOINT PUBLICATIONS

Most joint publications are available online: http://www.jcs.mil/doctrine.

JP 1, *Doctrine for the Armed Forces of the United States*, 25 March 2013.

JP 3-0, *Joint Operations*, 17 January 2017.

JP 3-05, *Special Operations*, 16 July 2014.

JP 3-05.1, *Unconventional Warfare*, 15 September 2015.

JP 3-17, *Air Mobility Operations*, 5 February 2019.

JP 3-24, *Counterinsurgency*, 25 April 2018.

JP 5-0, *Joint Planning*, 16 June 2017.

### DEPARTMENT OF THE ARMY PUBLICATIONS

Most Army doctrinal publications are available online: https://armypubs.army.mil.

ADP 1-01, *Doctrine Primer*, 2 September 2014.

ADP 2-0, *Intelligence*, 4 September 2018.

ADP 3-05, *Special Operations*, 29 January 2018.

ADP 5-0, *The Operations Process*, 17 May 2012.

ADP 7-0, *Training*, 29 August 2018.

ADRP 3-05, *Special Operations*, 29 January 2018.

ADRP 6-22, *Army Leadership*, 1 August 2012.

AR 40-562/BUMEDINST 6230.15B/AFI 48-110_IP/CG COMDTINST M6230.G, *Immunizations and Chemoprophylaxis for the Prevention of Infectious Diseases*, 7 October 2013.

AR 380-381, *Special Access Programs (SAPS) and Sensitive Activities*, 21 April 2004.

ATP 2-01.3, *Intelligence Preparation of the Battlefield*, 1 March 2019.

ATP 2-33.4, *Intelligence Analysis*, 18 August 2014.

ATP 3-01.81, *Counter-Unmanned Aircraft System Techniques*, 13 April 2017.

ATP 3-05.1, *Unconventional Warfare*, 6 September 2013.

ATP 3-05.40, *Special Operations Sustainment*, 3 May 2013.

ATP 3-05.71, *(U) Army Special Operations Forces Resistance and Escape (C)*, 26 February 2014.

ATP 3-06, *Urban Operations*, 7 December 2017.

ATP 3-07.5, *Stability Techniques*, 31 August 2012.

ATP 3-18.10, *Special Forces Air Operations*, 24 February 2016.

ATP 3-18.11/AFMAN 11-411(I)/NTTP 3-05.26M, *Special Forces Military Free-Fall Operations*, 24 October 2014.

ATP 3-18.12, *Special Forces Waterborne Operations*, 14 July 2016.

ATP 3-18.16, *(U) Preparation of the Environment (S//NF)*, 27 February 2018.

ATP 3-18.20, *(U) Advanced Special Operations Techniques (S//NF)*, 19 February 2016.

ATP 3-18.72, *(U) Special Forces Personnel Recovery (S//NF)*, 13 January 2016.

ATP 3-21.8, *Infantry Platoon and Squad*, 12 April 2016.

ATP 3-34.20, *Countering Explosive Hazards*, 21 January 2016.

ATP 3-90.8, *Combined Arms Countermobility Operations*, 17 September 2014.

ATP 4-48, *Aerial Delivery*, 21 December 2016.

ATP 5-19, *Risk Management*, 14 April 2014.

FM 3-04, *Army Aviation*, 29 July 2015.

FM 3-05, *Army Special Operations*, 9 January 2014.

FM 3-05.70, *Survival*, 17 May 2002.

FM 3-07, *Stability*, 2 June 2014.

FM 3-18, *Special Forces Operations*, 28 May 2014.

FM 3-24, *Insurgencies and Countering Insurgencies*, 13 May 2014.

FM 3-53, *Military Information Support Operations*, 4 January 2013.

FM 3-90-1, *Offense and Defense, Volume 1*, 22 March 2013.

FM 3-90-2, *Reconnaissance, Security, and Tactical Enabling Tasks, Volume 2*, 22 March 2013.

FM 6-0, *Commander and Staff Organization and Operations*, 5 May 2014.

FM 27-10, *The Law of Land Warfare*, 18 July 1956.

GTA 31-01-003, *Detachment Mission Planning Guide*, 30 July 2012.

GTA 41-01-005, *Religious Factors Analysis*, 27 February 2015.

TC 3-21.76, *Ranger Handbook*, 26 April 2017.

TC 7-100.3, *Irregular Opposing Forces*, 17 January 2014.

TC 18-01.1, *Unconventional Warfare Mission Planning Guide for the Special Forces Operational Detachment—Alpha Level*, 27 October 2016.

TC 18-01.2, *Unconventional Warfare Mission Planning Guide for the Special Forces Operational Detachment—Bravo Level*, 22 November 2016.

TC 18-01.3, *Unconventional Warfare Mission Planning Guide for the Special Forces Operational Detachment—Charlie Level*, 22 November 2016.

TC 18-02, *Special Forces Advisor Guide*, 31 August 2016.

TC 18-11, *Special Forces Military Free-Fall and Double-Bag Static Line Operations*, 30 April 2015.

TC 18-32, *Special Forces Sniper Training and Employment*, 10 August 2017.

## OBSOLETE PUBLICATIONS

This section contains references to obsolete historical doctrine. The Archival and Special Collections in the Combined Arms Research Library on Fort Leavenworth in Kansas contains copies. These publications are obsolete doctrine publications referenced for citations only.

FM, *Basic Field Manual: Strategic Services (Provisional)*, 1 December 1943 (obsolete).

FM 2, *Morale Operations Field Manual: Strategic Services (Provisional)*, 26 January 1943 (obsolete).

FM 3-05.201, *Special Forces Unconventional Warfare Operations*, 30 April 2003 (obsolete).

FM 4, *Special Operations Field Manual: Strategic Services (Provisional)*, 23 February 1944 (obsolete).

FM 5, *Secret Intelligence Field Manual: Strategic Services (Provisional)*, 22 March 1944 (obsolete).

FM 6, *Operational Groups Field Manual: Strategic Services (Provisional)*, 25 April 1944 (obsolete).

FM 31-20, *Special Forces Operational Techniques*, 12 February 1971 (obsolete).

FM 31-20, *Doctrine for Special Forces Operations*, 20 April 1990 (obsolete).

FM 31-20-2, *Unconventional Warfare Tactics, Techniques, and Procedures for Special Forces (ID)*, 7 December 1998 (obsolete).

FM 31-21, *Organization and Conduct of Guerrilla Warfare*, 5 October 1951 (obsolete).

FM 31-21, *Guerrilla Warfare and Special Forces Operations*, 8 May 1958 (obsolete).

FM 31-21, *Guerrilla Warfare and Special Forces Operations*, 29 September 1961 (obsolete).

FM 31-21, *Special Forces Operations—U.S. Army Doctrine*, 5 June 1965 (obsolete).

FM 34-3, *Intelligence Analysis*, 15 March 1990 (obsolete).

FM 100-20, *Military Operations in Low Intensity Conflict*, 5 December 1990 (obsolete).

ST 31-201, *Special Forces Operations*, November 1978 (obsolete).

## UNITED STATES LAW

Unless noted otherwise, most acts and public laws are available online: http://uscode.house.gov.

Article 15, Uniform Code of Military Justice, http://www.ucmj.us/

DOD Law of War Manual, https://dod.defense.gov/Portals/1/Documents/pubs/DoD%20Law%20of%20War%20Manual%20-%20June%202015%20Updated%20Dec%202016.pdf?ver=2016-12-13-172036-190

Executive Order 12333, *United States Intelligence Activities*. https://www.archives.gov/federal-register/codification/executive-order/12333.html

Executive Order 13284, *Amendment of Executive Orders, and Other Actions, in Connection with the Establishment of Homeland Security*. https://www.hsdl.org/?abstract&did=623

Executive Order 13355, *Strengthened Management of the Intelligence Community*. https://www.hsdl.org/?abstract&did=449323

Executive Order 13470, *Further Amendments to Executive Order 12333, United States Intelligence Activities*. https://www.hsdl.org/?abstract&did=487886

Geneva Conventions, http://www.loc.gov/rr/frd/Military_Law/pdf/GC_1949-I.pdf

Hague Conventions, https://www.hcch.net/en/instruments/conventions

Title 10, United States Code, *Armed Forces*.
Section 167, *Unified Combatant Command for Special Operations Forces*.

Title 50, United States Code, *War and National Defense*.
Section 3093, *Presidential Approval and Reporting of Covert Actions*.

## OTHER PUBLICATIONS

ACP-125(G), *Communications Instructions-Radiotelephone Procedures*, 28 November 2016. http://www.k1chr.org/ACP%20125%20%28G%29%20Radio%20Telephone%20Procedures%20NOV%202016.pdf

Al-Zawahiri, Ayman, *Knights Under the Prophet's Banner*, 2001.

Aristotle and W.D. Ross, *Aristotle's Metaphysics*, Oxford [England]: Clarendon Press, 1981.

Azzam, Abdullah, *The Defence of the Muslim Lands*, 1 February 2002.

Borsykowsky, Risa, The Ayalon Institute: Kibbutzim Hill Rehovot, https://www.jewishgiftplace.com/Ayalon-Institute.html

Caraccia, Marco J., *Guerrilla Logistics*, U.S. Army War College, 8 April 1966.

Cheng, Steve, Navigating Ambiguity: The Contentious World of Nonstandard Logistics (Unpublished Essay), 2016.

Chief of Staff of the Army, Strategic Studies Group, *Megacities and the United States Army: Preparing for a Complex and Uncertain Future*, June 2014.

Combat Aviation Brigade, *Army Aviation Handbook* (FOUO), 20 April 2018. (A copy may be requested by contacting the U.S. Army Aviation Center of Excellence at this website: https://www.rucker.army.mil/info/contact/index.html.)

Dach, Von H., *Total Resistance*, Paladin Press, 1965.

Fiske, John, *The American Revolution, Volume II*.

Ganser, Daniele, *NATO's Secret Armies: Operation GLADIO and Terrorism in Western Europe*, 2005.

Giap, Vo Nguyen, *Peoples' War, Peoples' Army*, Praeger Publishers, 1962.

Giap, Vo Nguyen, *The Military Art of People's War*, Monthly Review Press, 1970.

Griffith, Samuel B., *Strategy, Tactics, and Logistics in Revolutionary War*, Translation of Mao Tse-Tung's Introduction to On Guerrilla Warfare, Praeger Publishers, 1961.

Guevara, Ernest (Che), *Guerrilla Warfare: A Method*.

Heilbrunn, Otto, *Partisan Warfare*, 1962.

Irwin, Will, *Historical Case Study: The Jedburghs and Unconventional Warfare in World War II*, Unconventional Warfare Symposium, Joint Base Lewis, McCord, WA, 25–26 September 2014

Lawrence, T.E., *The Seven Pillars of Wisdom*, 1926.

Lenin, Vladimir Ilyich, *What is to be Done?* Collected Works, Vol. 5. 1961.

Levy, "Yank" Bert, *Guerrilla Warfare*, 1964.

Lewis, S.J., *Jedburgh Team Operations in Support of the 12th Army Group*, 1944.

Malik, S.K., *The Quranic Concept of Power*

Marighella, Carlos, *Mini-Manual of the Urban Guerrilla*, 1969.

*Merriam-Webster Online Dictionary*, Retrieved from https://www.merriam-webster.com/ on 22 March 2018.

Paine, Thomas, *Dissertation on First Principles of Government*, 1791.

Qutb, Sayyid, *Milestones*, 1964.

Rousseau, Jean Jacques, *The Social Contract*, 1762.

SAV SER SUP 7, *Special Operations Forces Signal Operating Instructions Supplemental Instructions*, no date.

Sharp, Gene, *The Politics of Nonviolent Action*, Boston: Porter Sargent, 1973.

Stankiewicz, Czeslaw, *Polish Underground Veteran*. Interview with the Author, 1993.

Tanham, George K., *The Belgian Underground Movement, 1941–1944*.

Tanham, George K., *Communist Revolutionary Warfare*, 1968.

Taruc, Luis, *Born of the People*. Colonel Napoleon Valeriano: Speech at the Counterinsurgency Officer's Course, Special Warfare School, Fort Bragg, NC, 1963.

Tayacan, CIA, *Psychological Operations in Guerrilla Warfare*, 18 October 1984.

Tse-Tung, Mao, *On Protracted War*, 1938.

Tse-Tung, Mao, *Policy for the Work in the Liberated Areas for 1946*, Selected Works, Vol. IV. Red Thrust Star, 1995 and 1996.

Tse-Tung, Mao, *The Role of the Chinese Communist Party in the National War*, Selected Works, Vol. II, 1938.

Tse-Tung, Mao, *The Tasks for 1945*, 1944.

Tso-Peng, Li, Commentary on Mao's *On Protracted War*

U.S. Department of State, *U.S. Landmine Policy*, https://www.state.gov/t/pm/wra/c11735.htm, September 2014.

USAJFKSWCS, *Psychological Operations Commandant, Primer–Military Information Support Operations Support in Unconventional Warfare*, 28 May 2015.

USASOC, *Understanding Stages of Resistance, Assessing Revolutionary and Insurgent Strategies*, September 2016.

USSOCOM Directive 525-5, *(U) Advanced Special Operations Techniques (S//NF)*, 14 November 2013.

USSOCOM Directive 525-16, *(U) Preparation of the Environment (S//NF)*, 14 November 2013.

USSOCOM Directive 525-89, *(U) Unconventional Warfare (S//NF)*, 31 May 2012.

USSOCOM Publication, *U.S. Government Support to Resistance Framework (DRAFT) Version 0.33*, 11 January 2017.

USSOCOM Publication 4-0, *Logistics Support to Special Operations Forces*, 6 February 2013.

> (*Note:* For organizations that require USAJFKSWCS, USASOC, or USSOCOM publications listed above, please send a request to Commander, U.S. Army Special Operations Center of Excellence, USAJFKSWCS, ATTN: AOJK-SFD, 3004 Ardennes Street, Stop A, Fort Bragg, NC 23810-9610.)

Yousaf, Mohammed, *The Bear Trap: Afghanistan's Untold Story*, 1992.

# PRESCRIBED FORMS

This section contains no entries.

# REFERENCED FORMS

## DEPARTMENT OF THE ARMY FORMS

DA forms are available at the Army Publishing Directorate website: https://armypubs.army.mil.

DA Form 2028, *Recommended Changes to Publications and Blank Forms.*

## DEPARTMENT OF DEFENSE FORMS

DD forms are available at the Executive Services Directorate website: http://www.esd.whs.mil/DD/.

DD Form 3007, *Hasty Protective Row Minefield Record.*

This page intentionally left blank.

# Index

Entries are by paragraph number.

## A

administrative considerations, 4-68, T-1

area assessment, 3-41, 4-69, 4-94, 4-100, 4-101, 4-111, L-34, M-20, V-11, W-66, X-2, X-3, Y-18, Y-21, Y-27, Y-31, Y-36, Y-41

area command, 2-71, 2-92, 2-93, 2-94, 4-24, 4-105, 4-121, 5-6, 5-141, 5-150, 5-211, M-4, O-1, O-6, O-12, O-23, Q-5, Q-8, Q-11, Q-19, S-5, S-37, T-2, T-5, T-13, T-14, T-15, V-16, V-17, V-18, V-19, V-20, V-24, V-25, V-36, V-37, V-41, V-46, V-59, V-64, V-70, V-71, V-76, V-82, V-220, X-25, X-31

area complex, 2-42, 2-52, 2-55, 2-87, 2-105, 2-106, 4-120, 4-121, 5-23, 5-33, 6-52, M-5, S-1 through S-3, S-5 through S-7, S-11, S-24, S-25, S-33, S-36, S-45, T-2, V-22

area study, 4-34, 4-111, H-1, J-2, V-99, W-66, X-1, X-3

auxiliary, 2-39, 2-40, 2-41, 2-43, 2-44, 2-46, 2-49, 2-52, 2-54, 2-55, 2-68 through 2-70, 2-73 through 2-82, 2-90, 2-96, 2-106, 3-9, 3-30, 4-120, 5-139, 5-146, 5-162, 5-168, 5-182, 5-184, 5-198, 5-201, 5-218, F-3, F-5, L-3, L-10, L-34, Q-19, S-20, S-22, S-35, S-40, S-42, T-4, V-12, V-25, V-26 through V-28, V-34, V-40, V-46, V-47, V-50, V-64, V-66 through V-69, V-82, V-84, V-121, V-256, X-33, X-40, Y-18 through Y-20, Y-37

## B

base population, 2-42, 2-87, 2-108

## C

caching, N-1, N-3, N-5 through N-7, N-12, N-45, N-60, N-61, N-73, N-75, N-80, N-83, N-84, N-86, N-87, N-98, N-109, N-116, N-117, N-118,

N-120, N-121, N-123, N-125, T-6, V-138, W-34

Civil Affairs Operations, 3-29, 5-220, 6-19

coerce, 1-21 through 1-23, 1-28, 1-29, 3-4, 3-9, 3-46, 3-58, 4-123, 4-128, 5-100, C-114, C-155, C-164, P-10, W-72, W-152

combat employment of guerrilla forces, 4-70, 4-126, V-72

communications, 1-24, 1-25, 2-20, 2-50, 2-69, 2-75, 2-105, 2-106, 3-13, 3-14, 3-23, 3-30, 3-50, 4-12, 4-19, 4-36, 4-40, 4-61, 4-71, 4-73, 4-96, 5-11, 5-26, 5-38, 5-51, 5-55, 5-68, 5-109, 5-110, 5-118, 5-135, 5-140, 5-142, 5-144, 5-175 through 5-178, 5-184, 5-191, 5-193, 5-194, 5-196, 5-197, 5-206, 5-209, 5-212, 5-216, 5-218, C-16, C-18, C-23, C-34, C-36, C-39, C-40, C-50, C-125, C-127 through C-129, C-140, C-143, F-2, K-3, K-13, K-16, L-1 through L-4, L-8, L-13 through L-19, L-21 through L-32, L-34 through L-36, M-3, M-5, M-13, M-15 through M-18, M-21, N-54, N-119, O-2, O-5, O-16, O-21, O-23, O-26, Q-13, Q-14, S-2, S-7, S-14, S-22, S-31, S-34, S-47, V-11, V-20, V-22, V-26, V-31, V-38, V-51 through V-53, V-59, V-72, V-81, V-90, V-123, V-126, V-129, V-136, V-158, V-169, V-177, V-239, V-251, W-5, W-21, W-33, W-38, W-61, W-62, W-74, W-75, W-77 through W-80, W-83, W-84, W-87, W-99, W-116, W-140, W-166, X-31, X-41, X-43, X-44, Y-37 through Y-40, Z-9, Z-10, Z-12, Z-14, Z-17, Z-23, Z-24

conventional forces support, 6-31, 6-32

critical cadres, 2-101, 2-103, 4-121

## D

disrupt, 1-1, 1-21, 1-22, 1-24, 1-25, 1-28, 3-4, 3-9, 3-46, 3-58, 4-123, 4-128, 5-100, 5-110, 5-163, 5-173, 5-174, 5-178, 5-197, 6-32, C-164, D-6, S-26, V-55, V-118, V-219, W-72, W-99, W-120, Z-2, Z-6

## F

feasibility assessment, 1-18, 3-17, 3-24, 3-27, 3-28, 3-32, 3-33, 4-9 through 4-14, 4-17, 4-19, 4-83, 4-86, 4-90, 4-130, J-1, J-2

## G

government-in-exile, 2-23, 2-34, 2-41, 2-83, 2-87, 2-113, 2-120 through 2-122, 3-12, 3-30, 4-118, 5-213, 6-22, M-22, M-27, T-15

guerrilla force(s), 2-39, 2-44, 2-46, 2-48, 2-58, 2-59, 2-65, 2-68, 2-69, 2-81, 3-9, 4-126, 4-127, 5-31, 5-33, 5-132, 5-139 through 5-142, 5-146, 5-150, 5-160, 5-177, 5-185, 5-186, 5-188, 5-189, 5-194, 5-199, 5-202, 5-205, 5-207 through 5-209, 5-213, 6-26, M-8, Q-7, Q-8, Q-10, Q-19, S-5, S-7, S-11, S-35, S-44, S-45, T-3, T-12, T-17, V-1 through V-3, V-10, V-16, V-17, V-24, V-28, V-29, V-31, V-42, V-44, V-50, V-52, V-56, V-57, V-60, V-61, V-66, V-72, V-84, V-86, V-90, V-121, V-144, V-155 through V-157, V-218, V-221, V-226, V-227, V-255, V-257, V-259, V-260, V-261, V-290, V-293, V-295, X-24, X-37, X-38, X-42, X-47, X-51, Y-5, Y-7, Y-10, Y-14, Y-16

guerrilla periods of instruction, 4-75, Q-10, U-1, U-2

guerrilla warfare, 2-11, 2-30, 2-32, 2-40, 2-58, 2-61, 2-65 through 2-67, 2-92, 3-16, 3-20, 3-42, 3-45, 3-46, 5-1, 5-23, 5-73, 5-131 through

This page intentionally left blank.

ATP 3-18.1
21 March 2019

By Order of the Secretary of the Army:

**MARK A. MILLEY**
*General, United States Army*
*Chief of Staff*

Official:

**KATHLEEN S. MILLER**
*Administrative Assistant*
*to the Secretary of the Army*
1908706

**DISTRIBUTION:**
Distributed in electronic media only(EMO).

www.ingramcontent.com/pod-product-compliance
Lightning Source LLC
Chambersburg PA
CBHW052106020426

42335CB00021B/2669